Practical Geophysics II

for the Exploration Geologist

Practical Geophysics II

for the Exploration Geologist

Compiled by Richard Van Blaricom

Northwest Mining Association, 10 N. Post, Ste. 414, Spokane, WA 99201-0772

 Published 1992.
Printed in the United States of America.
ISBN: 0-931986-05-2.

PREFACE TO THE SECOND EDITION

The popularity of the First Edition in 1980, and its reprint in 1988, made it clear to me that the short course and resulting volume were needed by practitioners, students and teachers, and economic geologists as well. The authors of the addenda in this volume represent a far smaller group of practicing mineral exploration geophysicists due largely to the shrinking amount of mineral exploration in North America. As the world changes to new political and economic relationships, geophysics will increase in importance as a highly developed adjunct to the discovery and development of the minerals critical to growth.

Technology innovations alone would call for an update of this important volume. While basic theory is essentially the same, its application has been greatly enhanced by new measurement, application and interpretation techniques, not the least of which are new computer applications. On behalf of the authors, Northwest Mining Association and myself, we are certain you will find this second edition a valuable source for practical answers to geophysical exploration questions.

Richard (Dutch) Van Blaricom
Geophysicist

Summer 1992

PREFACE TO THE REPRINT OF THE FIRST EDITION

August, 1988

PRACTICAL GEOPHYSICS was first printed in 1980 following the very successful short course presented that year by the Association. It has been an important volume for the student, teacher, practitioner and consultant. As the inventory became smaller, the Association began the process of how to reprint the volume at a reasonable cost while maintaining the quality and appearance of the original volume.

This reprinting meets my goal for quality and cost. I am pleased that we can continue to provide this useful reference for those interested in application of geophysics to the finding of orebodies and to the better understanding of the earth's crust. Our aim is to publish a revised edition of this volume based on the proceedings of the 1989 geophysics course, hopefully to be available within a year after the course.

The success of this volume is attributable to the time and effort of the experts whose work is presented here, as well as to the members and staff who compiled this work. I thank them all for their commitment to excellence.

Richard (Dutch) Van Blaricom
Geophysicist

PREFACE TO THE FIRST EDITION

The search for minerals and the lure of mineral wealth have sparked the imaginations of many. The excitement and romance of the Old West are probably one reason we are in the exploration business . . . a business that contributes much to our nation's economy. Thus we not only enjoy our work, but also accomplish a great deal of good. Through the years, the discovery of new ore deposits supports the idea that the "easy" mines have been found. We must, with some regret, lay aside the "burro and the pick" and use IP equipment, magnetometers, and computer models to find deeper and larger deposits which so far have evaded detection from the geologist using classical exploration techniques.

In the competitive world of exploration, we must use every new tool and every new trick to gain advantage in discovering mineral deposits that will allow us to maintain our present standard of living. Effectively and intelligently applying geophysical methods to an exploration endeavor is certainly one of the last frontiers of exploration. The prospector with his pick and roving eye was able to probe but a few inches into the crust of the earth to find metal deposits. With the effective use of exploration geophysics, we are now able to probe several thousands of feet into the earth to locate valuable mineral commodities.

Using geophysical methods in exploration has another important advantage: in most cases, the surface is undisturbed—significant in light of environmental concerns. Thus, we are able to go into an area, do some deep exploration, come to a definitive conclusion, and leave the surface undisturbed.

In addition, geophysics can maximize the effective use of exploration dollars. Geophysics helps the geologist see the third dimension, thereby minimizing physical exploration costs. If the exploration geologist understands some of these geophysical techniques, he may be more inclined to apply geophysics in his next exploration endeavor.

The authors' efforts and contributions are gratefully acknowledged, as well as their companies' willingness to present this information. I believe their efforts resulted in substantial contributions which will certainly aid in the application of practical geophysics. I am personally grateful and very appreciative of having the opportunity to be associated with the production of this book.

"Dutch" Van Blaircom

INTRODUCTION

This is a manual of applying practical geophysics. Its purpose is to aid and guide geologists to help them more intelligently use geophysics in their exploration endeavors. This manual should also be useful in an undergraduate exploration geophysics course. Authors have concentrated on mining problems, but other applications are touched on lightly.

The underlying premise for using geophysics, as well as any other means of exploration, is that the commodity being sought has properties substantially different from the properties of the host rock. As geophysicists, we are mainly concerned with the physical properties of the country rock versus the physical properties of the ore-bearing rock. Physical properties we are generally interested in are color, density, magnetic susceptibility, seismic velocity, radioactivity, resistivity, induced polarization effect, thermal conductivity, temperature, and a few others.

Differences between country rock and the ore or the commodity being sought can be established by insight or by physical measurement. One determines which physical properties are more varied between the country rock and the ore, and then selects the appropriate technique to measure and detect these properties. If there are no substantial differences in the physical properties of the ore rock and host rock, then geophysics is probably not the answer. The amount of scatter in the data also has to considered. We may be able to accurately measure small differences in parameters, but if the data are scattered, the differences being measured may be obscured.

The following guidelines can help an individual select appropriate geophysical techniques. The most common and also the least scientific approach is to try all the techniques on a known condition. That is, the geologist or geophysicist can use a known deposit as a test case. He runs every survey that he can possibly think of over the area to see which techniques are most useful. Extremely effective, this method is expensive and time consuming. In-situ measurements are more accurate, more diagnostic, and more definitive of a particular condition that the following two approaches.

The next most commonly used technique is to list all the physical parameters of the commodity being sought and all the physical parameters of the host rock. One can estimate which of these physical properties will differ most between the ore and the host rock. Appropriate geophysical techniques having been selected, one can go to a test area and try them over a known condition.

Another approach is to test small samples of the ore and host rock under laboratory conditions. This is easily done if one can acquire core or surface samples of ore and country rock in a particular locality. One can measure all the parameters on these small samples, determine which technique would be most useful, and proceed with the appropriate survey. A major limitation to this approach is that one tests very small numbers of samples which may not be representative of the bulk parameters of the rock in question. Thus, a large number of samples should be tested to gain a valid statistical sample. Another limitation of the sample testing technique is that the samples are usually solid and one cannot measure what effect fractures or faults may have on the data.

Geophysics can be used in an exploration program in direct detection, indirect detection, and as an aid in geologic mapping. Direct detection is measuring the parameter associated with the commodity being sought. One example is the use of induced polarization (IP) to detect sulfide deposits. The measured IP effect is attributed to sulfides being sought. There is definitely a positive correlation between percent sulfides and measured IP effect. Other examples of direct detection are the use of a magnetometer to help locate a magnetite deposit, and the use of a gravimeter in the detection of massive sulfide deposits. Massive sulfides are extremely dense. Thus, we usually have a definite, well-defined gravity high over massive sulfides, which is attributed to the sulfides themselves.

Indirect detection, on the other hand, uses the principle of association. For example, we look for a commodity, i.e. sphalerite which has no IP effect, but we use IP to find pyrite, and we hope there is a good positive correlation between the measured IP effect associated with the pyrite and the sphalerite being sought. Another example is the use of IP effect to find gold deposits. Again, we look at pyrite and hope locate porphyry copper deposits. Many times, magnetics help locate hydrothermal alteration (magnetic lows) and this definitely aids in porphyry copper exploration. Other examples would be the use of magnetics in the exploration for asbestos. Asbestos is not in itself magnetic, but often asbestos is associated with increased percentages in magnetite formed during serpentinization. There is a relationship between the magnetic susceptibility, degree of serpentinization, and occurrences of asbestos.

Many times, geophysical techniques are used not only to find deposits per se, but also to provide information which helps describe the geologic setting. Such data aid in determining if a particular location is a "permissive geologic area." For example, in the Basin and Range Province, gravity can be used to help establish depth to bedrock and the approximate location of the pediment fault. (One probably would not look for porphyry copper deposits in the middle of a deep basin). In addition, magnetic studies can be useful in determining signatures which some believe are extremely useful in locating favorable exploration areas for both porphyry copper and massive sulfide deposits.

CONTENTS

CHAPTER 1
RADIOMETRICS
GEOLOGICAL APPLICATIONS FOR PORTABLE GAMMA RAY SPECTROMETERS

The purpose of this chapter is to provide the nongeophysicist with an understanding of the complexities, the strengths, and the weaknesses of gamma ray spectrometry. This will aid the explorationist in making better use of the available tools to solve geological problems. Airborne and drill hole surveys will be dealt with briefly in order to consider the full range of the source-detector geometry. However, the major emphasis will be on the use of portable ground spectrometers. Wherever possible, pictorial diagrams or line drawings have been used to illustrate mathematical or verbal descriptions. The line drawings represent simplifications of more complicated data or concepts and therefore should be regarded as expository devices and not as evidential information. For more detailed data, the reader may consult the references listed in the back of this chapter.

The purpose of gamma ray spectrometry is to provide information about the distribution of the three radioactive elements—uranium, thorium and potassium. These elements emit gamma radiation whose energy is characteristic of the emitting isotope. Therefore, by measuring the intensity and the energy of the gamma radiation at a particular geographic

Don A. Hansen is president of REXCON, INC., G-Cubed, Inc., and ScienTerra, Inc., (1980) three Spokane-based companies specializing in various aspects of exploration activity. He has developed new radiochemical techniques and equipment for uranium exploration. A primary interest has been the geologic interpretation of induced polarization magnetic and gamma ray spectrometer data. Dr. Hansen received an A.B. degree in physics at Brigham Young University followed by graduate studies in physics at the University of Washington and in geology at Brigham Young University. He holds a Ph.D. in geology from U.C.L.A. Subsequent exploration projects have taken him to Peru, Chile, Ecuador, Puerto Rico, central Africa, Australia and Canada. The work has involved exploration for ferrous and nonferrous metals, precious metals, uranium, petroleum and ground water. Dr. Hansen has been an invited lecturer at University of California at Berkeley, University of Toronto, and the People's Republic of China.

Don A. Hansen, Ph.D.
REXCON, INC.

location, theoretically one should be able to determine the abundance of the three elements in that location. The steps which must be taken in order to achieve such an end are given in this chapter.

Impediments to accurate measurement of the three radioactive elements come from a wide variety of sources. Probably no other geophysical method requires the consideration of so many variables as does gamma ray spectrometry. In addition to the normal variables of geometry and physical property contrasts, it is necessary to consider the size of the gamma ray source, the efficiency of the detector, the speed at which the detector moves, the effects of meteorological variables such as barometric pressure, air stability and the movement of major air masses. One must consider disequilibrium conditions in the uranium and thorium decay sequence and the peculiarities of the chemical behavior of uranium, thorium, and potassium in the measuring environment.

The benefits of determining the concentration of the three radioactive elements in the earth are many. Changes in the concentration of the radioactive isotopes accompany most major changes in lithologic composition. Variations in radioisotope concentrations can indicate primary geologic processes such as mineralizing solutions or metamorphic gradients. Variations can also characterize secondary geologic processes such as those of leaching and supergene alteration.

The term *gamma ray spectrometer* is applied to any device that separates gamma radiation into two or more energy components. Such devices comprise a detector and console. The detector, normally a sodium iodide crystal, absorbs the gamma radiation and converts it to a light flash or a scintillation. The light is received by a photomultiplier tube that converts the light flash into a voltage proportional to the intensity of the light flash. The console separates the voltage into a number of classes depending upon magnitude, thus providing an energy spectrum of the incident gamma rays.

It is generally true that the more sophisticated the spectrometer, the more accurate is the information obtained about the earth. The cost and effort are greater for accuracy. Indeed there is an almost linear relationship between (1) the accuracy of the gamma ray data and (2) the cost of the equipment and associated procedures required to render the data useful.

Natural radioactivity

Radioactivity refers to the process of decay or change in an atomic nucleus. By emitting radioactivity, the nucleus changes its state of charge or its state of mass or both.

Radioactive decay products

There are three common types of radioactive decay products. These are called alpha rays, beta rays, and gamma rays. The first two terms are in a sense misnomers; the alpha and beta rays have both mass and electrical charge and, therefore, are more properly called particles.

The *alpha particles* are helium nuclei. Each consists of four elementary particles: two protons and two neutrons. Soon after the expulsion of an alpha particle from a nucleus, the alpha particle acquires orbital electrons from the environment and becomes a neutral helium atom. Some helium atoms migrate to the surface of the earth and contribute to the helium background of the atmosphere; the remainder are bound in rock minerals and do not escape until liberated by decomposition of the mineral host. Alpha particles have a discrete energy, and the emitting isotope can be identified by measuring the characteristic energy of the alpha particle. Alpha rays come from the disintegration of uranium and thorium decay series.

Beta rays are ordinary electrons and, as such, bear a negative charge. They do not have a characteristic energy. They are accompanied by the emission of a neutrino; the neutrino carries the difference in energy between (1) that carried by the beta ray and (2) the total energy defect in the nucleus. Because alpha and beta rays possess both a mass and a charge, their penetration through earth materials is a few millimeters or less. Therefore, the detection of alpha and beta rays does not lend itself to remote sensing applications as does the detection of gamma rays. Figure 1.1 illustrates this relative penetrating power of the three radioactive decay products.

The third type of radioactive decay and most significant to this chapter is *gamma radiation*. Gamma radiation is simply a high energy form of electromagnetic radiation. Gamma rays have their origin in a nuclear decay process rather than an atomic electron decay process (x-rays) or a thermal electron decay process (light rays). Table 1.1 gives the wave length and the frequency, which is proportional to energy, of the electromagnetic spectrum.

Because gamma rays have no electrical charge and no mass, they are capable of penetrating a foot or more of earth-type materials. Each gamma-emitting isotope gives off one or more gamma rays of characteristic energy. The two properties of characteristic energy and penetrating power confer upon gamma rays their utility in geologic exploration. The remainder of this chapter will deal with the methods that are used to make gamma ray measurements more accurately reflect the concentration of uranium, thorium, and potassium.

Radioactive decay sequence

Table 1.2 presents the decay sequence for the three naturally occurring radioactive isotopes—uranium, thorium, and potassium. Each isotope is identified by a name and a number. The name identifies a specific chemical species, an atom with a unique nuclear charge. The number identifies the number of protons and neutrons in the nucleus of the atom.

It is geologically significant that there are geochemically mobile isotopes between the long-lived uranium or thorium end members and the isotope that emits the useful gamma ray energy. The isotopes, important in near-surface environments, are indicated in the last column of table 1.2.

Significant spectral peaks

Note that in table 1.2 there is a large number of *gamma ray lines* or energy peaks in both the uranium and thorium decay sequence. Only one line in each sequence is normally used in gamma ray spectrometry. In order to be of use in uranium prospecting applications, a line must

FIGURE 1.1
Relative penetrating power of radiation

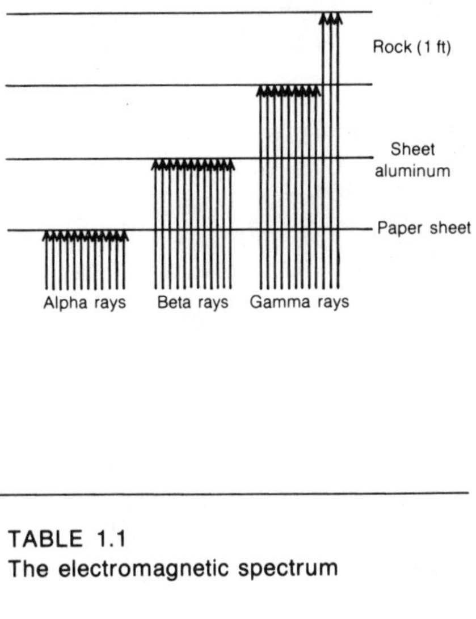

TABLE 1.1
The electromagnetic spectrum

Name of spectral division	Approximate frequency range Cycles per second
Direct Current and Magnetostatic	0
Power Line Frequency	60
Audio Frequency (AF)	$2 \times 10^1 - 2 \times 10^4$
Very Low Frequency (VLF)	$10^4 - 3 \times 10^4$
Radio Frequencies (RF)	$3 \times 10^4 - 3 \times 10^8$
Microwave (RADAR)	$3 \times 10^8 - 3 \times 10^{11}$
Infrared (IR)	$3 \times 10^{11} - 3 \times 10^{14}$
Visible and Ultraviolet (UV)	$3 \times 10^{14} - 3 \times 10^{16}$
X-Ray (X-R)	$3 \times 10^{16} - 3 \times 10^{19}$
Gamma Ray (γ-R)	$3 \times 10^{19} - 3 \times 10^{23}$
Cosmic Rays	3×10^{20} and greater

TABLE 1.2
Radioactive decay series of U-238, Th-232, K-40

Isotope-Mass No. (Symbol)	Obsolete Usage	Atomic Number (Z)	Li
URANIUM-238 SERIES			
(*1) Uranium-238	(U) Uranium I	92	4.5 ×
Thorium-234	(Th) Uranium X_1	90	24.1 d
Protactinium-234	(Pa) Uranium X_2	91	1.18 m
(*1) Uranium-234	(U) Uranium II	92	2.6 ×
(*1) Thorium-230	(Th) Ionium	90	8 × 1C
(*1) Radium-226	(Ra) Radium	88	1600 y
(*1) Radon-222	(Rn) Radon	86	3.82 da
Polonium-218	(Po) Radium A	84	3.05 m
Lead-214	(Pb) Radium B	82	26.8 m
(*1) Bismuth-214	(Bi) Radium C	83	17.9 m
Polonium-214	(Po) Radium C′	84	1.6 ×
Lead-210	(Pb) Radium D	82	19.4 yr
Bismuth-210	(Bi) Radium E	83	5.0 day
Polonium-210	(Po) Radium F	84	138.4 d
Lead-206	(Pb) Radium G	82	Stable
THORIUM SERIES			
(*1) Thorium-232	(Th) Thorium	90	1.4 × 1
(*1) Radium-228	(Ra) Mesothorium I	88	6.7 yr
(*1) Actinium-228	(Ac) Mesothorium II	89	6.1 hr
(*1) Thorium-228	(Th) Radiothorium	90	1.9 yr
Radium-224	(Ra) Thorium X	88	3.64 day
Radon-220	(Rn) Thoron	86	54.5 sec
Polonium-216	(Po) Thorium A	84	0.16 sec
Lead-212	(Pb) Thorium B	82	10.6 hr
Bismuth-212	(Bi) Thorium C	83	60.5 min
Polonium-212	(Po) Thorium C′	84	0.3 × 1C
(*1) Thallium-208	(Tl) Thorium C″	81	3.1 min
Lead-208	(Pb) Thorium D	82	Stable
POTASSIUM			
(*1) Potassium-40	(K) Potassium	19	1.31 × 1

* Footnotes:
(*1) Isotopes of particular geological or geochemical interest.
(*2) Half Life: The time required for a radioisotope to lose ½ of its activity.
(*3) Energy in million electron volts, Mev, (% yield) represents the number of gamma pho per 100 disintegrations of the particular isotope. Tl-208 adjusted for branching ratio from Bi-212.
(*4) Auto oxidation (loss of electrons) occurs at these states.
(*5) The Curie is the unit of activity and is defined as 3.7×10^{10} radioactive disintegrations per second.
(*6) General geochemical properties (M = Mobile, I = Immobile) of isotopes having geologic significance, in general those of long half life.
(*A) 2.43 (2), 2.20 (6), 2.12 (1), 1.85 (3), <u>1.76 (19)</u>, 1.73 (2), 1.51 (3), 1.42 (4), 1.38 (7), 1.28 (2), 1.24 (7), 1.16 (2), 1.12 (20).
(*B) 1.64 (13), 1.59 (12), 1.10, 1.04, 0.97 (18), 0.99 (25), 0.46 (3), 0.41 (2), 0.34 (11), 0.23, 0.18 (3), 0.13 (6), 0.11, 0.10, 0.08
(*C) 1.18 (1), 1.61 (3), 1.34 (2), 1.04 (2), 0.83 (8), 0.73 (10), 0.04 (1)
(*D) <u>2.62 (100)</u>, 0.86 (14), 0.76 (2), 0.58 (83), 0.51 (25), 0.28 (9), 0.25 (2)
The most geologically useful radiation energies are <u>underlined</u>.

first be isolated from adjacent lines in the thorium sequence and from the potassium 40 (K) energy peak. Second, the useful line should have a high percentage yield. The yield is the number of photons emitted for each 100 disintegrations of the particular isotope.

For example, only a few gamma ray peaks for the uranium 238 (U) sequence are above 10 percent in yield—thorium 234 (Th), uranium 234 (U), lead 214 (Pb) and bismuth 214 (Bi). Low energy lines—such as 0.09 Mev (million electron volts) from Th-234 and 0.05 Mev from U-234 are not useful because of a high gamma ray background at the low energy end of the spectrum. The high background is due to Compton scattering and is discussed later in this chapter.

Gamma Radiation (*3) :nergy in Mev, (% yield)	Valence States (*4)	Grams Curie (*5)	Geochemical Mobility Environment (*6) Oxi-dizing	Reduc-ing
	4*, 6	2.97×10^{6}	M	I
5), 0.06 (7), 0.03 (7)	4	4.3×10^{-5}	I	I
), 0.77 (1), 0.04 (3)	3, 4, 5	1.4×10^{-9}		
3)	3, 4, 6	1.64×10^{2}	M	I
	4	5.3×10^{1}	I	I
)	2	1.0	M	M
	0*	6.46×10^{-6}	M	M
	6	3.15×10^{-9}		
4), 0.24 (11), 0.29 (24), 0.05 (2)	2,4	3.03×10^{-8}		
	3, 5	2.23×10^{-8}		
	6	2.78×10^{-15}		
	2, 4	1.28×10^{-2}		
)	3, 5	8.0×10^{-6}		
	6	2.24×10^{-4}		
	2, 4			
24)	4	2.11×10^{9}	I	I
	2	1.0	M	M
	3, 4	1.4×10^{-4}		
	4	2.63×10^{-1}	I	I
	2	1.48×10^{-3}		
	0*	2.49×10^{-7}		
	6	7.08×10^{-10}		
5), 0.24 (82), 0.18 (1), 0.12 (2)	2.4	1.68×10^{-4}		
	3.5	1.60×10^{-5}		
	6	6.42×10^{-16}		
	1, 3	2.73×10^{-7}		
	2, 4			
(11)	1			

ement	Atomic No.	Val-ence	Remarks
ranium	92	+4	Essentially insoluble in water.
		+6	Soluble as sulphate, carbonate, or chloride in which U^{+6} exists as a complex divalent cation $(UO_2)^{+2}$.
horium	90	+4	Essentially insoluble in water.
adium	88	+2	Does not form minerals but may "co-crystallize" with insoluble Ba, Sr, Pb sulphates. Silicate minerals in alkaline or neutral environments tend to strongly absorb radium.
adon	86	0	One of the noble gases and forms no compounds.

The most useful line for the uranium sequence is that of Bi-214. It has an energy of 1.76 Mev and a total energy yield of 19 percent. The useful thorium sequence line is that of thallium 208 (Tl) having an energy of 2.62 Mev and a yield of 100 percent. Potassium has a single emission line at 1.46 Mev energy.

Interaction of gamma rays with matter

In space, a gamma ray can be considered as a small packet of wave energy traveling with the speed of light (300,000 kilometers per second) and having a frequency that is directly proportional to the energy of the radiation.

Gamma rays interact with matter in three principal modes: (1) Compton scattering, (2) photoelectric absorption, and (3) pair production. The relative importance of these effects in a lead shield is illustrated in figure 1.2. The photoelectric effect and pair production are relatively *unimportant* in the absorption of gamma rays in earth materials. The photoelectric effect *is* important in the detection of gamma rays in crystal scintillators.

FIGURE 1.2
Total absorption in lead

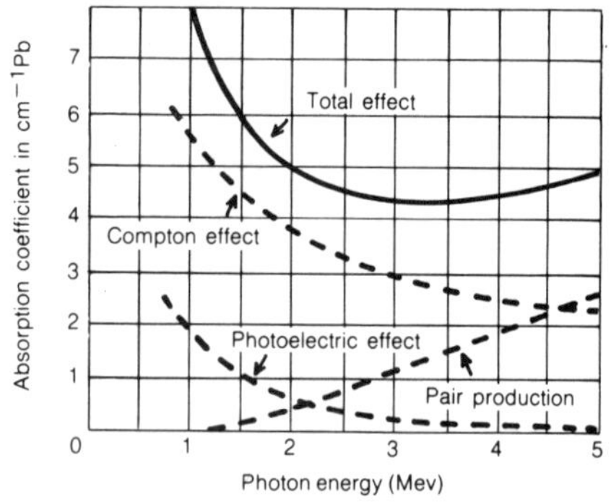

FIGURE 1.3
Interaction of gamma rays with the earth, the air and the detector

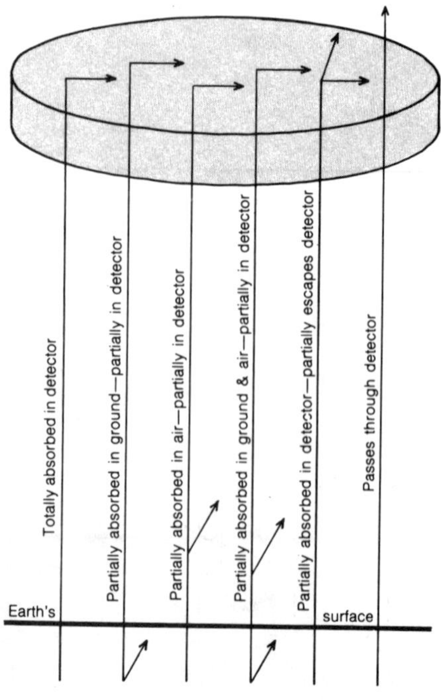

Compton scattering

Compton scattering is the name used to describe a process whereby a single high energy gamma ray photon produces many other gamma ray photons of lower energy by interaction with the earth or the detector.

The subject of Compton scattering merits considerable attention because a photon given off by Th-208 interacts with the environment to produce secondary photons having energy characteristic of Bi-214 and of K-40. In order to determine the isotopic abundance of uranium or potassium, one must correct the observed count rate in the uranium and potassium channels by an amount which is proportional to the number of counts in the thorium channel. Similarly, any counts originating from the uranium daughter, Bi-214, contribute to the total count in the potassium channel. This process of degradation of gamma ray energy produces a logarithmic increase in the low energy count rate—it is termed the *Compton tail* or the *Compton continuum.*

It is useful to think of Compton scattering as a collision between billiard balls. Depending on the angle of impact, the kinetic energy or the motion of one ball may be transferred totally or only partially to the second ball. In earth materials, the gamma ray photon acts as the billiard ball in motion; the orbital electrons of the material act as the target balls. The absorbing or the scattering power of the earth material is directly proportional to the density of the electrons in the material.

When a gamma ray collides with an electron, it transfers all or only part of its kinetic energy to the electron and either stops or changes its direction of travel. The probability of the gamma ray photon scoring a direct hit on an electron, and thereby transferring all of the energy, is quite small. Most interactions result in only a small change in direction and energy of the scattered gamma ray. After an initial collision with an electron, the gamma ray may undergo additional encounters each resulting in a decrease in energy. Figure 1.3 illustrates Compton scattering taking place in the earth, the air, and the detector. This collision process results in a relatively large number of photons of low energy, hence the exponential buildup in count rate shown in figure 1.4.

Photoelectric effect

In the detector crystal, because of its high density, the photoelectric effect is important. In this interaction, all of the gamma photon energy is transferred to an electron as kinetic energy. In the process of slowing down, the electron ionizes a short path length in the crystal with a resultant scintillation or flash of light. *The light flash intensity is proportional to energy of the incident gamma ray.* A photomultiplier tube detects the light flash and amplifies the signal for ultimate recording or display.

Cosmic rays

Cosmic rays are a nongeologic source of radiation. The earth's atmosphere is constantly under bombardment by charged particles—primarily protons and alpha particles—moving at a very high energy. The interaction of these light particles with the components of atmosphere results in secondary radiation consisting of various products such as mesons, electrons, and gamma ray photons. Many of the gamma rays so produced are of very high energy and their Compton scatter products constitute a noise source impressed upon the entire gamma ray spectrum.

Altitude and latitude effect of cosmic rays

Because the primary cosmic particles bear an electrical charge, there results a latitude dependence of the distribution of cosmic rays (illustrated in figure 1.5a). Both the charged primary particles and the secondary products are absorbed by collision with the constituents of the earth's atmosphere. The result is a marked increase in cosmic ray intensity with increasing elevation above sea level (illustrated in figure 1.5b). Cosmic rays contribute to the Compton-scattered gamma rays and, therefore, to the background radiation.

In ground surveys the percentage contribution of cosmic rays relative to ground sources is small and normally neglected. In precise airborne work, the Compton scattering products of cosmic ray interactions must be corrected for. A common method is to designate as cosmic background all events having an energy above the 2.62 channel of Tl-208. The Compton scattering contribution from these high energy photons is removed from all subsequent energy channels.

FIGURE 1.4
Example of a natural gamma ray spectrum obtained by multichannel spectrometer

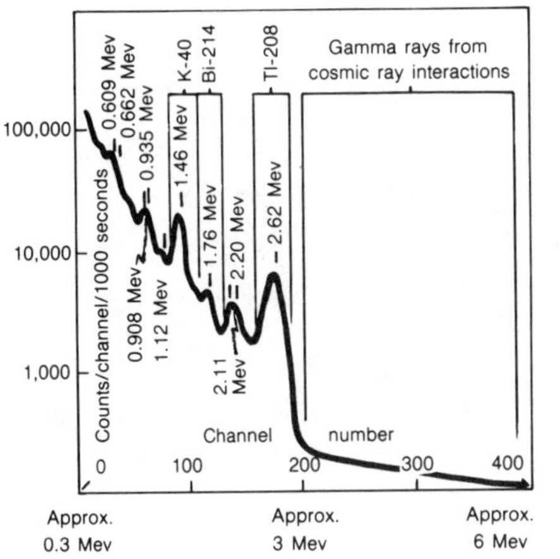

Radiation detection and instrumentation

There are many types of devices for the detection of gamma rays. Among these are electrometers, ionization chambers, Geiger tubes, scintillation crystals, semiconducting detectors, plastic phosphors and liquid scintillators. In this section, our purpose is to describe the use of spectrometers in exploration. Our discussion will be concerned primarily with the most widely applicable of detectors, the *thallium-activated sodium iodide* or cesium iodide crystals.

Sodium iodide crystals are used in preference to the other detectors for a variety of reasons. The high density and good transparency of the crystals result in high absorption of gamma rays and effective transmission of energy to the photomultiplier tubes. The sodium iodide can be grown and fabricated with relative ease as large volume detectors.

FIGURE 1.5a
Total cosmic ray intensity at sea level

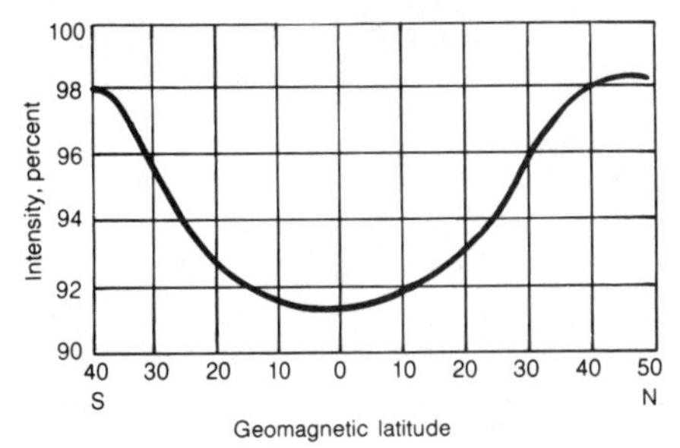

Geiger tubes versus scintillators

Geiger tubes have in the past been the dominant detectors used for field investigation of radiation from geologic sources. It is informative to compare some of the characteristics of scintillation crystals and Geiger tubes.

Geiger tubes have an advantage in low cost of the detector and associated electronics. They are rugged and can withstand field use with little impairment. They are sensitive to the high energy components of gamma radiation. However, Geiger tubes lack the characteristic of producing an output pulse that is proportional to the incoming gamma ray energy.

Scintillation crystals are sensitive to thermal and mechanical shock; they must be sealed against the ingress of water vapor. The associated electronics demand a high degree of sophistication. However, because of its greater overall sensitivity, the scintillation counter has to a large extent replaced the Geiger tubes for field reconnaissance.

FIGURE 1.5b
Total cosmic ray intensity vs elevation

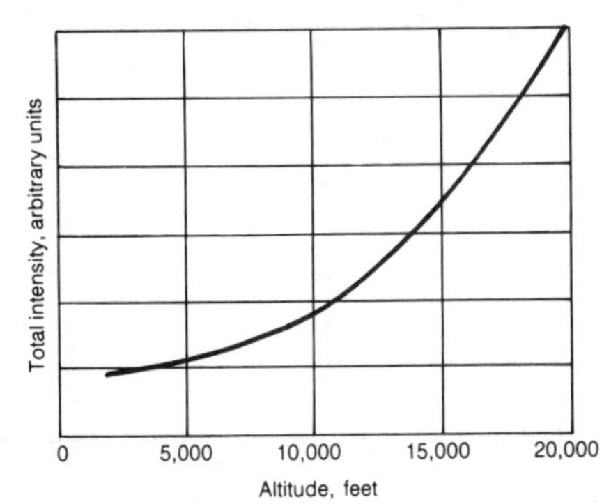

Scintillation counters differ from *spectrometers* in the method of processing the electronic signal from the photomultiplier tube. In the scintillator, the pulses are shaped and registered without regard to their voltage amplitude. In a spectrometer, the console electronics perform the function of classifying the voltage pulses into a number of classes which are proportional to the energy of the incident gamma ray. The single characteristic of *signal strength being proportional to gamma ray energy* requires that scintillation crystals rather than Geiger tubes be employed for spectrometer applications.

Solid state detectors

A third type of detector which will probably play an increasing role in exploration efforts in the future is the *solid state semiconducting detector.* The most suitable solid state detector at present is the *lithium-drifted germanium, Ge(Li), crystal.* These crystals have a very high resolving power, about 50 to 80 times that of thallium-activated sodium iodide, NaI(Tl), crystals.

The disadvantage of the Ge(Li) crystal is that it is difficult to grow and fabricate large crystals. Most crystals must also be maintained at liquid nitrogen temperatures in order to hold their effectiveness. Intrinsic detectors are being developed which do not require continual maintenance at cryogenic temperatures. Ge(Li) crystals offer substantial advantages in certain applications. These applications will probably increase in the future.

Integral and differential spectrometers

In the previous section of this chapter, we said that the signal strength in a scintillation crystal is proportional to the energy of the incoming gamma ray. A spectrometer preserves the proportionality while classifying the signals. If the signals are classified into a large number—about 100 to 1000—of small energy classes defined by both an upper and lower limit, the classes are called "channels." If the number of classes is small, 4 or 5, and each is defined by an upper and lower energy limit, the classes are said to be "windows." Instruments using these kinds of energy discrimination are called *differential spectrometers.*

If the signals are classified into a small number of classes each defined by only a lower energy limit, the instruments are called *integral spectrometers.*

Differential spectrometers offer advantages in precision of measurement, greater signal-to-noise ratio, and ease of processing the data. Integral spectrometers have an advantage in lower initial cost and higher count rate. The energy discrimination characteristics of the two types of instruments are illustrated in figure 1.6.

Drift stabilization

The energy stability of spectrometer windows or channels varies with changes in the power supply voltage and with changes in the amplification characteristics of photomultiplier tubes. Both power supplies and photomultiplier characteristics are dependent upon temperature. Steps taken to increase spectrometer stability include quality electronic design, thermal insulation of components, and cooling of power supplies. In addition, many spectrometers make use of *automatic gain control* wherein a radioactive calibration source of known energy emission is inserted permanently or periodically into the viewing range of the detector crystal. A feedback electronic circuit maintains the relationship between the spectrometer output and the known source energy. Assuming spectrometer linearity, all other channels or windows will also be stable with respect to gamma energy.

Calibration sources used in the past are americium 241 (Am), cesium 137 (Cs), and cobalt 57 (Co). For various reasons which will not be enumerated here, Cs-137, having an emission line at 0.662 Mev, is the most commonly used calibration source.

Temperature control

Temperature control in field equipment is essential. Thermal shock due to sharp temperature changes can crack detector crystals. Temperature changes also produce calibration changes by changing photomultiplier voltage. Equipment is carefully insulated, and drift correction circuits are incorporated to cancel the effects of temperature variations. In sophisticated airborne systems, heating coils are often placed in the

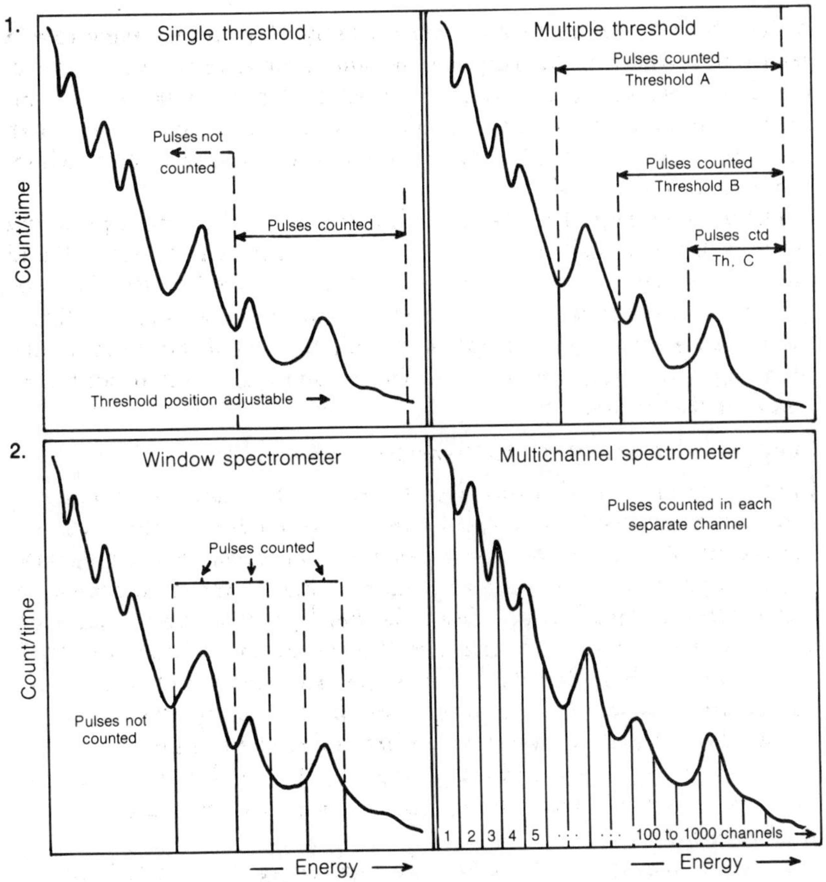

FIGURE 1.6
Energy discrimination characteristics of integral and differential spectrometers

1. **Integral spectrometers**
2. **Differential spectrometers**

detector housing in order to aid in maintaining a stable temperature environment for the crystals and photomultipliers.

Dead time

During the interval of time that a crystal registers a photoelectric interaction and the electron cascade proceeds through the photomultiplier tube, the system is unable to resolve simultaneous interactions. An electronic circuit measures the time that the detection system is dead (i.e., registering a pulse) and multiplies the registered counts by an appropriate number to bring them back to a "zero dead time" count. This is known as a *live time correction.* It is significant at very high count rates—above 10,000 counts per second.

Calibration of spectrometers

Gamma ray spectrometers must be calibrated in two essential modes. If a gamma ray spectrometer will be used as a geochemical prospecting or mapping tool, it must be calibrated in terms of the isotopic sensitivity (i.e., the conversion from counts per unit of time into isotopic concentration in parts per million or percent). Spectrometers must also be calibrated for the stripping constants that must be corrected for the Compton scattering contribution to the various isotopic channels.

Calibration may be done theoretically, and an approximation obtained. Due to variations in mounting configuration, the best calibrations are done empirically over natural or artificial standards.

Natural standards are selected in such a way that each of the separate standards is high in one radioactive element. The uranium, thorium and potassium content of each of the standards must be known, and the distribution of the elements must be determined or assumed to be uniform and large enough to satisfy the 2 pi or 4 pi geometry requirement. The practical problems of finding three geological units—intrusives, for example—that fulfill these requirements and are within reasonably accessible locations are great. Therefore, artificial sources are best for most circumstances. An exception may be in the calibration of airborne spectrometers where the mounting configuration makes a 2 pi geometrical requirement difficult to fulfill.

Artificial standards can be constructed as (1) rigid bodies of low activity concrete salted with radioactive materials or (2) loose material in a container.

The radioactivity of any standard, be it natural or artificial, depends on its emanation coefficient—the ease with which radon is liberated to the environment. Standards made from sealed containers may have higher radioactivity than rocks having the same uranium content because escape of radon is prohibited. After the manufacture of any standard, sufficient time must be allowed for radon to come into equilibrium with the parent members. Care must also be used in selecting uranium ores that are in equilibrium with the radium daughter.

Compton stripping ratio

In order to achieve a quantitative result, spectrometer data must be corrected for Compton scattering of gamma rays in the detector. The logarithmic increase of the spectrum in figure 1.4 is the result of Compton scattering. The smoothed version of the curve is called the *Compton continuum.* All photons can contribute to the lower energy portions of the spectrum when they interact with matter.

The numbers used to correct the count in any portion of the spectrum for the contribution from higher energy portions are called the *Compton correction coefficients* or *stripping ratios.* The stripping ratio is a function of the size and geometry of the detector and the geometrical relationship between the detector and the radiation source. Thus, it is necessary to use different stripping ratios for different crystal sizes, shielded and unshielded crystals, and for crystals used on the surface or in drill holes.

Sensitivity constants

The *sensitivity constants* relate the corrected count rate in any isotopic channel to the abundance of the isotope. The following expressions define the relationship between abundance and count rate:

$$\text{Th(ppm)} = C_{Th}(N_{Th} - S_1N_U - S_2N_{Cos} - b_{Th}), \tag{1.1}$$

$$\text{U(ppm)} = C_U(N_U - S_3N_{Th} - S_4N_{Cos} - b_U), \tag{1.2}$$

$$\text{K(\%)} = C_K(N_K - S_5N_{Th} - S_6N_U - S_7N_{Cos} - b_K). \tag{1.3}$$

Here C_{Th}, C_U and C_K are the sensitivity constants that relate the response to the abundance of the decaying isotope. The S_n coefficients are the Compton stripping ratios; b_{Th}, b_U, and b_K are the background counts in the various channels. The sensitivity constants and the stripping ratios must be determined by controlled experiment (1) for each crystal size and shape and (2) for each different geometrical relationship between the crystal and the radiation source. The factors N_{Th}, N_U, N_K, and N_{Cos} are the observed counts per unit of time in the Tl-208, Bi-214, K-40, and cosmic ray channels. The cosmic ray count is normally taken as that above Tl-208 energy window.

Note that in equation (1.1) above there is a stripping ratio for N_U to correct the thorium count. This is not a true Compton scattering ratio but is designed to correct for tails of Bi-214 lines at 2.43 and 2.20 Mev which might occur in a wide Tl-208 channel.

In surveys with hand portable instruments, the signal from the earth is frequently so strong compared with the background effects of atmospheric and cosmic sources that these may be considered second order corrections. The equations (1.1), (1.2), and (1.3) may then be written as follows:

$$\text{Th(ppm)} = C_{Th}(N_{Th} - b_{Th}), \tag{1.4}$$

$$\text{U(ppm)} = C_U(N_U - S_3N_{Th} - b_U), \tag{1.5}$$

$$\text{K(\%)} = C_K(N_K - S_5N_{Th} - S_6N_U - b_K), \tag{1.6}$$

where the Compton stripping ratios are:

$$S_3 = \frac{N_U - b_U}{N_{Th} - b_{Th}} = \frac{\text{Counts in U channel due to}}{\text{Counts in Th channel}}, \tag{1.7}$$

$$S_5 = \frac{N_K - b_K}{N_{Th} - b_{Th}} = \frac{\text{Counts in K channel due to}}{\text{Counts in Th channel}}, \tag{1.8}$$

$$S_6 = \frac{N_K - b_K}{N_U - b_U} = \frac{\text{Counts in K channel due to}}{\text{Counts in U channel}}. \tag{1.9}$$

Examples of sensitivity constants and stripping ratio

L. Lϕvborg (1973) has determined the sensitivity constants and stripping ratios for three sizes of crystals; two in the field, and one in the laboratory. His results are presented in table 1.3.

It is useful to note the variation in stripping ratio between the 2 × 2 in and 3 × 3 in crystals. The S_3 ratio, representing the thorium counts subtracted from the total count in the uranium channel, is 42 percent greater with the 2 × 2 in crystal than with the 3 × 3 in crystal. This can be interpreted as a lack of efficiency of the smaller crystal in detecting the higher energy gamma rays of thorium. Because of its greater volume (3.4 times) and greater efficiency, comparable counting statistics can be obtained about seven times faster with a 3 × 3 in crystal than a 2 × 2 in crystal. The small stripping ratios of the 5 × 5 in crystal are due to the laboratory geometry.

Drill hole probes can be calibrated readily. Suitable calibration standards for drill hole probes with acceptable boundary errors can be 1-2 ft in radius and 3-4 ft long with a central probe hole. The standard is constructed of concrete mixed with known amounts of uranium, thorium, and potassium in separate test standards. The sensitivity constants and the Compton stripping coefficients can thereby be determined.

The calibration of hand portable equipment by artificial standards makes use of test pads which approximate a half-space in dimension. Separate pads contain known amounts of potassium, thorium, and uranium. Such facilities are available under the auspices of the Canadian Geological Survey and are located at the Uplands Airport in Ottawa, Ontario. The United States Atomic Energy Commission also maintains test pits for drill hole probes and is presently planning test pads for calibration of portable equipment.

TABLE 1.3
Sensitivity constants and stripping ratios of three spectrometer crystals

NaI(Tl)	Sensitivity constant ppm/counts/min			Compton stripping ratio		
Size	C_{Th}	C_U	C_K	S_3	S_5	S_6
2 × 2 in (a)	0.63	0.16	0.019	1.7	2.3	1.5
3 × 3 in (a)	0.14	0.036	0.0048	1.2	1.6	1.3
5 × 5 in (b)	1.43	0.36	0.039	0.51	0.81	1.03

(a) In situ analysis of outcrops
(b) Laboratory analysis of 230 grams material

An excellent example of a standard for portable spectrometers ha been constructed in Denmark at the Research Establishment Risϕ. Th Risϕ calibration standards comprise four concrete slabs, each thre meters in diameter and one-half meter thick. They form a line o working platforms level with the ground. Three standards contain respectively about 200 ppm uranium, 150 ppm thorium, and 7 percen potassium. The fourth standard contains less than 1 ppm uranium, 3 ppm thorium and 1 percent potassium. This standard serves as a zero reference, thus eliminating the need for a separate background determination.

Example of calibration calculations

As an example of the use of test pads, assume the following results have been obtained by taking long observations—say 30 minutes—on each of four calibration pads at the Risϕ facility.

	Counts per minute		
Test pad	Thorium Tl-208	Uranium Bi-214	Potassium K-40
Background Th < 3 ppm U < 1 ppm K < 1%	10	15	40
Thorium Th = 150 ppm U < 1 ppm K < 1%	800	245	400
Uranium Th < 3 ppm U = 200 ppm K < 1%	20	2200	1100
Potassium Th < 3 ppm U < 1 ppm K = 7%	15	20	4000

To determine the Compton stripping, we use the expressions 1.7, 1.8, and 1.9:

$$S_3 = \frac{N_U - b_U}{N_{Th} - b_{Th}} = \frac{245 - 15}{800 - 10} = 0.42,$$

$$S_5 = \frac{N_K - b_K}{N_{Th} - b_{Th}} = \frac{400 - 40}{800 - 10} = 0.46,$$

$$S_6 = \frac{N_K - b_K}{N_U - b_U} = \frac{1100 - 40}{2200 - 15} = 0.48.$$

The sensitivity constants may now be computed from equations 1.4, 1.5, and 1.6:

$$C_{Th} = \frac{\text{Th(ppm)}}{N_{Th} - b_{Th}} = \frac{150}{790} = 0.19,$$

$$C_U = \frac{\text{U(ppm)}}{N_U - S_3 N_{Th} - b_U} = \frac{200}{2178} = 0.92,$$

$$C_K = \frac{\text{K\%}}{N_K - S_5 N_{Th} - S_6 N_U - b_K} = \frac{7}{3943} = 0.0056.$$

Airborne instruments

The establishment of a half-space geometry for airborne spectrometers is a still more difficult problem. The calibration can be accomplished, as at the Canadian facility at Ottawa, by taxiing the aircraft over the separate calibration pads. If the clearance between the detectors and the oad is relatively small, a satisfactory half-space approximation is obtained. The approximation to a half-space for large aircraft and high-mounted detectors becomes increasingly poor.

An alternative method for calibration of airborne equipment is to fly over large areas of known geological concentrations of the radioelements. Extensive sampling and ground radiometric work are required to establish the concentration values of the radioelements. The Canadian Geological Survey has a test area at Bancroft, Ontario, that has been extensively sampled by ground spectrometers and geochemical sampling; the radioactive isotope content is well known.

Digital recording of data

The type of equipment used for detecting and recording the gamma rays affects the accuracy of the survey. For maximum flexibility in data processing, digital recording on magnetic tapes is preferred. This permits "reflying" of the survey in a computer center. There, various weighting functions can be applied to the data in order to increase the statistical confidence level, and stripping can be accomplished automatically. Analog recording of raw or partially processed spectrometer data during survey flights permits immediate examination of results. It is recommended that analog recording supplement, rather than supplant, digital recording.

Radiation characteristics of geologic and cultural sources

The radiation characteristics of a rock or soil depend upon the initial abundance of the radioactive elements and upon the subsequent geologic history of the rock. Metamorphism and weathering play particularly important roles in modifying the radioelement distribution. The gamma radiation recorded by a spectrometer is indicative of potassium, uranium, and thorium only if the latter two elements are in equilibrium with their daughter products that emit the gamma rays.

In the case of both uranium and thorium, there are ten isotopes between the long-lived parent member and the isotope that emits the remotely sensed gamma ray. Any geologic conclusion based upon parent isotopic abundances derived from gamma spectral data must necessarily involve an assumption of equilibrium. Equilibrium is far from the normal case in the upper few feet of the earth, a region of intense biologic and chemical activity. For the foregoing reasons, it is good practice to think and to write the radiation sources as bismuth (Bi-214) for uranium and thallium (Tl-208) for thorium. Potassium (K-40) is the sole radioactive member, and no chemical separation problem exists.

Uranium and thorium in rocks

Much of the uranium and thorium in igneous rocks is concentrated in the common accessory minerals such as zircon, sphene, and apatite. Other highly radioactive minerals—such as monazite, allanite, pyrochlore, xenotime, uraninite, and thorite—are widespread but erratically distributed. In general, the uranium and thorium content of igneous rocks increases with the acidity.

Weathering of igneous, metamorphic and sedimentary rocks releases uranium which is later redeposited in many kinds of sediments. In the weathering environment, uranium is oxidized and released from soluble minerals. Other uranium- and thorium-bearing minerals such as zircon, monazite, xenotime, and thorite are resistant to chemical decomposition and mechanical disintegration. Zircon, the most abundant of these minerals, accumulates in placers and in the heavy mineral fraction of clastic sediments.

Potassium in rocks

The content of potassium increases with the acidity of the rocks. Dunites and serpentines are virtually free of potassium and the other alkaline metals. In general, the mica and feldspar-free igneous rocks have very low concentrations of potassium. The dispersion in values of potassium content for many rocks is quite high. For example, the range for basalts is from about 0.15 percent to more than 2 percent. Calculations of the mean potassium content of sedimentary rocks are almost meaningless, since the potassium content varies dramatically even in rocks from the same formation.

Table 1.4 gives the average values and ranges for the major rock types. Appendix 1A provides more extensive data for minerals and for the less common rocks.

Solubility of uranium and thorium

In the supergene environment, the behavior of uranium and thorium is markedly different. Thorium is distinguished in having but a single valence state, +4. Hence, its response to changes from reducing conditions to oxidizing conditions is nil. Thorium is present in natural waters primarily as a dispersion of thorium-bearing minerals. It should be noted, however, that radium 228 (a thorium daughter) can be mobilized in natural waters and displaced from the parent site. However, the half life of 6.7 years is so short that only very rapid geologic processes can cause a redistribution.

Uranium, on the other hand, responds to a near-surface oxidizing environment by changing from a +4 to a +6 valence state; it complexes with oxygen as a uranyl ion $(UO_2)^{+2}$ and forms a number of compounds soluble in a variety of surficial environments. The solubility of uranium can be significantly reduced in the near-surface zone by the formation of vanadates phosphates and organic complexes.

The radium 226 (Ra) isotope in the uranium decay sequence has a half life of 1620 years, and has time for greater movement than the 6.7 year thorium daughter.

Solubility of potassium

The major part of the potassium in soils, 90 to 98 percent, is bound up in the minerals fraction, predominantly in the feldspars and the micas. The remainder of the potassium is in the process of being liberated by weathering from mineral grains. When liberated, the potassium is either utilized in the metabolism of plants and soil organisms or is carried away in the ground water solution.

TABLE 1.4
Average radioelement content of rocks

	*K^{40} ppm	Th ppm	U ppm	U/Th	Th/K^{40}	U/K^{40}
Basaltic Rocks						
Average	0.8	4.0	1.0	.25	5.0	1.2
Range	0.2–2.0	0.5–10.0	0.2–4.0			
Granitic Rocks						
Average	3.0	12.0	3.0	.25	4.0	1.0
Range	2.0–6.0	1.0–25.0	1.0–7.0			
Shales						
Average	2.7	12.0	3.7	.31	4.5	1.4
Range	1.6–4.2	8.0–18.0	1.5–5.5			
Sandstones						
Average	1.1	1.7	0.5	.29	1.5	.46
Range	0.7–3.8	0.7– 2.0	0.2–0.6			
Carbonates						
Average	0.3	1.7	2.2	1.3	5.6	7.3
Range	0.0–2.0	0.1– 7.0	0.1–9.0			

*Chemical potassium contains 0.012 percent K-40.

Mobility of radon

There are three radon isotopes in the uranium and thorium decay series. These are listed in table 1.5. All are noble gases and hence are unconstrained by participation in chemical bonds with other elements.

TABLE 1.5
Isotopes of radon

Radon isotope	Name	Long-lived parent	Half-life
Ra-222	Radon	U-238	3.82 days
Ra-220	Thoron	Th-232	54.5 sec
Ra-219	Actinon	U-235	3.92 sec

These gases can migrate in the earth or atmosphere by two principal modes, diffusion and transport. The diffusion coefficient of most earth materials is so low that migration by diffusion is limited, to several tens of meters in the case of radon 222 (Rn). Thoron diffusion is substantially less because of its much shorter half-life. The diffusion of actinon is inconsequential because of the short half-life and the low initial abundance of the parent isotope U-235.

The second recognized mode of migration is by transport, either in air or in water. Underground water flow as a transport mechanism might be measured in distances ranging from a few feet per year to tens of feet per day. Contained radon would migrate at the same rate. The movement of air is extremely variable. Interstitial air in the earth moves at low rates in response to temperature or barometric pressure changes. Air above the ground is subject to very rapid movement in response to both local and regional climatic changes.

Emanating power of rocks

A frequently used term in radiation prospecting is the *emanating power* or the *coefficient of emanation* of a rock or mineral. When a Ra-226 atom disintegrates forming a Rn-222 atom and an alpha particle, the recoil of the disintegration propels the radon atom along a path length of roughly 3×10^{-6} cm before it is stopped. Therefore, if the mineral grain containing the radium atom is smaller than about one micron, the recoil radon will arrive in the interstitial space between mineral grains. That rock or mineral is then said to have a high emanating power. Of course, in a highly compacted, fine-grained rock, the recoil atom may find itself buried in an adjacent mineral grain resulting in little emanating power.

Depletion of radon

Figure 1.7 shows the depletion of radon by diffusion in saturated and unsaturated porous alluvium and in rock. In the unsaturated case, the near-surface radon may be depleted by diffusion for a depth of several feet. The absolute amount of depletion is a function of the diffusion coefficient and the emanating power of the mineral grains. If, for example, 25 percent of the recoiling radon atoms are liberated from mineral grains, the depletion pattern would be as in curve B. If the emanating power of the mineral grains is 50 percent, curve A might result. In the case of water-saturated porous alluvium, the diffusion is substantially reduced by the water, and the earth is depleted only to a depth of several inches. In a tight rock, the depletion depth is further reduced.

The combination of a porous, medium- and high-emanating power results in a depletion of Bi-214 from the upper layer of the earth. Because the surface layer acts as a shield for gamma rays from the deeper layers, the gamma activity from the surface of the earth may be greatly reduced through depletion of the near-surface radon.

FIGURE 1.7
Depletion of radon-222 in geologic sources

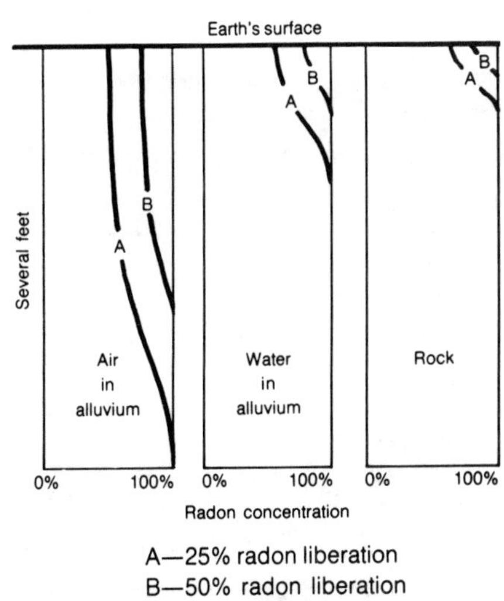

Effects of barometric pressure on radon

Figure 1.8 illustrates the effect of changing barometric pressure on the distribution of radon in soil air. Under stable atmospheric pressure conditions, either high or low, the atoms of radon assume a rather uniform distribution by diffusion, with the only perturbation resulting from depletion near the earth/air interface. During a period of increasing barometric pressure, the atmospheric air acts as a piston and forces the radon gas down through the pores of the earth. During a period of decreasing barometric pressure, the piston effect is reversed and the radon is drawn from the earth.

Atmospheric inversions

The uranium daughter (Ra-226) in the near-surface soils constantly emits a flux of radon to the atmosphere. Under conditions of temperature inversion, the atmospheric radon becomes trapped and accumulates in a low-lying layer of stagnant air. Such inversions are particularly troublesome for airborne radiometric surveying. The methods of dealing with this problem will be discussed later in this chapter.

Cultural radiation sources

Nongeological or cultural sources of radiation must be considered during radiation surveys. Cultural contamination is comprised of low-level activity in the detection equipment, of radium watch and instrument dials, and of some thorium-containing camera lenses. An important source of nongeologic signal is fallout from atomic explosions, whose products are carried into the upper atmosphere where they may be distributed over large areas of the earth. They constitute a source of background noise until the radioactive products decay below the levels of detection.

Additional sources of radiation that are not geological result from the interaction of man and geologic environment. These are wastes from radioactive disposal facilities and tailings from mines and mills. Tailings have been carried down drainage many miles from their original source and cause spurious anomalies.

Airborne survey procedures

Airborne survey procedures are especially important because of the costs involved in correcting erroneous procedures or reflying a project area. In planning a survey, it is important to understand as much as possible about the geology of the survey area, including lithology, strike and dip, possible mineralization or metamorphism, topography, character of overburden, and depth of weathering.

In addition to the geology, many other factors influence the accuracy and usefulness of the survey results. These factors may be in the ground, in the air, or in the detector system and the aircraft. Additional factors influencing usefulness of the survey are to be found (1) in the reduction procedures that are applied to the recorded data and (2) in the presentation of the data so that they can be considered in a geological context—not simply as an array of numbers.

The remainder of this section discusses the dominant factors to be considered in airborne surveys.

Radioelement concentration

The radioelement content of the ground influences the radiation intensity in direct proportion to the abundance of the isotope, if all other factors are assumed constant. The isotopic concentration is the information we wish to determine from a radiation survey.

FIGURE 1.8
Effect of changing barometric pressure on Radon-222 in soil air

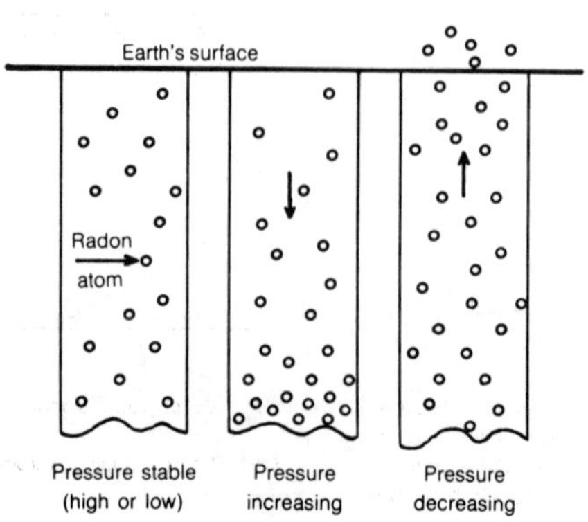

Thickness of radiation source

The thickness of the source also influences the intensity of radiation. For outcrop areas, rock thickness in excess of one foot is an infinitely thick source. Any additional thickness makes no contribution to the detector. For loosely consolidated soils, an *infinitely thick source might be 18 in. to 2 ft thick.* The effective absorption is modified by the moisture content of porous materials; the greater the moisture content, the greater the absorption.

Gamma ray absorption vs. energy

The absorption of gamma rays is also a function of their energy. A K-40 source having a primary energy line of 1.46 Mev is more readily absorbed and scattered than the gamma rays from a thorium source having an energy of 2.62 Mev.

Buried radiation source

If the radioactive source is buried by nonradioactive alluvium, water, or rock, the intensity of the source may be drastically diminished to the extent that it is undetectable. *One foot of rock, 18 in. to 2 ft of soil, or 3 ft of water effectively obscures underlying radiation sources, no matter how intense.* The important exception is that radon migration through overlying material can give evidence of buried sources.

Circle of influence

The radiation intensity is a function, among other things, of the horizontal extent and the thickness of the radiation source. In airborne surveys, an effectively infinite source might be 2000 ft wide, whereas in ground surveys a few tens of feet might constitute a source of effectively infinite horizontal extent. Figure 1.9 shows the circle of investigation from various altitudes of the aircraft above ground. For example, at a terrain clearance of 400 ft, 60 percent of the incoming data is from the circle with the radius of 460 ft. For an aircraft at 100 ft, 60 percent of the incoming information is from a circle with the radius of 180 ft.

FIGURE 1.9
Percent of total signal coming from circle below detector as a function of aircraft altitude

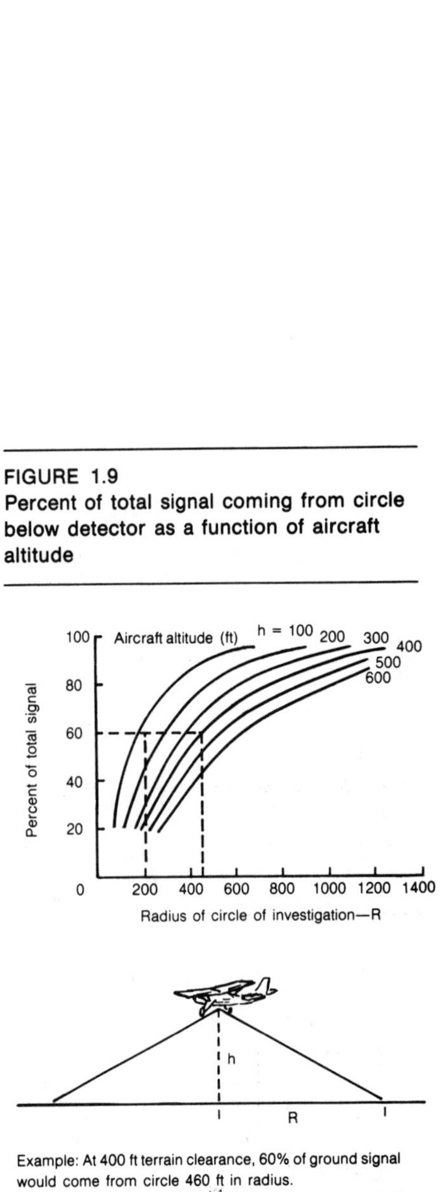

Example: At 400 ft terrain clearance, 60% of ground signal would come from circle 460 ft in radius.

At 100 ft terrain clearance, 60% of ground signal is from circle 180 ft in radius.

Effect of rainfall on radiation

The story of radon and rainfall is one of competing effects that can either enhance or subdue the radiation from the ground.

Radon escapes more freely from dry soil than from wet soil. Therefore rainfall can have the effect of inhibiting the escape of radon and thus *increasing* the intensity of the radiation from the ground. A persistent rain, however, can saturate the ground and drive the radon deeper into the ground, thus increasing the absorption due to the water content of the ground. This *decreases* the radiation from the ground.

If the rainfall is contaminated by atmospheric radon and radon decay products, these may completely override the other two effects and produce a strong *increase* in radiation from the ground. It has been observed that the recovery time after a rain is about three hours, corresponding to the decay rate of the radon daughters lead 214 (Pb) and Bi-214. The daughter products are probably attached to dust particles which are washed from the air during rainstorms.

Effect of snow on radiation

Snow and frozen ground can effectively block the normal flux of radon from the ground into the atmosphere. This has the effect of increasing the gamma emission of the ground because of the radon trapped beneath the snow. During airborne surveys, it has sometimes been observed that the snow-covered north slopes of the ridges exhibit a higher Bi-214 count than the sunny slopes on the south—presumably due to entrapment of radon under the snow. However, if the snow pack is very thick, say 10-20 ft, the inverse effect is observed as the thick snow cover absorbs the radiation. Figure 1.10 illustrates the effect of thick and thin snow cover, and other nongeologic topographic effects.

FIGURE 1.10
Nongeologic topographic effects

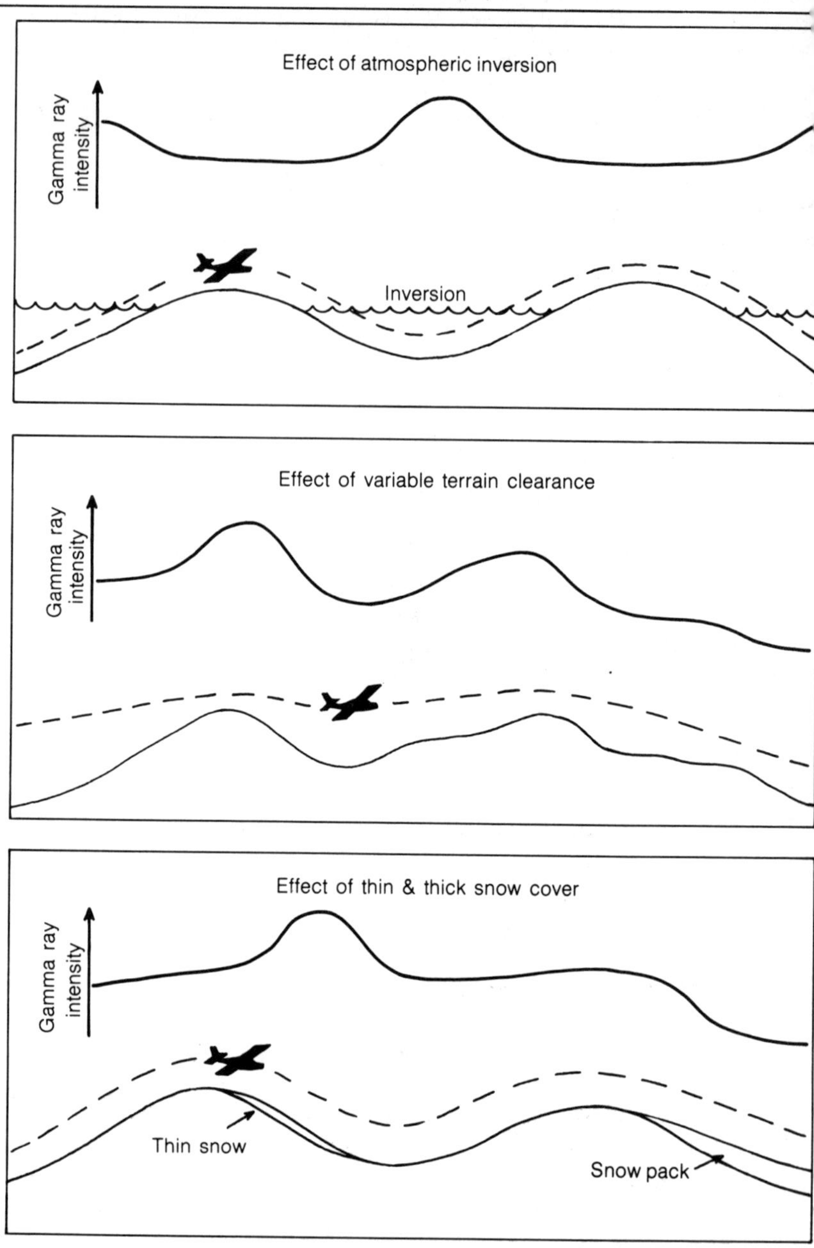

FIGURE 1.11
Sample of Bi-214 count recorded at ground station showing diurnal variation and rainfall effect

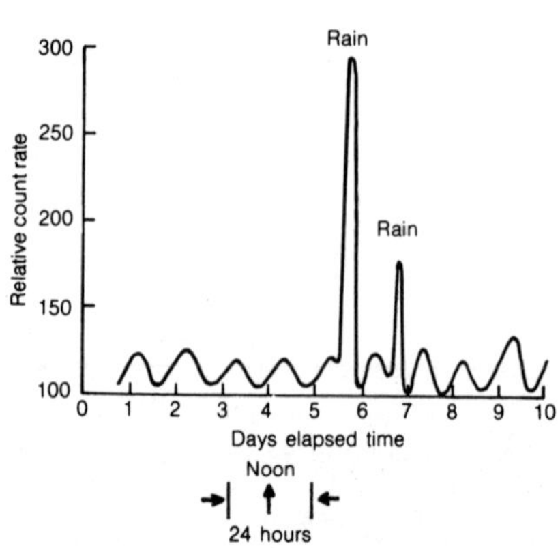

Radon and atmospheric inversions

During periods of wind or convective air movements, the radon escaping from the ground is mixed thoroughly with the air and forms a uniform background. However, during periods of temperature inversion, the escaping radon is retained in the blanket of cool air near the ground. The radon is continually accumulated until the inversion layer is broken up by winds or convective mixing of the air. For airborne surveys, an inversion layer then constitutes an inhomogeniety and a source of variable background signal that substantially reduces the accuracy of the surveys. It is necessary to wait until any atmospheric inversion has dissipated before making the tests for bismuth content of the air and flying survey profiles. Experience has shown that the atmospheric Bi-214 count can be up to 75 percent of the total bismuth count. This means the airborne component of the uranium daughter is as much as 300 percent of the component originating in the ground.

Figure 1.11 shows the diurnal modulation due to daily temperature inversions and the effect of rainfall which may result in a several hundred percent increase in bismuth count. Obviously, the inversion effect constitutes a very substantial source of extraneous signal and must be accounted for if precise survey results are desired.

Information about the presence or absence of inversions can frequently be obtained by calling the local airport or, more reliably, simply by making a test flight. The top of the inversion is normally quite visible as one slowly increases altitude. Figures 1.12a and 1.12b illustrate the effect of various detector locations with respect to the ground surface and atmospheric inversions.

Removing effects of atmospheric radon

The radon content of the air is a source of background noise and must be removed from the data. The effects of the radon content are commonly accounted for using one of several methods. One method uses a lead shield underneath one of the crystal detectors in the aircraft. The lead-shielded crystal is called a *2 pi crystal;* it records gamma radiation from the upper surface only. Additional detectors on the aircraft are unshielded and receive radiation from all directions. These are called *4 pi crystals.* The Bi-214 count from the 2 pi crystal is subtracted from that of the 4 pi crystals giving an effective Bi-214 count from the ground.

The effects of atmospheric Bi-214 may also be estimated by making periodic test flights over lakes. Three feet of water effectively shields any radiation from the earth, and at distances sufficiently far from shorelines there is no effective radiation from the side. The Bi-214 count rate is accumulated over the water. A linear interpolation is then made between each over-water reading, and interpolated values are subtracted on a time base from the survey data.

Alternatively, one may fly at high altitudes, say 3000 ft above the ground, where the influence of the ground is negligible. Any Bi-214 count observed is then attributed to the atmosphere and is subtracted from subsequent flight data. It must be kept in mind that the effects of cosmic rays increase at high elevations.

Effect of surveying speed

Other factors which influence the precision of radiometric measurements are those related to the properties of the detection system and the vehicle which transports it. The count observed during any specified period of time in any particular radiation environment is directly proportional to the volume, within certain limits, of the crystal detector. In airborne surveying the maximum count that may be obtained from any section of flight line is determined by the minimum aircraft speed and crystal volume. The precision of measurement is related to the count.

For example, if the aircraft speed is 250 feet per second and the average net count in the thorium channel is 4 counts per second, then the thorium count over 1000 ft of flight line will be 16 counts. The fractional standard deviation is from equation (1.12).

$$\sigma = \frac{1}{(16)^{1/2}} = \pm 0.25 = \pm 25\%$$

By increasing the crystal volume by a factor of 6.7, we would obtain 100 counts over the same 1000-ft interval, and increase the confidence level to ±10 percent. Note the very large increase in crystal volumes; hence, the cost required to reduce the error (standard deviation) from 25 percent to 10 percent.

There are other options to decrease the standard deviation such as slower aircraft speed and lower terrain clearance. If the same 1000 ft of profile were flown by helicopter in 18 seconds instead of 4 seconds the

FIGURE 1.12a
Contributors to total count rate at various detector locations

Sources of radiation at aircraft detector

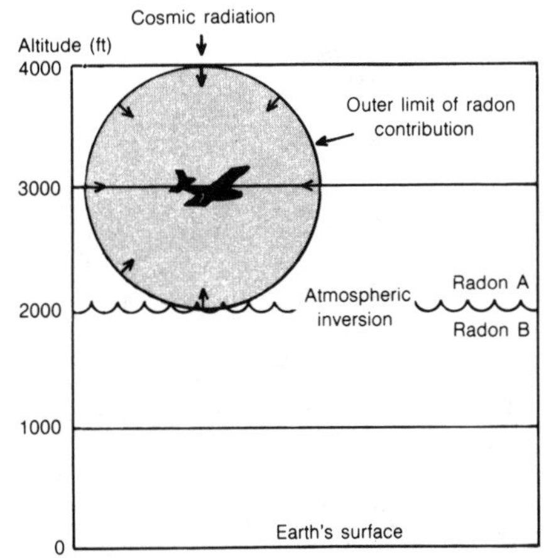

Cosmic at 3000 ft
Radon A
Aircraft background
Ground (0)

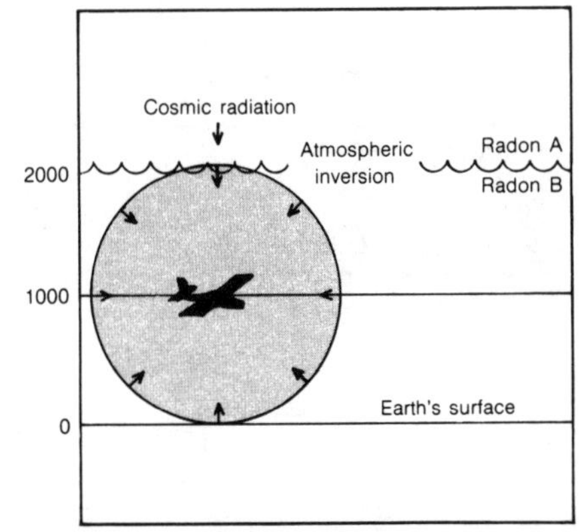

Cosmic at 1000 ft
Radon B
Aircraft background
Ground (0)

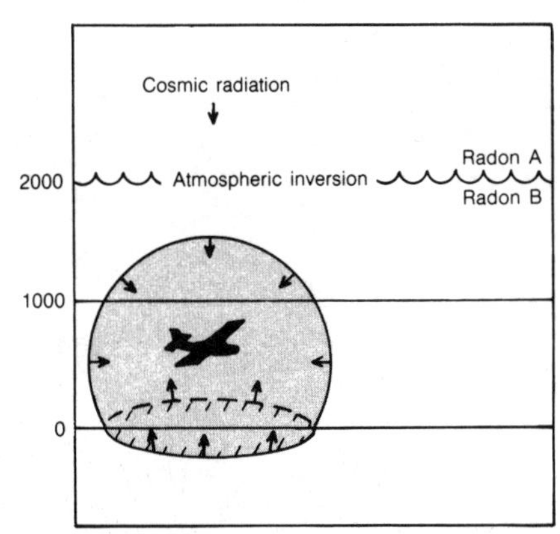

Cosmic at 300 ft
Radon B (less truncation effect)
Aircraft background
Ground (truncated area)

FIGURE 1.12b
Contributors to total count rate at various detector locations

Sources of radiation at detector

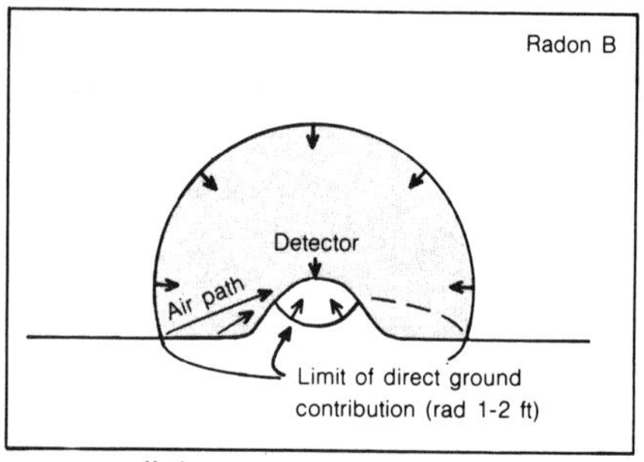

—on small rise—
Cosmic (0)
Radon B
Detector vehicle background
Ground (direct & air path)

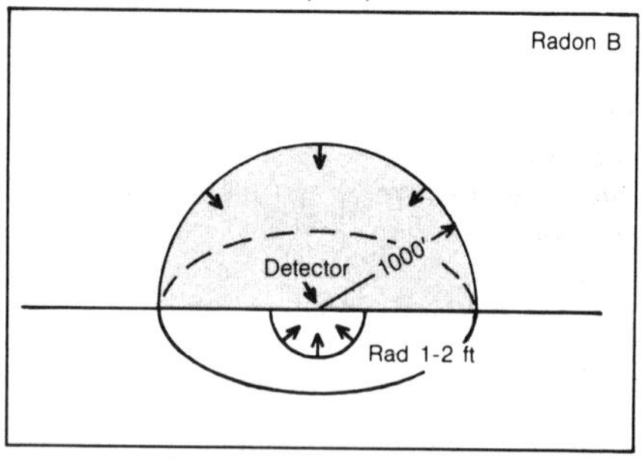

—at ground level—
Cosmic (0)
Radon B
Detector vehicle background
Ground-hemisphere (rad. 102 ft)

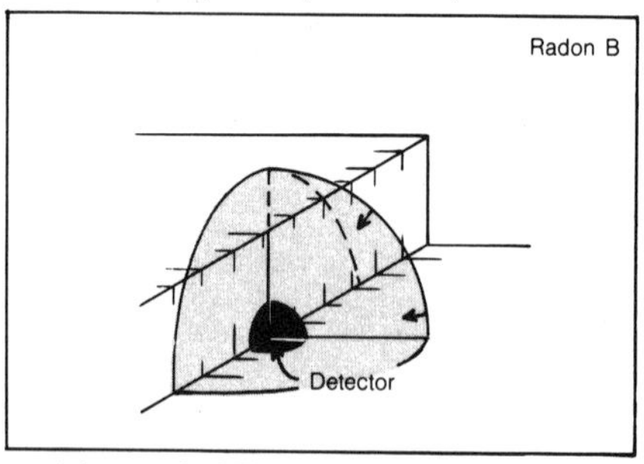

—at base of cliff—
Cosmic (0)
Radon B (½ hemisphere)
Detector vehicle background
Ground—1½ hemispheres (rad. 1-2 ft)

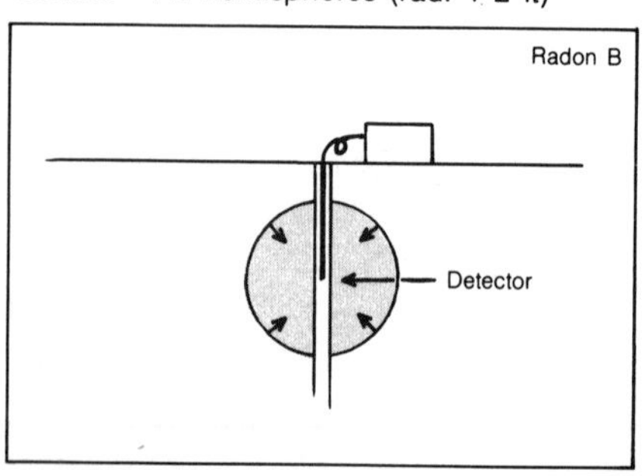

—in drill hole—
Cosmic (0)
Atmosphere (0)
Detector background
Ground-sphere (rad. 1-2 ft)

count would be 72. The confidence level would be ±12 percent. Because the gamma ray signal from the ground falls off exponentially with altitude, the net count could be increased by flying lower. However, safety considerations set a lower limit to the reduction of terrain clearance. Figure 1.13 shows the falloff of gamma ray intensity as a function of altitude for both large and small sources.

Background of instrument and aircraft

Background radioactivity of the detection system and the aircraft can be a significant source of extraneous signal. Sources of background are minute amounts of radioactive substances in the structure of the instrument and of the aircraft. Particular attention must be paid to radium instrument dials and clocks and to camera lenses that may be used for the aerial spotting camera. Many high quality camera lenses contain thorium oxide as a clarifying agent and therefore must be shielded or replaced when used with radiation equipment.

Terrain clearance effects

Topographic terrain clearance effects may add substantially to the uncertainties of airborne surveying. In a uniform geologic environment, topographic highs result in positive radiation response in uncorrected data. Various mathematical methods are employed for reducing the effects of terrain clearance. The corrections are based upon the assumptions (1) that the source is a uniform half-space, a two-dimensional body, or a point source; and (2) that changes in density of the air and changes in elevation above sea level during the course of the survey are negligible. Figure 1.10 illustrates some of the nongeologic topographic effects encountered in airborne radiometric surveys.

Flightline direction

There are times when it is advisable to contradict the conventional wisdom of flying survey profiles at right angles to the geologic strike.

As an example, suppose that the outcrop width of a geologically significiant formation was only 200 ft. This would represent about one second of data accumulation in most aircraft operations. Unless the source was very strong, there would be a very small chance of obtaining statistically significant data. By crossing the geologic strike at a small angle—about 20-30 degrees—the traverse time over the same true width can be increased to 2 or 3 seconds.

Topographic considerations may also dictate flight line direction. If steep topography prevents maintenance of low ground clearance when crossing strike, it may be desirable to lay out the flight lines to best maintain ground clearance and ignore the geologic strike. In terms of aircraft performance, the rate of climb in feet per mile is far more significant than the normal aircraft specification of feet per minute. A survey should be designed so that terrain clearances are not in excess of 500-600 ft, or kept to an absolute minimum. Above that altitude, the ground signal is normally too small to have statistical validity even with the best systems available today.

In airborne surveying, probably the dominant factor is terrain clearance. There are many advantages in maintaining minimum elevation. The earth signal is stronger because of the exponential falloff of the atmospheric attenuation. The atmospheric radon problem is, therefore, proportionally less. Ground resolution of small lithologic features is substantially increased.

Test strip

In every flight area, it is advisable to select a test strip—approximately 5-6 mi long. This should be along a readily recognizable route such as a fence line or a road for ease of navigation. Test flights should be made daily to verify operation of the entire system. For high resolution spectrometer work, the test strip should be flown at approximately 200-ft increments in altitude from 200-1000 ft. This is particularly useful in monitoring the aircraft background radioactivity and the absorption coefficients used in the computer reduction program. Flights over the same path at different altitudes, when fully corrected and reduced, should yield spectrometer profiles that are the same within the statistical noise envelope.

It has been found that sometimes the aircraft background changes. The exact cause is not known; however, it might be due to electrostatic pickup of airborne radon or radon decay products.

Ground monitor

A ground station consisting of an upward looking 2 pi crystal provides an efficient means of monitoring the atmospheric radon levels in the area of a survey. It is not sufficient for providing information to correct the airborne survey data.

Integration time

Integration time refers to the amount of time that pulses are counted in a spectrometer before dumping into the recording system, either digital or analog. In airborne survey work, the standard integration time is frequently one second. This is compatible with magnetic tape recording of data for later computer processing. At the computer stage, the data from incremental time units can be added or weighting functions can be applied in order to improve the statistical reliability of the data. Some airborne systems operate on a fixed-distance base (variable time). The latter require conversion of the data to time base in order to apply the sensitivity coefficients and convert to isotopic concentration.

FIGURE 1.13
Decrease of gamma ray intensity with altitude

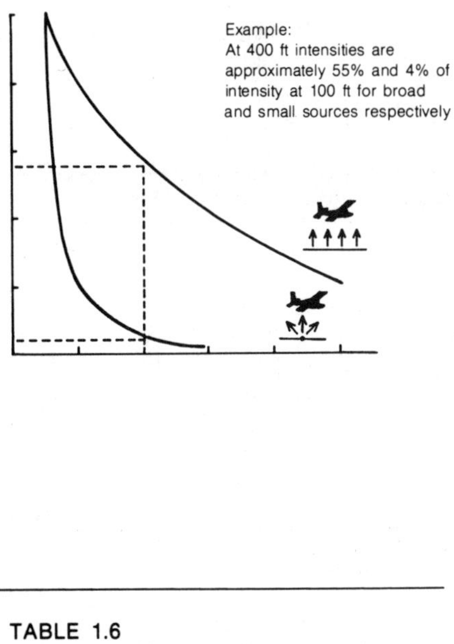

Radioactive fallout

Fallout products of thermonuclear fission can at times make a significant contribution to the total gamma ray flux. There are more than 100 radionuclides produced in fission and in the neutron reactions accompanying fission. Some of the more prominent products are listed in table 1.6. Note that there are few gamma emissions in the high energy region, particularly among the long-lived products. Most of the contribution is in the low-energy range.

Airborne systems—general

This section is intended to provide examples of instructive field experience with gamma spectrometers. There is a definite dearth of published material on spectrometric surveys. This is attributed to the

TABLE 1.6
Prominent Fission Products

Fission Product	Half-life	Beta Energy MEV	Gamma Energy MEV
Sr-89	50.5 days	1.50	—
Sr-90	27.7 years	0.60	—
Y-90	64.2 hours	2.26	1.75
Y-91	57.5 days	1.53	1.19
Zr-95	65 days	0.39	0.73
Nb-95	35 days	0.16	0.76
Ru-103	40 days	0.2	0.56
Ru-106	1.00 years	0.03	—
Rh-106	130 minutes	0.7	1.56, 1.23, 1.07, 0.80, 0.74
I-131	8.08 days	0.60	0.37
Cs-137	26.6 years	0.5, 0.8	0.66
Ba-140	12.8 days	1.0, 0.4	0.5
La-140	40 hours	2.2, 1.7	1.6, 2.3
Ce-144	285 days	0.3, 0.18	0.13, 0.08

fact that accurate, high sensitivity equipment has not been readily available. The Geological Survey of Canada, under Dr. Arthur Darnley, director of Resource Geophysics and Geochemistry, has developed a high sensitivity system utilizing twelve 9 × 4-in. (3000 in^3) crystals mounted in a Skyvan aircraft. In the United States, comparable crystal volumes are being flown by Geodata International, Inc., and Geophoto, Inc., of Dallas, Texas. The latter systems have lead-shielded crystals for the removal of airborne radon effects. Similar systems are reported under development by other parties.

System selection

Frequently, topography is the most significant factor in selection of the system. A high sensitivity spectrometer system may weigh several thousand pounds and require use of a twin-engine aircraft such as the DC-3. Such an aircraft could not maintain the required terrain clearance in mountainous areas. The choice might then be a light fixed-wing aircraft used in the gross count mode to outline higher radioactivity regions, to be followed by a helicopter-borne high sensitivity spectrometer to provide data on isotopic composition of the sources.

Various methods of estimating the cost effectiveness of radiometric systems have been devised. However, the choice of system often is determined by the elimination of certain classes of systems; final selection of a specific system is then made on the basis of cost and availability. In general, one should select the system that will provide the smallest percentage standard deviation per unit of distance on the ground. This creates a prejudice toward large detector volume, slow aircraft speed, and small terrain clearance. There are, of course, many other factors to be considered such as corrections for Compton scattering and for aircraft altitude. Table 1.7 lists the various characteristics of systems classified by crystal volume, and by gross count or spectrometer mode of operation.

TABLE 1.7
Characteristics of airborne systems

	High Sensitivity Spectrometers Fixed Wing	High Sensitivity Spectrometers Helicopter	Medium Sensitivity Spectrometers Fixed Wing	Spectrometer Fixed Wing Helicopter	Total Count Fixed Wing
Detector Volume, in^3	+3000	300 to 1000	500 to 1000	300 to 500	100 to 200
Aircraft Performance					
Speed mph	120	30–60	100–150	100–150	100–150
Terrain clearance	400	100–300	200–500	200–500	200–500
Rate of climb, ft/mile	Poor	Excellent	Good	Good	Good
Operational range—hrs	5–12	2–4	4–6	4–6	4–6
Instrumentation					
Correct for altitude	+	+	+	+	0
Correct for cosmic	+	+	0	0	0
Correct for aircraft background	+	+	0	0	0
Correct for atmospheric Bi-214	+	+	0	0	0
Correct for air density	+	+	0	0	0
Correct for Compton scatter	+	+	+	0	0
Internal stabilization	+	+	+	0	0
Digital recording	+	+	0	0	0
Analogue recording	+	+	+	+	+
Flight path recovery	+	+	+	+	+
Data Presentation					
Stacked profiles by isotope, or ratio	+	+	0	0	0
Contour data by isotope, or ratio	+	+	0	0	0
In flight analogue	+	+	+	+	+
Most Useful for					
Airborne geochemistry and geologic mapping	+	+	0	0	0
Finding localized ore outcrops	0	+	0	+	0
Finding weak broad halos	+	+	0	0	0
Recon for high radioactivity areas	+	+	0	+	0

Advantages of high sensitivity

There are valid geological reasons for the growing emphasis on large crystal volume in gamma ray prospecting. During the early days of uranium exploration, there were no well-established geological-geochemical models to guide the search for ore. However, at present the geological-geochemical characteristics of uranium ore provinces are much better understood; and, in some parts of the world, the most favorable areas have been well covered with Geiger counters and airborne scintillators. Therefore, there are definite advantages with the use of more sensitive techniques to identify favorable environments and to map low-intensity alteration or dispersion halos.

Computer processing

While manual calculation is satisfactory for processing ground spectrometer data, the use of computer-processing is almost mandatory in airborne work because of the large number of data points involved. Each data point must be processed for Compton stripping; and altitude and terrain clearance corrections must be applied. By computer techniques, the statistical validity of the data can be manipulated using various methods of combining adjacent data values. The use of the computer-processed data in producing isotopic ratios is a particularly valuable procedure because isotopic ratios are commonly more diagnostic of geological variations than are the absolute values of the isotopic abundances.

Lack of line-to-line correlation

Quite unlike the potential field methods of gravity and magnetics, one does not often find good line-to-line correlation even though line-to-line correlation of the geology may be excellent. The depth of investigation of the radiometric method is limited to the upper foot or so of the earth. Thus, local inhomogenieties in transported overburden due to wind or stream channels can obscure a radioactive source. Uranium under oxidizing conditions is especially prone to movement from the surface environment. Thorium, on the other hand, is normally found in the fine-grained resistates and potassium normally stays with the essential minerals—the feldspars and micas. The result of inhomogenieties in transported overburden and the relatively facile movement of uranium is a lack of line-to-line correlation, particularly for mineralized zones and narrow geologic sources.

Use of ratios

In many geologic environments the radioactive isotopes vary sympathetically in abundance; in particular, the uranium and thorium. This observation is so general that it is frequently worth geologic examination of areas where the uranium/thorium ratio varies substantially from normal. Particularly in uranium exploration, the uranium/thorium ratio has been found to increase near uranium deposits.

Example of limitations of airborne systems

The results of flying the high sensitivity mapping system in a *favorable* situation (above normal radioactivity and low atmospheric background) and in an *unfavorable* situation (below normal radioactivity and high atmospheric background) are shown in table 1.8, taken from a Geological Survey of Canada publication.

Table 1.8 yields the very important conclusion that, even with a 3000-cu in. detector volume in a favorable signal-to-background situation, one standard deviation is still in excess of ±25 percent during a 2.5 second data accumulation period. During that period, the Skyvan aircraft would travel approximately 440 ft. This represents the maximum ground resolution. Of course the standard deviation could be improved for any fixed crystal volume by flying lower and/or slower or by increasing the integration time. The first two constitute safety hazards, and a longer integration time destroys spatial resolution.

TABLE 1.8
Flight tests of a high sensitivity GSC system

Channel	Gross Count	K	U	Th	U/Th
Sample interval	0.5 sec	2.55s	2.55s	2.55s	ratio
Favorable situation: high radioactivity, low background					
Atmospheric background	220	95	40	38	—
Corrected count rate	1000	400	50	150	0.33
Standard deviation	±35.0	±25.2	±13.6	±13.9	0.09
Percent SD	± 3.5	± 6.3	±27.2	± 9.3	28.5
Unfavorable situation: low radioactivity, high background					
Atmospheric background	325	130	70	38	—
Corrected count rate	400	150	20	30	0.66
Standard deviation	27	18.7	11	8.5	0.4
Percent SD	7	12.5	55	28	62.5

Ground survey procedures

There are substantial differences in the factors that influence the accuracy of ground and drill-hole surveys relative to airborne surveys. The differences are largely the result of (1) the increase in ground signal strength and resolution with decreased altitudes (in the case of surface surveys) and (2) complete shielding of surface-related factors (in the case of drill-hole surveys). Table 1.9 summarizes the relative importance of the various factors in airborne surface and drill-hole surveys.

Station vs. continuous surveys

In ground surveys there are two general modes of operation. One is the continuous recording of data from a moving vehicle. Data can be monitored by an instrument operator, or a permanent record can be made by a chart or tape recorder. The second mode is station-by-station survey by hand-held or back-pack mounted equipment.

TABLE 1.9
Factors influencing accuracy of gamma ray spectrometer measurements

Factor	Airborne	Surface	Drill Hole
Abundance of isotopes in ground	M	M	M
Airborne Bi-214	M	m	0
Atmospheric inversions (inhomogenieties in Bi-214)	M	m	0
Radon depletion near surface	M	m	0
Recent rainfall effect (washout of airborne Bi-214)	M	M	0
Moisture content of ground	m	m	0
Snow and frozen ground	M	M	0
Nuclear fallout	M	m	0
Cosmic rays	M	m	0
Cultural effects (dials, watches, mining activity)	m	m	0
Vehicle speed	M	M	M
Crystal volume	M	M	M
Detector elevation	M	m	0
Topography	M	m	0

Major factor = M
Minor factor = m
No significance = 0

The truck-mounted mode permits rapid coverage but suffers lack of access in many areas. Traverses along roadways always offer the hazard of contamination or masking (1) by foreign materials in the roadbed or (2) from radioactive material accidentally dropped from ore trucks. Even when roadbeds are constructed of native materials, the removal of soil from the borrow pit can produce distortions in the gamma field by exposing deeper horizons or by masking when water filled.

The area of influence of the detector may be increased by mounting it on top of the vehicle. This subdues the effect of inhomogenieties of roadbeds, borrow pits. Needless to say, one should provide electronic or manual means of recording check points, cultural features and geologic features encountered along the profile. Changes in roadbed are particularly important—for example, changes from surfaced to unsurfaced roads, change in the character of the fill, roadcuts.

Detector elevation

Where should an instrument be placed with respect to the ground surface? If data are desired on a small outcrop or lithologic unit a few feet or so in lateral dimension, the optimum location is on the outcrop. Several closely spaced readings should be taken in order to estimate the geologic variations. For a larger lithologic unit, the instrument should be raised to a height slightly above positive irregularities within a few tens of feet of the detector (see figure 1.14). For still larger geologic units, a detector height of 3-5 ft (e.g., back-pack mounted sensor) is more appropriate. The back-pack sensor has the added substantial advantage of permitting hands-free operation. In using truck-mounted equipment, certain advantages accrue by mounting the detector on top of the car. The sample size is large, and contribution of foreign road surfacing materials is reduced relative to the undisturbed materials at roadside.

Background of instrument and vehicle

In ground traversing by vehicle and foot, contamination problems are possible. A vehicle may be contaminated by rock or ore specimens in the truck. Frequently, it is too easy to forget the small button of radioactive material that accompanies the instrument as a calibration source. It can provide a variable source of background error if carried by the operator or too near a truck installation.

Survey closure

For accurate ground surveys, it is advisable to run traverses in closed loops, starting and ending each traverse at exactly the same spot. Closure errors can be caused by instrumental drift, changes in atmospheric radon content or change in the emissivity or absorption characteristics of the ground caused by rainfall.

Effective sample size

The effective area and volume sampled by a spectrometer depend upon several factors including the height of the detector above the surface, energy of the gamma ray, absorption coefficient of the ground, and whether or not the detector is shielded or collimated. The sample size for an instrument on a flat rock surface has been defined in a manner such that the effective sample contributes only about 60 percent of the detected gamma rays. Figure 1.15 illustrates the effective sample configuration from an uncollimated detector centered 5 cm above a flat rock surface. The size and weight of the effective sample are given in table 1.10 for 1.76 and 2.62 Mev. Corresponding absorption coefficient values were 0.123 cm^{-1} and 0.111 cm^{-1} respectively.

If we considered a sample size that would contribute 90 percent of the detected rays instead of 60 percent, the sample volumes would be approximately ten times as great.

FIGURE 1.14
Location of detector with respect to local ground roughness

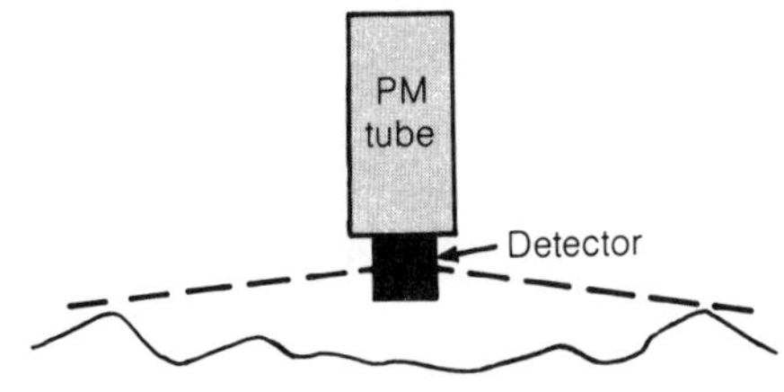

FIGURE 1.15
Effective sample configuration with detector 5 cm above flat rock surface

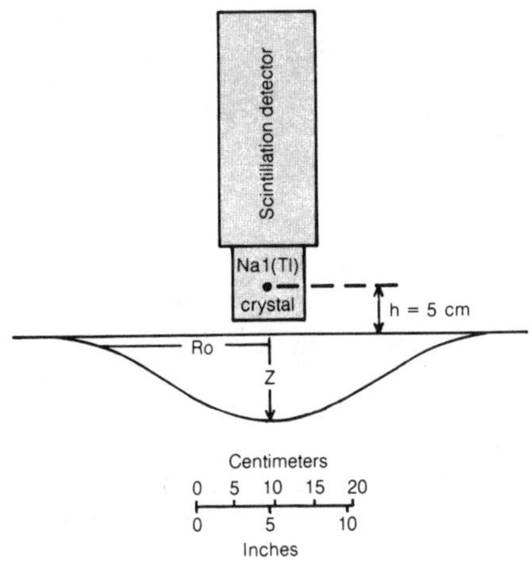

TABLE 1.10
Mass, radius and depth of effective sample obtained with detector centered 5 cm above a flat rock surface

Isotope	Kg/lbs	Ro Cm/in	Z Cm/in
Bi-214	37/81	40/16	12/4.7
Tl-208	49/110	42/17	14/5.5

In situ analysis for uranium, thorium and potassium

Effective use of the gamma ray spectrometer as an instrument for analysis of uranium, thorium, and potassium content of the earth depends upon verifying or assuming that secular equilibrium exists in the uranium and thorium series. It is normally assumed that the K-40/K-39 ratio is constant in nature. Successful assaying depends upon an appreciation by the operator of the geometric requirements for good data. The count rate observed at any particular site is strongly influenced by the solid angle, or the circle of influence, "seen" by the detector.

Most field gamma ray spectrometers are calibrated for a 2 pi solid angle. In flat terrain, an approximate 2 pi angle can be readily attained. In irregular terrain, the detector should be held at an elevation slightly above the level of local irregularities. In exceedingly rough terrain, the 2 pi solid angle is difficult to estimate. In such cases, it is desirable to use a collimator.

A *collimator* consists of a 2-3 in. thick cylinder of lead surrounding the detector crystal. The collimator substantially reduces the circle of influence of gamma rays, and therefore the effect of local topographic irregularities. Separate stripping and sensitivity constants must be used with a collimator. Also, the collimator adds substantially to the total weight of the spectrometer system and may be several times the weight of the rest of the equipment.

Station spacing

Station spacing in ground radiometric surveying, as in any other geophysical technique, is dependent upon the desired geological resolution. Therefore, station spacing might be from 10-1000 ft depending upon the geological resolution desired.

Integration time

The integration time in spectrometer surveying is determined by the degree of accuracy required in the results. By reference to equations 1.12 and 1.13 and figure 1.16, one may determine the standard deviation of any particular observation and may prolong the integration time to achieve any desired precision in the measurement. For example, using a 3×3 in. crystal in a back-pack mounted position, one normally would obtain quite satisfactory field results by a 3- to 5-minute integration time. This observation may be taken in one spot while traversing or while walking in a circular or zigzag path, depending upon one's philosophy about sampling that particular environment.

Use of ratios

In considering field data, it is frequently useful to present data in terms of isotope ratios as well as in terms of absolute abundances of isotopes. For example, if data are taken at the base of a cliff or on a point of topographic elevation, the stripping ratios and the sensitivity constants determined by a half-space are not strictly applicable. By using ratios, the effect of erroneous stripping ratios is substantially reduced.

Laboratory determination of hand specimens

Hand specimens must be crushed and homogenized before they can be accurately analyzed by gamma ray spectrometry. The homogenized sample is put in a standard-size container, and the container is carefully sealed to prevent contamination of the spectrometer. Assaying of the specimen is accomplished by a long period count in a shielded environment. Shielding is normally 10 cm of low radioactivity lead or 15 cm of low activity iron.

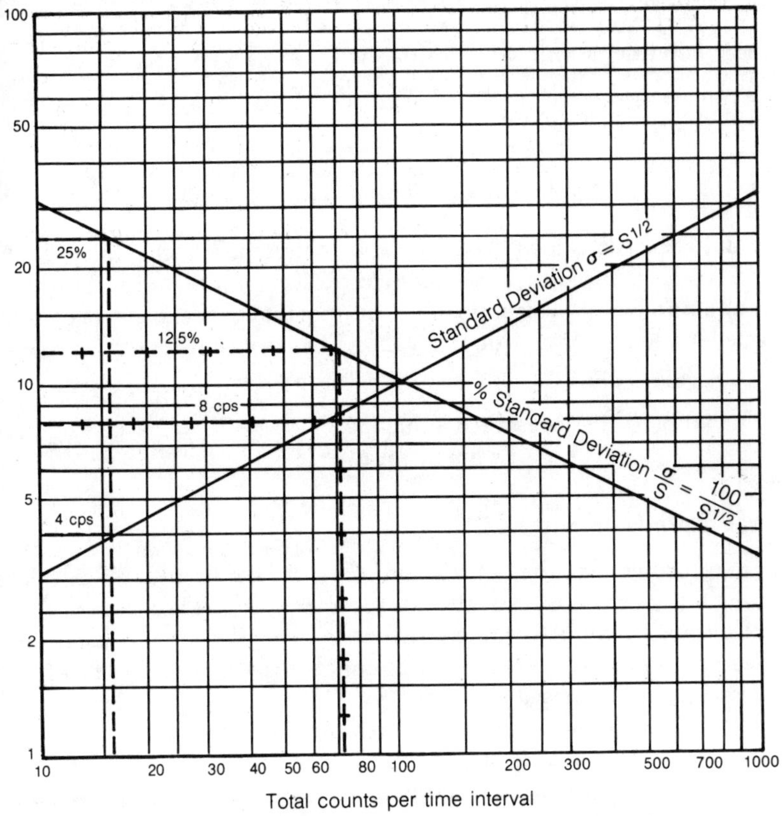

FIGURE 1.16
Curves for finding standard deviation (estimation of error)

The gamma ray spectrum observed in the laboratory is essentially different from that observed in the field. In the field, the gamma spectrum is strongly influenced by the interaction of gamma rays emitted within the rocks. The interaction results in a strong Compton continuum that is absent in laboratory work where the sample is of a restricted volume.

The crushing and homogenizing process liberates the radon contained in the rock and therefore much of its gamma radioactivity. There are two methods of correcting for the loss of radon. One is to store the sample for two weeks in a sealed container so that the radon buildup is back to the normal state. The second method is to make the reading at two separate times after the sample container is sealed. The final value is obtained by extrapolation from the two determinations on the basis of the known half-life of radon.

Data reduction

Errors in radioactivity measurements

It is exceedingly important to have a firm idea as to the reliability of radioactivity survey data. A wrong assumption of the probable error or the accuracy of a radioactivity measurement can lead to useless expenditure of time and funds in follow-up of false anomalies. There are many errors and uncertainties in radiometric work. Some errors are inherent in the measurement of radigactivity; they are entirely due to the nature of the radioactive disintegration process. Still other uncertainties are introduced by the interaction of the radiation with matter between the source and detector, and by independent radiation sources in the intervening distance.

Systematic and random errors

The errors from all sources may be classified as systematic errors or random errors. The effect of systematic errors can be minimized by careful procedures and use of proper calibration and stripping constants. Even without systematic errors, a set of measurements under supposedly identical conditions will yield values with some variation. When the variations are considered from a statistical probability viewpoint, the results of the measurements can be expressed with an assigned probability of error. The error is expressed as follows:

$$\text{Error (E)} = \text{Measured value (M)} - \text{True value (T)} \qquad (1.10)$$

In reality, the True value (T) can never be determined with certainty. In the real world, the best approximation of the True value is the arithmetic mean of a number of measurements, but only if systematic errors can be corrected or reduced to a negligible value. The error or deviation in terms of measurable quantities may be expressed as:

$$\text{Deviation (D)} = \text{Measured value (M)} - \text{Arithmetic mean } (\bar{S}) \qquad (1.11)$$

FIGURE 1.17
Illustration of precise, accurate, & biased

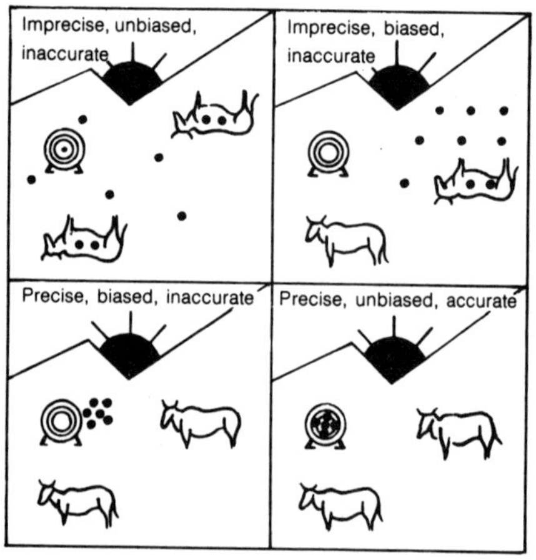

Accuracy, precision and bias

The magnitude of the Deviation (D) is the measure of the reproducibility or precision of the measurements. There is a great deal of confusion about the use of the terms precision, accuracy, and bias in physical measurements. For a measurement to be *accurate*, it must be both precise and unbiased.

Precision is a measure of the agreement among a group of individual measurements. Precision is strongly influenced by random decay. The precision may be measured in terms of the deviation of an individual count rate from the mean count rate.

Bias is influenced primarily by the systematic errors such as voltage changes or wrong stripping constants that introduce constant errors into the data. Figure 1.17 illustrates the meaning of the terms accurate, precise and biased.

Gaussian distribution

It has been experimentally verified by physicists that the radioactive disintegration process is truly random so long as the total number of atoms disintegrating during the observation period is negligibly small compared with the total number of atoms present. The statistical law that describes random processes is the Poisson-distribution law. Because the Poisson-distribution law approximates the normal Gaussian distribution as a limit, the mathematical descriptions applicable to the Gaussian or normal law are used. Figure 1.18 illustrates the relationship between the normal curve and the standard deviation.

FIGURE 1.18
Normal distribution function approximates Poisson distribution laws

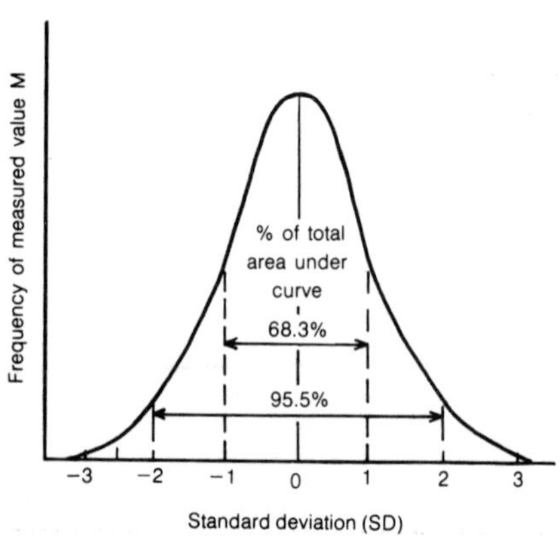

Standard deviation

The standard deviation, abbreviated S.D. or σ, is the most commonly used of several indices of the precision or error in radiometric data. The standard deviation indicates how reliable a measurement or a set of measurements is. If S is the sum of all gamma rays recorded during a specified interval of time, the Standard Deviation is given by:

$$\text{Standard Deviation } \sigma = S^{1/2} \qquad (1.12)$$

We may also write the standard deviation in fractional form as:

$$\frac{\sigma}{S} = \frac{1}{S^{1/2}} = \frac{100}{S^{1/2}} \text{ percent} \qquad (1.13)$$

Substantial field experience indicates that isotopic ratios are frequently more useful as geologic indicators than are the isotopic abundance values. If we wish to compute the standard deviation of the ratio $\frac{S_U}{S_{Th}}$, we use the expression:

$$\sigma_{U/Th} = \frac{S_U}{S_{Th}} \times \left(\frac{\sigma^2{}_U}{S^2{}_U} + \frac{\sigma^2{}_{Th}}{S^2{}_{Th}}\right)^{1/2} \quad (1.14)$$

It is important to note that we can compute the precision of measurement even though the count is measured only over a single interval of time. The solution of foregoing equations for standard deviation and percent standard deviation can be obtained directly from figure 1.16. It can readily be seen that as the total number of counts increases, the uncertainty in the result decreases.

As an example of the use of figure 1.16, assume that the number of counts in the Bi-214 energy window is 16 counts per minute. From the curves, we find that the standard deviation is ±4 counts per minute or 25 percent. This means that if we took additional readings at the same site, we would expect (see figure 1.18) that 32 percent of the readings would be more than 20 counts per minute or less than 12 counts per minute.

Reduction of error

If we desire to reduce the error (standard deviation) by one-half to 12.5 percent, we must increase the count rate by four times, to 64 counts per minute. Other things being equal, this could be accomplished by increasing the crystal volume by four times or by increasing the observation time by a factor of four.

The first alternative increases instrumental cost, size and weight. The second is practical only in ground surveys, as aircraft do not permit that much variation in ground speed. Therefore, in evaluating radiometric data, large percentage changes in small count rates should be regarded with great suspicion. They may be merely a manifestation of the random nature of a radioactive decay process. Figure 1.19 is an illustration of the response of a spectrometer at a particular sample site as a function of time.

FIGURE 1.19
Gamma ray spectrometer response at a single location as a function of time

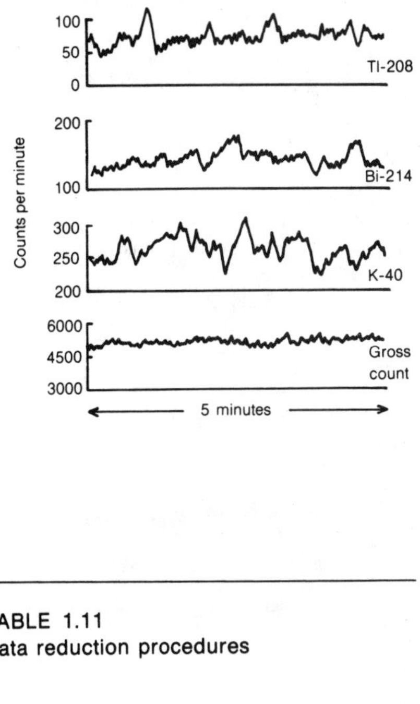

Correction procedures

The spectrometer data, once accumulated, may be subject to a variety of correction procedures. The choice of procedures to be applied depends upon the desired accuracy of the result and the type of data acquired—airborne, ground, or drill hole. Table 1.11 summarizes the kinds of correction procedures that are applied to accurate data to obtain isotopic abundance values as an end result.

Compton stripping

Correction of spectrometer data for Compton scattering is essential in order to obtain a reduced count rate proportional to the isotopic abundance. In one field example, the potassium channel count rate was

TABLE 1.11
Data reduction procedures

Procedure	Airborne	Ground	Drill Hole
Compton stripping	M	M	M
Background	M	M	M
Sensitivity constant	M	M	M
Atmospheric radon	M	m	0
Altitude	M	0	0
Live time	m	m	M
Cosmic ray	m	0	0

M = Major importance
m = Minor importance
0 = Not applicable

FIGURE 1.20
Airborne gamma ray spectrometer profiles and geologic section
Southeast Texas, U.S.A.

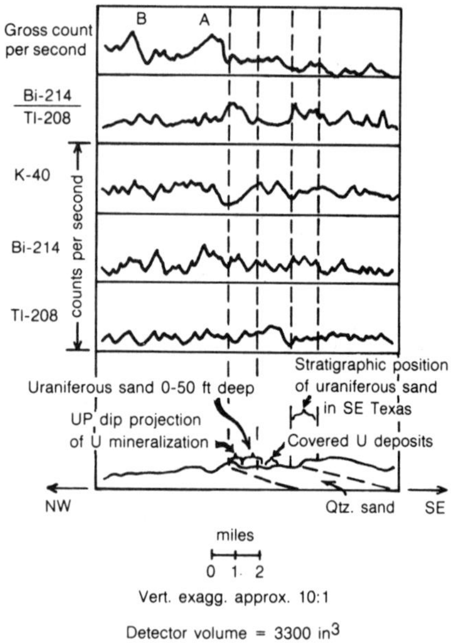

FIGURE 1.21
Airborne gamma ray spectrometer profiles and geologic section
Southeast Texas, U.S.A.

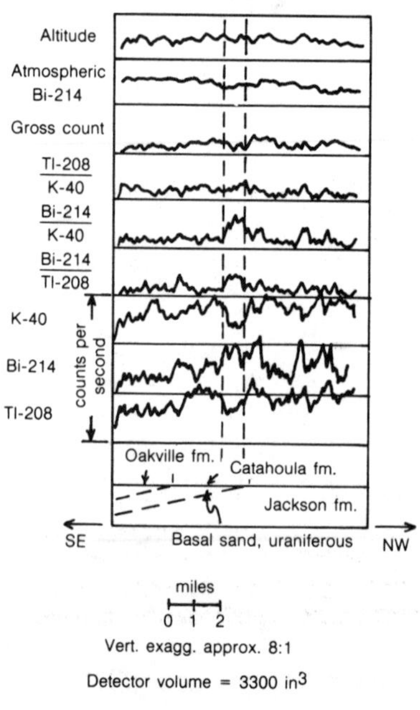

reduced by more than 90 percent by stripping the Compton scattered photons. This was due to an abnormally high thorium content of the sediments.

The stripping corrections are readily applied to digitally recorded data. Data from hand portable equipment must normally be reduced manually by application of equations (1.4), (1.5), and (1.6).

Background

The term *background* has a confusing variety of definitions as commonly used in radiometric work.

In airborne surveys, background is considered the sum of the counts due to the structural elements of the aircraft and the instrumentation package. If the airborne equipment has no provision for monitoring and correcting for the time-varying cosmic ray contribution, that too must be included in background. Likewise, if atmospheric radon cannot be monitored and compensated, the airborne Bi-214 and its Compton scatter products become simply part of background.

In ground and drill hole surveys, the contribution from the instrumental package, atmospheric radon, and cosmic rays is frequently small compared with the ground signal. They may, therefore, safely be considered as background. The exception is in areas of very weak ground activity during periods of strong atmospheric inversion. The atmospheric component may then be a significant portion of the total signal.

Background may be determined by observations over water. However, in most areas this is not a practical solution. Alternatively, readings can be taken with the detector protected by shielding materials, e.g., low activity lead or iron. The most frequent procedure is to simply define as background the lowest reliable count rate in the survey. Thus defined, the background includes a minimum geologic contribution. Therefore, any quantitative determinations of radio isotope content will be in error by an amount proportional to the isotope content of the background area.

If gamma spectrometry is applied to the determination of narrow structures or collected samples, background is dramatically changed. Almost all natural radioactivity becomes background and the signal-to-noise ratio is correspondingly reduced.

Background may be reduced by the use of collimating shields. These add considerably to the weight of the equipment. For these reasons, in situ determinations are considerably more reliable than field assay of hand specimens.

Sensitivity constants

Sensitivity constants enable the conversion of counts per unit of time into isotopic abundance data—and here are strong arguments for doing so. Surveys from various instruments can be related and compared. In addition, the data from the discipline of isotope geochemistry, also given in abundance units, becomes immediately comparable.

The subject of sensitivity constants has been covered more fully in the section on radiation detection and instrumentation.

Atmospheric radon

Correction for atmospheric radon is essential for high sensitivity airborne surveys; it is normally of minor importance in surface work and makes no contribution to drill hole logging signals. The reader should refer to the section on airborne survey procedures for a full discussion of atmospheric radon.

Altitude correction

There are a variety of altitude corrections for airborne surveying. They vary in the geometric assumptions and the manner in which the absorption coefficient is treated.

The assumed geometry of the radioactive source can be that of (1) a point source, (2) a slab source, or, most commonly, (3) a half-space. Mathematical expressions are then derived to express the decrease in spectral intensity with increasing altitude.

A second type of altitude correction is called *Weiner Filtering*. In this procedure, a computer program looks for positive correlations between radiometric response and topography, and subdues that response. The defect of Weiner Filtering is that positive correlations between geology and topography are also subdued.

The amount of absorption suffered by the gamma rays in the air between depends upon the distance traveled, the energy of the gamma ray, and the density of the air. Thus, the 1.76 Mev rays from Bi-214 are attenuated more rapidly than the 2.62 Mev rays from Tl-208. However, there is also a buildup effect because of the Compton scattering interactions that take place before the gamma rays enter the detector, i.e., in the ground and in the air. These scattering products enter the detector and result in a buildup of photons in lower energy channels. The buildup is not removed by applying normal Compton correction, and must be compensated for in the altitude absorptional coefficient.

The expression for altitude correction has the form $S = S_0 e^{\bar{\mu}h}$. Here, S is the observed channel count rate at height h, S_0 is the count rate at zero altitude and $\bar{\mu}$ is the empirically determined attenuation coefficient. It is separately determined for each energy channel. Determination is made by flying the test profile at a series of altitudes that cover the range expected in a survey, for example, 200, 400, 600, 800, and 1000 ft. The count is accumulated along the entire test profile in order to obtain a large total count. The values of the attenuation coefficients are then adjusted so that the altitude-corrected values of S are all the same.

Live time correction

The correction for excessively high count rates is a procedure that is essential primarily in logging or assaying for radioactive ores. The correction must be made electronically and is one of the instrument design considerations.

Cosmic ray correction

Cosmic ray interactions with the components of the earth's atmosphere produce gamma rays that stimulate the detector and produce Compton scattering products. For a more complete discussion of cosmic rays, please refer to the section on radiation detection and instrumentation.

FIGURE 1.22
Airborne gamma ray spectrometer profiles and geologic section
Southeast Texas, U.S.A.

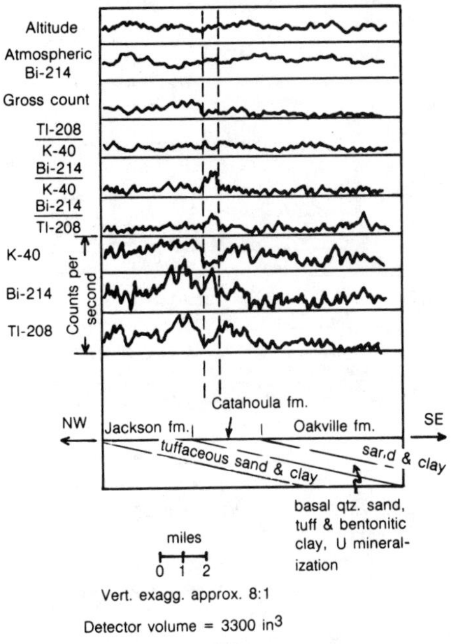

FIGURE 1.23
Airborne gamma ray spectrometer response
Southeast Texas, U.S.A.

Tl-208

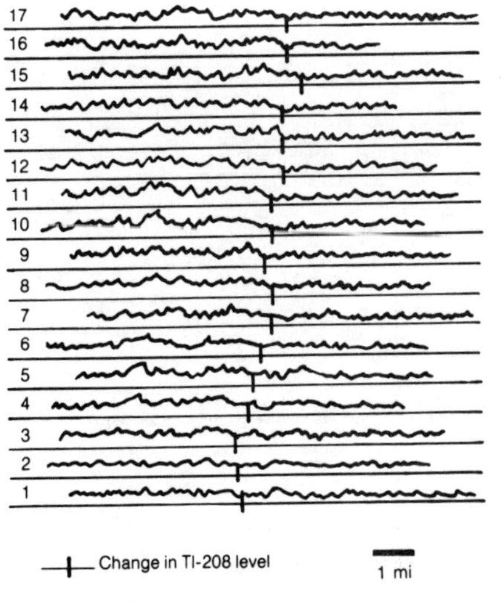

Examples of surveys

Published case histories of gamma spectrometer surveys are not abundant. However, it is hoped that the following examples will be of value when thought is given to the geological use of spectrometer data.

Use of isotopic ratios

Ratios of elements can be used as diagnostic parameters of lithologic units and of special geologic conditions. Figure 1.20 is an illustration of such a case. It represents a slightly smoothed version of an airborne spectrometer profile over a known uranium occurrence of the roll-front type in southeast Texas.

There are several notable features in the records. The gross count record shows two high values, A and B, northwest of the known mineralization. The Bi/Tl ratio shows high values over quartz sand that hosts a local uranium deposit. The K-40 curve indicates a minor low in the region of each ratio high; the Tl-208 profile shows similar low features. The low K-40 has been correlated with sandstone of low feldspar content, and the low Tl-208 has been related to low clay content of the sands.

Figures 1.21 through 1.27 illustrate two modes of data presentation for an area in southeast Texas. All data were taken at a nominal 400 ft terrain clearance with a system having a detector volume of about 3300 cu in.

Figures 1.21 and 1.22 show the intensities of the bismuth, thallium and potassium radiation and the intensity ratios on two lines from an extensive survey. Also shown are the radar altimeter profile, the atmospheric Bi-214 count averaged for 21 seconds, and the gross count. A distinctive high in the ratios bismuth/thalium, and bismuth/potassium shows good correlation with the outcrop area of a relatively clean uraniferous quartz sand. The K-40 and Tl-208 profiles show relative lows in the same area, while Bi-214 shows a relative high. The Bi-214 high is not the most intense on either profile, and the gross count shows a relative low. The features having the best correlation and the most unique response are the two Bi-214 ratios. A diagrammatic geologic section is sketched below the profiles.

Figures 1.23 through 1.27 show data from the same area in southeast Texas. Here the data are presented as stacked profiles of the same isotope. Line spacing is a nominal one-quarter mile. Figure 1.23 represents the Tl-208 radiation. The change in background level has been marked on each profile. The change correlates very closely with the mapped position of the Catahoula-Oakville formational contact. Figure 1.24 shows the Bi-214 profiles; the highest intensities appear on the left side of the map area in the Jackson formation. Figure 1.25 illustrates the intensity of K-40 radiation; there are no distinctive features having good line-to-line continuity. Figure 1.26 shows the ratio K-40/Tl-208. Lines 1 through 4 have been omitted in order to expand the line spacing and avoid overlap of the curves. A series of strong and consistent highs appear southeast of the Catahoula-Oakville contact; these indicate the basal sand in the Oakville formation. Figure 1.27, displaying the ratio Bi-214/Tl-208, shows a series of highs at the Catahoula-boakville contact, a favorable ore environment.

FIGURE 1.24
Airborne gamma ray spectrometer response
Southeast Texas, U.S.A.

Bi-214

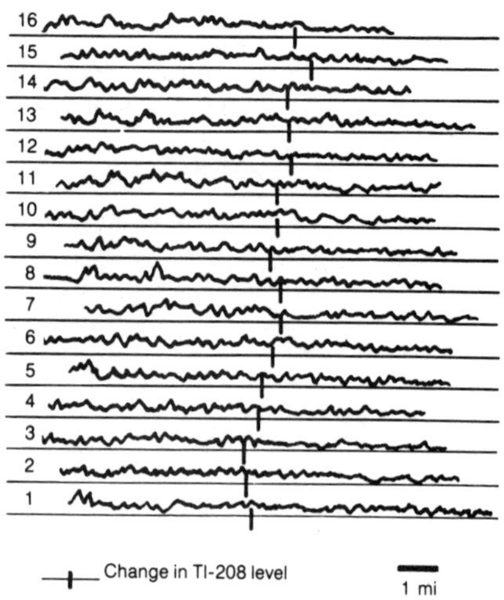

FIGURE 1.25
Airborne gamma ray spectrometer response
Southeast Texas, U.S.A.

K-40

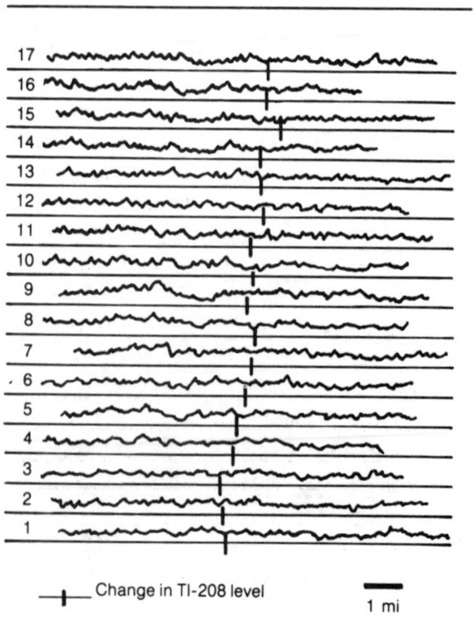

Other exploration applications

Gamma ray spectrometer surveys have played a role in exploration for a wide variety of ores and minerals. Some ores have associated potassium alteration halos, for example, porphyry copper deposits. The high potassium content can be used as an exploration guide in searching for potassium alteration features.

Uranium is a constant associate of phosphate deposits; hence, spectrometer surveying can be used in mapping and in searching for phosphate beds. Coal normally bears a very low content of the radioactive minerals, and, therefore, coal seams can be identified on drill logs and in outcrop areas by radiation lows. Conversely, the shale overlying coal beds sometimes exhibits a relatively strong radioactivity. In particular, the clinker horizons adjacent to burned-out coal seams frequently exhibit a strong radioactive response which can be mapped from the ground or from the air.

The literature has reported many occurrences of radiation halos over hydrocarbon deposits. There appears to be increasing agreement that halos actually exist, but, as yet, there is not a generally accepted explanation of their origin.

Heavy metal placer deposits can sometimes be mapped and identified by their associated zircon, sphene, and monazite.

Other applications of gamma ray spectrometry include tracer techniques. Radioactive components are introduced into geologic environments such as ground water systems, and their movement is ecorded by detecting the radioactive tracer.

Measurements of snow cover have also been made by measuring the ttenuation of natural or artificial sources of known intensity. The ttenuation due to snow cover is a function of the water content of the snow.

Geologic mapping

The applications of gamma spectrometer techniques to geologic mapping are many and varied. Potential applications can be considered on the basis of a knowledge of the general radioactivity of rock types and of the geochemical properties of the elements. Geologic mapping by radioactivity has been applied in the following kinds of situations:

- It has been found that some intrusive rocks have high radioactivity zones near the margins of the body. Therefore, it is possible to identify marginal faces of some intrusive rocks by relative radioactivity highs.
- Lithologic mapping has been effective in deeply weathered areas; for example, shales juxtaposed on limestones where residual weathering has left evidence of original rock composition in radioelement concentrations.
- Faults sometimes exhibit a radioactive high due to increased permeability allowing access to migrating radon. Other faults exhibit a radioactive low because of increased leaching of radioactive minerals along fault zones.

In considering gamma spectrometry work, as in any geochemical method, an orientation survey is desirable. There is no way to consistently predict the effectiveness of a survey solely on the basis of known lithology and structure. Thus portable spectrometer surveys are a useful but not infallible method of assessing the probable effectiveness of airborne surveys.

Don Hansen wrote this chapter while under contract to GeoMetrics of Sunnyvale, CA. It is with their permission that this condensation appears.

FIGURE 1.26
Airborne gamma ray spectrometer response
Southeast Texas, U.S.A.

K-40 / Tl-208

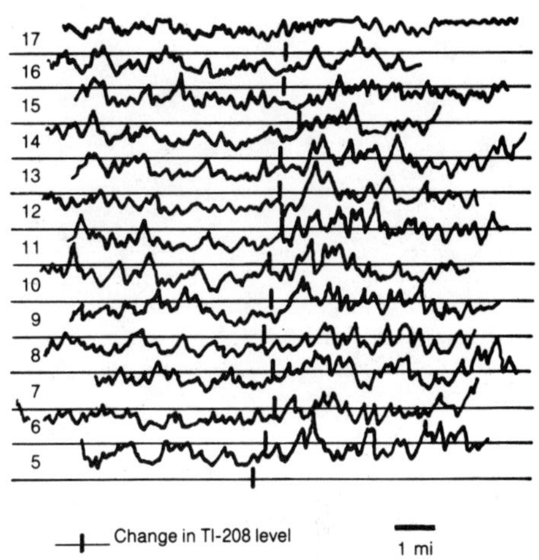

FIGURE 1.27
Airborne gamma ray spectrometer response
Southeast Texas, U.S.A.

Bi-214 / Tl-208

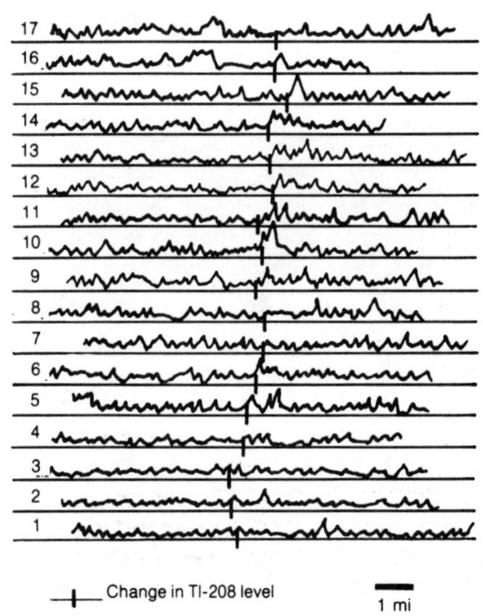

Appendix 1A

Units of Measure

The following units are commonly used in radiometric mensuration.

Curie

Radioactivity is measured in units of *curies*. The curie is defined as 3.7×10^{10} radioactive disintegrations per second. A curie is a measure of the source strength. A curie of radium is approximately one gram, a curie of radon 222 is 0.66 cubic millimeters at standard temperature and pressure, and a curie of uranium 238 is about 3000 kilograms (3.3 tons). When a radioactive decay series is in equilibrium, the activity of each radioactive daughter isotope, measured in curies, is exactly the same.

Roentgen

It is a frequently overlooked fact that radiation produces effects only through absorption. If a gamma ray passes through matter with no loss of energy, it produces no effect. Only those photons which interact with the detector lose energy and produce an observable effect. Therefore, the unit of radioactivity of the source, the curie, does not necessarily describe the amount of energy absorbed by a mass or by the detector. The quantity of radiation absorbed per unit volume or mass of material is the roentgen. A roentgen of gamma radiation is that which will produce 1.6×10^{12} ion pairs per gram of air resulting in the appearance of 83 ergs of heat energy per gram of air.

Scintillometers and Geiger counters are frequently calibrated in terms of milliroentgens (mR) per hour. Careful measurements have shown that one gram of radium in equilibrium with its decay products and enclosed in 0.5 millimeters of platinum will produce a gamma ray exposure of 8.4 roentgens per hour at a distance of one centimeter. If the radium is assumed to be a point source, the intensity at other distances can be calculated by the inverse square law. The National Bureau of Standards maintains sets of radiation calibration standards that serve as a primary facility for calibrating other standards.

Electron volts

The unit of gamma ray energy is the electron-volt. It represents the energy acquired by an electron when accelerated by a potential difference of one volt. The energy is commonly expressed as million electron volts (Mev) or thousand electron volts (Kev).

Crystal volume

The volume of a crystal detector is of first order importance in assessing the data collecting capability of spectrometers. The following table gives the volume in inches and centimeters of some of the standard-size crystals.

Crystal size (in)	Volume (in^3)	Volume (cm^3)	Volume relative to 2 × 2
2 × 2	6.3	103	1
3 × 3	21	340	3.3
5 × 5	98	1600	15
6 × 4	113	1800	17
9 × 4	254	4100	40
11½ × 4	415	6700	65

Appendix 1B

Radio Element Content of Rocks and Minerals

The following table gives the range and average content of rocks and minerals. The data are from various sources believed to be reliable; in some cases where data are scarce, the possible values have been indicated. (Compiled by John G. Roylance.)

Natural Concentrations of Uranium, Thorium, and Potassium

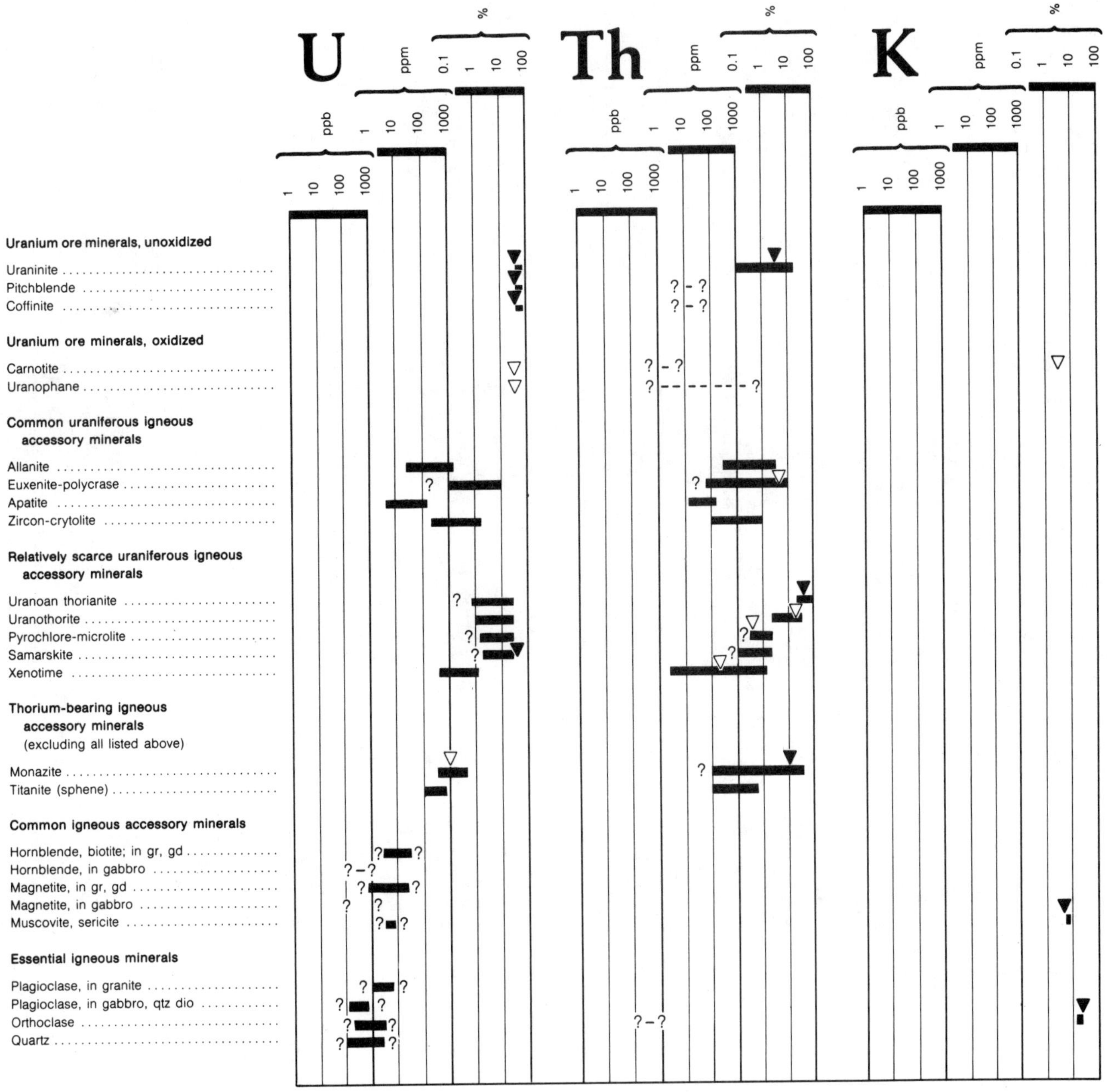

SYMBOLS:
Average concentration— ▼
Possible average concentration, poor data— ▽
Range of concentration— ▬
Possible range of concentration— ? – ?

Natural Concentrations of Uranium, Thorium, and Potassium

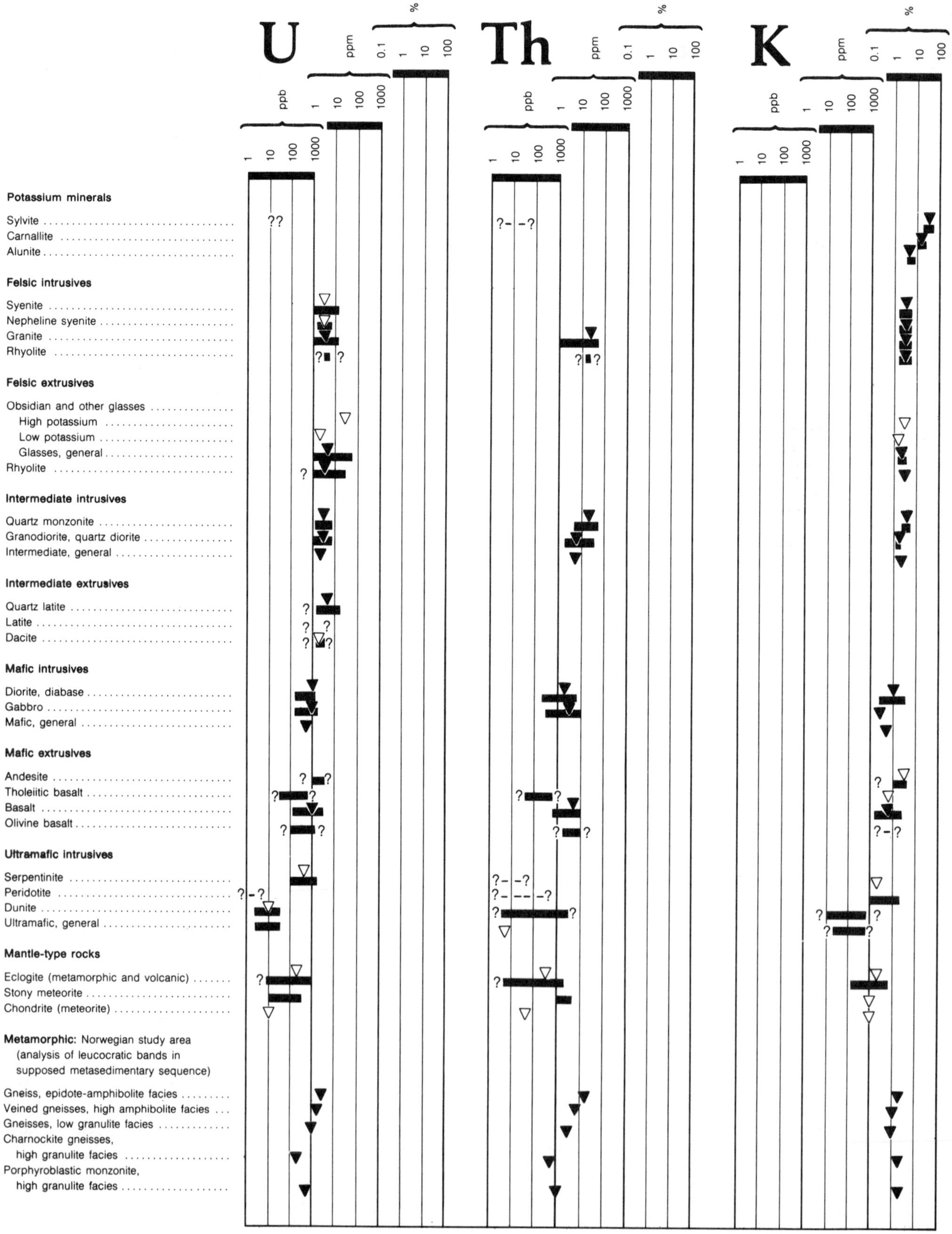

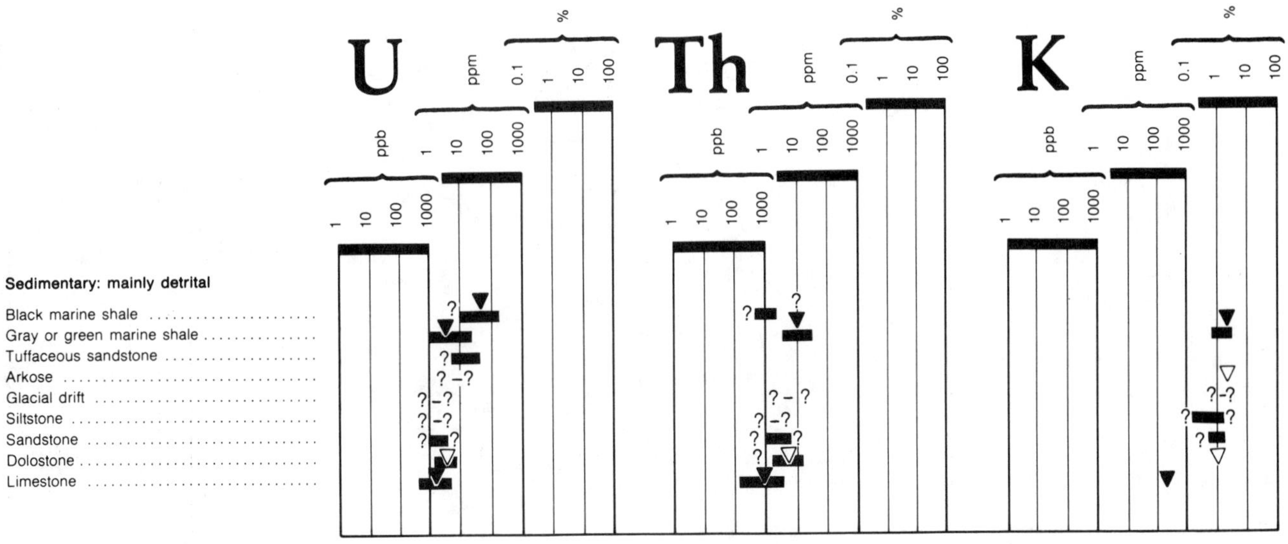

1) 99.3% is U^{238}, which yields daughter Bi^{214} that is measured by gamma ray spectrometer as an indication of total U present in sample. U^{238} half-life = 4.5×10^9 yr. 1% U = 1.2% U_3O_8.

2) In an average granite, Th^{232} is about 4×10^{12} times more abundant than Th^{230}, a U^{238} daughter with a half-life of 80,000 years and the only geologically significant Th isotope besides Th^{232}. Th^{232} half-life = 1.4×10^{10} yr. Spectrometer measures radiation of $T1^{208}$, a daughter of Th^{232}, to indicate Th^{232} concentration in sample.

3) 0.012% of total K is K^{40}, 11% of which yields gamma radiation upon decay. K^{40} half-life = 1.3×10^9 yr.

SYMBOLS:
Average concentration— ▼
Possible average concentration, poor data— ▽
Range of concentration— ▬
Possible range of concentration— ?—?

Selected References (Chapter 1)

Adams, J. A. S., and Gasparini, P. 1970. *Gamma-ray spectrometry of rocks.* Holland: Elsevier.

Adams, J. A. S., and Louder, W. M., eds. 1964. *The natural radiation environment.* Chicago: Univ. of Chicago Press.

Bowie, S. H. U., Davis, M., and Ostle, D. 1972. *Uranium prospecting handbook.* London: Institute of Mining and Metallurgy.

Duval, J., and Cook, M. 1971. *Development of remote methods for obtaining soil information and location of construction materials using gamma-ray signatures for Project Tathemis.* Rice University: Dept. of Geology.

Faul, H., ed. 1954. *Nuclear geology.* New York: Wiley.

International Atomic Energy Agency. 1969. *Nuclear techniques and mineral resources.* Vienna, Austria.

International Atomic Energy Agency. 1973. *Uranium exploration methods.* Vienna, Austria.

Lapp, R. E., and Andrews, H. L. 1948. *Nuclear radiation physics.* New Jersey: Prentice-Hall.

Løborg, L. 1973. The future development in the uses of gamma-ray spectrometry for uranium prospecting on the ground. In *Uranium exploration methods,* International Atomic Energy Agency.

Rankama, Kalervo. 1963. *Progress in isotope geology.* New York: Wiley.

Sakakura, A. Y. 1957. *Scattered gamma rays from thick uranium sources.* U.S. Geological Survey Bulletin 1052-A.

Figure Credits (Chapter 1)

Figure 1.5a and b
Reprinted with permission of Henry Faul, *Nuclear geology,* 1954.

Figure 1.9
Reprinted with permission of Duval and Cook, Dept. of Geology, Rice University, 1971.

Figure 1.13
Reprinted with permission of Henry Faul, *Nuclear geology,* 1954.

CHAPTER 2
ELECTRICAL: IP AND RESISTIVITY

GROUNDED ELECTRICAL METHODS IN GEOPHYSICAL EXPLORATION

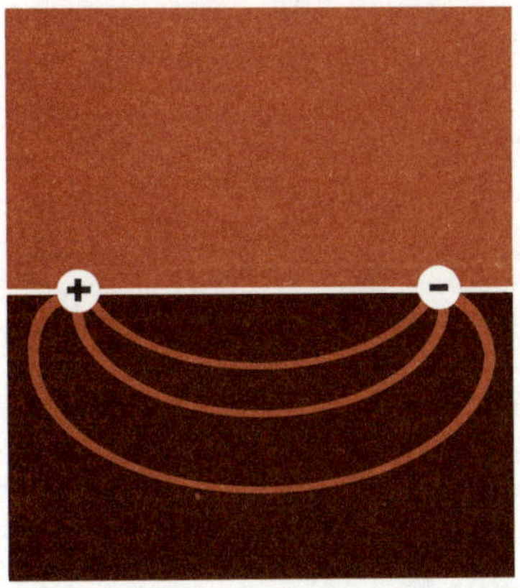

Resistivities of earth materials

For almost all electrical geophysical methods, the true, or, more scientifically, the specific resistivity of the rocks is of interest. The true resistivity of a rock unit is a bulk property of the rock itself—it is the resistance that the rock unit has to the passage of electrical current. The true resistivity of a rock unit is defined as being equal to the resistance of a unit cube of the rock. Since voltage = current × resistance, the resistivity of a rock unit can be measured as shown in figure 2.1.

$$\text{resistivity} = \frac{\text{voltage/L}}{\text{current/A}} = \frac{\text{volts/meters}}{\text{amperes/meters}^2} = \text{ohm-meters}$$

or

$$= \text{ohm-feet, etc.}$$

When current flow is introduced into the surface of a half-space of uniform resistivity, the current flow is that shown in figure 2.2. In this situation, the voltage at a point or distance (r) from the current electrode is

$$V = \frac{\rho I}{2\pi r} \text{ for voltage equal to zero at infinity}$$

or

$$\rho = \text{resistivity} = \frac{V}{I} \times 2\pi r$$

FIGURE 2.1
Resistivity measurement of a cube

Resistivity measurement of a cube of side length L using two current and potential measuring electrodes.

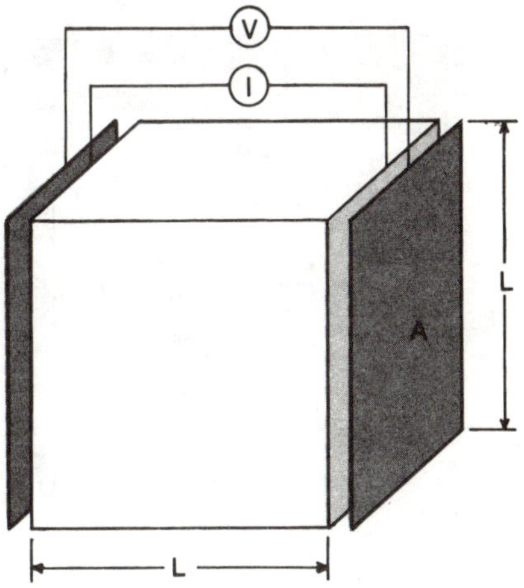

Philip G. Hallof, president of Phoenix Geophysics Limited of Willowdale, Ontario, is one of the early contributors to the development of induced polarization. Dr. Hallof took a B.S. degree in geophysics from Massachusetts Institute of Technology, and completed his Ph.D. work at MIT on the variable frequency IP method. Upon completion of his Ph.D., he was employed by and ultimately became president of McPhar Geophysics Limited where he instituted the first commercial use of the variable frequency IP method in contract surveys as well as in the manufacture and sale of survey instruments. At McPhar and later at Phoenix Geophysics, Dr. Hallof's work has included the use of IP and other electrical methods in almost all mining areas of the world including Russia, Eastern Europe and mainland China.

Philip G. Hallof, Ph.D.
Phoenix Geophysics Limited

FIGURE 2.2
Point current electrode on surface of a conducting solid

Section through point current electrode on surface of a conducting solid, showing how potential V is related to resistivity ρ, current I, and distance from electrode.

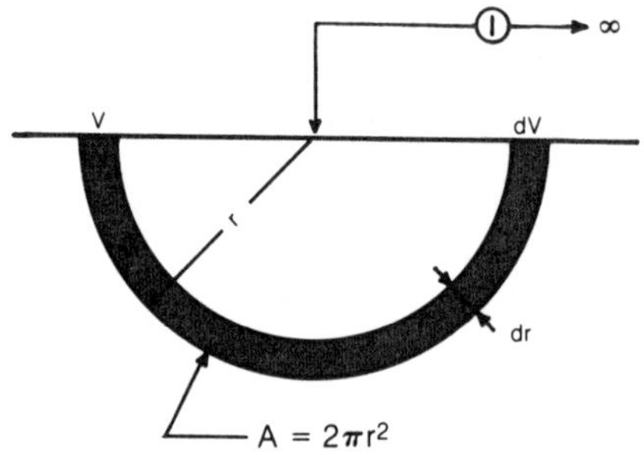

FIGURE 2.3
Range of typical resistivity values of general rock types

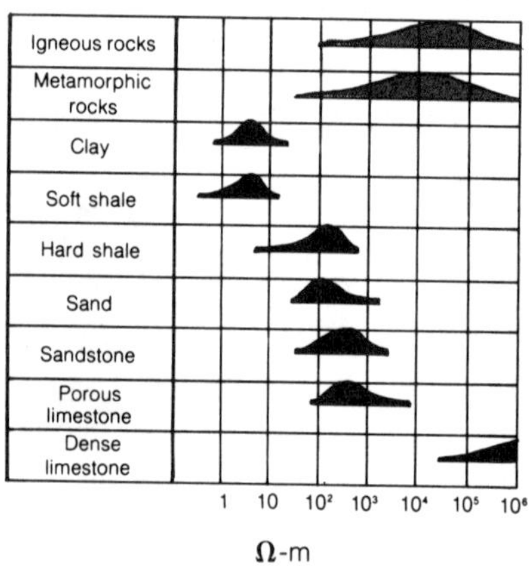

Therefore, the true resistivity of a uniform half-space can be determined by a four-electrode measurement with one current electrode and one voltage electrode at infinity.

Almost all rock-forming minerals are effective insulators; they will not conduct electricity in the normal sense. Therefore, all rocks, if absolutely dry, will not conduct electricity. In the earth, the pore spaces of rocks are usually filled with water. The resistivity of a rock is directly related to the porosity of the rock. Tables 2.1 to 2.5 and figure 2.3 will show that increased porosity due to grain size, fracturing, and alteration will result in a lower true resistivity for the rock unit.

Most groundwaters are moderate in resistivity value, and therefore the statement about porosity usually holds. However, the resistivity of the fluids in the pore spaces of a rock also determines the true resistivity of the rock unit. Increased salinity of a fluid decreases its resistivity so that rocks filled with saline solutions have lower resistivities than those filled with fresh water.

The tables also list some rather uncommon rock-forming minerals that are very good-to-excellent conductors, i.e., they have very low true resistivities. These are the metallic luster minerals, mostly sulfides. These minerals are opaque in thin-section because they have free electrons in their crystal lattice. They conduct electricity in the same way, and almost as well as, the pure metals.

One also notices the crystal form of carbon called graphite is a semiconductor material that has a low true resistivity. Other types of pure carbon, ranging from coal to diamonds, do not conduct.

In general then, the use of the resistivity method in geophysics will depend upon a change in the porosity of the rocks in the subsurface, a variation of the salinity of the fluids filling the rocks in the subsurface, or the presence of some of the specific minerals that are excellent conductors of electricity.

Exploration using the resistivity method

It has been shown that the true resistivity value in any rock type is dependent upon the porosity of the rock and the salinity (or resistivity) of the solution filling the pore spaces. This situation arises because most rock forming minerals are insulators and will not permit current flow. A few minerals (mostly sulfides) (1) have metallic luster, (2) have free electrons in their crystal structure, and (3) are opaque in thin section; therefore, they conduct electricity much better than rocks.

The resistivity method can then be used in exploration within rocks in situations where the geophysicist is looking for some structure that has a different resistivity than surrounding rocks. This could be a fault or shear zone, a zone of increased alteration, a rock unit that is more porous than the surrounding rocks, the presence of basement rocks beneath a more porous, less resistive overburden layer, or a conducting sulfide ore zone.

Making a useful resistivity measurement requires the use of four electrodes. Current is applied to the earth through two electrodes and the potential across the ground is measured using two other electrodes. If only two electrodes are used, the contact conditions largely determine the resistivity measured. This is entirely unsatisfactory.

Since the earth resistivity is almost always a variable value, the voltage/current relationship measured will vary across the surface of the earth. The parameter measured and plotted is almost always the *apparent resistivity* (ρ_a). This is the true resistivity value that a uniform half-space would need to create the voltage measured given the electrode placement and current level used. The values of apparent resistivity measured are then interpreted relative to variations in electrode placement, distances between electrodes, etc., to give the electrical parameters and geometry of the earth.

Figure 2.4 shows the three regions of potential distribution that are commonly used in resistivity exploration. The actual placement of the four electrodes and the calculation of the apparent resistivity value for V and I are shown in figure 2.5. Each of these electrode configurations, and there are many more, has advantages and disadvantages.

In the use of the resistivity measurement, the geophysicist is usually looking for lateral variations in the earth's resistivity or changes in the earth's resistivity with depth. The theoretical results (figures 2.6 and 2.7) show the profiles measured with three different electrode configurations over a buried sphere. The dipole-dipole configuration can be seen to give the sharpest and largest magnitude anomaly over the sphere. This is not surprising since the dipole-dipole measurement is essentially a measurement of the lateral second derivative of the potential from a simple current source. This characteristic of the measurement should yield the maximum sensitivity to lateral variations.

If earth resistivity varies only with depth, the layered earth model applies. It is not important where the electrodes are placed, but how far apart they are on the surface. A greater electrode separation gives a greater depth of detection. The theoretical results for a two-layer earth (figure 2.8) show how the apparent resistivity varies with electrode separation for this configuration. In this situation, the second derivative nature of the dipole-dipole measurement is of no particular advantage and, indeed, the dipole-dipole measurements respond somewhat more slowly to the high resistivity basement as the electrode interval is increased.

One of the most widespread uses for the resistivity method is the layered media case. Until recently, this was the only geometry in which

FIGURE 2.4
Current distribution

Three basic regions about two current electrodes where potential measurements are made.

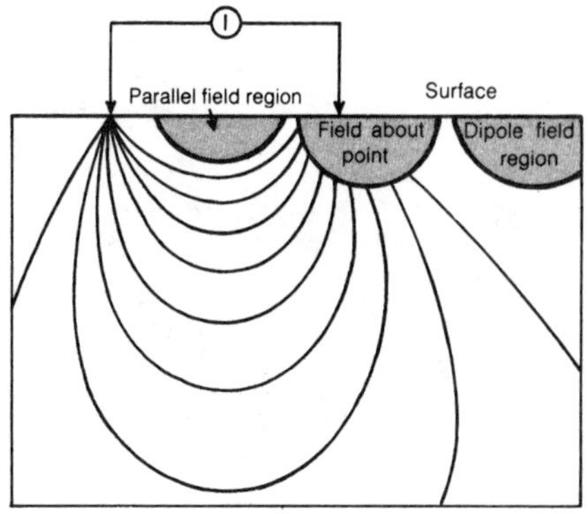

FIGURE 2.5
Common resistivity arrays

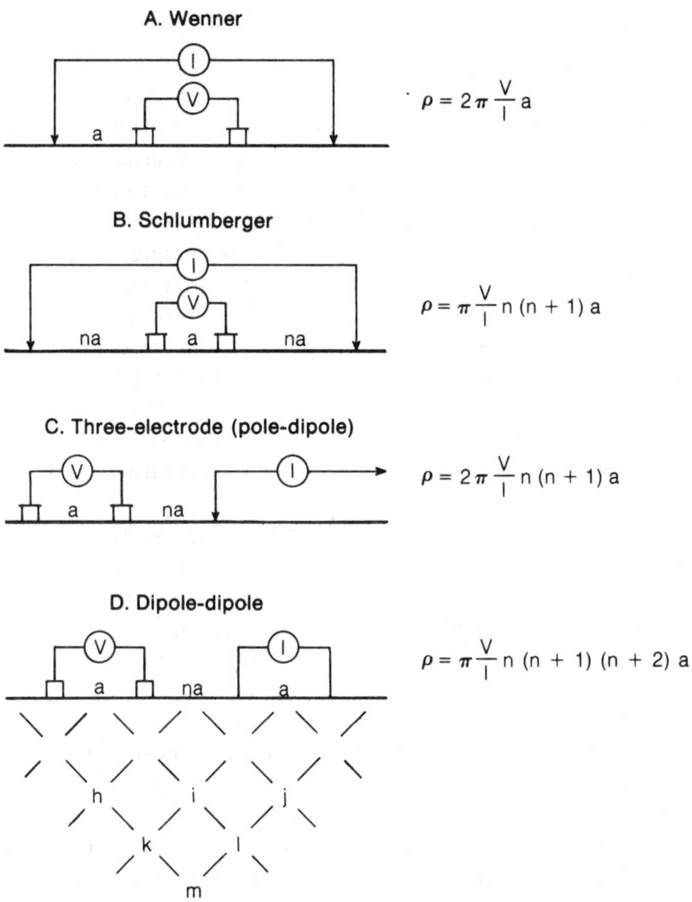

FIGURE 2.6
Comparison of depth penetration

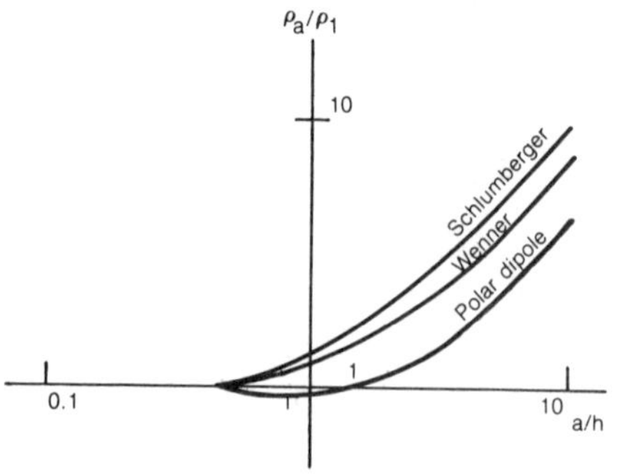

theoretical results were available. This type of problem could include (1) groundwater problems, (2) depth to basement problems, (3) geothermal problems, and (4) other problems in which the resistivity of the earth can be assumed to vary only with depth.

The two sets of theoretical curves in figures 2.8 and 2.9 are for the two-layer earth and the Wenner configuration and the dipole-dipole configuration. They show that for very small separations the apparent resistivity is equal to the true resistivity value in the upper layer. As the electrode separations are increased, the apparent resistivity value smoothly approaches the true resistivity value in the basement layer. As shown in figure 2.10, matching of the field curves to a family of theoretical curves has been the accepted method of quantitative interpretation. As will be discussed later in this chapter, direct inversion of field data using the digital computer is now possible.

The resistivity method has been employed in past years to locate sulfide ore zones. The success has been limited. Although the ores may have a large resistivity contrast with the host rocks, the size of the sulfide zones is usually small relative to other earth structures which, although smaller in resistivity contrast, have much greater volumes. The resistivity anomalies from various ore zones will be studied in later discussions of IP field results.

A few of the difficulties in the use of the resistivity method are demonstrated by the theoretical scale modeling results in figures 2.11 through 2.16. These data show that even for resistivity contrasts of 10/1 to 25/1 the anomaly from a source at zero depth disappears very quickly as it is moved to some depth. Further, the introduction of an overburden layer of even a small contrast will almost completely mask an anomaly from a source at depth. These difficulties in interpretation are almost always further compounded by the fact that the basement rocks themselves are variable in thickness and resistivity.

FIGURE 2.7
Comparison of resolution

Resistivity profiles over a buried sphere, giving a comparison of the resolution of three commonly used electrode arrays.

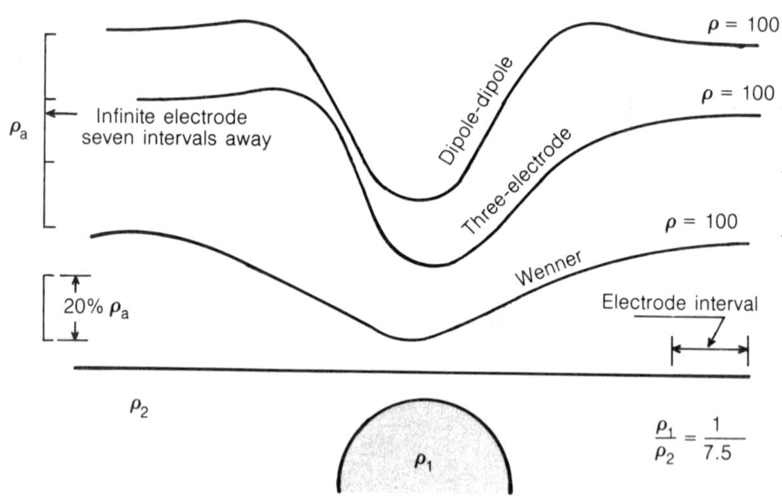

The figure on this page is referred to in the preceding section *Exploration using the resistivity method.*

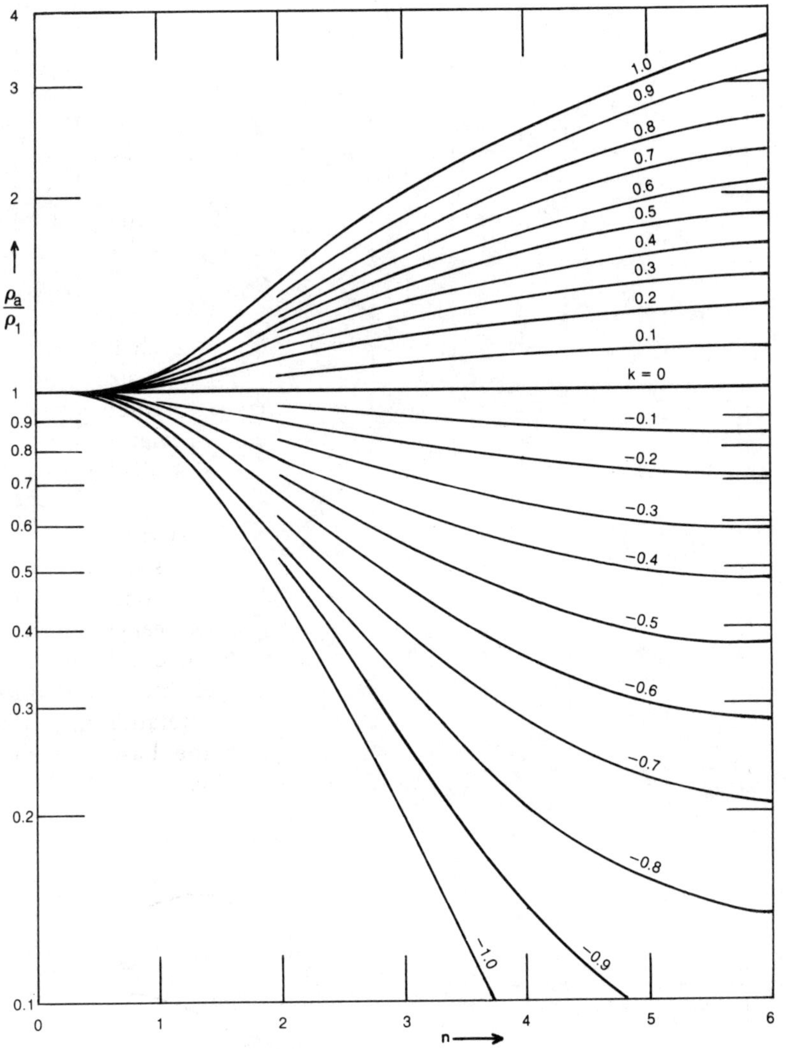

FIGURE 2.8
Two-layer resistivity curves, dipole-dipole array, for $\frac{D}{a} = 1.0$

$$k = \frac{\rho_2 - \rho_1}{\rho_2 + \rho_1}$$

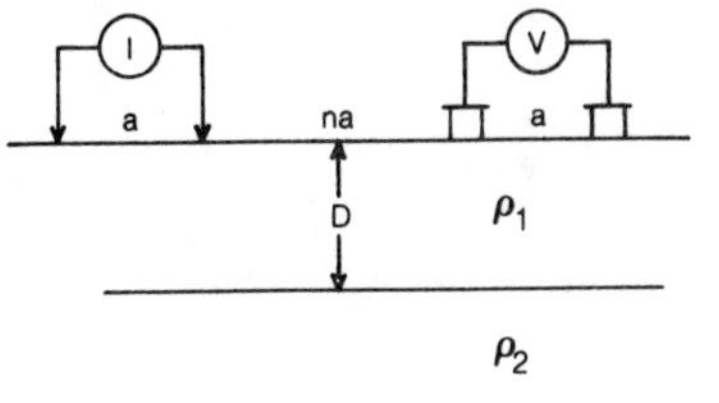

The figures on this page are referred to in the preceding section *Exploration using the resistivity method.*

FIGURE 2.9
Two-layer resistivity curve, Wenner array

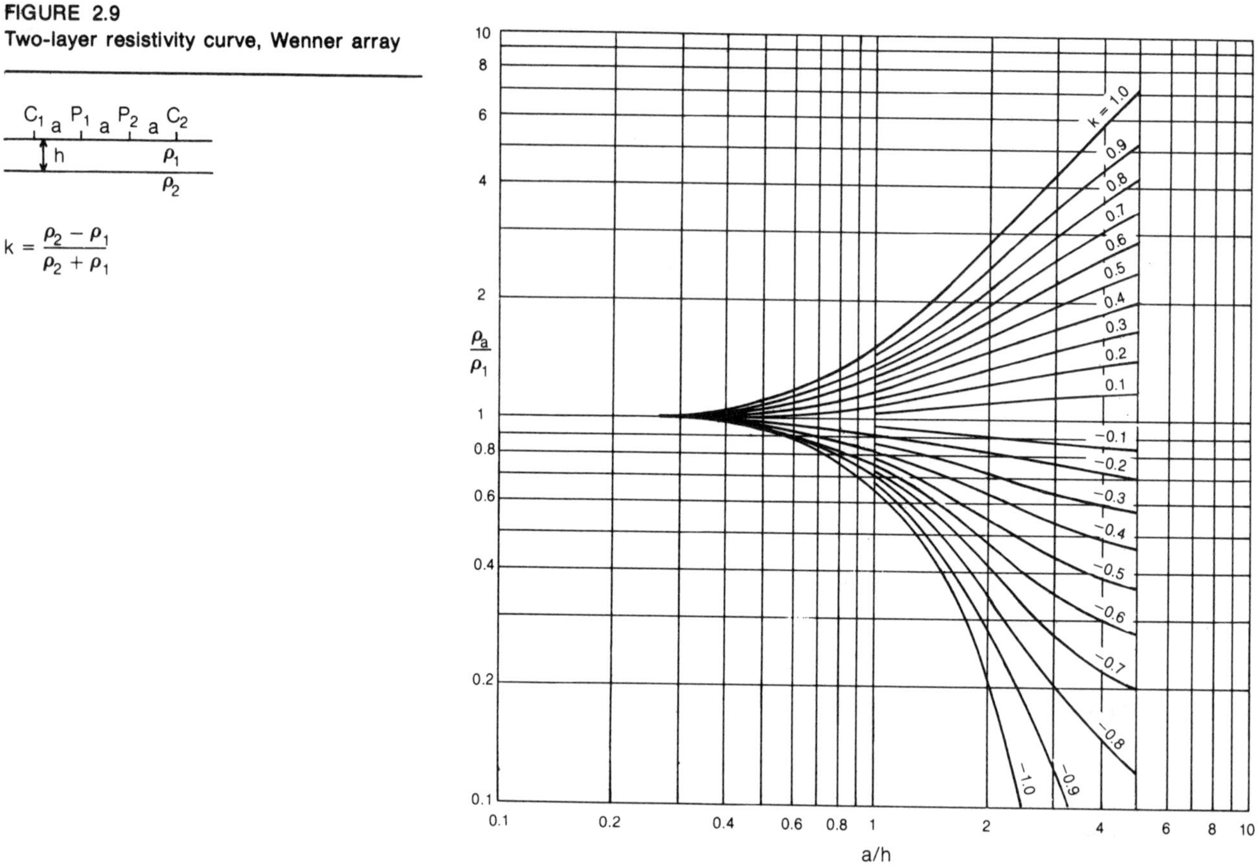

FIGURE 2.10
Interpretation using three-layer curves, Wenner array

The interpretation of a profile of apparent resistivity in the vicinity of Fond-du-Lac, Wisconsin.

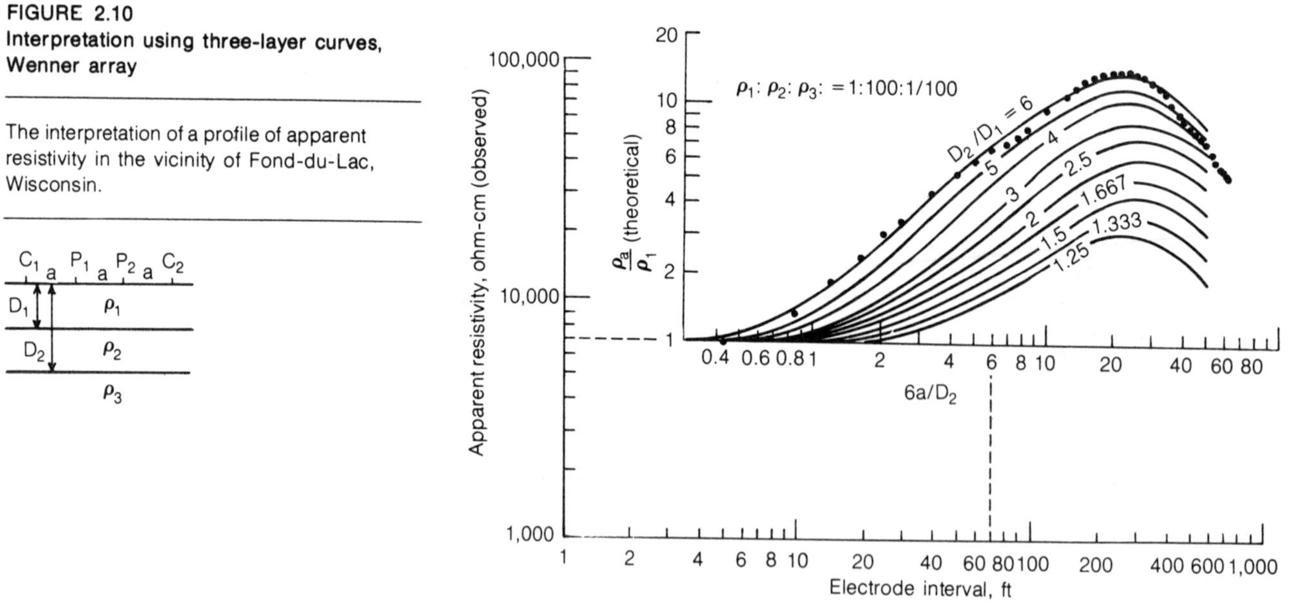

The figures on this page are referred to in the preceding section *Exploration using the resistivity method.*

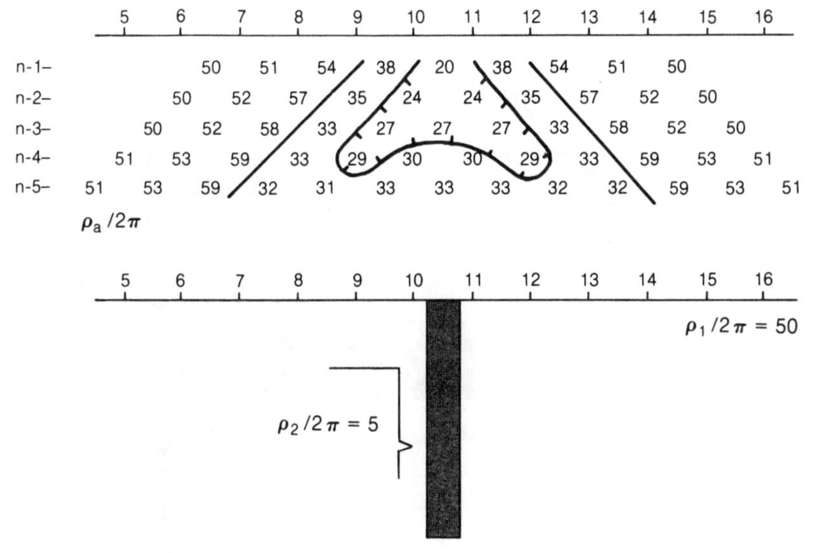

FIGURE 2.11
Scale model case

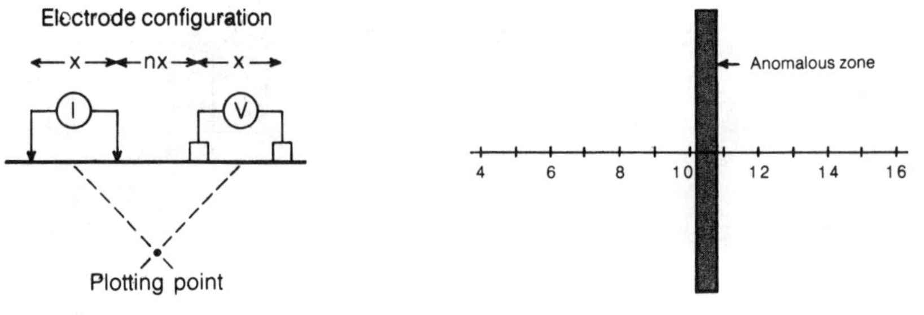

Plan view of anomalous zone and survey line

Electrode configuration and plan view for figures 2.11 and 2.12

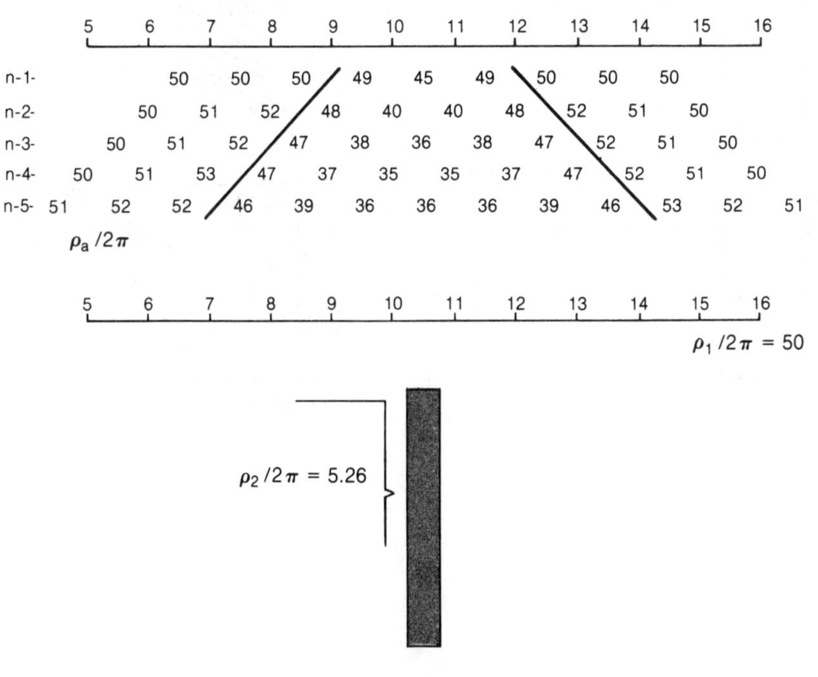

FIGURE 2.12
Scale model case

The figures on this page are referred to in the preceding section *Exploration using the resistivity method.*

FIGURE 2.13
Scale model case

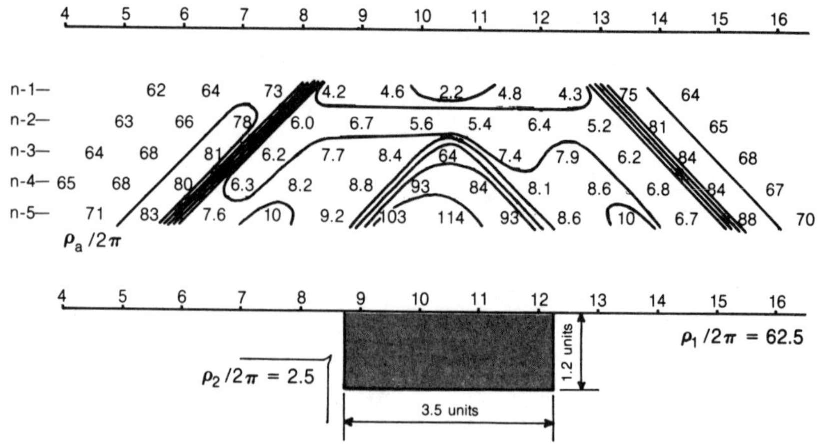

Electrode configuration and plan view for figures 2.13 and 2.14

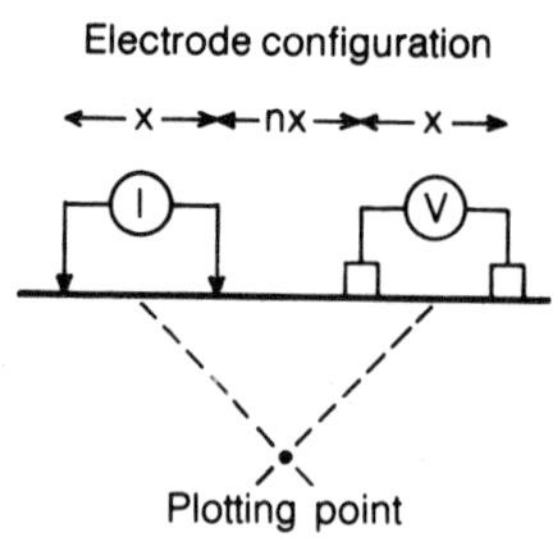

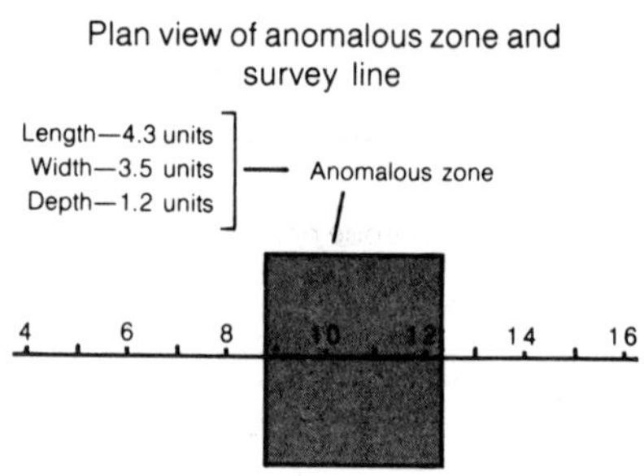

FIGURE 2.14
Scale model case

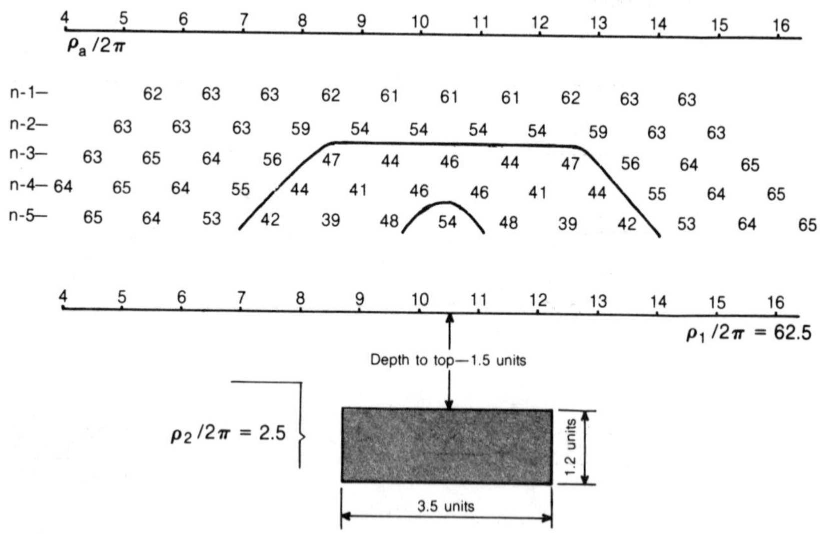

The figures on this page are referred to in the preceding section *Exploration using the resistivity method.*

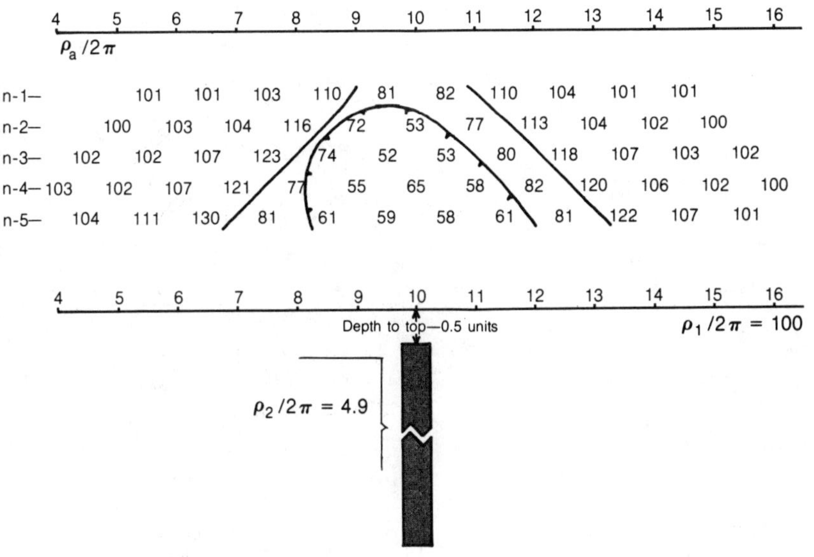

FIGURE 2.15
Scale model case

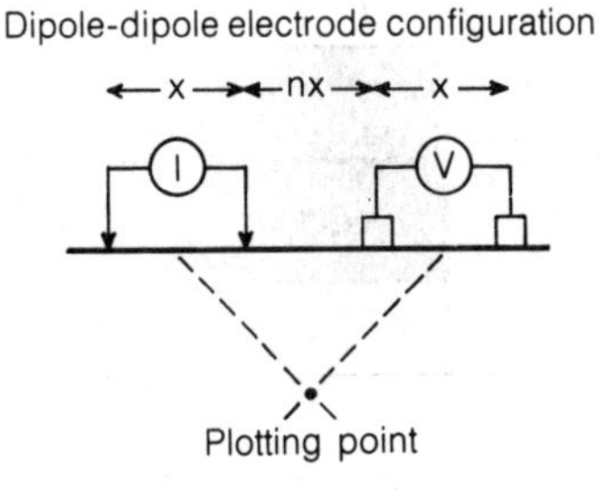

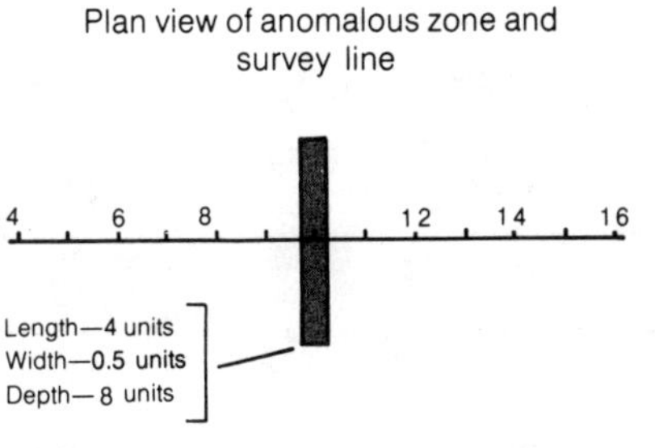

Electrode configuration and plan view for figures 2.15 and 2.16

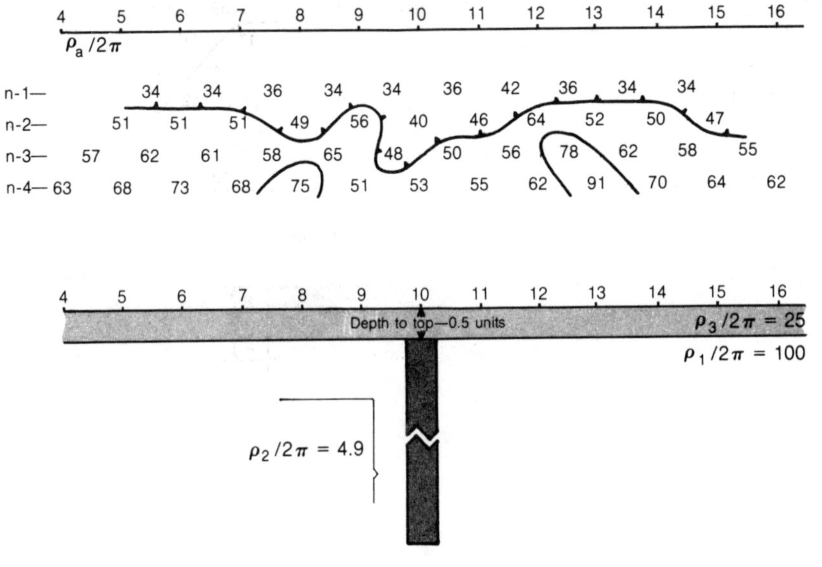

FIGURE 2.16
Scale model case

The table on this page is referred to in the preceding section *Resistivities of earth materials.*

TABLE 2.1
Resistivities of igneous and metamorphic rocks

Rock type	Resistivity range (Ω-m)
Granite	3×10^2 - 10^6
Granite porphyry	4.5×10^3 (wet) - 1.3×10^6 (dry)
Feldspar porphyry	4×10^3 (wet)
Albite	3×10^2 (wet) - 3.3×10^3 (dry)
Syenite	10^2 - 10^6
Diorite	10^4 - 10^5
Diorite porphyry	1.9×10^3 (wet) - 2.8×10^4 (dry)
Porphyrite	10 - 5×10^4 (wet) - 3.3×10^3 (dry)
Carbonatized porphyry	2.5×10^3 (wet) - 6×10^4 (dry)
Quartz porphyry	3×10^2 - 9×10^5
Quartz diorite	2×10^4 - 2×10^6 (wet) - 1.8×10^5 (dry)
Porphyry (various)	60 - 10^4
Dacite	2×10^4 (wet)
Andesite	4.5×10^4 (wet) - 1.7×10^2 (dry)
Diabase porphyry	10^3 (wet) - 1.7×10^5 (dry)
Diabase (various)	20 - 5×10^7
Lavas	10^2 - 5×10^4
Gabbro	10^3 - 10^6
Basalt	10 - 1.3×10^7 (dry)
Olivine norite	10^3 - 6×10^4 (wet)
Peridotite	3×10^3 (wet) - 6.5×10^3 (dry)
Hornfels	8×10^3 (wet) - 6×10^7 (dry)
Schists (calcareous and mica)	20 - 10^4
Tuffs	2×10^3 (wet) - 10^5 (dry)
Graphite schist	10 - 10^2
Slates (various)	6×10^2 - 4×10^7
Gneiss (various)	6.8×10^4 (wet) - 3×10^6 (dry)
Marble	10^2 - 2.5×10^8 (dry)
Skarn	2.5×10^2 (wet) - 2.5×10^8 (dry)
Quartzites (various)	10 - 2×10^8

The table on this page is referred to in the preceding section *Resistivities of earth materials.*

TABLE 2.2
Resistivities of Typical Rock Materials

Material	Frequency at measurement, cps (if not d-c)	Resistivity, ohm-cm
MINERALS		
Galena	...	0.5—5.0
Pyrite	...	0.1
Magnetite	...	0.6—1.0
Graphite	...	0.03
Rock salt (impure)	...	3×10^3—5×10^5
Serpentine	...	2×10^4
Siderite	...	7×10^3
Igneous rocks		
Granite	...	10^8
Granite	16	5×10^5
Diorite	...	10^6
Gabbro	...	10^7—1.4×10^9
Diabase	...	3.1×10^5
Metamorphic rocks		
Garnet gneiss	...	2×10^7
Mica schist	16	1.3×10^5
Biotite gneiss	...	10^8—6×10^8
Slate	...	6.4×10^4—6.5×10^6
Sedimentary rocks		
Chattanooga shale	50	2×10^3—1.4×10^5
Michigan shale	60	2×10^5
Calumet and Hecla conglomerates	60	2×10^5—1.3×10^6
Muschelkalk sandstone	16	7×10^3
Ferruginous sandstone	...	7×10^5
Muschelkalk limestone	16	1.8×10^4
Marl	...	7×10^3
Glacial till	...	5×10^4
Oil sand	...	4×10^2—2.2×10^4

TABLE 2.3
Resistivities of sediments

Rock type	Resistivity range (Ω-m)
Consolidated shales	20—2×10^3
Argillites	10—8×10^2
Conglomerates	2×10^3—10^4
Sandstones	1—6.4×10^8
Limestones	50—10^7
Dolomite	3.5×10^2—5×10^3
Unconsolidated wet clay	20
Marls	3—70
Clays	1—100
Alluvium and sands	10—800
Oil sands	4—800

The table on this page is referred to in the preceding section *Resistivities of earth materials.*

TABLE 2.4
Variation of rock resistivity with water content

Rock	% H_2O	(Ω-m)
Siltstone	0.54	1.5×10^4
Siltstone	0.44	8.4×10^6
Siltstone	0.38	5.6×10^8
Coarse grain SS	0.39	9.6×10^5
Coarse grain SS	0.18	10^8
Medium grain SS	1.0	4.2×10^3
Medium grain SS	1.67	3.2×10^6
Medium grain SS	0.1	1.4×10^8
Graywacke SS	1.16	4.7×10^3
Graywacke SS	0.45	5.8×10^4
Arkosic SS	1.26	10^3
Arkosic SS	1.0	1.4×10^3
Organic limestone	11	0.6×10^3
Dolomite	2	5.3×10^3
Dolomite	1.3	6×10^3
Dolomite	0.96	8×10^3
Peridotite	0.1	3×10^3
Peridotite	0.03	2×10^4
Peridotite	0.016	10^6
Peridotite	0	1.8×10^7
Pyrophyllite	0.76	6×10^6
Pyrophyllite	0.72	5×10^7
Pyrophyllite	0.7	2×10^8
Pyrophyllite	0	10^{11}
Granite	0.31	4.4×10^3
Granite	0.19	1.8×10^6
Granite	0.06	1.3×10^8
Granite	0	10^{10}
Diorite	0.02	5.8×10^5
Diorite	0	6×10^6
Basalt	0.95	4×10^4
Basalt	0.49	9×10^5
Basalt	0.26	3×10^7
Basalt	0	1.3×10^8
Olivine-pyrox.	0.028	2×10^4
Olivine-pyrox.	0.014	4×10^5
Olivine-pyrox.	0	5.6×10^7

The tables on this page are referred to in the preceding section *Resistivities of earth materials.*

TABLE 2.5
Resistivities of minerals (zero frequency)

		Reported resistivities (Ω-m)	
		Low	High
METALLIC MINERALS			
Copper		1.2×10^{-8}	3.0×10^{-7}
Gold			2.4×10^{-8}
Graphite	C, parallel to base plane	2.8×10^{-5}	
SEMICONDUCTING MINERALS			
Argentite	Ag_2S	8.0×10^{-1}	2.0×10^{-3}
Arsenopyrite	FeAsS	2.0×10^{-5}	15
Bismuthinite	Bi_2S_3	3.0	12
Bornite	$Fe_2S_3(3-5) \times Cu_2S$	1.6×10^{-6}	6.0×10^{-3}
Cassiterite	SnO_2	4.5×10^{-4}	$1.0 \times 10^{+4}$
Chalcocite	Cu_2S	8.0×10^{-5}	1.0×10^{-4}
Chalcopyrite	$Fe_2S_3 \times Cu_2S$	3.0×10^{-5}	2.0×10^{-1}
Cobaltite	CoAsS	1.0×10^{-5}	1.3×10^{-1}
Covellite	CuS	3.0×10^{-7}	8.3×10^{-5}
Cuprite	Cu_2O	10	50
Enargite	Cu_3AsS_4	2.0×10^{-4}	9.0×10^{-1}
Galena	PbS	6.8×10^{-6}	5.8×10^{-1}
Hauerite	MnS_2	10	20
Ilmenite	$FeTiO_3$	1.0×10^{-3}	4.0
Jamesonite	$Pb_4FeSb_6S_{14}$	2.0×10^{-2}	1.5×10^{-1}
Magnetite	Fe_3O_4	9.3×10^{-5}	$1.0 \times 10^{+4}$
Manganite	$MnO \times OH$	1.8×10^{-2}	5.0×10^{-1}
Marcasite	FeS_2	1.0×10^{-3}	1.5×10^{-1}
Melaconite	CuO	$6.0 \times 10^{+3}$	
Metacinnabarite	4 (HgS)	2.0×10^{-6}	1.0×10^{-3}
Millerite	NiS	2.0×10^{-7}	4.0×10^{-7}
Molybdenite	MoS_2	1.2×10^{-2}	$1.0 \times 10^{+6}$
Nicollite	NiAs	1.1×10^{-7}	2.0×10^{-6}
Pentlandite	$(Fe, Ni)_9S_8$	1.0×10^{-6}	1.1×10^{-5}
Psilomelane	$kMnO \times MnO, nH_2O$	8.8×10^{-2}	$6.0 \times 10^{+3}$
Pyrite	FeS_2	1.0×10^{-5}	6.0×10^{-1}
Pyrolusite	MnO_2	7.2×10^{-3}	31.8 ± 5.3
Pyrrhotite	Fe_7S_8	2.0×10^{-4}	1.6×10^{-4}
Rutile	TiO_2	29	$9.1 \times 10^{+2}$
Skutterudite	$CoAs_3$	1.1×10^{-6}	1.6×10^{-4}
Smaltite	$CoAs_2$	1.1×10^{-6}	1.2×10^{-5}
Sphalerite	ZnS	1.8×10^{-2}	$4.0 \times 10^{+4}$
Tetrahedrite	Cu_3SbS_3	3.0×10^{-1}	$3.0 \times 10^{+4}$
Uraninite	UO_2	1.5	$2.0 \times 10^{+2}$

Geothermal exploration using the resistivity method

The resistivity method has not been of reliable, consistent assistance in the search for sulfide deposits. The resistivity method, however, has been of continuing use in the exploration for geothermal energy.

The resistivity of a rock unit decreases with increased porosity in the rock and increased salinity in the fluids in the pore volumes. The resistivity of a solution in a rock will decrease further as the temperature increases.

Therefore, a large, porous, hot saline geothermal cell, which may have a volume of several cubic kilometers, often presents an inviting target for a resistivity survey. The surface temperature data and the surface resistivity data shown in figure 2.17 are from an area in New Zealand where geothermal energy has been tapped at a very shallow depth. Low resistivities correlate well with higher temperatures.

The results shown in figure 2.18 are part of an exploration survey for geothermal sources on Java. The zone of very low resistivities (labelled zone B) correlates with a known fault zone that is marked by hot springs. The deeper zone of low resistivities [anomalous for larger volumes of (N)] marked zone A is obviously of the same character, but it has no surface manifestation.

An interesting area for geothermal exploration in the United States has been the Mesa Area in Imperial County, California. A rough correlation of several types of geophysical data is shown in figure 2.19. The zone of low resistivities at depth is outlined on three lines using the dipole-dipole configuration and x = 2000 ft; n = 1, 2, 3, 4, 5.

The resistivity results from test line no. 1 are shown in figure 2.20. All of the sediments in the Imperial Valley have a low resistivity. The geothermal cell, confirmed by several drill holes, is seen to have an extremely low resistivity level.

FIGURE 2.17
Geophysical survey
Broadlands Area, New Zealand

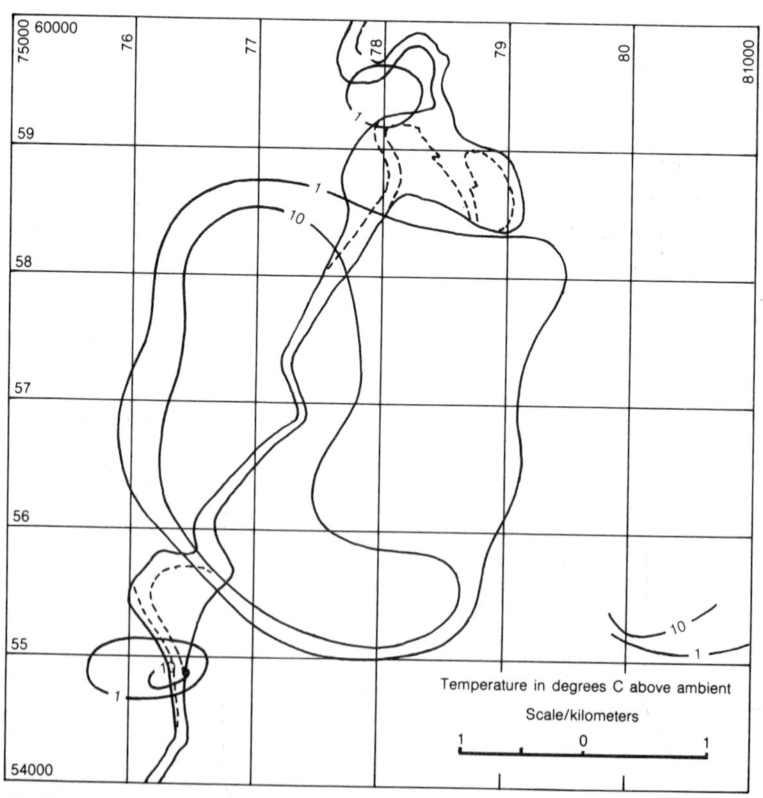

A. Temperature at 15 m depth

The figures on this page are referred to in the preceding section *Geothermal exploration using the resistivity method.*

Figure 2.17, continued

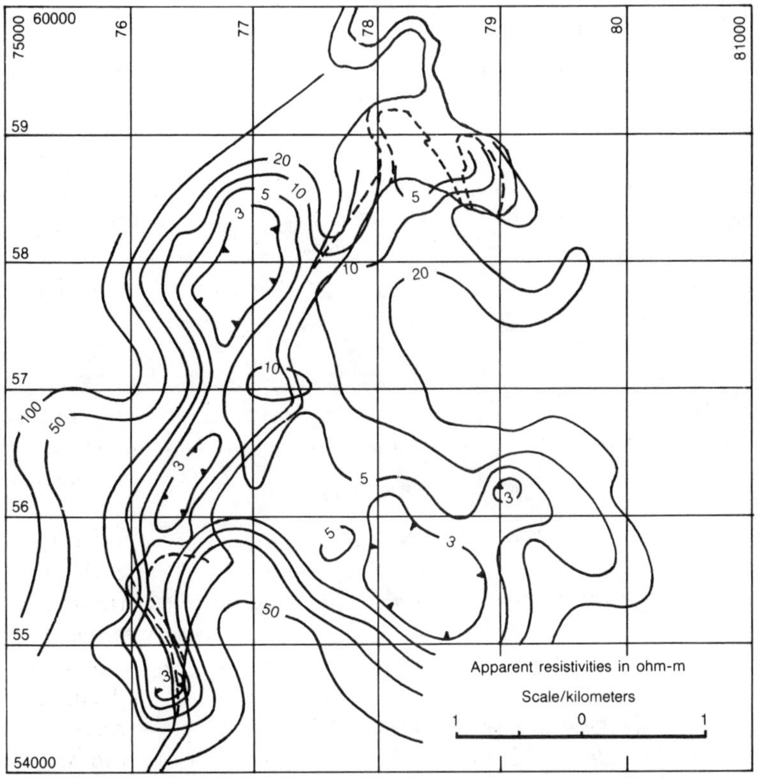

B. Apparent resistivity survey using Wenner configuration, A = 180m

FIGURE 2.18
Apparent resistivity survey, Dieng Plateau Area, Java, Indonesia

Pseudo section plotting method along Dieng-Bator Road

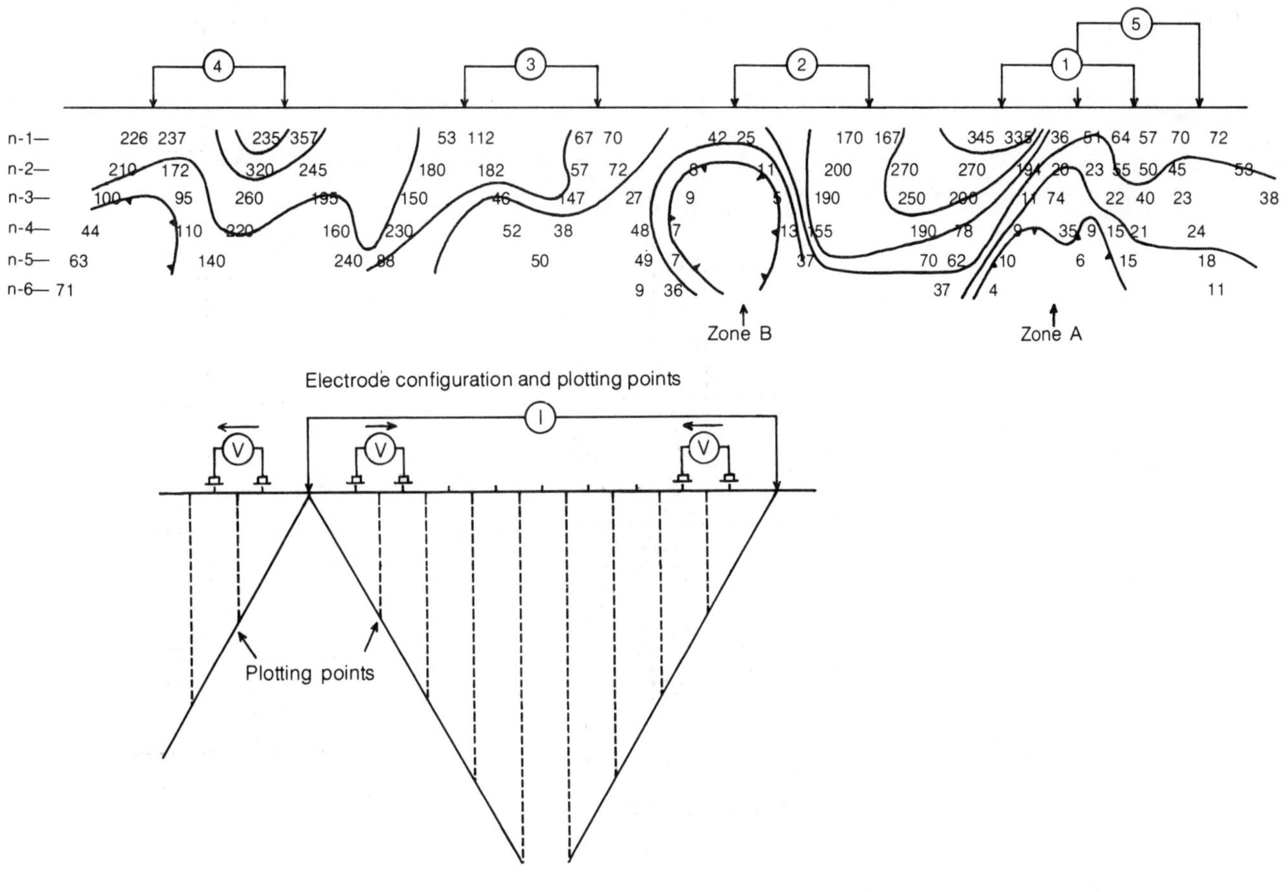

The figures on this page are referred to in the preceding section *Geothermal exploration using the resistivity method.*

FIGURE 2.19
Mesa area, Imperial County, California

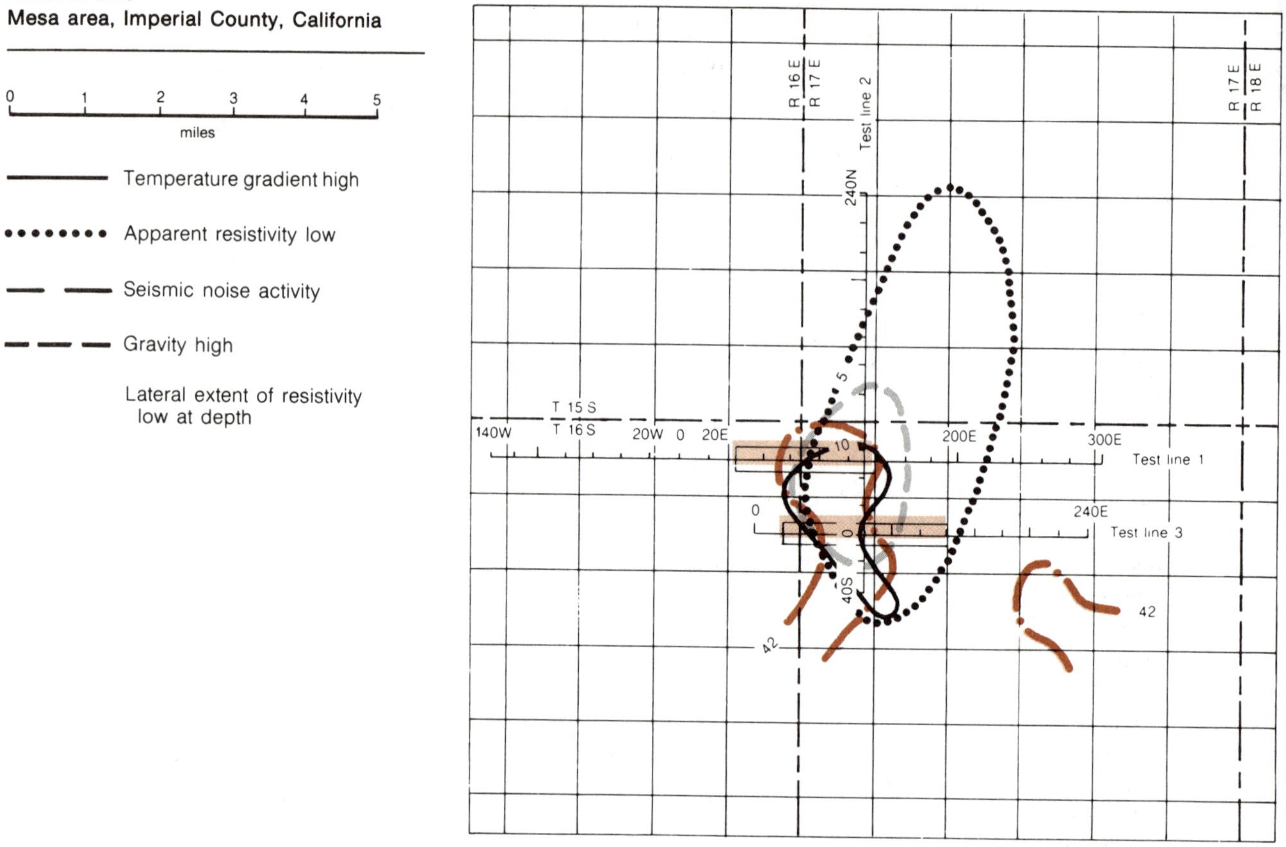

FIGURE 2.20
East Mesa geothermal anomaly, Imperial County, California
Test line 1

Frequency—0.125 Hz
Values corrected for inductive coupling
$\rho/2\pi$ in ohm-ft = (ρ in ohm-m) × 0.52

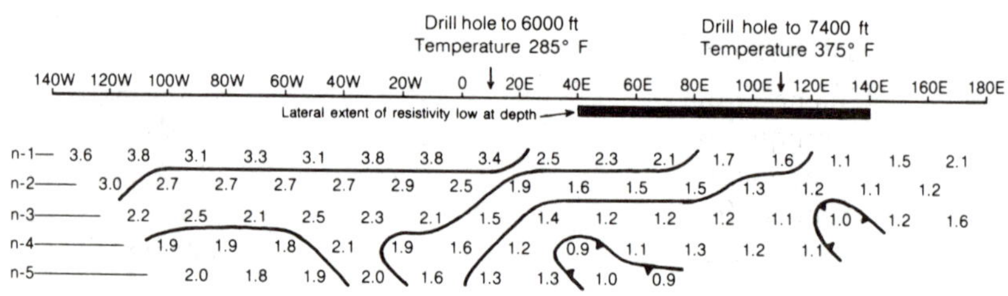

Electrode configuration

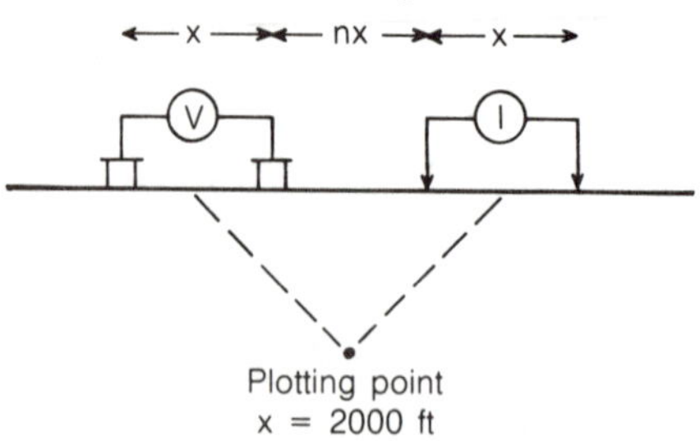

Equipotential or mise-à-la-masse method

The equipotential or mise-à-la-masse method is a modification of the resistivity method that is especially useful when it is possible to make contact directly to one or more points on a zone of massive sulfide mineralization. It is also very helpful if two or more locations of massive mineralization are known. With this rather simple technique, the geophysicist can trace the extension of a known zone of massive mineralization or determine which of several zones are interconnected.

Figures 2.21 and 2.22 indicate the basic theory. Potential theory requires that an excellent conductor maintaining current flow must have the same potential throughout its entire volume. If a current electrode from a high voltage current source is inserted into a good conductor (with the other current electrode at infinity), the entire outside surface of the sulfide zone will have the same high potential.

The potentials in the area of interest are determined by use of a suitable voltmeter, a long wire, and two potential electrodes. One potential electrode is placed at infinity and is established as the zero potential level.

The field results in figures 2.23 through 2.26 are from a survey at York Harbour, Newfoundland. In this case, the sulfide was exposed in underground workings. It was desired to know where the zones most closely approached the surface and if the H-1 zone and the H-2 zone were the lower portions of a single zone near surface.

The concentric potential distribution on the surface above a single current electrode is shown in figure 2.23. The center of the potential pattern is immediately above the current electrode at depth.

This was not the case when the current electrode was placed first in the H-1 zone and then in the H-2 zone. In both cases the center of the surface potential distribution is considerably to the east of the underground position of the mineralization. Further, since almost exactly the same potential distribution was measured for both locations for the current electrode at depth, both zone H-1 and zone H-2 must be part of a single zone of sulfide mineralization that has its most shallow position beneath the center of the surface potential pattern.

Self-potential method

Just as the name implies, the self-potential method is an electrical exploration technique based upon natural direct currents flowing in the ground. Recent research has suggested there may be more than one natural mechanism for these potentials. The potentials arising from the chemical reactions in oxidizing massive sulfides (figure 2.27) are the best known.

The method can be used (1) as profiles along lines (figure 2.28), or (2) as contours of plan data from several lines.

With the use of any one of several types of high impedance dc voltmeters, the measurement is quite simple. One nonpolarizing electrode is placed at infinity to establish a zero potential level. The potential (relative to the remote electrode) over the entire area of interest is then established using a long wire, a second nonpolarizing electrode and the voltmeter. As shown on the sketch drawings, zones of massive sulfide mineralization that extend into the oxidizing zone are usually marked by definite voltage lows.

The figures on this page are referred to in the preceding section *Equipotential or mise-à-la-masse method.*

FIGURE 2.21
Distortion of the equipotentials

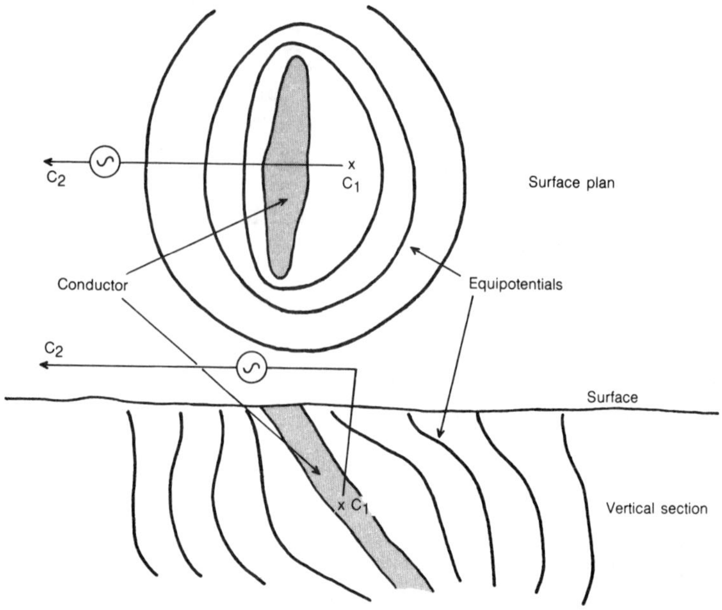

FIGURE 2.22
Mise-à-la-masse method used at York Harbour, Newfoundland

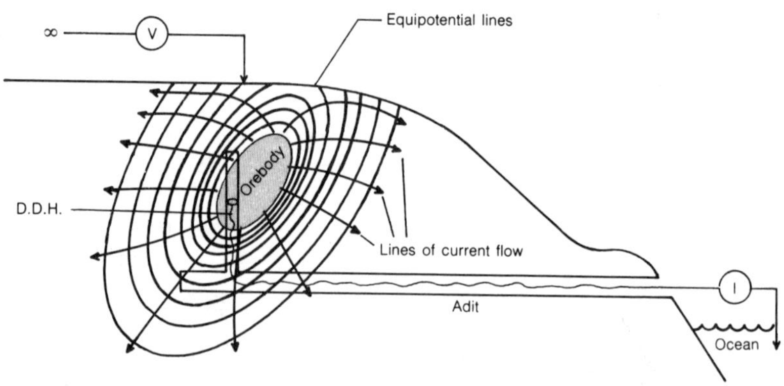

The figures on this page are referred to in the preceding section *Equipotential or mise-à-la-masse method.*

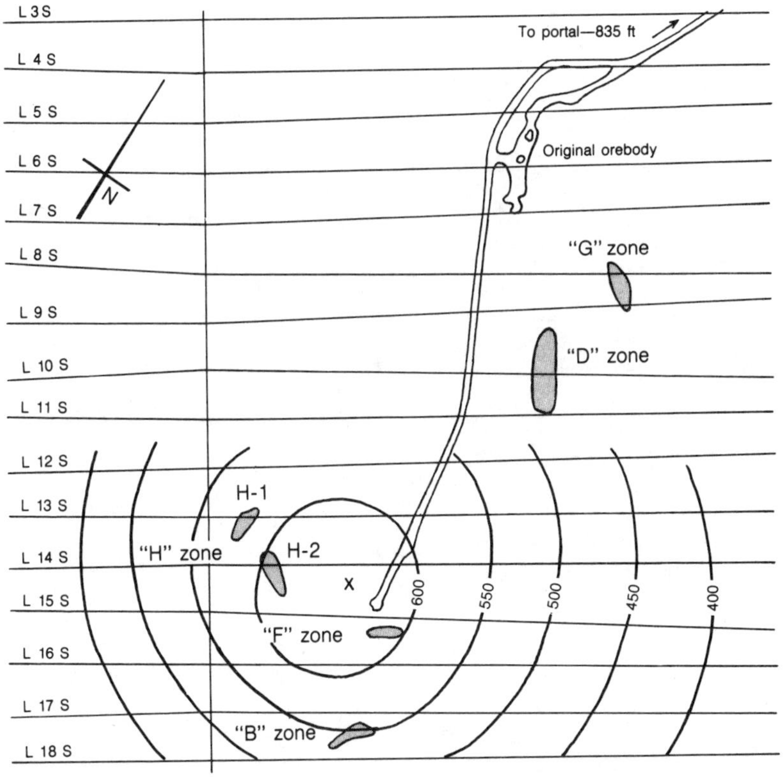

FIGURE 2.23
Potential pattern from current source in test position

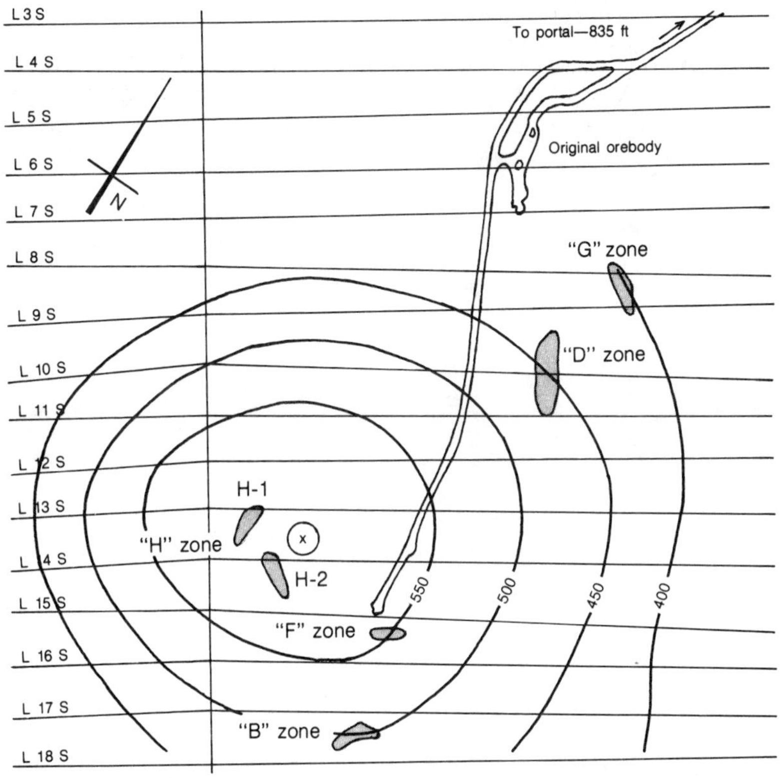

FIGURE 2.24
Potential pattern from current source in H-1 zone

The figures on this page are referred to in the preceding section *Equipotential or mise-à-la-masse method.*

FIGURE 2.25
Potential pattern from current source in H-2 zone

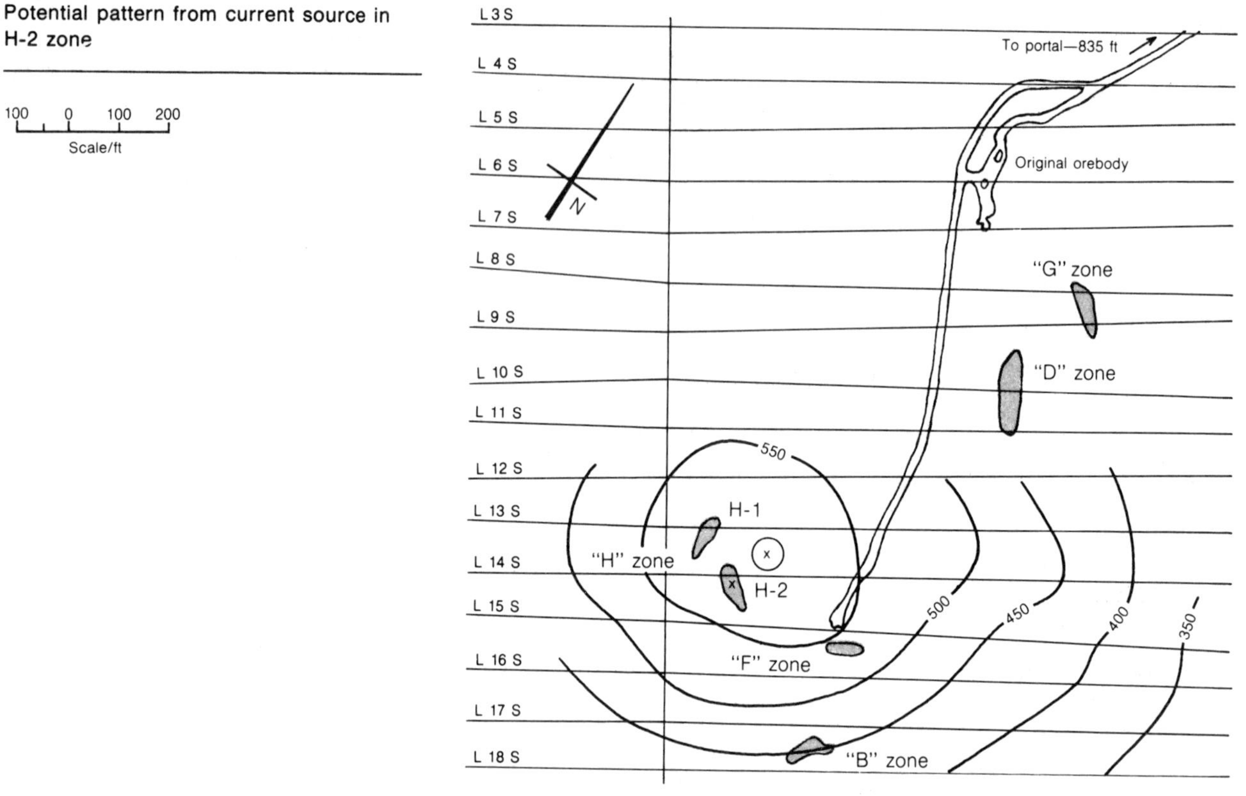

FIGURE 2.26
Potential profile along line 13-S for current electrode in H-1 zone

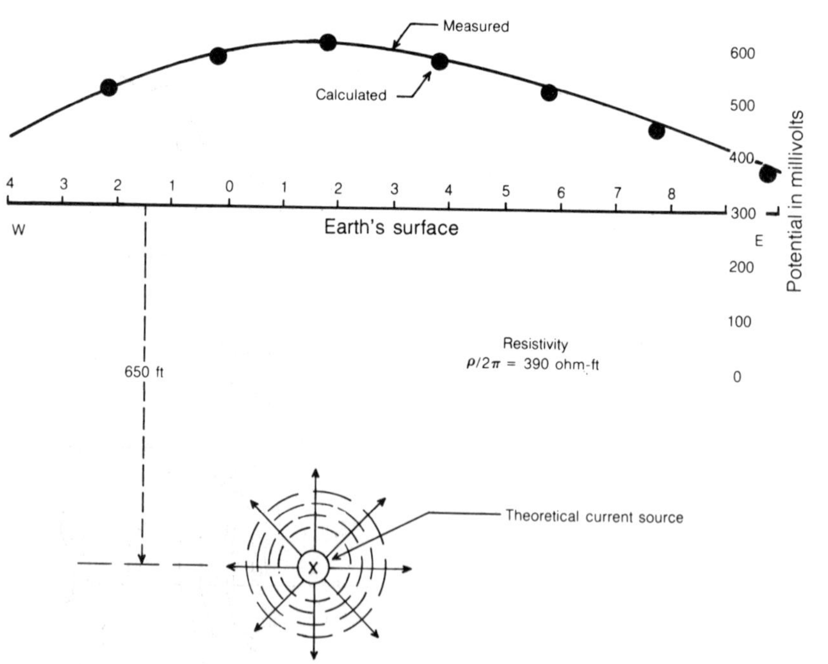

The figures on this page are referred to in the preceding section *Equipotential or mise-à-la-masse method.*

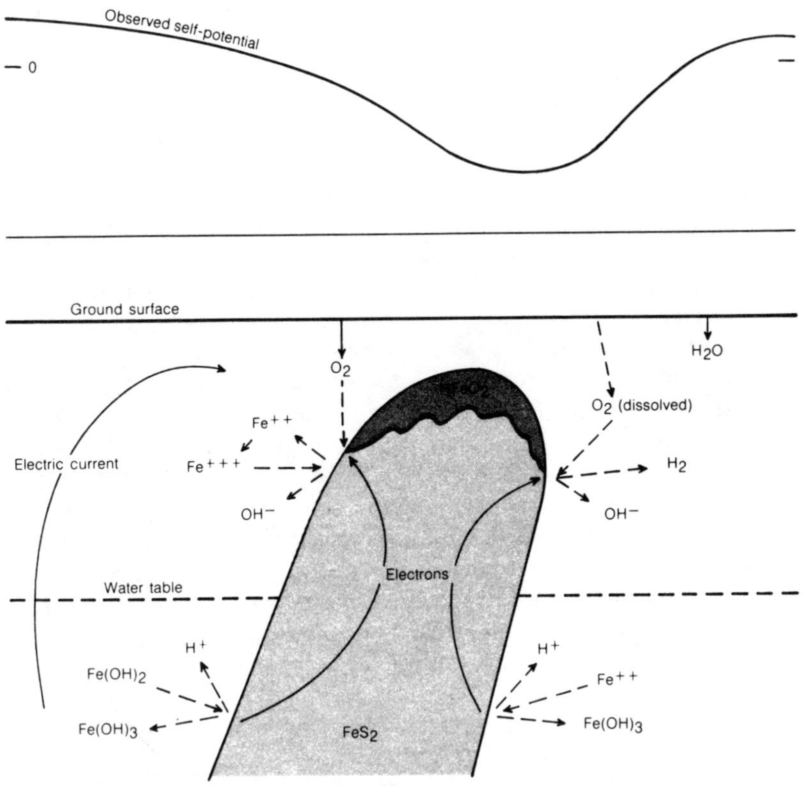

FIGURE 2.27
Idealized SP case

Self-potentials near an oxidizing massive sulfide body showing current flow lines and chemical reactions.

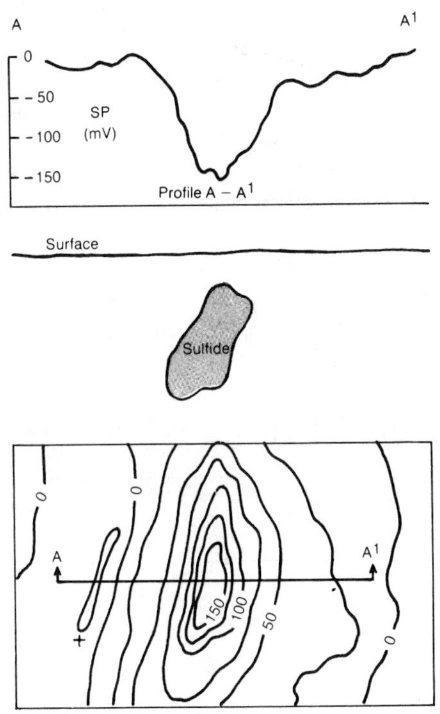

FIGURE 2.28
Idealized SP case

Typical SP profile and contours over sulfide body.

FIGURE 2.29
Sketch showing overvoltage effect

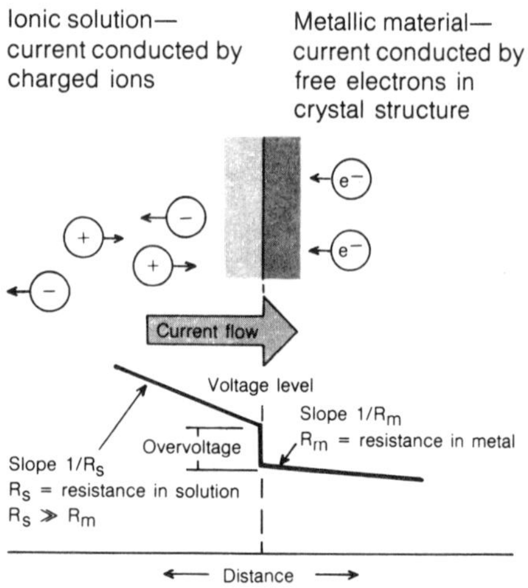

FIGURE 2.30
Frequency spectra for electrode polarization

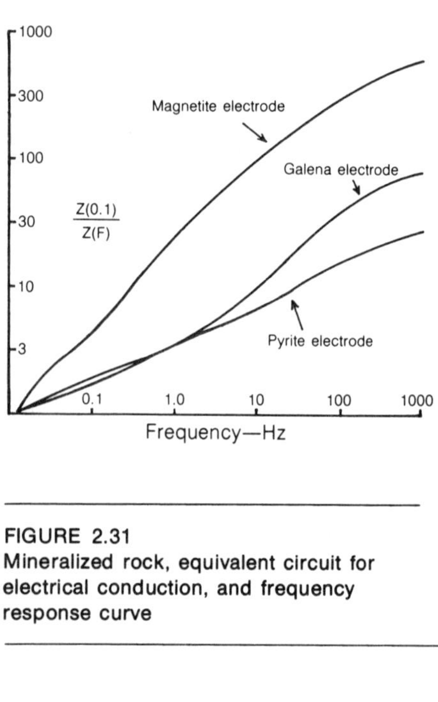

FIGURE 2.31
Mineralized rock, equivalent circuit for electrical conduction, and frequency response curve

General theory of the IP effect

As discussed later, recent research has produced a fairly complex description of the IP phenomenon. At this point a fairly simple explanation will suffice. This chapter has shown that in almost all situations current flow in the earth is carried in the solutions filling the pore spaces of the rocks. The current flow is actually maintained by charged ions in the solutions.

The IP effect is created when this *ionic* current flow is converted to *electronic* current flow at the surface of metallic minerals that are present and in contact with the fluids in the pore spaces. A simple explanation of the IP effect can be seen in figures 2.29 and 2.30.

Solid-state physicists have known for many years of the existence of the *over-voltage effect* at the surface between a metal and an ionic solution. This voltage drop is over and above the I × R voltage drop in the materials to either side of the surface. An ion that approaches the surface during current flow, but which does not possess enough electrical energy to overcome this *over-voltage* cannot take an electron from the crystal lattice of the metal, or give an electron to the crystal lattice of the metal. In this situation, the charged ion remains at the surface, and its electrical charge decreases the current flow through the surface.

As time passes, this induced charge increases the polarization at the surface and the current flow continues to decrease. If the current flow is terminated, these induced polarization charges will return to their normal positions under the influence of their own electromotive forces. This transient flow of charged ions will be measured as a voltage that (1) exists after the applied voltage and current are terminated, but (2) decays to zero rapidly.

It can be seen that if a voltage is applied across the interface, the current flow will immediately begin to decrease, i.e., the resistance of the surface will increase with time. Therefore, the average resistance of the surface will decrease if the applied voltage and current are reversed in direction more and more rapidly.

These two simple descriptions outline the time domain (or pulse transient) IP method and the frequency domain (or variable frequency) method. In the first method the geophysicist looks for portions of the earth where current flow is maintained for a short time after the applied current is terminated. In the second method the geophysicist tries to locate portions of the earth where the resistivity of the rocks decreases as the frequency of the applied current is increased.

The sketches in figures 2.31 through 2.33 show some additional idealized curves for resistivity vs. frequency and also some equivalent electrical circuits that can be used to simply approximate the electrical effects at a metal-solution interface.

The simple mathematical formulation in figure 2.34 shows that the chargeability parameter (m) measured in the time domain method is exactly equivalent to the frequency effect parameter (fe) used in the

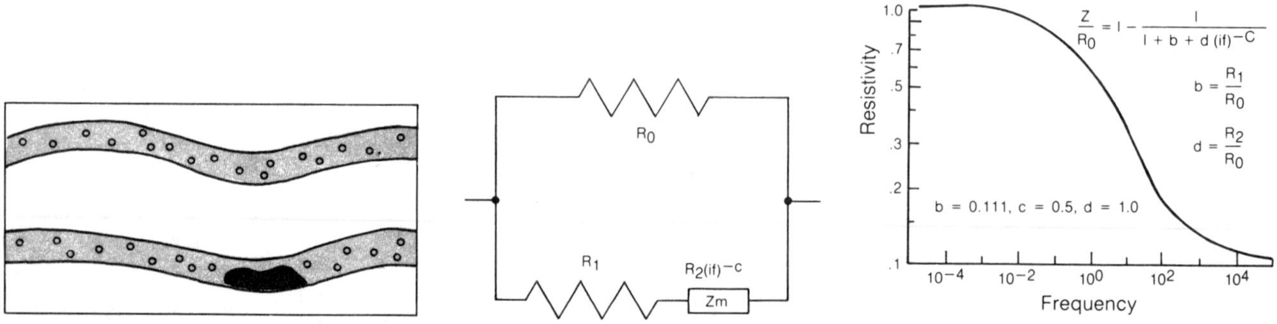

A. Mineralized rock
B. Equivalent circuit
C. Frequency response

variable frequency method of IP exploration. It has been known for some time that the mechanism that controlled the IP effect would also lead to a phase shift between the current flowing through the rock and the voltage across a region containing metallic mineralization. Recently, the phase lag ϕ of the voltage, with respect to the applied current, is being used in IP exploration. This phase shift (ϕ) is equivalent to (m) and (fe).

Figure 2.35 shows that the chargeability and frequency effect measured at the surface has exactly the same relationship to the chargeability and frequency effect in the mineralized zone at depth. It can be seen that all of the IP measurement techniques used in induced polarization exploration (time domain, variable frequency, phase IP) are exactly equivalent.

Figure 2.35 also shows the definition of the metal factor (mf) parameter. Unlike the chargeability and the frequency effect, this parameter has dimensions. The metal factor has the dimensions of conductivity, the inverse of resistivity. In situations in which the earth resistivity varies widely, the use of this parameter sometimes gives more easily interpretable anomalies.

FIGURE 2.32
Log-log plot for conductivity vs. frequency in mineralized rock

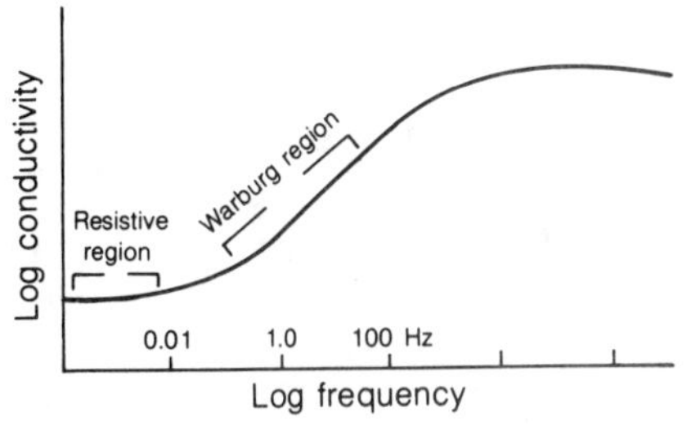

FIGURE 2.34
Time domain and frequency domain by LaPlace transform theory

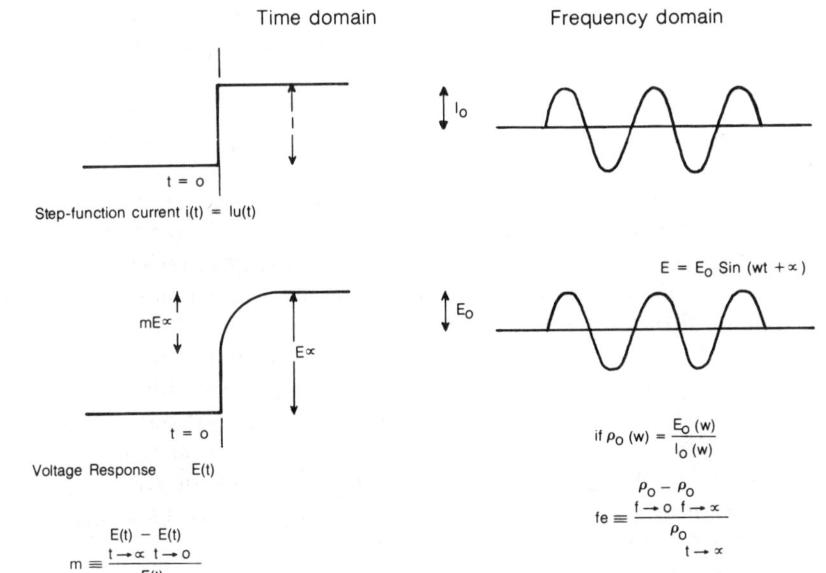

$$m \equiv \frac{E(t)_{t\to\infty} - E(t)_{t\to o}}{E(t)_{t\to\infty}}$$

$$fe \equiv \frac{\rho_{o\,f\to o} - \rho_{o\,f\to\infty}}{\rho_{o\,t\to\infty}}$$

$$m = \frac{fe}{I + fe} \cong fe$$

FIGURE 2.33
Equivalent circuit for electrical conduction in mineralized rocks

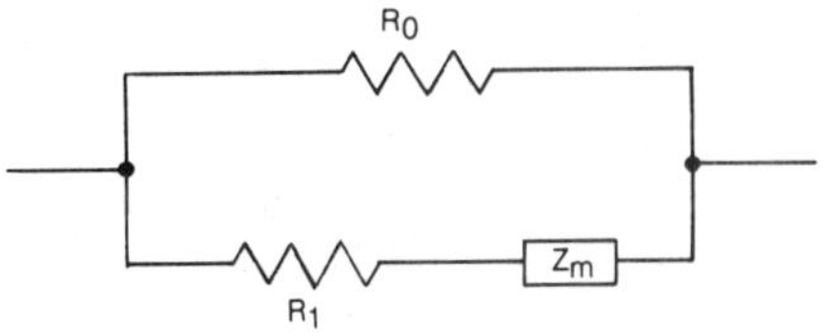

$$Z_m = \frac{R_m}{f^{1/4}}$$

FIGURE 2.35
Chargeability and frequency effect measurement

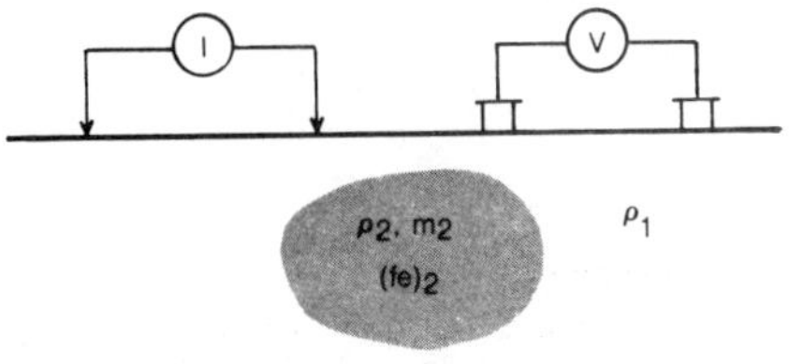

$$m_a \simeq m_2\, \rho_2 \times \frac{1}{\rho_a}\frac{\partial \rho_a}{\partial \rho_2} \qquad (fe)_a = (fe)_2\, \rho_2 \times \frac{1}{\rho_a}\frac{\partial \rho_a}{\partial \rho_2}$$

$$mf \equiv \frac{(fe) \times 100}{\rho_{dc}/2\pi} \times 1000 = \frac{[\rho_{dc}/\rho_{ac} - 1] \times 100}{\rho_{dc}/2\pi} \times 1000$$

$$= \left[\frac{1}{\rho_{ac}} - \frac{1}{\rho_{dc}}\right] \times 2\pi \times 10^5 = [\sigma_{ac} - \sigma_{dc}] \times 2\pi \times 10^5$$

FIGURE 2.36
Electrode configurations for IP surveys

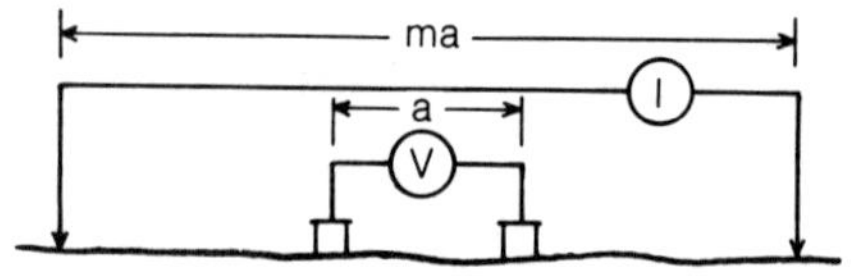

A. Gradient (Schlumberger) array m > 15

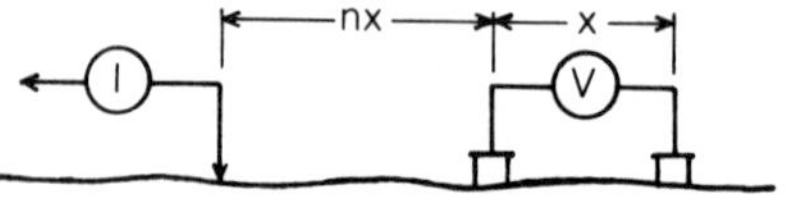

B. Pole-dipole array n = 1,2,3,4, etc.
For n = 1, special case is three array

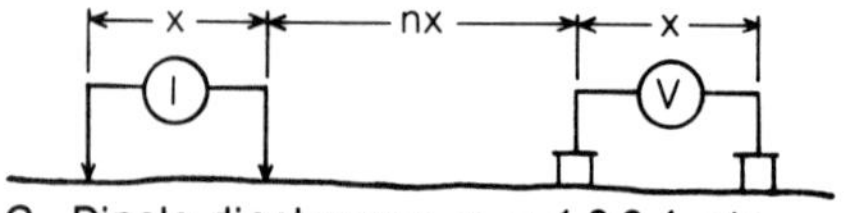

C. Dipole-dipole array n = 1,2,3,4, etc.

Electrode configurations and data plotting techniques

When the resistivity exploration method was discussed, it was shown in figures 2.4 through 2.7 that there were several different ways to place the four electrodes on the earth. The various configurations have inherent advantages in terms of labor required, amount of wire required, depth detection for large sources, and detectability of small, local sources. The three configurations most commonly used to make induced polarization and resistivity measurements are shown in figure 2.36.

Before the advent of modern, solid state electronics, there was one other overriding consideration. Whatever the geophysicist's preference, it was necessary to create a large enough voltage at the potential electrodes to make a satisfactory measurement in the presence of whatever natural electrical noise was present. This was determined by the sensitivity and noise-rejection capabilities of the voltmeter and the power and current capabilities of the transmitter. As shown in figure 2.37, the voltage produced for an ampere of current varies considerably from one configuration to another.

For most electrode configurations, the IP and resistivity results are plotted as profiles along the measurement line. The dipole-dipole measurements for x = 100 ft, n = 2 from a flat-lying ore zone in southeast Missouri are shown in figure 2.38. This presentation could be made with measurements using other electrode configurations; however, as will be shown, the apparent effects measured may be different.

For the dipole-dipole configuration, once the value of the electrode interval (x) is decided upon, the geophysicist can, and usually does, make measurements with several values of (n). The field data in figures 2.39 and 2.40 show values measured for two values of (n). In both cases, the effect of the depth to the top of the source can be seen in that the apparent effects measured increase for the larger values of (n).

For more than two values of (n), the profile method of presentation becomes confused, particularly because we wish to show the resistivity data at the same time. A two-dimensional (pseudo-section) format has been developed to present dipole-dipole data with several values of (n).

If the results are plotted in the two-dimensional array shown in figures 2.40 and 2.41, the effects of lateral variations in the earth's parameters can be easily separated from the effects of vertical variations in the earth's parameters. The lateral position of the apparent values is plotted in relationship to where the electrodes were placed when that particular measurement was made. The distance of the plotted apparent value from the measurement line is related to the separation between the electrodes when the measurement was made.

The 45-degree angle used to plot the data is entirely arbitrary. The pseudo-section plots are contoured, and the resulting anomalous patterns can be recognized as being caused by a particular geometry of source, and/or correlated from line to line. However, the contoured data plots are *most emphatically not* meant to represent sections of the electrical parameters of the subsurface. The relationship between the plotted, contoured pseudo-section and the actual section of the electrical parameter is a very complex mapping function. The pseudo-section data plots are merely a convenient method for showing all of the apparent results along a given line in one presentation.

In figures 2.42 and 2.43 the effect of using different electrode configurations can be studied, as well as the advantages of the pseudo-section plotting technique over the profile method. All of the data are from a relatively narrow, vertical source that is buried under an overburden layer that is 0.5 units in thickness.

In the resistivity data, the gradient array shows a higher resistivity value and is therefore seeing more of the basement; however, there is no indication of the buried conductor. The pole-dipole and the dipole-dipole configuration give some evidence of the conductor at depth for the IP measurement. The gradient array gives no evidence of the vertical source; the dipole-dipole measurements give a recognizable anomaly. These data confirm the data from figures 2.6 and 2.7. The dipole-dipole measurement, which is a lateral second derivative measurement, is the most sensitive to relatively small, lateral variations in the parameters of the earth.

The psuedo-section plot in figure 2.43 should be studied to understand the advantages of this method of presentation over the profile method. The presence of the conductive overburden is clearly shown. The depth to the top of the source, its position, and the fact that the width is limited, are all easily interpreted from the data.

Field data from a flat-lying source are shown in figures 2.44 and 2.45. For this position of a shallow source, there is much less difference between the gradient array profiles and the dipole-dipole profiles. It seems fairly evident that the pseudo-section data plot is the most useful.

The dipole-dipole configuration is the best to use. It can be easily employed in the field because the lengths of wire are relatively short. In field work, electrode intervals have been used that range from x = 10 ft to x = 2000 ft; and n = 1 alone to n = 1, 2, 3, 4, 5, 6.

The disadvantage of the dipole-dipole configuration is the relatively small resistive coupling (figure 2.37). With modern electronics, the design of IP equipment with higher power capabilities is practical. This is particularly true for variable frequency equipment or time domain equipment that uses coherent filters (signal stacking) since noise rejection is better and current requirements are less.

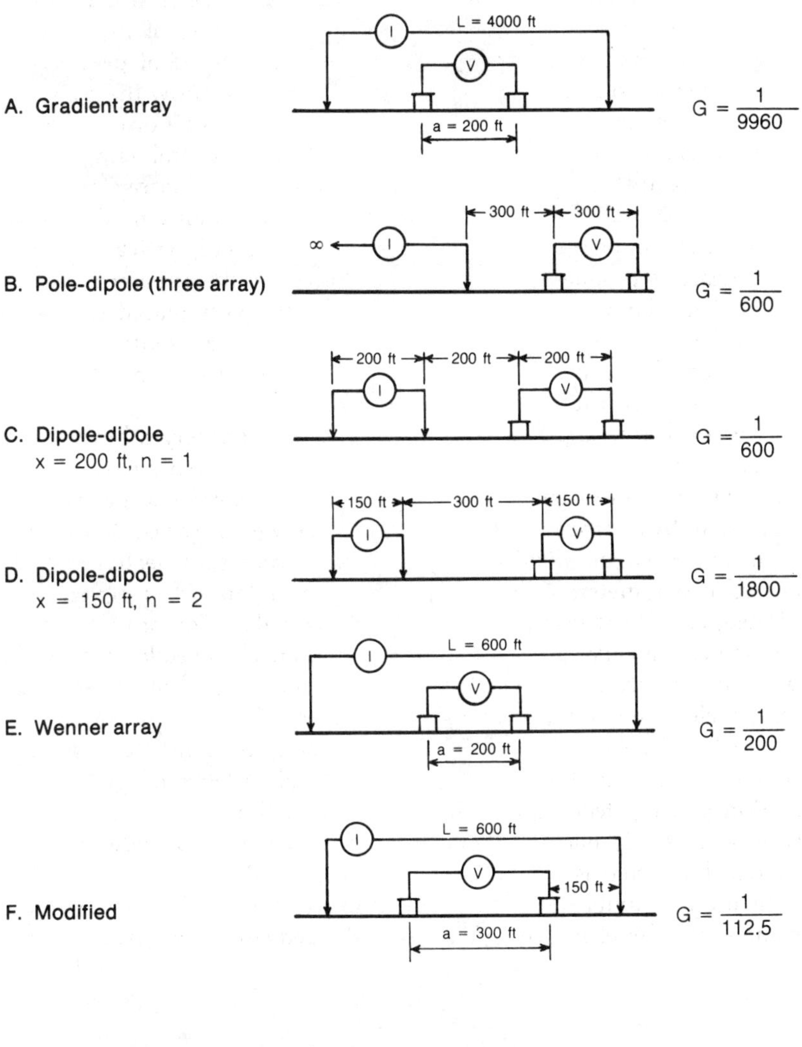

FIGURE 2.37
Electrode configuration and voltage

The figures on this page are referred to in the preceding section *Electrode configurations and data plotting techniques.*

FIGURE 2.38
Induced polarization survey over a southeast Missouri lead deposit

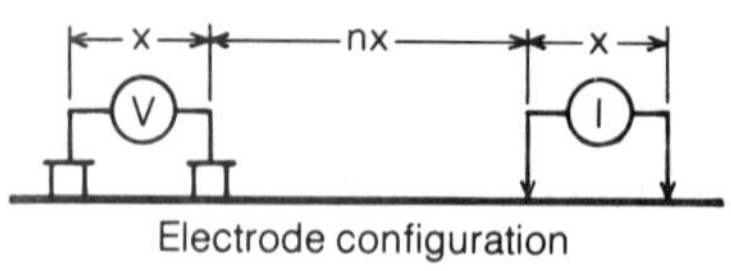

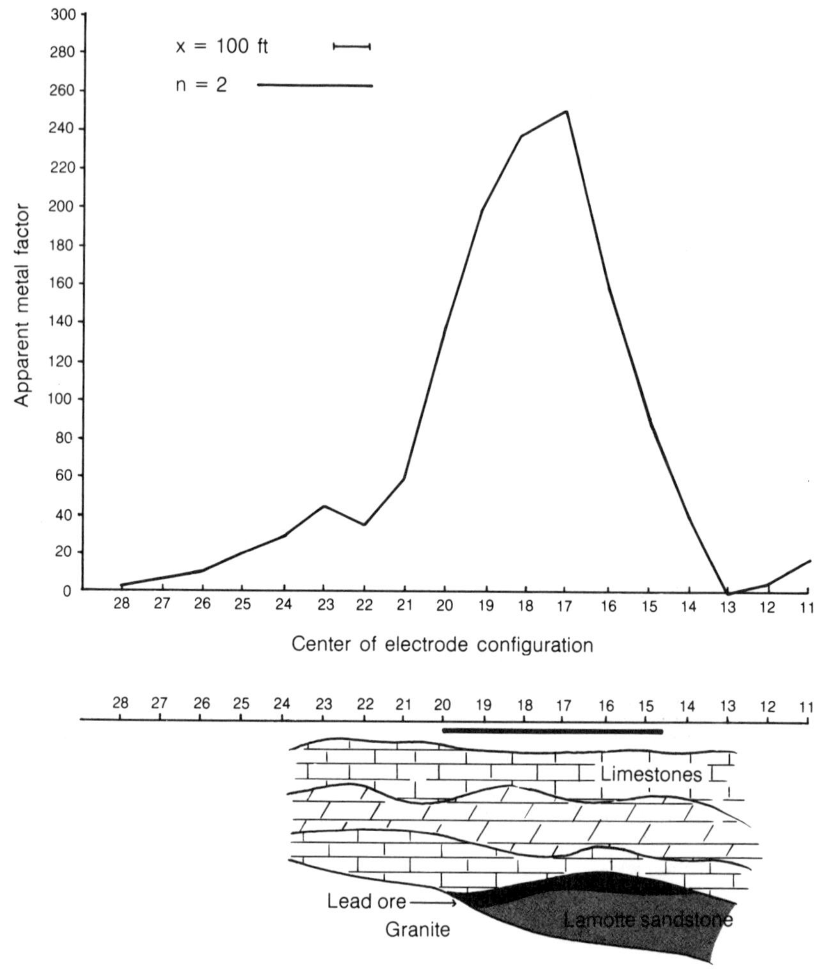

FIGURE 2.39
Induced polarization survey over vertical sulfide orebody

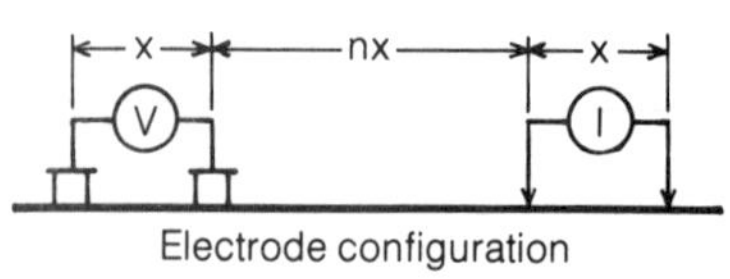

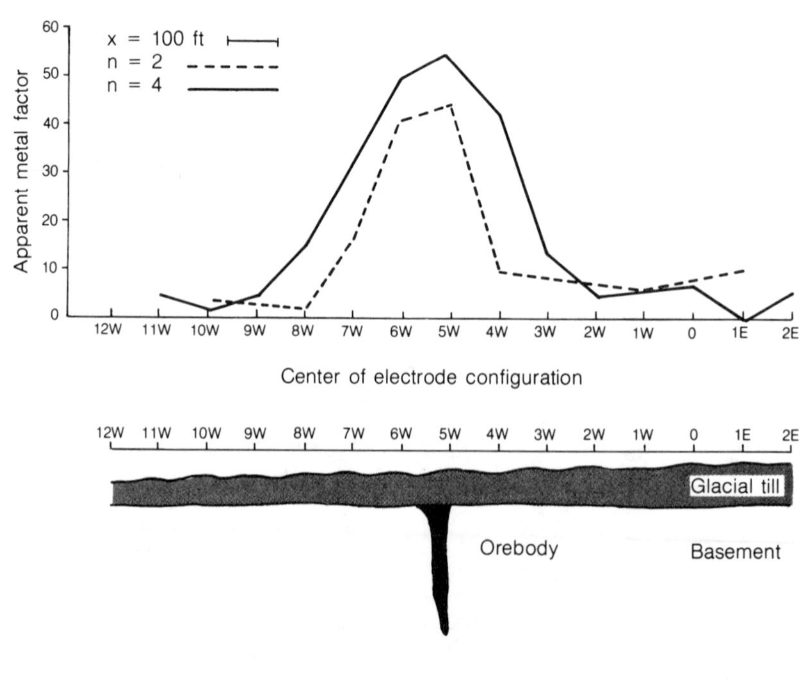

The figures on this page are referred to in the preceding section *Electrode configurations and data plotting techniques.*

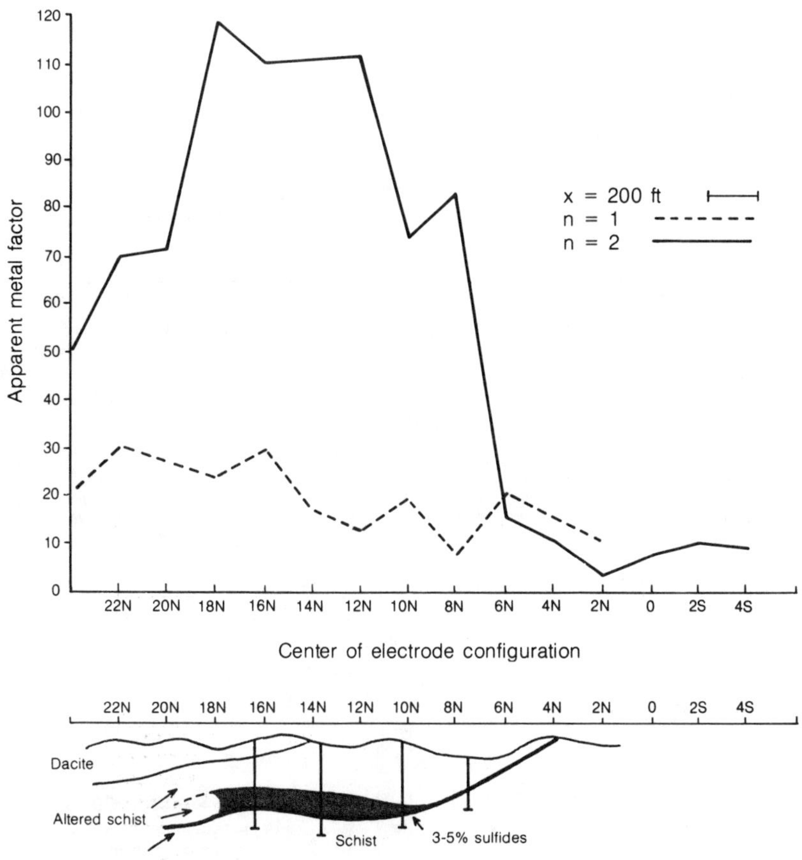

FIGURE 2.40
Induced polarization survey over Cactus Orebody

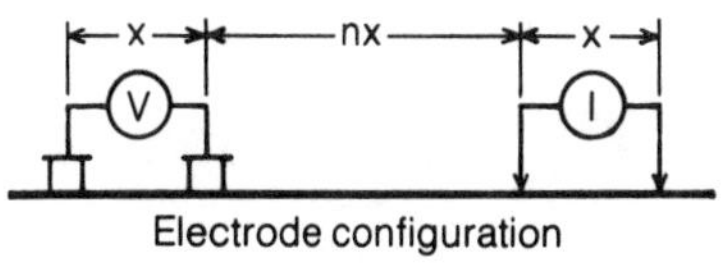

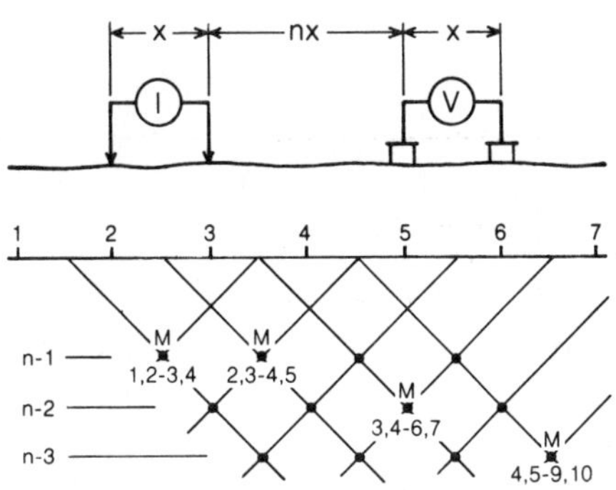

FIGURE 2.41
Dipole-dipole plotting method

The figure on this page is referred to in the preceding section Electrode configurations and data plotting techniques.

FIGURE 2.42
Theoretical induced polarization and resistivity studies

Vertical source with overburden

Dipole-dipole array, x = 1 unit, n = 3
Pole-dipole array, x = 1 unit, n = 3
Gradient array, a = 1 unit

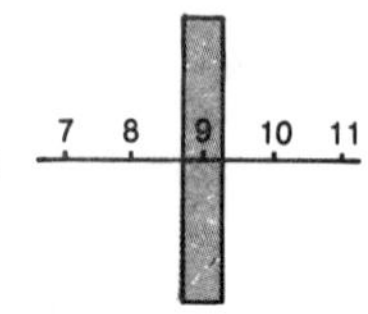

Plan view of anomalous zone and survey line

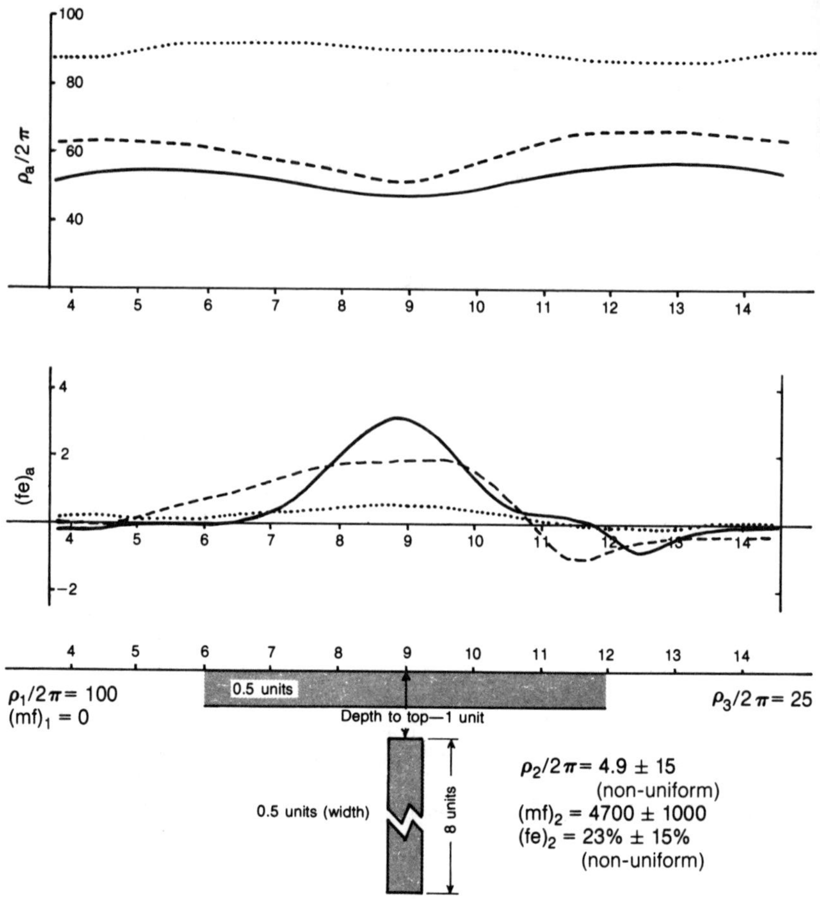

The figure on this page is referred to in the preceding section *Electrode configurations and data plotting techniques.*

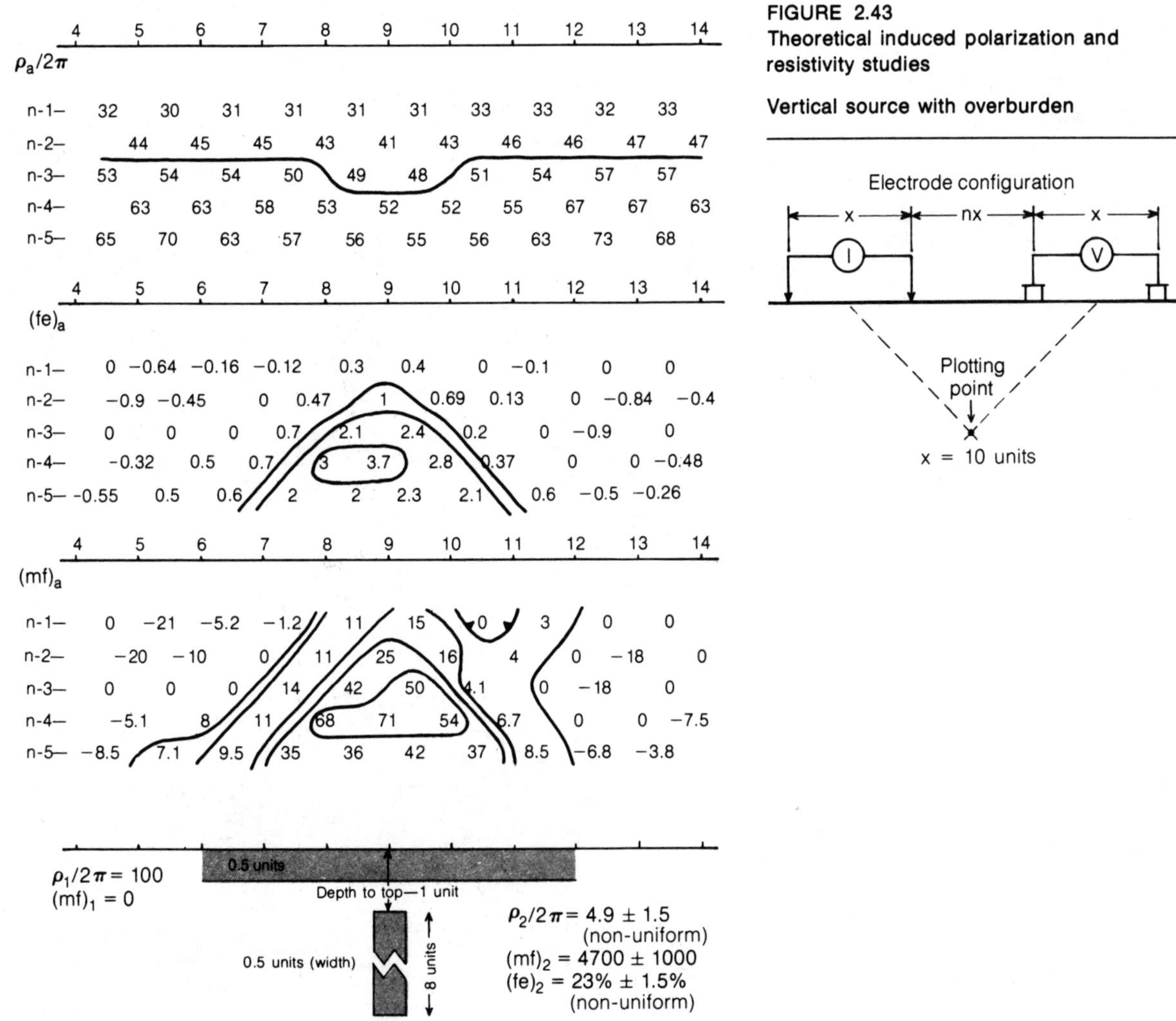

FIGURE 2.43
Theoretical induced polarization and resistivity studies

Vertical source with overburden

The figure on this page is referred to in the preceding section *Electrode configurations and data plotting techniques.*

FIGURE 2.44
Geophysical data and drilling results from Pine Point type mineralization

Pyramid Orebody
Line-375W

—— Dipole-dipole array, x – 200 ft, n = 1
- - - - Gradient array, a = 100 ft

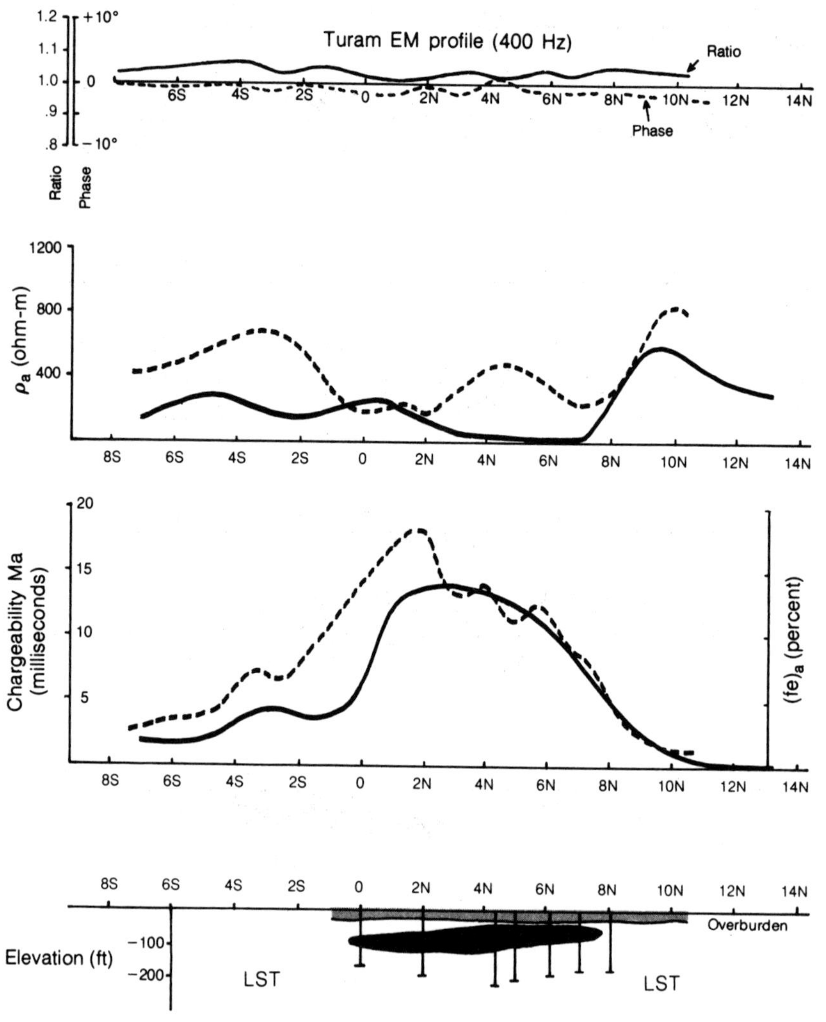

The figure on this page is referred to in the preceding section *Electrode configurations and data plotting techniques.*

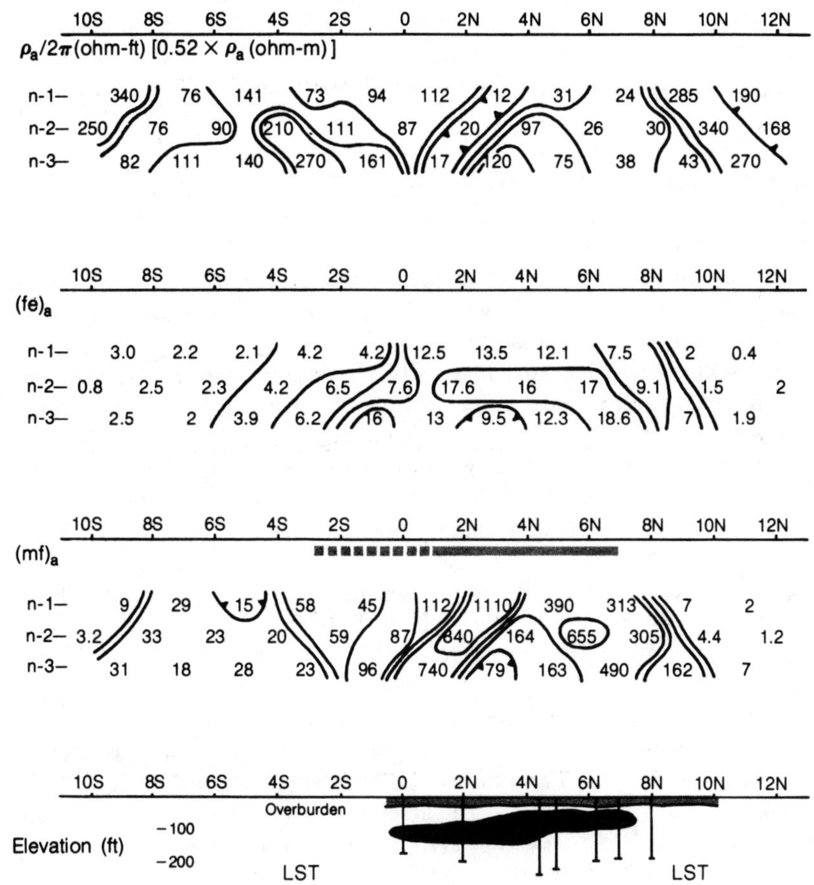

FIGURE 2.45
Induced polarization and drilling results from Pine Point type mineralization

Pyramid Orebody
Line-375W

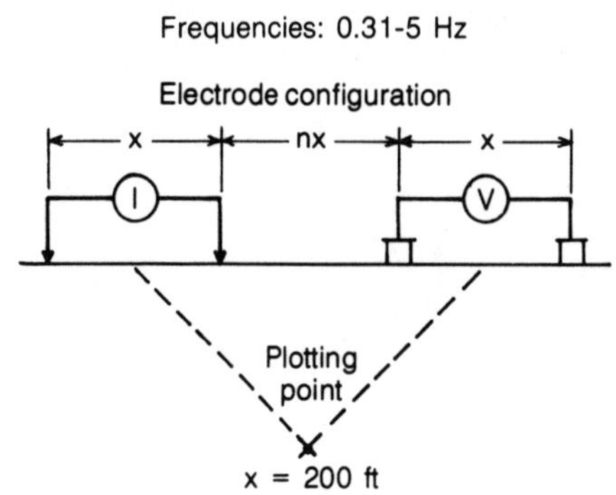

General information concerning exploration with IP method

Everyone who has tried to employ geophysical exploration techniques in the field knows there is a big difference between knowing something about the causes of a particular geophysical effect and being able to effectively use the method to find orebodies. Before examining the usefulness of the IP method in the search for metallic mineralization, the reader must comprehend some general information.

True IP effect vs. apparent measured values

In the discussion of the general theory of the resistivity of rocks and the general theory of the induced polarization effect, the *true* values within the rocks themselves were studied. That is, (1) the true resistivity value within a large geothermal cell or (2) the true IP effect (frequency effect, chargeability or phase shift) within the disseminated sulfide zone or the lead-zinc ore zone was discussed.

In making field measurements during an exploration program, the geophysicist rarely has the opportunity to measure these *true* values. The source being located is relatively small and at some depth. Only the apparent effect at the surface can be measured; the apparent value has been reduced or diluted by the size of the volume of the earth that is affecting the measurement being made. The apparent effect being measured will be influenced by the size, depth, and attitude of the source as well as the length of the electrode interval being used and the exact position of the electrodes themselves. Further, it is obvious that the electrical parameters of the surrounding rocks will also affect the apparent values measured from a source with a given true effect.

The scale modelling IP and resistivity results shown in figures 2.46 through 2.49 can be used to study the difference between the true IP effect in a given source and the apparent effect measured at the surface. In figures 2.46 and 2.47, the true effects in the source are the same for both figures. It has a true resistivity of about 4.9 and a true frequency effect of about 23 percent. It can be seen that none of the apparent effects measured approach these true effects in magnitude.

For the source in a vertical position, the lowest apparent resistivity is about 70, and the highest frequency effect is about 5.5 percent; due to the depth to the top of the relatively narrow source, the maximum effects are measured for n = 5. For the flat-lying source, there is a greater volume of anomalous material near the surface. The lowest resistivity (47) and the highest frequency effect (13.7 percent) are measured for n = 4.

The theoretical results shown in figures 2.48 and 2.49 show that even for a shallow source (maximum apparent anomaly for n = 1), the apparent effects measured are much less than the true effects in the source. These results also confirm that the apparent effects measured increase if the true effects are increased. This is the usual situation.

True IP effects and sulfide concentration

If the source of the IP anomaly is very broad and shallow (relative to the electrode interval used), the apparent effects measured will be approximately equal to the true effects in the source. If this geometry is encountered, the apparent effects will be relatively uniform throughout the anomalous pattern.

The field data in figures 2.50 through 2.52 show anomalous patterns in which the apparent effects are obviously approximately equal to the true effects. In all three cases the resistivity value is in the range 40 to 50, the frequency effect is 7-15 percent and the metal factor is 150 to 300. Numerous drill holes in the second two locations and mining operations at Gaspe copper have given an excellent estimate of the total sulfide mineral concentration present.

It can be seen that the largest IP effects were measured at Mt. Morgan and the sulfide concentrations were only 1-2 percent. At Gaspe and in New Mexico, the IP effects are less, but appreciably more sulfide mineralization is present. Such characteristics as grain size and interconnection determine the true IP effect to be expected from a given concentration of sulfide mineralization. Therefore, the interpretation of an IP anomaly cannot be rigorous.

IP anomalies are usually smooth

It has been shown that for a large zone of disseminated sulfide mineralization the IP effect can be treated as a property of the rock volume. The effects are averaged and smooth, because of the smooth nature of solutions to the equations that govern electrical potentials and current flow.

This smooth nature of potentials and the resulting averaged, smooth anomalies that are measured can make it difficult to interpret and test the source of a given IP anomaly. The field results shown in figure 2.53 were measured with $x = 200$ ft and $x = 100$ ft over the same IP source. The pattern suggests a single, simple source at depth (maximum values for $n > 1$).

In fact, later drilling shows that the source is quite complex. The averaging effect of the IP measurement results in the very smooth anomaly, even for $x = 100$ ft. Drilling to test the source of this kind of anomaly can, therefore, be a hazardous and frustrating experience.

IP anomalies and "ore"

Not all IP anomalies, even strong, definite ones, are ore. The results in figures 2.54 and 2.55 are from the Chibougamau, Quebec area. One anomaly is due to an orebody; the other is not.

Inductive coupling and measurements with large electrode intervals

A complete mathematical development of the effects of inductive coupling is beyond the scope of this treatise. However, some attention must be given to the effect and when it can be expected. The theoretical results in figure 2.56 show that in the search for large volumes of disseminated porphyry copper mineralization at depth, the apparent effects measured increase with increasing values of (n). If a large source is at a depth of 1000–2000 ft, it is necessary to use electrode intervals (x) of 1000–1500 ft and $n = 1, 2, 3, 4, 5, 6$.

If the resistivity level in the area of the survey is low to moderate, the equations that govern electromagnetic phenomena will begin to be important. As explained previously, the basic assumption is always that only the potential equations are important. In this situation, i.e., large electrode intervals and low resistivity level, apparent effects from inductive coupling that appear to be due to sources of true IP effect at depth can be measured (figures 2.57 and 2.58).

Until recently, it was necessary to choose the specifications of the survey to avoid inductive coupling effects; this was not always possible. There is now some hope that coupling effects can be removed from field data.

The figure on this page is referred to in the preceding section *General information concerning exploration with IP method.*

FIGURE 2.46
Theoretical induced polarization and resistivity studies

Vertical source without overburden

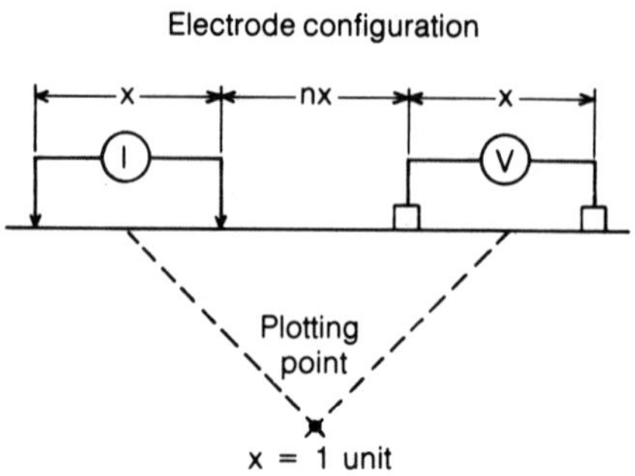

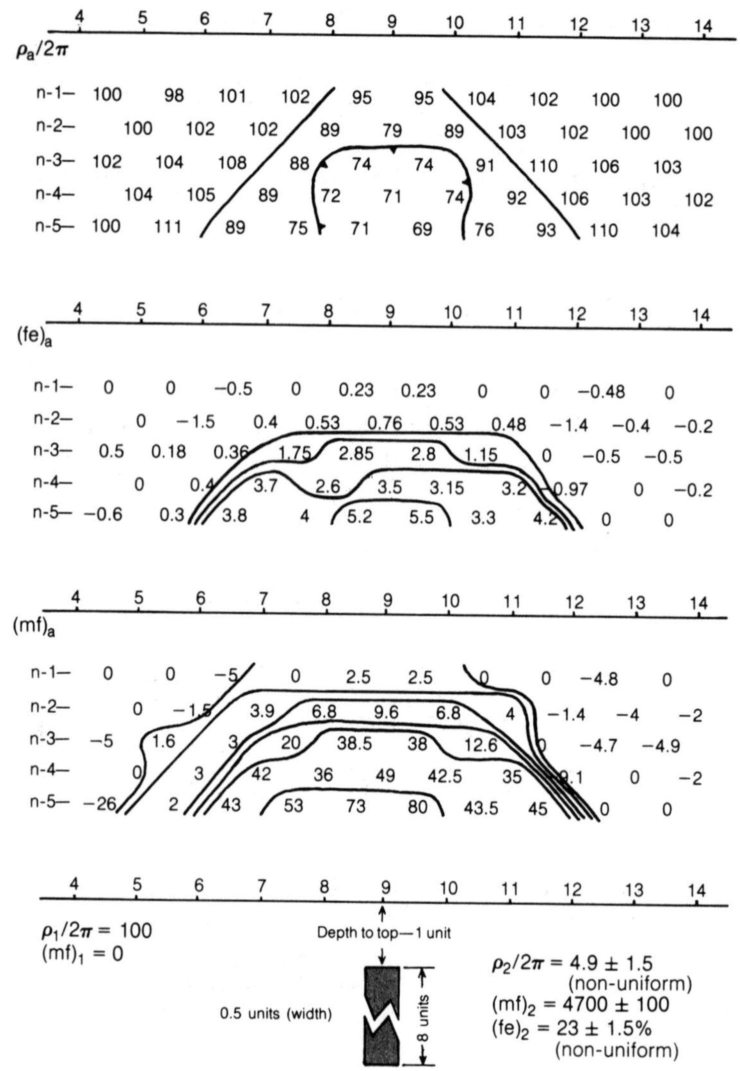

The figure on this page is referred to in the preceding section *General information concerning exploration with IP method.*

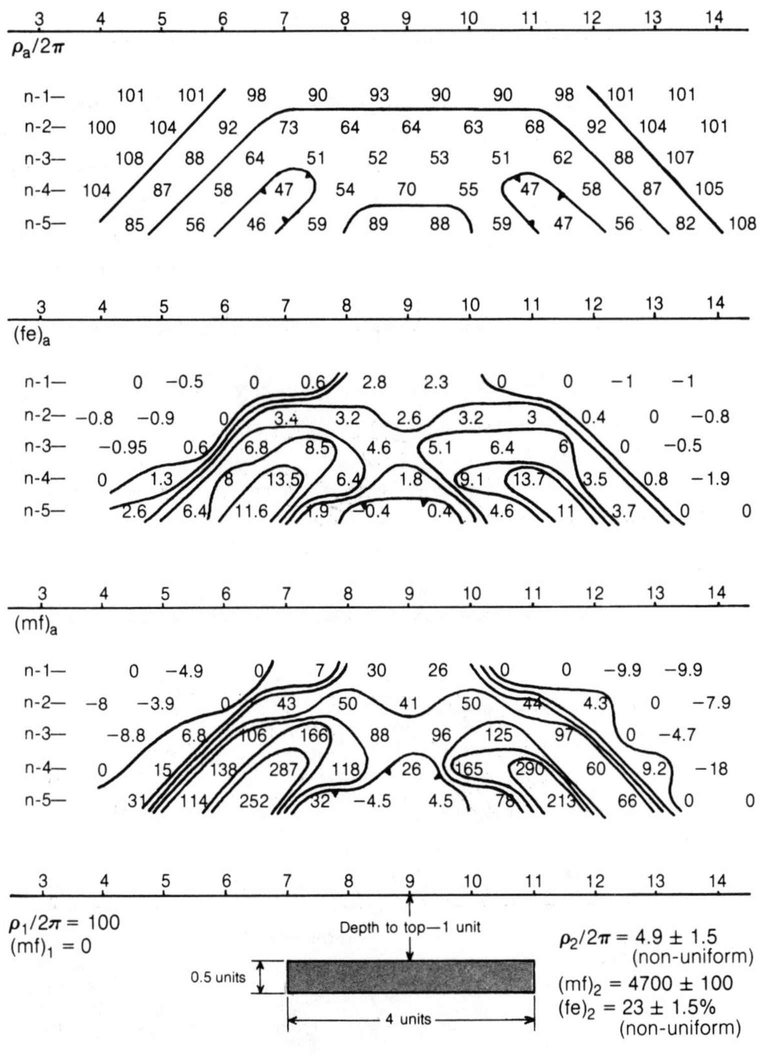

FIGURE 2.47
Theoretical induced polarization and resistivity studies

Horizontal source without overburden

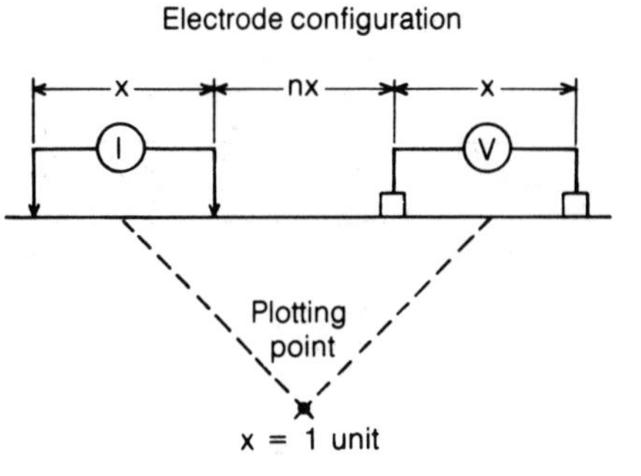

The figures on this page are referred to in the preceding section *General information concerning exploration with IP method.*

FIGURE 2.48
Theoretical induced polarization and resistivity studies

Calculated cases, dipole-dipole configuration

Electrode configuration

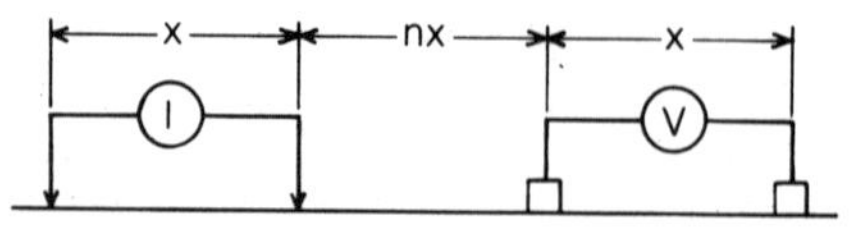

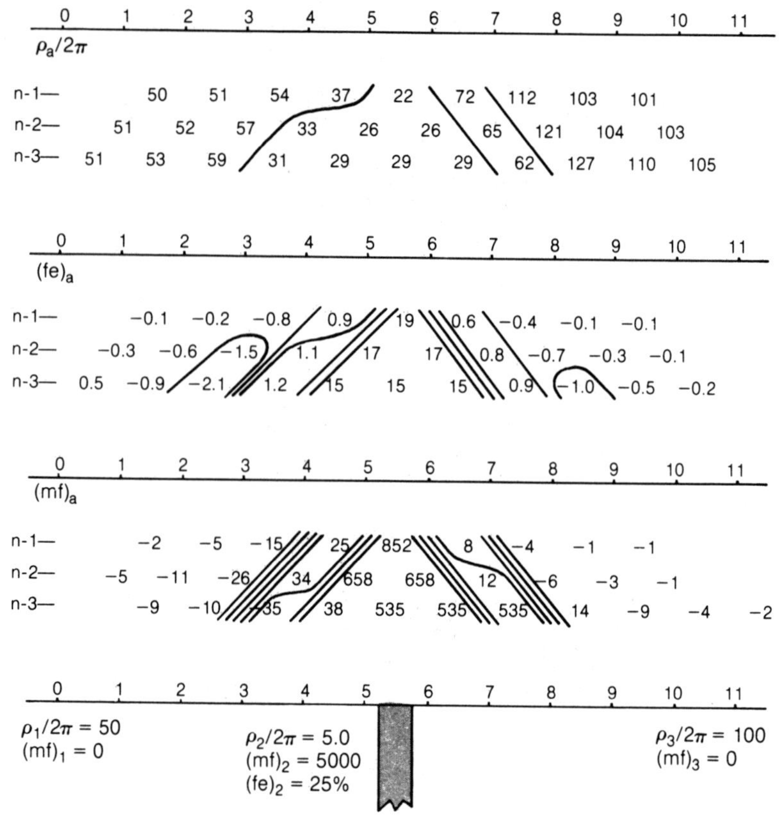

FIGURE 2.49
Theoretical induced polarization and resistivity studies

Calculated cases, dipole-dipole configuration

Electrode configuration

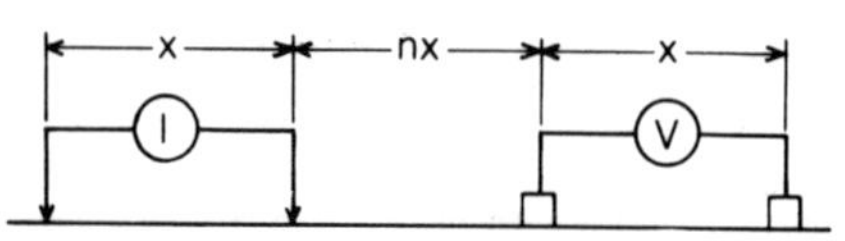

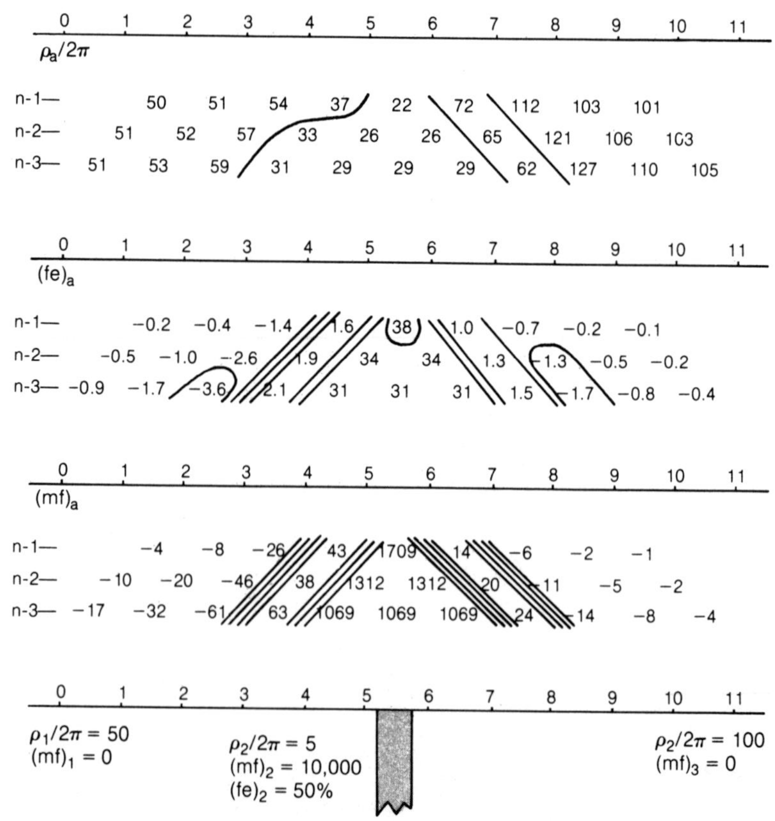

The figure on this page is referred to in the preceding section *General information concerning exploration with IP method.*

FIGURE 2.50
Induced polarization and drilling results from Copper Mountain Orebody, Gaspe Area, Quebec

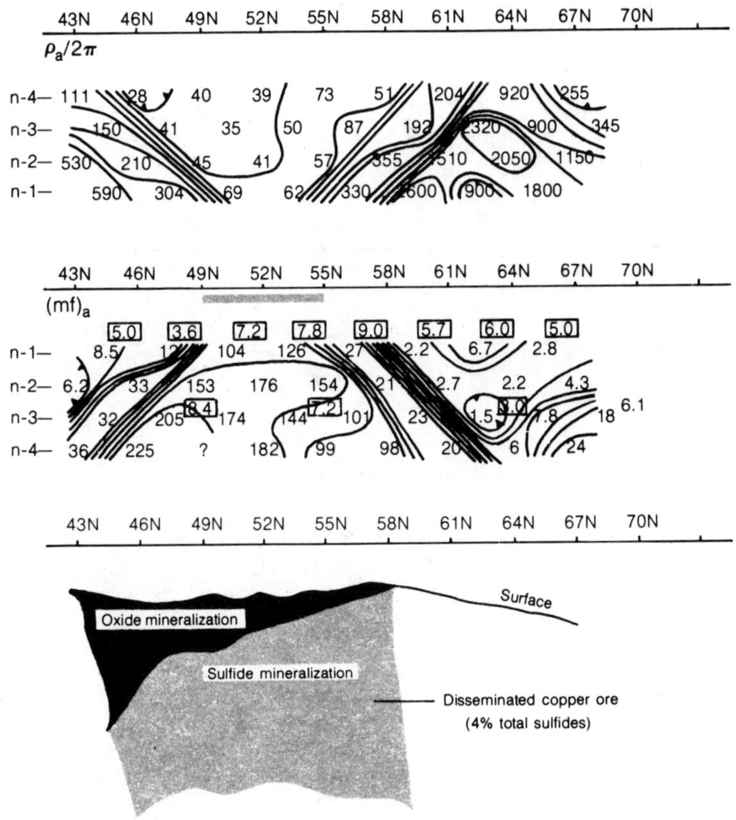

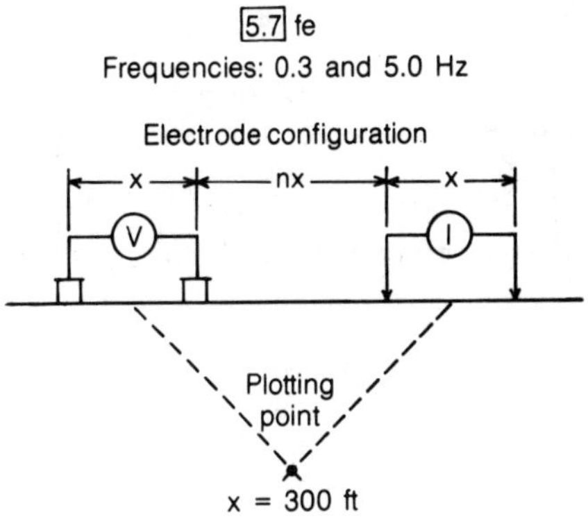

The figure on this page is referred to in the preceding section *General information concerning exploration with IP method.*

FIGURE 2.51
Induced polarization and drilling results from Western New Mexico, U.S.A.

Line-40W

Frequencies: 0.31-2.5 Hz

Electrode configuration

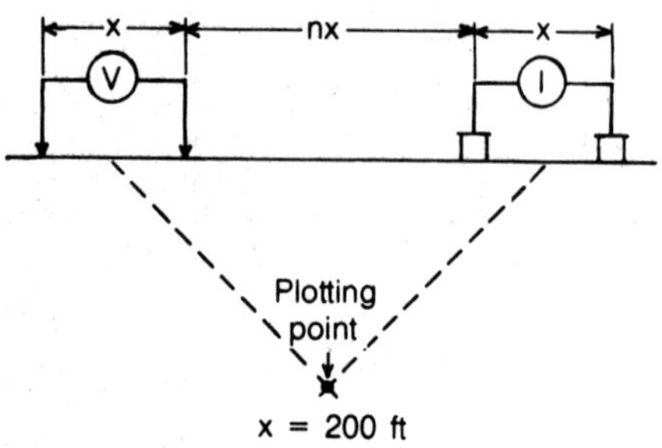

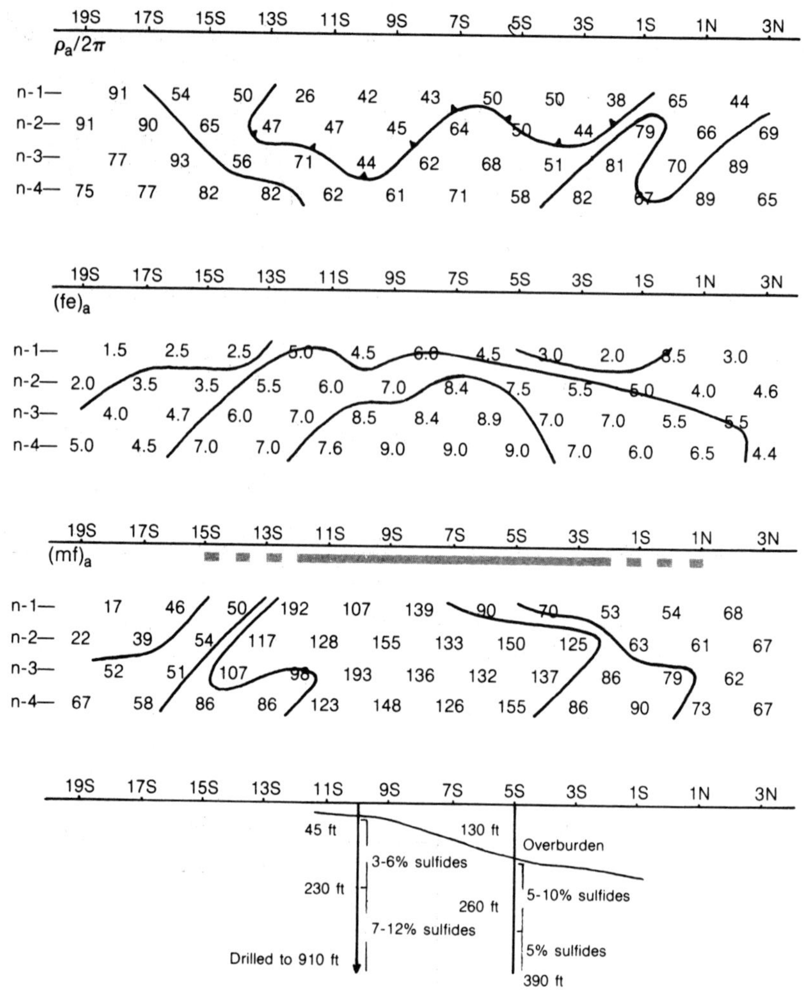

The figure on this page is referred to in the preceding section *General information concerning exploration with IP method.*

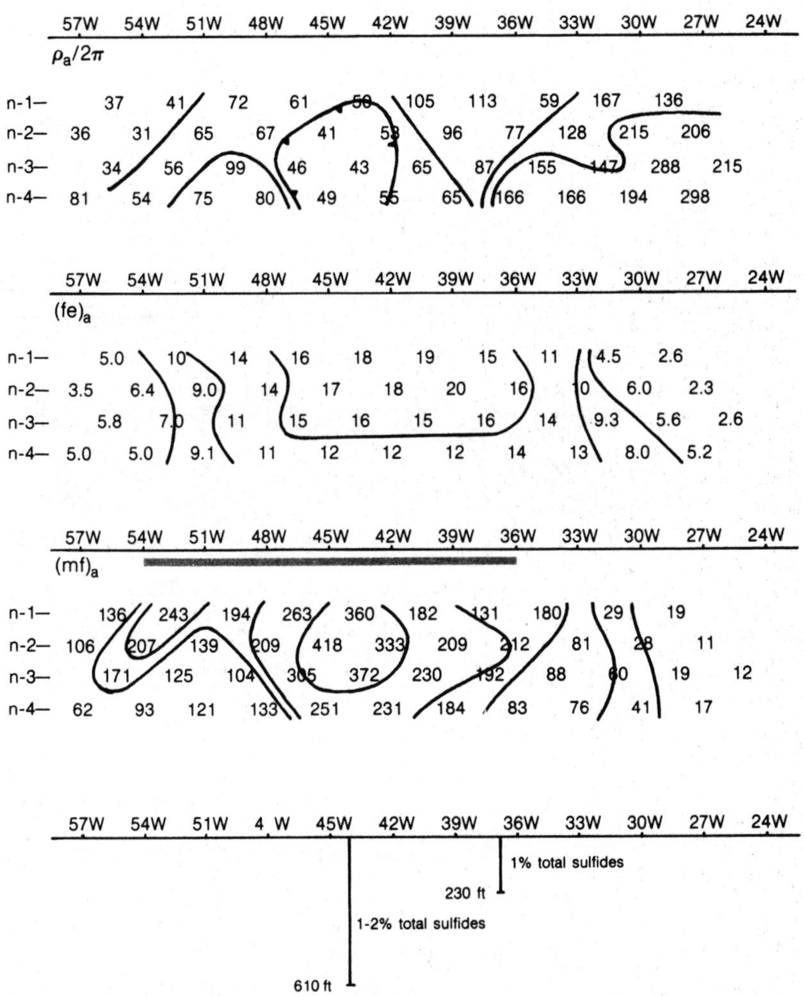

FIGURE 2.52
Induced polarization and drilling results from Moonmera Area, Mt. Morgan, Queensland, Australia

Line-B

Frequencies: 0.31-5 Hz

Electrode configuration

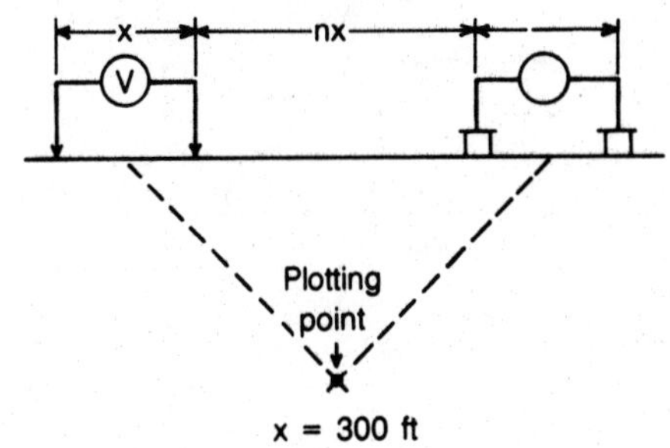

The figures on this page are referred to in the preceding section *General information concerning exploration with IP method.*

FIGURE 2.53
Induced polarization and drilling results from prospectors airways property, Joutel Township, Quebec

Line-2W

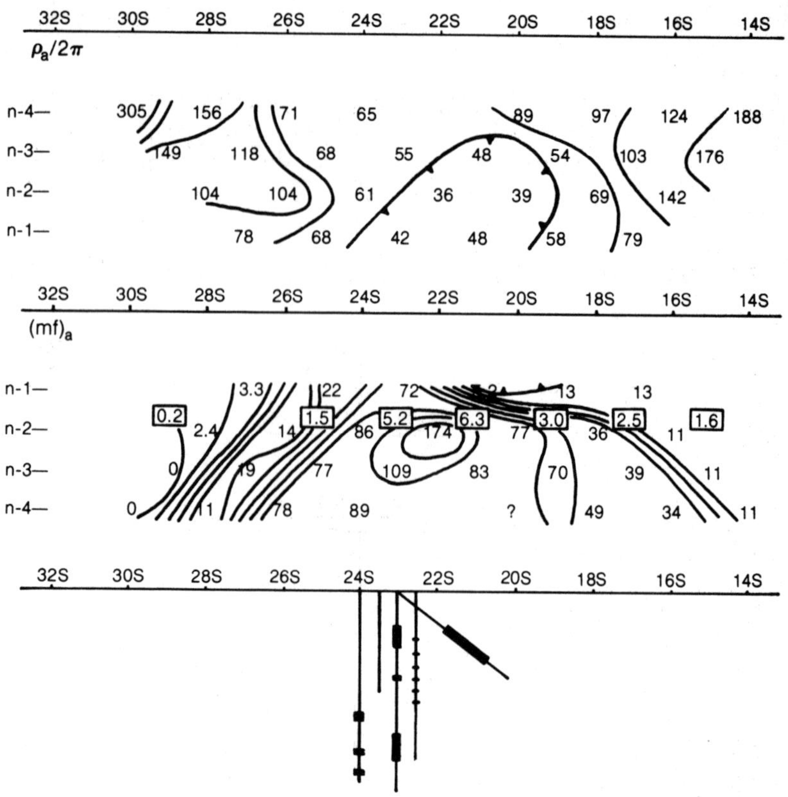

[0.2] fe
Frequencies: 0.3 and 5.0 Hz

Electrode configuration

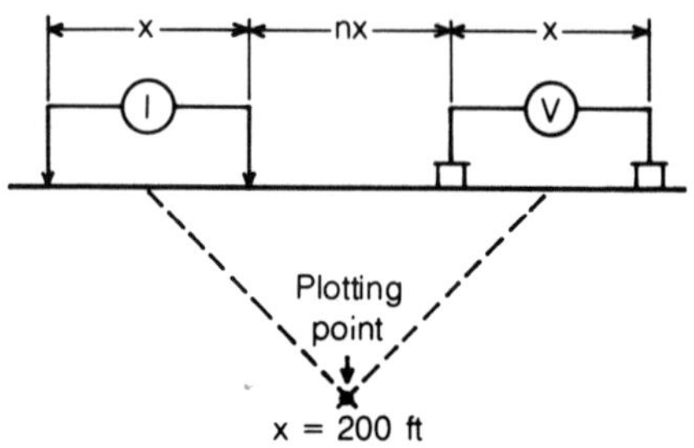

FIGURE 2.54
Induced polarization on Lake Chibougamau, Quebec, Canada

Line-12W

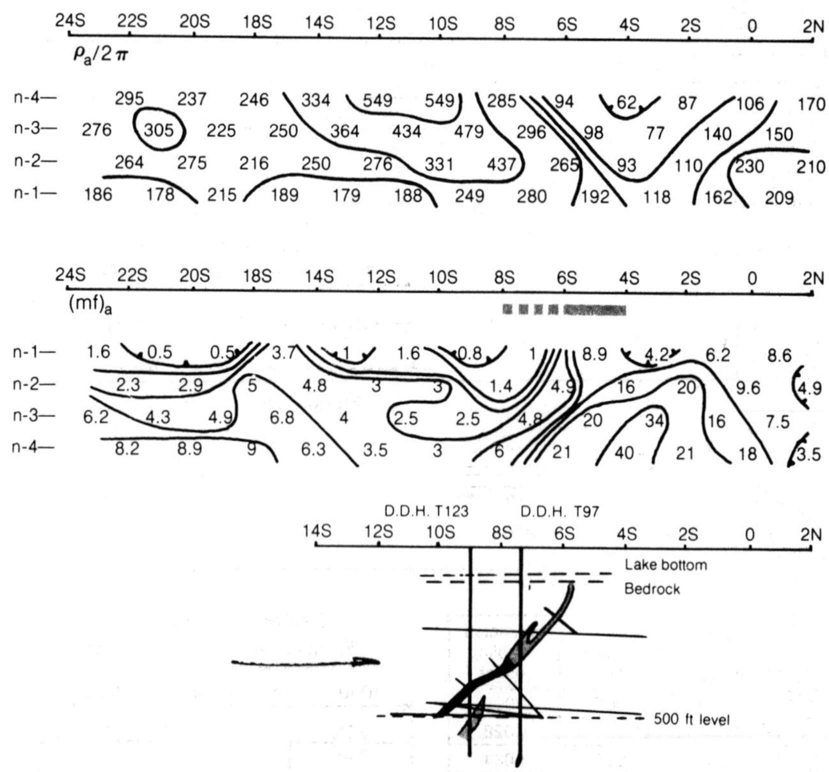

■ > 15% sulfide

▨ < 15% sulfide

Electrode configuration

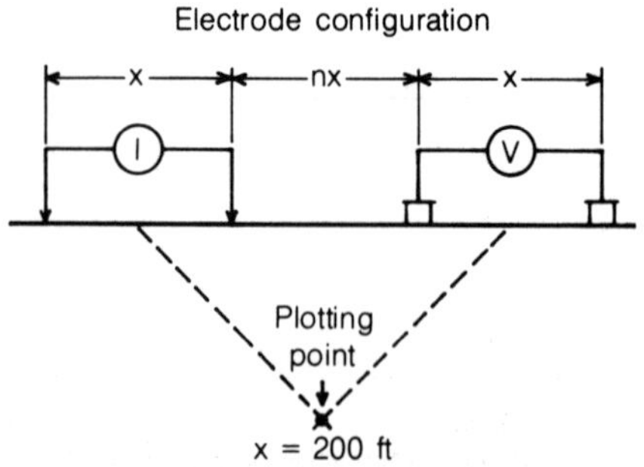

The figures on this page are referred to in the preceding section *General information concerning exploration with IP method.*

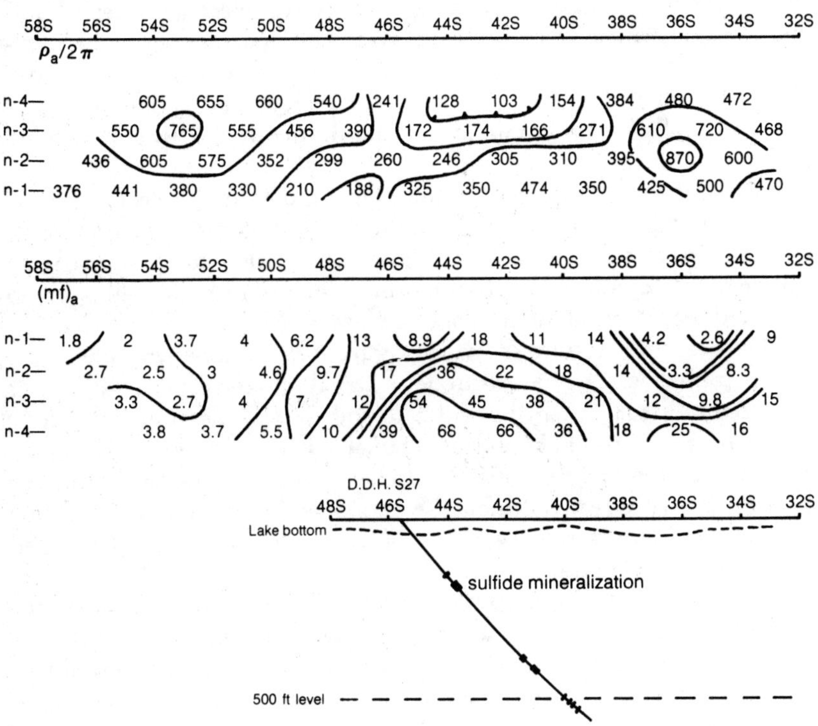

FIGURE 2.55
Drilling results on Lake Chibougamau, Quebec, Canada

Line-72W

n	$\rho_a/2\pi$	$(mf)_a$	ρDC/ρAC
1	23.3	137	1.032
2	12.4	830	1.103
3	13.5	881	1.119
4	16.3	711	1.116
5	19.1	586	1.112

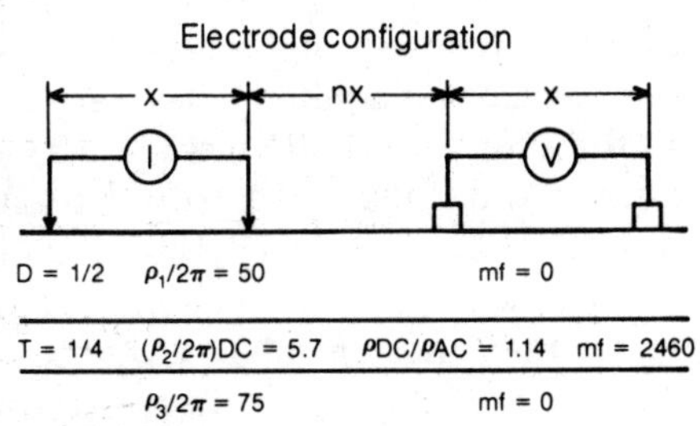

FIGURE 2.56
Theoretical induced polarization and resistivity studies

Calculated cases, dipole-dipole configuration

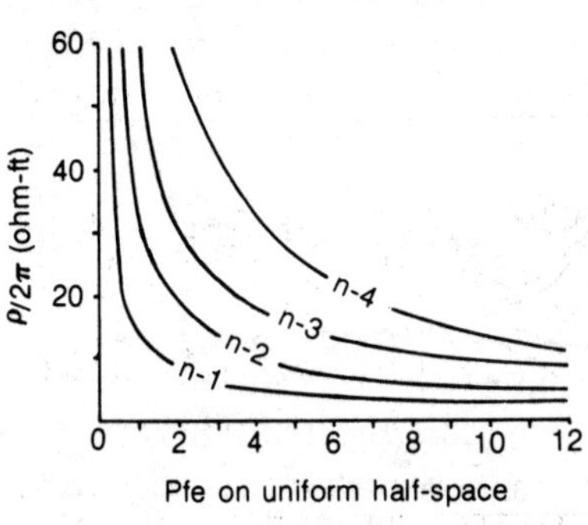

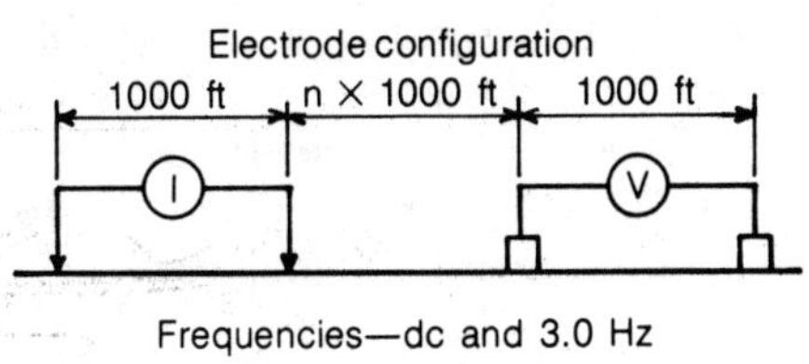

FIGURE 2.57
Expected IP effects from inductive coupling

n	ρdc/2π	Exact coupling effect	Estimated coupling effect
1	43.9	1.004	1.002
2	16.6	1.032	1.028
3	11.8	1.096	1.084
4	10.8	1.183	1.155

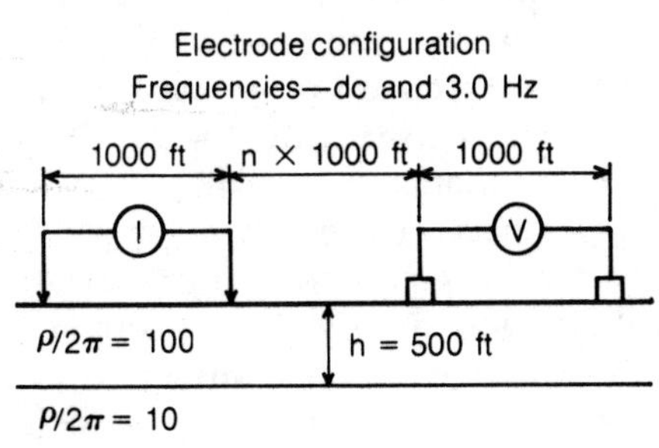

FIGURE 2.58
Expected frequency effects from inductive coupling on layered earth

Choice of electrode interval

The various electrode configurations that can be selected for induced polarization and resistivity surveys have already been discussed. An even more important decision is the choice of the electrode interval; it is this decision which determines the overall scale of the survey. An improper choice of electrode interval can result in an apparent anomaly that is too small to be interpreted from background variations, or no anomaly at all. It is important to consider the depth of detection required and minimum target sought.

This author's experience is that the dipole-dipole electrode configuration is generally the most useful to the exploration geophysicist; therefore, with a few exceptions, most of the field examples discussed will be dipole-dipole data. However, the conclusions made will apply to any of the configurations in common use.

The theoretical results shown in figure 2.59 demonstrate one of the major problems. If the electrode interval chosen for a survey is appreciably greater than the width of the source, the apparent effects from a narrow zone of concentrated metallic mineralization will be much the same as the apparent effects from a source of greater width, but smaller true IP effect. Further, the anomalous pattern and magnitude will be much the same regardless of where the source is positioned between the electrodes.

In a general way, if the source is relatively shallow, the closer the electrode interval is to the width of the source the greater will be the apparent IP effects. That is, there is less unmineralized rock averaged into the measurement, and the apparent effect measured is more nearly equal to the true effect.

For many sources in many geologic environments, the choice of electrode intervals is not critical. The anomalous pattern and particularly the anomalous magnitude may vary as the electrode interval is changed, but the anomaly is still present. The measurements with both $x = 200$ ft and $x = 100$ ft (figures 2.60 and 2.61) over a shallow zone of zinc-lead mineralization indicate a definite anomaly. Because the source is only 25 ft deep, the apparent effects measured with $x = 100$ ft are appreciably greater than those measured with $x = 200$ ft.

If the shallow source has less width, the result of changing the electrode interval by a factor of two can be even more dramatic. For a narrow zone of copper ore in the Canadian shield, even with some depth to the top (figures 2.62 and 2.63), the pattern measured with $x = 100$ ft can be quite different from that measured with $x = 50$ ft.

In some cases, a particular source in a particular environment can be detected with widely varying electrode intervals; however, the anomalous pattern will look very different. Measurements with $x = 200$ ft and $x = 25$ ft both detected the ore zone at Pine Point (figures 2.64 and 2.65). Because the scale of the measurements was so different, however, the anomalous patterns are totally different.

If the mineralized zone has a more complicated shape, the measurements with different electrode intervals may detect different portions of the source. This is what happened in the discovery of the Lakeshore orebody (figures 2.66 through 2.69). The old pit at 525E on line 500N contained secondary copper minerals that were apparently the weathered remnant of a near-vertical zone of copper-bearing sulfides. The measurements with $x = 200$ ft, and particularly with $x = 300$ ft, do show an anomaly suggesting a narrow source at depth under the pit.

Later measurements with $x = 500$ ft and $x = 1000$ ft indicated a much larger volume of metallic mineralization at a still greater depth. Deeper drilling located a significant copper orebody.

Measurements with shorter electrode intervals can be expected to find only the shallow sources. These can be either the most important, or the least important, from an economic viewpoint. The a = 200 ft, three-array survey over the Lornex orezone was completed first (figure 2.70). Drilling confirmed that the source was pyrite with small amounts of copper. It was only later, after measurements with a = 400 ft and a = 800 ft, that it was realized that deeper sources might exist to the west. Later drilling located the source at greater depth that was ore grade.

Although it is always possible to "look to greater depth" by increasing the electrode interval, it must be remembered that measurements with large electrode intervals average a large volume of rock into each measurement. The apparent anomaly from a small, possibly important source may be missed.

In an extreme case, an important ore deposit can be missed. The reconnaissance survey using x = 500 ft was planned to locate an extension of the Broken Hill orebody with a maximum depth of detection (figures 2.71 through 2.73). No thought was given to smaller volumes of ore that might still be important. The very weak anomaly centered at 5+00W on line 25N (the reconnaissance lines were 2000 ft apart) was ignored for several years. Finally, detailed measurements with x = 300 ft and x = 100 ft, and also measurements on more closely spaced lines, outlined an anomaly zone that was found to be caused by 7 million tons of typical Broken Hill zinc-silver-lead ore. The ore was obviously at a very shallow depth.

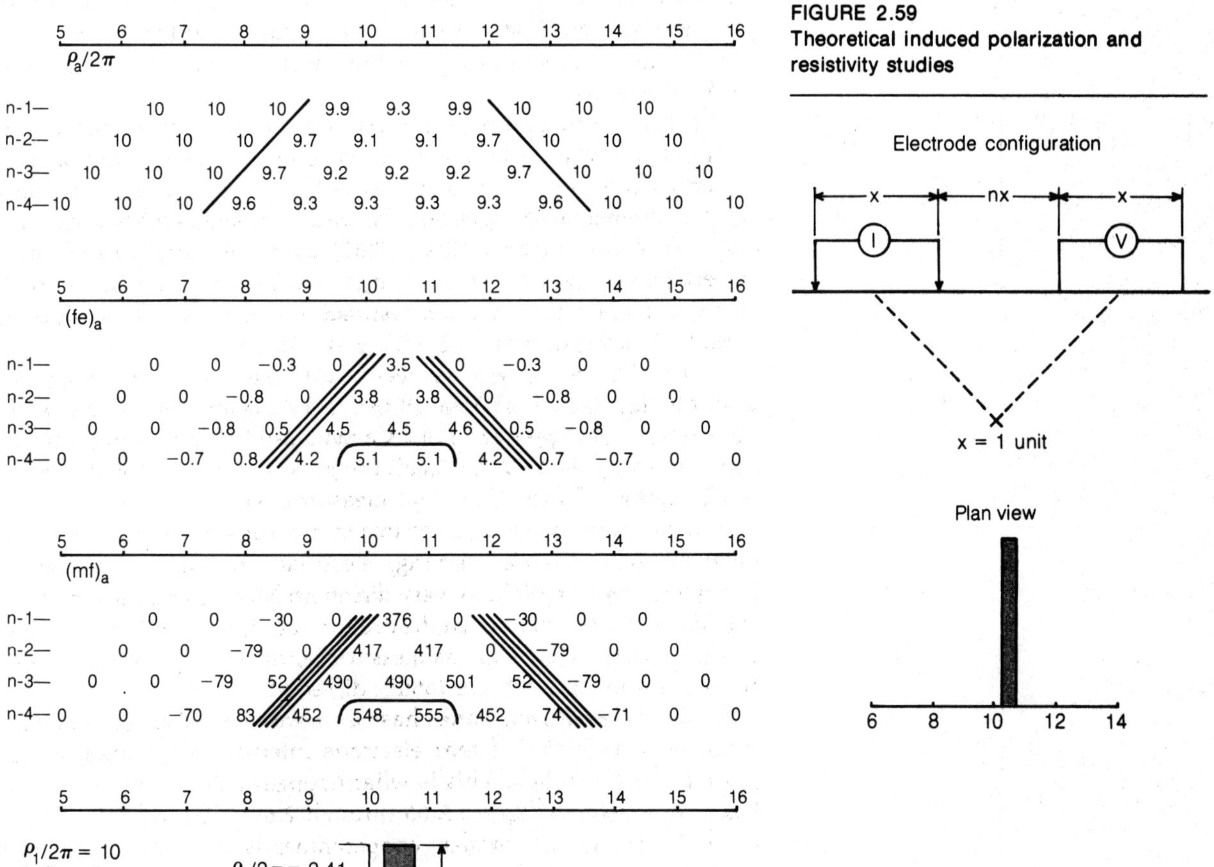

FIGURE 2.59
Theoretical induced polarization and resistivity studies

The figures on this page are referred to in the preceding section *Choice of electrode interval.*

FIGURE 2.60
Induced polarization and drilling results from Pine Point type mineralization

N-38A orebody, line-P3.0E

Frequencies: 0.31-5 Hz

Electrode configuration

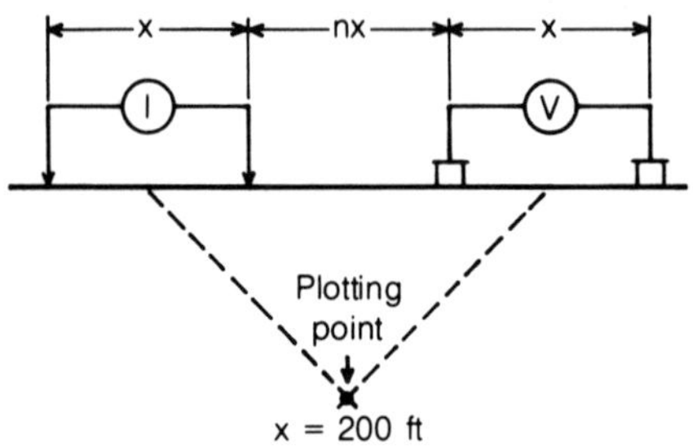

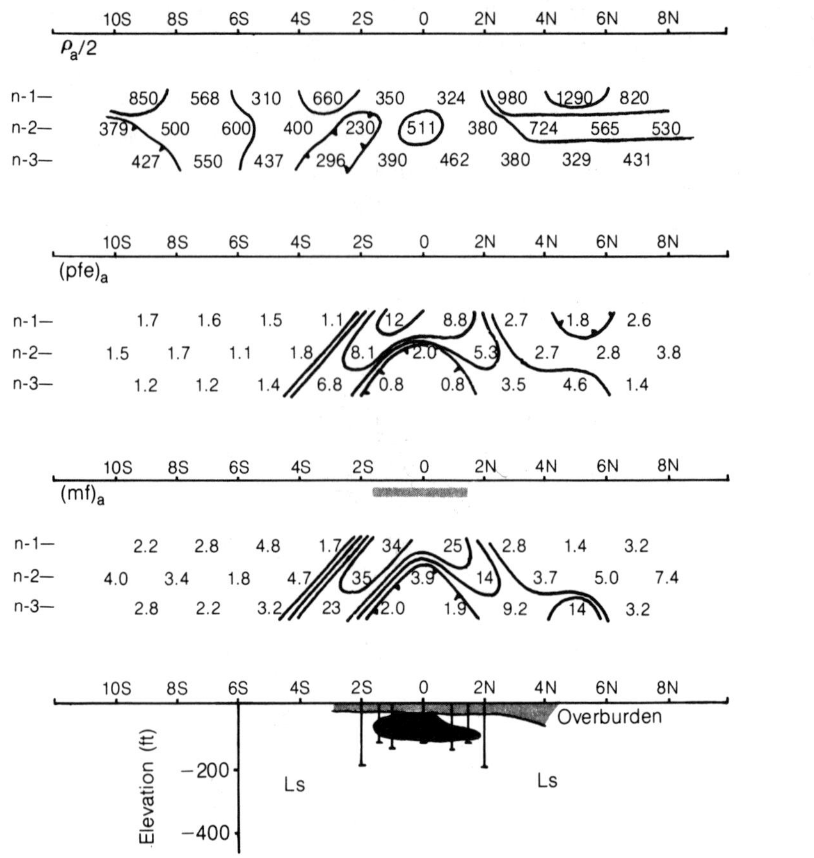

FIGURE 2.61
Induced polarization and drilling results from Pine Point type mineralization

N-38A orebody, line-P3.0E

Frequencies: 0.31-5 Hz

Electrode configuration

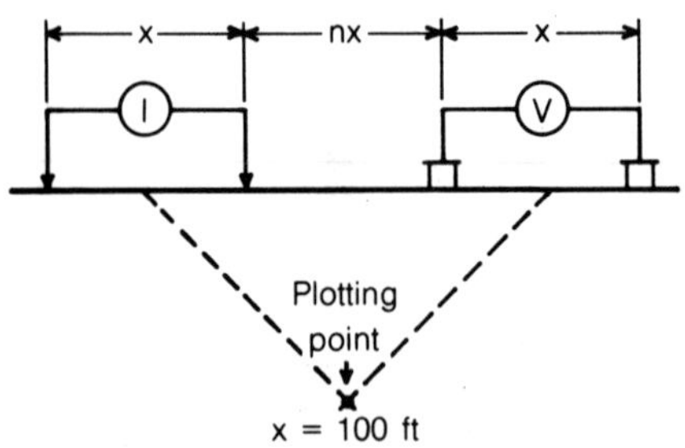

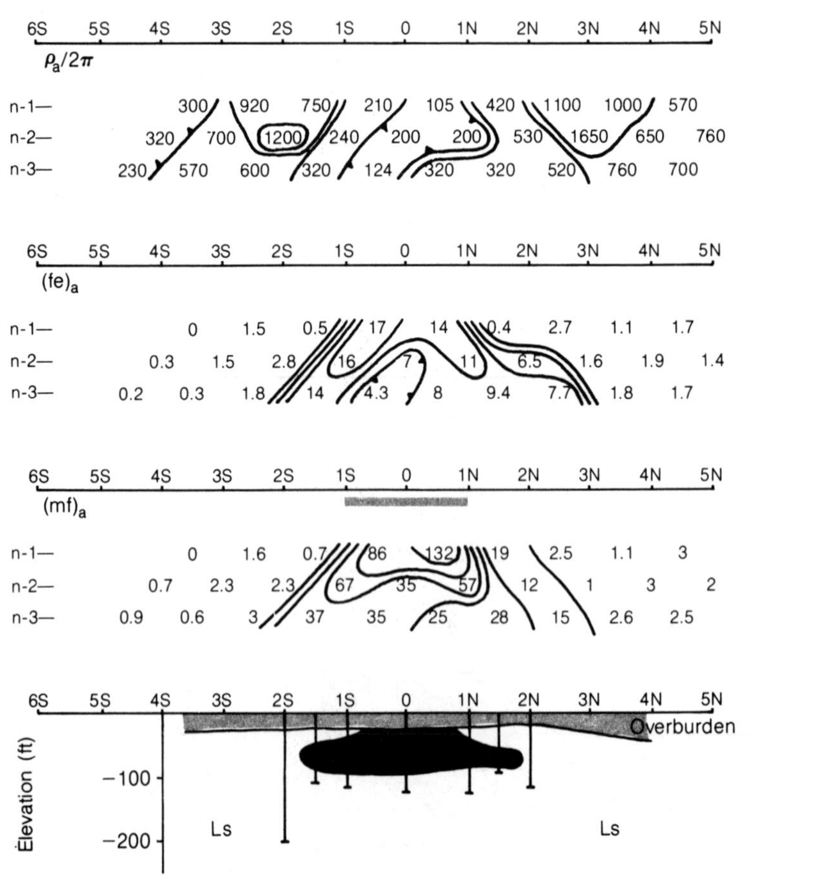

The figures on this page are referred to in the preceding section *Choice of electrode interval.*

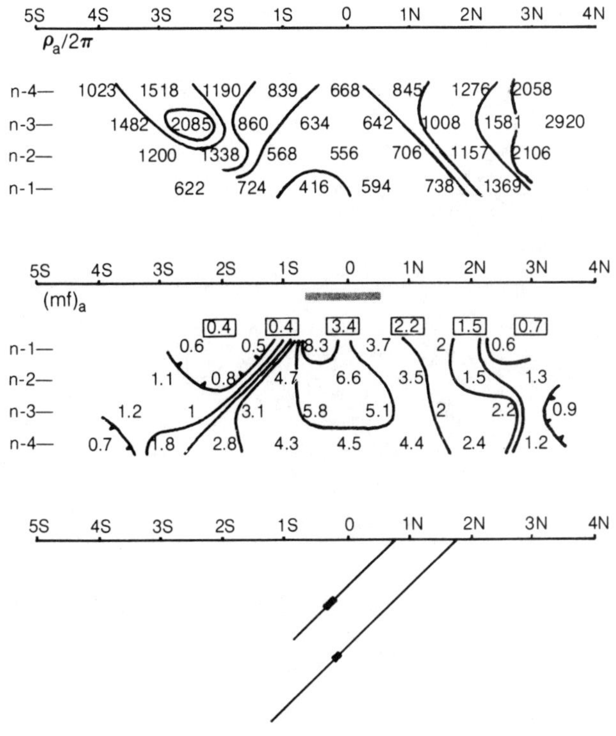

FIGURE 2.62
Induced polarization and drilling results from Ethel Copper Mines property, James Township, Ontario

Line-2W

3.4 fe
Frequencies—0.3 and 5.0 Hz

Electrode configuration

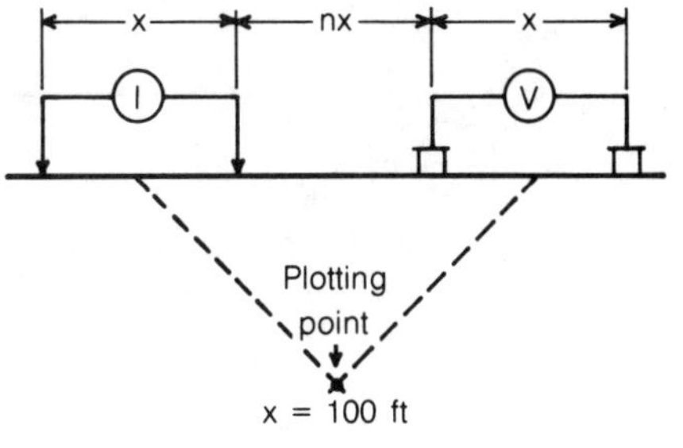

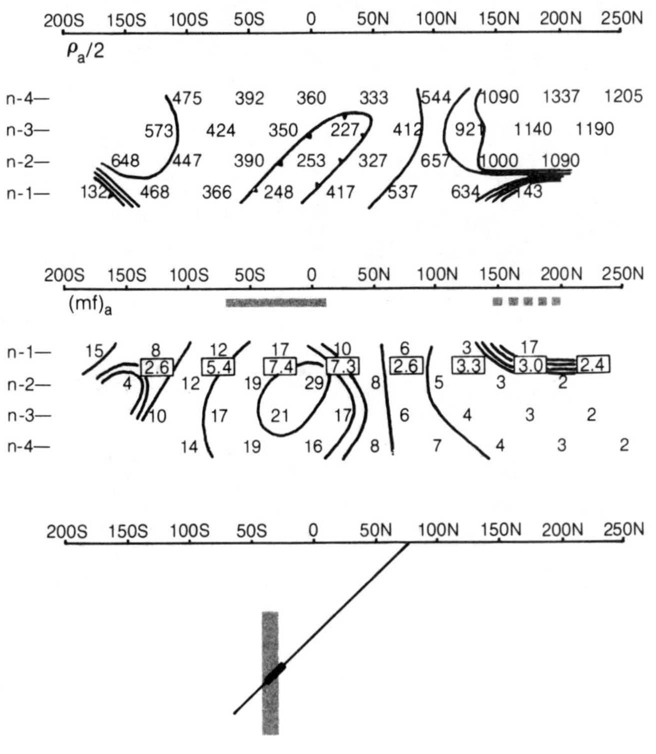

FIGURE 2.63
Induced polarization and drilling results from Ethel Copper Mines property, James Township, Ontario

Line-2W

2.6 fe
Frequencies—0.3 and 5.0 Hz

Electrode configuration

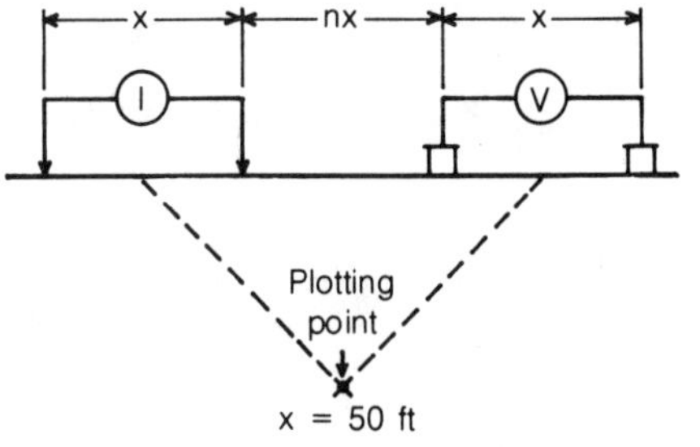

The figure on this page is referred to in the preceding section *Choice of electrode interval.*

FIGURE 2.64
Induced polarization and drilling results from Pine Point type mineralization

K-62 ore zone, line-T4.25E

Frequencies: 0.31-5 Hz

Electrode configuration

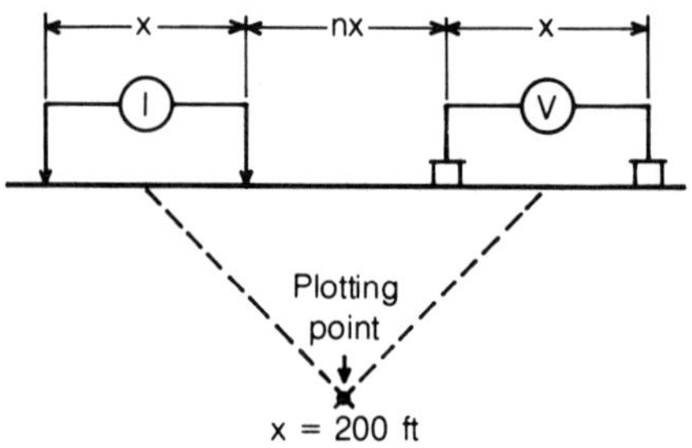

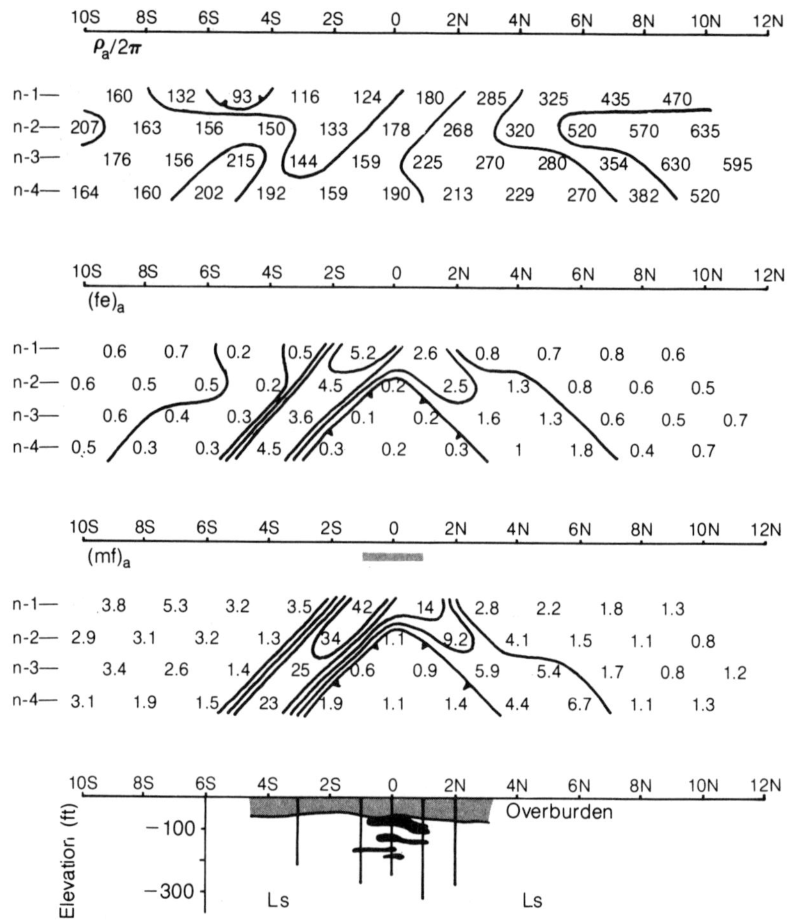

The figure on this page is referred to in the preceding section *Choice of electrode interval.*

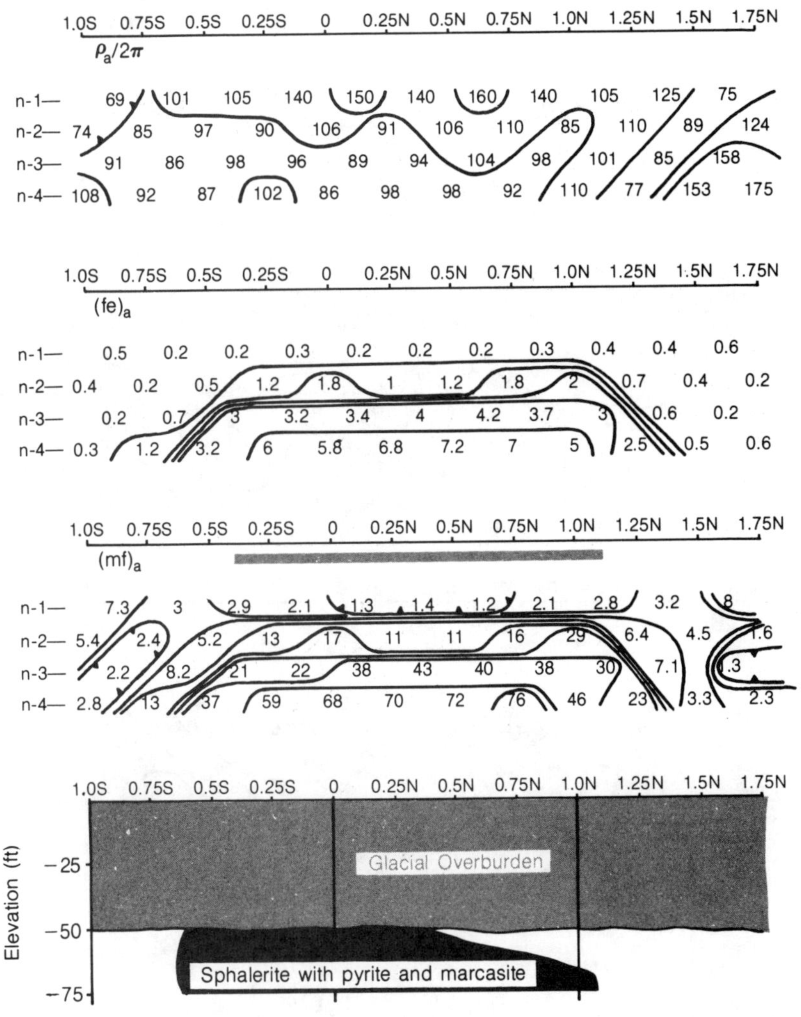

FIGURE 2.65
Induced polarization and drilling results from Pine Point type mineralization

K-62 ore zone, line-T4.25E

Frequencies: 0.31-5 Hz

Electrode configuration

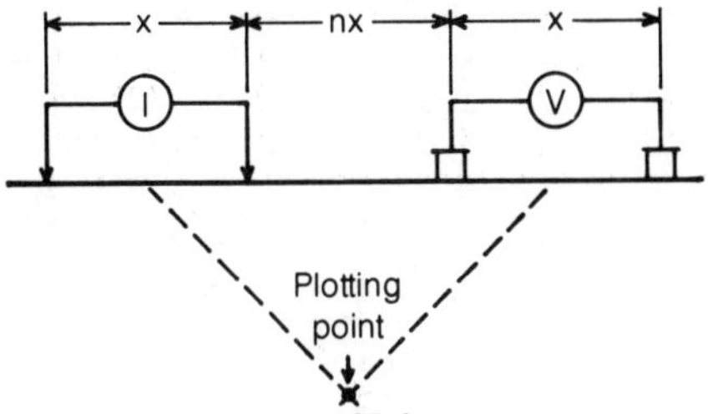

x = 25 ft

The figure on this page is referred to in the preceding section *Choice of electrode interval.*

FIGURE 2.66
Induced polarization and drilling results from Lakeshore orebody, Pinal County, Arizona

Line-500N (1966)

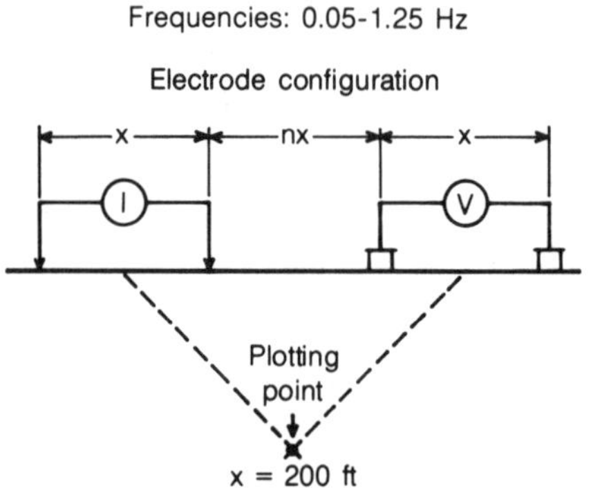

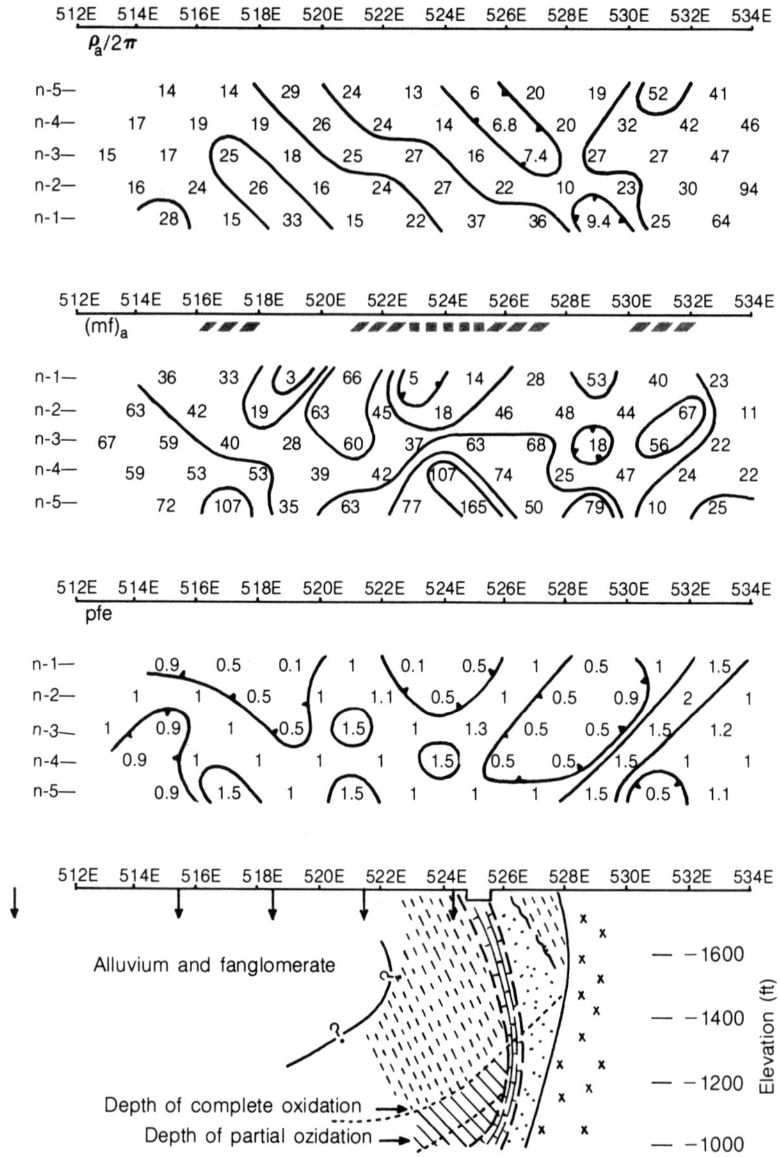

The figure on this page is referred to in the preceding section *Choice of electrode interval.*

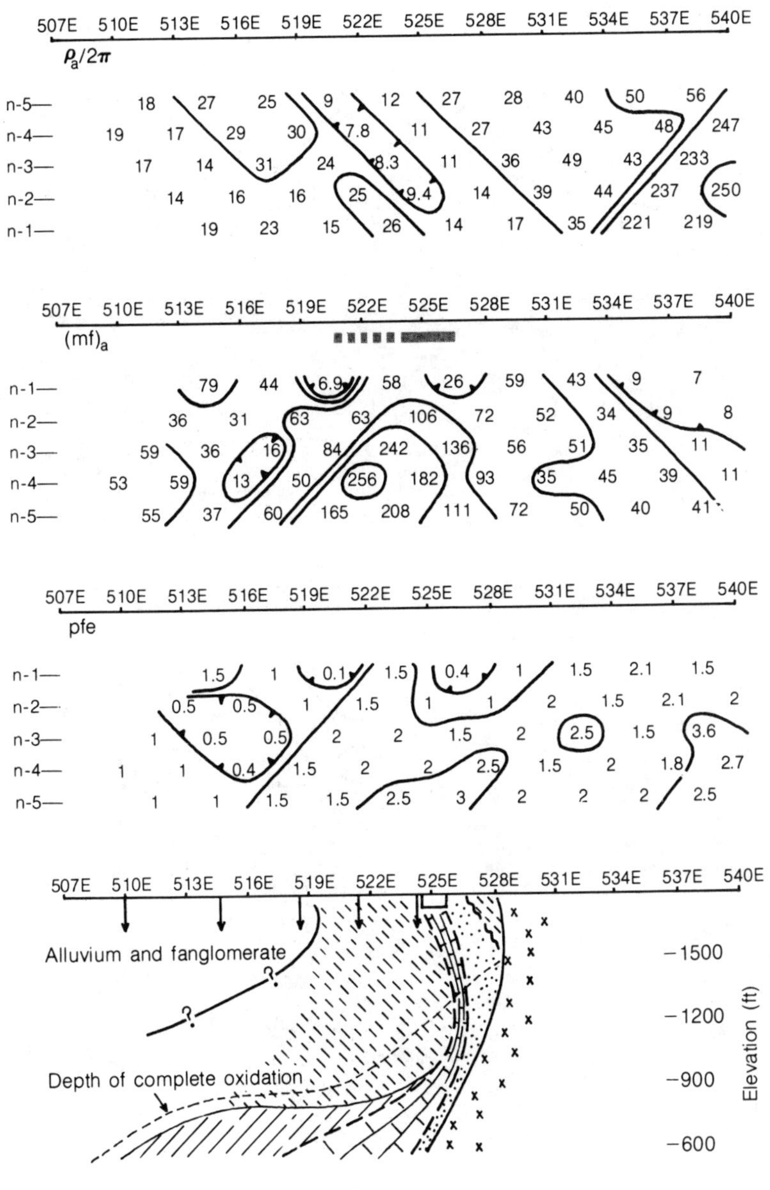

FIGURE 2.67
Induced polarization and drilling results from Lakeshore orebody, Pinal County, Arizona

Line-500N (1966)

Frequencies: 0.05-1.25 Hz

Electrode configuration

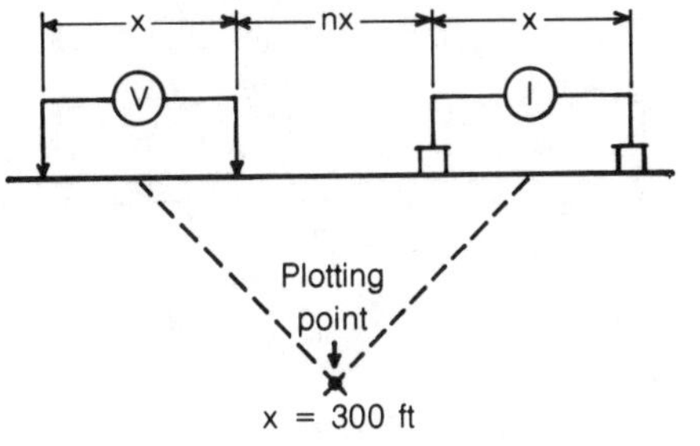

The figure on this page is referred to in the preceding section *Choice of electrode interval.*

FIGURE 2.68
Induced polarization and drilling results from Lakeshore orebody, Pinal County, Arizona

Line-498N (1967)

Frequencies: 0.125-1.25 Hz

Electrode configuration

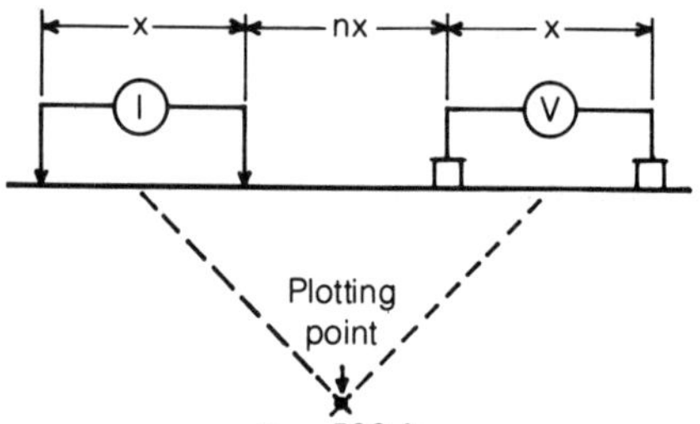

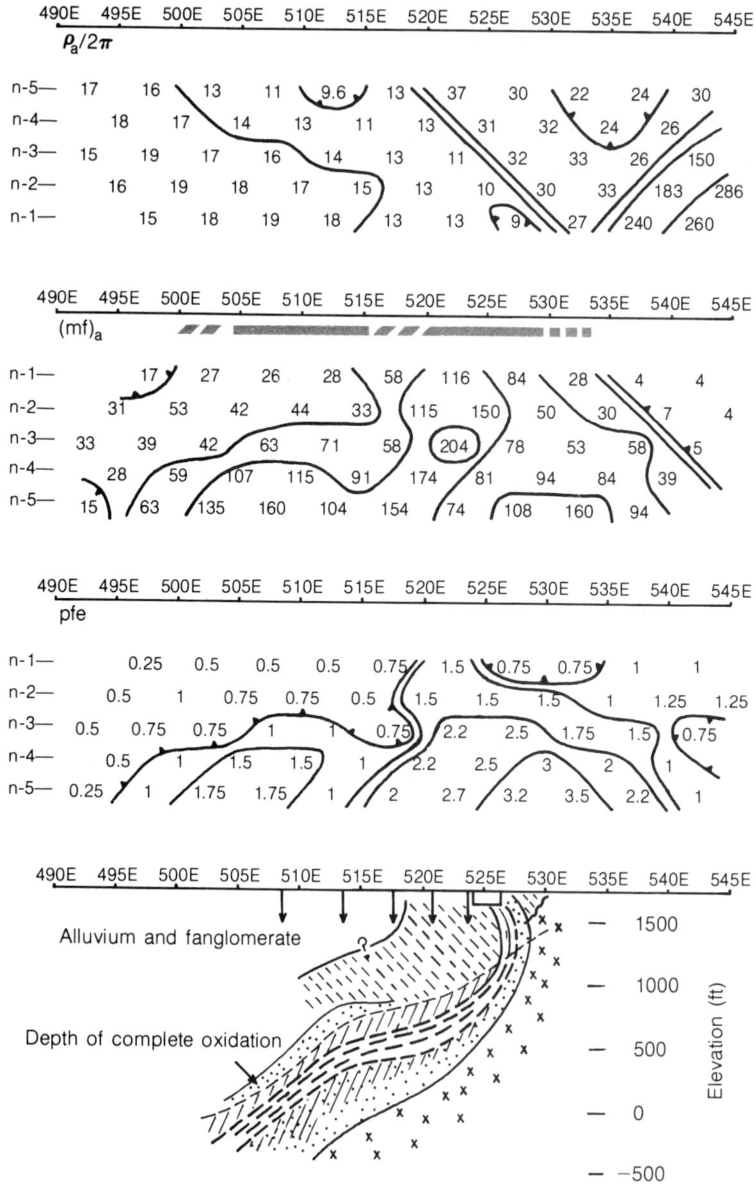

The figure on this page is referred to in the preceding section *Choice of electrode interval.*

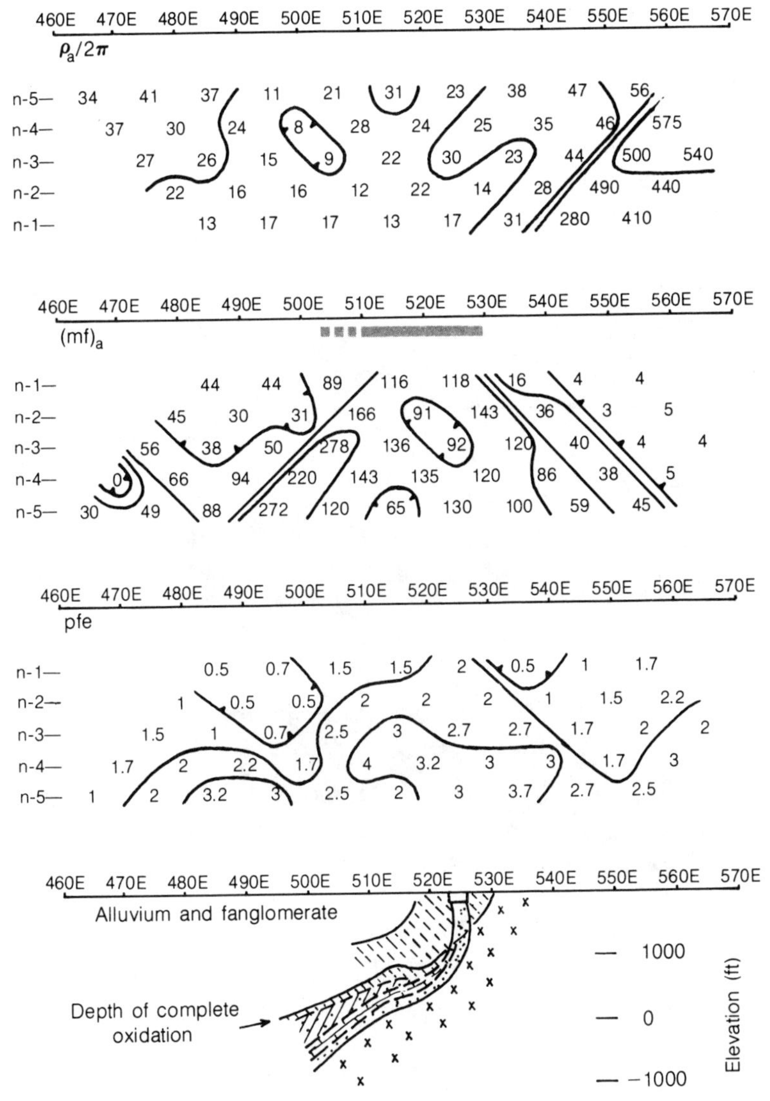

FIGURE 2.69
Induced polarization and drilling results from Lakeshore orebody, Pinal County, Arizona

Line-498N (1967)

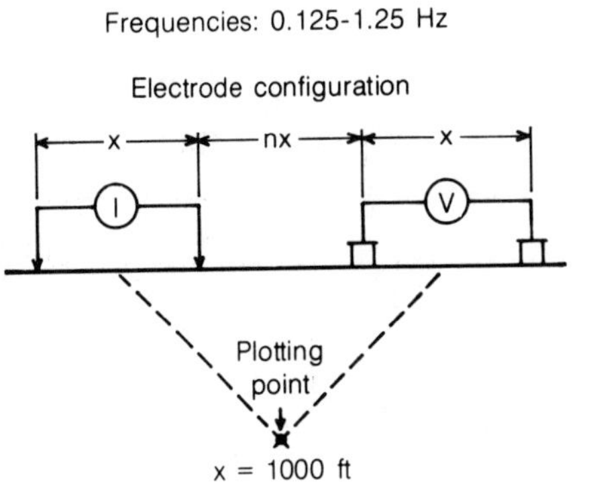

The figure on this page is referred to in the preceding section *Choice of electrode interval.*

FIGURE 2.70
Induced polarization and geologic section from British Columbia disseminated sulfides

Lornex Mining Corp.

Line-15N

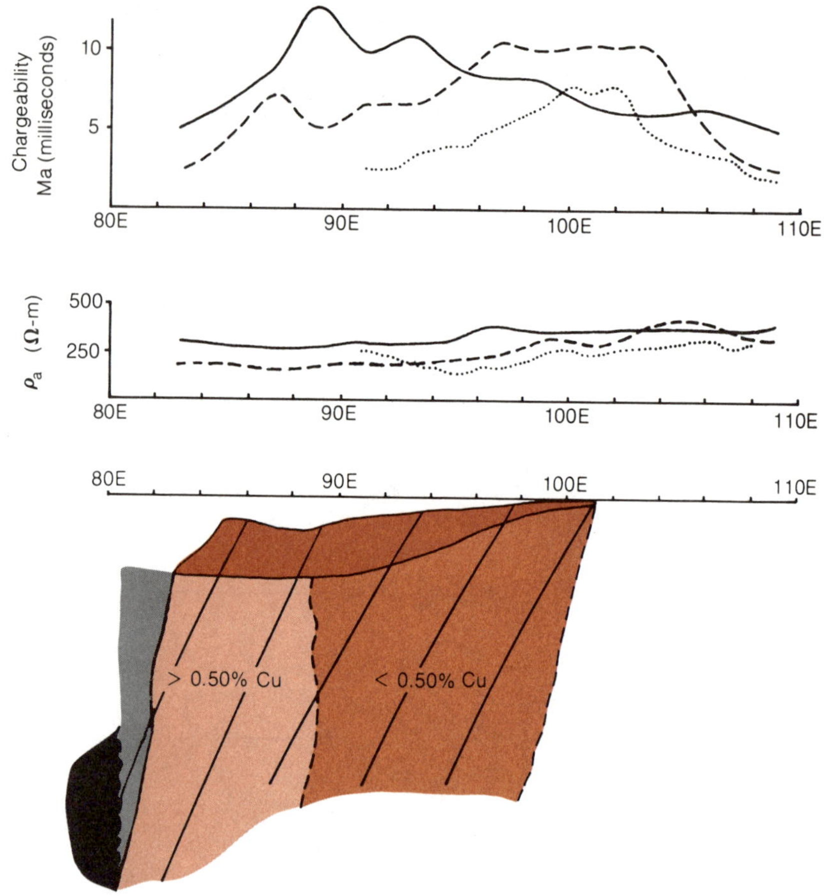

The figure on this page is referred to in the preceding section *Choice of electrode interval.*

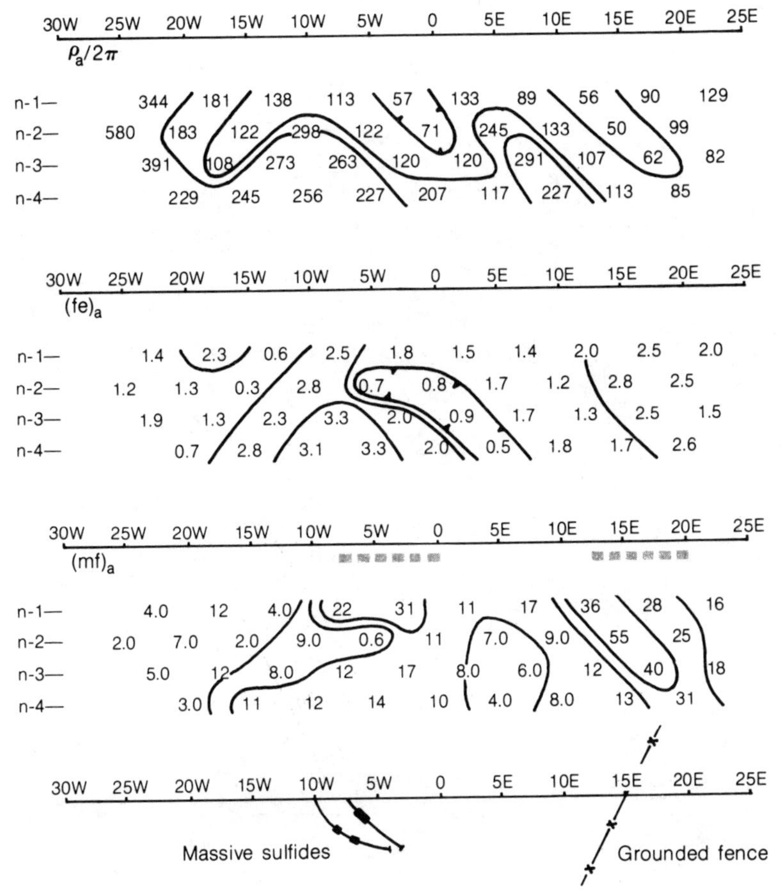

FIGURE 2.71
Induced polarization and drilling results from Broken Hill area, New South Wales

Line-25N

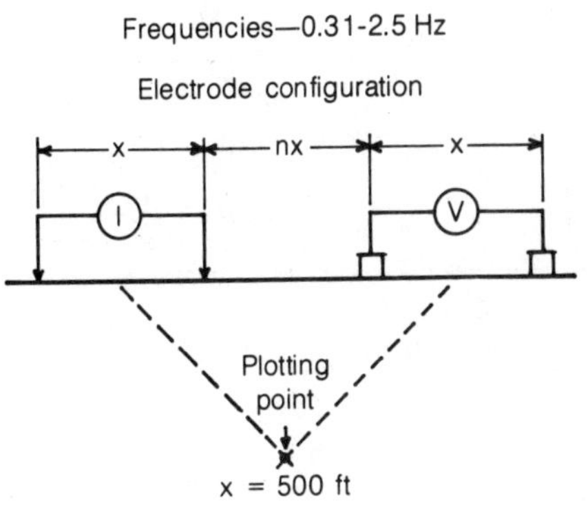

The figure on this page is referred to in the preceding section *Choice of electrode interval.*

FIGURE 2.72
Induced polarization and drilling results from Broken Hill area, New South Wales

Line-25N

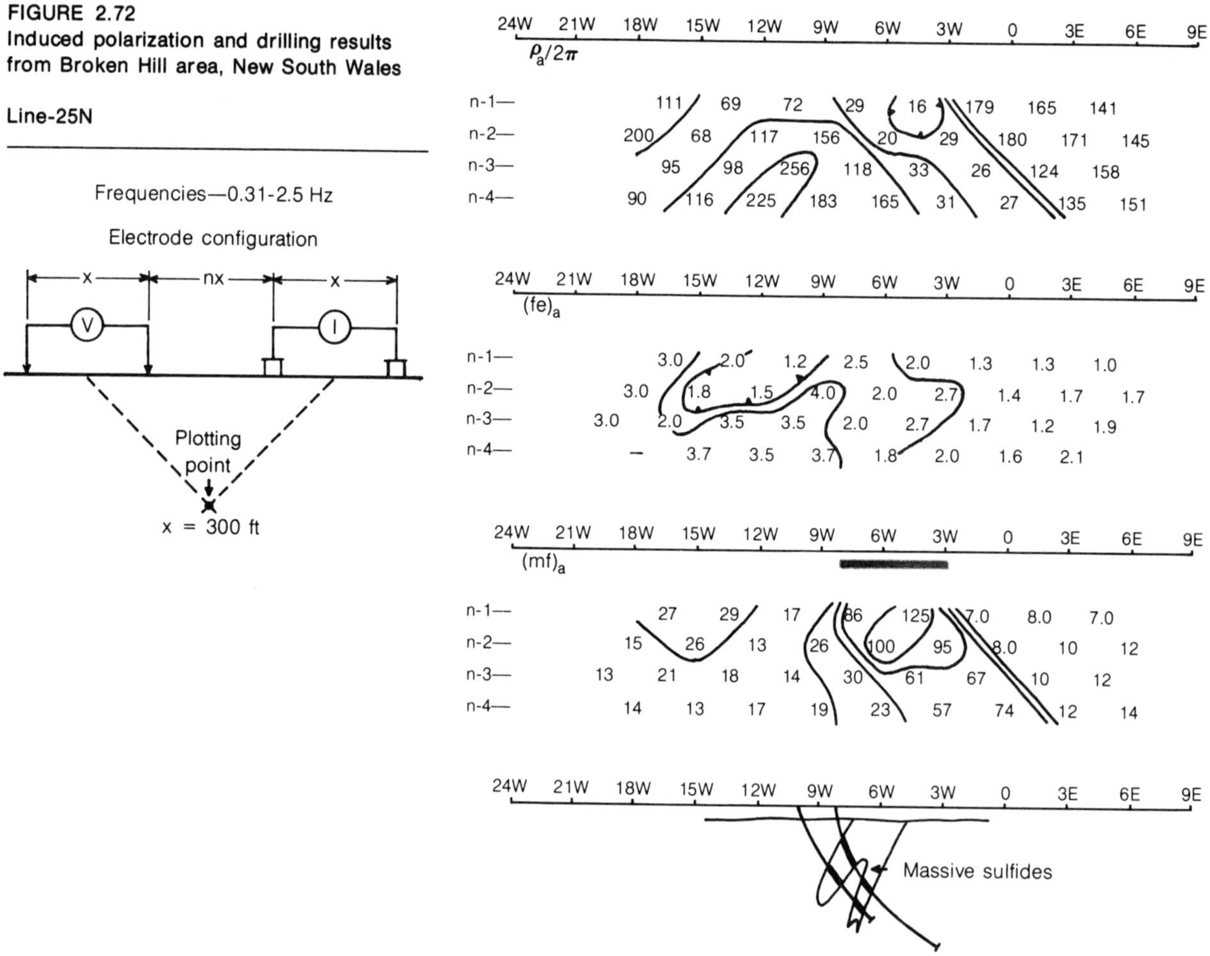

The figure on this page is referred to in the preceding section *Choice of electrode interval.*

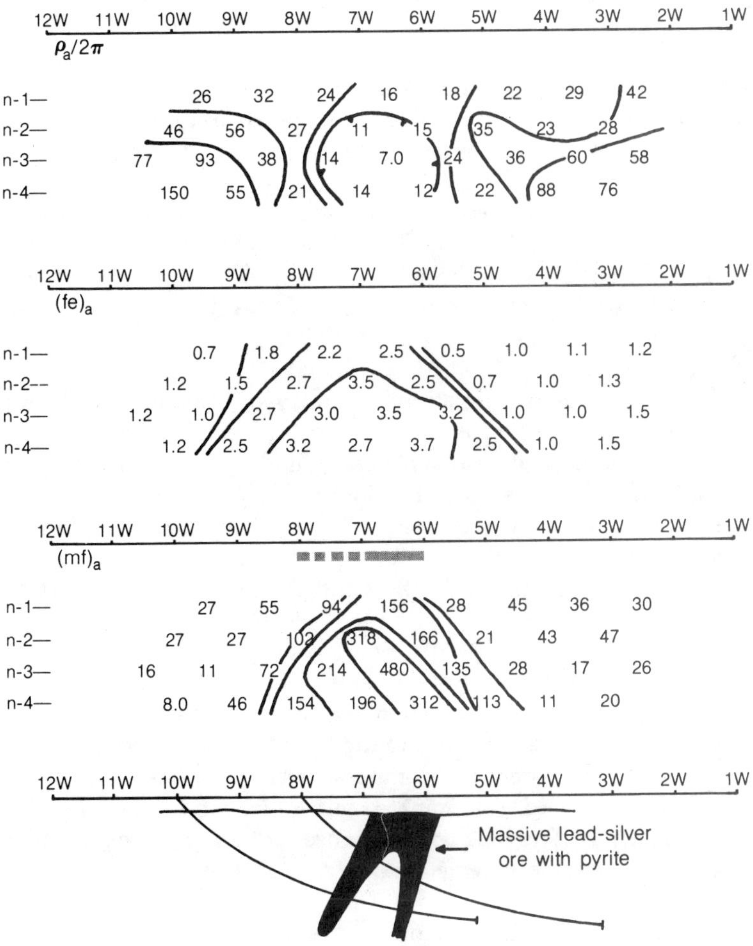

FIGURE 2.73
Induced polarization and drilling results from Broken Hill area, New South Wales

Line-25N

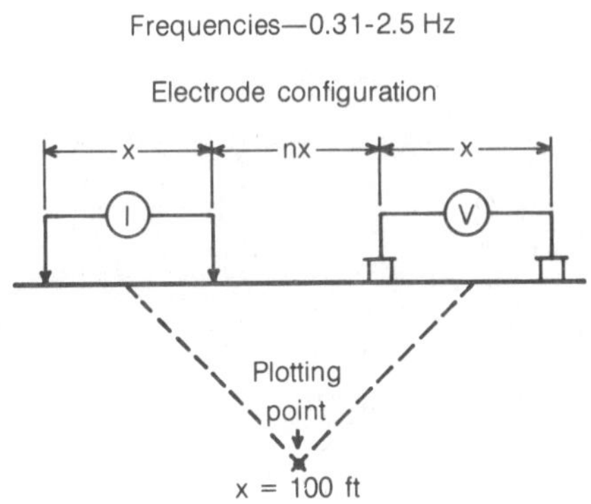

Use of IP in the search for disseminated sulfide mineralization

As previously discussed, the addition of a few percent of metallic mineralization to an otherwise normal, or a badly altered, rock will probably not change its true resistivity very much. However, even concentrations of 0.5-1.0 percent sulfides can give rise to appreciable IP effects. Therefore, it is not surprising that in the period 1960 to 1970, when EM methods could be used to find massive sulfide deposits and when disseminated, porphyry copper deposits were in great demand, the IP method was first developed to search for disseminated, metallic mineralization.

The data shown in figure 2.74 were measured in 1959 with the use of frequencies of dc and 5.0 Hz. There is enough veinlet sulfide present to lower the resistivity of the altered intrusive. The IP effects from the shallow deposit are extremely large. The geometry and anomalous pattern indicate that the apparent effects are nearly equal to the true effects.

There is almost no resistivity anomaly in the data shown in figure 2.75. The apparent effects are probably equal to the true effects. However, the anomaly is much weaker, and there is only 1.0-1.5 percent metallic minerals present. It is ore-grade because only copper and molybdenum minerals are present. The anomaly that located the Brenda orebody was interpretable only because the surrounding rocks contained almost no metallic mineralization and the background IP effects were very low.

The results in figure 2.76 are perhaps the first variable frequency IP data measured in the southwest United States. The measurements at dc-5.0 Hz were made in 1958. There was 200-400 ft of rock overlying the Cactus deposit, and the depth to the top is clearly shown by the results.

If the zone of disseminated mineralization is extensive and the survey is properly specified, disseminated mineralization can be located at considerable depth. The field data in figure 2.77 are from the Cordilleran in Argentina. The measurements with x − 200 m located a source at a depth of about 200 m.

During the 1960s, numerous zones of disseminated mineralization that contained copper (loosely described as porphyry copper) were found in British Columbia. One of the first was the Stikine River Deposit (figure 2.78). The anomaly is quite distinct in the low-background environment; however, as in many deposits, only a portion of the broad zone of mineralization located contained enough copper to be ore grade.

Through the years, other types of disseminated mineralization that were not of the porphyry copper type became important. The data in figure 2.79 located a high-grade, disseminated copper deposit that did become a mine. The data in figure 2.80 located a deposit of copper-silver mineralization that did not become a mine.

Use of IP to locate strata-bound deposits

Soon after IP began to be used regularly in the exploration for porphyry copper deposits, another large class of ore zones came under study. The strata-bound deposits, usually in limestones, containing zinc, lead and silver in varying concentrations did not respond to the electromagnetic methods that were in wide use. Sphalerite is a nonconductor and galena is a poor conductor. The conducting minerals pyrite and/or marcasite were usually present in some concentrations, but rarely in a quantity which would result in an EM anomaly. The field results previously shown in figures 2.44 and 2.45 demonstrated that, even from the very conductive Pyramid deposit at Pine Point, there was no recognizable EM anomaly.

Field data from the southeast Missouri lead belt, the Wisconsin lead-zinc field, Ireland, and the Pine Point area identify strata-bound sulfide deposits.

There are often only very weak concentrations of metallic minerals present (in the east Tennessee area there is reportedly no galena, pyrite, or marcasite). The surrounding limestones, however, are usually the ideal host rock within which to work because the background effects are very low. Therefore, with careful work, the geophysicist can often find relatively small zones at relatively great burial depths.

Use of IP to locate narrow sources

The proper choice of electrode intervals for exploration with the IP method was discussed in the section "Choice of electrode interval." The point was made that narrow zones of mineralization, if they are shallow, can be detected by the use of short electrode intervals.

The zones of mineralization located by the field data in figures 2.81 through 2.86 had widths that varied from 1-15 ft. The mineralization was of interest to the geologists who originated the surveys because it contained high-grade silver, high-grade copper, high-grade copper-zinc or high-grade zinc. In all of the cases shown, except that from Broken Hill, New South Wales, the source is indicated to be shallow, i.e., the maximum apparent effects were measured for n = 1. As was explained previously, the sources could be better located and evaluated by the use of detailed measurements with shorter electrode intervals.

This was done in many of the examples presented here, and detail with shorter electrode intervals is always worthwhile to better evaluate a shallow, narrow source. At least one line across an anomalous zone that indicates a shallow, narrow source should be selected for detail. Detailed measurements should be repeated with shorter electrode intervals; the electrode interval should be reduced in this way until the maximum apparent effects are measured for n = 2 or n = 3. The additional cost of the further IP work is always repaid by the requirement for fewer and shorter drill holes.

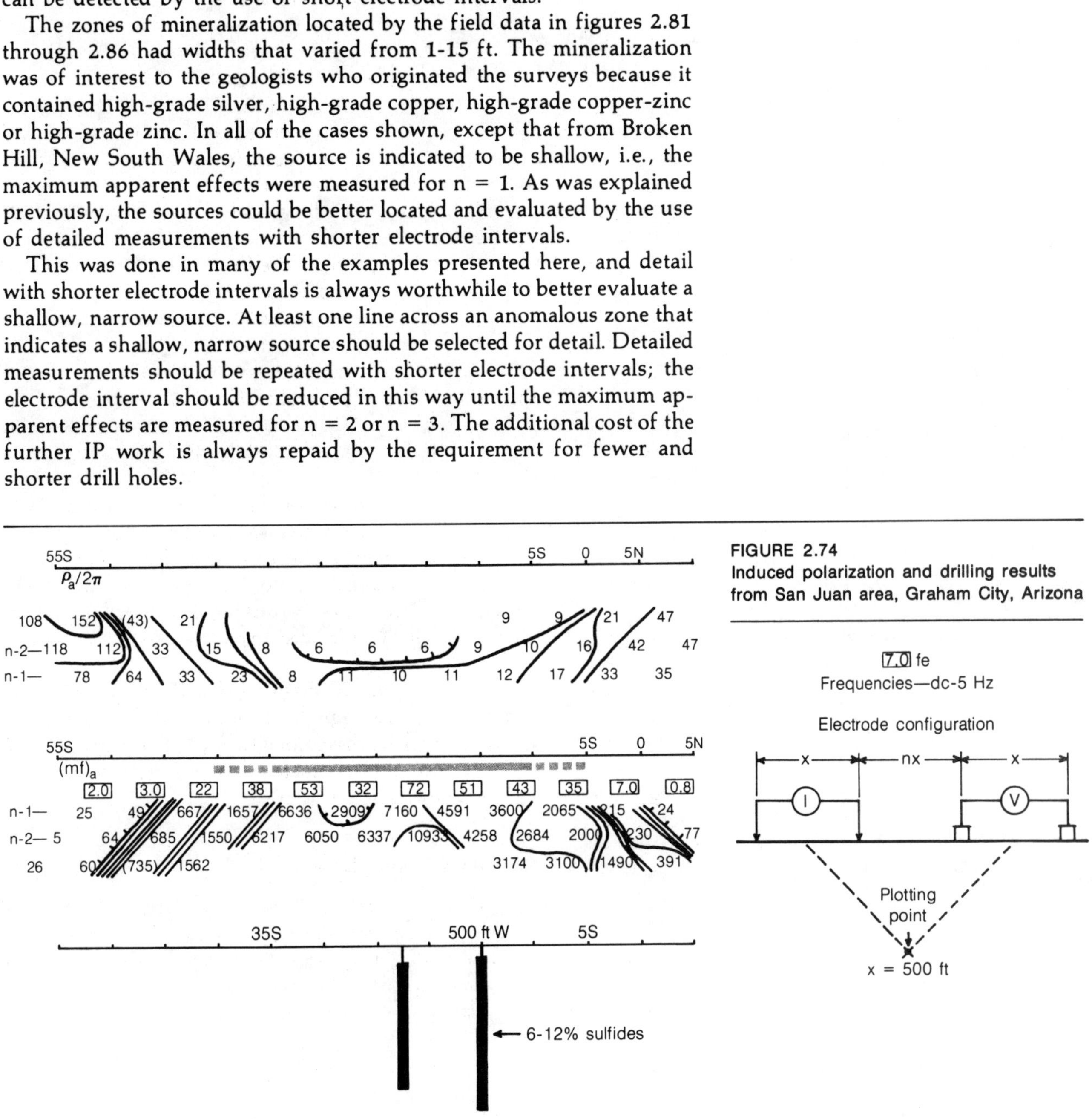

FIGURE 2.74
Induced polarization and drilling results from San Juan area, Graham City, Arizona

The figures on this page are referred to in the preceding section *Use of IP in the search for disseminated sulfide mineralization.*

FIGURE 2.75
Induced polarization and drilling results from zone of disseminated mineralization, Brenda area, Peachland, B.C.

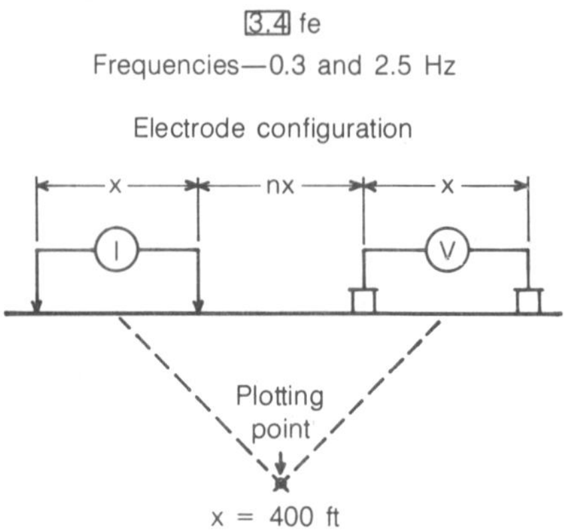

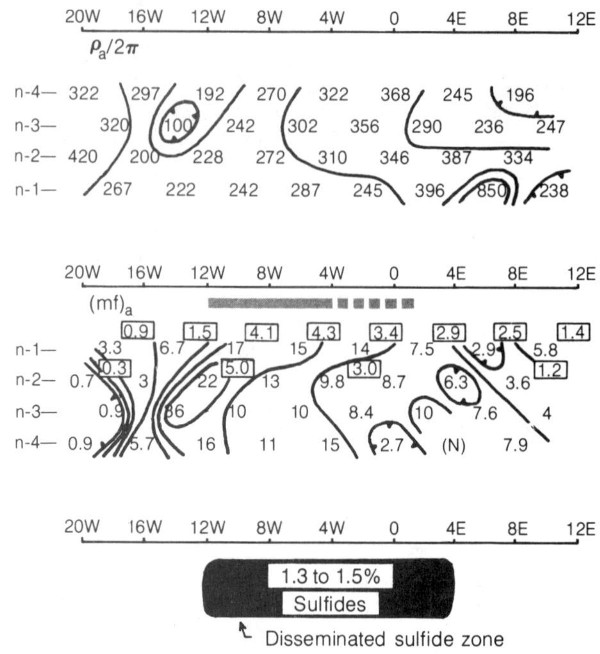

FIGURE 2.76
Induced polarization and drilling results from Cactus area, Pinal City, Arizona

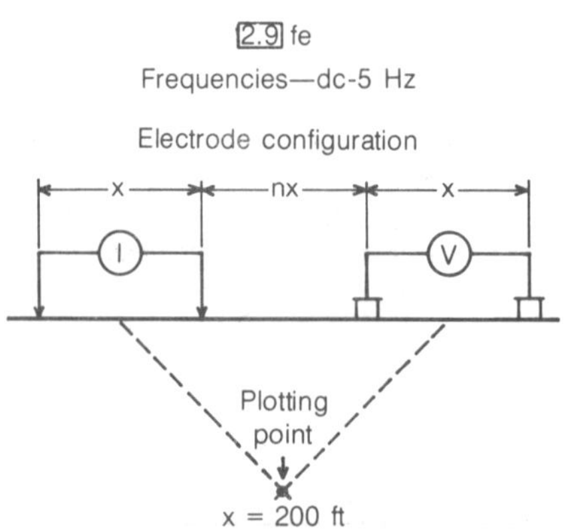

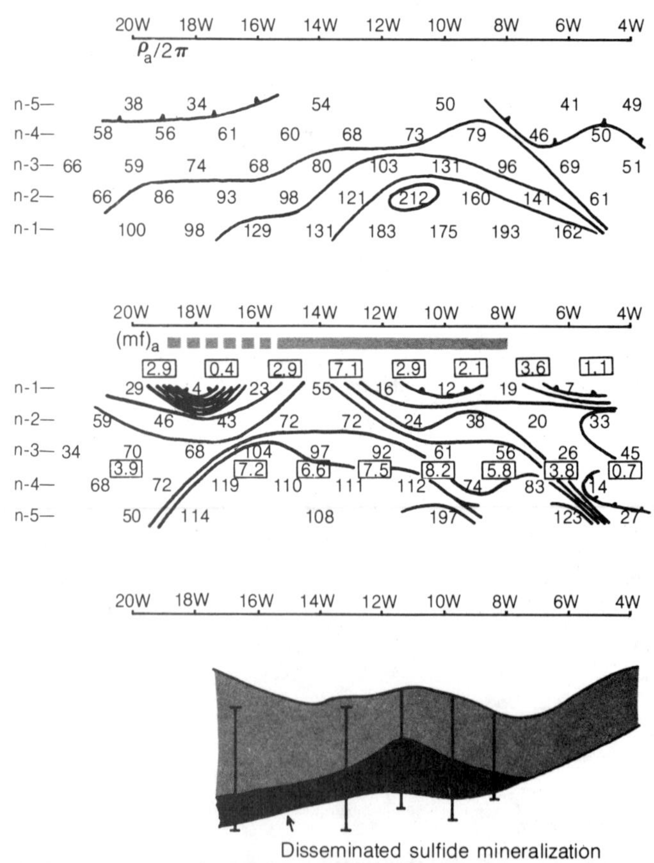

The figures on this page are referred to in the preceding section *Use of IP in the search for disseminated sulfide mineralization.*

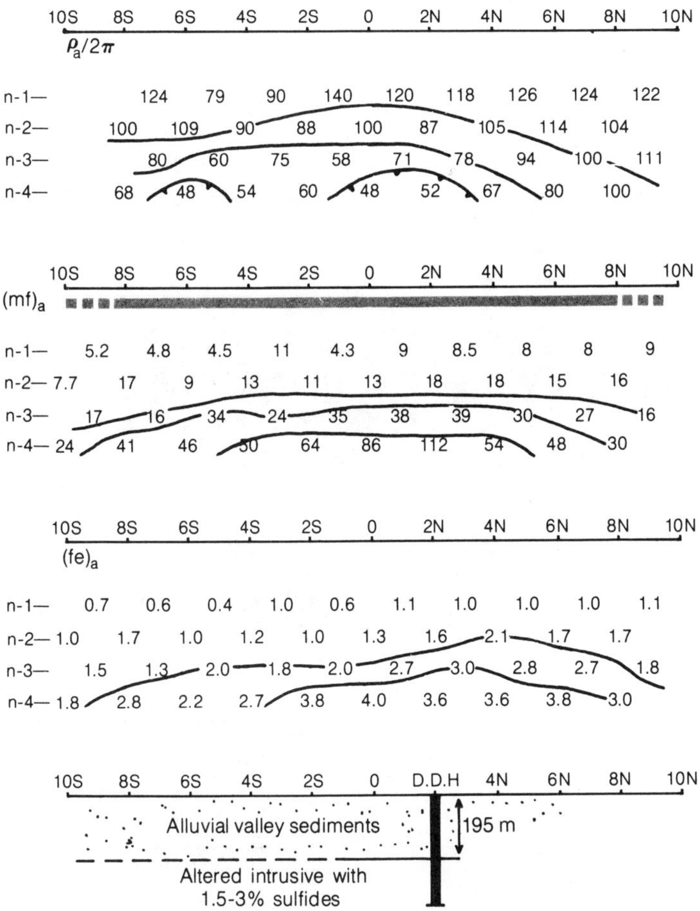

FIGURE 2.77
Induced polarization results from deep source Yalguarez Valley, Mendoza Province, Argentina

Line-8

Frequencies—0.31-2.5 Hz

Electrode configuration

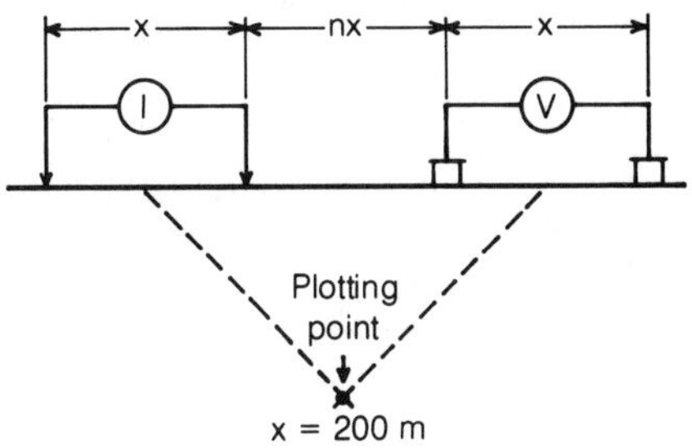

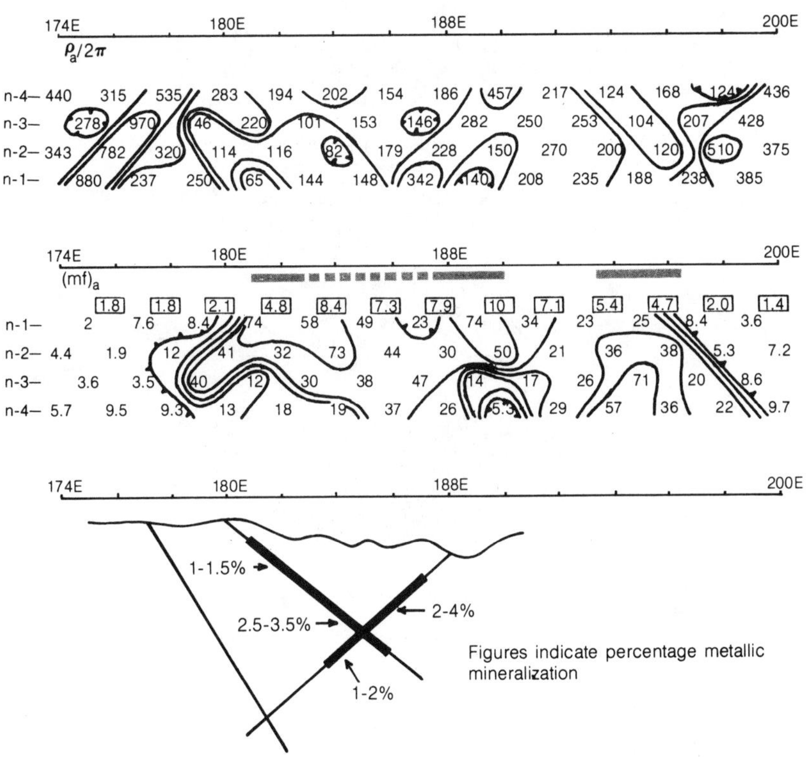

FIGURE 2.78
Induced polarization and drilling results from zone of disseminated mineralization, Stikine River area, Northwestern British Columbia

[5.4] fe

Frequencies—0.3 and 5.0 Hz

Electrode configuration

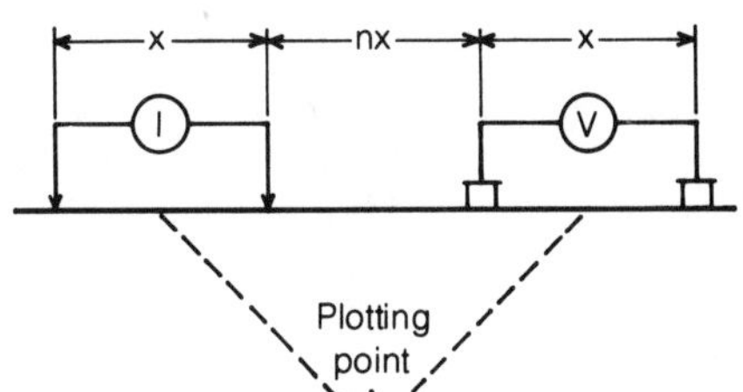

The figures on this page are referred to in the preceding section Use of IP in the search for disseminated sulfide mineralization.

FIGURE 2.79
Induced polarization and drilling results from copper-bearing mineralization in breccia zone, Tribag Property, Batchawana area, Ontario

2.7 fe

Frequencies—0.3 and 5.0 Hz

Electrode configuration

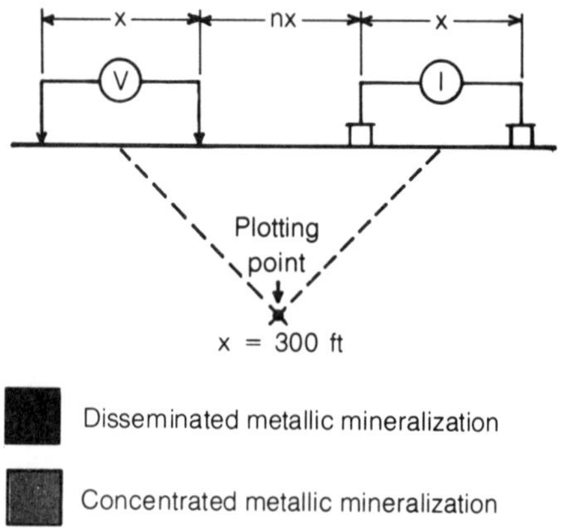

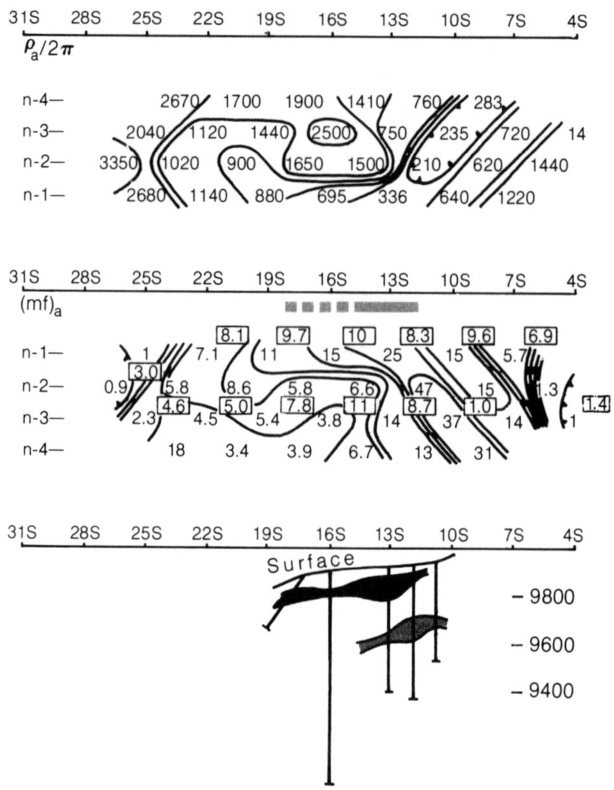

FIGURE 2.80
Induced polarization and drilling results from zone of disseminated mineralization, Medford, Oregon

5.2 fe

Frequencies—0.3 and 2.5 Hz

Electrode configuration

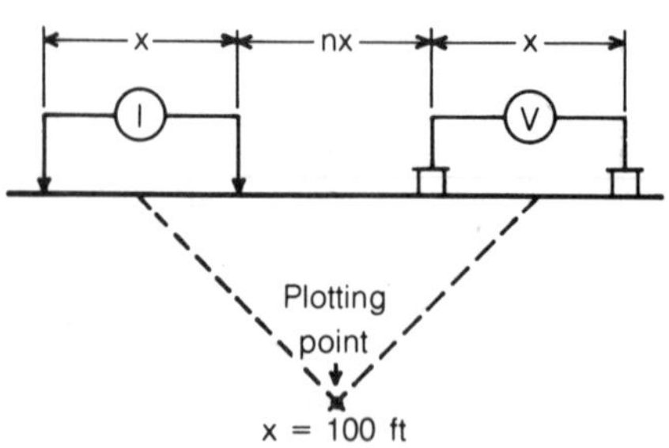

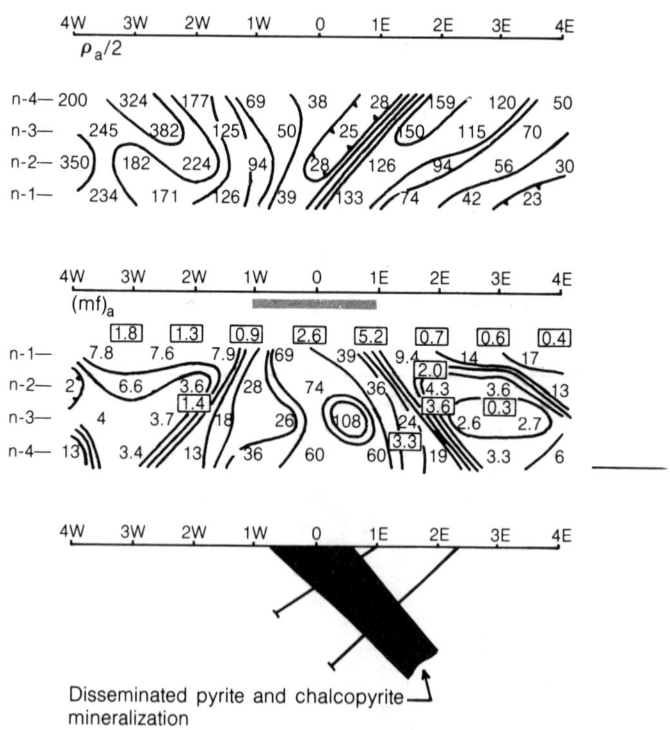

The figures on this page are referred to in the preceding section *Use of IP to locate narrow sources.*

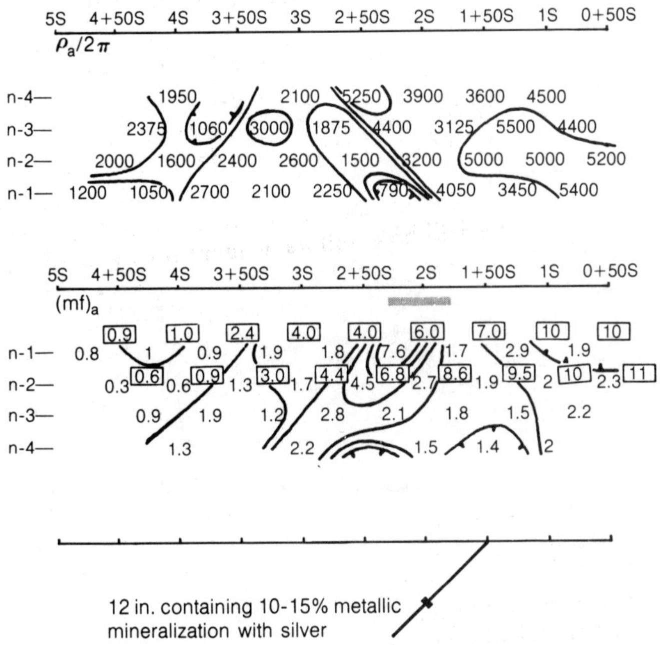

FIGURE 2.81
Induced polarization and drilling results from narrow silver-bearing vein, Cobalt area, Ontario

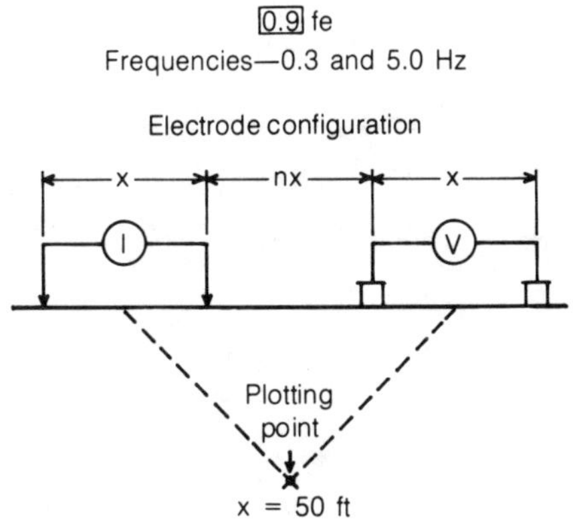

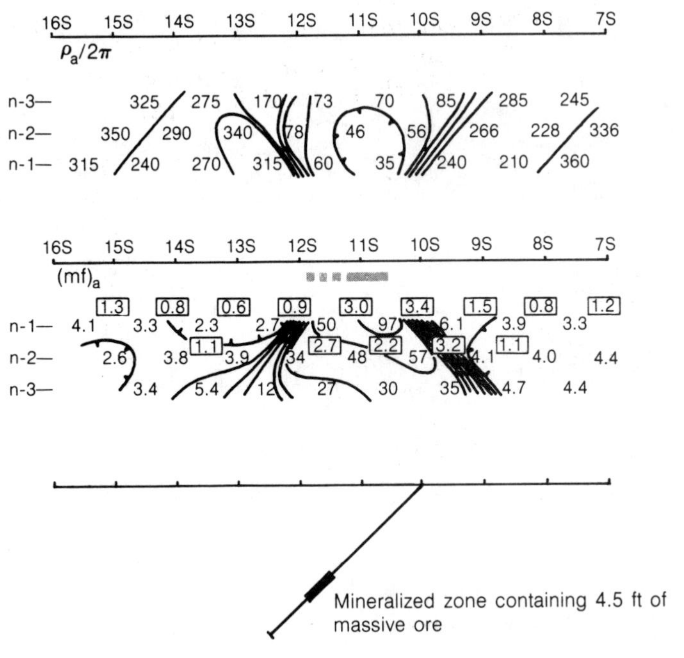

FIGURE 2.82
Induced polarization and drilling results from massive sulfide ore zone, Nigadoo River Mines, Bathurst area, New Brunswick

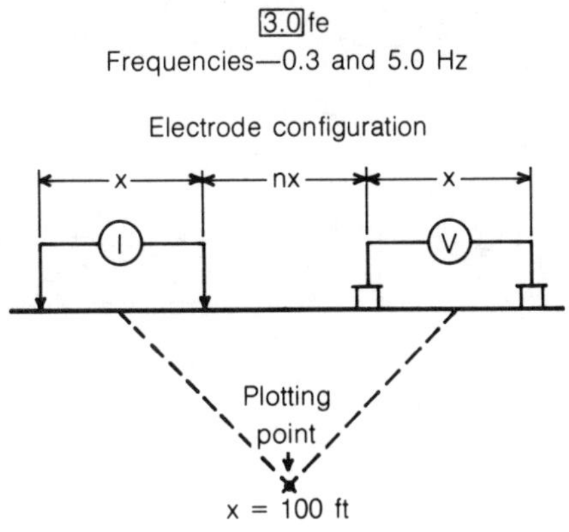

The figures on this page are referred to in the preceding section *Use of IP to locate narrow sources.*

FIGURE 2.83
Induced polarization and drilling results from massive sulfide ore zone, Heath Steele Mines, Newcastle area, New Brunswick

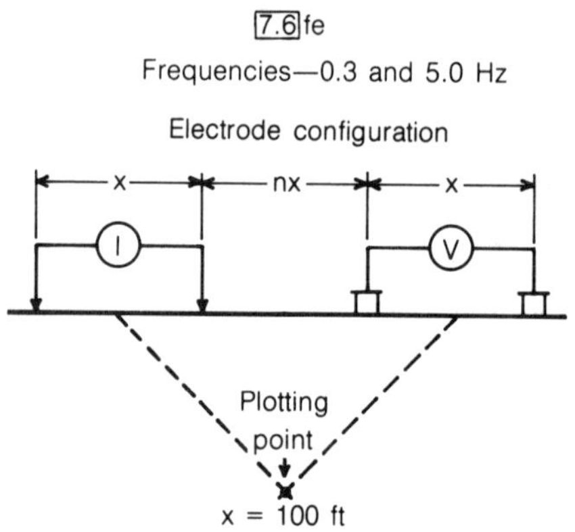

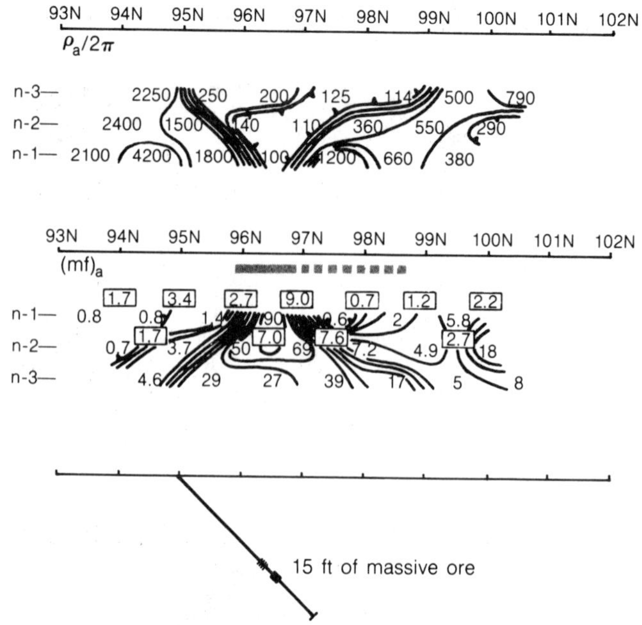

FIGURE 2.84
Induced polarization and drilling results from shallow zone of massive mineralization, Manitouwadge area, Ontario

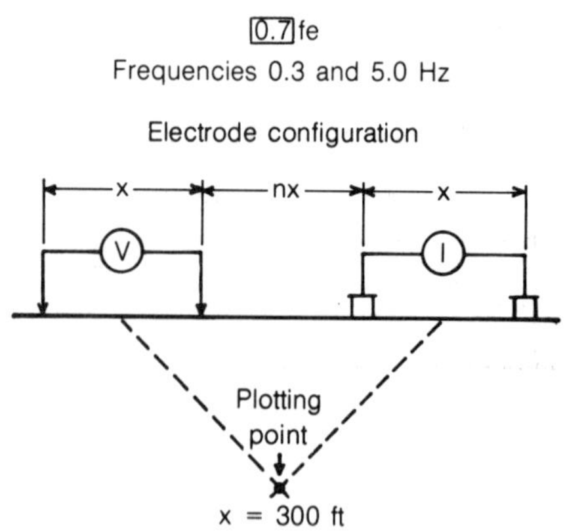

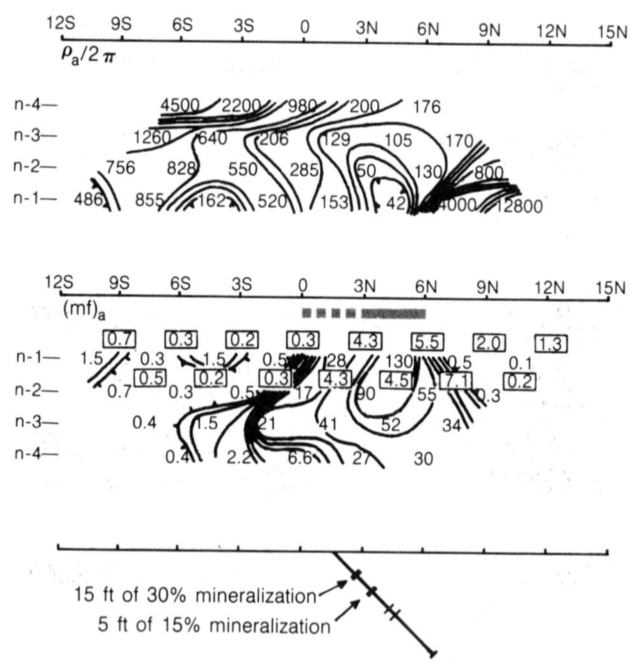

The figures on this page are referred to in the preceding section *Use of IP to locate narrow sources.*

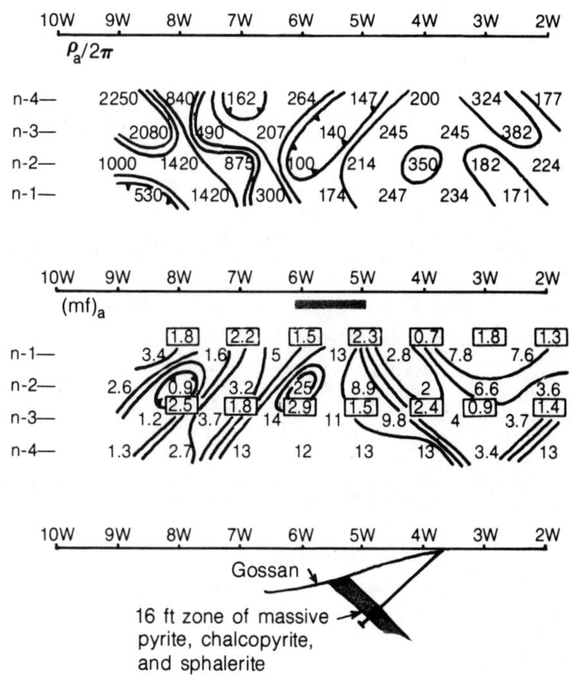

FIGURE 2.85
Induced polarization and drilling results from narrow zone of massive mineralization, Medford, Oregon

[1.5] fe
Frequencies—0.3 and 2.5 Hz

Electrode configuration

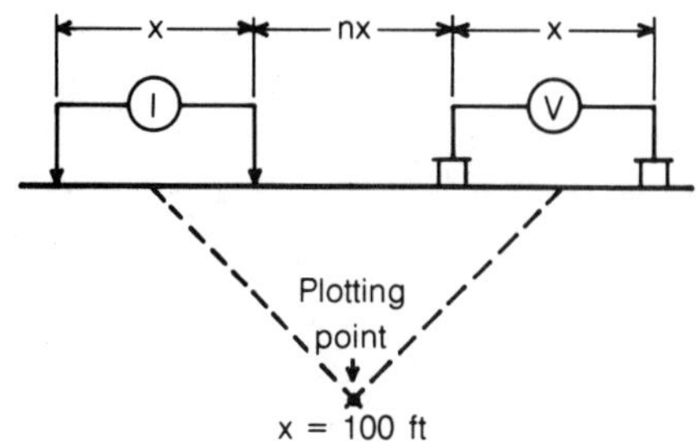

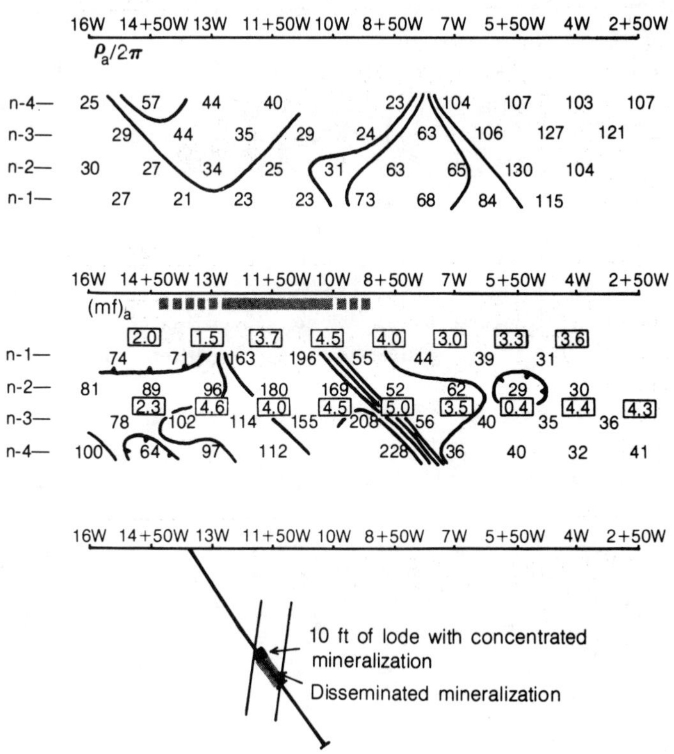

FIGURE 2.86
Induced polarization and drilling results from mineralized lode band, Broken Hill, N.S.W. Australia

[1.5] fe
Frequencies—0.3 and 2.5 Hz

Electrode configuration

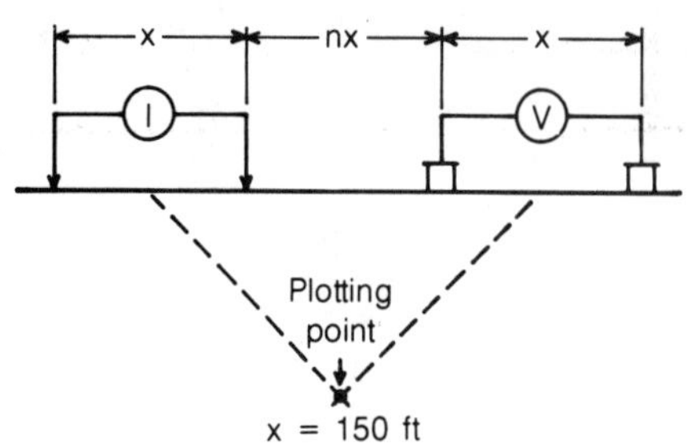

Use of IP in exploration in areas of high surface conductivity

Exploration in areas of deep, complete surface weathering has always been a difficult problem for geologists and geophysicists. These areas are usually tropical, or arid to semi-arid. Even when transported alluvium cover is not present, the geologist's job is a difficult one. Complete weathering reduces outcrops of widely varying rock types to very uniform-appearing rubble. Geochemistry may or may not be useful depending upon the chemistry of the area, the local plumbing, and the presence or absence of transported material.

Geophysicists have become accustomed to using electrical methods in the search for sulfide ores. However, the weathered layer is always very porous and often contains saline solutions. Such a weathered layer always has much lower resistivities than the basement rocks from which it was derived. In many areas, the conductivity of the surface layer approaches that of sea water. In this environment, it is not surprising that electromagnetic and resistivity methods have been found to be of little or, more frequently, no assistance in exploration.

The weathered layer is not uniform, but contains an infinite number of resistivity variations. Every crack and shear in the basement rock can contain conductive saline solutions. The sulfide mineralization has been converted to oxides so that there is usually a considerable depth to the top of the zone of mineralization. It is not so much that no anomalies are present from the sulfide mineralization at depth, but that the weak anomalies measured are lost among the large variations in the EM results that are present due to the surface conducting layer.

The field results in figure 2.87 are from Cyprus. Most of the non-mountainous portion of the island has resistivities as low as sea water. The results shown here were measured over a known orebody that was being prepared for mining.

Much of Australia is deeply weathered, and the field results in figure 2.88 found a high-grade copper orebody in South Australia. The IP anomaly is quite definite and large in magnitude. The apparent resistivities are low magnitude and variable. The lowest resitivities are within the narrow conductor at 1+00E; this feature correlates with a major fault structure that can be traced for tens of miles.

One of the lowest resistivity areas in Australia (indeed, in the world) is the old Shield surface in Western Australia. This very old surface is deeply weathered and very saline. The apparent resistivities are very low, and the sulfide zones are usually weathered to depths of 100 ft or more.

The field results shown in figures 2.89 and 2.90 located nickel ore zones in this low resistivity environment. The apparent resistivities at Nepean are lower than those at Scotia; they are about equal to those in sea water.

The results in figure 2.91 did not locate an orebody. The data are included because the massive sulfide zone centered at 4+00W represents the discovery that was made, using electrical methods, in the lowest resistivity environment of which the author has knowledge.

The IP method can be used to locate metallic mineralization in areas of high conductivity. However, planning is necessary and special care must be taken. The low apparent resistivities will result in very low voltages at the potential electrodes. These low voltages must be detected in the presence of what is often considerable electrical noise. High-powered current sources and very stable, sensitive receiver-voltmeters must be used. Finally, inductive coupling effects can be expected to distort IP measurements made in areas of low resistivities.

The figures on this page are referred to in the preceding section Use of IP in exploration in areas of high surface conductivity.

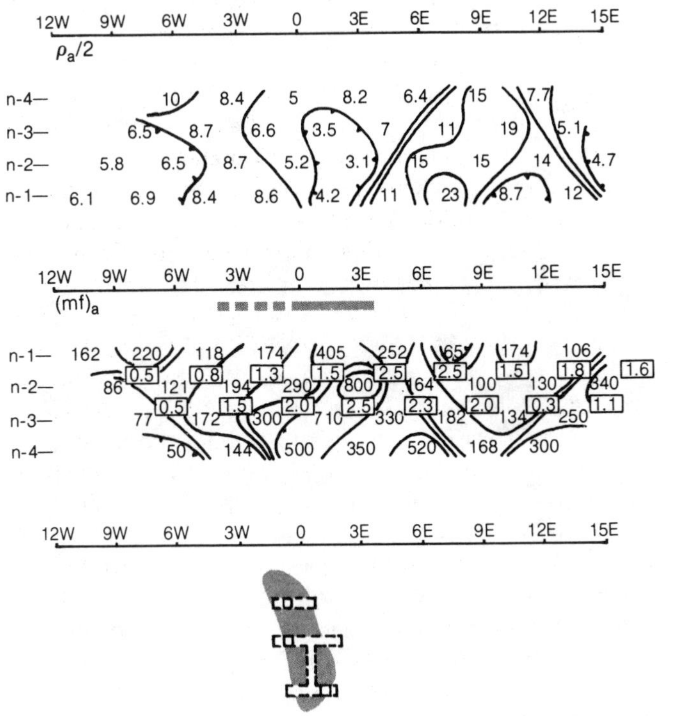

FIGURE 2.87
Induced polarization and drilling results from massive pyrite-chalcopyrite zone, Cyprus

2.5 fe
Frequencies—0.1 and 1.25 Hz

Electrode configuration

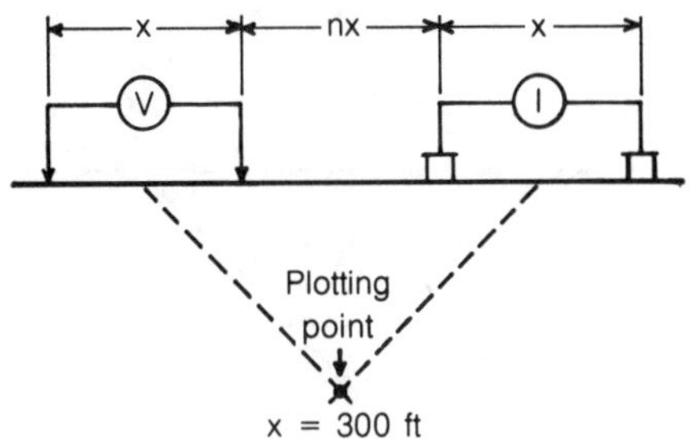

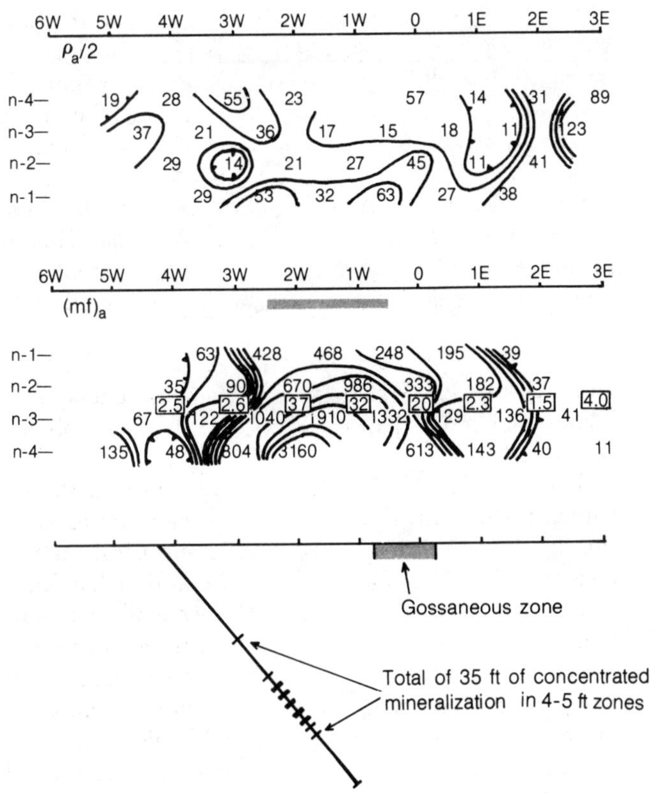

FIGURE 2.88
Induced polarization and drilling results from concentrated mineralization, oxidized at surface, South Australia

32 fe
Frequencies—dc and 1.25 Hz

Electrode configuration

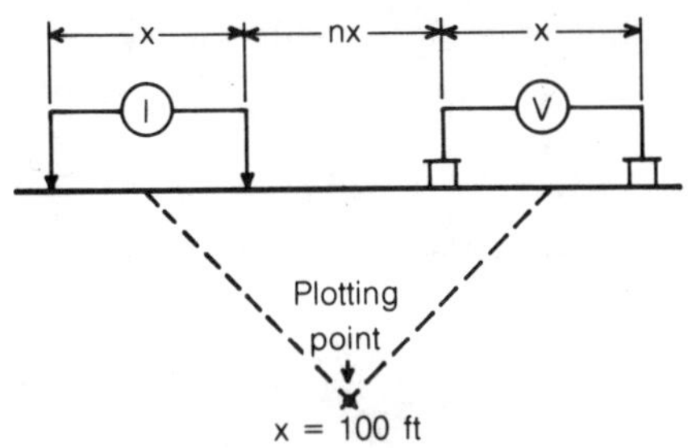

The figures on this page are referred to in the preceding section *Use of IP in exploration in areas of high surface conductivity.*

FIGURE 2.89
Induced polarization and drilling results from narrow zone of massive mineralization, Scotia Grid, Kalgoorlie, West Australia

0.5 fe
Frequencies—0.1 and 1.25 Hz

Electrode configuration

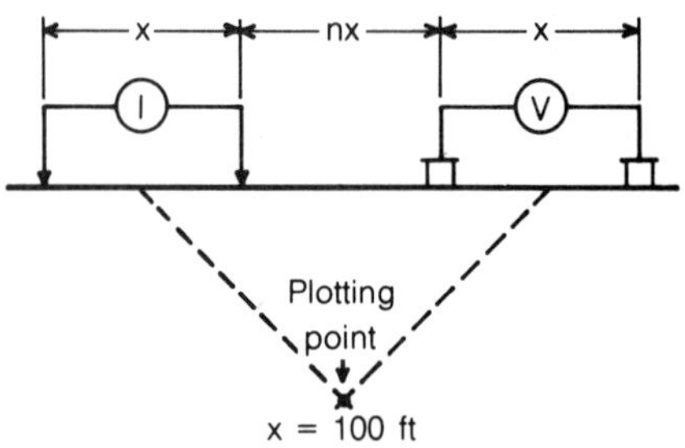

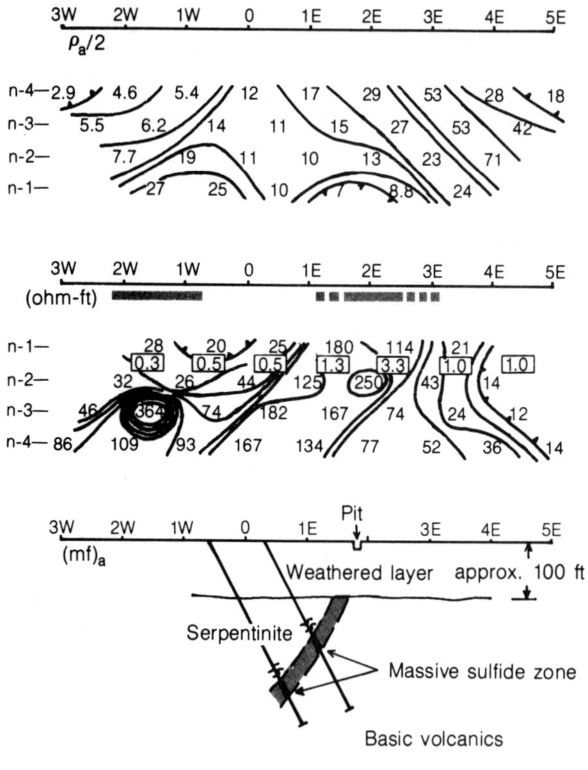

FIGURE 2.90
Induced polarization and drilling results from narrow zone of massive mineralization, Nepean area, West Australia

3.0 fe
Frequencies—0.1 and 1.25 Hz

Electrode configuration

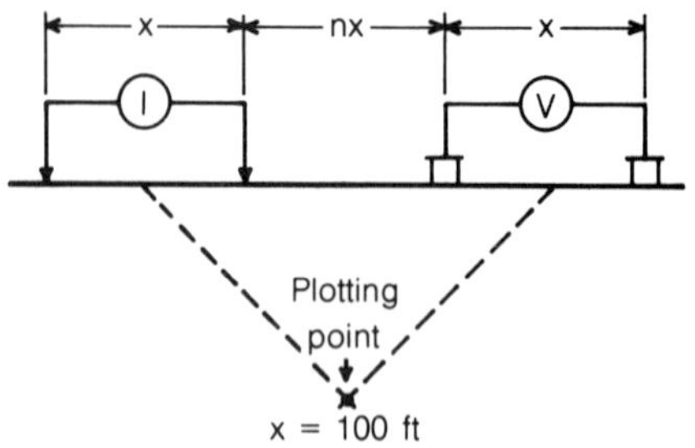

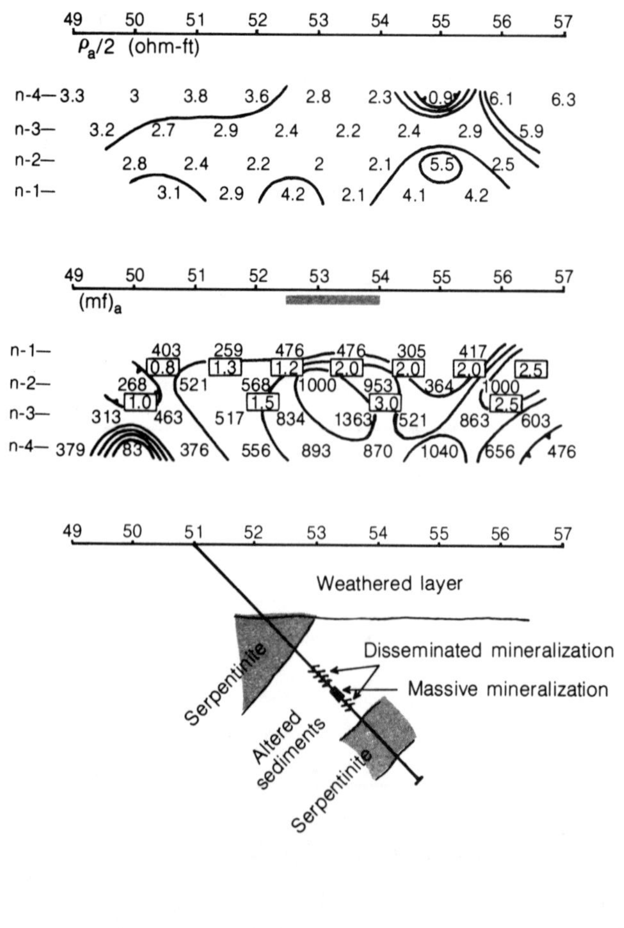

The figure on this page is referred to in the preceding section *Use of IP in exploration in areas of high surface conductivity.*

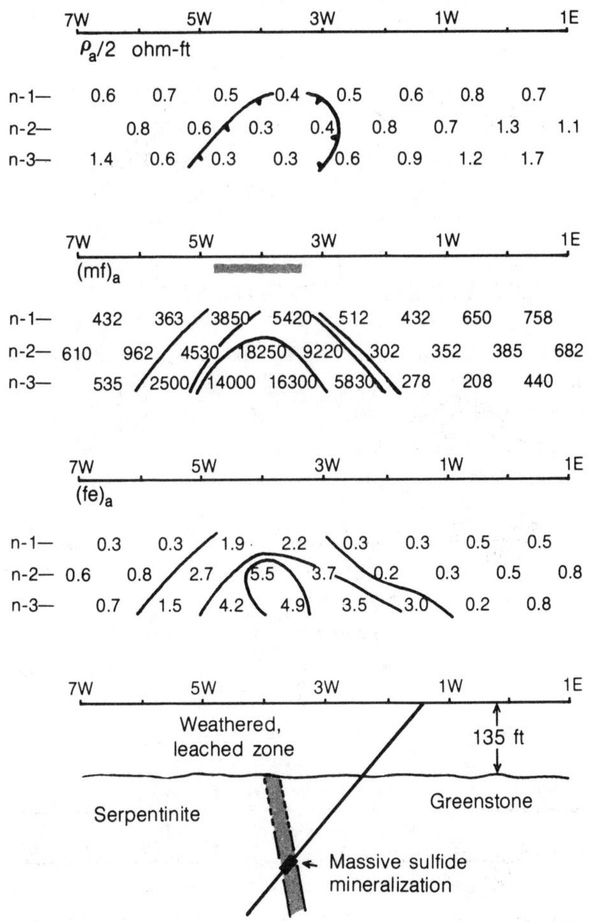

FIGURE 2.91
Induced polarization results from deep source, Kalgoorlie area, West Australia

Line-175N

Frequencies—0.31-2.5 Hz

Electrode configuration

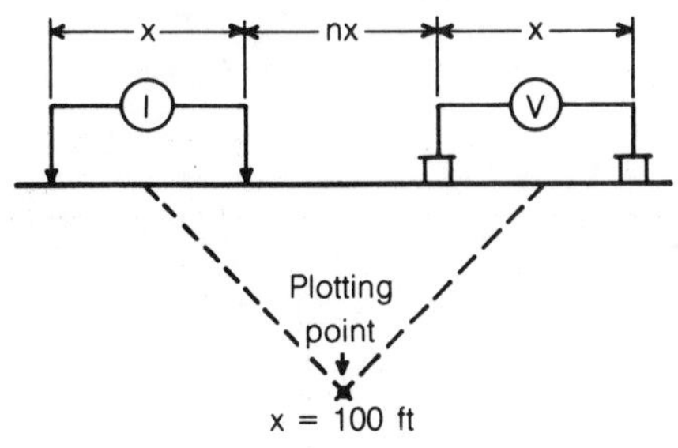

Comparison of IP and EM results

In the period from 1945 to 1975, the electromagnetic method (in its various forms) was the primary exploration method used in the search for massive sulfide mineralization. This was particularly true in the shield areas in Canada and Scandinavia. Systems were developed for use in the air and on the ground. A great many zones of massive sulfide mineralization were located, and a significant number of these were found to be orebodies. Through the years, it became common practice to abandon areas that contained no electromagnetic responses. It was only in areas immediately adjacent to known ore deposits that drilling was done which was not guided by electromagnetic data.

Beginning in the mid-1960s, experiments with the use of the IP method in the search for massive sulfide ore began. As shown in figure 2.92, a zone of massive sulfide mineralization that gave a definite EM anomaly could be expected to give a strong IP response. It was also learned that some electromagnetic anomalies, presumably from a purely ionic conductor such as a shear zone, would give EM and sometimes even resistivity anomalies (figure 2.93). However, because no metallic mineralization is present, there are no correlating IP responses.

From this use for the IP method in the search for massive sulfide zones, progression was made to the use of the IP method in the search for concentrated sulfide mineralization that cannot be detected with electromagnetic methods.

The electromagnetic, IP and resistivity field data in figures 2.94 through 2.100 are from various areas in North America. There are no complications due to deep weathering, highly conductive overburden, etc.; the zones of massive sulfide mineralization might normally be expected to give EM anomalies, but they do not.

In all cases shown, a resistivity low correlates with the IP anomaly. Therefore, it is obvious that some resistivity contrast exists between the ore and the country rock.

It may be that the resistivity contrast for some ore mineralization is not as great as geophysicists have let themselves believe. If EM anomalies have been drilled almost exclusively, it would not be surprising if all the orebodies found were conductive enough to be detected by EM methods. Sphalerite, for example, is a nonconductor and galena is only a poor conductor. Mineralization containing pyrrhotite tends to be a good conductor because the usual nature of the mineral is to develop large, interconnected crystals. Pyrite usually has a smaller grain-size and is more equidimensional.

It must be remembered that all of the mineralization shown in these data had enough of a resistivity contrast to indicate a resistivity anomaly. Even for the deep source at Timmins (figure 2.94), the deep apparent resistivities over the source are lower than those to the east and west.

The depth to the top of the source may have reduced or eliminated the EM anomaly at Timmins (figure 2.94), at Bathurst (figure 2.95), and at Flin Flon (figure 2.96) since the IP and resistivity anomaly was detected only for the larger values of (n). However, at Noranda (figures 2.97 and 2.98), at Newcastle (figure 2.99) and in Wisconsin (figure 2.100), the effects were measured for $n = 1$ indicating the source is shallow.

It is probable that another factor affecting the electromagnetic response is the shape and attitude of the source. The EM techniques respond best to a sheet-like conductor with considerable depth extent and great strike length. The IP measurements satisfy different physical equations and show recognizable anomalies from zones with relatively short strike lengths and even vertical, pipe-like sources.

The use of EM methods is also made extremely difficult in areas of a thick, relatively conductive upper layer. This layer may be due to conductive overburden such as marine sediments or even old lake-bed sediments; it could also be due to deep weathering that leaves a porous, saline, residual soil.

In many of these situations IP can be used successfully since it is unaffected, to a large degree, by the ionic conductivity of the surface layer.

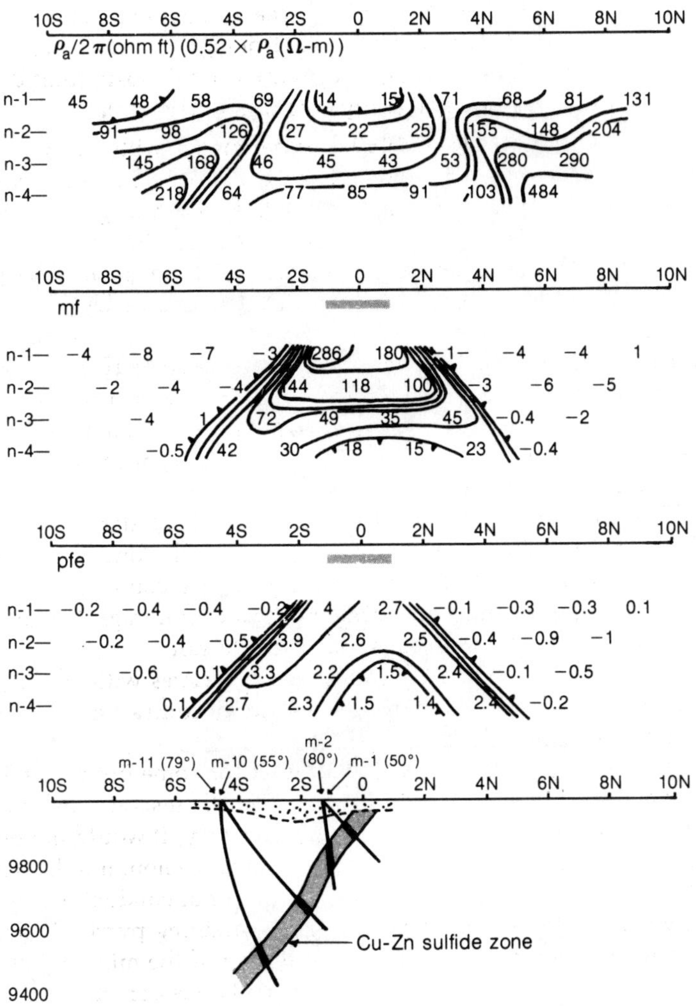

FIGURE 2.92
Iso-Copperfields test, Hebecourt Township, Noranda area, Quebec, Canada

Line No.-"0"

200 ft dipole-dipole electrode configuration
Frequencies—0.3 and 5.0 Hz

The figure on this page is referred to in the preceding section *Comparison of IP and EM results.*

Figure 2.92, continued

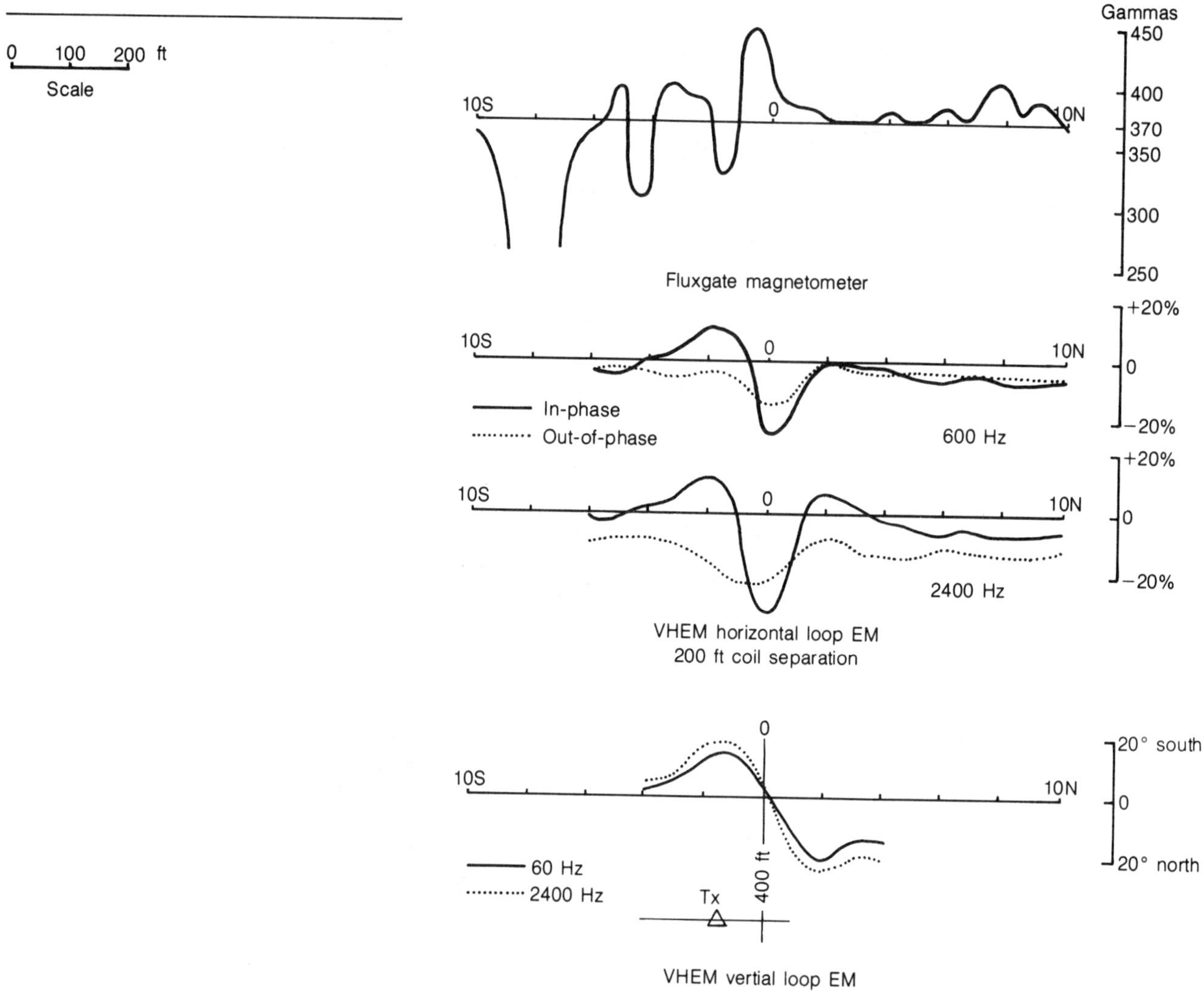

The figures on this page are referred to in the preceding section *Comparison of IP and EM results.*

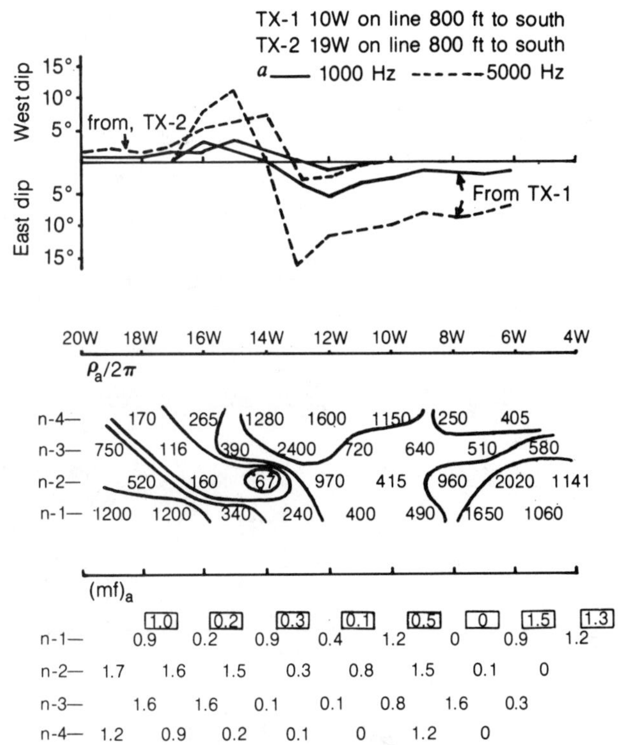

FIGURE 2.93
EM and IP results from Timmins area, Ontario, Canada

1.0 fe
Frequencies—0.3 and 5.0 Hz
x = 200 ft

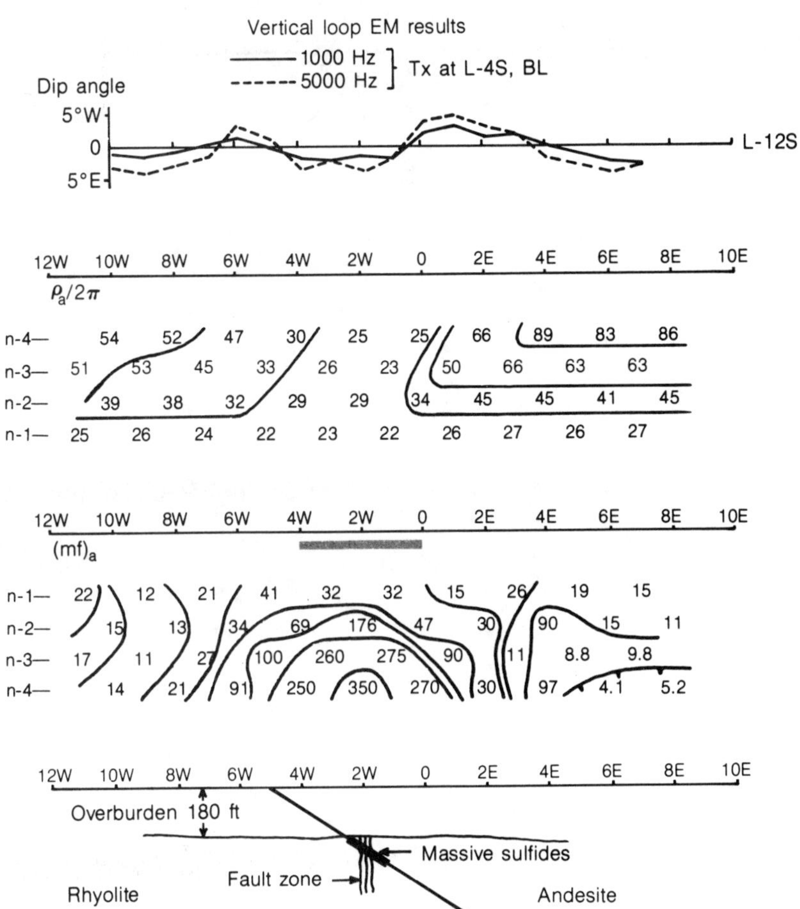

FIGURE 2.94
Comparison of EM methods and IP results from Timmins area, Ontario, Canada

Line-12S

Frequencies—0.31 and 2.5 Hz

Electrode configuration

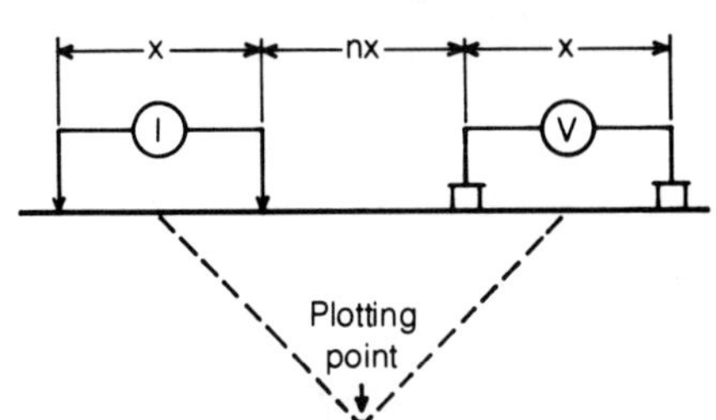

x = 200 ft

The figure on this page is referred to in the preceding section *Comparison of IP and EM results.*

FIGURE 2.95
Comparison of EM methods and IP results from Bathurst area, New Brunswick, Canada

Line-28E

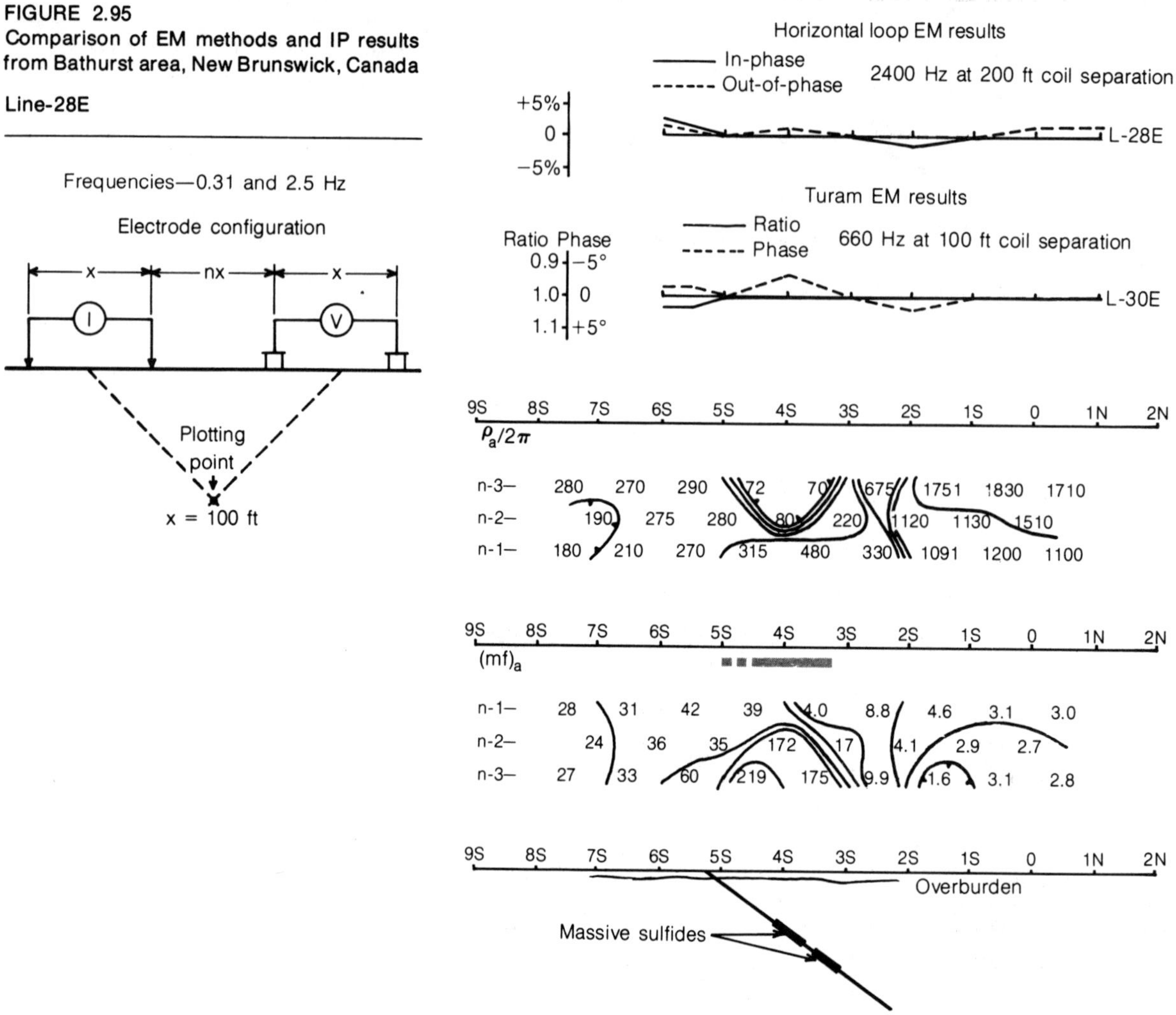

The figures on this page are referred to in the preceding section *Comparison of IP and EM results.*

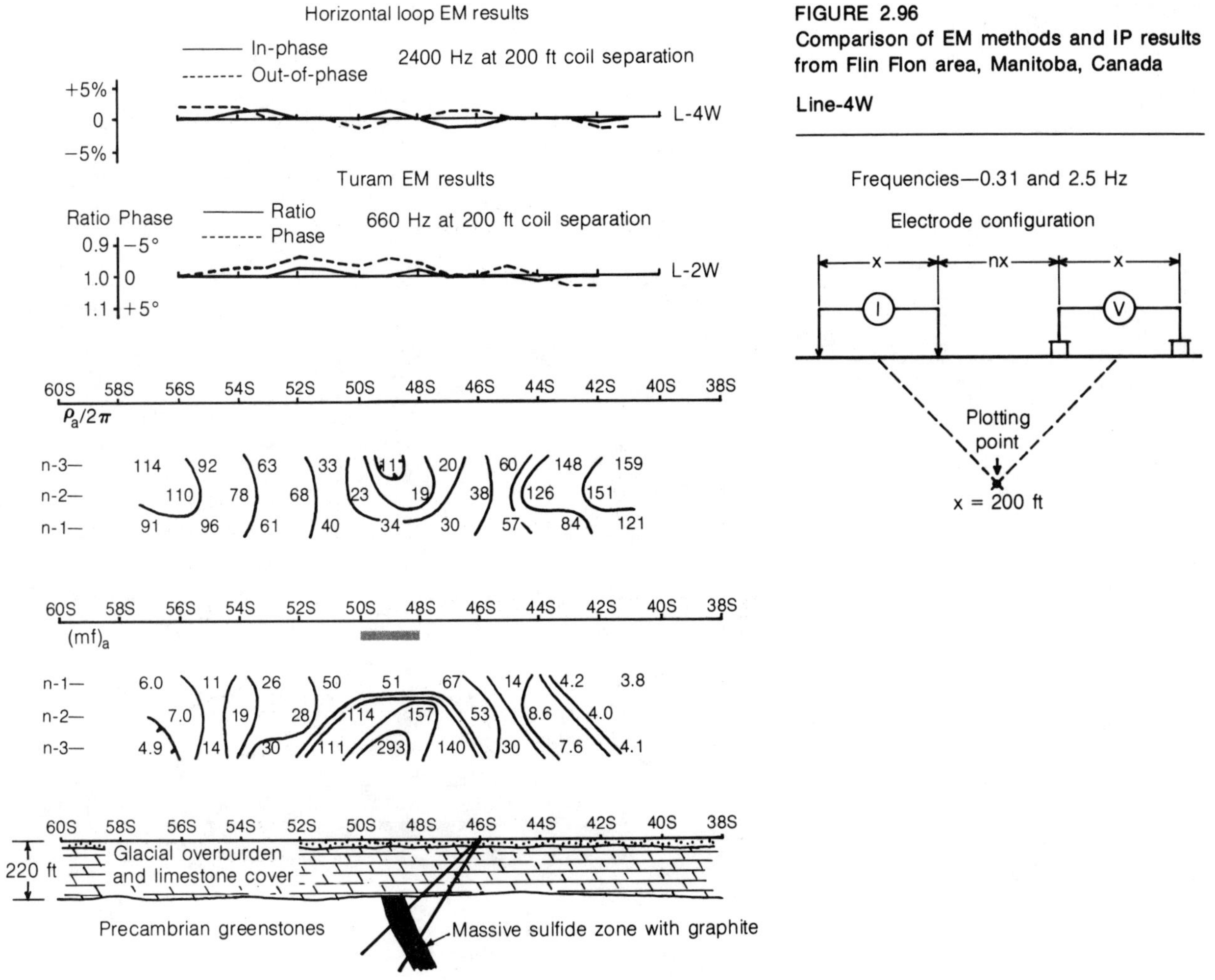

FIGURE 2.96
Comparison of EM methods and IP results from Flin Flon area, Manitoba, Canada

Line-4W

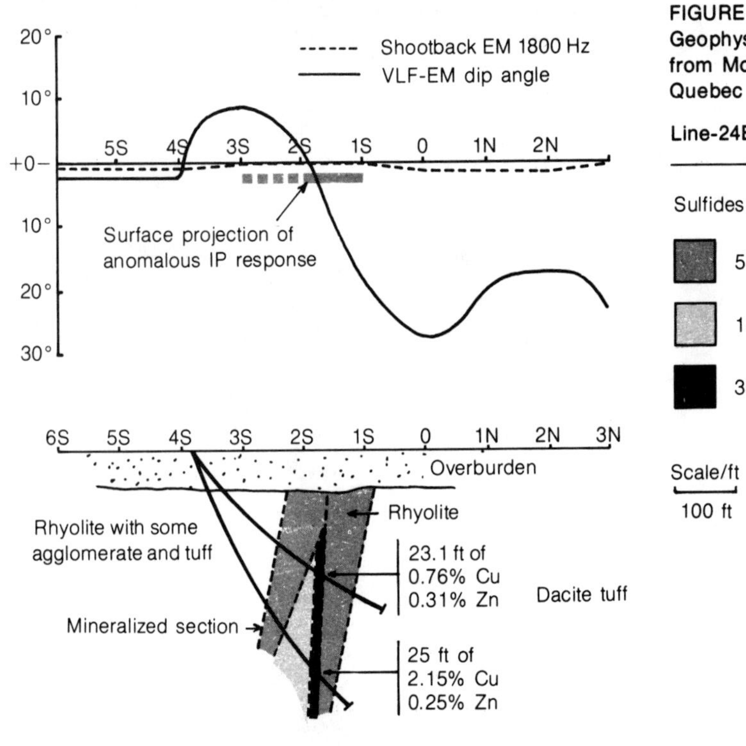

FIGURE 2.97
Geophysical data and geologic section from Mobrun Extension, Noranda area, Quebec

Line-24E

The figure on this page is referred to in the preceding section *Comparison of IP and EM results.*

FIGURE 2.98
IP and geologic data from Mobrun Extension, Noranda area, Quebec, Canada

Line-24E

Frequencies—0.31 and 5 Hz

Electrode configuration

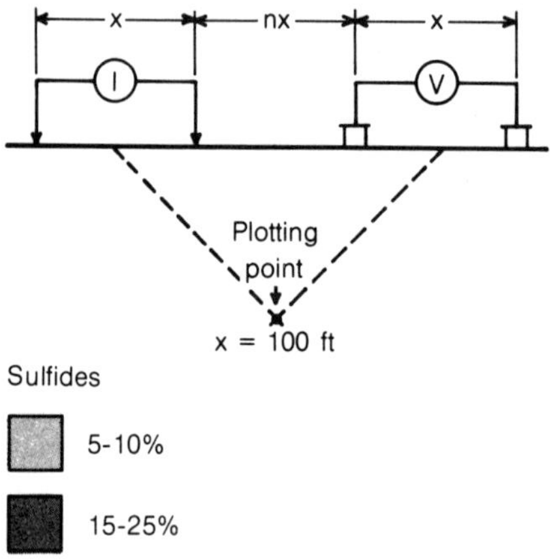

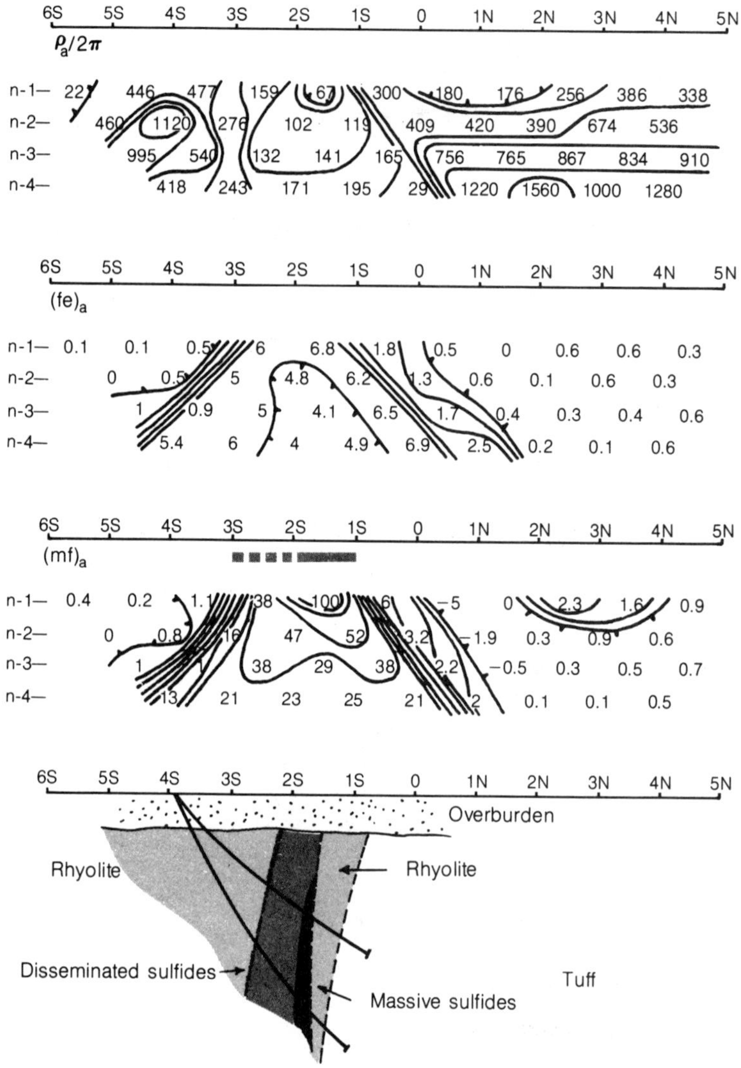

The figures on this page are referred to in the preceding section *Comparison of IP and EM results.*

FIGURE 2.99
Comparison of EM methods and IP results from Newcastle area, New Brunswick, Canada

Line-196W

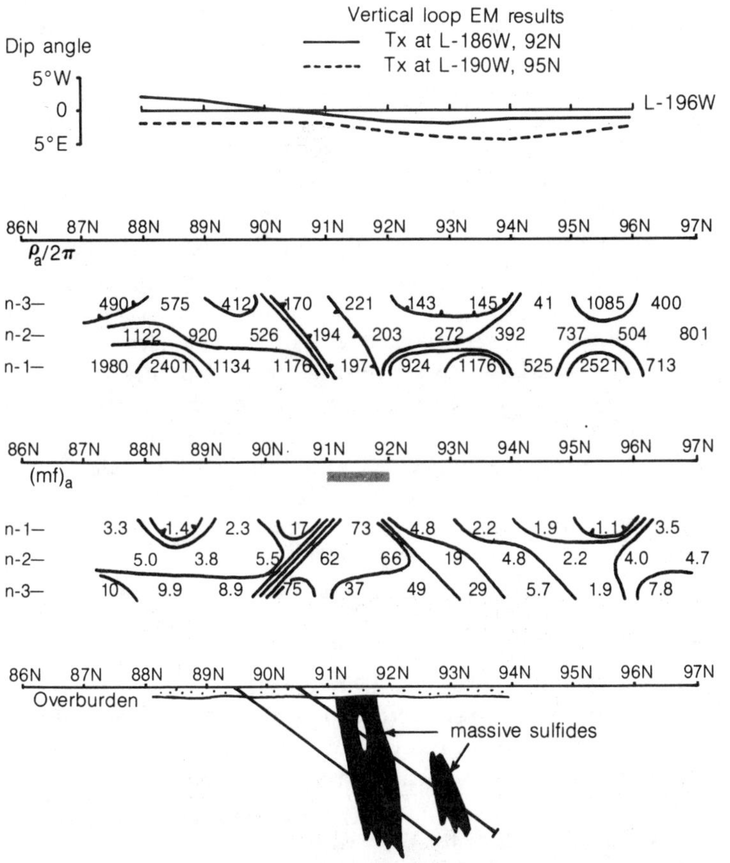

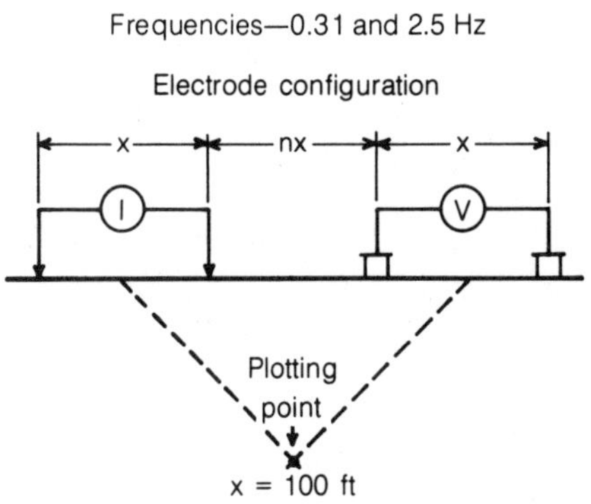

FIGURE 2.100
Comparison of EM methods and IP results from Linden area, Wisconsin, USA

Line-2E

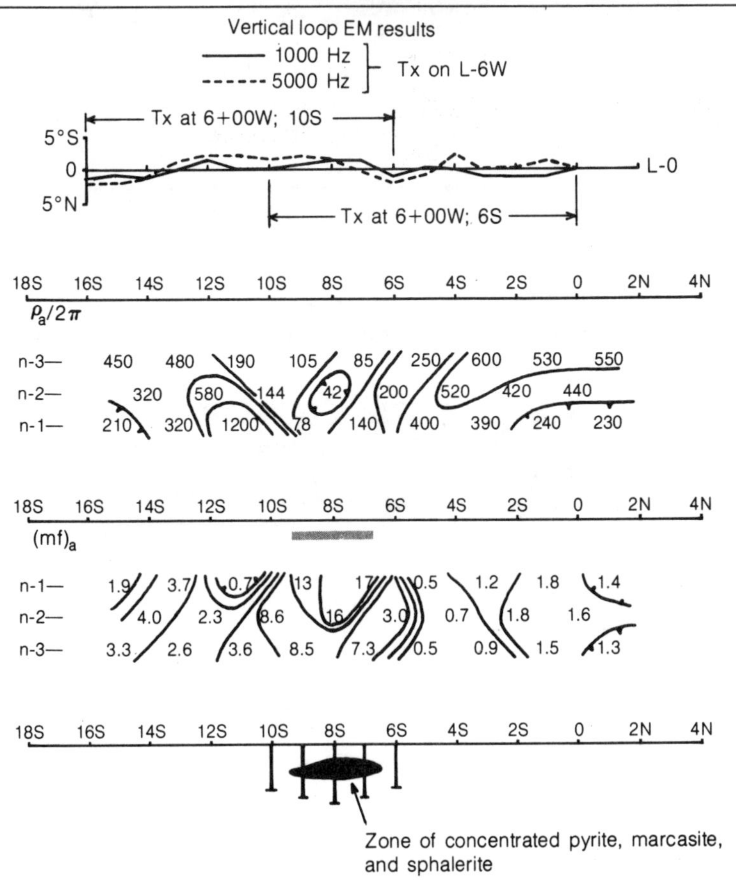

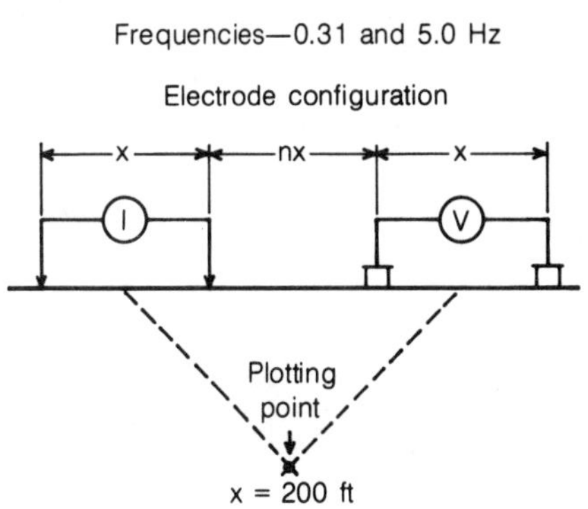

Use of the IP method for exploration in drill holes

The successful use of geophysical techniques in drill holes has long been the dream of geologists and geophysicists alike. As the search for ore goes to greater and greater depths and the cost of drilling increases, it becomes even more important to make as much use as possible from the deep drill hole.

The use of magnetics, electromagnetics, and even gravity has become possible in recent years if the hole has a large enough diameter. Resistivity and IP measurements have also become commonplace in logging techniques to investigate the walls of the hole.

Using the drill hole to do exploration with the IP method presents some special problems. First, the drill hole is relatively small, the water in the drill hole (at least in hard rock areas) has a high resistivity, and the fresh rock surface inside the drill hole usually has a very high resistivity. In this situation a current electrode in the hole sees a very high impedance. It is difficult to establish a significant current flow.

Further, the capacitive coupling between the current wire and the water in the drill hole will have about the same impedance to a non-dc signal as the resistive impedance at the current electrode. Finally, if the geophysicist tries to establish current wires and electrodes in the same drill hole with potential wires and electrodes, capacitive coupling between the wires will become significant. In short, it is extremely difficult to make satisfactory measurements with all the wires in a single hole.

The scale modeling results in figures 2.101 through 2.108 and the field data in figures 2.109 and 2.110 demonstrate procedures that can be used to explore the area surrounding a single drill hole. As the scale modeling shows, the technique works much better if the bottom of the hole is beneath the unknown source of mineralization. This should be remembered by geologists who are planning drilling programs in which they would like to use this procedure.

If more than one hole is present, particularly if a *fence* of holes has been drilled to test a region of particular importance, there are other electrode configurations that can be used. Some of them are quite effective at detecting small sources that exist between, or immediately beneath, a group of closely spaced drill holes.

FIGURE 2.101
Voltages measured in drill hole technique

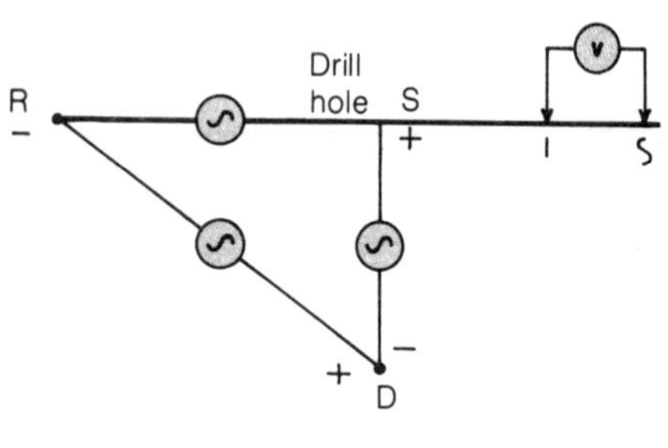

Dipole—dipole   Pole—dipole

$$(5)\quad \frac{V_{12}}{I_{S\text{-}R}} = \frac{\rho_{S\text{-}R}}{2\pi}\left(\frac{1}{GF}\right)_{S\text{-}R} = \frac{-\rho_R}{2\pi}\left(\frac{1}{GF}\right)_R + \frac{\rho_S}{2\pi}\left(\frac{1}{GF}\right)_S$$

$$(6)\quad \frac{V_{12}}{I_{D\text{-}R}} = \frac{\rho_{D\text{-}R}}{2\pi}\left(\frac{1}{GF}\right)_{D\text{-}R} = \frac{-\rho_R}{2\pi}\left(\frac{1}{GF}\right)_R + \frac{\rho_D}{2\pi}\left(\frac{1}{GF}\right)_D$$

$$(7)\quad \frac{V_{12}}{I_{S\text{-}D}} = \frac{\rho_{S\text{-}D}}{2\pi}\left(\frac{1}{GF}\right)_{S\text{-}D} = \frac{-\rho_D}{2\pi}\left(\frac{1}{GF}\right)_D + \frac{\rho_S}{2\pi}\left(\frac{1}{GF}\right)_S$$

From the above expressions,

$$(8)\quad \frac{V_{12}}{I_{S\text{-}R}} = \frac{V_{12}}{I_{D\text{-}R}} + \frac{V_{12}}{I_{S\text{-}D}}$$

It can also be shown that

$$(9)\quad \left(\frac{1}{GF}\right)_{S\text{-}R} = \left(\frac{1}{GF}\right)_{D\text{-}R} + \left(\frac{1}{GF}\right)_{S\text{-}D}$$

The figures on this page are referred to in the preceding section *Use of the IP method for exploration in drill holes.*

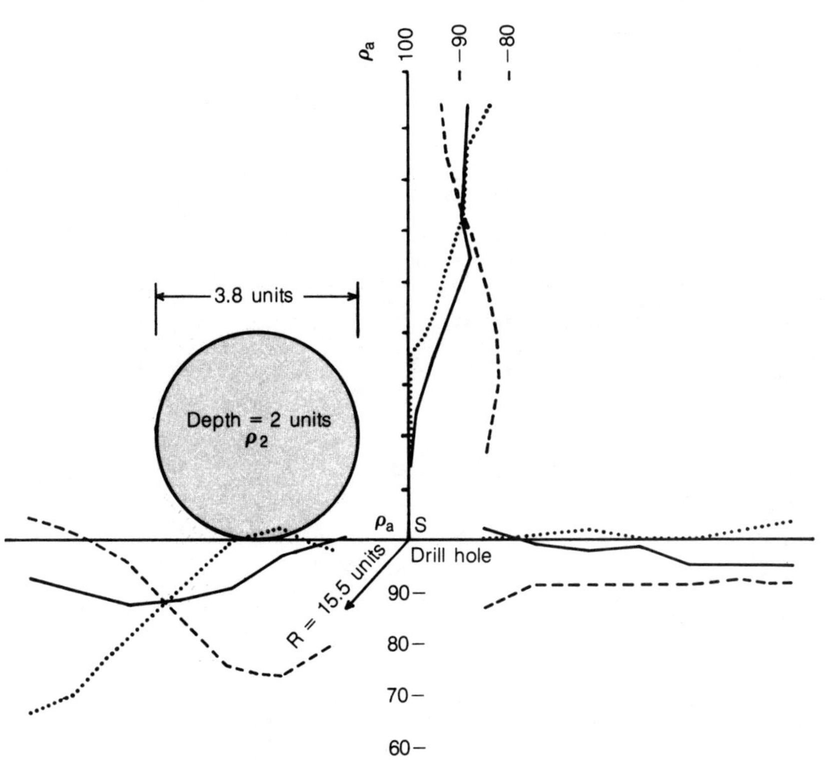

FIGURE 2.102
Downhole resistivity

Scale model cases

$\rho_1/2\pi = 100$

$\rho_2/2\pi = 7.7$

—— S-R
----- D-R
·········· S-D

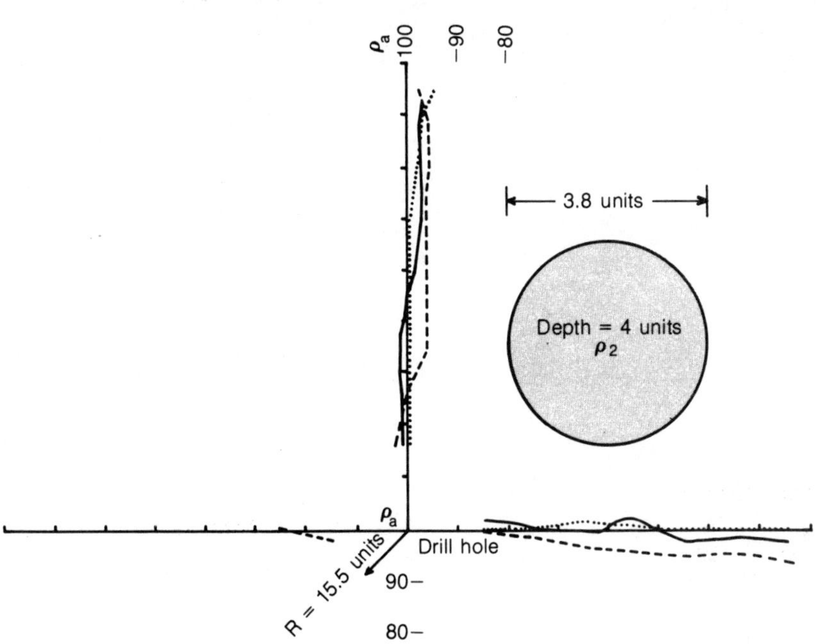

FIGURE 2.103
Downhole resistivity

Scale model cases

$\rho_1/2\pi = 100$

$\rho_2/2\pi = 7.7$

—— S-R
----- D-R
·········· S-D

The figure on this page is referred to in the preceding section *Use of the IP method for exploration in drill holes.*

FIGURE 2.104
Downhole induced polarization and resistivity

Scale model cases

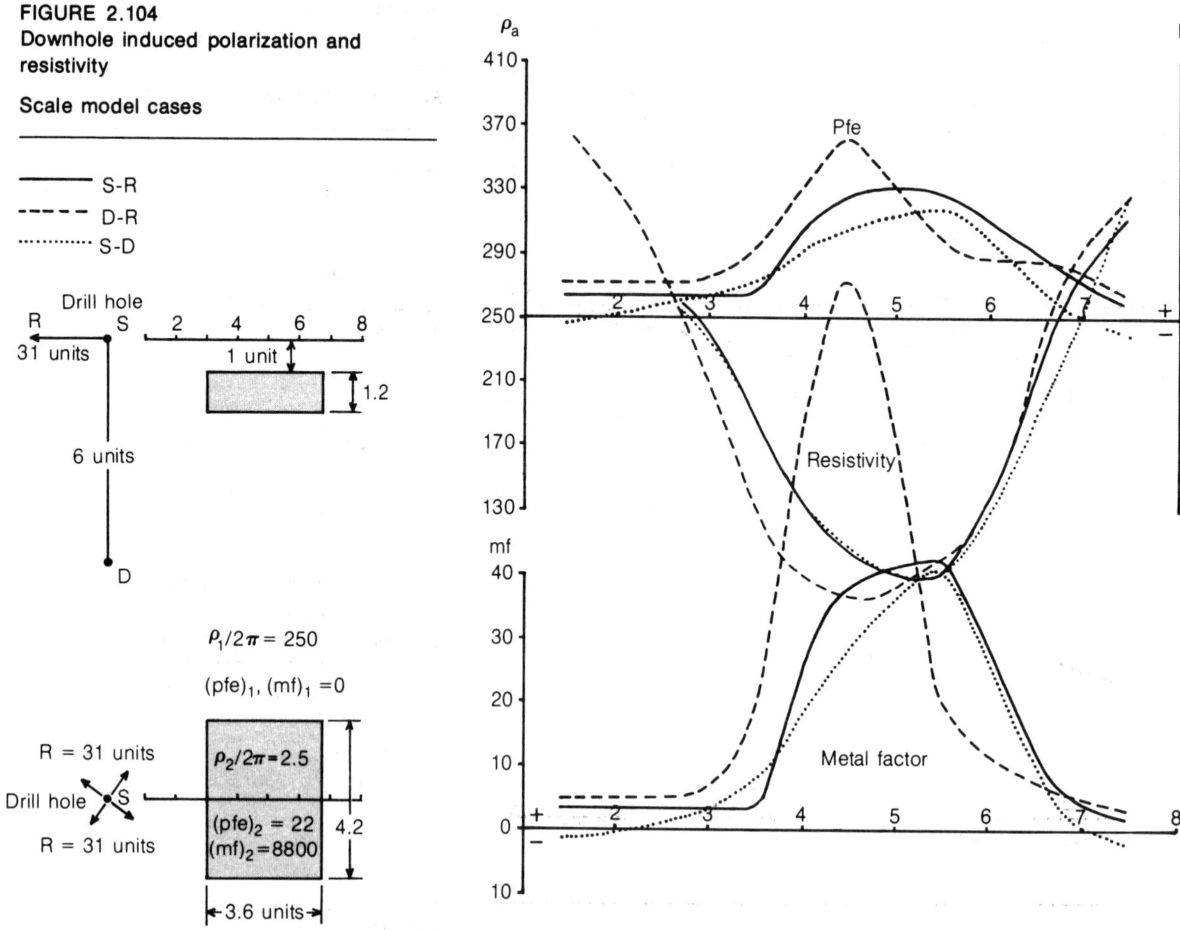

The figures on this page are referred to in the preceding section *Use of the IP method for exploration in drill holes.*

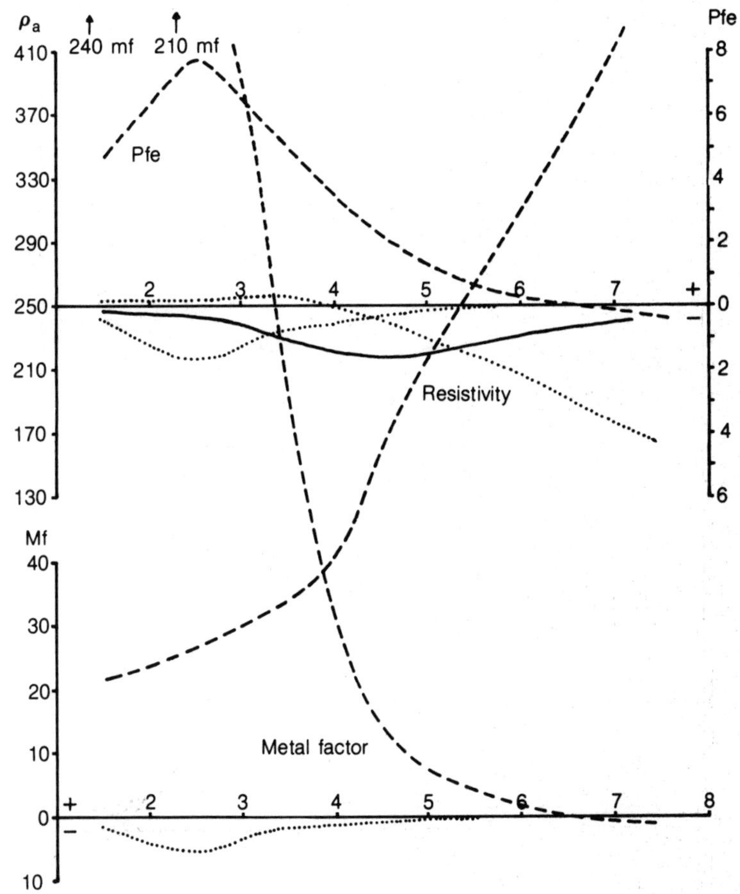

FIGURE 2.105
Downhole induced polarization and resistivity
Scale model cases

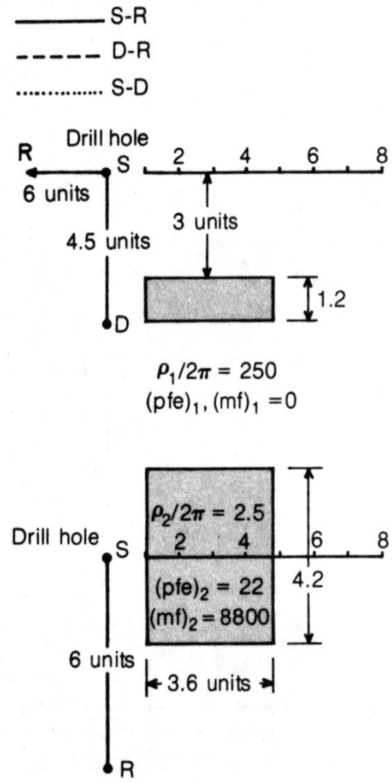

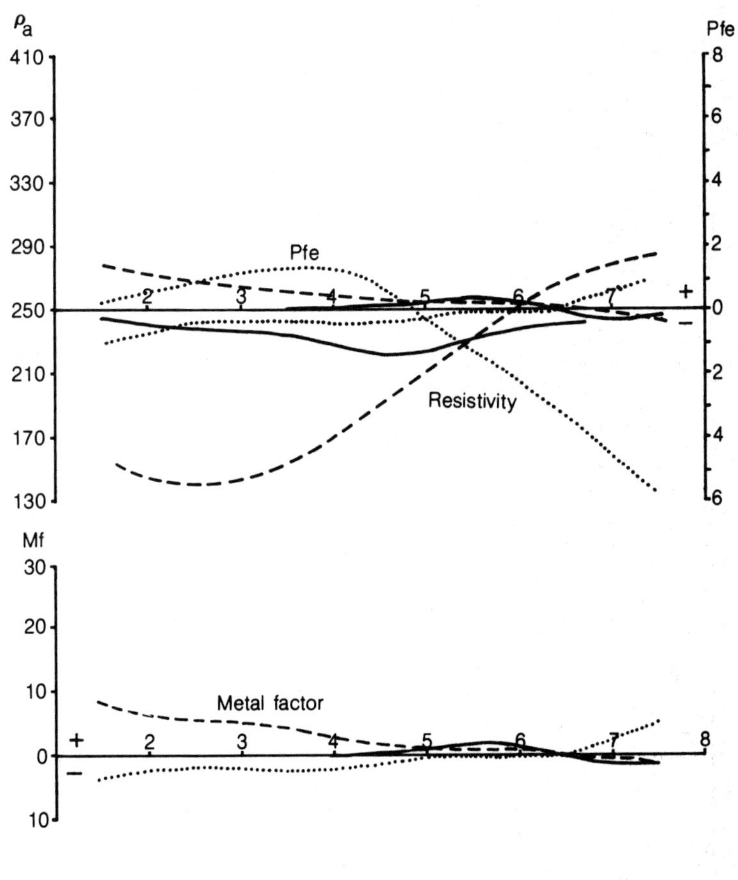

FIGURE 2.106
Downhole induced polarization and resistivity
Scale model cases

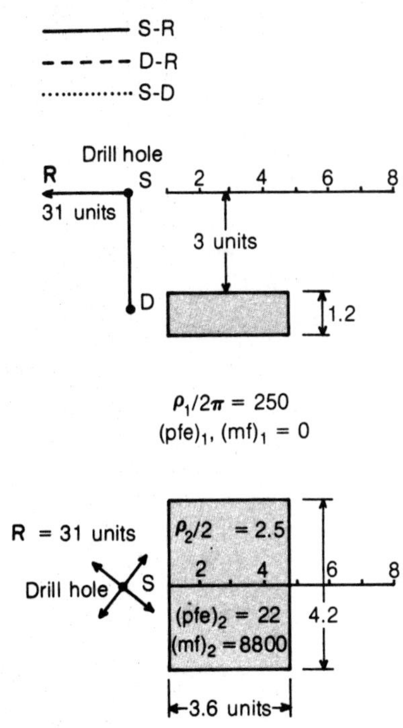

The figure on this page is referred to in the preceding section *Use of the IP method for exploration in drill holes.*

FIGURE 2.107
Downhole induced polarization and resitivity

Scale model cases

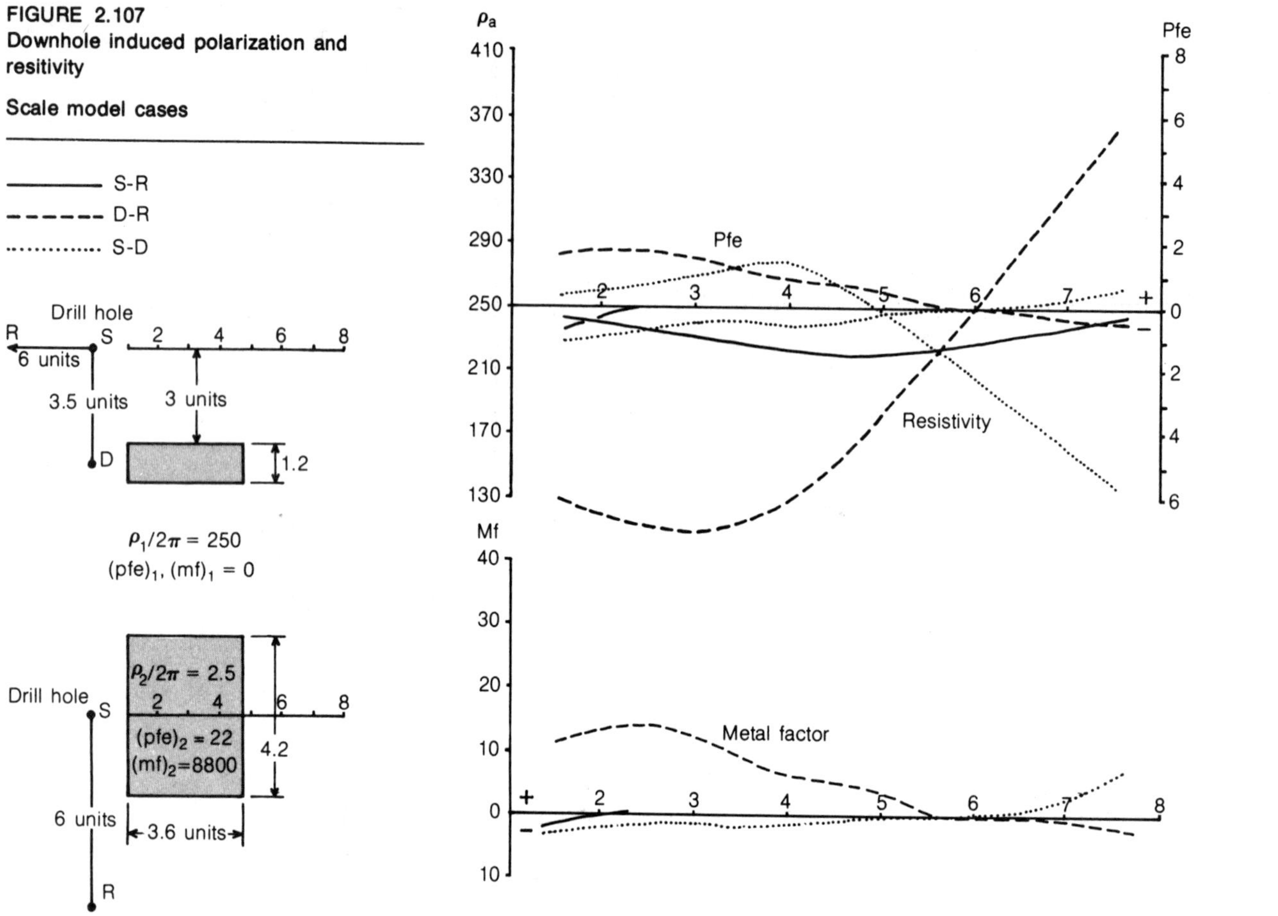

The figures on this page are referred to in the preceding section *Use of the IP method for exploration in drill holes.*

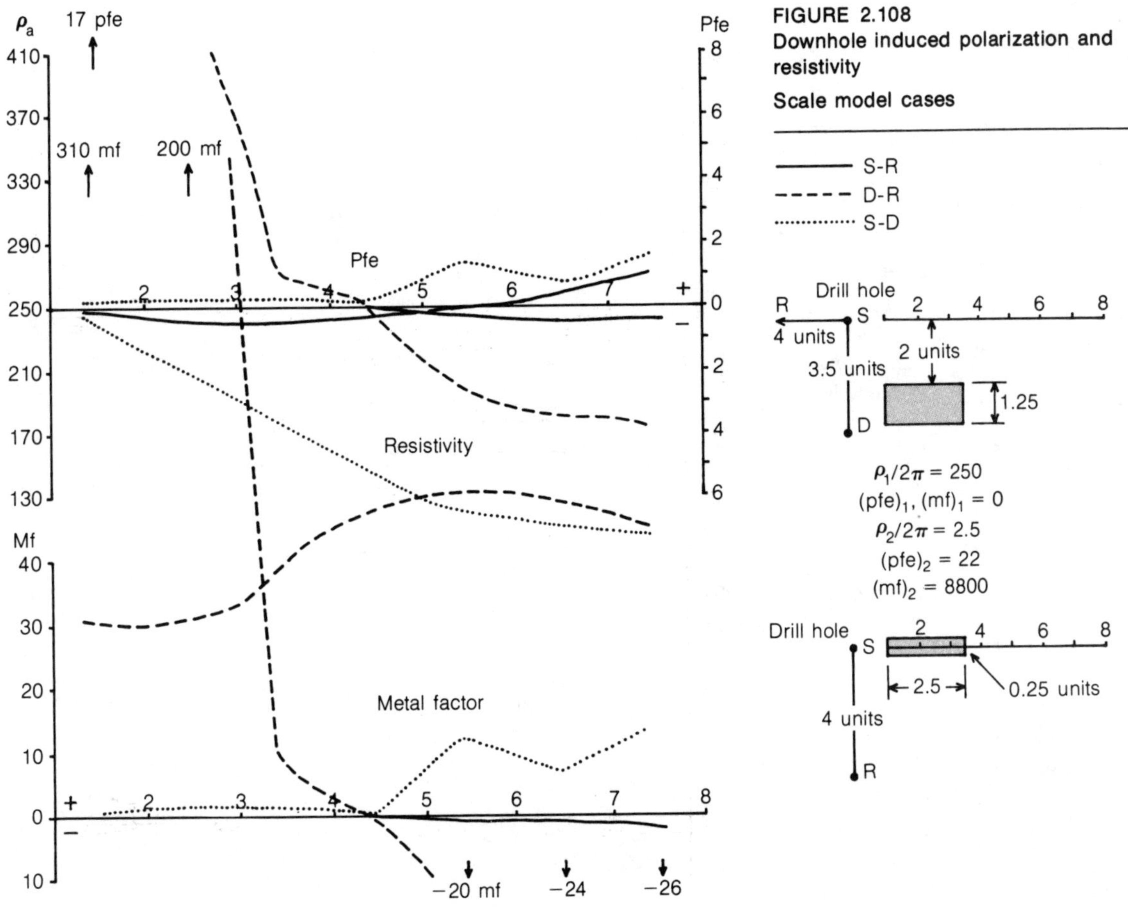

FIGURE 2.108
Downhole induced polarization and resistivity
Scale model cases

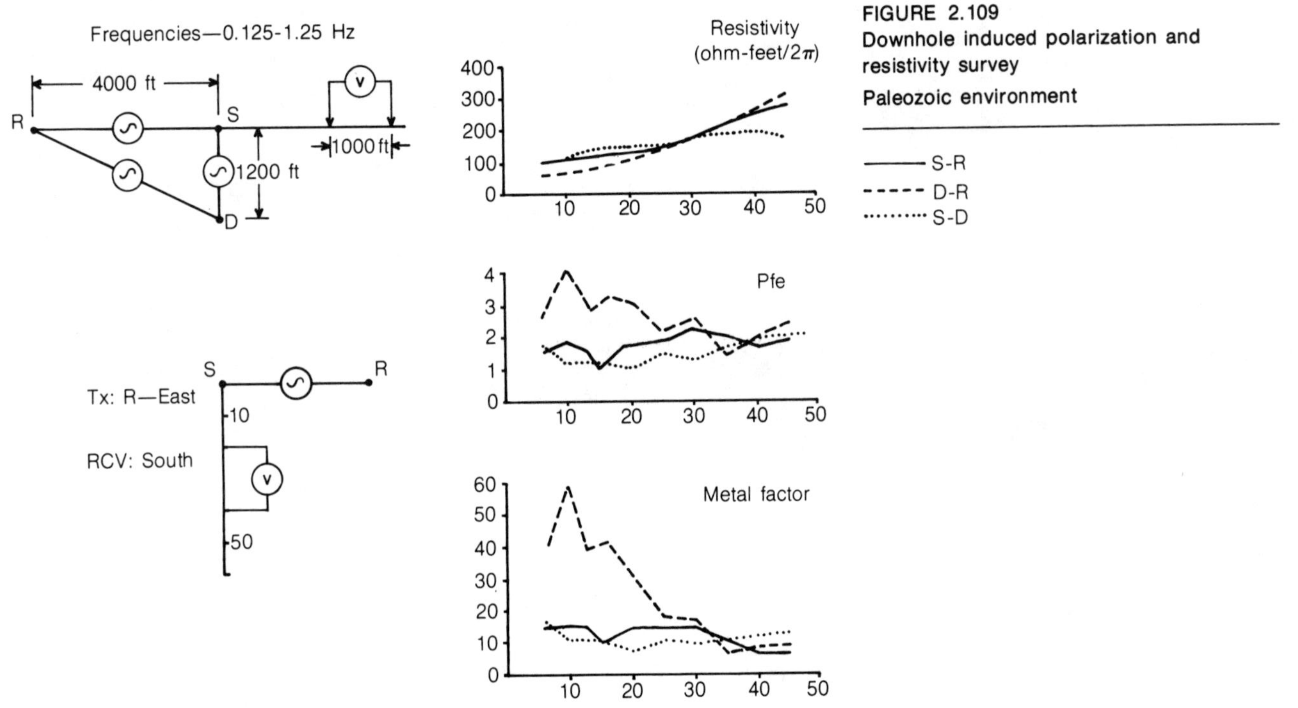

FIGURE 2.109
Downhole induced polarization and resistivity survey
Paleozoic environment

The figure on this page is referred to in the preceding section *Use of the IP method for exploration in drill holes.*

FIGURE 2.110
Downhole induced polarization and resistivity survey
Porphyry copper environment

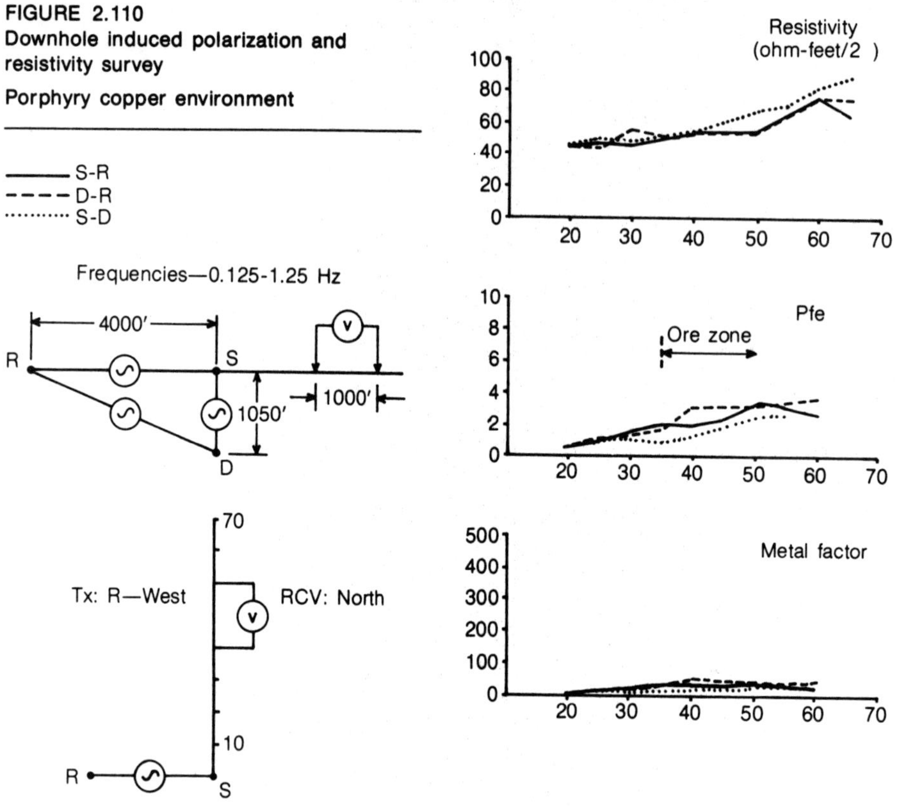

Use of the digital computer to interpret IP and resistivity data

It is obvious that the digital computer has altered all phases of modern life. This is true for the mining industry as well. At operating mines, the computer is used for operations ranging from inventory control to tonnage and grade evaluations and even to the planning of internal, day-to-day instructions for the operation of open-pit mines.

The high-speed calculation capability and almost infinite storage capacity of the modern computer have made it more and more useful to the exploration geophysicist. Initially, the computer was used to record, store, and analyze large volumes of data. Later, the geophysicist learned to use the computer to generate *type curves.* In this process, the geophysicist would calculate the magnetic or the gravity profile expected from a certain earth geometry. This is termed the *forward problem.* The mathematical expressions for these calculations were fairly well known; few cases had been calculated because each required considerable time and labor. The computer changed that, and great volumes of *type curves* were calculated in a few years.

As the theoretical geophysicist became more adept at using the computer and approximate mathematical techniques, the forward problems that could be solved became more useful to the exploration geophysicist in the interpretation of field data. It was also found that the forward problem could be used in the planning of surveys.

The three forward problem solutions shown in figures 2.111, 2.112 and 2.113 were calculated to help plan a reconnaissance IP survey at Anvil in the Yukon. The electrical parameters used for the source are those measured in the open-pit mines in the area. The forward problem solutions show that the use of x = 1200, n = 1, 2, 3, 4 could be expected to locate the typical source to a depth of 2000 ft. At depths of 2400 ft or greater the measurement for x = 1200 ft, n = 4 would probably not be detectably anomalous.

Because of the high resistivity host rocks in this area, inductive coupling would not be a problem in this survey.

The inverse problem

As more and more *forward problems* were solved by the theoretical geophysicist using digital techniques to evaluate complex mathematical expressions, the possibility of solving the *inverse problem* came under study. In this procedure, the geophysicist begins with a set of field data and the forward problem solution for the geophysical data he has under investigation. He proceeds from one forward problem solution to another until the computer tells him that it has determined the parameters of the earth that give a forward problem solution that is the *best fit* to the field data being investigated.

The theory of this automatic method to interpret geophysical data has been known for some time. However, computer programs designed to implement the theory tended to be unstable. The cost was very high due to the large number of forward problem solutions that had to be calculated for each inverse problem case.

In recent years, William Pelton of Phoenix Geophysics Limited has investigated very fast, low-cost forward problems and has developed an inversion program for geophysical data using the *ridge regression* technique. This has reduced the cost and increased the stability of inverting geophysical data even for geophysical methods (such as induced polarization, resistivity and electromagnetics) in which the forward problem solutions are quite complex. This new approach to the problem has resulted in reduced costs for the computer time to invert geophysical problems. The inversion examples presented here have computer time costs that are in the range of $3-10. Thus, for the first time, the computer charges to invert geophysical field data are appreciably less than the cost for the geophysicist's time to analyze and report on the result of the inversion.

One-dimensional resistivity problems

For the resistivity problem, the potential equations must be solved with an applied source, i.e., the current source applied to the earth's surface. The simplest problem that can be studied is the one-dimensional problem of the layered earth.

There are many different electrode configurations used by geophysicists to make resistivity measurements. Examples will be shown of the inversion of one-dimensional resistivity data measured with use of the Schlumberger electrode configuration and the dipole-dipole electrode configuration.

The Schlumberger resistivity data shown in figures 2.114 and 2.115 are from a large survey in Brazil. The purpose was to trace a gravel aquifer that had been located in a few widely spaced drill holes. The drill hole at location VES #35B had not penetrated to the basement; however, it had located the gravel aquifer at a depth of 11.3-25.3 m. The known depths to the various geologic formations were used as known values to invert the resistivity data points shown. As shown in figure 2.114, an excellent *fit* was obtained with the use of a six layer earth. The fresh water aquifer was found to have a true resistivity value of 107 ohm-m between lower resistivity rock units.

The resistivity measurements at location VES #30 were inverted with the use of the resistivity values from the known case as initial guesses. As shown in figure 2.115, an excellent inversion was possible here also. A somewhat simpler earth is given by the computer, and the gravel aquifer is shown to be at about the same depth; however, the thickness is much less.

The dipole-dipole data for the layered earth shown in figure 2.116 also gave an almost perfect fit for the data. The distance between the electrodes was not great enough to accurately determine the bedrock resistivity. The standard deviations of the expected errors for the resistivity and thickness of the glacial layer were 4.6 percent and 6.1 percent respectively.

Two-dimensional resistivity problems

For two-dimensional resistivity problems, the dipole-dipole electrode configuration is the most useful in the separation between vertical changes in the earth's resistivity and the lateral variations in resistivity. The computer inversion of dipole-dipole pseudo-sections is now used routinely.

The results shown in figure 2.117 are from a geothermal program in a volcanically active area in Kenya. The area of the survey is covered by an unknown thickness of recent lava flows. The resistivity low centered at depth at station 15 to station 16 is clearly evident. The interpretation problem was (1) to estimate the thickness of lava flows on top of the zone and (2) to determine if the resistivity in the source of the anomaly was low enough to be an active geothermal zone.

The agreement between the observed data and the calculated data pseudo-section is quite good. The center of the conductive zone is accurately determined to be at station 15 to station 16; the depth to the top is reasonably well established at approximately 375 m.

The true resistivity value within the source was poorly defined by the inversion; however, the value was clearly established as being less than 1.0 ohm-m. This could be an excellent geothermal source.

The two-dimensional resistivity inversion program permits the geophysicist to simultaneously invert two sets of data for the same method. This has been done with the data shown in figures 2.118 and 2.119. Here resistivity data measured at the Iso-Copperfields deposit in

Quebec are shown using x = 200 ft and x = 100 ft. The double set of data puts a considerable constraint on the inversion solution. The agreement between the measured data and the calculated data is very good.

The simultaneous inversion gives a depth of overburden of about 45 ft, a width for the conductor of about 60 ft, and the three resistivity values shown. The resistivity contrast between the bedrock and the conductor is about 3600/1.

Induced polarization field data

The dipole-dipole induced polarization measurement is somewhat more complex than the resistivity measurement. The inversion program has been developed to interpret the IP *pseudo-section.* The results shown in figure 2.120 are from a porphyry copper exploration program carried out in Mendoza Province, Argentina. The survey was carried out using x = 200 m and n = 1, 2, 3, 4. The resistivity results showed little contrast between the alluvial valley gravels and the bedrock.

There was, however, a distinct IP anomaly at depth, centered east of the base line. The inversion of the IP results gives a large source with an IP effect of ten percent. The depth to the top of the source is determined to be slightly greater than 200 m; this is a little more than the depth to bedrock in the single drill hole that had been previously completed.

The versatility of this inversion program is demonstrated by the fact that it can be used to simultaneously invert several sets of geophysical field data to obtain the *best fit* source for all the sets. As shown previously, this can be data with two different electrode intervals. It can also be two completely different types of geophysical data taken on the same survey line. For instance, it is possible to simultaneously invert magnetic data, gravity data, and resistivity data over a zone of massive sulfides and to determine the location and size of the source that gives the best fit, as well as the magnetic susceptibility, density, and conductivity.

One of the most useful applications is to simultaneously invert IP and resistivity field data. The results shown in figures 2.121 and 2.122 were measured over a typical, shallow lead-zinc ore zone in the Pine Point area. The measurements were made using x = 200 ft and n = 1, 2, 3.

The simultaneous inversion of both sets of data gives a source that agrees well with the known size of the ore zone; the shallow depth is also accurately predicted. The resistivity contrast between the lead-zinc ore and the limestones is only about two to one; the frequency effect in the ore is 14.5 percent, compared to about 1.9 percent in the limestones.

The cost of this type of data inversion is now small enough that it can be used by field geophysicists in their day-to-day interpretation of field data.

The observed resistivity and frequency effect data from a property in a Greenstone area in Quebec are shown in figure 2.123. The completely interpreted solution to the problem, with the calculated results to be expected from the specific earth geometry and parameters chosen by the computer, is shown in figure 2.124. The source is quite reasonable, and the data fit is good.

The observed data in figure 2.125 and the solution and calculated data in figure 2.126 are from a highly conductive environment such as Cyprus or Western Australia. Again, the geometry and parameters of the source are reasonable and the data fit is excellent.

Both of these sources have been tested by drilling, and the computer inversions were found to be accurate.

The figures on this page are referred to in the preceding section *Use of the digital computer to interpret IP and resistivity data.*

FIGURE 2.111
Induced polarization and resistivity
Forward problem calculation

Dipole—dipole
Electrode configuration
x = 1200 ft

Source:
Width—1500 ft
Thickness—600 ft
Length > 6000 ft

ρ_1 – 1.0 Ω-m
$(fe)_1$ – 20%

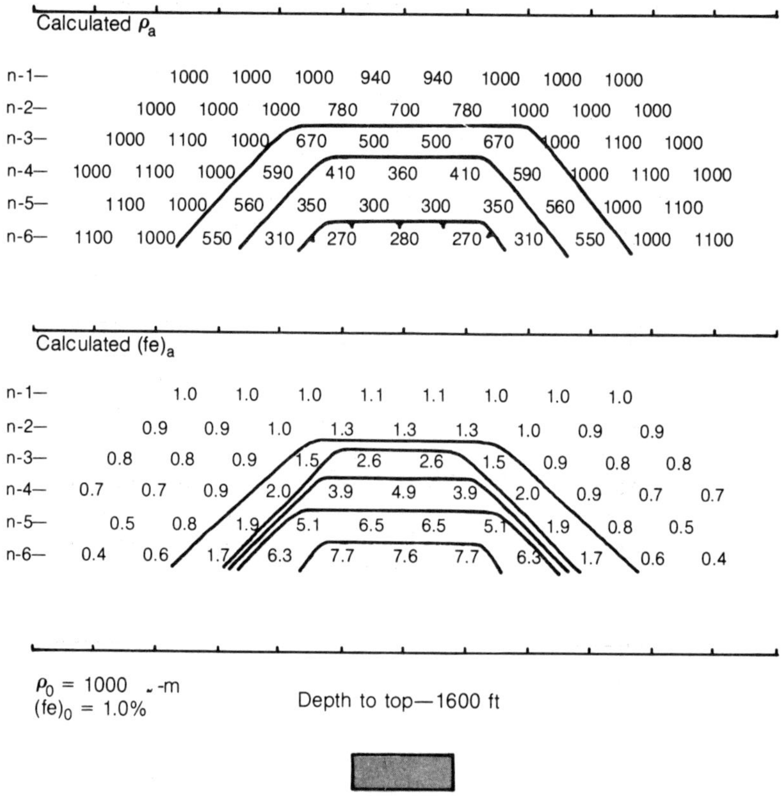

FIGURE 2.112
Induced polarization and resistivity
Forward problem calculation

Dipole—dipole
Electrode configuration
x = 1200 ft

Source:
Width—1500 ft
Thickness—600 ft
Length > 6000 ft

ρ_1 – 1.0 Ω-m
$(fe)_1$ = 20%

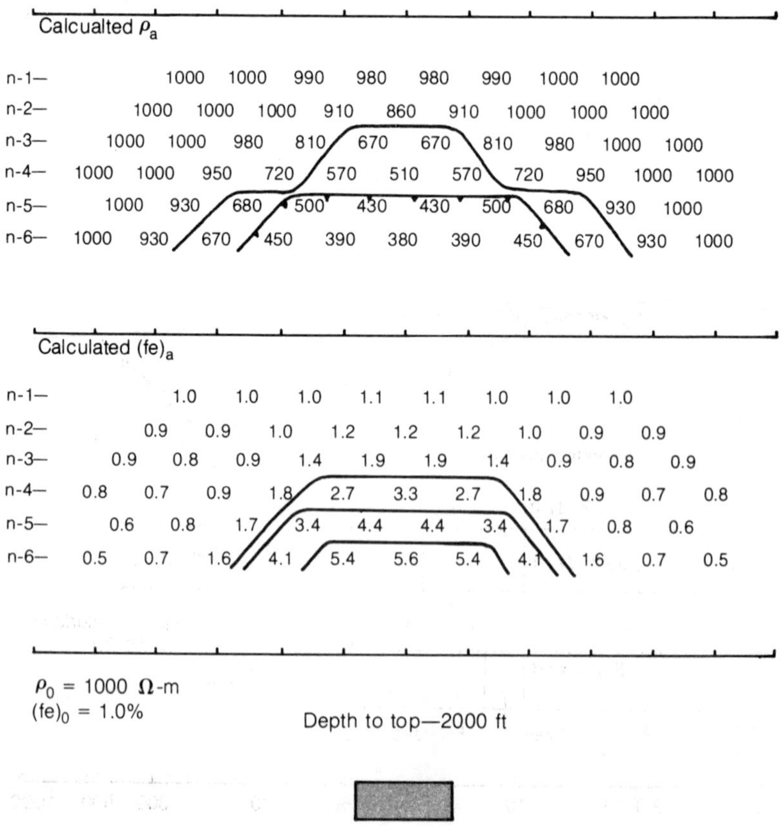

The figures on this page are referred to in the preceding section *Use of the digital computer to interpret IP and resistivity data.*

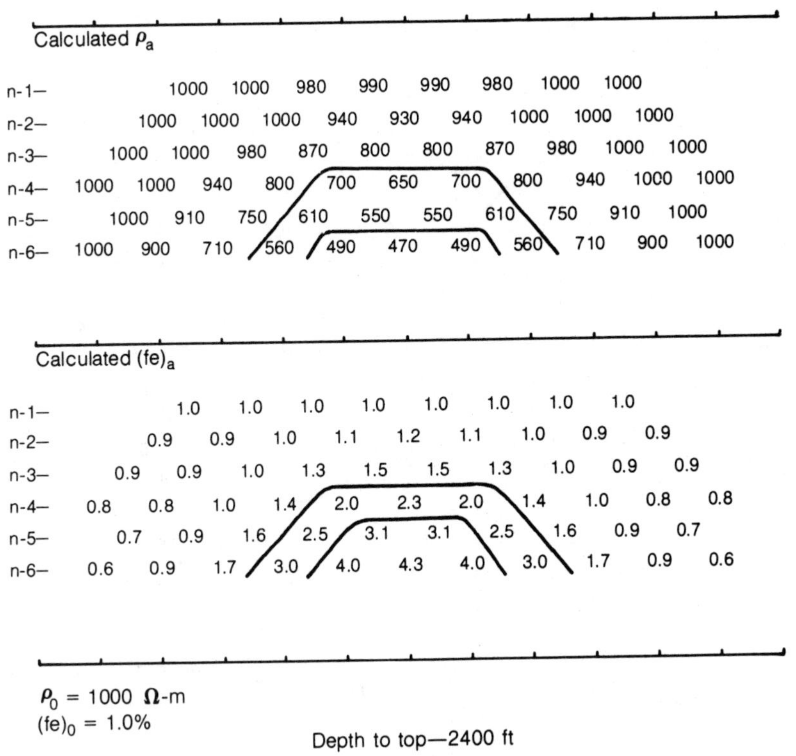

FIGURE 2.113
Induced polarization and resistivity

Forward problem calculation

Dipole—dipole
Electrode configuration
x = 1200 ft

Source:
Width—1500 ft
Thickness—600 ft
Length > 6000 ft

ρ_1 – 1.0 Ω-m
$(fe)_1$ – 20%

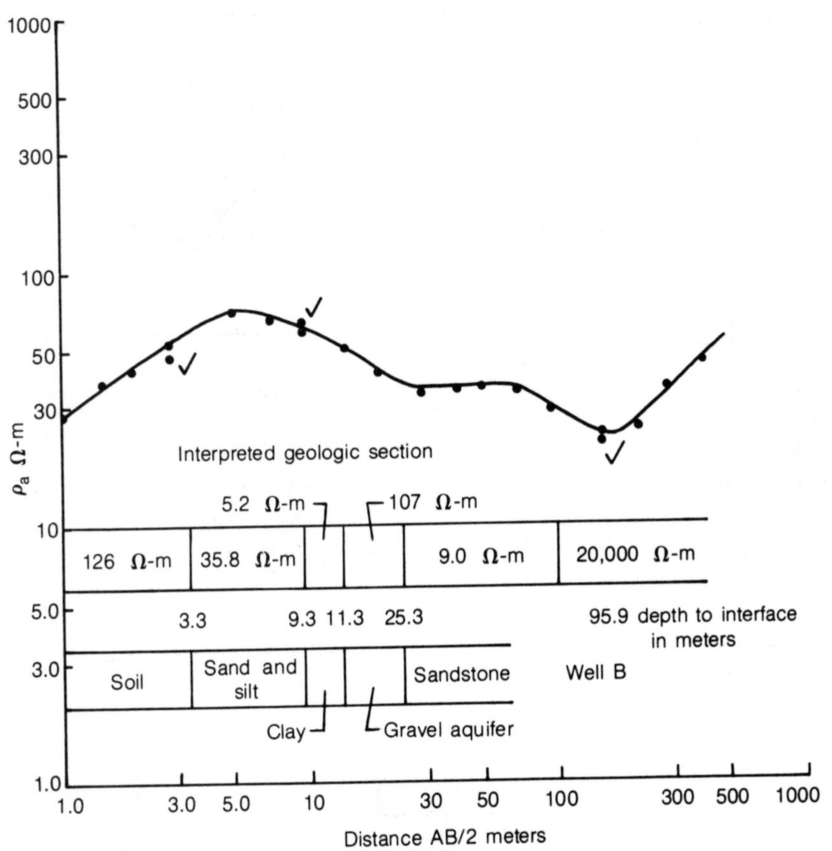

FIGURE 2.114
Resistivity data inversion
Apodi Valley, Brazil

Location VES #35 B

• Measured value
— Calculated profile
√ Change in MN distance
χ^2 error = 4.1%

Schlumberger electrode configuration

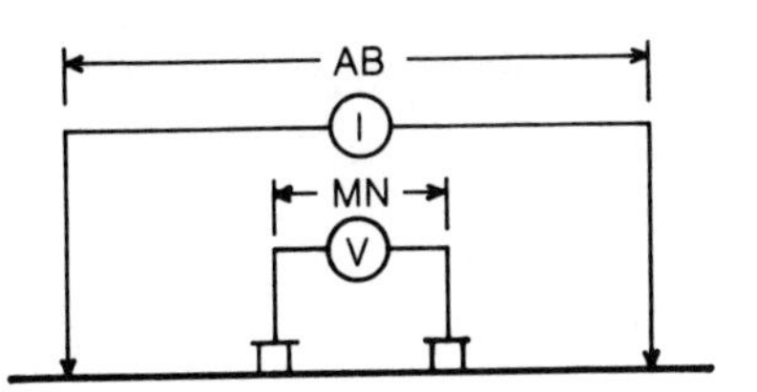

The figures on this page are referred to in the preceding section *Use of the digital computer to interpret IP and resistivity data.*

FIGURE 2.115
Resistivity data inversion, Apodi Valley, Brazil

Location VES #30

- • Measured value
- — calculated profile
- ✓ Change in MN distance

χ^2 error = 9.0%

Schlumberger electrode configuration

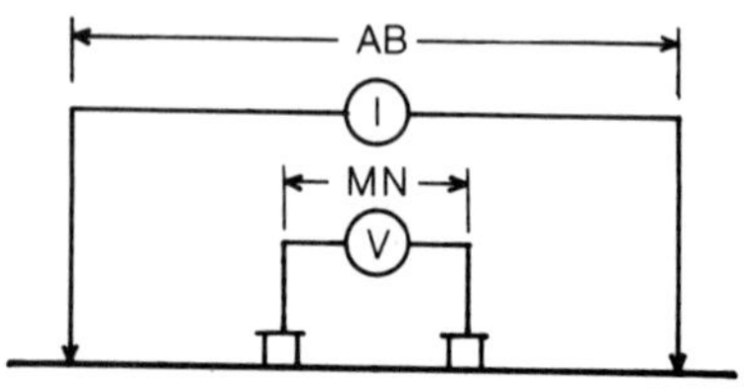

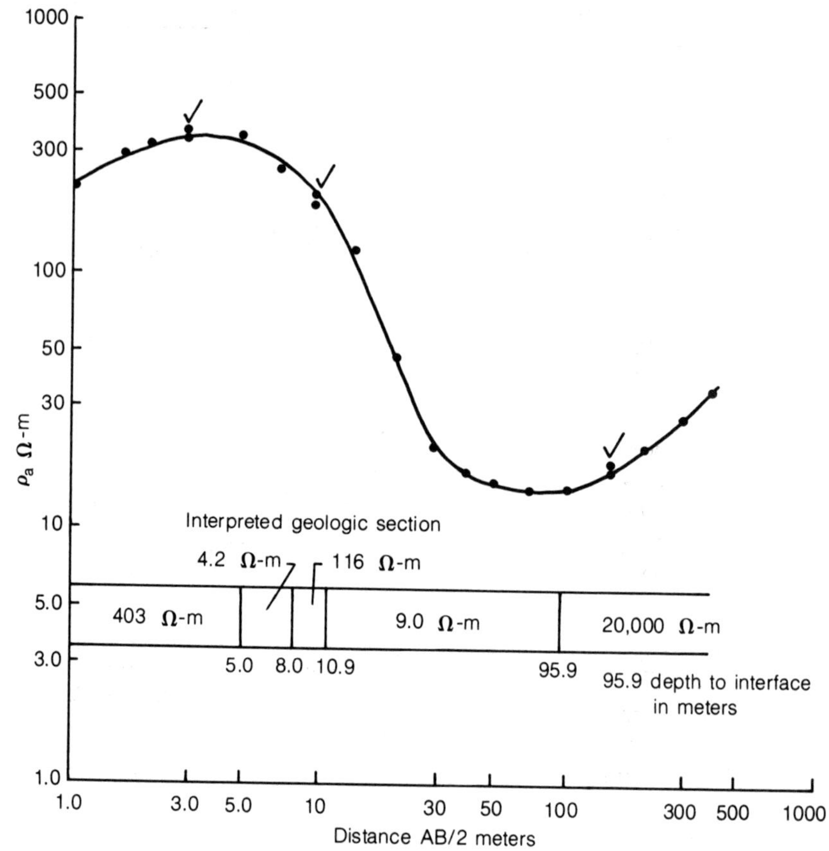

FIGURE 2.116
Resistivity data inversion, Northern Ontario

Thickness of overburden determination

Dipole—dipole
Electrode configuration
x = 200 ft

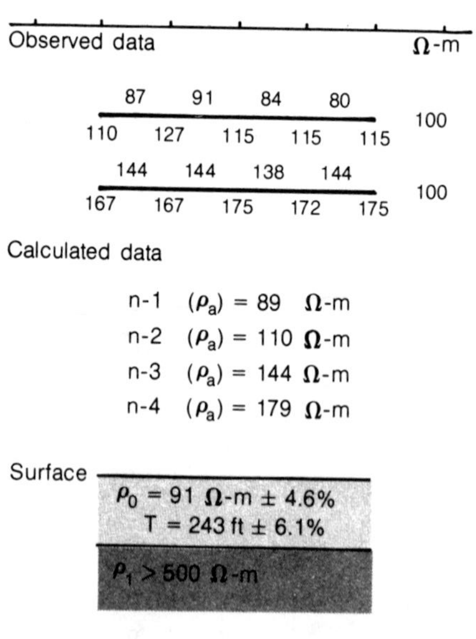

The figure on this page is referred to in the preceding section *Use of the digital computer to interpret IP and resistivity data.*

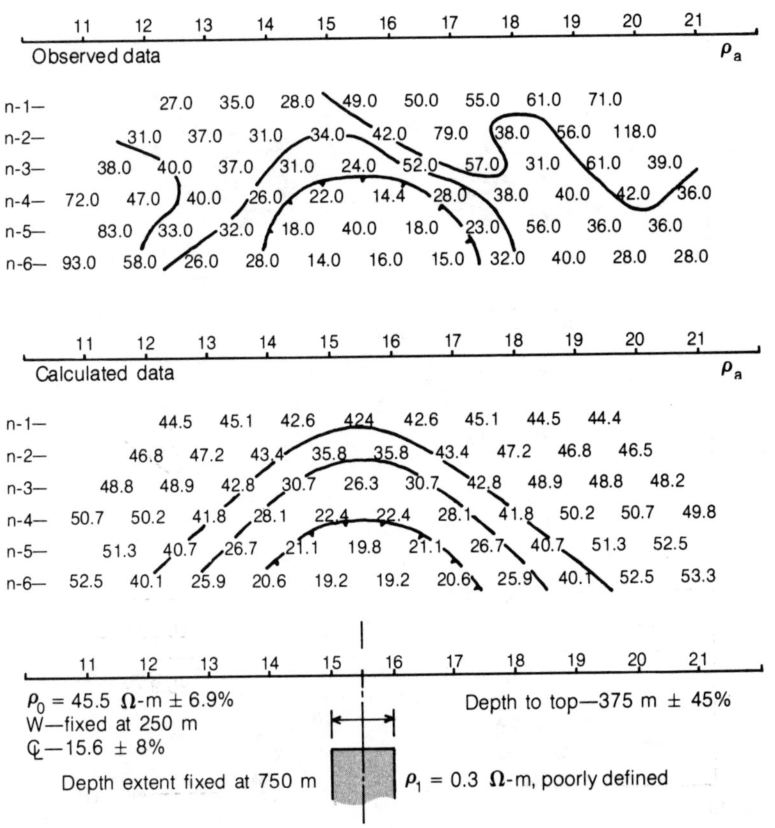

FIGURE 2.117
Resistivity data inversion, Kenya

Geothermal survey

Dipole—dipole
Electrode configuration
x = 250 m

The figure on this page is referred to in the preceding section *Use of the digital computer to interpret IP and resistivity data.*

FIGURE 2.118
Resistivity data inversion, Iso-Copperfields deposit, Quebec, Canada

x = 200 ft

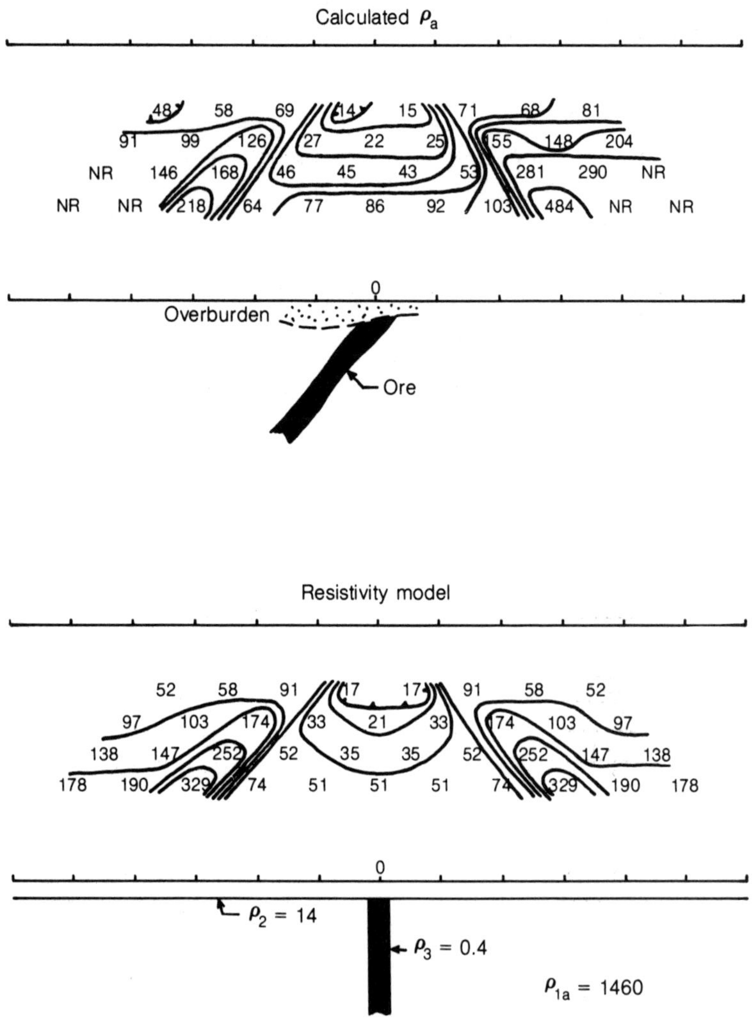

The figure on this page is referred to in the preceding section *Use of the digital computer to interpret IP and resistivity data.*

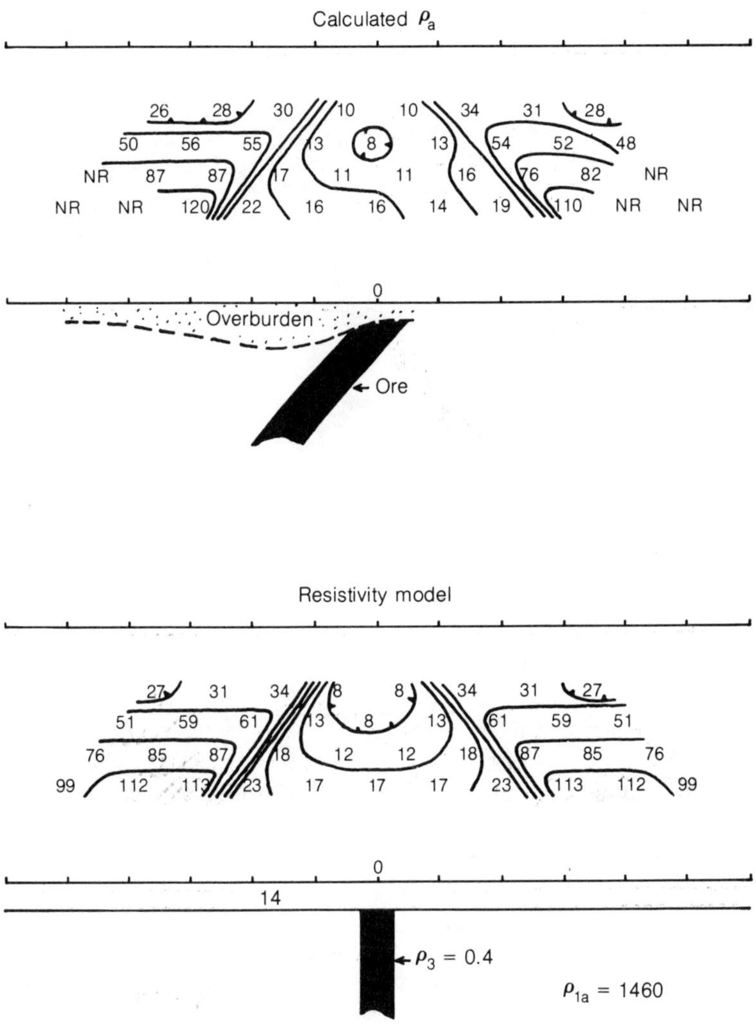

FIGURE 2.119
Resistivity data inversion, Iso-Copperfields deposit, Quebec, Canada

x = 100 ft

The figure on this page is referred to in the preceding section *Use of the digital computer to interpret IP and resistivity data.*

FIGURE 2.120
IP data inversion, Mendoza Province, Argentina

x = 200 m

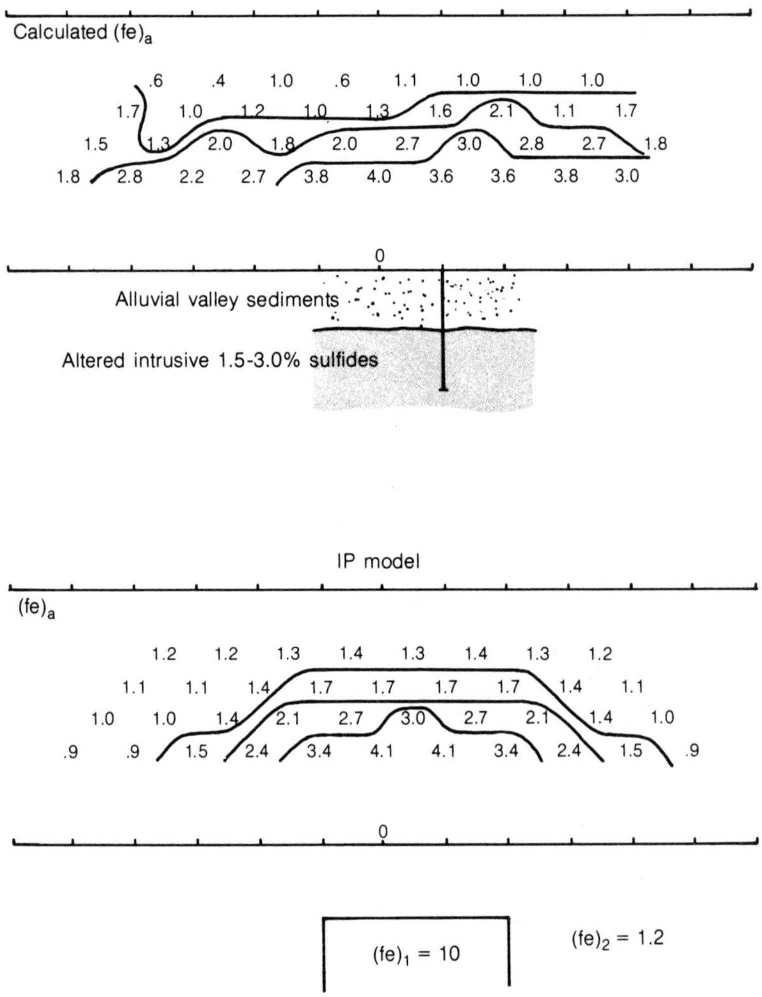

The figure on this page is referred to in the preceding section *Use of the digital computer to interpret IP and resistivity data.*

FIGURE 2.121
Simultaneous IP and resistivity field data inversion, Pine Point area

N-38A

x = 200 ft

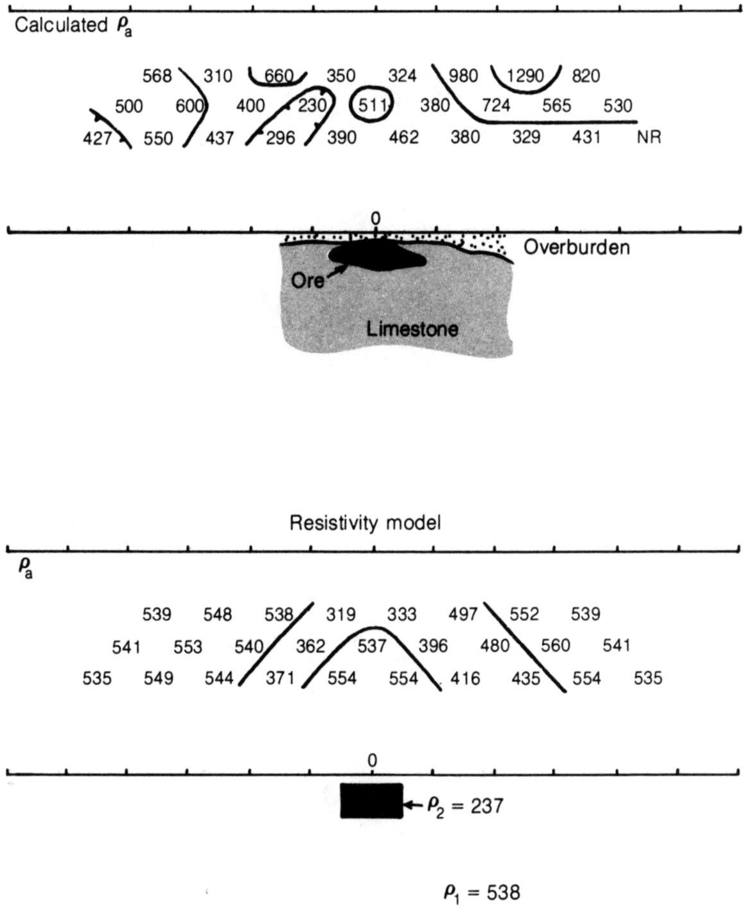

The figures on this page are referred to in the preceding section *Use of the digital computer to interpret IP and resistivity data.*

FIGURE 2.122
Simultaneous IP and resistivity field data inversion, Pine Point area

N-38A

x = 200 ft

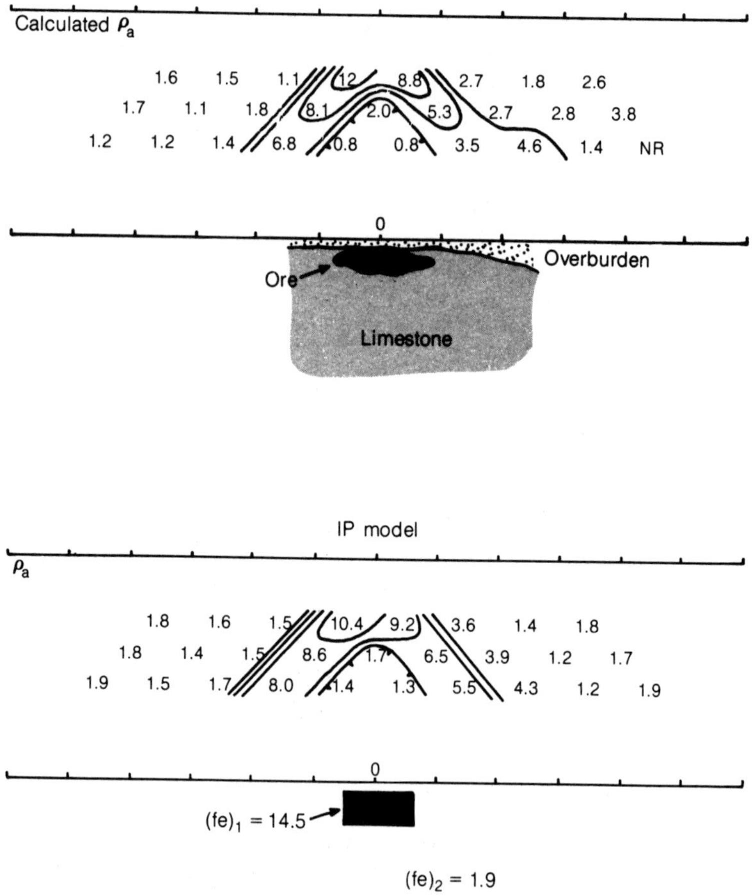

FIGURE 2.123
Induced polarization and resistivity data, Iso-Copperfields

Line—0+00

Computer interpretation of geophysical data
Canadian Shield massive sulfide deposit

Overlain by conductive overburden

Determine: horizontal position
Depth to top
Width
Res. of source
Res. of overburden
IP of source

Dipole-dipole array

x = 100 ft

100 ft

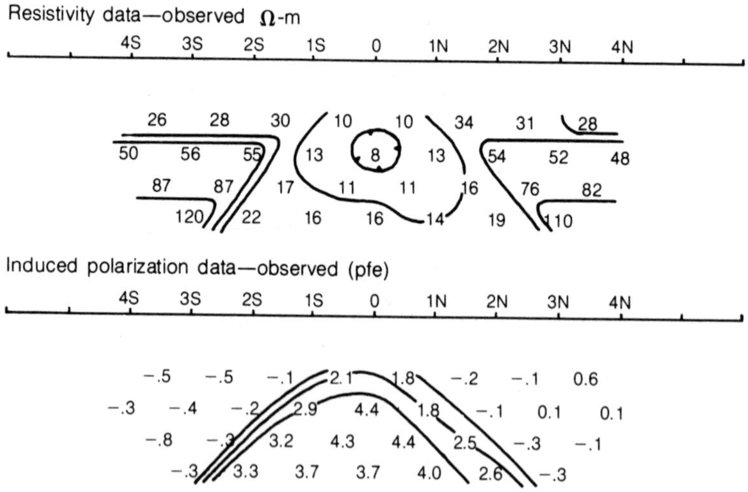

The figures on this page are referred to in the preceding section *Use of the digital computer to interpret IP and resistivity data.*

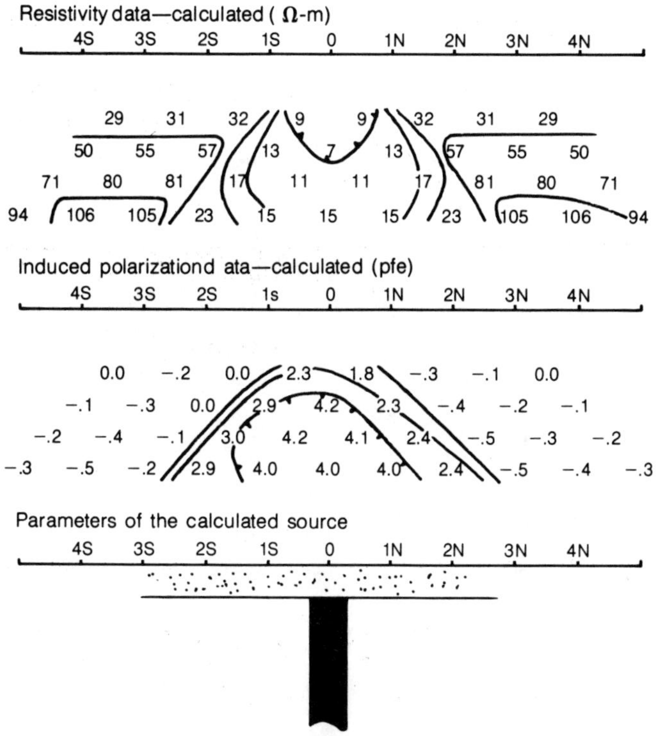

FIGURE 2.124
Solution and calculated data, Greenstone area, Quebec

Center at stn 0+13S ± 0.3%

Depth of overburden—51 ft ± 3%

Width—60 ft ± 20%

Res. of source—0.4 ω-m ± 63%

Res. of overburden—15 ω-m ± 5%

IP of source—7.5% ± 1%

Depth extent greater than 200 ft

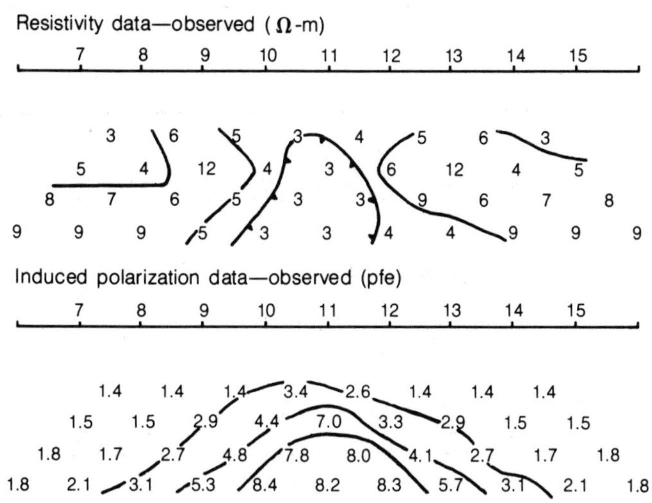

FIGURE 2.125
Induced polarization and resistivity data, deeply buried source in very conductive environment

Computer interpretation of geophysical data

Determine position, depth, and physical properties of source

Dipole-dipole array

x = 25 m

25 m

The figure on this page is referred to in the preceding section *Use of the digital computer to interpret IP and resistivity data.*

FIGURE 2.126
Solution and calculated data, Greenstone area, Quebec

Depth to top—23 m ± 11%

Res. of source—0.3 -m ± 45%

Res. of host—6.5 ω-m ± 5%

IP of source—23 ± 16%

IP of host—2.3 ± 11%

Center at stn 10+90 ± 1%

Depth extent fixed at > 50 m

Width fixed at 25 m

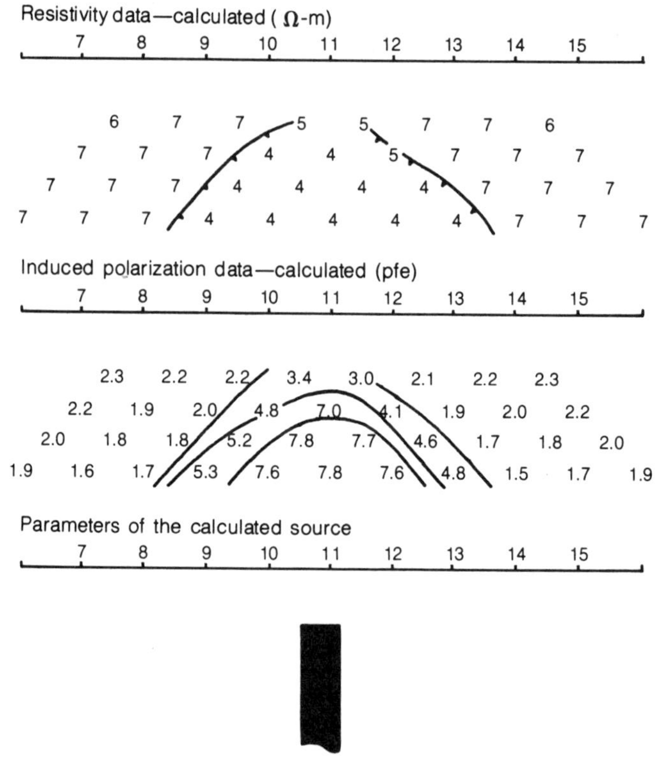

Selected References (Chapter 2)

Auston, J. S. 1969. Uchi Lake—a geophysical case history. Presented at Annual Meeting of Society of Exploration Geophysicists, 1969, Calgary, Alberta, Canada.

Bhattacharyya, B. K., and Morrison, H. F. 1963. Some theoretical aspects of electrode polarization in rocks. *Geophysical Prospecting* II: 176-96.

Cannon, R. W., Thornton, J. M., Rotherham, D. C. 1972. Induced polarization and resistivity in the Gibralter area, B.C. Presented at Annual Meeting of American Institute of Mining and Metallurgical Engineers, February 1972, San Francisco, California.

Caven, R. J. 1976. Geophysical surveying at the Umex Thierry deposit. *Canadian Mining Journal* 97, No. 10: 34-43.

Dey, A., and Morrison, H. F. 1973. Electromagnetic coupling in frequency and time domain induced polarization surveys over a multilayered earth. *Geophysics* 38: 380.

Donohoo, H. V., Podolsky, G., and Clayton, R. H. 1970. Early geophysical prospecting at Kidd Creek mine. *American Mining Congress Journal* 56: 44-53.

Dowsett, J. S. 1967. Geophysical exploration methods for nickel. *Mining and Groundwater Geophysics,* Geological Survey of Canada Economic Geology Report No. 26.

Fountain, D. K. 1968. The application of the induced polarization method at Brenda mines, B.C. *Canadian Institute of Mining and Metallurgy* 61, No. 670: 153-7.

Fraser, D. C., Keevil, N. B., and Ward, S. H. 1964. Conductivity spectra of rocks from the Craigmont ore environment. *Geophysics* 29: 832-47.

Grant, F. S., and West, G. F. 1965. *Interpretation theory in applied geophysics.* New York: McGraw-Hill.

Hallof, P. G. 1960. The uses of induced polarization in mining exploration. *Transactions of the AIME (Mining)* 217: 319-27.

________________. 1963. A study of the usefulness of the various parameters employed in the variable frequency IP method. Presented at the 33rd Annual International Meeting of the Society of Exploration Geophysicists.

________________. 1964. A comparison of the various parameters employed in the variable frequency induced polarization method. *Geophysics* 29: 425-34.

________________. 1965. The proper choice of frequencies for induced polarization measurements. Presented at the Annual Meeting of the Society of Exploration Geophysicists, November 1965, Dallas, Texas.

________________. 1966. Geophysical results from the Orchan Mines Ltd. property in the Mattagami area of Quebec. *Mining Geophysics* I: 157-71.

________________. 1966. Induced polarization and resistivity results from the Cactus deposit, Miami, Arizona. *Mining Geophysics* I: 313-16.

________________. 1967a. An appraisal of the variable frequency IP method after twelve years of application. Presented at the University of California, Berkeley; Symposium on Induced Polarization, February 1967.

———. 1967b. An appraisal of the variable frequency IP method after twelve years of application. In *Proceedings of the Symposium on Induced Polarization.* Berkeley: University of California.

———. 1967c. *IP spectral response: a geophysical case history of Cavendish Township, Ontario, Section 9.* Toronto: McPhar Geophysics Ltd.

———. 1971. A geophysical case history of the Lakeshore orebody. *Geophysics* 36: 1232-49.

Halverson, M.O. 1967. A look at dispersion and induced polarization using the phase angle concept. In *Proceedings of the Symposium on Induced Polarization.* Berkeley: University of California.

Hansen, D. A., and Barr, D. A. 1966. Exploration case history of a disseminated copper deposit. *Mining Geophysics* I: 306-12.

Hauck, Anthony M. 1970. A reconnaissance downhole induced polarization and resistivity survey method. Presented at Annual Meeting of Society of Exploration Geophysicists.

Heiland, C. A. 1946. *Geophysical exploration.* New York: Prentice-Hall.

Hendrick, D. M., and Fountain D. K. 1970. Induced polarization as an exploration tool, Noranda area, Quebec. Presented at Annual Meeting of Canadian Institute of Mining & Metallurgy.

Keller, G. V., and Frischknecht, F. C. 1966. *Electrical methods in geophysical prospecting.* London: Pergamon Press.

Lacy, R. J., and Morrison, B. C. 1966. Case history of integrated geophysical methods at Mission deposit, Arizona. *Mining Geophysics* I: 321-25.

Marshall, D. J., and Madden, T. R. 1959. Induced polarization, a study of its causes. *Geophysics* 34: 790-816.

Millett, F. B. 1967. Electromagnetic coupling of colinear dipoles on a uniform half-space. *Mining Geophysics* II.

Parasnis, D. S. 1966. *Mining Geophysics.* Amsterdam: Elsevier.

Paterson, N. R. 1966. Mattagami Lake mines—a discovery by geophysics. *Mining Geophysics* I: 185-96.

Pelton, W. H., and Hallof, P. G. 1971. *The applied potential method in the search for massive sulfides at York Harbour, Newfoundland.* Ontario, Canada: Phoenix Geophysics, Ltd.

Pelton, W. H., and Hallof, P. G. 1977. *New techniques in the interpretation of induced polarization, resistivity, and magnetic data.* Ontario, Canada: Phoenix Geophysics, Ltd.

Pelton, W. H., and Smith, P. K. 1976. Mapping porphyry copper deposits in the Philippines with IP. *Geophysics* 41: 106-22.

Rijo, L. and others. 1977. Interpretation of apparent resistivity data from Apodi Valley, Rio Grande Do Norte, Brazil. *Geophysics* 42.

Seigel, H. O. 1959. Mathematical formulation and type curves for induced polarization. *Geophysics* 24: 547-63.

———. 1962. Induced polarization and its role in mineral exploration. *Transactions of Canadian Institute of Mining and Metallurgy 1962* 65: 151-58.

Sumner, J. S. 1976. *Principles of induced polarization for geophysical prospecting.* Amsterdam: Elsevier.

—————. 1977. *Induced polarization for exploration geologists and geophysicists.* From Symposium at Department of Geosciences, Univ. of Arizona, Tucson, Arizona.

Van Voorhis, G. D., Nelson, P. H., and Drake, T. L. 1973. Complex resistivity spectra of porphyry copper mineralization. *Geophysics* 38: 49.

Wait, J. R., ed. 1959. *Overvoltage research and geophysical applications.* London: Pergamon Press.

Ward, S. H., and Fraser, D. C. 1967. Conduction of electricity in rocks. *Mining Geophysics* II.

Zonge, K. L., Sauck, W. A., and Sumner, J. S. 1972. Comparison of time, frequency, and phase measurements in induced polarization. *Geophysical Prospecting* 20: 626.

Figure Credits (Chapter 2)

Figures 2.1 & 2.2
Reprinted with permission of J. S. Sumner, *Principles of induced polarization for geophysical prospecting,* 1976.

Figure 2.3
Reprinted with permission of G. V. Keller and F. C. Frishknecht, *Electrical methods in geophysical prospecting,* 1966.

Figures 2.4-2.9
Reprinted with permission of J. S. Sumner, *Principles of induced polarization for geophysical prospecting,* 1976.

Figure 2.10
Reprinted with permission of G. V. Keller and F. C. Frishknecht, *Electrical methods in geophysical prospecting,* 1966.

Figures 2.21-2.26
Reprinted with permission of Wm. Pelton and P. G. Hallof, *The applied potential method in the search for massive sulfides at York Harbour, Newfoundland,* 1971.

Figures 2.27 & 2.28
Reprinted with permission of J. S. Sumner, *Principles of induced polarization for geophysical prospecting,* 1976.

Figure 2.31
Reprinted wtih permission of D. J. Marshall and T. R. Madden, Induced polarization, a study of its causes. *Geophysics* 34: 790-816, 1959.

Figure 2.44
Reprinted with permission of H. O. Seigel, Induced polarization and its role in mineral exploration. *Transactions at Canadian Institute of Mining and Metallurgy* (1962) 65: 151-58.

Figures 2.66-2.69
From P. G. Hallof, A geophysical case history of the Lakeshore orebody. *Geophysics* 36: 1232-49, 1971.

Figure 2.92
Reprinted with permission of D. M. Hendrick and D. K. Fountain, Induced polarization as an exploration tool, Noranda Area, Quebec. Canadian Institute of Mining and Metallurgy Annual Meeting, 1970.

Figures 2.94-2.96
From P. G. Hallof, An Appraisal of the variable frequency IP method after twelve years of application. UC-Berkeley Symposium on IP, Feb. 1967.

Figures 2.97 & 2.98
Reprinted with permission of D. M. Hendrick and D. K. Fountain. Induced polarization as an exploration tool, Noranda Area, Quebec. Canadian Institute of Mining and Metallurgy Annual Meeting, 1970.

Figure 2.99
From P. G. Hallof, An appraisal of the variable frequency IP method after twelve years of application. UC-Berkeley Symposium on IP, Feb. 1967.

Figures 2.101-2.110
Reprinted with permission from Anthony M. Hauck III, A reconnaissance downhole induced polarization and resistivity survey method. S.E.G. Annual Meeting, 1970.

Figures 2.111-2.113
Reprinted with permission from Wm. Pelton and P. G. Hallof, *New techniques in the interpretation of induced polarization, resistivity, and magnetic data,* Phoenix Geophysics, Ltd., Ontario, Canada 1977.

Figures 2.114 & 2.115
Reprinted with permission of L. Rijo and others, Interpretation of apparent resistivity data from Apodi Valley, Rio Grande Do Norte, Brazil. *Geophysics* 42, 1977.

Figures 2.118-2.122, 2.127-2.137
Reprinted with permission of W. H. Pelton, *Interpretation of induced polarization and resistivity data,* Phoenix Geophysics, Ltd., Ontario, Canada, 1977.

Figure 2.138
Reprinted with permission of J. S. Sumner, ed., *Induced polarization for exploration geologists and geophysicists.* From Symposium at Department of Geosciences, Univ. of Arizona, 1977.

CHAPTER 2 ADDENDUM ELECTRICAL

RESISTIVITY AND INDUCED POLARIZATION

THE SPECTRAL INDUCED POLARIZATION METHOD

As shown in figure 2.1, there are three different techniques to measure the IP effects. The older methods were the "variable frequency" and the "time-domain" IP techniques. In recent years, the Phase IP method has been perfected. Here the presence of induced polarization is detected as a phaselag between the voltage measured across the earth and the applied current. Therefore, a measurement at a single frequency is sufficient to complete an IP measurement.

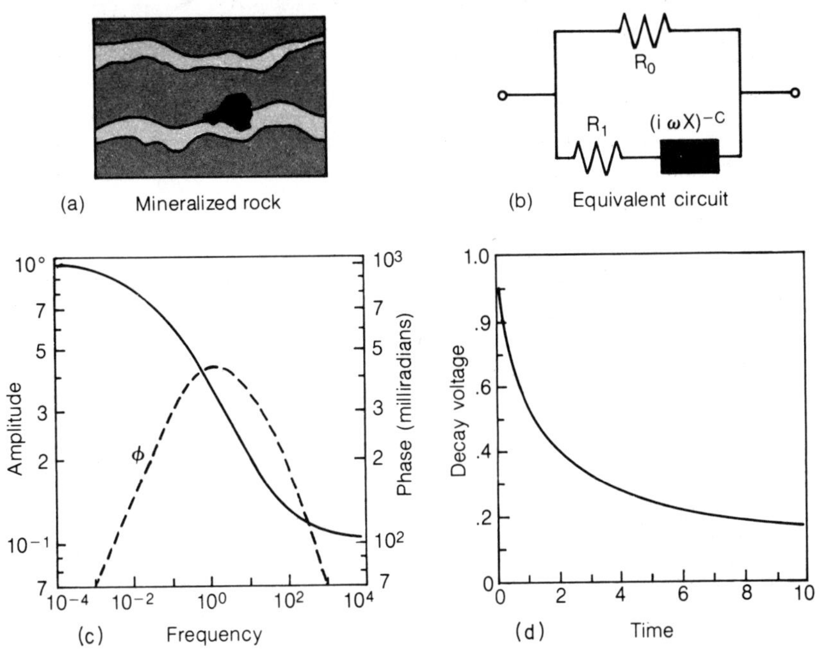

FIGURE 2.1
Equivalent circuit of mineralized rock and response

In the Spectral Induced Polarization method, measurements of phase and magnitude are made over a wide range of frequencies (see figure 2.2). Field survey measurements are usually limited to five or six orders of magnitude of frequency. In a typical survey, the measurement frequencies might range from 1/32 ($1/2^5$) Hz to 1024 (2^{10}) Hz. This, of course, can also be done in the "time domain" by using long current "on" times and looking at the decay curve over a long decay period.

Spectral IP Measurements

As we shall see, the purpose of the spectral IP measurement is to determine the electrical response (transfer function) of the subsurface for a particular electrode placement over a wide frequency range. Since small changes in the response can be important, it is necessary to make accurate measurements. There are three sources of "noise" and "distortion" that must be overcome; telluric current noise, current electrode variations and inductive coupling distortions. The first two must be overcome to make an accurate measurement. The third must be recognized in the data and removed, if present, before the spectral plot is interpreted. Telluric current electrical noise is present in the earth at all frequencies; however, it becomes a greater problem at very low frequencies. We have overcome this serious problem by using "digital coherent filters" to remove unwanted, low-frequency noise. Coherent digital filters require extremely accurate phase control between the current source and the receiver. This can be done using either a hard wire link, FM signals, or coherent clocks.

FIGURE 2.2
Cole-Cole relaxation model

$$Z(\omega) = R_0\left[1 - m\left(1 - \frac{1}{1 + (i\omega\tau)^c}\right)\right]$$

$R_0 = 1.0$ $\tau = .01$ $c = .25$

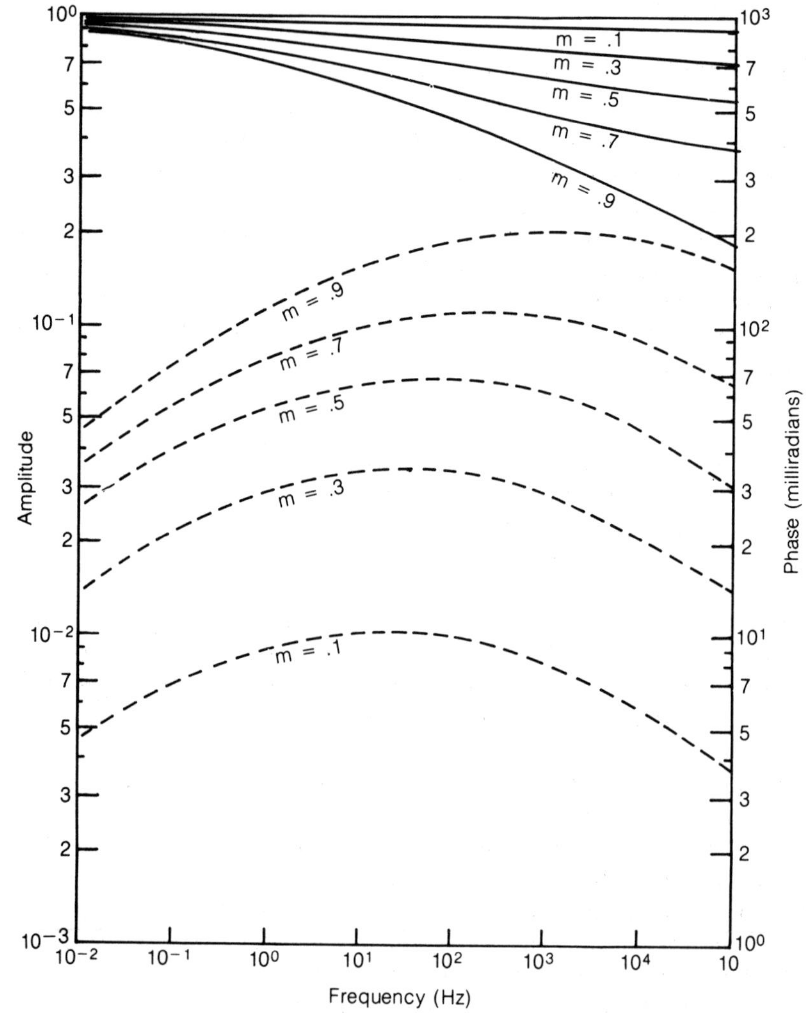

As electrical currents pass through current electrodes, severe changes can take place with the passage of time; the magnitude changes if current flow can change; and the phase shift of applied current. These factors create variations in the waveform of the applied current. Accurate spectral IP measurements cannot be made unless these variations are determined and compensation is made.

Electromagnetic effects can be expected to distort spectral IP measurements because of the interaction of the currents flowing in the wire and because of interaction between the wires and the currents flowing in the ground. These "inductive coupling" distortions increase in magnitude with increases in the electrode interval (a), the separation between receiver and transmitter (n), the conductivity of the subsurface (σ), and the frequency of the measurement (f). Inductive coupling is generally a problem in spectral IP measurements only at high frequencies (Dey and Morrison, 1978; Hohmann, 1973 and 1975). Fortunately, the frequency dependence of these electromagnetic effects is different from the expected frequency dependence of the IP effect. By continuing the spectral IP measurement to higher frequencies, it is possible to identify the magnitude and frequency dependence of the inductive coupling effects for a particular electrode placement and earth geometry. Computer inversion techniques have been developed to separate the electromagnetic distortions from the entire spectral response.

The Spectral IP Response

If accurate measurements are made of the magnitude and phase of the earth response over a wide frequency range, and if the distortions from inductive coupling are properly removed, the spectral IP response for a particular electrode placement can be accurately determined. All of the work to date confirms that the Cole-Cole Response Equation shown in figure 2.2 will adequately describe all spectral responses that have been measured (Pelton et al, 1978; Hallof and Klein, 1982). Many other individuals, have examined this, including Wait, Zonge and Seigel.

Four parameters can be used to describe the response over the entire frequency range:

- R_0 - the resistivity at zero frequency
- m - the dimensionless IP parameter (chargeability)
- τ - the time constant of the IP response
- c - the exponent of the frequency term

The frequency (Fc), the "critical frequency", can often be used to describe a spectral response. It is the specific frequency at which the maximum phase shift is measured. For the Cole-Cole Response Equation, the critical frequency (Fc) can be described in terms of the other parameters. Thus, the critical frequency (Fc) is inversely proportional to the time constant (τ). If (Fc) occurs at high frequencies, the time constant is small. If (Fc) occurs at low frequencies, the time constant is large.

For any given polarizable rock, the values of (R_0) and (m) will depend upon the porosity of the rock itself and the surface area of metallic minerals exposed to electrolyte in conductive pore spaces. Increasing the metallic mineral will decrease (R_0) and increase (m). Research efforts and field experience show that (τ) and (c) are much more dependent upon the "texture" of the metallic mineralization present than upon the type of metallic minerals present. There are two factors that describe this "texture".

- the grain size for each population of polarization particles that is causing the IP effects measured,

- the number of populations of metallic particles present as well as the specific distribution of grain sizes within each population of polarizable particles.

Grain Size And Time Constant

Field work and research indicates that the time constant (τ) of the measured IP effect is directly dependent upon the grain size of the metallic particles that are the source of the IP effect. For fine-grained mineralization, (Fc) is at high frequencies and (τ) is small. For coarse-grained mineralization, (Fc) is at low frequencies and (τ) is large. Dr. J. Wong studied this relationship in his doctoral thesis (Wong, 1979). Figures 2.3 and 2.4 show some of the theoretical and experimental results that confirm this direct relationship. Other experimental data and extensive field measurements are available to support this conclusion (Hallof et al, 1979, Hallof and Klein, 1982). The first use of spectral IP data in defining the "texture" of metallic mineralization in the subsurface relates the time constant (τ) of the IP effect measured to the grain size of the mineralization. After the inductive coupling distortions are removed from each spectral plot, the resulting data are interpreted to determine the four parameters (R_0, m, τ, c) that give the Cole-Cole dispersion curve that best "fits" the spectral IP data.

FIGURE 2.3
Theoretical results for variation in grain size of polarizable particle, Wang 1979

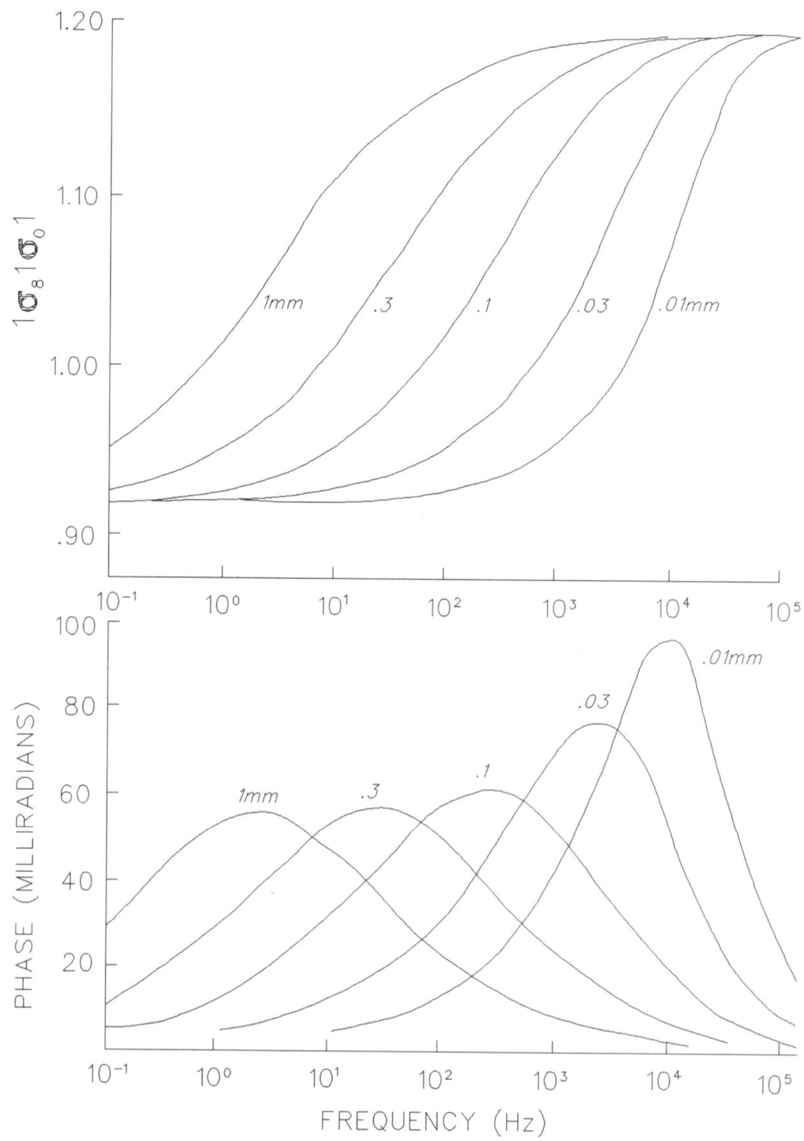

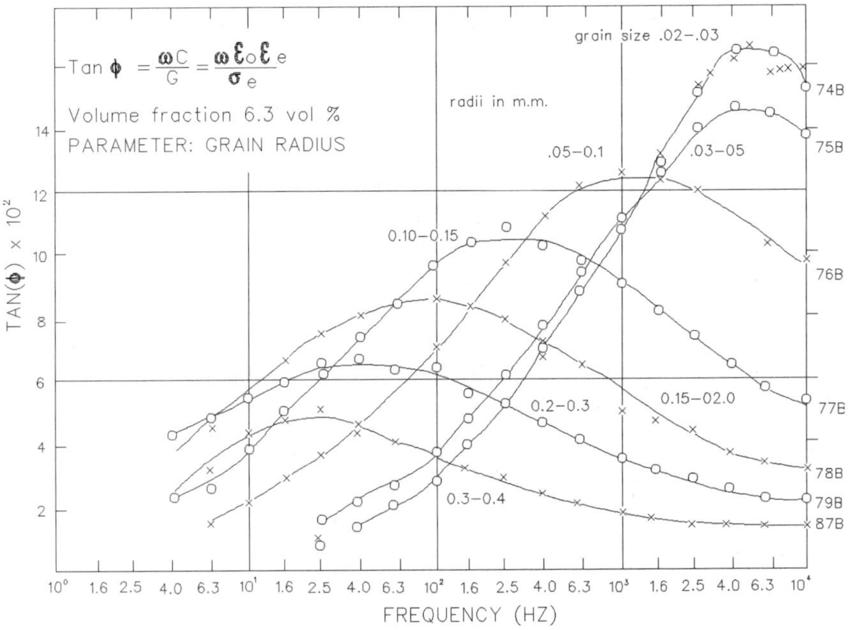

FIGURE 2.4
Variations of response with grain size

In almost all of the thousands of examples now available, the Cole-Cole equation gives an excellent "fit" to the measured data; therefore, the four spectral parameters are all that are necessary to describe the response. A change in the time constant (τ) obtained by this procedure reflects a change in the grain size of the polarizable particles.

Grain Size Distributions And Frequency Exponent

The value of (c) determines the slope of the phase-versus-frequency, log-log plot, for both high and low frequencies. As shown in figures 2.4 and 2.5, the curves are symmetric. For small values of (c), the phase-versus-frequency curves are quite "flat." For larger values of

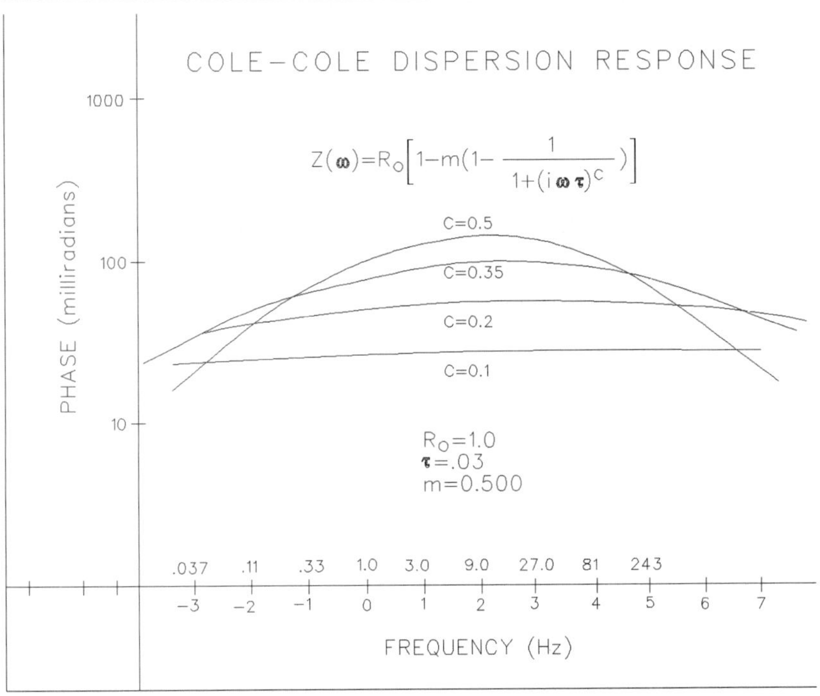

FIGURE 2.5
Spectral plots with variations in "c"

(c), the slopes become more pronounced and the phase-versus-frequency curves become more "peaked." For larger (c), the value of (Fc) is easily determined; for very small (c), the value of (Fc) may not be so obvious.

Field measurements have produced spectral curves of all shapes. Computer inversion has given curves from as flat as (c = 0.10) to as peaked as (c = 0.50). Theoretical work shows that (c = 0.50) would be the maximum theoretical value to be expected.

Dr. J. Wong's work can be extended to include the case for metallic particle shapes other than spheres. The result is the same. If a single, unique grain size is present, the phase-versus-frequency curve is very peaked with a value (c = 0.50), as shown in figure 2.6. The position of the critical frequency (Fc) is determined by the grain size; the shape of the phase-versus-frequency spectral plot is determined by the distribution of grain size.

FIGURE 2.6
Theoretical spectral plot of single grain size

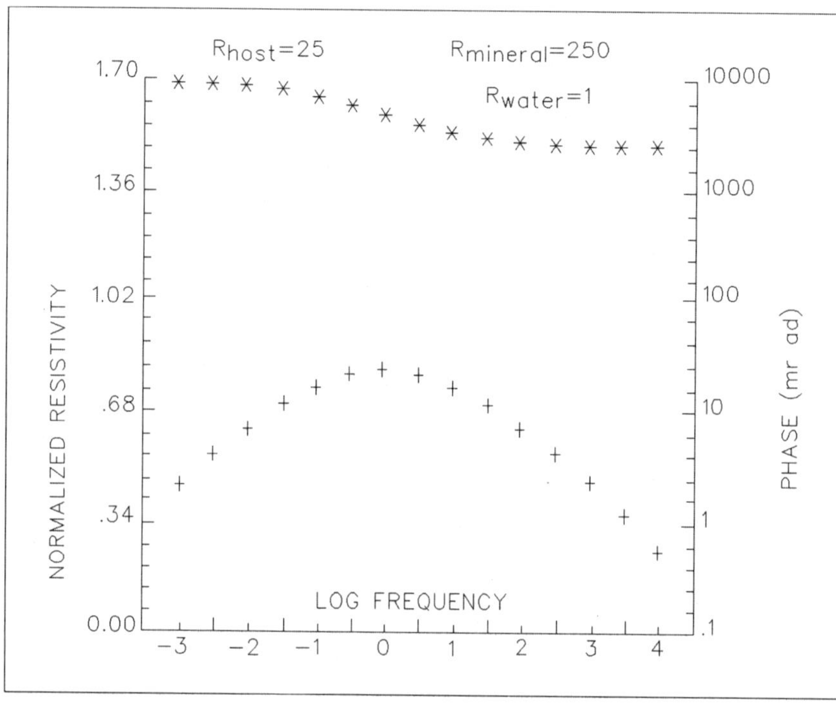

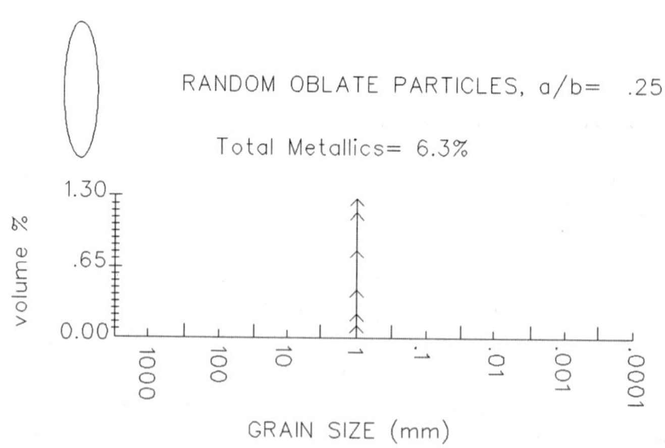

The spectral curve shown in figure 2.7 is the theoretical spectral plot of a normal distribution of grain size; the mean grain size for the normal distribution shown is smaller than the unique grain size used in figure 2.6. The small mean grain size results in an (Fc) value at a higher frequency than for figure 2.6; this is the expected result.

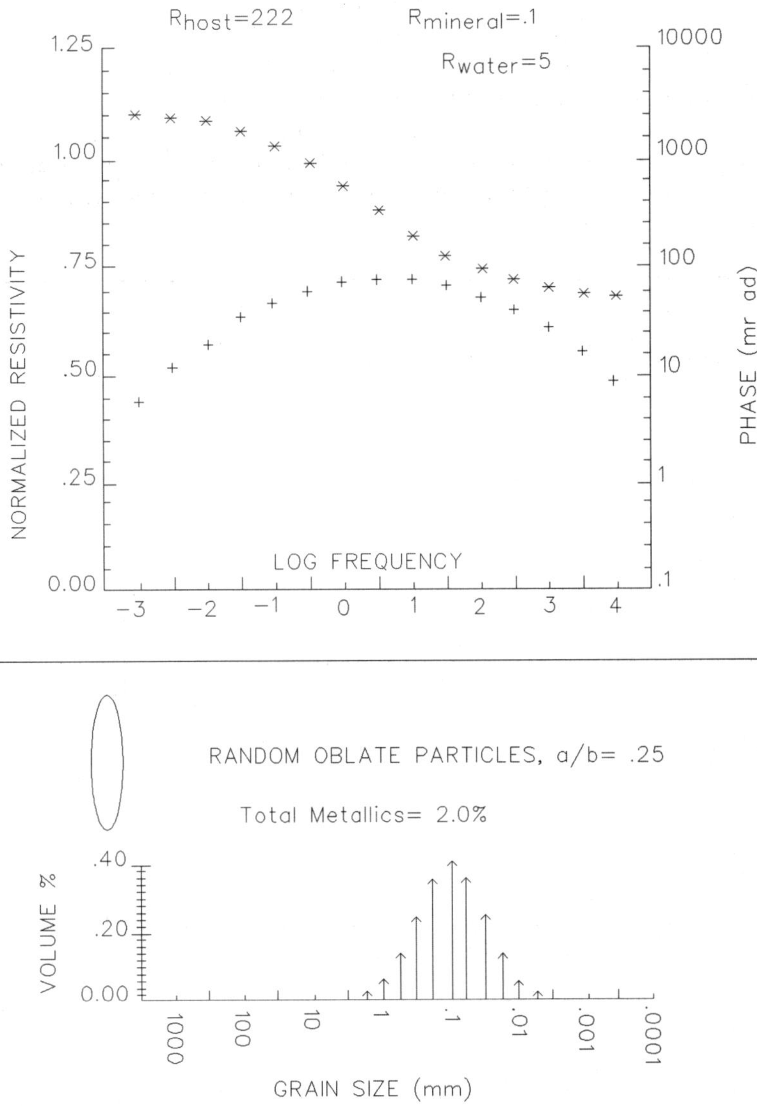

FIGURE 2.7
Theoretical spectral plot of "normal" grain size distribution

The broader distribution of the grain size population, however, results in a smaller value for (c). The phase-versus-frequency spectral curve is less peaked. This result agrees with field experience. If a broad, non-unique distribution of grain size is present, the (c) value obtained by the inversion of the spectral curve is in the range 0.25 to 0.35. For most spectral IP surveys, the values of (c) fall into this range.

For some zones of metallic mineralization, there may be two populations of metallic particles present. If the grain sizes for these two populations are quite different, the spectral curve would be expected to look something like the theoretical curve shown in figure 2.8. The presence of the two distinct grain sizes can be seen in the double peak for the phase-versus-frequency curve.

FIGURE 2.8
Theoretical spectral plot of two grain sizes "B:-modul" response

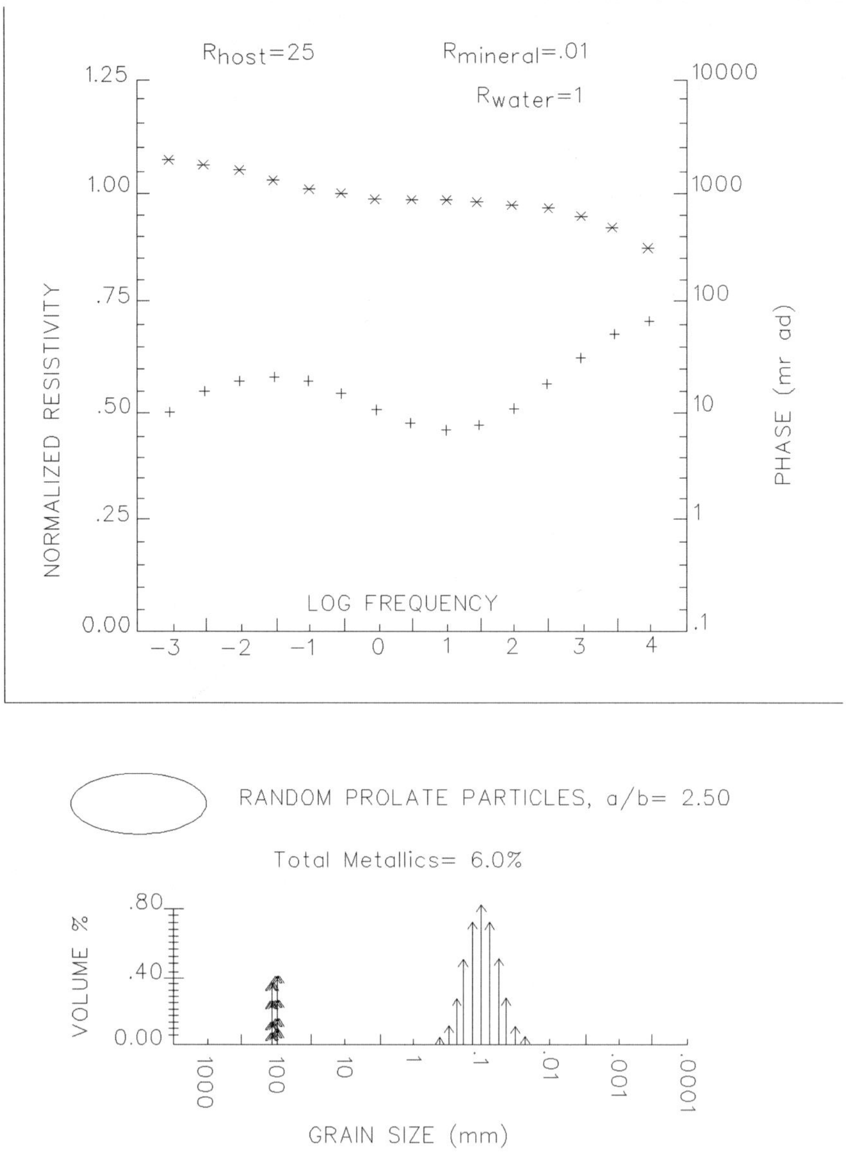

In more usual situations, however, there are two overlapping distributions of grain size or a continuous distribution over a wide range of grain sizes. The phase-versus-frequency curve for this situation is that shown in figure 2.9. The value for (c) is small, in the range from 0.100 to 0.200. The spectral curve is very "flat." The magnitude of the phase shifts measured is low; however, there are phase shifts measured at all frequencies.

Keep in mind that when the spectral curve has this "flat" character, the spectral parameters (τ and c) are poorly determined by computer inversion. If some estimate of the time constant (τ) values is available from other investigations, these values can be "fixed" in the inversion. These solutions are expected to be non-unique, but they will result in some estimate of the spatial distribution of the grain size populations that have been modeled.

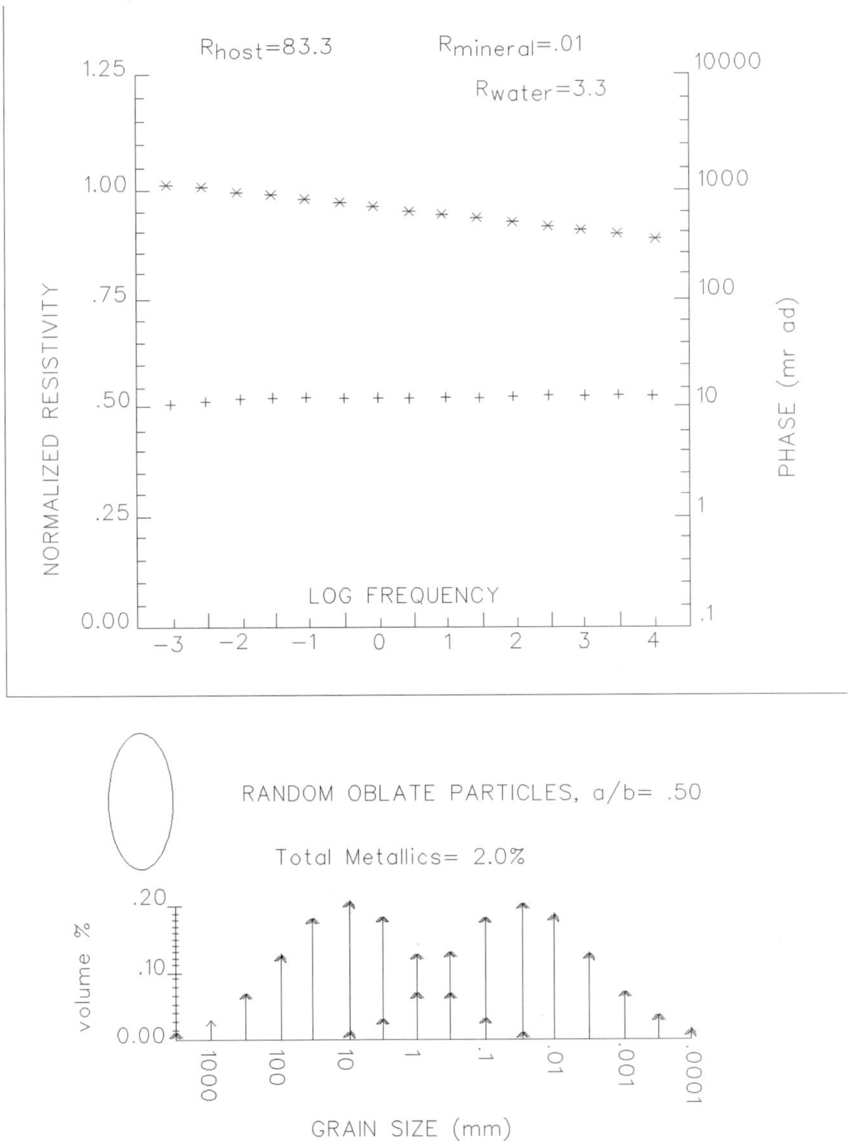

FIGURE 2.9
Theoretical spectral plots, two grain size with overlapping response parameters

Spectral Examples From Field Work

All of the "texture" characteristics described can be measured during a spectral induced polarization survey. The examples shown here are from a program of test surveys within base metal sulphide mines in Ontario. They have been chosen because the electrode intervals used were very small (x = 1.0 meter or x = 2.0 meters). For this reason, the inductive coupling effects encountered were usually small.

The results shown in figure 2.10 are from a zinc-silver ore zone in the Mattabi Mine near Ignace, Ontario. All of the measurements within this sulphide zone gave essentially the same spectral parameters. The value of $c \leq 0.5$ suggests that the entire zone reacts as a single, unique grain size (see figure 2.6). Further, the time constant value $\tau = 0.24$ seconds indicates that the unique grain size indicated is quite small.

FIGURE 2.10
Typical spectral plot from field data massive sulfides at Mattabi mines

MASSIVE ZINC-SILVER ORE
ETDF PROJ. MATTABI MINE IGNACE550-1040 SILL. DR. 1m 3/82 L373 D=8 N=1 X= 2m

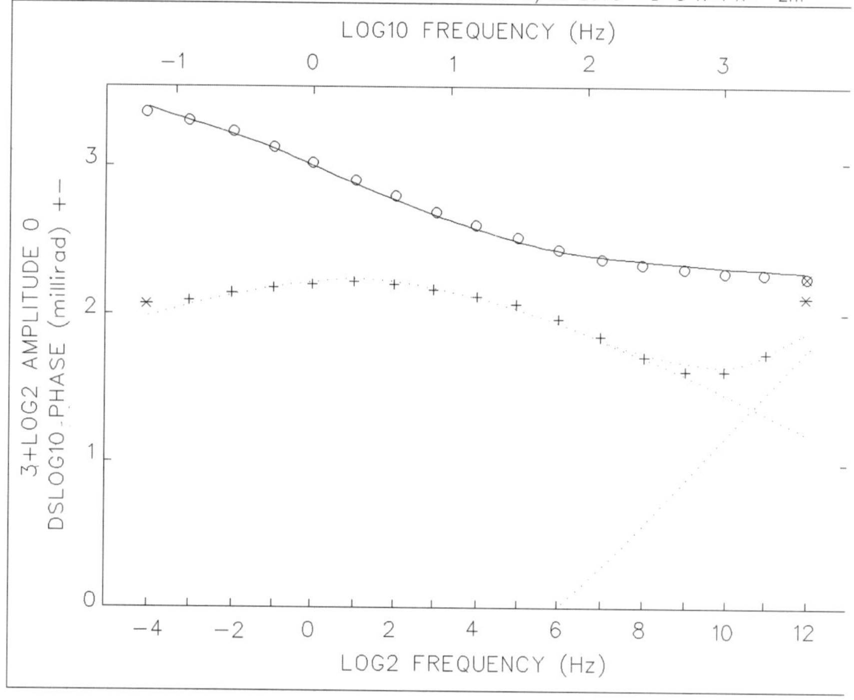

CRL: Number of dispersions= 2
C2=1 Fixed

Iter	lambda	Rchsq	R0	M1	T1	C1	M2	T2
0	1.E−02	.00129	1.500	.601	2.9E−01	.471	.700	4.9E−06
1	1.E−02	.00062	1.480	.598	2.6E−01	.465	.579	4.1E−06
2	1.E−03	.00061	1.471	.595	2.4E−01	.470	.562	4.1E−06
3	1.E−04	.00061	1.470	.595	2.4E−01	.471	.481	4.8E−06
Pct Std Deviations			1.8	1.5	13.1	2.7	9999.9	9999.9
Correlation Matrix			1.000					
			.931	1.000				
			.918	.852	1.000			
			−.746	−.807	−.803	1.000		
			.053	.224	.031	−.150	1.000	
			−.057	−.228	−.033	.155	−1.000	1.000

Apparent resistivity measured at 1 Hz is .220

Apparent resistivity calculated from inductive coupling is 1.307

This confirms all experience with massive, volcanogenic complex sulphide ores. Most of these zones have time constants that are less than one second. Only rarely is encountered a spectral plot from a massive sulphide ore that has a time constant greater than 10 seconds.

The exceptions to this experience are the veinlet chalcopyrite (stringer copper) zones that often occur in the footwall of massive, volcanogenic sulphide zones and zones of massive pyrrhotite. These two types of sulphide mineralization usually have a large grain size texture; the time constant (τ) values are often greater than 10 seconds.

The spectral plot shown in figure 2.11 was measured using the dipole-dipole electrode configuration, with x = 2 meters. Electromagnetic effects (inductive coupling) can be seen in the phase data, but only at the highest frequencies used.

The spectral plot shown in figure 2.11 has the more usual shape with c = 0.25; this zone of disseminated sulphide mineralization is indicated to have a normal distribution of grain size about some preferred value. The time constant (τ) value is less than one second.

FIGURE 2.11
Spectral plot with some EM coupling

DISSEMINATED COPPER–ZINC ORE

ETDF PROJ. LYON LAKE MINE 320LVL CREEK ZONE 2m 3/16/82 L372 D=9 N=3 X= 2m

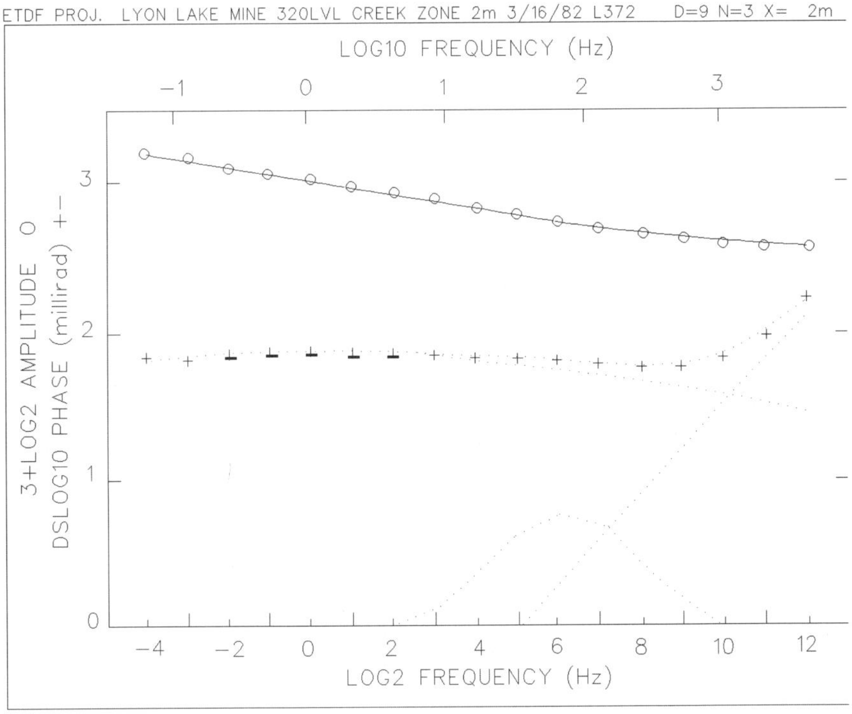

CRL: Number of dispersions=3
C1=.25 C2=1 C3=1 fixed

Iter	Lambda	Rchsq	R0	M1	T1	M2	T2	M3	T3
0	1.E−02	.00014	1.464	.535	7.6E−01	.011	2.3E−03	.971	5.1E−06
1	1.E−02	.00014	1.464	.534	7.6E−01	.012	2.3E−03	.983	5.1E−06
Pct Std Deviations			.9	.7	19.4	47.5	37.6	71.0	72.9

Correlation Matrix

1.000						
.285	1.000					
.890	−.104	1.000				
.551	−.527	.778	1.000			
−.090	−.202	−.057	.055	1.000		
−.217	.170	−.241	−.161	−.084	1.000	
.237	−.183	.268	.184	.093	−.999	1.000

Apparent resistivity measured at 1 Hz is 5.810

Apparent resistivity calculated from inductive coupling is 2.962

The resistivity and the IP effect are typical of those encountered for disseminated sulphides.

The spectral plot shown in figure 2.12 is much the same as that shown in figure 2.7. The value of c = 0.25 indicates a normal distribution grain size. The value of τ = 870 seconds, however, indicates a large grain size.

The spectral plots show on on figure 2.12 and figure 2.13 indicate the result of overlapping, or broad distributions of grain size and time constant.

If the spectral curve shown in figure 2.12 is inverted using one Cole-Cole dispersion, the value for (c) is very small. If the value for (c) for two dispersions is "fixed" before the computer inversion is attempted, it is possible to determine an estimate of the IP effects and the time constants that could overlap and result in the spectral curve that was measured.

FIGURE 2.12
Spectral plots with broad distribution of grain size

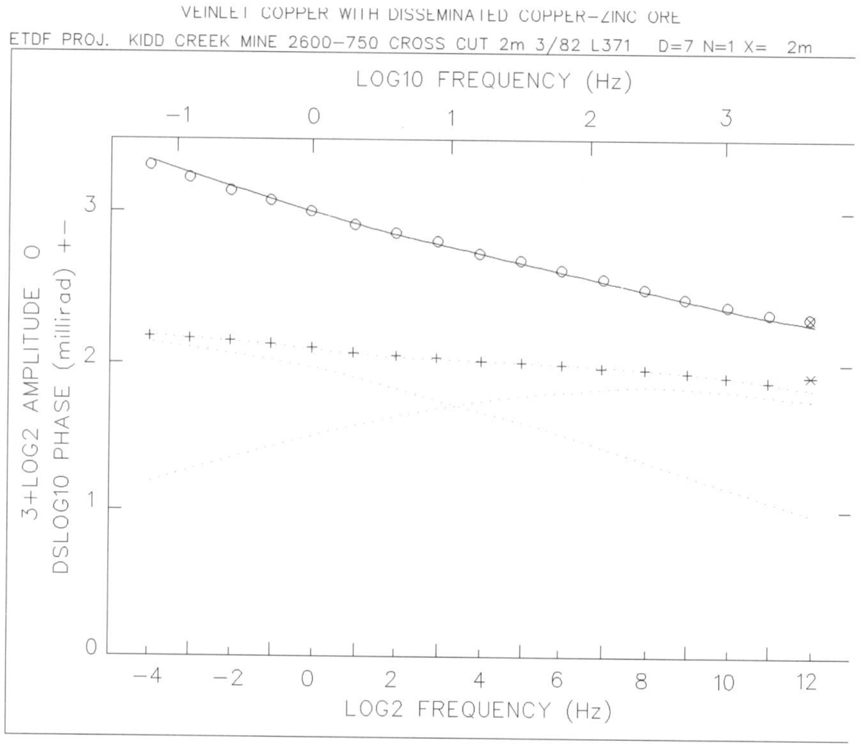

CRL: Number of dispersions=2
C1=.325 C2=.325 fixed

Iter	Lambda	Rchsq	R0	M1	T1	M2	T2
0	1.E-02	.00027	2.717	.686	6.8E+01	.403	1.1E-03
1	1.E-02	.00027	2.726	.686	7.1E+01	.408	1.1E-03
2	1.E-03	.00026	2.809	.692	8.7E+01	.412	1.2E-03
3	1.E-03	.00026	2.876	.698	1.0E+02	.414	1.3E-03
Pct Std Deviations			10.9	3.8	98.3	3.3	46.4

Correlation Matrix					
	1.000				
	.979	1.000			
	.991	.947	1.000		
	.008	.703	.857	1.000	
	.713	.567	.774	.778	1.000

Apparent resistivity measured at 1 Hz is 21.60

Apparent resistivity calculated from inductive coupling is .000

FIGURE 2.13
Spectral plot with broad distribution of grain size

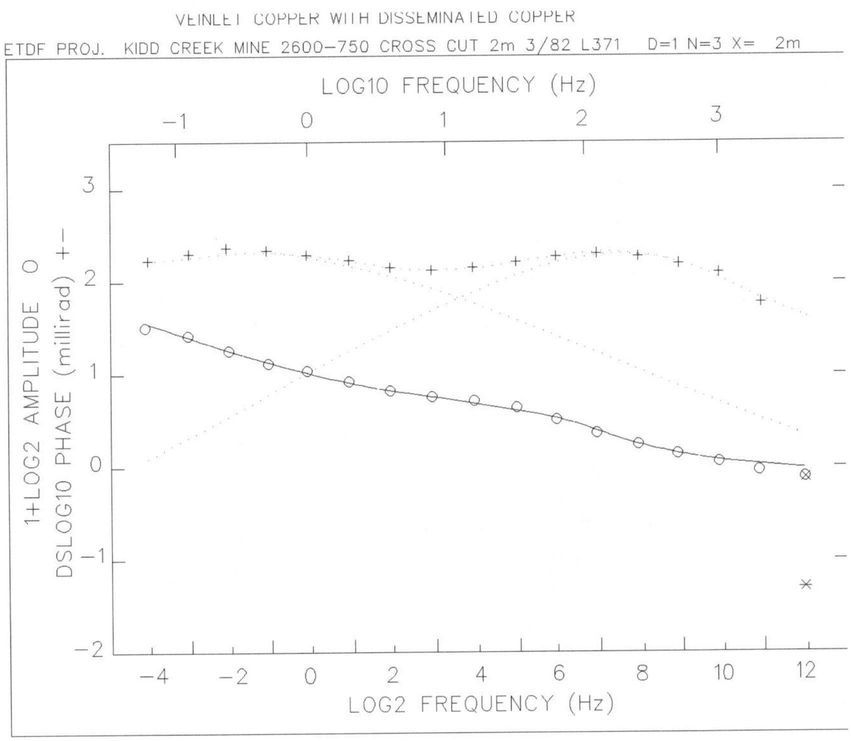

CRL: Number of dispersions=2

Iter	Lambda	Rchsq	R0	M1	T1	C1	M2	T2	C2
0	1.E−02	.00109	1.746	.550	1.0E+00	.610	.405	1.5E−03	.765
1	1.E−02	.00095	1.761	.550	1.1E+00	.610	.401	1.5E−03	.769
2	1.E−03	.00095	1.766	.551	1.1E+00	.607	.400	1.5E−03	.769

	R0	M1	T1	C1	M2	T2	C2
Pct Std Deviations	3.9	5.1	19.3	7.7	4.9	12.8	4.4
Correlation Matrix	1.000						
	.930	1.000					
	.918	.785	1.000				
	−.840	−.938	−.750	1.000			
	−.594	−.824	−.400	.814	1.000		
	−.512	−.752	−.307	.726	.862	1.000	
	.393	.605	.209	−.560	−.784	−.762	1.000

Apparent resistivity measured at 1 Hz is 1.610

Apparent resistivity calculated from inductive coupling is .010

CONCLUSIONS

With the necessary equipment to make accurate induced polarization measurements over a wide frequency range and the necessary analysis software and computer software, it is possible to produce a spectral IP response curve for each dipole-dipole pair occupied during a spectral IP survey. In addition to permitting anomalies to be detected in difficult areas, these results often permit some interpretation of changing "textures" within the source of an IP anomaly. If conditions are proper, it is sometimes possible to speculate upon the source of an IP anomaly based upon the "texture" interpreted from the spectral IP results.

The determination of the time constant (τ) of the IP effect measured and the exponent (c) of the frequency dependence of the IP impedance lead to the interpreted "texture" of the source and especially changes in the "texture." These parameters describe the character of the spectral IP effect with regard to grain size.

It is possible to make some determinations as to the grain size of the metallic particles causing the IP effect detected. Also, it is possible to determine something about the distribution of grain sizes within the grain size population as well as the number of grain-size populations that may be present.

Knowledge of the above variations can often be of assistance in mineral exploration. Some of the problems in which spectral IP data are useful are as follows:

- The texture of a large zone of "porphyry copper"-type mineralization can vary considerably. In particular, it is possible to separate regions that contain primarily veinlet mineralization from those that are primarily disseminated.

- It is often possible to separate sources that are massive sulphide mineralization from those that contain concentrated graphite. In tests in mining areas from Bathurst, New Brunswick, to Faro, Yukon Territory, graphite deposits invariably have a low resistivity, a high IP effect and a time constant (τ) in excess of 10 seconds to 100 seconds. The graphite sources have a large grain size. On the other hand, massive volcanogenic sulphide sources have a small grain size, usually with a time-constant value less than 10 seconds. Some care must be exercised, however — concentrated pyrrhotite zones and the veinlet chalcopyrite (stringer copper) zones in the footwall of volcanogenic sulphide deposits usually have a very large grain size.

- Spectral IP checks across a long zone of sulphide or oxide iron formation will locate positions where a variation in texture may reflect an increase in base metal sulphide content.

- Often a change in texture can be detected when gold, copper or zinc is present within an otherwise barren zone of pyrite.

A considerable volume of data is available to substantiate the applications described above.

SPECTRAL IP RESULTS — SOURCES WITH LARGE (τ) VALUES

The texture of a metallic mineral deposit can be interpreted from the values of (τ) and (c) obtained by the inversion of spectral IP data. Further, field data from New Brunswick to the Yukon indicates that graphitic material can be distinguished from massive volcanogenic deposits on the basis of texture.

The geologic processes that convert carbonaceous acid fragmentals into highly conductive, graphite tuffs appear to always result in a very large grain size for the conductive graphite particles. On the other hand, even very massive volcanogenic sulphide deposits appear to be fine grained. The time constants (τ) for graphite deposits are very large; for volcanogenic sulphide deposits, they are small.

Considerable spectral IP work underground at Sudbury, Ontario, indicates that massive pyrrhotite mineralization can be expected to have large time constant (τ) values. The crystal habit of pyrrhotite results in long, thin particles. Field data from three deposits in Ontario help to evaluate this circumstance:

- a typical graphitic tuff zone
- a typical "pyrrhotite dike" from Shield volcanic rocks
- a "Geco-type" sulphide ore deposit containing pyrrhotite.

Graphitic Tuff — Schreiber, Ontario

The geologic staff of the Ontario Geological Survey chose a graphitic tuff band exposed in a road cut west of Schreiber, Ontario, as being typical of the many similar geologic occurrences that are known in the Canadian Shield. There is a small amount of rubble in the road cut, but the graphitic material is present at a very shallow depth. A 13-current dipole survey was completed, using X = 1.0 meters.

The results (figure 2.14), when plotted in pseudosection format, show the extremely conductive, highly polarizable source to be expected from a massive graphitic tuff. A few noisy readings could not be measured successfully. Most of the data, even those for very low apparent resistivities, however, gave quite accurate spectral plots.

For all of the measurements made, the phase angles measured are increasing as the frequency is decreased; even at the lowest frequency of (2.0) Hz. As shown on the examples, the computer inversion of these spectral plots results in very large time constant (τ) values.

The pseudosection plots for (c_1) and (τ_1) show the expected result for a source containing a normal population distribution of very large grain size, polarizable particles. The (τ_1) values are variable but large in magnitude (100 seconds to 100,000 seconds), and the (c_1) values are in the range 0.25 to 0.35.

These results confirm those measured elsewhere from graphitic sources.

FIGURE 2.14
Pseudosection showing response of graphitic tuff

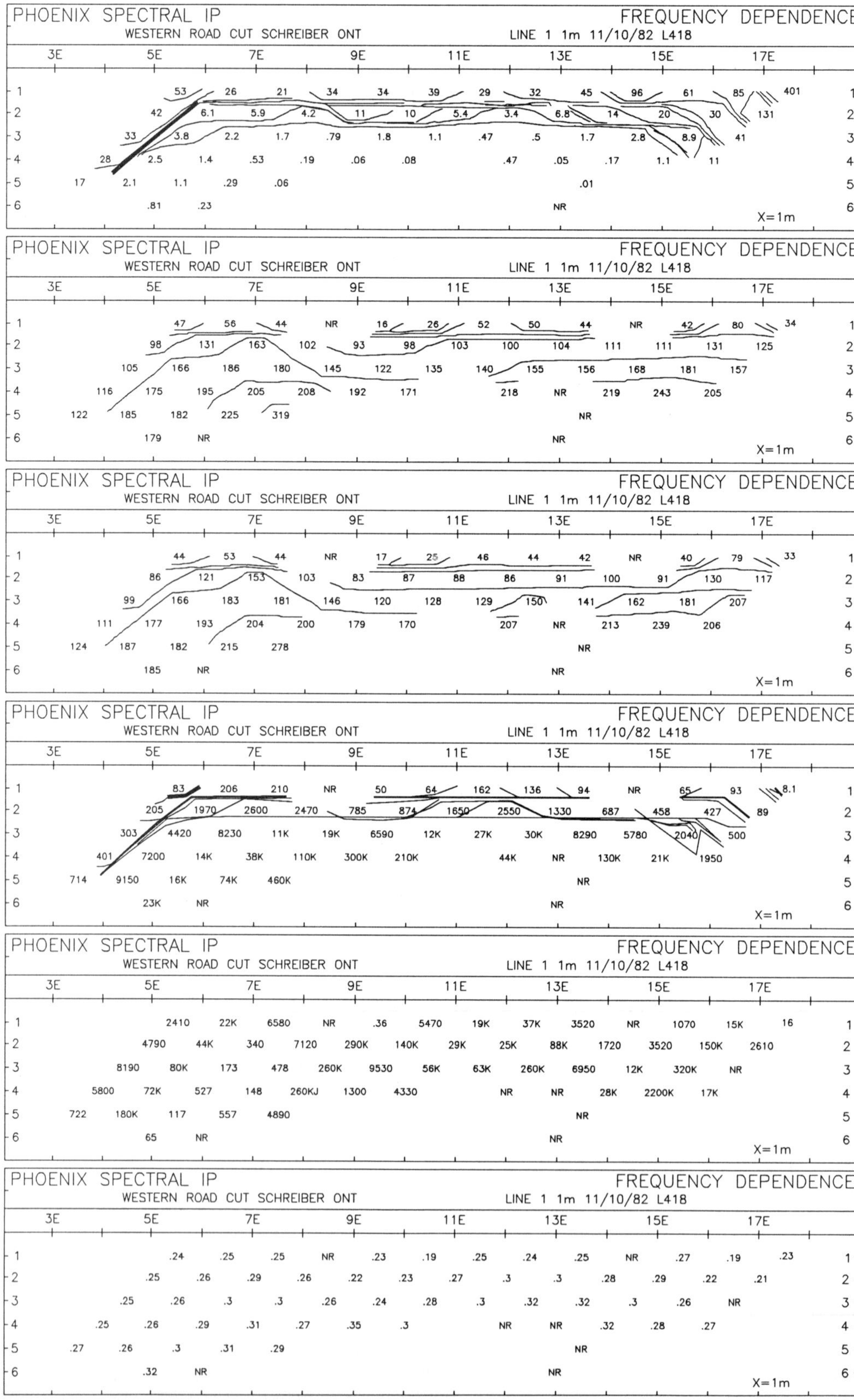

Massive Pyrrhotite in Canadian Shield Greenstones

There are numerous occurrences of massive, long, pyrrhotite "dikes" in the Canadian Shield. These zones are often classified as "sulphide iron formation." They are excellent electromagnetic conductors and give rise to sharp, strong magnetic highs. The geologic staff of Kidd Creek Mines Ltd. has made available data from a massive, pyrrhotite zone on the Kidd Creek Mines Ltd. property near Timmins, Ontario.

Line 155E (figure 2.15) passes over the zone near its center. A magnetic anomaly of more than 5,000 gammas correlates with the strong electromagnetic anomaly associated with the zone. Since there is a depth of 20 m to 30 m of glacial overburden overlying the zone, it was necessary to use an electrode interval of X = 20 m for the spectral IP survey. A 16-current dipole spectral IP survey was completed on Line 155E.

The apparent resistivity pseudosection shows a very strong, somewhat complex anomalous pattern that correlates exactly with the electromagnetic and magnetic anomalies. There is a much weaker resistivity low at the extreme northern end of the data plot.

There is a strong, definite phase-IP anomaly that correlates exactly with the apparent resistivity low. The anomalous pattern for the phase values indicates a relatively narrow, tabular source. There is a much weaker IP anomaly that correlates with the very weak apparent resistivity anomaly at the north end of the line.

The time constant (τ_1) values for these two very different IP anomalies are of greatly different magnitudes. The weak IP anomaly centered at 11620N has (τ_1) values that are all less than five seconds. There is an indistinct EM anomaly here and no correlating magnetic high. The holes drilled to test this anomaly intersected disseminated pyrite with traces of sphalerite.

The very strong IP anomaly, centered at 11420N to 11440N, has (τ_1) values that are very large. Within the very strong resistivity low, the signals were at a low level and noisy. Therefore, some of the spectral data plots are uncertain. Even though some of the (τ_1) values are poorly determined, however, it is clear that all of the values are greater than 100 seconds (or perhaps even greater than 1,000 seconds). This range of time-constant (τ_1) values for massive pyrrhotite agrees very well with those determined underground in the nickel mines at Sudbury, Ontario.

FIGURE 2.15
Pseudosection with a massive sulfide ore with pyrrhotite

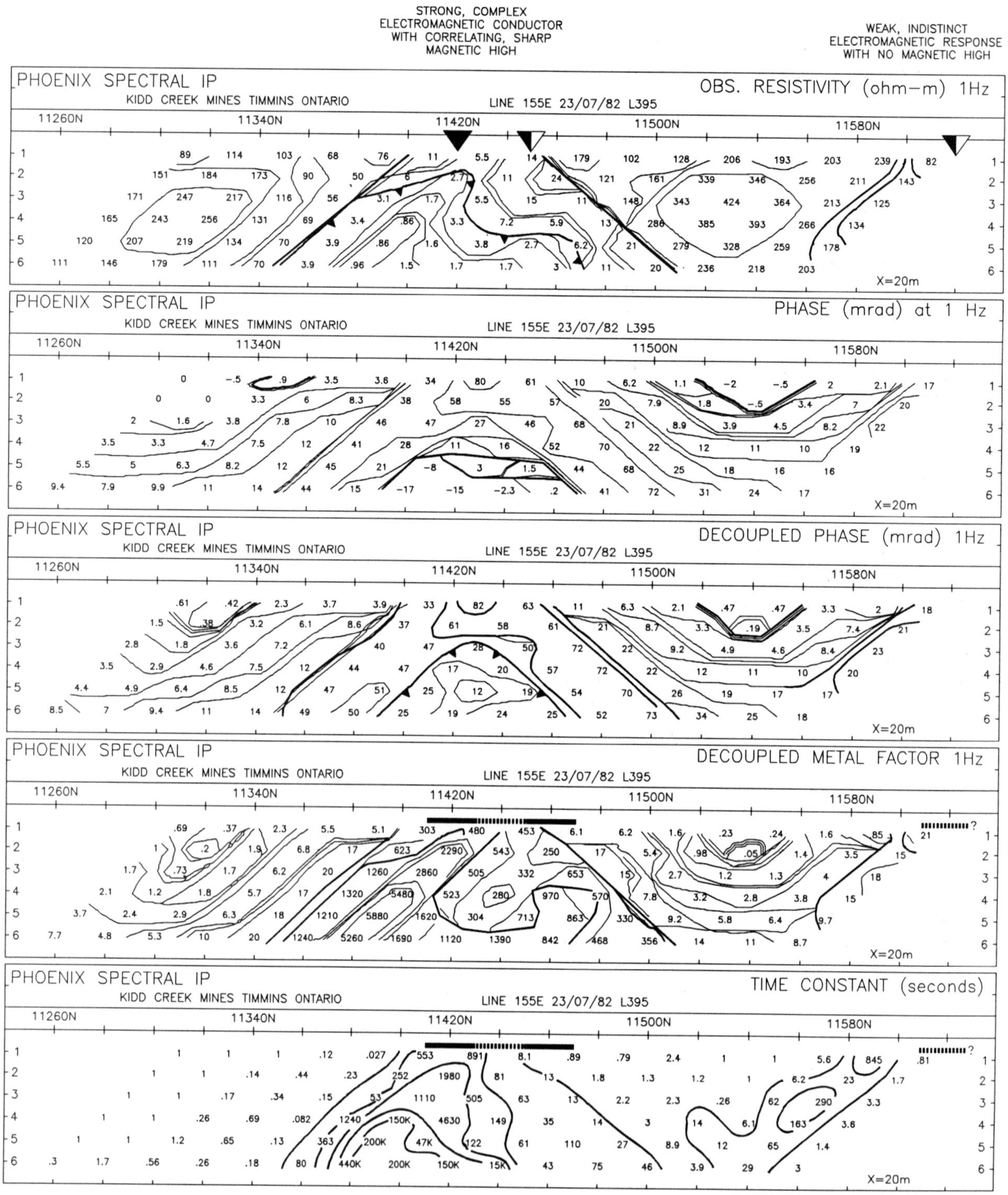

Massive Sulphide Ore Zones Containing Pyrrhotite (Geco Type)

The volcanogenic ore zones that have been previously examined did not contain pyrrhotite — consequently, they were non-magnetic. Numerous orebodies in the Canadian Shield do contain pyrrhotite, however. The largest and best known is probably the Geco Ore Body in western Ontario. The Kam-Kotia ore zones, near Timmins, Ontario, have much the same metallic mineral assemblage.

Main Metals — Cu, Zn

Location — Robb Twp.; claims P.12403-12405, P.12339-12343, P. 12353, P. 12356.

Reference: ODM map 53c.

Geology —Andesite flows and tuffs contain lenses of rhyolite tuffs and breccias. The strike is N55W, and the rock sequence faces N and dips about 75N. The orebodies occur as numerous irregular lenses of massive sulphides with associated quartz-sulphide stockworks. Massive sulphide zones consist of 80 percent to 95 percent pyrite, pyrrhotite, chalcopyrite and sphalerite, and the stockworks contain copper mineralization. On surface, the main ore lens was 350 ft in length, up to 200 ft in width and persisted from surface to beyond the 150-ft level.

Economic Features — Ore reserves, December 1967 (Canadian Mines Handbook, 1968-1969, p. 182) were 3,480,000 tons including positive and broken ore of 2,100,000 tons averaging 2.38 percent copper and 4 percent zinc, 150,000 tons averaging 1 percent copper and 1.9 percent zinc, and 480,000 tons averaging 0.36 percent copper and 4.6 percent zinc; probable ore of 100,000 tons averaging 1.5 percent copper and 1.6 percent zinc; and possible ore of 650,000 tons averaging 1.5 percent copper and 2.51 zinc.

A 10-current dipole spectral IP survey has been completed in the bottom of the inactive open-pit mine belonging to Kam-Kotia Mines Ltd. (figure 2.16). An electrode interval of 2 m was used for the spectral IP survey. A lens of massive sulphide ore is exposed in the bottom of the pit.

The pseudosection profiles (figure 2.16) show a very conductive, highly polarizable source. The time constants (τ_1) for the mixed sulphide source are larger than expected for massive, volcanogenic sulphide ore zones. They are generally smaller than for the massive pyrrhotite "dike" surveyed at Timmins, however, and appreciably smaller than for the massive graphite zone near Schreiber, Ontario. The spectral plots are typical of all those recorded.

For many dipole pairs, the (c_1) has a value less than c_1= 0.250. This low value for (c_1) indicates a broadly dispersed grain size population or two overlapping grain size populations. This could be the result of the overlapping spectral characteristics of the pyrrhotite mineralization (large grain size) and the base-metal sulphide mineralization (small grain size).

For instance, the spectral plot for D3, N3 can be re-processed (figure 2.17), fixing the parameters for the indicated coupling dispersion at the values already determined. Two dispersions have been used for the IP effect. The "fit" to the measured data is excellent, using two values for the time constant (500 sec. and 0.05 sec.). The two sets of parameters for the IP dispersion would be fairly typical for normal distributions of grain sizes for pyrrhotite and base metal sulphides.

FIGURE 2.16
Pseudosection with a massive sulfide ore with pyrhotite

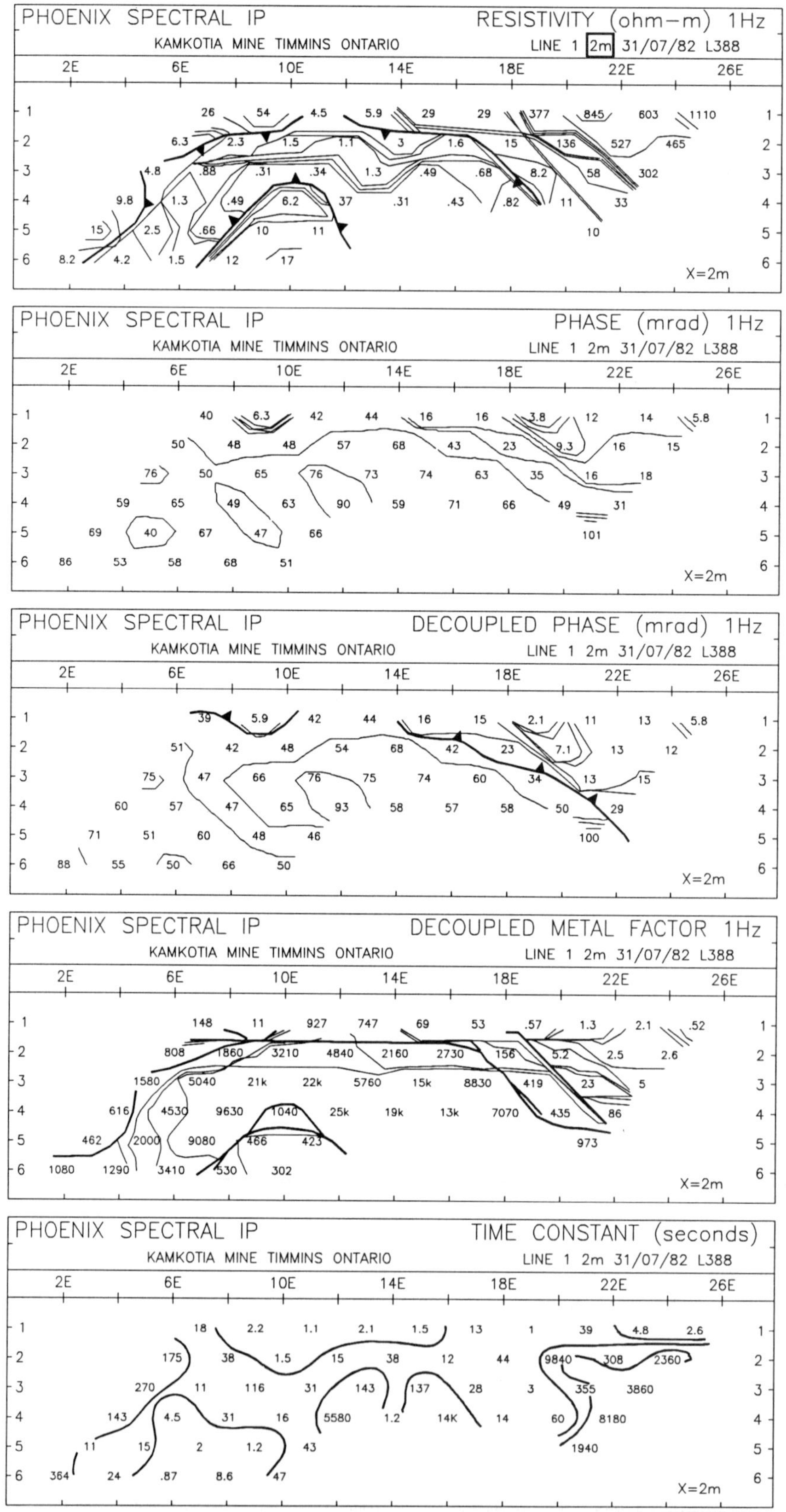

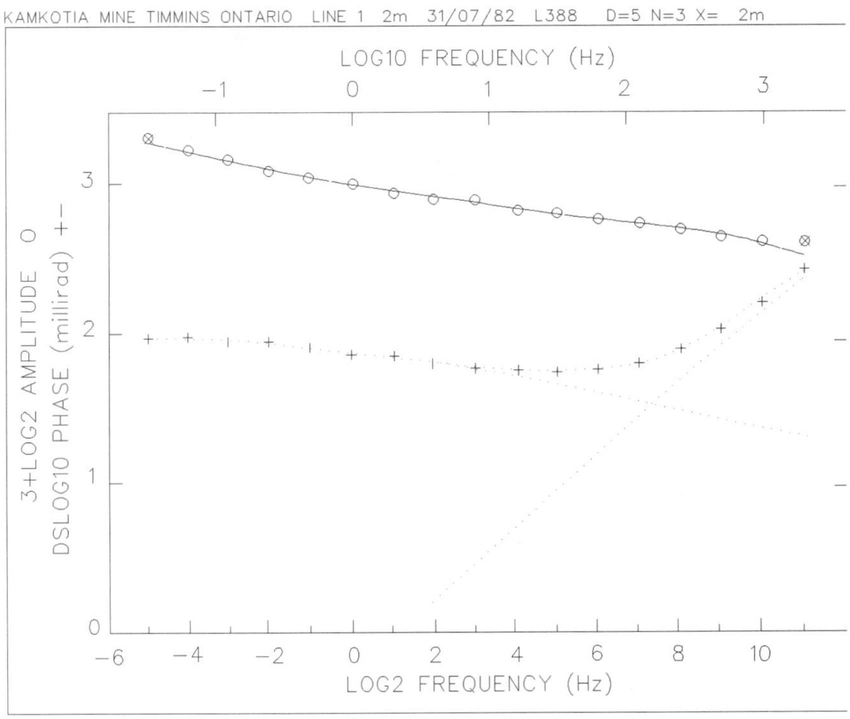

FIGURE 2.17
Spectral plots with M2 fixed

CRL: Number of dispersions=2
M2=1 fixed

Iter	Lambda	Rchsq	R0	M1	T1	C1	T2	C2
0	1.E−02	.00013	2.206	.657	1.4E+02	.232	1.7E−05	.825
1	1.E−02	.00013	2.209	.657	1.4E+02	.232	1.7E−05	.825

Pct Std Deviations	11.9	7.1	243.0	8.7	4.2	3.0
Correlation Matrix	1.000					
	.998	1.000				
	.999	.998	1.000			
	−.972	−.983	−.977	1.000		
	.174	.199	.173	−.235	1.000	
	.621	.658	.623	−.718	.730	1.000

Apparent resistivity measured at 1 Hz is 1.300

Apparent resistivity calculated from inductive coupling is .460

This observed variability of grain size, and therefore time constant, for sulphide ore zones containing pyrrhotite is apparently fairly typical.

UNDERGROUND SPECTRAL IP MEASUREMENTS AT SUDBURY, ONTARIO

An extensive program of spectral induced polarization measurements has been completed within the mines and mineralization at Sudbury, Ontario. The measurements were made with short electrode intervals (1.0 m and 2.0 m). The Geological Survey of Canada and the various mining companies operating in Sudbury were very helpful in this work.

A few thousand spectral IP measurements (for a particular dipole-dipole pair) were made. Several pseudosections were measured in the stopes and workings that were made available. In all but a few cases, one or two Cole-Cole expressions quite accurately described the results. This approximation quite accurately "fits" all of the data. Four parameters are all that are necessary.

The results shown on figure 2.18 show how well the Cole-Cole equation fits the data. The measurement was made in mineralization that the geologists described as disseminated and the magnitude of the measured IP effects in the same zone varies considerably; the apparent resistivities also vary considerably. The time constant determinations made in the area are much the same as shown. The value of C1 = 0.546 shown on figure 2.18 implies that we are detecting the effects from a single, unique grain size (see figure 2.6). As we have seen, the time constant usually measured from pyrrhotite mineralization is large. The measurements on figure 2.19 are perhaps more typical.

FIGURE 2.18
Spectrical plot of disseminated mineralization single grain size

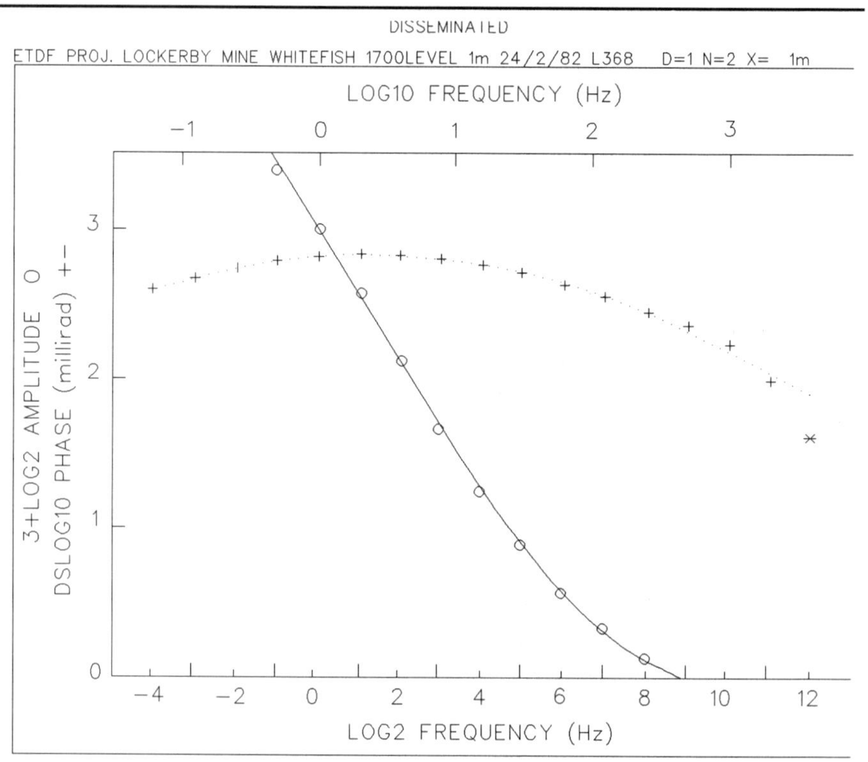

CRL: Number of dispersions=1

Iter	Lambda	Rchsq	R0	M1	T1	C1
0	1.E−02	.00056	4.690	.979	2.3E+00	.546
1	1.E+04	.00056	4.690	.979	2.3E+00	.546
Pct Std Deviations			5.7	.1	14.6	.9

Correlation Matrix

1.000			
.961	1.000		
.969	.950	1.000	
−.566	−.660	−.731	1.000

Apparent resistivity measured at 1 Hz is .210

Apparent resistivity calculated from inductive coupling is 0.000

MASSIVE

ETDF PROJ. LOCKERBY MINE WHITEFISH 1700LEVEL 1m 24/07/82 L368 D=9 N=4 X= 1m

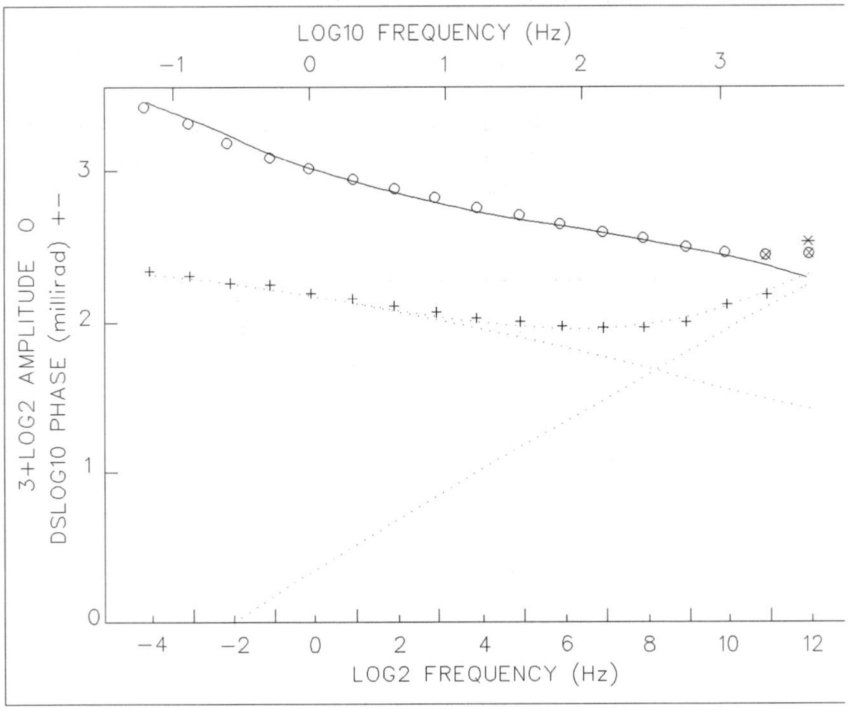

CRL: Number of dispersions=2

Iter	Lambda	Rchsq	R0	M1	T1	C1	T2	C2
0	1.E−02	.00040	7.931	.917	7.8E+03	.273	2.8E−06	.535
1	1.E−02	.00040	8.013	.918	8.2E+03	.273	2.8E−06	.535
Pct Std Deviations			77.9	5.9	2123.7	9.3	22.9	12.4

Correlation Matrix					
1.000					
.999	1.000				
.996	.999	1.000			
−.900	−.917	−.933	1.000		
.161	.189	.193	−.352	1.000	
.444	.479	.492	−.697	.853	1.000

Apparent resistivity measured at 1 Hz is 5.840

Apparent resistivity calculated from inductive coupling is .159

FIGURE 2.19
Spectral plot disseminated mineralization normal distribution of grain size

If more than one grain size (time constant) is present within the source (for instance, a normal distribution of grain sizes), the value for c will be less (see figure 2.2). The value of 0.256 shown in figure 2.20 is much more typical of that measured with short electrode intervals. In figure 2.20, the distortions from inductive coupling can be seen at the higher frequencies (these data are from the underground workings at the Fraser Mine).

FIGURE 2.20
Spectral plot normal distribution of grain size with EM coupling

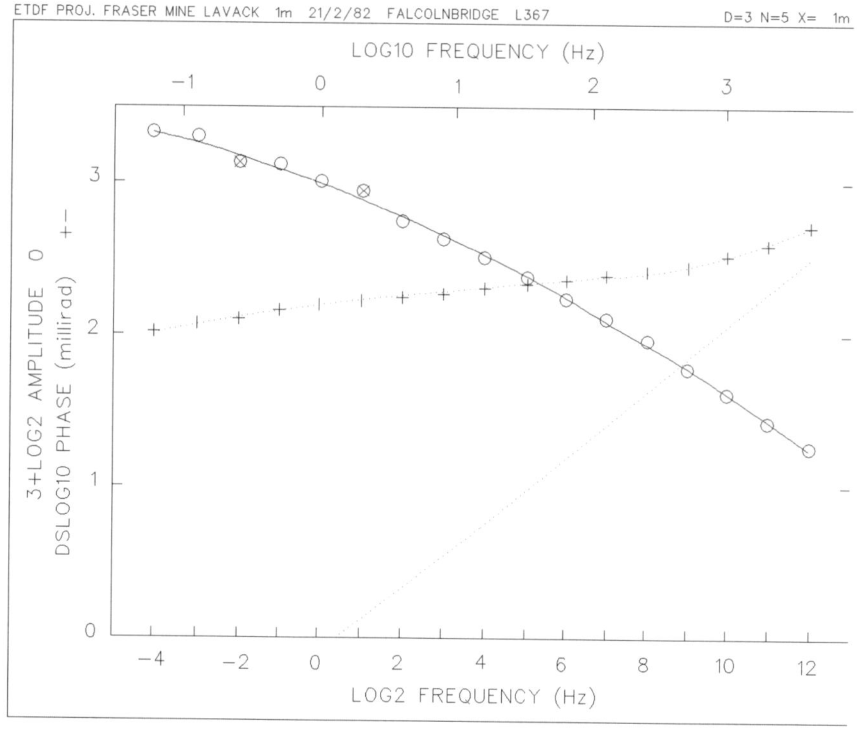

CRL: Number of dispersions= 2
M2=1 fixed

Iter	Lambda	Rchsq	R0	M1	T1	C1	T2	C2
0	1.E−02	.00021	1.703	.911	9.1E−02	.256	3.4E−07	.575
1	1.E+00	.00021	1.703	.911	9.1E−02	.256	3.4E−07	.575
Pct Std Deviations			3.0	4.2	17.6	7.3	35.5	14.1
Correlation Matrix			1.000					
			.853	1.000				
			.055	−.463	1.000			
			−.919	−.970	.275	1.000		
			.232	.539	−.633	−.378	1.000	
			.649	.908	−.636	−.797	.829	1.000

Apparent resistivity measured at 1 Hz is 6.610

Apparent resistivity calculated from inductive coupling is 2.762

FIGURE 2.21
Determination of grain size at a porphyry copper deposit in Nevada at the Kennedy Grid at Granite Mountain area in Pershing County base map

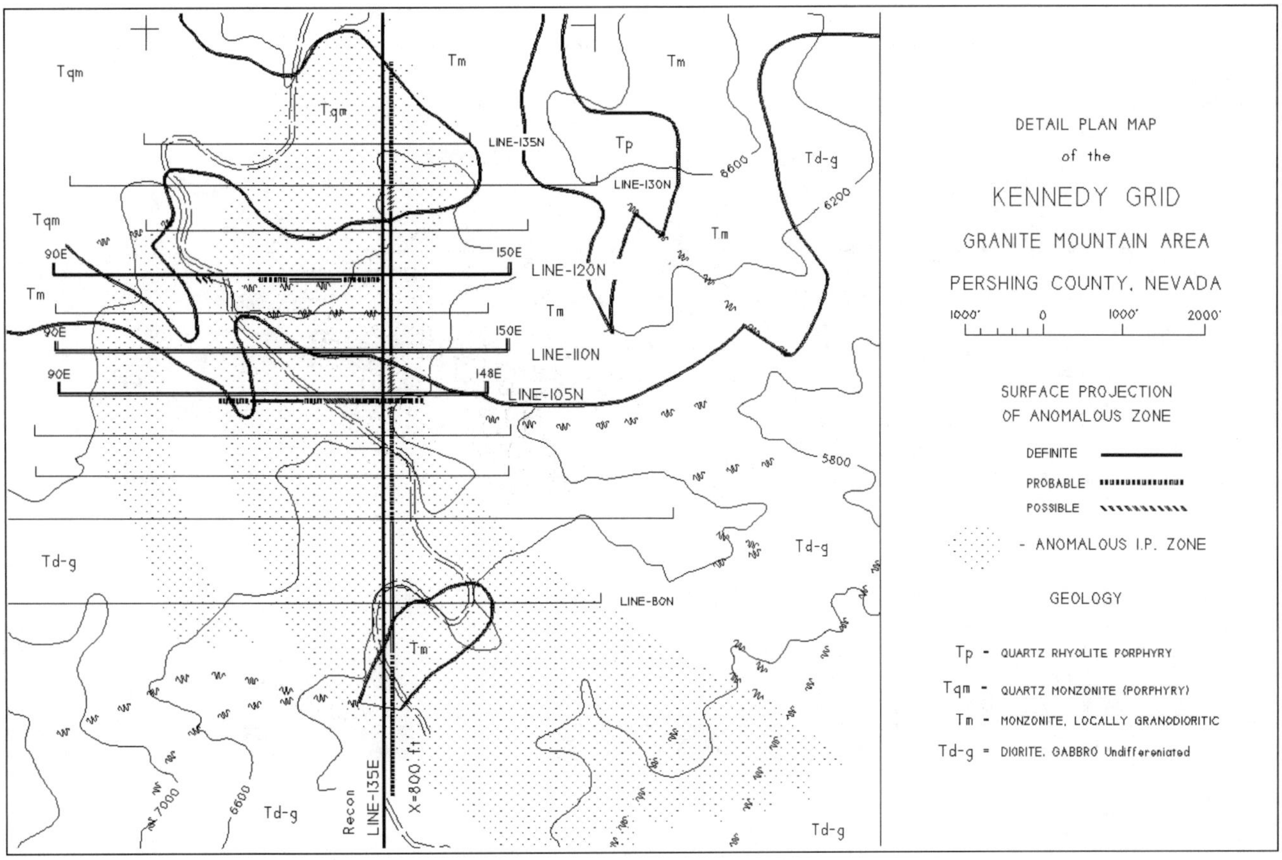

At the Kennedy Grid located in the Granite Mountain Area of Pershing County, Nevada, a large porphyry copper system was detected by an IP survey. At a later date, a long line was surveyed with Phase IP on line 110N with 300-ft electrode intervals. The frequencies used were 0.11, 0.33, 1.0, 3.0 and 9.0 Hz. The mini-spectra data are shown on figure 2.22. The line is actually three times as long as that shown; only the center one-third of the line of data is shown.

FIGURE 2.22
Mini-spectral data

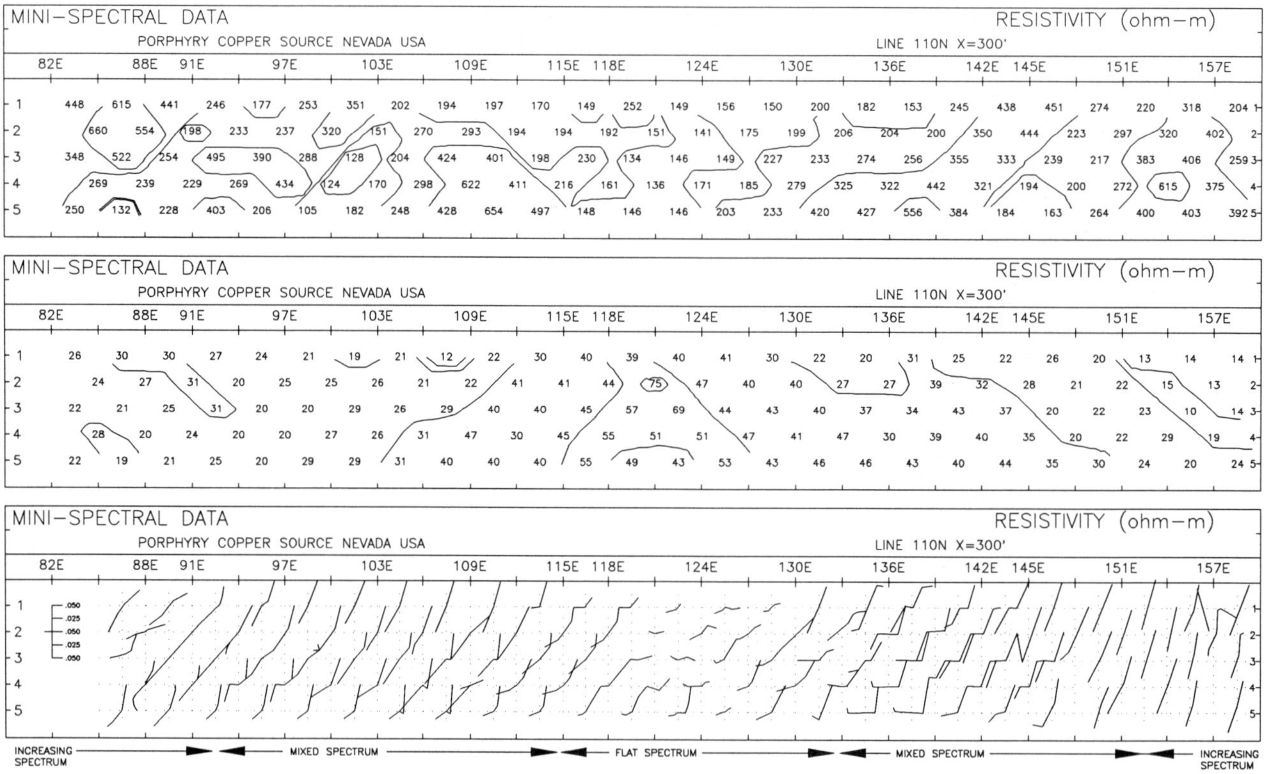

For most of the line surveyed, the mini-spectra were seen to be as "increasing spectra." This was regardless of the magnitude of the IP effects measured. In the center of the line, however, the shape of the mini-spectra changed.

They were characterized as being "flat spectra" in that area. It should be noted that there is no resistivity anomaly on any part of the measured data.

At a later date, the measurements on line 110N were repeated with a spectral IP system; the complete spectrum of frequencies were used. As shown on the data plot, the measured values were essentially the same. The values of the time constant varied considerably (see figure 2.23). In the interval from 115E to 121E, the time constants are larger than on the remainder of the line. This is exactly the same portion of the line at which the "flat spectra" were measured. Later drilling intersected veinlet mineralization (containing both molybdenum and gold). Disseminated pyrite was intersected to one side, while disseminated metallic magnetite lay on the other side.

FIGURE 2.23
S. I. P., Pseudosection with time constraints plotted

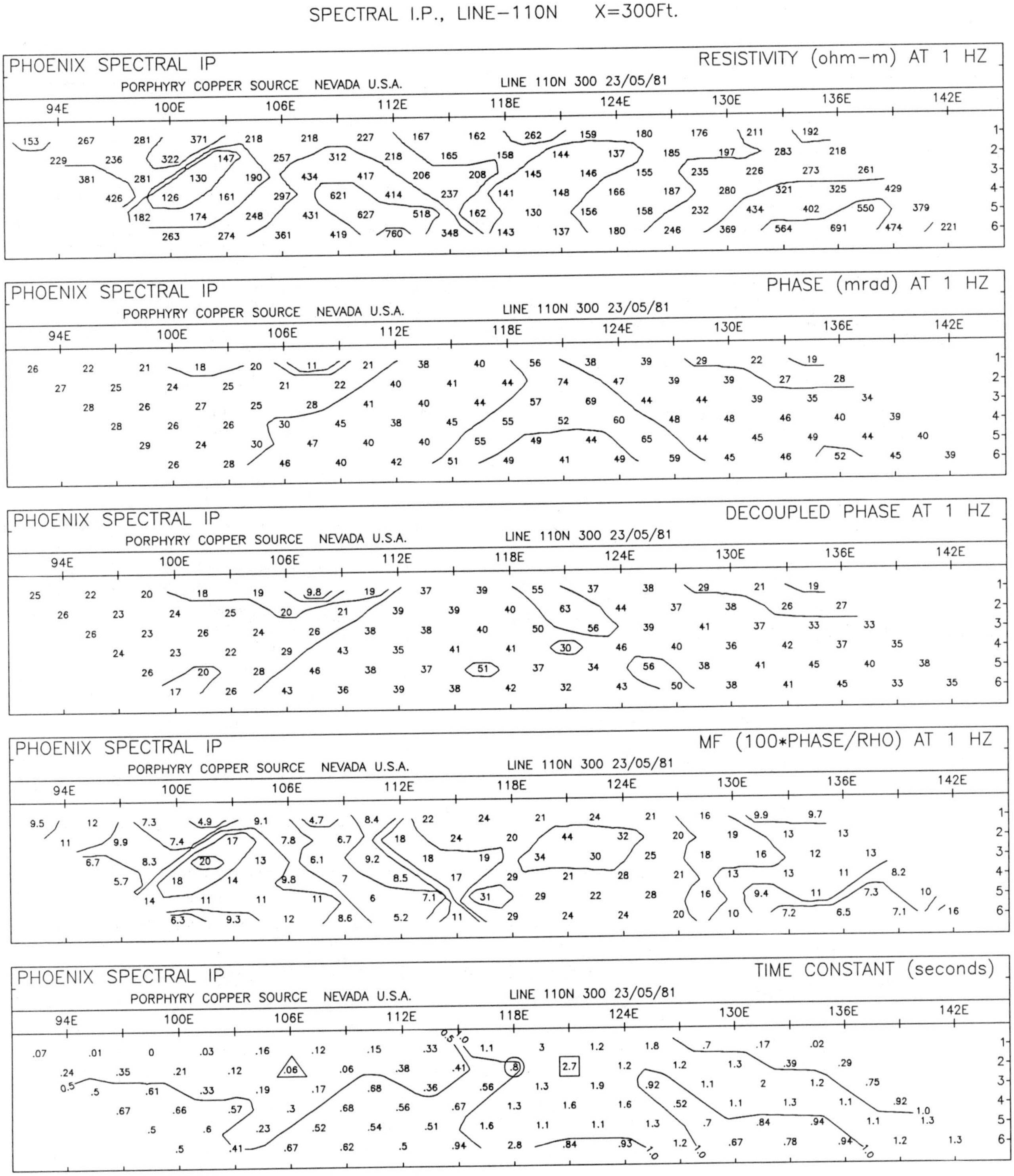

Three examples of the spectral IP plots are included on figures 2.24, 2.25 and 2.26. The variations of the time constant and the shape of the mini-spectrum curves can be seen on these plots. The location of the data plots is identified by symbols on the pseudosection.

FIGURE 2.24
Spectral plot from porphyry copper system

PORPHYRY COPPER SOURCE NEVADA U.S.A. LINE 110N 300' 23/05/81 D=6 N=2 X= 91m

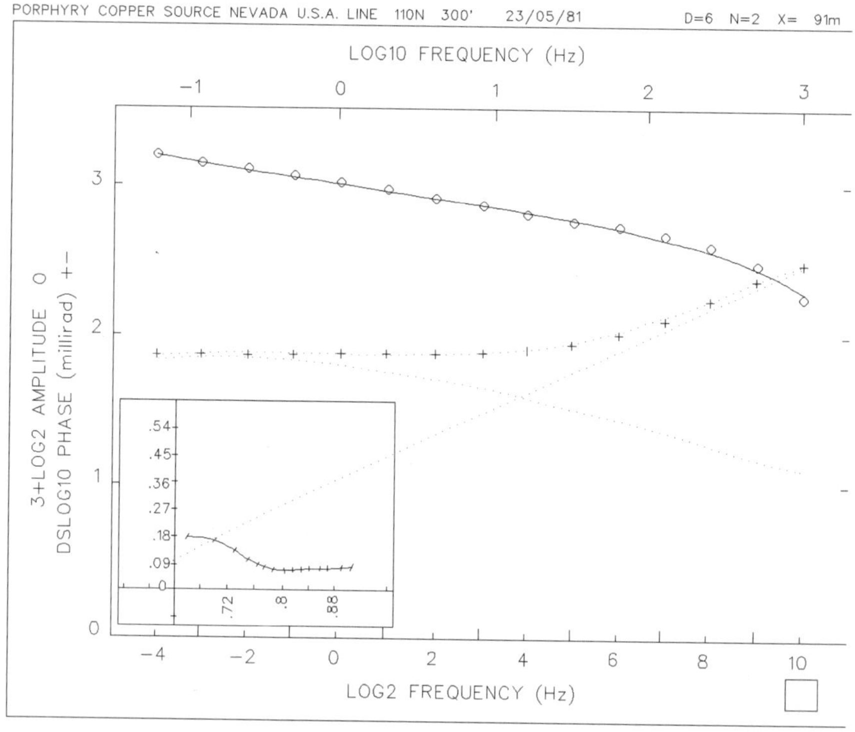

CRL: Number of dispersions=2
M2=1 fixed

T=2.7 sec.

Iter	Lambda	Rchsq	R0	M1	T1	C1	T2	C2
0	1.E−02	.02679	1.400	.400	1.0E+00	.250	1.0E−05	.600
1	1.E−02	.00066	1.352	.389	1.2E+00	.361	1.0E−05	.493
2	1.E−03	.00026	1.407	.400	2.0E+00	.345	1.0E−05	.469
3	1.E−04	.00025	1.440	.416	2.6E+00	.331	9.7E−06	.467
4	1.E−05	.00025	1.441	.414	2.7E+00	.333	9.7E−06	.465

	R0	M1	T1	C1	T2	C2
Pct Std Deviations	5.9	14.0	78.5	16.8	28.1	4.9
Correlation Matrix	1.000					
	.968	1.000				
	.958	.863	1.000			
	−.959	−.993	−.858	1.000		
	−.953	−.925	−.912	.930	1.000	
	.378	.573	.146	−.548	−.250	1.000

Apparent resistivity measured at 1 Hz is 143.7

Apparent resistivity calculated from inductive coupling is 54.40

FIGURE 2.25
Spectral plot from porphyry system

PORPHYRY COPPER SOURCE NEVADA U.S.A. LINE 110N 300 23/05/81 D=7 N=2 X= 91m

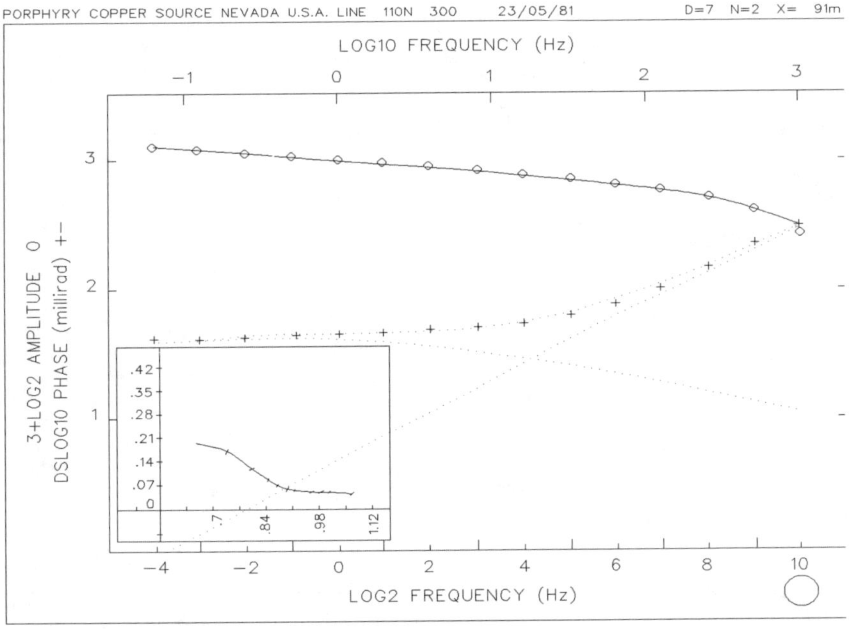

CRL: Number of dispersions=2 T=0.8 sec.
M2=1 fixed

Iter	Lambda	Rchsq	R0	M1	T1	C1	T2	C2
0	1.E−02	.11220	1.300	.200	1.0E+00	.250	6.0E−05	1.000
1	1.E−02	.00217	1.195	.324	8.0E−01	.285	2.8E−05	.692
2	1.E−03	.00037	1.193	.274	7.0E−01	.329	2.1E−05	.595
3	1.E−04	.00035	1.210	.289	7.9E−01	.304	2.0E−05	.599
4	1.E−05	.00035	1.205	.282	8.0E−01	.313	2.1E−05	.595

	R0	M1	T1	C1	T2	C2
Pct Std Deviations	3.4	17.8	49.4	20.8	14.3	4.6
Correlation Matrix	1.000					
	.954	1.000				
	.715	.495	1.000			
	−.958	−.993	−.529	1.000		
	−.859	−.813	−.632	.837	1.000	
	.444	.650	−.171	−.596	−.147	1.000

Apparent resistivity measured at 1 Hz is 198.1

Apparent resistivity calculated from inductive coupling is 102.6

FIGURE 2.26
Spectral plot from porphyry system

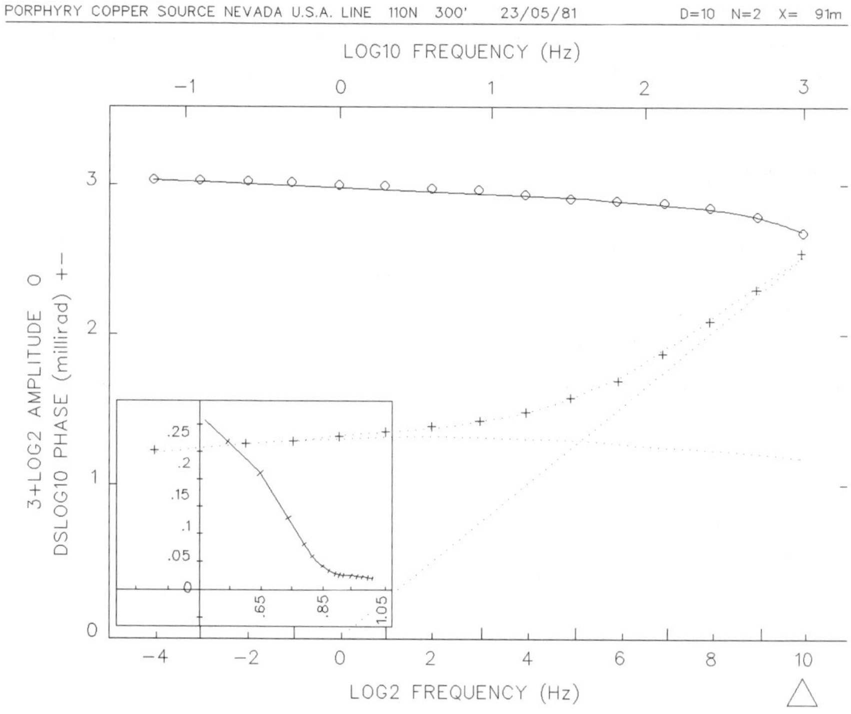

CRL: Number of dispersions=2
M2=1 fixed

T=0.057 sec.

Iter	Lambda	Rchsq	R0	M1	T1	C1	T2	C2
0	1.E-02	.00018	1.094	.202	9.5E-02	.233	3.9E-05	.824
1	1.E-02	.00005	1.101	.200	9.2E-02	.238	3.9E-05	.830
2	1.E-03	.00005	1.102	.206	7.7E-02	.232	3.9E-05	.836
3	1.E-04	.00005	1.106	.219	6.0E-02	.219	3.9E-05	.842
4	1.E-05	.00005	1.106	.219	5.7E-02	.219	3.9E-05	.842

Pct Std Deviations	.8	10.3	45.6	10.5	2.7	1.2
Correlation Matrix	1.000					
	.936	1.000				
	-.716	-.904	1.000			
	-.961	-.992	.855	1.000		
	-.681	-.608	.434	.646	1.000	
	.417	.631	-.764	-.558	.139	1.000

Apparent resistivity measured at 1 Hz is 312.5

Apparent resistivity calculated from inductive coupling is 240.8

REMOVING INDUCTIVE COUPLING DISTORTION USING SPECTRAL IP

The Elura massive sulfide deposit — located near Cobar, New South Wales — is a vertical, pipe-shaped source within a sedimentary host. The economic metals are lead, zinc and silver. Enough pyrrhotite occurs with the ore for the deposit to have been originally discovered by a magnetic survey. It can also be detected by a well-controlled gravity survey.

Line 50,800N was used for the spectral IP test survey. The measurements over a broad frequency range (at Elura .125 hz to 128 hz) make it possible to, among other things, remove inductive coupling effects from the measurements made for each dipole pair. On this line, the ore zone is centered just west of station 2500E; this would position the top of the "pipe-like" source between electrode No. 3 and electrode No. 4 on the spectral IP survey. The depth to the top of the metallic mineralization is approximately 100 m, resulting from complete weathering of the surface layer. The weathered surface layer has a resistivity value of about 10 ohm-m. At a depth of 200 m, the near-vertical, pipe-like zone of mineralization has a width along the line of approximately 80 m; it has a total horizontal extent perpendicular to the line of about 200 m.

The Field Data

The results shown on figure 2.27 are from Line 50,800N at Elura, using X = 100 m and n = 1, 2, 3, 4, 5, 6. The computer printout has east to the left, with electrode No. 4 at station 2500E on Line 50,800E. We have shown the apparent resistivity data and the measured phase shift pseudosection for 3.0 Hz, 1.0 Hz, 0.33 Hz and 0.11 Hz.

FIGURE 2.27
Pseudosection with inductive coupling

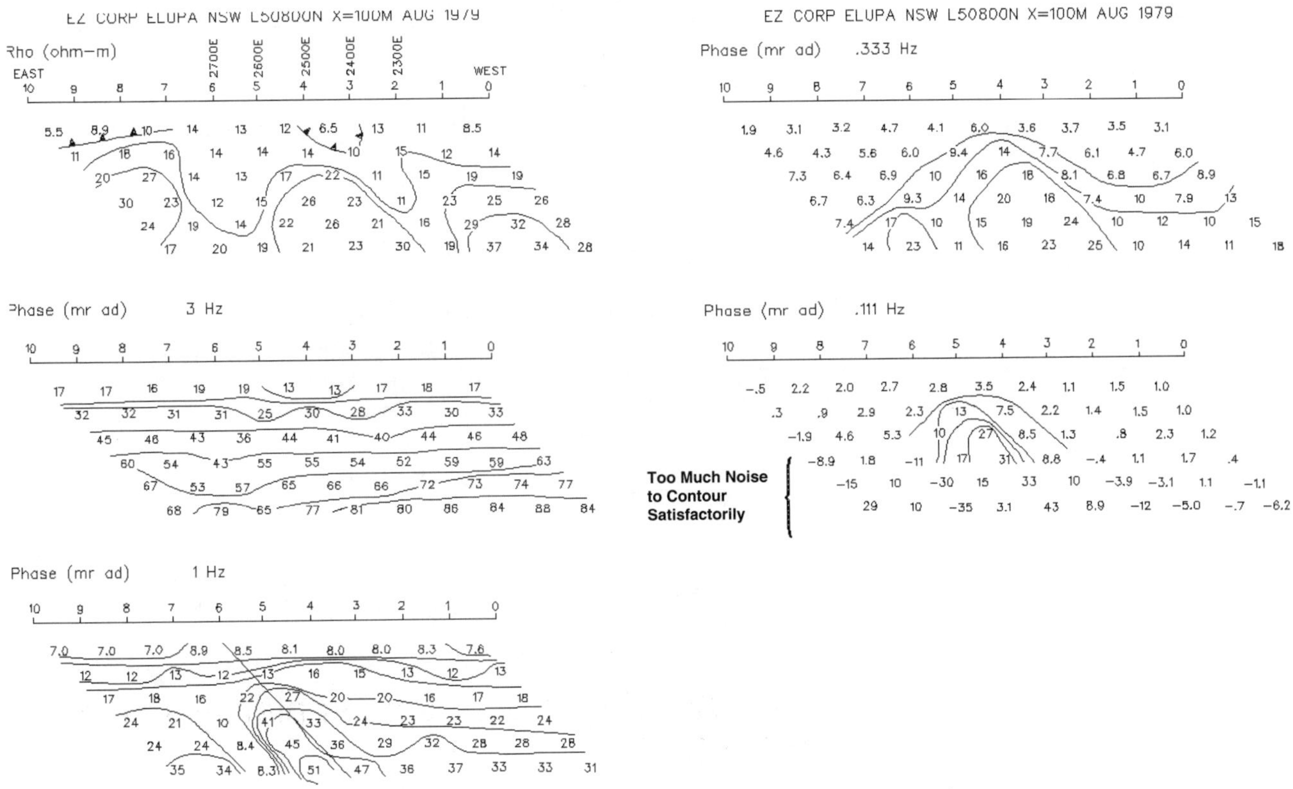

The apparent resistivity measurements show a typical pattern for variable deep weathering. The lowest value of apparent resistivity does not correlate with the known ore zone which is centered at approximately electrode No. 4. The results from a one-dimentional (1-D) computer inversion of the resistivity for a one-layer case gives an inverted resistivity for the first layer of 10.1 ohm m ± 18 percent with a thickness of 102 m ± 26 percent; with the homogenious half space being 204 ohm-meters ± 318 percent. Notice the error in the half space (bottom layer) is larger than the other parameters.

The pseudosections for the measured phase shifts are of considerable interest. For the higher frequencies (particularly at 3.0 Hz), there is a definite increase in phase shift with increased electrode separation. This is the expected result for inductive coupling from colinear dipoles on a uniform earth or a layered earth.

For the lowest frequency (0.11 Hz), it can be seen that only the measurements for n = 1, 2, 3 and possibly n = 4 can be considered reliable. For the larger values of (n), we will see that the telluric noise (at about 0.05 Hz) results in high standard deviations for the phase shift measurements.

The most useful measured data appears to be that at 0.33 Hz. The inductive coupling effects are perhaps distorting the data somewhat. There is clearly an anomaly at depth, however, in the interval from electrode No. 3 to electrode No. 5.

Inversion of Spectral IP Data for Each Electrode Set

For each position of the four electrodes, the computer will print the results of the frequency scan. As outlined earlier, computer inversion can be used to separate the inductive effects from the phase shifts resulting from the IP effects.

The basic assumption is that both the IP transfer function and the inductive coupling function can be approximated by a Cole-Cole dispersion having the form:

$$SZ = R_o \left[1 - m_1\left(1 - \frac{1}{1+(i\omega\tau_1)^{c_1}}\right)\right]\left[1 - m_2\left(1 - \frac{1}{1+(i\omega\tau_2)^{c_2}}\right)\right]$$

For a nearly uniform earth, the value (m_2) for inductive coupling is expected to be near 1.0. The value for (c_2) would be in the interval 0.80 to 1.00. The value of (c_1) would be in the range 0.15 to 0.35.

The interpretation consists of a computer inversion using non-linear regression to obtain the "best-fit" parameters for the two dispersions is shown in figure 2.28 (in this case, after four iterations). The mean predicted value for each parameter is shown, along with the standard deviation for that parameter, in percentages. The data points, the IP curve, the inductive coupling curve and the combined curve are all shown.

Figure 2.29 shows non-anomalous data inversion examples to one side of the anomaly. It can be seen that in both cases the inductive coupling curve can be used to describe about all of the effect measured. The (m_1) values are not well determined, but the expected value is small. The examples shown in figure 2.30 is from a dipole pairs within the anomalous pattern. In this example, the data from the noisy, low-frequency measurements (0.11 Hz) were weighted to zero for the inversion. The spectral measurements deviate substantially from the inductive coupling curve. A considerable magnitude for the chargeability (m) parameter is necessary to fit the data, and the standard deviations are small. These measurements for n = 4 are quite accurate and reliable.

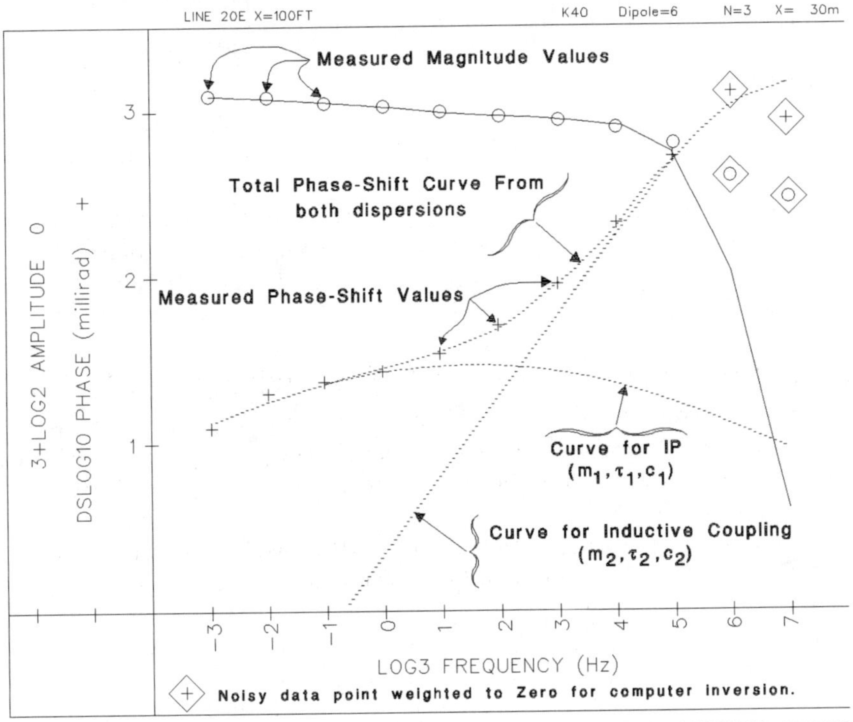

				IP EFFECT			INDUCTIVE COUPLING EFFECT		
Iter	Lambda	Rchsq	R0	M1	T1	C1	M2	T2	C2
0	1.E−02	.03408	1.089	.221	4.0E−02	.329	1.000	1.1E−04	1.000
1	1.E−02	.00168	1.063	.196	2.5E−02	.373	1.000	3.6E−04	1.000
2	1.E−03	.00068	1.072	.186	3.1E−02	.355	1.000	3.3E−04	1.000
3	1.E−04	.00067	1.071	.174	4.3E−02	.373	1.000	3.4E−04	1.000
4	1.E−05	.00066	1.071	.172	4.4E−02	.376	1.000	3.4E−04	1.000
Pct Std Deviations			.8	16.4	84.2	13.5	0.0	5.5	0.0
Correlation Matrix			1.0000						
			.5173	1.0000					
			−.4289	−.9715	1.0000				
			−.5302	−.9438	.8971	1.0000			
			0.0000	0.0000	0.0000	0.0000	0.0000		
			−.2093	−.6596	.6657	.5348	0.0000	1.0000	
			0.0000	0.0000	0.0000	0.0000	0.0000	0.0000	0.0000

Apparent Resistivity Measured at 1 Hz is 78

Apparent Resistivity Calculated from Inductive Coupling is 10

FIGURE 2.28
Inversion on spectral plots removal at coupling for anomalous reading

FIGURE 2.29
Inversion on spectral plots removal of coupling from non-anomolous reading

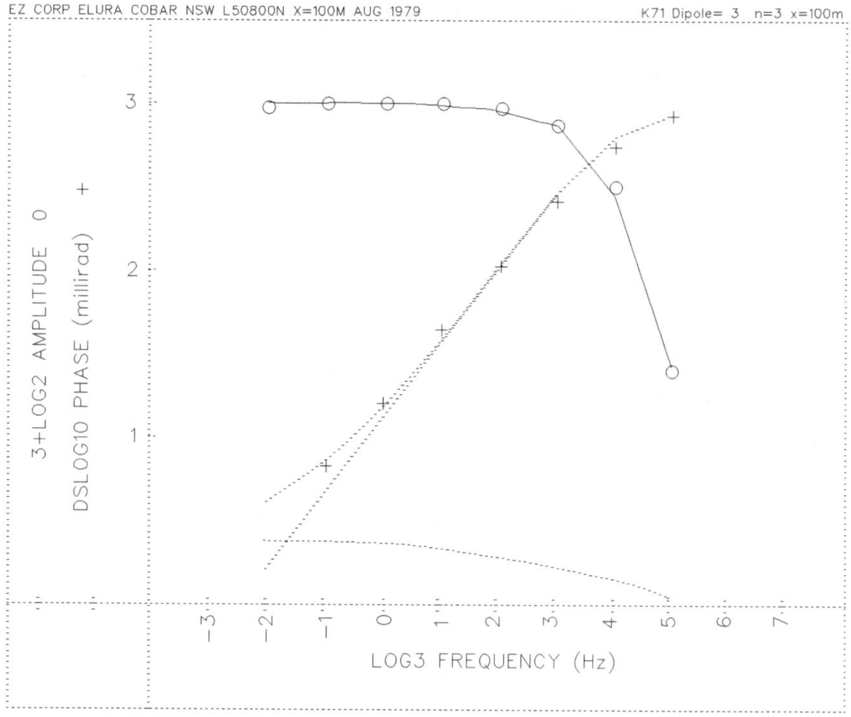

CRG: Neg= 0

T1=1 C1=.25 fixed

Iter	Lambda	Rchsq	R0	M1	M2	T2	C2
0	1.E-02	.00298	1.015	.018	.943	1.9E-03	.950
1	1.E-02	.00162	1.006	.020	.860	2.0E-03	.948
2	1.E-03	.00150	1.006	.023	.870	2.0E-03	.954
3	1.E-04	.00150	1.006	.024	.866	2.0E-03	.956
Pct Std Deviations			1.2	47.2	4.0	5.1	2.0

Correlation Matrix

1.000				
.347	1.000			
−.297	−.518	1.000		
.383	.087	−.728	1.000	
.293	.724	−.599	.355	1.000

Apparent Resistivity Measured at 1 Hz is 15.40

Apparent Resistivity Calculated from Inductive Coupling is 18.65

Interpreted Results

When the computer inversion is completed for each data point on the measured pseudosection, the computer can print out several types of interpreted data plots.

The results shown could be the phase shifts that would be measured if the effects predicted for inductive coupling at each frequency are subtracted from the measured values. These "decoupled" phase shifts are the interpreted magnitudes of the IP effects that would be measured in the absence of the inductive coupling phenomenon.

The final computer output from the computer interpretation of the spectral IP pseudosections is shown on figure 2.31. The apparent resistivities are those shown previously. The second pseudosection shows the interpreted value of (m_1) for each pair of dipoles. The anomaly centered below 2500E (electrode No. 4) is quite definite. There may be other IP effects at greater depth to the east. The complex pattern could also be partially explained, however, if the measurements for dipole No. 9 are assumed to be in error.

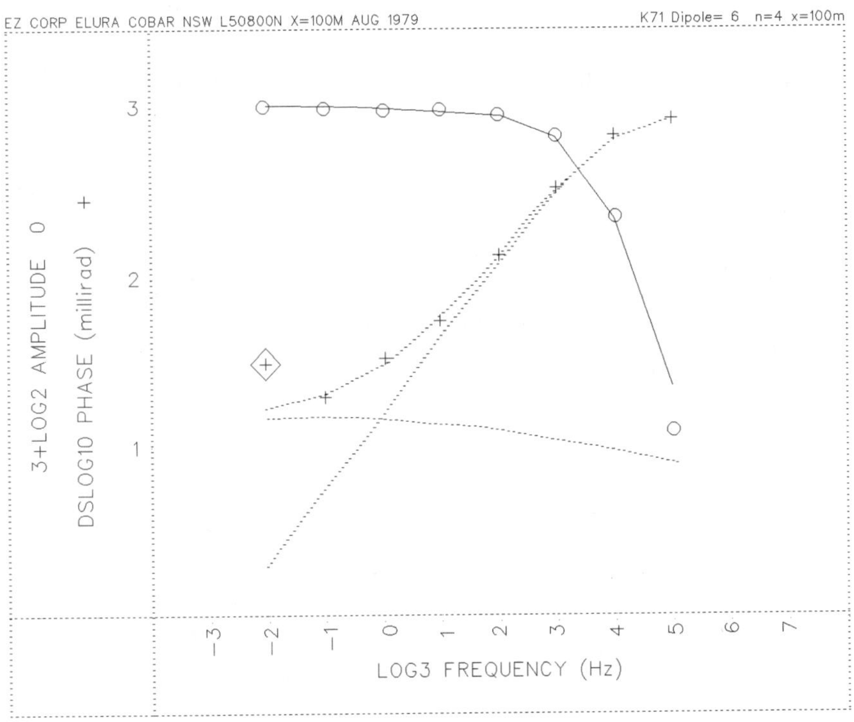

FIGURE 2.30
Spectral plot removal of coupling from noisy anomalous readings

CRG: Neg= 0

T1=1 C1=.25 fixed

Iter	Lambda	Rchsq	R0	M1	M2	T2	C2
0	1.E-02	.00132	1.106	.156	.762	2.5E-03	1.000
1	1.E-02	.00042	1.103	.143	.773	2.3E-03	.945
2	1.E-03	.00040	1.101	.139	.782	2.2E-03	.936
3	1.E-04	.00040	1.100	.138	.782	2.2E-03	.936
Pct Std Deviations			.8	7.7	2.6	3.8	2.1
Correlation Matrix			1.000				
			.679	1.000			
			-.523	-.738	1.000		
			.328	.305	-.647	1.000	
			.406	.706	-.735	.678	1.000

Apparent Resistivity Measured at 1 Hz is 25.65

Apparent Resistivity Calculated from Inductive Coupling is 28.00

The bottom pseudosection on figure 2.31 is plotted to give some feeling for the reliability of the interpretation. The value plotted is the standard deviation in the value for m_1 that has been interpreted and plotted above. In this case, the standard deviation is in the units of percentages.

CONCLUSIONS

The spectral IP measurements on Line 50,800N at Elura and the computer interpretation of the results reveal a definite anomaly, at depth, centered beneath station 2500E. This correlates quite well with the position of the known "pipe-like" ore zone. The depth of weathering of about 100 m is predicted by the resistivity results and would account for the depth to the top of the source indicated by the measurements with X = 100 m.

FIGURE 2.31
Pseudosection with m_1 plots

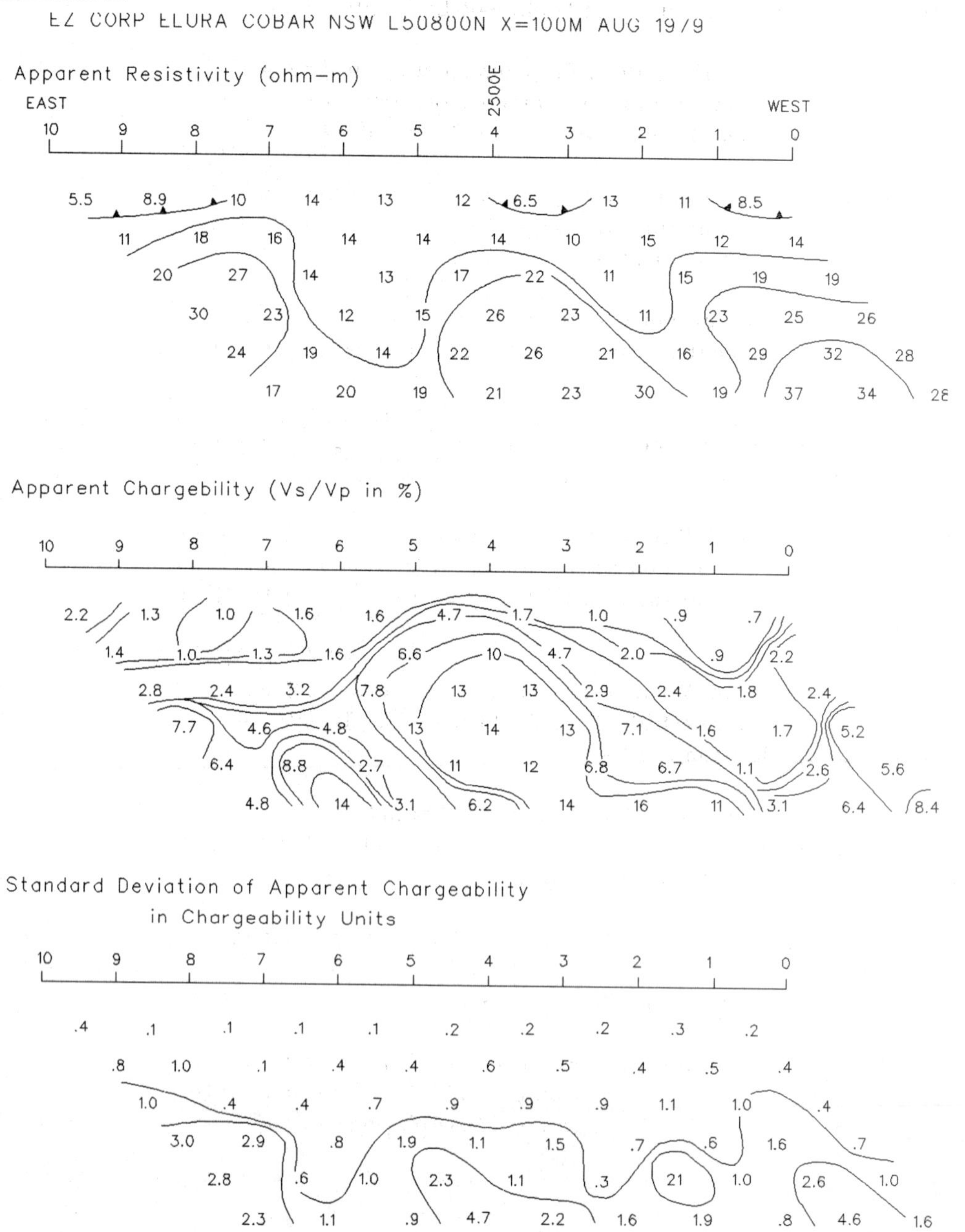

ACKNOWLEGDEMENT

This chapter could not have been prepared for presentation without the help and permission of the following companies:

Bachelor Lake Gold Mines Ltd.
Barrick Resources Corp.
B. P.-Selco Explorations Canada
Homestake Mineral Development JCo.
Kellar Lake Gold Mines Ltd.
Orequest Consultants Ltd.
Noranda Exploration Co. Ltd.
Teck Corporation

REFERENCES

Dewitt, G.W., 1978, Parametric studies of induced polarization spectra: M.S. Thesis, University of Utah, Salt Lake City, Utah.

Dey, A., and Morrison, H.F., 1973, Electromagnetic coupling in frequency and time-domain induced polarization surveys over a multilayered earth: Geophysics, Vol. 38, No. 2., pp. 380-405.

Grissemann, C., 1971, Examination of the frequency-dependent conductivity of ore-containing rock on artifical models. Scien. Rep. No. 21, Electr. Lab., Univ. of Innsbruck, Austria.

Hallof, P.G., 1965, The proper choice of frequencies for induced polarization measurements: Presented at the S.E.G. Annual Meeting, Dallas, Texas.

1974, the IP phase measurement and inductive coupling: Geophysics, Vpl. 39, pp. 650-665.

Hallof, P.G., Cartwright, P.A., and Pelton, W.H., 1979, The use of the Phoenix IPV-2 phase IP receiver for discrimination between sulphides and graphite: Presented at the S.E.G. Annual Meeting, New Orleans, Louisiana.

Hallof, P.G., and Pelton, W.H., 1980, The removal of Inductive Coupling Effects from Spectral IP Data: Presented at the S.E.G. Annual Meeting, Houston, Texas.

Hallof, P.G., and Klein, J.D., 1982, Electrical Parameters of Volcanogenic Mineral Deposits in Ontario: Presented at the S.E.G. 52nd Annual International Meeting, Dallas, Texas.

Hallof, P.G., and Klein, J.D., 1982, Electrical Parameters of Nickel-Copper Ores at Sudbury, Ontario: Presented at the S.E.G. 52nd Annual International Meeting, Dallas, Texas.

Hallof, P.G., and Yamashita, M., 1990, The Use of the IP to Locate Gold-bearing Sulphide Mineralization, pp. 227-279, Induced Polarization, published by S.E.G., Investigations in Geophysics, No. 4.

Hallof, P.G., 1983, "An Introduction to the Use of the Spectral Induced Polarization Method": Privately printed, Phoenix Geophysics Ltd.

Hallof, P.G., 1982, "Chapter 2: Grounded Electrical Methods in Geophysical Exploration": Practical Geophysics, pp. 39-151, compiled by Northwest Mining Association.

Hallof, P.G., 1990, "Reconnaissance and Detailed Geophysical Results, Granite Mountain Area, Pershing County, Nevada": Induced Polarization, pp. 325-354, published by S.E.G., Vol. 6 of Investigation in Geophysics.

Hohmann, G.W., 1973, Electromagnetic coupling between grounded wires at the surface of a two-layer earth: Geophysics Vol. 38, pp. 894.

1975, Three-dimensional induced polarization and electromagnetic coupling: Geophysics, Vol. 40, pp. 309-324.

1977, Numerical IP Modelling: Induced Polarization for Exploration Geologists and Geophysicists, The University of Arizona, March 14-16, 1977.

Paterson, N.R. and Hallof, P.G., 1991, "Geophysics Exploration for Gold," pp. 360-398, chapter 12, Gold Metalology and Exploration, edited by R. P. Foster, University of Southampton

Pelton, W.H., Ward, S.H., Hallof, P.G., Sill, W.R., and Nelson, P.H., 1978, Mineral discrimination and removal of inductive coupling with multi-frequency IP: Geophysics, Vol. 43, pp. 598-609.

Wong, J., 1979, An electrical model of the induced polarization phenomenon in disseminated sulphide ores: Geophysics, Vol. 44, pp. 1245-1265.

CHAPTER 3
GRAVITY

THE LAND GRAVITY EXPLORATION METHOD

Author's note: *The reader who is already familiar with the gravity method or who is only interested in applications may jump to the case history section.*

"Gravitation, n. The tendency of all bodies to approach one another with a strength proportioned to the quantity of matter they contain - the quantity of matter they contain being ascertained by the strength of their tendency to approach one another."

From The Devil's Dictionary by Ambrose Bierce.

Robert E. West, PhD, PE Geophysicist

Robert E. West is a geophysical consultant to the mining industry. Dr. West has more than 25 years experience in geophysical exploration for minerals, petroleum and groundwater. He has used gravity, magnetic, induced polarization and resistivity methods extensively in North and South America as a consultant and with EXXON Co. USA and Mining Geophysical Surveys, Inc. His first love is gravity and magnetic data interpretation. Dr. West has received a BS degree in physics from the University of Idaho and MS and Ph.D. degrees in geophysics from the University of Arizona.

The gravity method detects the gravitational field of lateral density changes in rocks. For the method to be useful, a density change of exploration interest must occur in the rocks. For example, the density change between a massive sulfide orebody and its host rock may cause a gravity anomaly that can lead to the detection of the sulfides (figure 3.1).

FIGURE 3.1
Surface gravity anomaly over a massive sulfide deposit

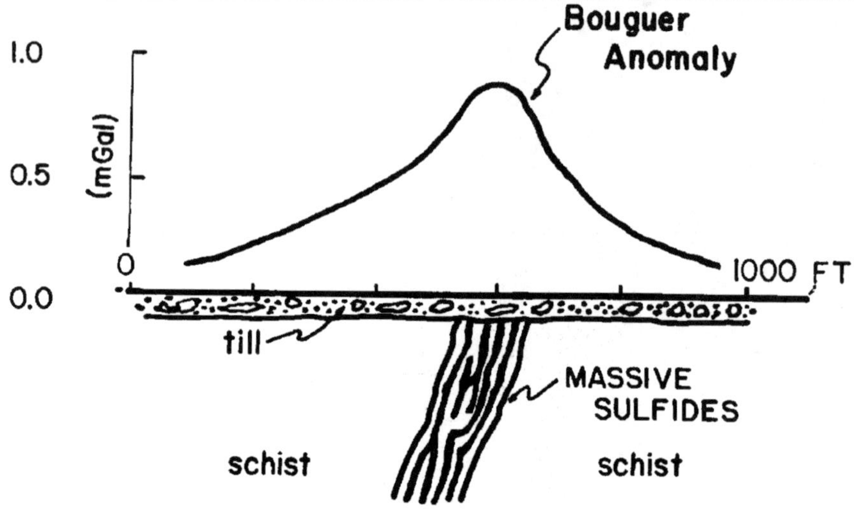

FIGURE 3.2
Surface gravity anomaly over a salt dome

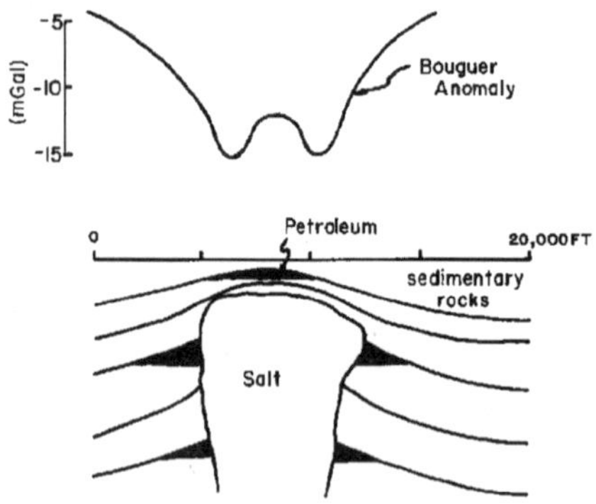

The density change between a salt dome and the sedimentary rocks that it has penetrated or uplifted can lead to the detection of the salt. This may, in turn, lead to the discovery of petroleum located in structural traps in the sedimentary rocks adjacent to the salt dome (figure 3.2).

The gravity method is adaptable to the study of large or small areas and can be used for reconnaissance or detailed exploration problems. Gravity surveys on land are relatively easy to carry out. Costs of land gravity surveys are, in general, quite reasonable. If accurate station elevations are required, however, the cost of a gravity survey can double. Gravity measurements in rugged terrain are complicated by terrain effects which make data reduction and interpretation more difficult.

Interpretation of gravity data is complicated by ambiguity. The gravity effect of many different mass distributions can match a single gravity anomaly. Insight about the geology leads the interpreter to a geologically reasonable mass distribution. This is the key to interpretation of gravity data. As a better understanding of the geology of the survey area is gained, interpretation of gravity data usually improves and becomes more valuable.

Exploration gravity surveys can be conducted on land, on the ocean floor, from ships, in boreholes and from helicopters. The discussions that follow will be limited to gravity anomalies on the continents.

THE BASICS

An introduction to the theory of gravitational fields, measurement of the earth's gravity, corrections to gravity measurements, gravity anomalies, planning a gravity survey and interpretation are covered in this chapter. The geophysical texts listed in the reference section provide in-depth coverage of these topics.

Gravitational Fields

Potential theory derives equations for the exact calculation of the gravity field of simple and complex mass distributions. The results are used in gravity data reduction and interpretation. Other topics in potential theory that are used in exploration not covered in this chapter include filtering of the gravitational field (e.g., upward and downward continuation and derivatives) and calculation of the anomalous mass.

Theory

The gravitational field **g** (symbols in the text in boldface indicate a vector, while an arrow over the symbol indicates a vector in the figures and equations; a list of symbols is included at the end of this chapter) at a point in space is defined by the gravitational force per unit mass exerted on small "test mass" m′ placed at the point (figure 3.3).

$$\vec{g} = \frac{\vec{F}}{m'} \tag{3.1}$$

FIGURE 3.3
The gravitational field

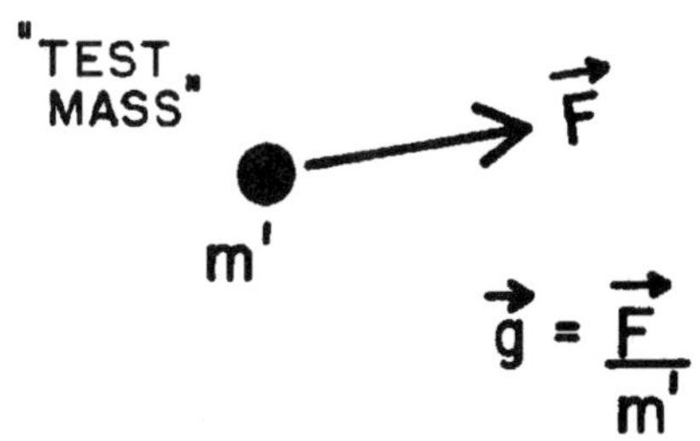

The gravitational force exerted on m′ by another point mass m is given by Newton's law of universal gravitation (figure 3.4).

$$\vec{F} = \frac{G\,m\,m'}{r^2}\frac{\vec{r}}{r} \tag{3.2}$$

G is the universal gravitation constant 6.670 X 10^{-8} dyne cm^2/gm^2. The mass of m and m′ is in grams, r is the distance between m and m′ in centimeters, and **r**/r is a unit vector along the line between m′ and m and directed from m′ to m (i.e., the force on m′ is directed toward m). The magnitude of *F* is

$$F = \frac{G\,m\,m'}{r^2} \tag{3.3}$$

The gravitational field resulting from the point mass m is

$$\vec{g} = \frac{\vec{F}}{m'} \tag{3.4}$$

$$\vec{g} = \frac{G\,m}{r^2}\frac{\vec{r}}{r} \tag{3.5}$$

FIGURE 3.4
The gravitational force and field due to point mass m

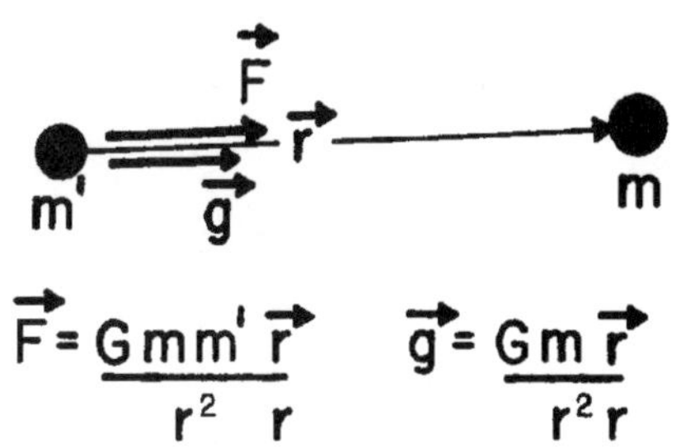

The units of the gravitational field are force per unit mass or dynes per gram. A dyne per gram, which is equivalent to a centimeter per square second, is called a Gal after Galileo. The common unit in gravity surveys is a milligal, or 1/1000 of a Gal. Another unit that is used occasionally is the gravity unit which is 1/10 of a milligal.

If the only force that acts on the test mass m′ in figure 3.4 is the gravitational force, then the gravitational field *g* is equal to the acceleration **a** of the test mass m′ with respect to the location of m, if the mass of m is much greater than the mass of m′. This follows from Newton's second law of motion, *F* = m′ **a**. The gravitational field of the earth has been **incorrectly** called the acceleration of gravity, because the earth is generally much more massive than any test mass used to measure its gravitational field. The acceleration of gravity refers to the acceleration of a body that is in free fall in the earth's gravity field. Acceleration is a property of the motion of a body. The field refers to the gravitational force acting on the body. The gravitational field and the acceleration may or may not be equal, but physically they are different concepts. If the test mass m′ is removed from the point in space, the gravitational field still exists at the point, i.e., the field has a physical significance of its own in the same way an electric or magnetic field has physical significance.

FIGURE 3.5
The gravitational field of a sphere with mass m, volume v, and uniform density m/v and a point mass m at the center of the sphere are equal

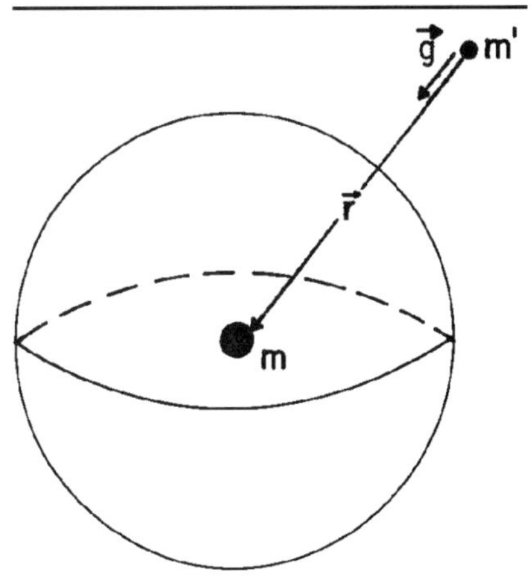

The magnitude of the gravitational field of the earth at its surface can be calculated approximately from equation 3.5 if we assume that the earth is a uniformly dense sphere. Newton proved that a uniformly dense sphere or a sphere whose mass distribution has spheri-

FIGURE 3.6
The field of two point masses m and M

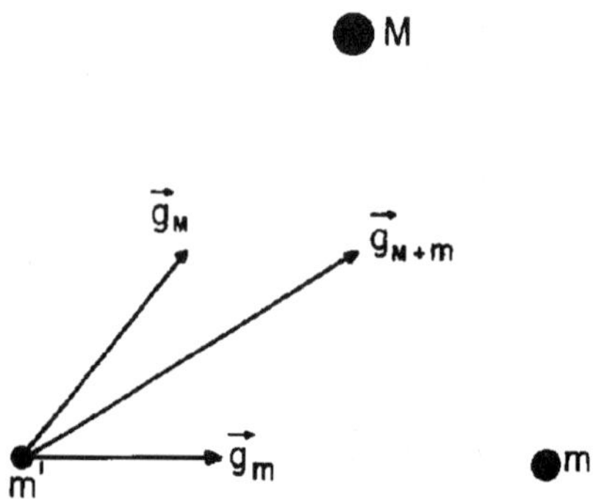

FIGURE 3.7
The gravitational field g and gravity effect g_z of a finite sized body

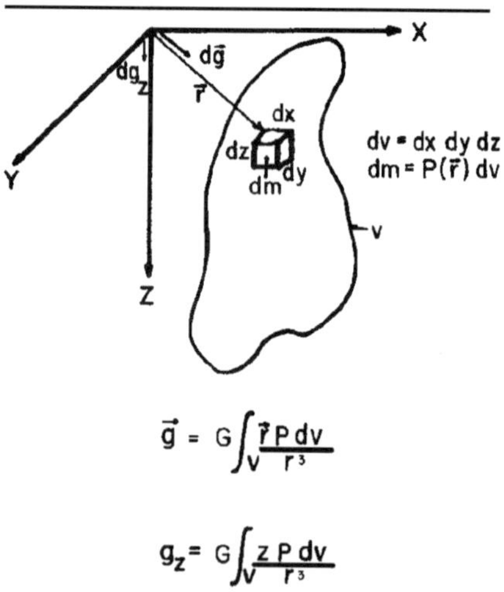

cal symmetry has the same gravitational field as a point mass with the same total mass as the sphere located at the center of the sphere (figure 3.5).

Using the mass of the earth 5.967 x 10^{27} gm and the equatorial radius 6378 x 10^5 cm, the field is 979.868 Gal. Using the polar radius 6356 x 10^5 cm, the field is 986.663 Gal. This approximation indicates that the field of the earth will change several Gals because of the shape of the earth.

The principle of superposition applies to gravitational fields. The field at a point in space due to a number of point masses is the vector sum of their individual fields (figure 3.6).

To deal mathematically with mass distributions of finite size, *g* and m in equation 3.5 are replaced with differentials d*g* and dm.

$$d\vec{g} = \frac{G\,\vec{r}\,dm}{r^3} \tag{3.6}$$

$$dm = P\,dv \tag{3.7}$$

$$d\vec{g} = \frac{G\,\vec{r}\,P\,dv}{r^3} \tag{3.8}$$

P is the density at the point given by the position vector *r* and dv is the volume element (figure 3.7).

For reasons that will be explained in the section on interpretation, g_z, the vertical component of *g*, is calculated for models to compare with gravity data rather than *g*. The vertical component is given by

$$dg_z = \frac{G\,z\,P\,dv}{r^3} \tag{3.9}$$

where z is the z component of **r**. Integration of equation 3.9 is carried out over the total volume v of the mass distribution to give g_z, the gravity effect. If the gravity effect of a model that is geologically reasonable agrees with the measured gravity anomaly, more confidence can be placed in the model.

The integration of equation 3.9 is only possible for a body with a simple geometry and density distribution (figure 3.8).

Relations are developed in gravitational potential theory that can make the determination of the field a less-complex process than is indicated by equations 3.8 and 3.9 (the gravitational potential and Gauss' law). Also, equations are developed for calculating the gravitational field of a body with an arbitrary shape. These equations are used to develop algorithms for use in computer programs.

Figure 3.8 gives examples of equations for the gravity effect of bodies with simple geometry and constant density. The term "two-dimensional body" that is used for some of the diagrams in figure 3.8 means that the body has infinite length into and out of the paper, i.e., in the y direction. A two-dimensional body is used to calculate the approximate gravity effect of a geologic feature that has a long strike length.

Figure 3.9 shows the approximation of the shape of a two-dimensional body with a polygon. The gravity effect of the polygon is easily calculated by computer using an algorithm introduced by Talwani and others (1959). The technique is often called the Talwani method.

Figure 3.10 shows the use of a number of two-dimensional rectangles whose bottom surfaces are adjusted up and down by computer until the gravity effect of all the rectangles matches the gravity anomaly measured over an alluvial basin. Computer methods have also been developed for calculating the gravity effect of three-dimensional bodies.

FIGURE 3.8
Gravity effect g_z in milligals for bodies with simple geometry and constant density. g_z is calculated at z = 0. The density P is in gm/cm³. The x direction is perpendicular to the strike of two-dimensional bodies. x = 0 is over the center of bodies that have finite size. The upper equation is for distances in kilofeet. The lower equation is for distances in kilometers

BODY | **COORDINATE SYSTEM** | **g_z in milligals**

1. Infinite horizontal slab

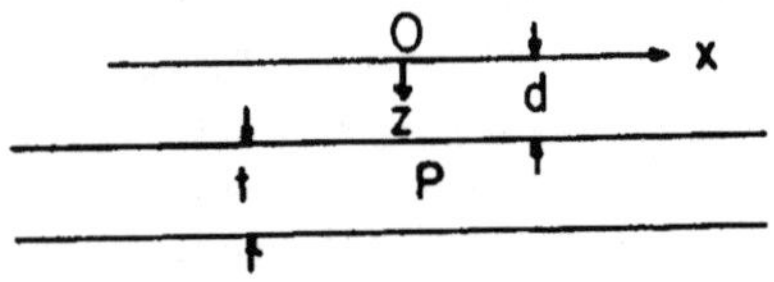

$$g_z = 12.77\ P\ t$$
$$g_z = 41.91\ P\ t$$

2. Thin semi-infinite horizontal slab

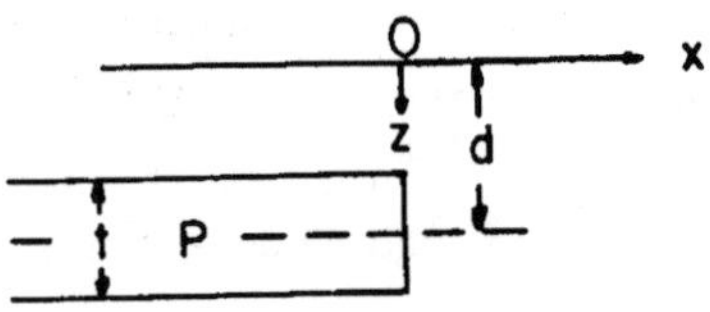

$$g_z = 4.066\ P\ t\ [\frac{pi}{2} - \tan^{-1}(\frac{x}{d})]$$
$$g_z = 13.34\ P\ t\ [\frac{pi}{2} - \tan^{-1}(\frac{x}{d})]$$

3. Sphere

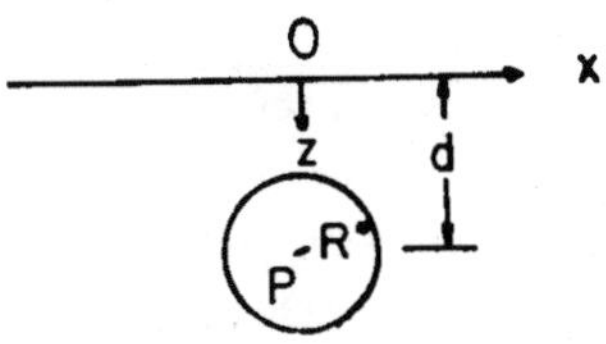

$$g_z = \frac{8.52\ P\ R^3}{d^2\ [1 + (\frac{x^2}{d^2})]^{\frac{3}{2}}}$$
$$g_z = \frac{27.94\ P\ R^3}{d^2\ [1 + (\frac{x^2}{d^2})]^{\frac{3}{2}}}$$

4. Two-dimensional horizontal cylinder

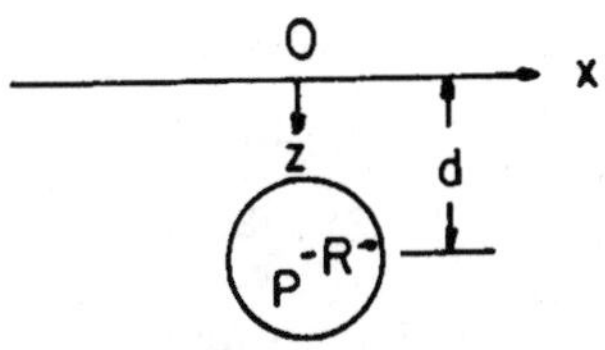

$$g_z = \frac{12.77\ P\ R^2}{d\ [1 + (\frac{x^2}{d^2})]}$$
$$g_z = \frac{41.91\ P\ R^2}{d\ [1 + (\frac{x^2}{d^2})]}$$

5. Vertical cylinder at x = 0

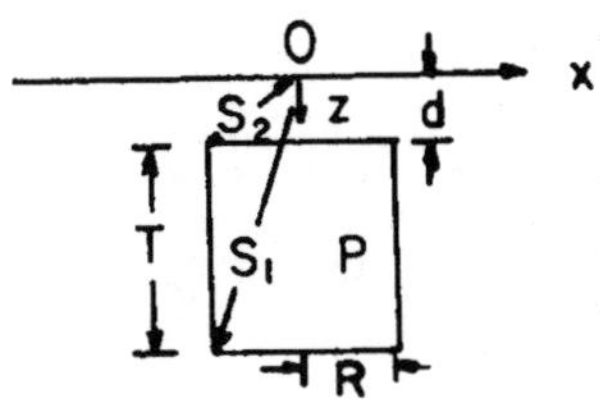

$$g_z = 12.77\ P\ (T - S_1 + S_2)$$
$$g_z = 41.91\ P\ (T - S_1 + S_2)$$

6. Vertical cylinder, top at z = 0

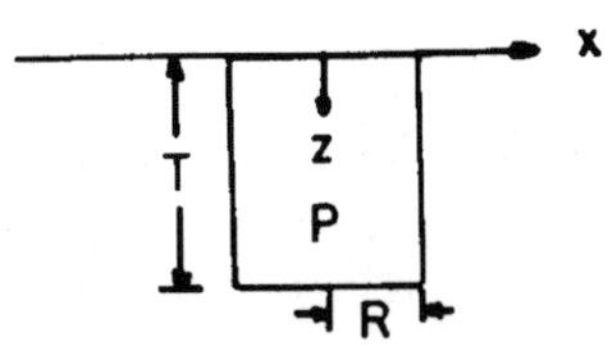

$$g_z = 3.193\ P\ \frac{R}{x}\ \{[T^2 + (x-R)^2]^{\frac{1}{2}} - [T^2 + (x+R)^2]^{\frac{1}{2}} - 2R\}$$
$$g_z = 10.48\ P\ \frac{R}{x}\ \{[T^2 + (x-R)^2]^{\frac{1}{2}} - [T^2 + (x+R)^2]^{\frac{1}{2}} - 2R\}$$

7. Two-dimensional dike. T>>B

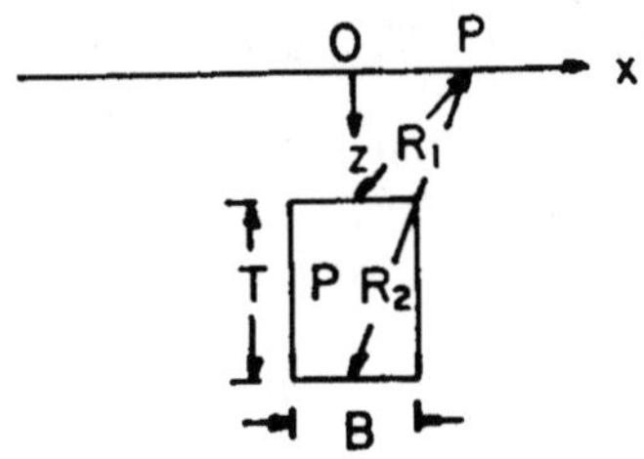

$$g_z = 4.066\ P\ B\ \ln\frac{R_2}{R_1}$$
$$g_z = 13.34\ P\ B\ \ln\frac{R_2}{R_1}$$

FIGURE 3.9
Gravity effect of a polygon calculated at Z=0

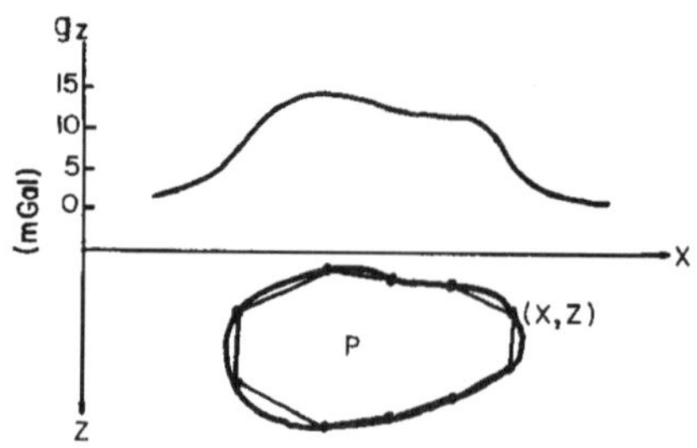

FIGURE 3.10
Iterative gravity model of an alluvial basin calculated at Z=0

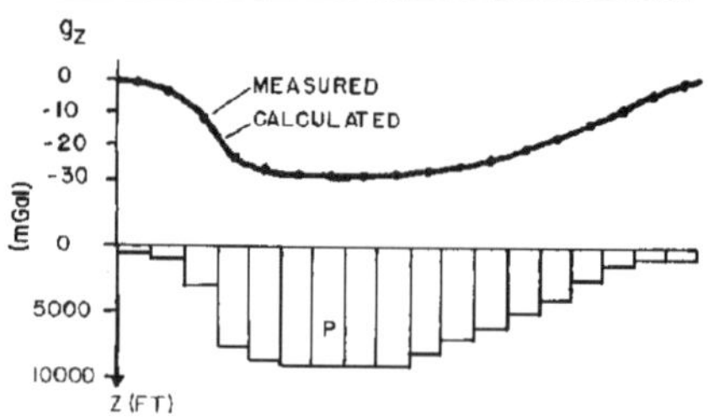

FIGURE 3.11
Definition of gravity g_e

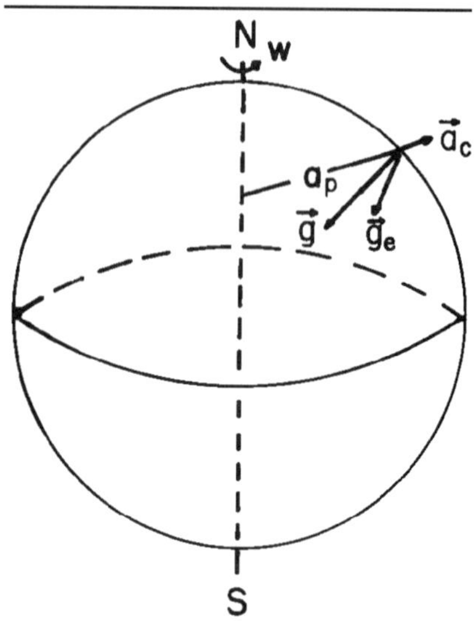

Ambiguity

The calculation of the gravity effect of a body with a known shape and density is unique and can be made as accurately as the computer has time to spend. Unfortunately, the "inverse problem", i.e., the determination of the mass distribution of a body from its gravity anomaly, is not unique. The problem occurs when the mass element dm is split into two parts, the density distribution P and the volume element dv in equation 3.7. When the gravity effect of the model is matched with a measured gravity anomaly, we determine two unknowns, the density and the geometry, from one known, the gravity anomaly. There is an infinite number of possible solutions. Geologic insight is necessary to choose one that is reasonable.

Gravity Measurements

The measurement of gravity on the earth's surface is the subject of this section. Gravity is defined and an explanation of how it is measured is given.

Gravity

Gravity is usually defined as a force. It is convenient, however, to define gravity as a field instead.

Gravity g_e (figure 3.11) is defined as the resultant of the gravitational field g and the centrifugal acceleration due to the earth's rotation a_c

$$\vec{g}_e = \vec{g} + \vec{a}_c \qquad (3.10)$$

A measurement of the magnitude of $\mathbf{g}_e$, g_e is called a gravity measurement.

The magnitude of the centrifugal acceleration of the earth is given by $-a_pW^2$. a_p is the length of a straight line from the earth's rotational axis to a point p on the surface. The line is perpendicular to the rotational axis. W is the earth's angular velocity (7.29212×10^{-5} radians/sec). Since a_p varies from 6378×10^5 cm at the equator to 0 cm at the pole, the centrifugal acceleration varies from -3.392 Gal at the equator to 0 Gal at the pole. Therefore, we can expect a variation in the earth's gravity of about 3 Gal because of centrifugal acceleration.

The direction of the gravity vector g_e is that of a plumb line, i.e., vertical. A surface perpendicular to g_e is called an equipotential or level surface. The equipotential surface at mean sea level is called the geoid.

Absolute Gravity Measurements

The first gravity measurements were probably made by determining the time it takes for a ball to roll a specific distance, from a standing start, down an inclined plane. Galileo made observations with an inclined plane and balls made out of different materials. He concluded that two objects dropped from the same height at the same instant with differing mass and substance would strike the ground at the same instant. This is called the principal of equivalence. The principal is accepted in classical mechanics and general relativity. It has recently been questioned and is discussed further in the case history section.

More accurate measurements of gravity can be made with a pendulum, a ruler and an accurate clock. Gravity is measured by determining the period of the pendulum and its length. The most accurate measurements of gravity are made by measuring the acceleration of a freely falling body in a vacuum.

The Gravity Meter

Precise measurements of absolute gravity can now be made to about 1/100 of a milligal. These measurements are difficult to carry out routinely in the field. Fortunately, changes in the value of gravity are easy to measure with a gravity meter, and this instrument is routinely used to carry out gravity surveys with precision near 0.1 milligal. The field procedure is relatively simple, but it requires a careful instrument operator. Since the field procedure is best discussed with the instrument in hand, it will not be given here.

A gravity meter is somewhat analogous to an extremely sensitive spring balance. If the balance is moved from place to place on the earth's surface and the weight of the same mass is measured, the weight of the mass will change because the gravitational field changes. The change in length of the spring is proportional to the change in gravity from place to place. The gravity meter reading is actually recorded from a dial. The readings taken from the dial are converted to milligals using calibration constants determined by the manufacturer.

The LaCoste and Romberg gravity meter is commonly used for exploration gravity surveys. Its design began in a geophysics class at the University of Texas in 1932 (LaCoste, 1988). The design is based on a long period vertical seismometer with a zero length spring (figure 3.12).

FIGURE 3.12
LaCoste 8 Romberg gravity meter suspension

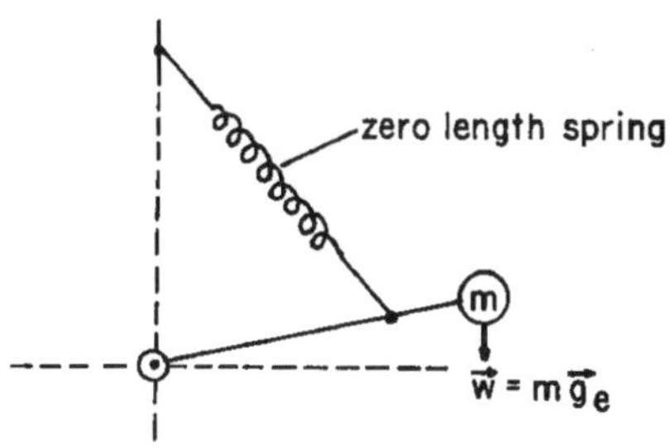

A zero length spring has a length that is zero when the stretching force is zero. An example of a zero length spring is a wire that is wound in a spiral that lies in a plane. The zero length spring gives a theoretically infinite period for the seismometer suspension. It also gives a theoretically infinite displacement sensitivity for a gravity meter. This means that a small change in the weight w resulting from a change in gravity g_e gives a large change in the position of the mass m. Since the gravity meter is also a seismometer, it detects earthquakes. Operation of the meter, therefore, must be suspended during a major earthquake.

Gravity Measurements With A Gravity Meter

The value of gravity at a field station is found by taking the first gravity measurement at a base station where the value of gravity, g_e, is known. Next, gravity measurements are made at field stations. A final gravity measurement is made at a base station. The magnitude of gravity is found by adding the change in gravity from the beginning base station to the field station to the value of gravity at the beginning base station.

Base stations have been established throughout the world with gravity meters. These bases are tied to the site of an absolute gravity measurement in Potsdam, East Germany, made with six different pendulums in 1906. This datum is referred to as the Potsdam datum. More recent absolute measurements of gravity indicated that the Potsdam datum is about 14 milligals too high. Correction for this error has led to a second datum referred to as International Gravity Standardization Network 1971 or IGSN 71. Prior to IGSN 71, the Potsdam gravity datum was used.

Gravity data available from different government agencies may be on different datums. For example, The California Division of Mines and Geology uses the Potsdam datum while the Nevada Bureau of Mines and Geology has used both the Potsdam and IGSN 71 datums. The main advantage of placing gravity data on a datum is that data from state and federal agencies can be added to the field data set. Since large amounts of data are available in the public domain, it is advisable to tie the survey to a datum.

Corrections To Gravity Meter Measurements

If gravity meter readings are taken at the same location over a period of time, the meter reading changes. This is caused by tides and instrument drift. Tides are the gravitational field resulting from extraterrestrial bodies and the gravity effect of the distortion of the earth that is caused by this field. Instrument drift is the change in meter reading caused by changes in the gravity meter suspension system. During the course of a gravity survey, meter readings are taken at gravity base stations to determine the change caused by tides and instrument drift. Drift corrections are made by assuming that instrument drift is linear with time over the time period between base station readings.

If the gravity meter has a high drift rate, i.e., greater than 0.1 mGal/hr, base station readings should be taken every two hours.

If the base station readings are taken often enough, such as every two hours, tides are approximately linear with time and both instrument drift and tidal effects can be corrected using the drift correction. If the gravity meter has a low drift rate, i.e., less than 0.01 mGal/hr, the time between base station readings can be longer, and tidal gravity effects can be removed using theoretical tide calculations.

The sun and moon are the only extraterrestrial bodies that are large enough and near enough to have an appreciable effect on gravity surveys. Theoretical tide corrections are expressed as a function of time, latitude, longitude and elevation. In the middle latitudes, the maximum change from tidal effects over a time period of about six hours is 0.3 mGal. If the gravity anomaly under investigation has a large amplitude, tide corrections can be ignored.

Calculating g_e From Gravity Meter Measurements

Data that are collected for each gravity station include the meter dial reading, date, time and station location (i.e., latitude, longitude and elevation). The survey is run in loops that include meter readings at the beginning base station, field stations and the final base station.

Gravity measurements made with a gravity meter require the following corrections:

- The meter dial reading is converted to milligals using constants supplied by the manufacturer.
- Theoretical tide corrections are added to the meter dial reading.
- Linear (with time) drift corrections are made to the meter dial reading.
- The change in gravity from the base station to the field station is added to the gravity value of the base station to give the gravity value of the field station.

Assume that two different base stations are used, a beginning and an ending base station. Gravity at a field station g_{ef} is equal to the change in gravity from the beginning base station to the field station plus the gravity value at the beginning base station g_{e1}.

$$g_{ef} = (dr_f \ x\ k_f + tc_f + dc_f) - (dr_1 \ x\ k_1 + tc_1) + g_{e1} \qquad (3.11)$$

where dr is the dial reading, k is the meter constant, tc is the tide correction and dc is the drift correction. The subscript f refers to the field station and the subscript 1 refers to the beginning base station.

The drift correction at the beginning base station, d_{c1}, is zero.

In general, the drift correction is given by

$$dc_f = (t_f - t_1) \times \frac{(g_{e2} - g_{e1}) - (q_2 - q_1)}{(t_2 - t_1)} \tag{3.12}$$

where t is the time and g_e is the gravity value. The subscripts f, 1, and 2 refer to the field station and first and second base stations. The values of q_1 and q_2 are given by

$$q_1 = dr_1 \times k_1 + tc_1 \tag{3.13}$$

$$q_2 = dr_2 \times k_2 + tc_2 \tag{3.14}$$

where dr_1, dr_2, k_1, k_2, tc_1 and tc_2 are the dial reading, dial constant and tide correction for the first and second base stations. The instrument drift rate is given by

$$Dr = -\frac{(g_{e2} - g_{e1}) - (q_2 - q_1)}{(t_2 - t_1)}$$

The term dial constant is somewhat contradictory since the constant may only hold for a limited range of dial readings depending on the kind of gravity meter used. Outside this range, a different dial constant must be used.

The theoretical tide correction can be calculated by subroutines in gravity data reduction programs. If no tide correction is made, tc in the above equations is zero and the tide correction is included in the values of dr_1 and dr_2 in equations 3.13 and 3.14.

The elapsed time between base station readings, $t_2 - t_1$, should be two hours or less when tide corrections are included in the drift correction unless tidal effects are small enough to be ignored.

Corrections To Gravity

Tidal gravity effects have been removed by reduction of the gravity meter data to g_{ef}. Gravity, g_{ef}, includes the following components:

- The gravity field of the whole mass of the earth from the crust to the core. This component includes the gravity effect of the earth's nonspherical shape.
- The centrifugal acceleration caused by the earth's rotation.

Our only interest lies in the gravity field of density changes in the upper part of the earth's crust. We now consider the removal of other unwanted components included in g_{ef}.

Normal Gravity

We assume that the earth's field can be split into two components, a normal gravity g_o and a small disturbing field. The normal gravity includes the large-scale components of the earth's field caused by rotation, nonspherical shape and density variations that are only a function of the distance from the center of the earth. The small disturbing field includes the field of non-uniform lateral density changes in the crust and upper mantle.

The normal gravity is given by the equation for the gravity field of a rotating ellipsoid of revolution (figure 3.13). The ellipsoid must give a good representation of the earth's shape and approximate the shape of the geoid. The choice of an ellipsoid is based on direct measurement of the meridian arcs of the earth, theoretical consideration of the form of a rotating fluid in hydrostatic equilibrium and an analysis of first order gravity variations over the earth's surface.

FIGURE 3.13
A rotating ellipsoid of revolution

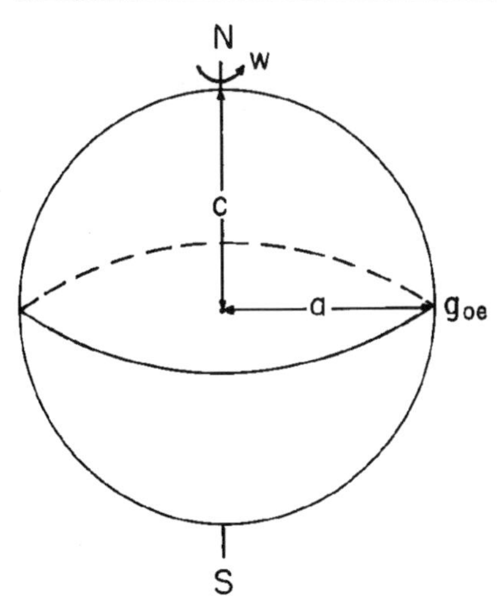

The gravity field of the rotating ellipsoid is completely determined by the following parameters:

- The equatorial radius a.
- The flattening f, where f is given by (a - c)/a, and c is the polar radius.
- The angular velocity W.
- The normal value of gravity at the equator g_{oe}.

More accurate measurement of these parameters has led to the definition of different ellipsoids.

The 1930 International Gravity Equation is based on the International Ellipsoid. The equation is

$$g_{o30} = 978049\ [1 + 0.0052884 \sin^2(L) - 0.0000059 \sin^2(2L)] \quad (3.15)$$

where L is the latitude and g_o is in milligals. A more recent version of the equation is the 1967 Geodetic Reference System Equation (GRS 67) given by

$$g_{o67} = 978031.85\ [1 + 0.005278895 \sin^2(L) + 0.000023462 \sin^4(L)] \quad (3.16)$$

Changes in the normal gravity equation resulting from changes in the definition of the ellipsoid are important because they change the value of gravity anomalies that are calculated from the equation. The 1930 International Gravity Formula is used with the Potsdam base station datum while GRS 67 is used with the IGSN 71 datum.

FIGURE 3.14
Correction of the normal gravity g_o from the geoid to a gravity station at elevation h and latitude L

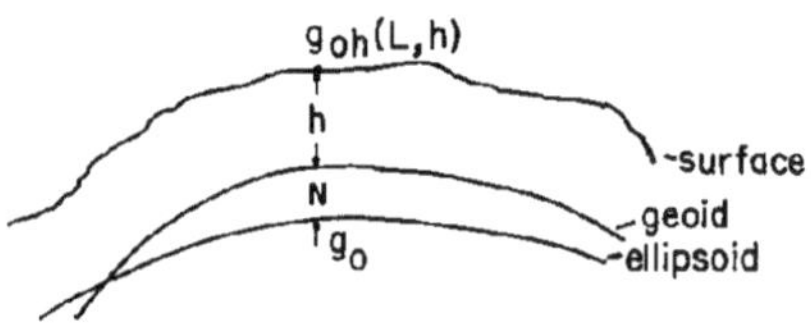

Gravity Anomalies

Gravity anomalies are calculated by correcting g_o (figure 3.14), the normal gravity, from the geoid (sea level) to the location of the field station (g_{oh} at latitude L and elevation h) and comparing it with the measured gravity g_{ef} of the station. The gravity anomaly can be thought of as a comparison of the measured and theoretical values of gravity. It is calculated by subtracting the corrected value of g_o from g_{ef}. This subtraction also removes the gravity effect of the ellipsoid, i.e., the effect of the earth's shape and rotation. Different corrections can be applied to g_o, giving different kinds of gravity anomalies.

The traditional approach to calculating gravity anomalies has been to correct the measured gravity g_{ef} to the geoid. This led to the mistaken belief that the calculated gravity anomalies were reduced to a sea level elevation datum. Correcting g_o to the station location indicates that the gravity anomaly is calculated at the level of the station (Ervin, 1977). If the gravity anomalies are needed on a constant elevation datum, they must be continued upward or downward using methods developed in potential theory.

The normal gravity g_o has been corrected in this text to the level of the station location. Depending on how the equation for the gravity correction is defined, this approach may change the sign of the equation relative to equations given in other texts. Gravity anomalies will have the same sign and magnitude regardless of the approach used to derive them.

The ellipsoid used to calculate the normal gravity only approximates the geoid (figure 3.14); the use of elevation h above (+) or below (-) the geoid to correct g_o to the station elevation causes an error in the corrections to g_o. The difference in elevation of the geoid and the ellipsoid (the geoidal undulation N) is relatively constant over small distances and the error causes a shift of the gravity anomalies that is also relatively constant over small areas. Since we are only interested in changes in the gravity anomalies, the constant error has no effect on the interpretation of data from exploration gravity surveys. These surveys are usually carried out over relatively small areas.

Free Air Correction

The free air correction to g_o (figure 3.15) accounts for the decrease in g_o with an increase in elevation above sea level.

It is approximated by (dg_o/dh) h, the normal gradient of gravity multiplied by the station elevation. If small variations in dg_o/dh with latitude are ignored, the free air correction in milligals is:

$$Fc = -0.3086\ h \quad (h \text{ in meters}) \tag{3.17}$$

$$Fc = -0.09406\ h \quad (h \text{ in feet}) \tag{3.18}$$

$$g_{ohf} = g_o + Fc \tag{3.19}$$

The expression (dg_o/dh) h is the first order term in the Taylor's series expansion of $g_{oh} - g_o$ as a function of h. If the second order term (h^2) is included in the free air correction, this can change gravity anomalies by several milligals at higher elevations. The free air correction to g_o is negative for stations above sea level and positive for stations below sea level.

FIGURE 3.15
Free air correction to g_o

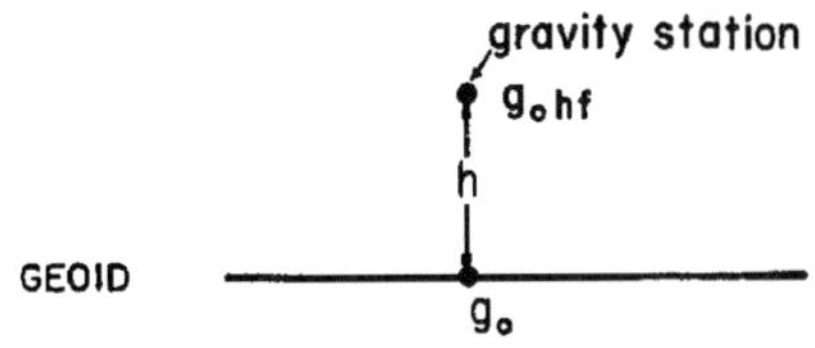

Bouguer Correction

This correction is named for Pierre Bouguer, who led the French Royal Academy expedition to measure the length of a degree of latitude in Peru in 1735. If the gravity station is assumed to lie on an infinite horizontal plane and rock of density P is placed below the plane, the normal gravity at the station is increased (figure 3.16).

The amount of increase is given by the equation for the gravity effect of an infinite horizontal slab (figure 3.8, body 1) of thickness h and density P. The Bouguer correction in milligals is

$$Bc = 0.04191\ P\ h \quad (h \text{ in meters, } P \text{ in gm/cm}^3) \tag{3.20}$$

$$Bc = 0.01277\ P\ h \quad (h \text{ in feet, } P \text{ in gm/cm}^3) \tag{3.21}$$

$$g_{ohsb} = g_o + Fc + Bc \tag{3.22}$$

FIGURE 3.16
Bouguer correction to g_o

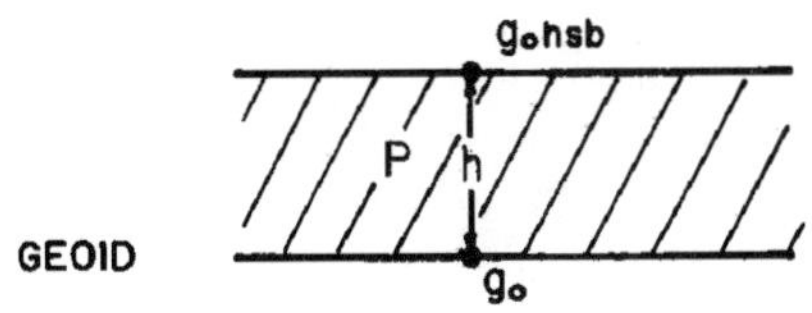

The amount of increase depends on the density that is chosen for the slab. The density that is used should be the average density of the terrain that surrounds the station. This density is used because the primary purpose of the Bouguer correction in exploration geophysics is that of a first order terrain correction.

A commonly chosen density is 2.67 gm/cm^3. This density is assumed to be the average density of crustal rocks. It is used by most government agencies in calculating Bouguer anomalies. For this density, the correction in milligals is

$$Bc = 0.1119\ h \quad (h \text{ in meters}) \tag{3.23}$$

$$Bc = 0.03411\ h \quad (h \text{ in feet}) \tag{3.24}$$

The correction to g_o is positive if h is above sea level and negative if h is below sea level.

In some cases, the free air and Bouguer corrections for a specific density are combined into a single correction called the elevation correction.

Curvature Correction

The infinite horizontal slab used in the Bouguer correction can be corrected to the curvature of the earth. This correction is a function of elevation. It changes sign and can have a magnitude of several milligals. The curvature correction Cc in milligals is

$$Cc = 1.464 \times 10^{-3}\,h - 3.533 \times 10^{-7}\,h^2 + 4.485 \times 10^{-10}\,h^3 \quad (h \text{ in meters}) \quad (3.25)$$

$$Cc = 4.462 \times 10^{-4}\,h - 3.282 \times 10^{-8}\,h^2 + 1.270 \times 10^{-15}\,h^3 \quad (h \text{ in feet}) \quad (3.26)$$

FIGURE 3.17
Terrain correction to g_o

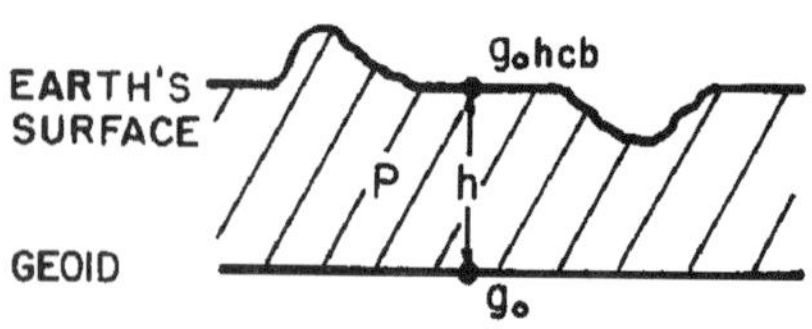

Terrain Correction

The terrain correction Tc corrects for the deviation of the actual terrain from that of an infinite horizontal slab and depends on the average density of the terrain that surrounds the station (figure 3.17). The correction requires information about the terrain in the form of topographic maps or in digital form.

Templates have been developed to make terrain corrections using topographic maps. More recently, large amounts of digital terrain data have become available. These data have made terrain corrections calculated by computer feasible. Terrain corrections calculated by computer are suitable for all terrain that surrounds a gravity station except the inner few kilometers next to the station. Digital terrain corrections may also include a correction for the earth's curvature. The terrain correction to g_o is almost always negative for stations above sea level. Terrain corrections in rugged terrain can be tens of milligals. If the gravity anomaly of interest is in rugged terrain and has a small amplitude, terrain corrections will be required. Gravity measurements near cliffs and on top of hills may give terrain effects that cannot be accurately calculated using the available topographic information.

The terrain correction and curvature correction are sometimes combined into a single correction. The terrain correction is calculated as a positive value. Since it decreases g_o,

$$g_{ohcb} = g_o + Fc + Bc + Cc - Tc \quad (3.27)$$

Isostatic Correction

The isostatic correction Ic assumes an isostatic model to calculate the gravity effect of the compensating mass deficiency below sea level. For example, the Airy-Heiskanen model of isostasy (figure 3.18) is assumed for the Isostatic Residual Gravity Map of the United States.

On continents, "roots" that lie beneath a normal sea-level crustal thickness of Ds compensate for the topography. The isostatic correction depends on the thickness Ds, the density of the topographic load P, the density change across the base of the crust DP, and on topographic information that is used to determine the geometry of the compensating mass. The values used for the Isostatic Residual Gravity Map of the United States are 30 km for Ds, 2.67 gm/cm^3 for P, and 0.35 gm/cm^3 for DP. The correction can be made using templates and topographic maps or by computer using digital elevation information.

In figure 3.18, the isostatic correction accounts for the gravity effect of the compensating mass below Ds (crosshatched). The isostatic correction is a positive number, and the effect is subtracted from g_o to account for the low-density roots.

$$g_{ohi} = g_o + Fc + Bc + Cc - Tc - Ic \quad (3.28)$$

Gravity Anomalies

Gravity anomalies are calculated by subtracting the corrected value of g_o (i.e., g_{oh}) from g_{ef}, the measured value of gravity.

Free Air Anomaly

$$FA = g_{ef} - g_{ohf} \quad (3.29)$$

g_{ohf} is given by equation 3.19.

Simple Bouguer Anomaly

$$SBA = g_{ef} - g_{ohsb} \tag{3.30}$$

g_{ohsb} is given by equation 3.22.

Complete Bouguer Anomaly

$$CBA = g_{ef} - g_{ohcb} \tag{3.31}$$

g_{ohcb} is given by equation 3.27.

Isostatic Anomaly

$$IA = g_{ef} - g_{ohi} \tag{3.32}$$

g_{ohi} is given by equation 3.28.

Gravity anomaly profiles calculated with these equations will be discussed in the case history section.

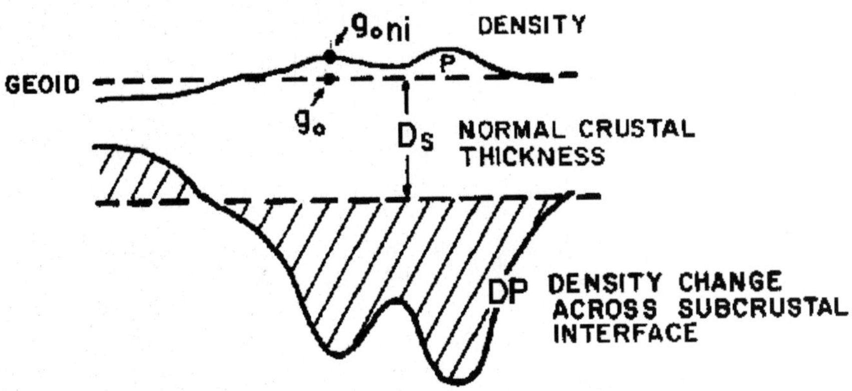

FIGURE 3.18
Isostatic correction to g_0 using the Airy - Heiskanen model of isostasy

Planning A Gravity Survey

In general, land gravity surveys conducted with modern gravity meters require relatively routine field operations. A well-trained gravity crew is a necessity. Accurate station locations are required. In particular, accurate station elevations may be needed. If data are collected in rugged terrain, terrain corrections will probably be required.

Base Stations

Gravity surveys should be tied to base stations. Base station locations can be obtained from:

Gravity Services Branch
Defense Mapping Agency Aerospace Center
3200 South 2nd Street
St. Louis, MO 63118-3399
(314) 263-8410

Field Stations

Gravity station locations depend on access to the survey area. If good access is available, the field station pattern and density will depend on the purpose of the survey and on the amplitude, shape and maximum gravity gradient of anomalies that are expected to occur in the survey area. Available gravity maps in the area of interest can help in the determination of the station pattern. Good station coverage may require a second trip to the field area after the first set of data is reduced and inspected.

Data Accuracy

The measurement of g_{ef} (equation 3.11) is normally precise to within 0.1 to 0.2 mGal. The error in g_o, the normal gravity (equation 3.15 or 3.16), depends on the accuracy of the latitude of the station. If the error in north-south horizontal position is 100 ft, the error in g_o is 0.02 mGal. If the error in north-south horizontal position is 100 m, the error in g_o is 0.08 mGal. The error in the combined free air and Bouguer correction for a density of 2.67 gm/cm^3 is about 0.06 mGal/ft. If gravity anomalies with errors of around 0.1 mGal are required, elevations should be accurate to the nearest foot.

Terrain effects can be estimated using templates described in the references. If possible, rugged terrain should be avoided.

If large gravity anomalies are expected, elevations interpolated from topographic maps may be adequate for stations that are accurately located on the maps, particularly in flat areas. In general, it is best to survey the station elevation to the nearest foot.

Interpretation

The goal of gravity data interpretation is to provide a new picture of the subsurface geologic conditions that will contribute to solving the exploration problem at hand. This new geologic picture may be extremely oversimplified or cloudy. It should give insight, however, that was lacking prior to the gravity survey.

First, this section discusses some important concepts in gravity data interpretation. Next, different approaches to interpretation are described. Finally, a discussion is given for an idealized interpretation using two different approaches.

Ambiguity In Interpretation

Ambiguity gives insight about the gravity method. In the field, we measure the gravity field of the mass distribution of exploration interest along with the gravity fields of many other mass distributions. In the office, we attempt to separate the gravity anomaly of interest from those caused by the "rest of the world" (the regional-residual separation), and we attempt to come up with a model of the mass distribution that is geologically reasonable and has a gravity effect that matches the residual gravity anomaly.

Or, we create a model of the whole world (at least from the upper mantle to the surface) and calculate the gravity effect of the model. The gravity effect of the entire model includes the gravity effect of the part of the model that we are interested in, along with the gravity effect of many parts of the model that we are not interested in. We design the model so that its gravity effect matches the gravity anomaly.

Ambiguity occurs in both approaches. Since there is no unique solution in gravity interpretation, geologic insight must be used to eliminate solutions that are unreasonable and help determine a model that approximates the buried subsurface geology.

Gravity Anomaly

The use of the term "anomaly" needs some clarification. The free air, Bouguer and isostatic residual anomalies are differences between the measured and theoretical (normal gravity) values of gravity, which explains the use of the term "anomaly". The term is also used to refer to a particular gravity feature that occurs on a map or profile which differs from the surrounding gravity values. For example, a gravity high that may be associated with a buried bedrock ridge could be of exploration interest. If a series of highs and lows occurs across a gravity map, it is not always a simple task to decide which is anomalous, the high or the low.

Gravity Effect

Consider the ideal situation shown in figure 3.19. A single anomalous source occurs beneath a flat earth. The source causes an anomalous gravity field, **DG**, that is superimposed on the earth's gravity, $\mathbf{g_{ef}}$. At x1, the field resulting from the anomalous mass is negligible, because the distance to the anomalous mass is large. At x2 the gravity $\mathbf{g_{ef}}$ is the resultant of $\mathbf{g_{ef}}$ at x1 and the anomalous field **DG**. The change in the magnitude of $\mathbf{g_{ef}}$ from x1 to x2, Dg_{ef}, is approximately equal to the vertical component (in the direction of $\mathbf{g_{ef}}$ at x2) of the anomalous field, DGz, at x2.

The magnitude of the anomalous field in figure 3.19, DG, is extremely small relative to the earth's gravity (about 0.01 percent to 0.0001 percent of the earth's gravity). The field of the anomalous mass does not change the vertical direction appreciably (unless you are performing an accurate land survey over a large area) and, for all practical purposes, the earth's field, g_{ef}, is in a constant z direction. This is why the vertical component of the gravitational field of a model, g_z (the gravity effect), is calculated to compare with the change in gravity, Dg_{ef} (the gravity anomaly).

In this situation, Dg_{ef} is determined from g_{ef}, the magnitude of gravity, for a flat earth. In real situations, Dg_{ef} is determined from the appropriate gravity anomalies or from the residual gravity which is determined in the regional-residual separation.

The terms "gravity anomaly" and "gravity effect" are used interchangeably in most text books. It is attempted here to restrict gravity effect to theoretical values of gravity that can be calculated (e.g. a model). The term gravity anomaly is restricted to real gravity features that have been measured in the field or are derived from field measurements (e.g. a residual gravity anomaly). It should be clear from the text whether the gravity effect or gravity anomaly is the subject of discussion.

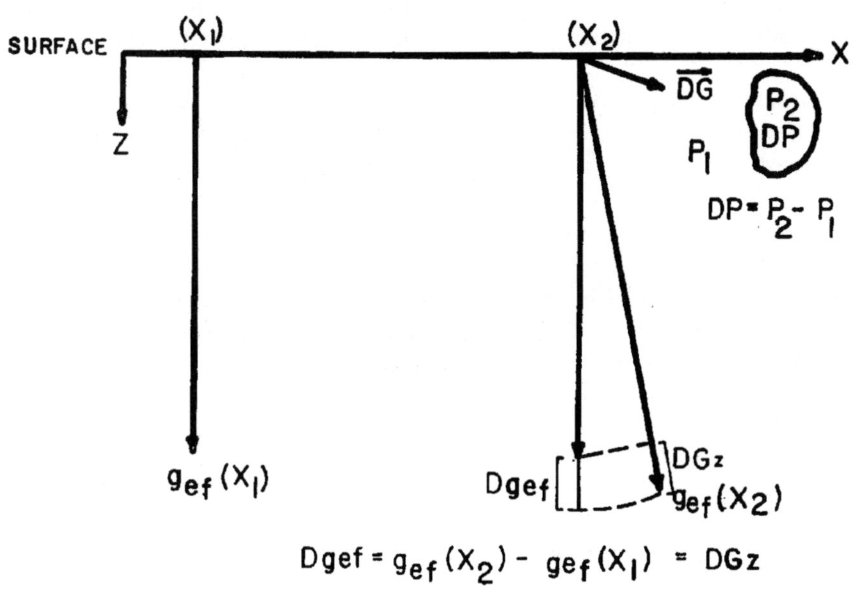

FIGURE 3.19
The anomaious gravity, the gravity change, and density contrast

Density Contrast

The density used to calculate the gravity effect of a model is not the density of the anomalous body, but the density contrast --- the difference in density between the body and its surroundings. In figure 3.19, the density contrast, DP, is given by

$$DP = P_2 - P_1 \tag{3.33}$$

where P_2 is the density of the anomalous body and P_1 is the density of the surroundings. In figures 3.8, 3.9 and 3.10, the density P should be replaced by the density contrast DP when these equations and figures are used to calculate the gravity effect of a model.

The density contrast for a model is estimated by considering the possible density of rocks that cause the gravity anomaly. This information may come from density measurements on rock samples, downhole gravity meter surveys, gamma gamma density logs, comparison of seismic velocity and density or from tables of rock density.

Many tables of rock density are available and can be found in the references listed at the end of this chapter. Generally speaking, the order of increasing density of rocks is:

- Unconsolidated clastic sediments
- Clastic sedimentary rocks
- Volcanic rocks
- Plutonic rocks and limestones.

Metamorphic equivalents of these rocks are usually more dense than the original rock. Mafic igneous rocks are generally more dense than felsic igneous rocks. Ultramafic rocks are more dense than mafic rocks. Exceptions to these generalizations, however, are not unusual. For example, the density of a compact basalt can exceed the density of most plutonic rocks while the density of highly vesicular basalt can approach the density of unconsolidated clastic sediments.

FIGURE 3.20
Gravity effect of structure vs intrabasement density contrasts at z=0

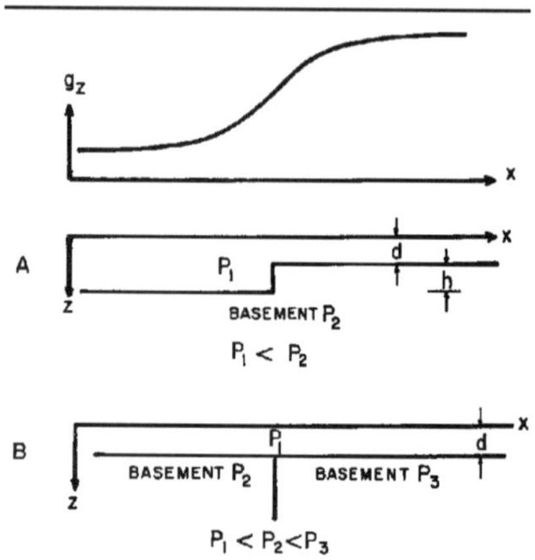

Structure Versus Intrabasement Density Contrasts

Figure 3.20 shows two models of the basement surface. In model A, basement rocks have been faulted, causing a lateral change in density. In model B, the density contrast occurs in the basement rocks.

If only a gravity profile is available, it would be difficult to tell whether the anomaly is caused by structure or by an intrabasement density contrast. This situation can occur in both petroleum and mineral exploration and has important implications for both.

If basement is made up of rocks with different densities separated by vertical boundaries that extend to depth, large gravity anomalies can occur. Watson and Brow (1966) have suggested that structures may occur in rocks that overlie intrabasement density changes. A somewhat similar situation occurs when old structures in Precambrian basement rocks have been reactivated and influence younger structures in the overlying sedimentary and volcanic rocks.

Depth Estimates

Figure 3.21 shows the gravity effect of three identical spheres buried at different depths (figure 3.8, body 3).

Figure 3.22 shows the gravity effect of three identical semi-infinite horizontal slabs buried at different depths (figure 3.8, body 2).

Both the spheres and the slabs have a density contrast of 0.1 gm\cm^3 greater than their surroundings. The diameter of the spheres and the thickness of the slabs are 1 km. The depths of the spheres and slab centers are 0.5, 1.0 and 1.5 km. The profiles go over the center of the spheres and are perpendicular to the edge of the slab. The model for the slab uses a thin plate located at the center of the slab to calculate the gravity effect. The thinner the slab relative to its depth of burial, the better this approximation works.

For the sphere, the amplitude of g_z and the horizontal gradient dg_z/dx increase as the depth to the center of the sphere decreases. For the semi-infinite horizontal slab, dg_z/dx also increases as the slab becomes shallow. Since the slab has infinite extent, the amplitude is the same for all three slabs, 4.191 mGal. Bott and Smith (1958) have shown that the maximum possible depth to the top of a source, d max, is given by the following equations.

$$d\ \max = \frac{0.86\ g_z\ \max}{\dfrac{dg_z\ \max}{dx\ \max}} \quad \textit{three-dimensional body} \tag{3.34}$$

$$d\ \max = \frac{0.64\ g_z\ \max}{dg_z/dx\ \max} \quad \textit{two-dimensional body} \tag{3.35}$$

These equations give the exact depth to a point mass and a horizontal line source. In general, the shallower the source, the "sharper" the anomaly, i.e., the greater the maximum gradient dg_z/dx max.

The most common method of making depth estimates is to compare a gravity anomaly with the gravity effect of a model. This method will be demonstrated in the case histories section.

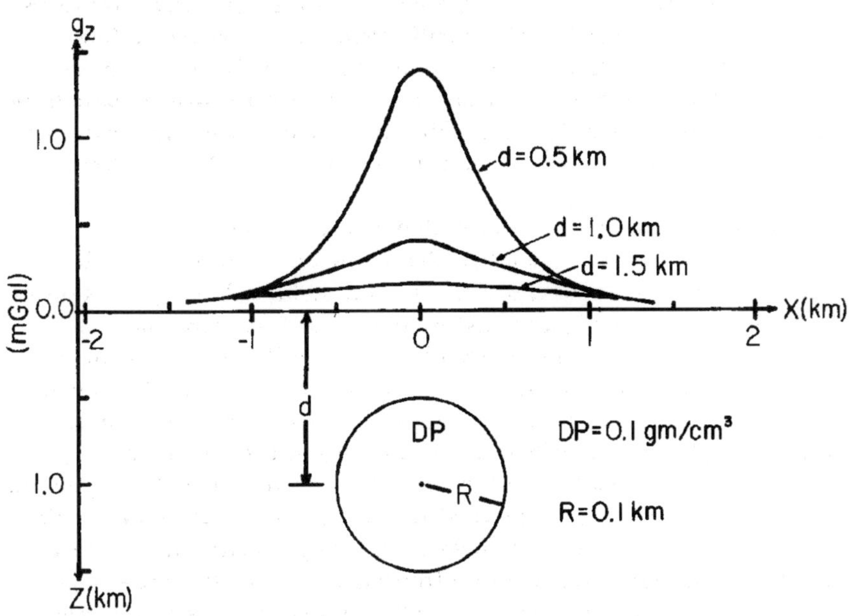

FIGURE 3.21
Gravity effect over a sphere buried at different depths calculated at Z=0

FIGURE 3.22
Gravity effect over a semi-infinite slab buried at different depths and calculated at Z=0

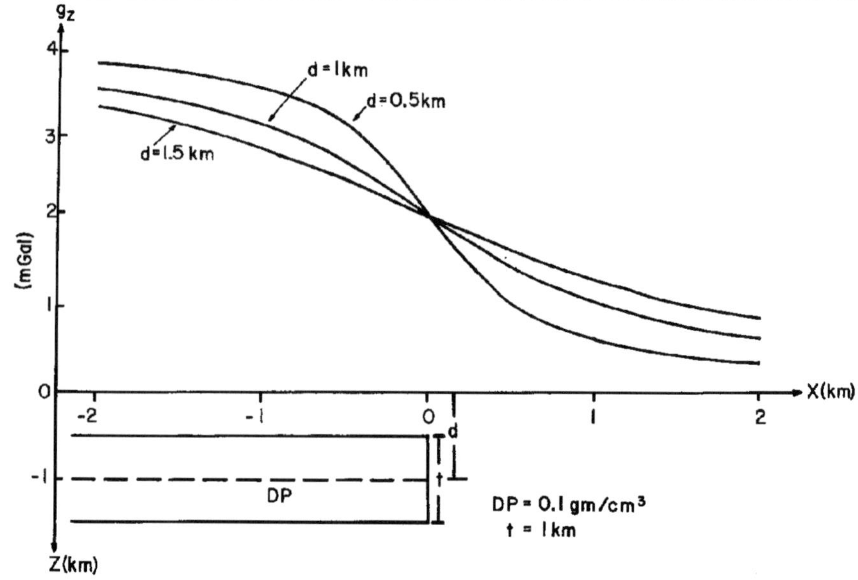

Gravity Anomaly Wavelength
The wavelength of a gravity anomaly is an expression that refers loosely to the relative horizontal extent and sharpness of a gravity anomaly. A precise definition of wavelength requires the use of Fourier analysis. The amplitude spectra of the sources in figures 3.21 and 3.22 would show a preponderance of short wavelength components for the shallow sources.

Vertical Contact Location
The horizontal location of a vertical contact that is the boundary between rocks with different densities is located at the position of the maximum horizontal gravity gradient. This is illustrated in figure 3.22. If the contact has a dip, the location of the maximum gravity gradient moves away from the contact. The distance it moves is small for all but shallow dips.

Regional-Residual Separation
The gravity anomaly of interest (the residual anomaly) can be separated from gravity anomalies caused by "the rest of the world" (the regional anomaly) so that it can be analyzed. As mentioned before, there is no method that allows for a unique regional-residual separation.

One method of separation is based on the wavelength of the anomaly. Shallow sources have short wavelengths, and deep sources have long wavelengths. Digital filtering can be used to remove either the short or long wavelength anomalies. Usually in exploration, we are interested in short wavelength anomalies caused by shallow sources.

Another approach to the regional-residual separation is to contour a "smoothed" regional gravity that is based on the interpreter's "preconceived" view of what the regional and residual gravity should look like. For example, a piercement salt dome is likely to have a gravity anomaly similar to a vertical cylinder, whereas gravity anomalies related to density changes caused by isostatic compensation should vary with the regional surface elevation. This information can be used to construct a regional gravity that removes gravity anomalies related to isostasy and other nearby salt domes.

Each method has its advantages. For example, if large amounts of data require analysis, digital filtering may be the only practical way to deal with the regional-residual separation. On the other hand, an interpreter can consider knowledge based on his or her experience in geologic interpretation of gravity data when the smoothing method is used.

It is not always necessary to perform a regional-residual separation. The area of interest can be modeled from the mantle to the crust, eliminating the need for the separation. It is also common to speculate about the source of a gravity anomaly without separating out the regional gravity.

Interpretation Philosophy

When an interpreter considers the cause of a gravity anomaly, his or her approach to explaining the anomaly depends on many factors. Interpretation philosophy varies with the interpreter. Some interpreters believe that the more quantitative and rigorous the interpretation, the more realistic it will be. For example, they may feel that including interpretations of other kinds of geophysical data (e.g., seismic and magnetic data) will bring them closer to the "unique solution". Other interpreters are less strict about their approach to interpretation. They believe that time is best spent considering the possible sources of a gravity anomaly in terms of the possible subsurface geology. All have logical arguments to support their approaches.

The best method depends on the situation. If good, hard evidence about density contrasts is available, rigor may be justified. On the other hand, when little is known about the geometry and density of the sources, rigor may be pointless. Speculation about the geologic cause of a gravity anomaly will still contribute ideas that can be tested.

Two Idealized Approaches To Gravity Interpretation

- The gravity survey is completed, data are reduced and several anomalies are found. Digital filtering is used to perform the regional-residual separation of the anomaly of interest from other anomalies. The anomaly of interest is modeled. The modeler uses information from other geophysical methods and density information from downhole gravity meter surveys. Based on the interpretation, drilling commences and the interpretation is confirmed.

- The gravity survey is completed, data are reduced and several anomalies are found. The interpreter considers the possible geologic sources of the anomalies and, based on these considerations, separates the anomaly of interest using smoothed contours for the regional gravity. Based on the same considerations and the interpreter's knowledge of rock densities, a simple model, e.g., a sphere, that gives a reasonable shape and density contrast to the suspected source of the gravity anomaly is used to estimate the depth to the source. Based on the interpretation, drilling commences and the interpretation is confirmed.

CASE HISTORIES

Case histories included in this section have been solicited from colleagues, summarized from the literature, or are my own. Each geophysicist has his or her own approach to interpretation. I believe that the reader will benefit from the different points of view expressed in these case histories.

First, corrections to gravity data, the free air, Bouguer and isostatic corrections, are discussed on a scale of hundreds of kilometers. Next, the measured gravity, free air anomaly and Bouguer anomaly are compared on a scale of thousands of feet. Gravity studies of depth to bedrock, the Boulder batholith and salt domes are discussed in some detail. Gravity studies of massive sulfide deposits, layered mafic intrusions and replacement orebodies are summarized. Finally, research on the "fifth force" is reviewed.

Comparison Of Simple Bouguer, Free Air And Isostatic Residual Gravity Anomalies

Figure 3.23 compares topography, refraction seismic data and gravity anomalies along a northeast-trending profile across Arizona. The profile follows a seismic refraction line (Warren, 1965). The free air and simple Bouguer gravity anomaly profiles are from West (1972). The isostatic anomaly profile is from the Isostatic Residual Gravity Map of the United States (Smith, 1985). The depth of the M-Discontinuity is based on Warren's interpretation of seismic refraction data.

FIGURE 3.23
Gravity and crustal profile across Arizona

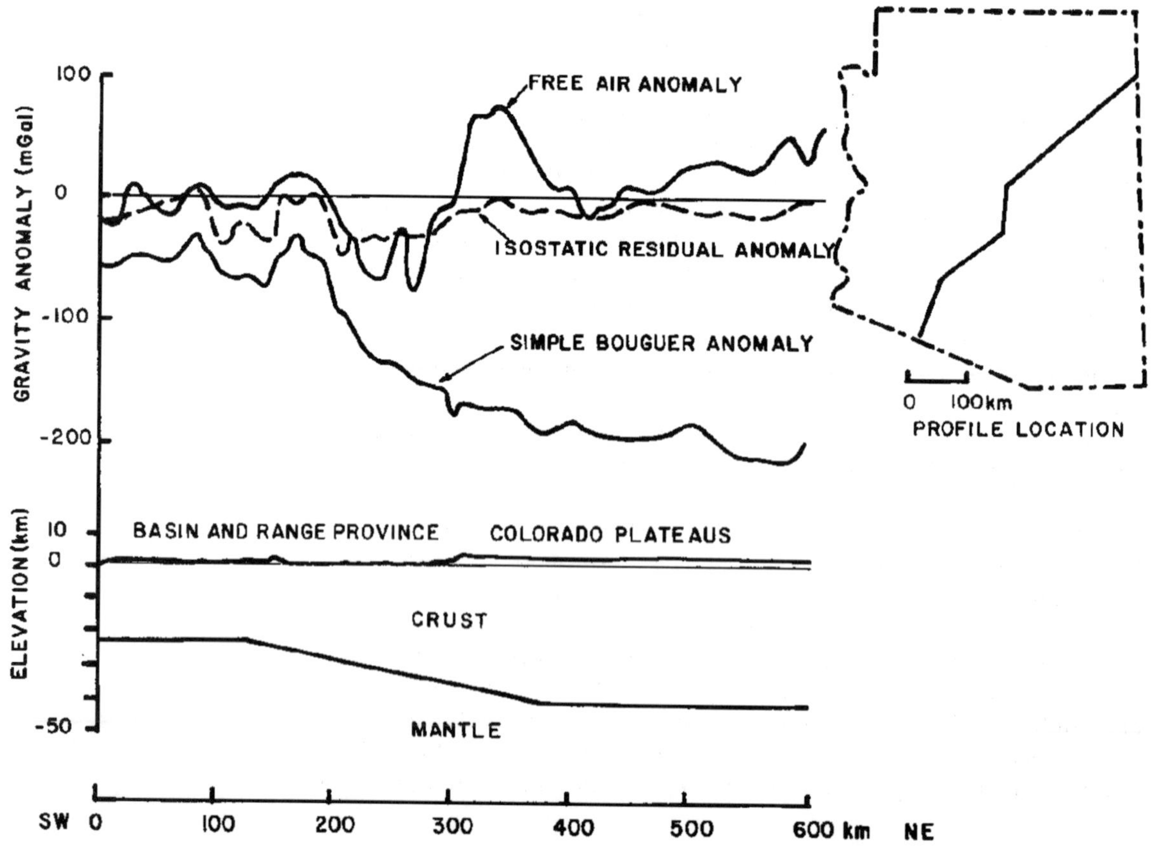

Free Air Anomaly

The free air correction corrects the earth's normal gravity for gravity changes caused by elevation differences of the stations where gravity measurements are made. The free air anomaly is calculated by subtracting the corrected normal gravity from the measured gravity. This removes the gravity effect of earth's rotation, shape and changes in elevation from the measured gravity. No correction is made for the gravity field of the earth's topography or for the gravity field of any compensating mass deficiency that maintains isostatic equilibrium. Although it is not obvious on figure 3.23, many of the short wavelength free air anomalies correlate with topography. This suggests that these anomalies are caused by terrain effects.

Over the full length of the profile, the free air anomaly oscillates about zero milligals. It has been pointed out that the free air anomaly should be near zero in areas where isostatic equilibrium occurs (e.g., Swick, 1942, p. 77). The reason for this is that the positive gravity effect of the topography should be cancelled by the negative gravity effect of the compensating mass (on continents, figure 3.18).

Figure 3.18 assumes "local compensation", i.e., the root forms a magnified mirror image of the surface topography. There is good evidence that local compensation does not occur, because the crust has enough strength that it can support a load. Areas that are $2^{\circ} \times 2^{\circ}$ or larger, however, are probably compensated (Woollard, 1969).

In general, free air gravity anomaly maps are not published for continental areas. The strong correlation of the free air gravity anomaly with topography makes them difficult to interpret. They are published for offshore areas. In these areas, the gravity effect of the topography of the ocean bottom has to be considered.

Simple Bouguer Anomaly

The Bouguer correction adds the gravity effect of an infinite horizontal slab of density P (2.67 gm/cm^3 was used for this profile) to the free air corrected normal gravity. The upper surface of the slab is at the gravity station and the lower surface of the slab is at the geoid. The simple Bouguer anomaly is calculated by subtracting the corrected normal gravity from the measured gravity. This correction and the free air correction remove the gravity effect of earth's rotation, shape, elevation changes and the gravity effect of the slab from the measured gravity. The Bouguer correction partially corrects the normal gravity for the earth's topography. No correction is made for the gravity field of the earth's topography that lies above or below the upper surface of the slab, and no correction is made for the gravity field of any compensating mass deficiency below the geoid that maintains isostatic equilibrium.

Some of the short wavelength free air anomalies that correlate with the topography are removed or reduced in amplitude by the Bouguer correction. For example, the large-amplitude free air anomaly at 340 km is almost completely removed. This anomaly correlates with Mogollon Mesa, a topographic high at the southern edge of the Colorado Plateau. This illustrates on a fairly large scale the effectiveness of the Bouguer correction in removing terrain effects.

The Bouguer correction removes the gravity effect of a homogeneous crust down to the geoid. The gravity effect of the compensating mass deficiency below the geoid remains. Consequently, Bouguer anomalies are usually negative on continents. There is good correlation between the long wavelength decrease of the simple Bouguer anomaly profile and the thickness of the crust as interpreted from seismic refraction data. There is an inverse correlation of the Bouguer anomaly profile and the overall (long wavelength) increase in elevation from the Basin and Range Province to the Colorado Plateau. The increased crustal thickness to the northeast causes a decrease in gravity in this direction. There is evidence that the low-density compensating mass extends beneath the crust into the upper mantle (Warren, 1969). Thus, while the Bouguer correction removes short wavelength terrain effects, there remains a long wavelength inverse correlation of the Bouguer anomaly with elevation that is believed to be caused by the gravity effect of the low-density compensating mass deficiency beneath the geoid.

In addition to the long wavelength decrease in the simple Bouguer anomaly, shorter wavelength anomalies remain that are caused by density inhomogeneities in the crust that have not been removed by the Bouguer correction. These anomalies and others that are not visible at this scale are the anomalies of interest in exploration geophysics.

Simple and complete Bouguer anomaly maps are commonly published for continental areas. The inverse correlation of the Bouguer anomaly with elevation should be kept in mind when Bouguer anomaly maps are used. For example, large-amplitude Bouguer gravity anomaly lows occur over the major topographic highs of the western United States such as the Rocky Mountains and the Sierra Nevada (Lyons and O'Hara, 1982) .

Isostatic Residual Anomaly

On continents, the gravity effect of the mass deficiency resulting from compensation of the earth's topography is reflected in the Bouguer gravity anomaly as an inverse correlation with elevation. If the geometry and density contrast of the compensating mass deficiency is known, its gravity effect can be calculated and removed from the Bouguer anomaly to enhance anomalies caused by density contrasts in the crust, the anomalies of interest in exploration. Unfortunately, little is known about the geometry and density contrast of the compensating mass deficiency, although interpretation of seismic data has provided some information.

One approach used to remove the gravity effect of the compensating mass is to calculate its gravity effect using a model based on isostatic theory. The model used for the Isostatic Residual Gravity Map of the United States is the Airy-Heiskanen model and is described in the first part of this chapter (Smith, 1985). The model used is highly idealized, and the crustal thickness compares rather poorly with crustal thickness determined from seismic data. Despite these limitations, Smith believes that the isostatic correction enhances gravity anomalies related to geologic features in the shallow crust.

The isostatic residual gravity profile on figure 3.23 is near zero mGal and the long wavelength inverse correlation of the Bouguer anomaly with elevation has been removed. Similarities and differences occur in the short wavelength anomalies shown on the Bouguer and isostatic residual anomaly profiles. The differences could be caused by the different data sets used, inaccurate plotting from the small-scale (1:2,500,000) isostatic residual gravity anomaly map, real differences in gravity caused by sources in the crust or the addition or removal of anomalies by the isostatic correction. Jachens and Griscom (1985) give examples of anomalies that are not easily recognized or are difficult to interpret quantitatively for California.

Summary

Free air gravity anomalies are not used in exploration because terrain effects obscure the gravity anomalies caused by density contrasts beneath the surface. Simple and complete Bouguer and isostatic residual gravity maps can be used for interpreting gravity anomalies caused by density contrasts beneath the surface that are of interest in exploration. The interpreter must be aware that terrain effects can influence simple Bouguer gravity anomalies.

Comparison Of Measured Gravity And Free Air And Simple Bouguer Gravity Anomalies

Figure 3.24 shows a plot of surface elevation (surveyed to the nearest foot), measured gravity, free air anomaly and simple Bouguer anomaly (2.67 gm/cm^3 reduction density) for an east/west gravity profile near Florence, Arizona. Since the profile is east/west, the normal gravity (equation 3.15) is constant. The trend of the hills and washes is northerly, perpendicular to the cross section. All of the gravity profiles decrease to the east. The discussion in the following paragraph disregards the long wavelength easterly gravity decrease. The short wavelength highs and lows are the subject of interest.

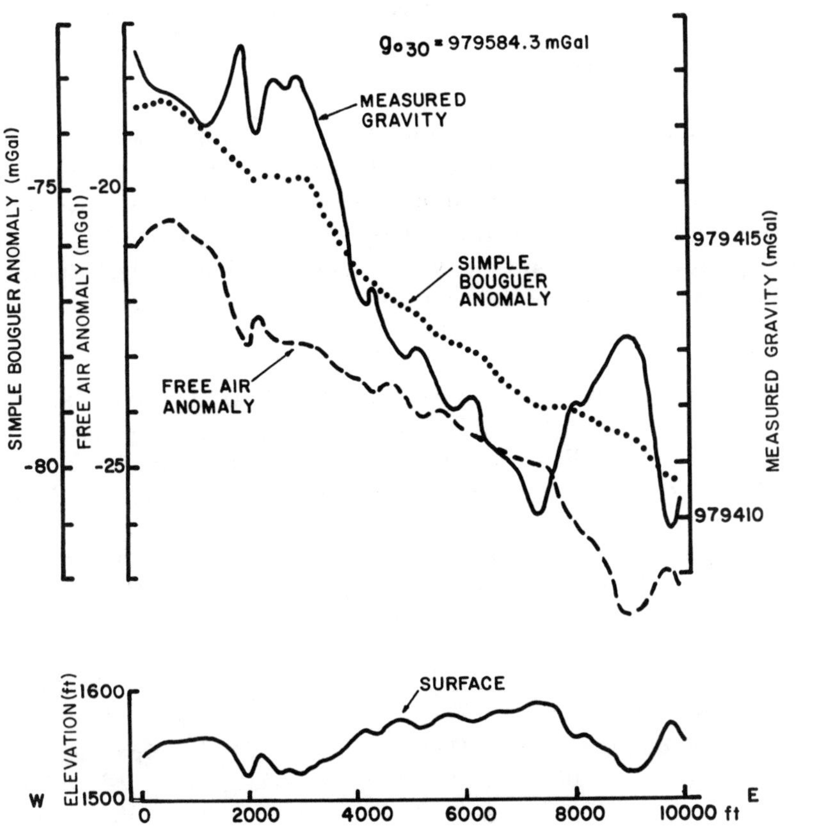

FIGURE 3.24
Gravity profile near Florence, Arizona

The measured gravity correlates inversely with topography. The free air anomaly shows a positive correlation with topography. There is very little correlation of the simple Bouguer anomaly with topography. This illustrates the effectiveness of the Bouguer correction in removing terrain effects from the measured gravity. Close inspection of the simple Bouguer anomaly profile reveals small-amplitude lows over some of the sharp peaks in the topography and, possibly, over the broad high area from 4000 ft to 8000 ft along the profile. In the simple Bouguer anomaly, terrain effects are almost always negative in sign. Consequently, a correlation of gravity lows with topographic highs suggests terrain effects that the Bouguer correction did not remove. When complete Bouguer anomalies correlate with topography, presumably the anomalies are caused by density contrasts that occur beneath the surface. It is also possible that errors occur in the terrain correction.

The long wavelength simple Bouguer gravity anomaly decrease to the east is caused by a decrease in the density of the rocks in this direction. Part of this decrease is probably caused by an increase in the thickness of the crust to the northeast (figure 3.23). The short wavelength decrease near 4000 ft may be caused by faulting that drops Precambrian and Late Cretaceous to Early Tertiary granites down to the east. It is also possible that this decrease is caused by a terrain effect.

Porphyry copper mineralization occurs in this area, and the purpose of the gravity survey was to attempt to determine the depth to granitic rocks which host the mineralization. Terrain corrections should have been made on the data in this area to see if the short wavelength decrease in the simple Bouguer anomaly profile at 4000 ft can be explained by terrain effects.

Depth To Bedrock

Estimates of the depth to bedrock in covered areas can be made using gravity surveys, because bedrock is usually more dense than the overlying cover. When mineralization is observed in outcrop or in drill holes that lie near a covered area, determination of the depth of the cover becomes important. If a groundwater supply is needed, determination of the depth or thickness of the basin fill deposits is equally important when the basin fill deposits are the groundwater aquifer.

Accurate quantitative estimates of the depth to bedrock with gravity data require an accurate regional-residual separation, knowledge of the density contrast between the basinfill deposits and bedrock, gravity models that reasonably represent the geologic situation in the area of interest, and relatively homogeneous bedrock and basinfill deposits. In most situations, it is not possible to meet these requirements, and relatively large errors can occur in the depth estimates. It is often possible, however, to give the relative depth to bedrock, i.e., one can say that the basin fill deposits are shallow here and deep there. This is a major contribution to most exploration situations.

Pediment Exploration Using Data With Large Station Spacings

Figure 3.25 shows a Bouguer gravity anomaly map in the vicinity of Florence, Arizona (Peterson, 1965). The density used for the Bouguer correction is 2.67 gm/cm^3 and the contour interval is 2 mGal. Gravity stations are shown by the black dots. The normal gravity was calculated from the 1930 International Gravity Formula. The Potsdam gravity datum was used. No information on how gravity station elevations were measured was given. Terrain corrections were made on stations that are near the mountains. The terrain corrections for stations in the valleys are assumed to be negligible.

Bedrock includes Precambrian metamorphic and granitic rocks that are intruded by Late Cretaceous to Early Tertiary (Laramide) granitic rocks. Bedrock and small areas of Tertiary basalts are surrounded by Tertiary basin fill deposits. Exploration for porphyry copper mineralization that is related to the Laramide granitic intrusions has been conducted in this area at various times in the past. Gravity surveys aided in the location of shallow bedrock in covered areas.

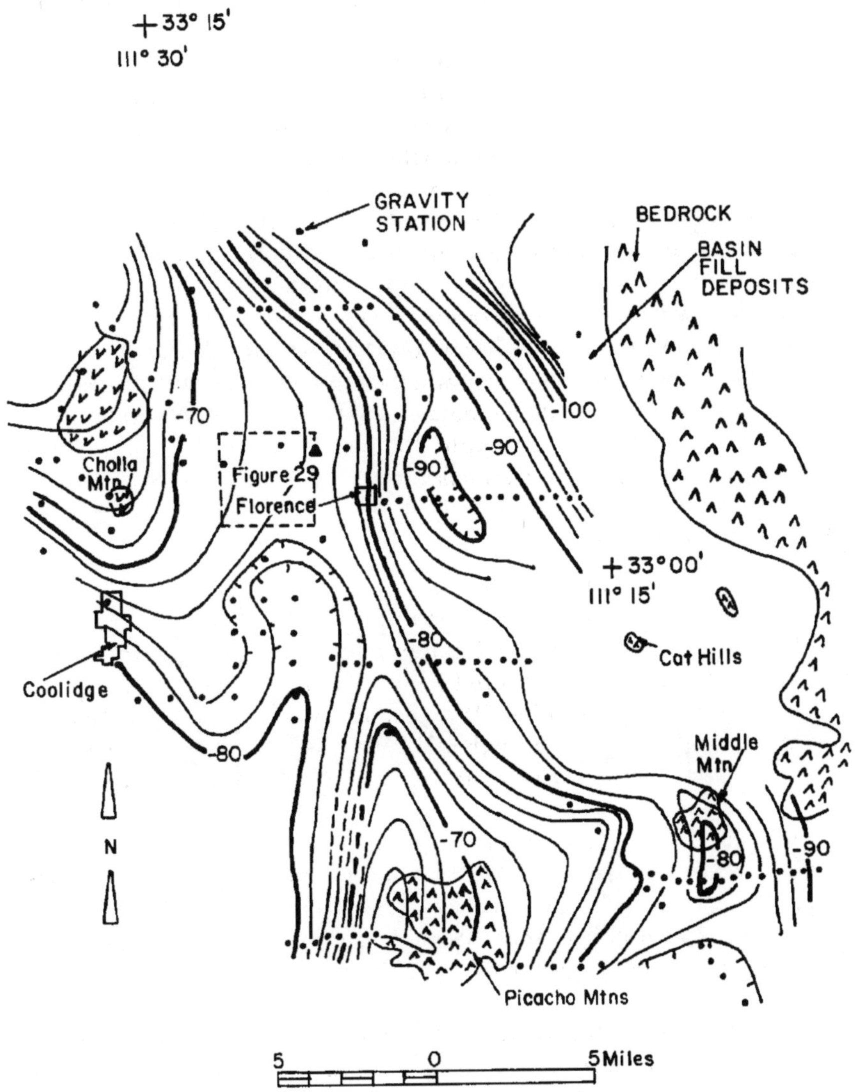

FIGURE 3.25
Bouguer gravity anomaly map of the Florence, Arizona, area

In general, the granitic rocks are more dense than the basin fill deposits, but exceptions to this generalization occur. For example, a petroleum exploration well encountered almost 6000 ft of anhydrite in the Picacho Basin west of the Picacho Mountains. The density of this rock was measured with a downhole gravity meter and it ranged from 2.55 to 2.67 gm/cm^3. It is not unusual for the density of bedrock to lie within this range.

This area (figure 3.25) lies in the Basin and Range Province, and the gravity map is typical of maps in this province. Gravity highs correlate with bedrock outcrops, and gravity lows correlate with basin fill deposits. When gravity highs extend into covered areas from bedrock, they may indicate that high-density granitic rocks form a pediment that extends beneath the basin fill deposits. Since bedrock is more dense than the basin fill deposits, it is assumed that the density contrast between these rocks accounts for the Bouguer gravity highs and lows. It is possible that intrabasement density changes also cause some of the gravity anomalies.

Superimposed on the Bouguer gravity highs and lows is the northeasterly decrease in the Bouguer anomalies caused by isostatic compensation of the higher topography to the northeast. Gravity values on bedrock give an indication of this regional gravity gradient. On bedrock north of Cholla Mountain, Bouguer anomalies are about -64 mGal. Bouguer anomalies on bedrock east of Florence are about -100 mGal (these gravity stations are not shown on figure 3.25).

A Bouguer gravity anomaly high extends north from the Picacho mountains into the area just west of Florence. This high may indicate that a shallow bedrock ridge extends north into the Florence area from the Picacho Mountains. Bouguer anomalies decrease to the east from the gravity high over bedrock north of Cholla Mountain. A rather sharp gravity decrease just east of bedrock may indicate that a north-trending fault drops bedrock down to the east. A similar feature occurs just east of Florence. These strong gravity gradients cannot be accounted for by the northeasterly regional decrease in gravity, and they must be caused by near-surface density contrasts.

Analysis of these widely spaced gravity data is summarized in figure 3.26. High-angle normal faults are plotted at the maximum gravity gradients to account for the lateral change in density that must occur. The change in density is caused by shallow, high-density bedrock on the upthrown side of the fault and thick, low-density alluvium on the downthrown side of the fault.

FIGURE 3.26
Interpretation of figure 3.25

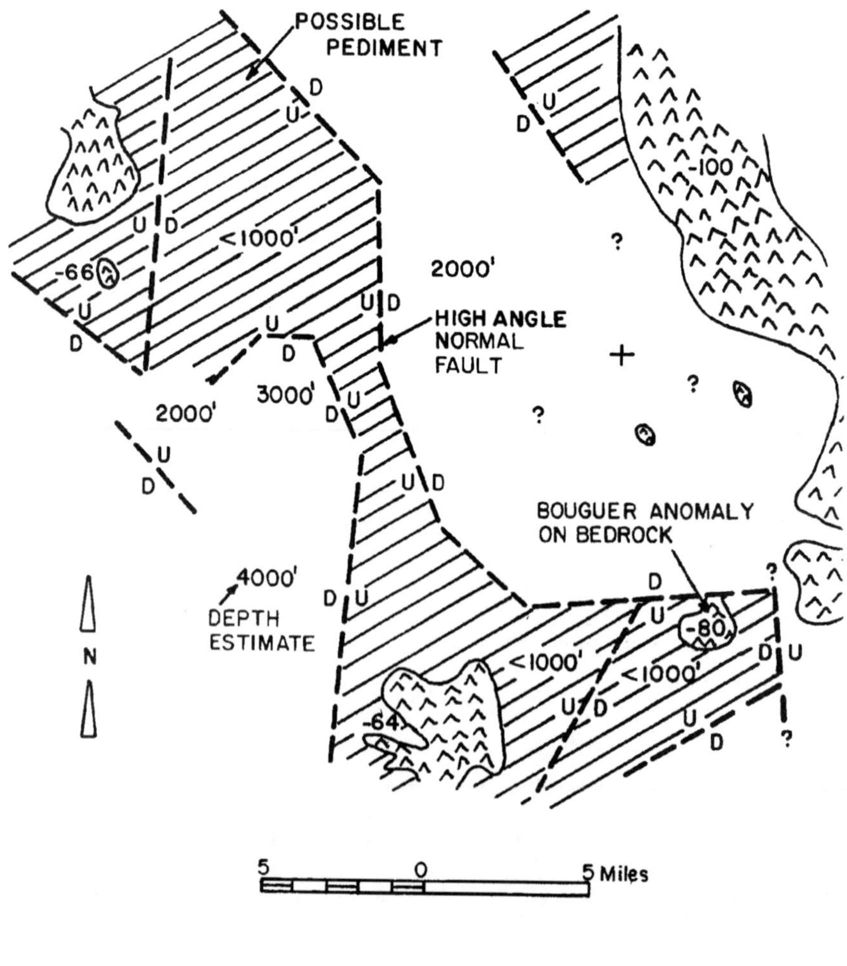

Depth estimates are also made for the basin fill deposits. The regional decrease in the Bouguer gravity to the east is assumed to be linear with distance from bedrock outcrop to bedrock outcrop. Using this assumption, the regional gravity can be calculated in covered areas. The residual gravity at any location in the covered area can be estimated by subtracting the calculated regional gravity from the Bouguer gravity. Depth estimates are made by assuming a reasonable density contrast between bedrock and the basin fill deposits (0.3 gm/cm^3) and using an appropriate model to calculate the depth of alluvium necessary to cause the residual gravity anomaly. An infinite horizontal slab is chosen as a model, because it is simple to use and because the width-to-depth ratio of the basins is large. This model is only appropriate for the center of a wide basin or a broad pediment.

Figure 3.26 suggests that reasonably large parts of the covered area near Florence may be shallow enough to explore for buried porphyry copper mineralization (when the price of copper makes such exploration feasible). The area with deeper basin fill deposits is the place to look for groundwater.

The interpretation method that was used is simple and straightforward. It can be used to evaluate large areas in a relatively short time. If there is interest in an area where pediments are suggested to occur, detailed gravity surveys can be carried out to further evaluate the area or a drill hole can give a direct test of the depth to bedrock.

Pediment Exploration With Detailed Gravity Data

Figure 3.27 shows a complete Bouguer gravity anomaly map on the west side of the Tucson Mountains west of Tucson, Arizona. The reduction density was 2.67 gm/cm^3. The 1930 International Gravity Formula was used to calculate the normal gravity. The Potsdam base station datum was used.

Three drill holes shown on figure 3.27 were drilled in the late '60s on the pediment on the west side of the Tucson Mountains. Their purpose was to explore for porphyry copper mineralization. The two easternmost drill holes encountered bedrock at depths of 300 ft to 400 ft. The westernmost drill hole bottomed in alluvium at a depth over 700 ft. City of Tucson water wells in the vicinity of Ryan Field have depths of several thousand feet and bottom in alluvium.

A U.S. Geological Survey gravity survey of the area was already available (Plouff, 1961). Stations in this area showed a strong northwest-trending gravity gradient that suggested that a large normal fault had dropped bedrock down to the west in Avra Valley. After the westernmost drill hole bottomed in alluvium, it was decided to attempt to pinpoint the location of the fault with detailed gravity survey lines A, B and C so that future drilling could be located on the pediment.

Figure 3.28 shows the results of line B. The position of the maximum horizontal gravity gradient for each line is plotted on figure 3.27 and is interpreted as a normal fault. Drilling east of this fault encountered shallow bedrock.

These quality data also allowed for interpretation of the fault dip, average depth to bedrock on the upthrown and downthrown side of the fault, and the density contrast between bedrock and the basin fill deposits using characteristic curves for a two-dimensional step model (Grant and West, 1965, p. 282). The results gave a dip of 60°, a depth to bedrock of 300 ft on the upthrown side of the fault, a depth to bedrock of 6000 ft on the downthrown side of the fault and a density contrast of 0.3 gm/cm^3.

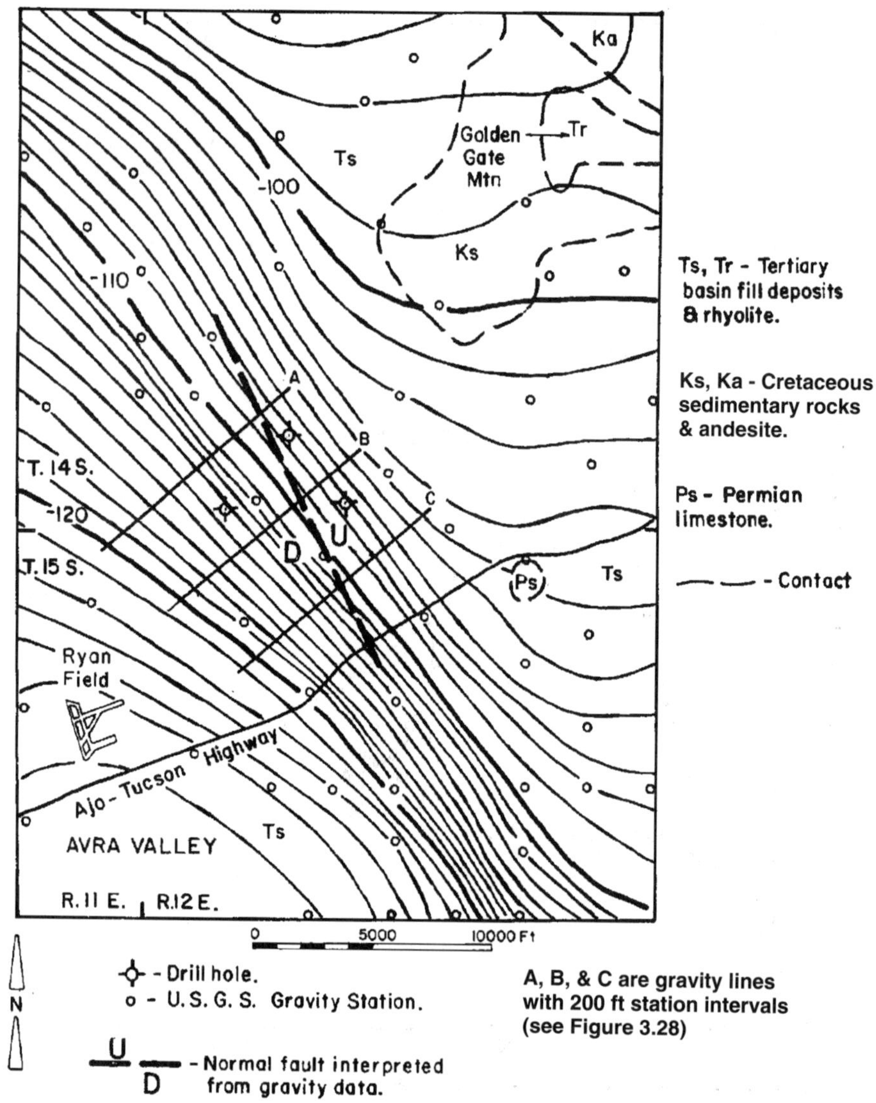

FIGURE 3.27
Gravity surveys west of the Tucson Mountains, Pima County, Arizona

Several lower-amplitude maximums occur in the horizontal gravity gradient on line B and also on lines A and C. These maximums may indicate that small displacement step faults occur on the pediment. They may also result from noise in the gravity data.

Estimates Of Depth To Bedrock On A Shallow Mineralized Pediment
A small outcrop of altered Laramide quartz monzonite occurs near Poston Butte (the triangle on figure 3.25), which is a remnant basalt flow. A water well south of Poston Butte encountered copper oxides at relatively shallow depths. Exploration drilling in this area in the early '70s defined a substantial area of buried porphyry copper mineralization at depths of about 400 ft. As exploration proceeded, two drill holes in this area encountered depths to bedrock of more than 1000 ft. At this point, a detailed gravity survey was run to locate new drill holes in areas with shallow cover.

Since drilling indicates that bedrock is relatively homogeneous Laramide quartz monzonite, the possibility that intrabasement density contrasts occur is decreased. Alteration and mineralization

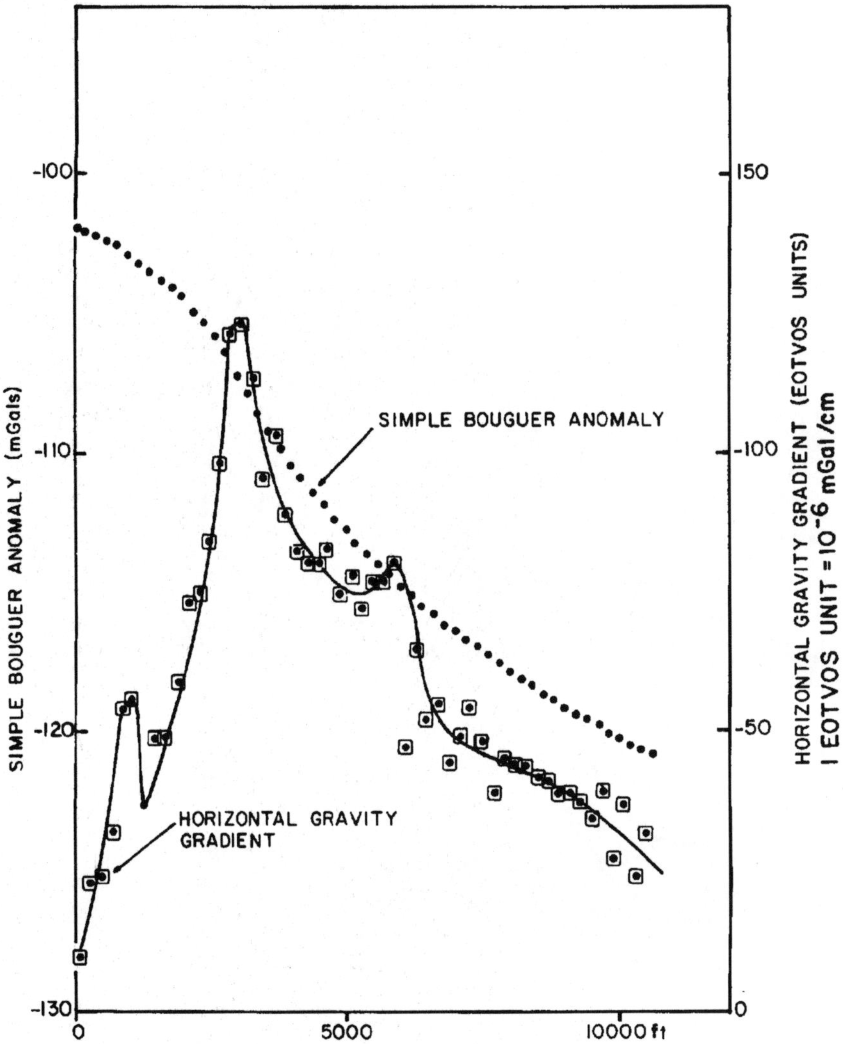

FIGURE 3.28
Gravity profiles for line B in figure 3.27

by the hydrothermal system, however, could have caused variations in the density of bedrock. No density measurements were made to test for possible changes in density of the quartz monzonite caused by alteration and mineralization. The average density of four core samples of quartz monzonite was 2.60 gm/cm^3. Figure 3.29 shows data from the detailed gravity survey in the area immediately southwest of Poston Butte. Also shown is the elevation of bedrock as determined by drilling. The gravity data reduction density is 2.67 gm/cm^3. The Potsdam base station datum was used, and the normal gravity was determined using the 1930 International Gravity Formula. Gravity station elevations were surveyed to the nearest foot. Terrain corrections were not made on the data, because the surrounding terrain is flat. With the exception of Poston Butte (elevation 1748 ft), the surface elevation on figure 3.29 only varies from 1440 ft to 1500 ft.

As shown on figure 3.29, several low-amplitude Bouguer anomaly highs and lows occur in the Poston Butte area. There is an easterly decrease in the Bouguer anomalies that is probably caused by the northeasterly regional decrease in gravity. Presumably, the low-amplitude gravity features are caused by the undulations of the bedrock topography beneath the basin fill deposits. The north-trending gravity low that correlates with the bedrock low just west of the center of figure 3.29 is encouraging. On the other hand, another north-trending low just east of the map center lies over a relatively flat bedrock surface.

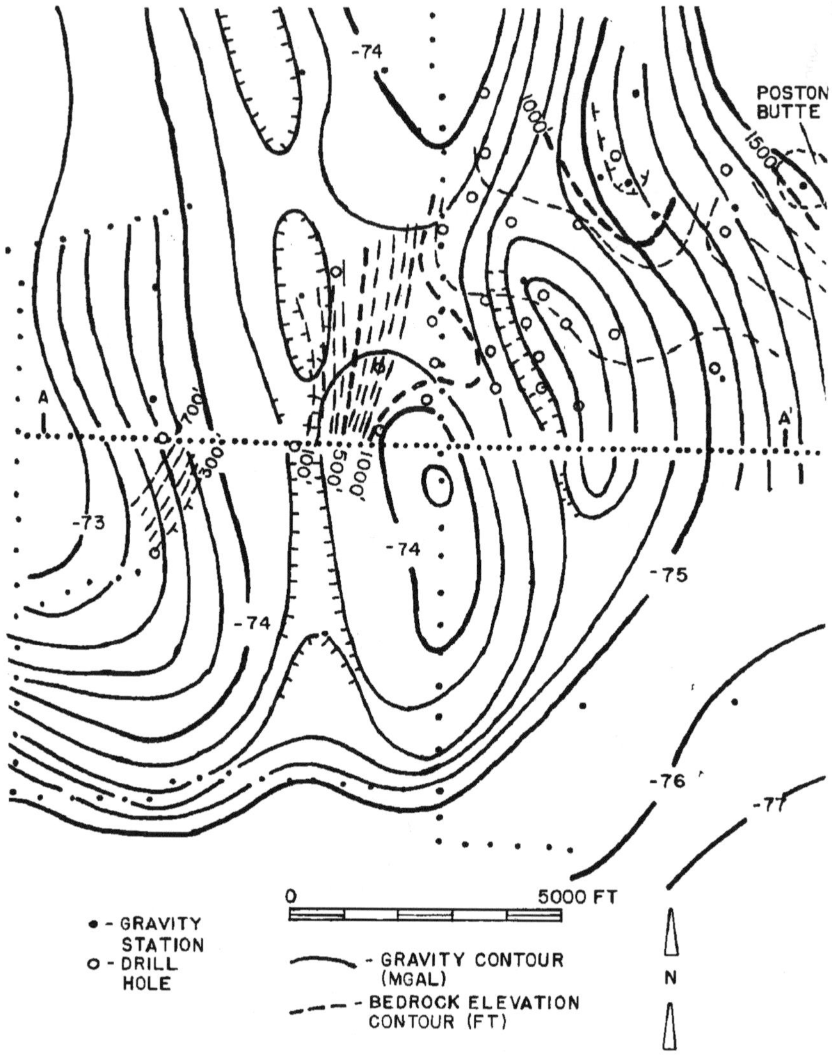

FIGURE 3.29
Comparison of Bouguer Anomalies and Bedrock Elevation near Florence, Arizona

Figure 3.30 shows Bouguer gravity anomaly profile AA′ located on figure 3.29. The Bouguer anomalies are constant, about -74 mGal at the location where the bedrock elevation decreases from about 1000 ft to 0 ft. This is caused by the easterly decrease in the regional gravity that has "flattened" the Bouguer anomalies. An estimated regional gravity decrease of about 2 milligals per mile is determined from Bouguer anomalies on bedrock north of Cholla Mountain (figure 3.25, -62 mGal) and east of Florence and north of Cat Hills (-100 mGal). This regional gravity gradient is too steep. A more appropriate regional gradient of 0.8 mGal/mi is shown on figure 3.30.

The residual gravity is calculated by subtracting the regional gravity from the Bouguer gravity. A low-amplitude, one-milligal gravity low occurs over the bedrock low. Another smaller gravity low occurs further east where there is no bedrock low. On the east end of the line, another residual gravity low occurs. This is the beginning of a major decrease in gravity that is believed to be caused by a north-trending normal fault that drops bedrock down to the east, just east of the city of Florence (figures 3.25 and 3.26).

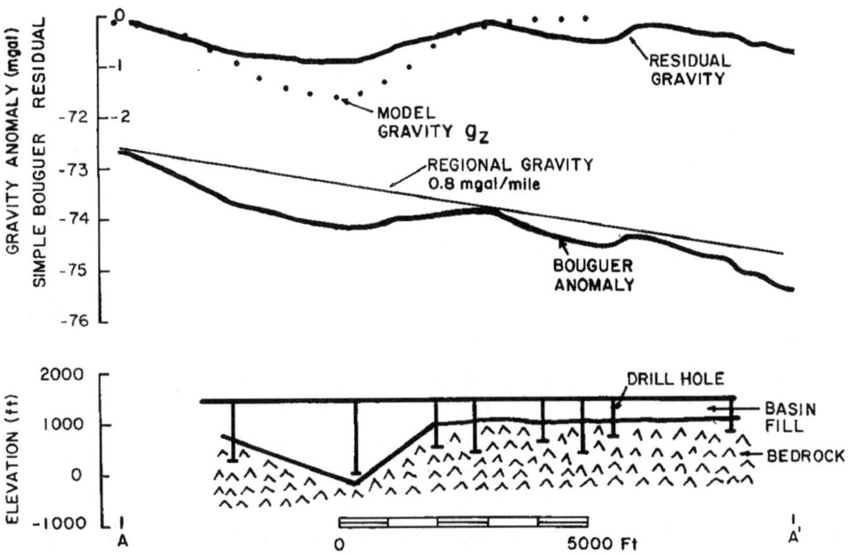

FIGURE 3.30
Profile AA' on figure 3.29

The gravity effect, g_z, at the surface, of a two-dimensional gravity model of the bedrock low was calculated. High-density, Mid-Tertiary conglomerate fills the bedrock low. Density measurements on eight core samples of the conglomerate averaged 2.40 gm/cm^3. A density contrast of 0.20 gm/cm^3 was used for the model, since the density of samples of quartz monzonite averaged 2.60 gm/cm^3. The basin fill deposits that lie above the bedrock trough (above an elevation of 1000 ft) were disregarded in the model calculations, since their gravity effect is a constant value (they are assumed to form an infinite horizontal slab with a thickness of about 400 ft).

The model gravity effect has an amplitude that is about 0.6 mGal greater that the residual gravity. Despite the difference, it is obvious that the gravity measurements have detected the bedrock low. The Bouguer anomalies can be used as a guide for the drilling, because the gravity lows probably indicate greater bedrock depths. The fact that a low-amplitude residual gravity low occurs where there is no increase in bedrock depth indicates that some caution must be used in evaluating areas for drilling based on the gravity data. Quantitative depth estimates can also be made by more sophisticated modeling of the gravity data, but even more caution should be used with this approach.

The most important requirement for successful use of the gravity method in this area would be to obtain more gravity stations so that the location of the gravity anomalies can be more accurately defined.

Gold Exploration On Pediments Of The Basin And Range Province

All of the previous examples are for porphyry copper mineralization in Arizona. These examples apply equally well to pediment exploration throughout the Basin and Range Province, with one exception: low-density, nonwelded silicic ash flow tuffs that can host precious metal ore deposits. Porphyry copper mineralization is Laramide in age and premineral rocks, including volcanic rocks, are relatively dense. Gravity exploration for pediments made up of these rocks is successful, because a significant density contrast exists between the host rocks and basin fill deposits that cover the pediments.

Precious metal mineralization in the Basin and Range province is often Tertiary in age. The density contrast between the host rocks and the basin fill deposits varies from several tenths of a gram per cubic centimeter for precious metal deposits that are hosted in marine sedimentary or plutonic rocks to virtually zero for precious metal deposits that are hosted in nonwelded ash flow tuffs. Analysis of density log information from 34 boreholes at Yucca Flat, Nevada, showed no statistical difference in density between Miocene silicic ash flow tuffs and basin fill deposits (Ferguson and others, 1988). The fact that large-amplitude Bouguer gravity anomaly lows can occur over Tertiary volcanic rocks exposed in the ranges of the Basin and Range Province indicates that these rocks can have a low density relative to the underlying "bedrock."

Unfortunately, the low density of nonwelded Tertiary ash flow tuffs may limit the use of the gravity method in determining the location of pediments formed by these rocks. If the volcanic host rocks form a relatively thin cover over bedrock, however, gravity surveys may still be useful in locating pediments. Also, as the welding of the tuff increases, so will its density and so will the possibility of detecting the change in density between the tuff and the basin fill deposits.

Groundwater Exploration In A Small Shallow Basin

A water supply was required for a mine in northern Chile. A search for groundwater was carried out in basin fill deposits near the mine. Gravity data were collected to guide the test drilling.

Figure 3.31 shows a Bouguer gravity anomaly map over one of the prospective areas. Bedrock includes Jurassic intrusive and volcanic rocks.

An arbitrary base station datum was used for the gravity survey. The normal gravity that was used is unknown. Elevations were surveyed to the nearest 0.1 m. The density for the Bouguer reduction is 2.20 gm/cm^3. The terrain is rugged in some of the areas where bedrock crops out, and terrain effects have probably influenced the data. No terrain corrections were made on the data, because topographic maps adequate for calculating these corrections were not available.

FIGURE 3.31
Gravity survey in Northern Chile

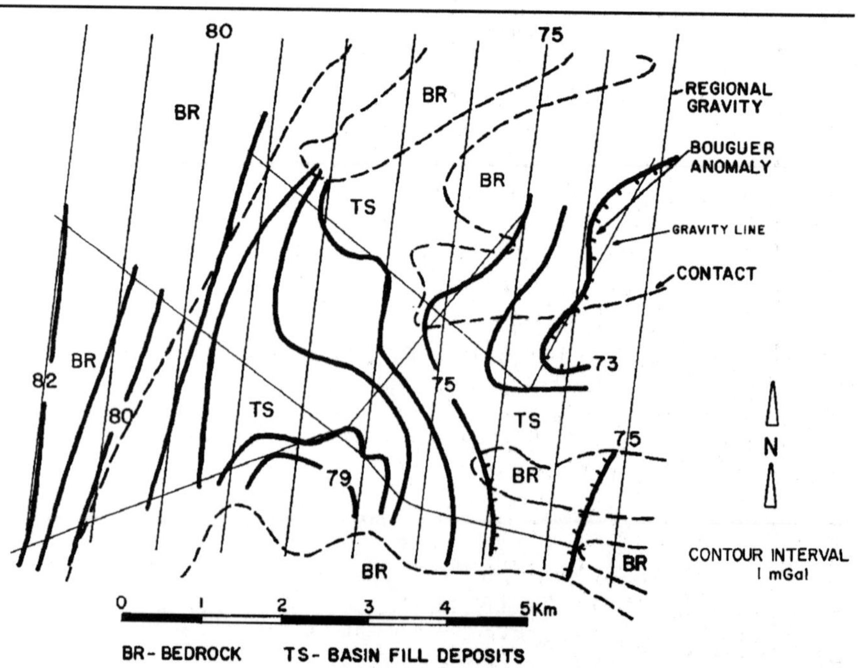

The purpose of the gravity survey was to determine where the volume of basin fill deposits is greatest so that test wells could be located in these areas. There is no obvious Bouguer gravity anomaly low over the basin fill deposits (figure 3.31). Inspection of the Bouguer gravity anomalies on bedrock indicate that the regional gravity anomalies decrease to the east. An easterly decreasing planar gravity was estimated from the Bouguer anomalies and is shown on figure 3.31. The regional gravity was subtracted from the Bouguer gravity, and the residual gravity is shown in figure 3.32.

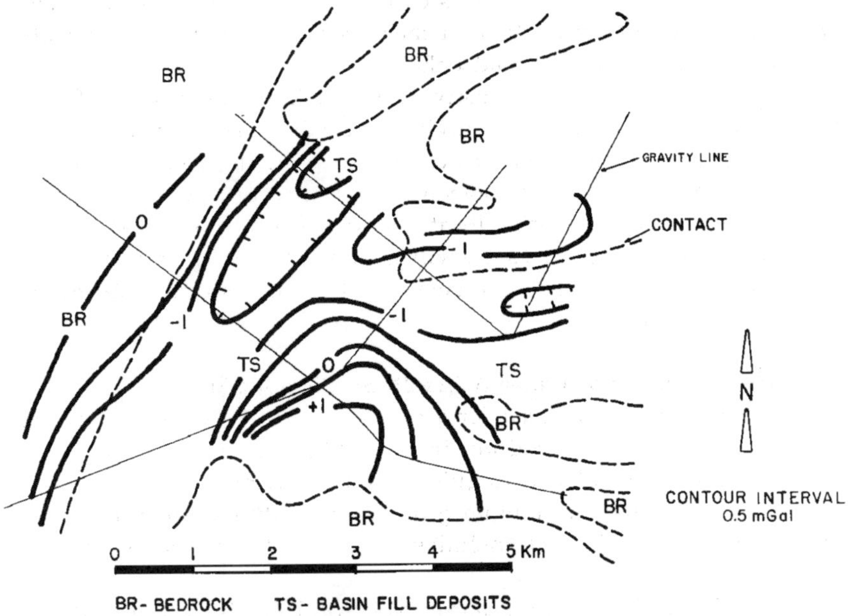

FIGURE 3.32
Residual gravity anomaly in Northern Chile

A northeast-trending residual gravity low suggests that the largest volume of basin fill deposits lies in a structurally controlled basin that has the same trend as the western contact between bedrock and the basin fill deposits. A strong gravity gradient associated with this contact suggests that it is fault-controlled. A residual gravity high extends to the northeast from bedrock in the south-central part of figure 3.32 and indicates that bedrock is probably shallow in this area. A narrow east/west-trending gravity low on the east side of figure 3.32 suggests that a narrow basin occurs here.

Subsequent test drilling verified the interpretation given above. Significant thicknesses of basin fill deposits were located along the residual gravity lows. Unfortunately, not all test wells intersected saturated basin fill deposits. Dipole-dipole resistivity surveys were used to locate areas where the basin fill deposits were saturated, and an adequate groundwater supply was found for the mine.

Discussion Of Gravity, Magnetics And Geology Of The Boulder Batholith, Montana

This discussion was written by Edgar O. McAlister, Consulting Mining Geophysicist, 524 Highway 208, Yerington, NV, 702/463-3599.

The Boulder batholith, located in central western Montana, is a crudely elliptical intrusive about 64 mi long and 30 mi across (figure 3.33; Klepper and others, 1971; Knopf, 1963). It invades rocks varying in age from lower Precambrian through Mesozoic. The Elkhorn Mountain volcanic series of late Cretaceous age is intruded, and the remains of this series occurs as roof pendants and remnants on the margins of the batholith. The Elkhorn volcanics are sufficiently allied in time and space to the Boulder batholith to suggest that they may be an early part of the batholith (Robinson and others, 1968; Becraft and others, 1963).

A late igneous event is the Lowland Creek series, consisting of flows and dykes of quartz latite, Oligocene in age (37-24 m.y.). Lowland Creek volcanics cover a large portion of the western batholith in a belt extending from west of Silver Bow to Basin (included as Czr on figure 3.33) while a dyke series and flow remnants follow the central core of the batholith from Basin past Wickes Corbin (not all shown on figure 3.33). The Lowland Creek event is so much later than the main batholith (78 -70 m.y.) that it is not considered to be part of the cooling history but seems more likely to be a resurgence from a deeper source following the older batholith emplacement fracture system.

FIGURE 3.33
Generalized geologic map of the Boulder Batholith modified from Klepper & others, 1971

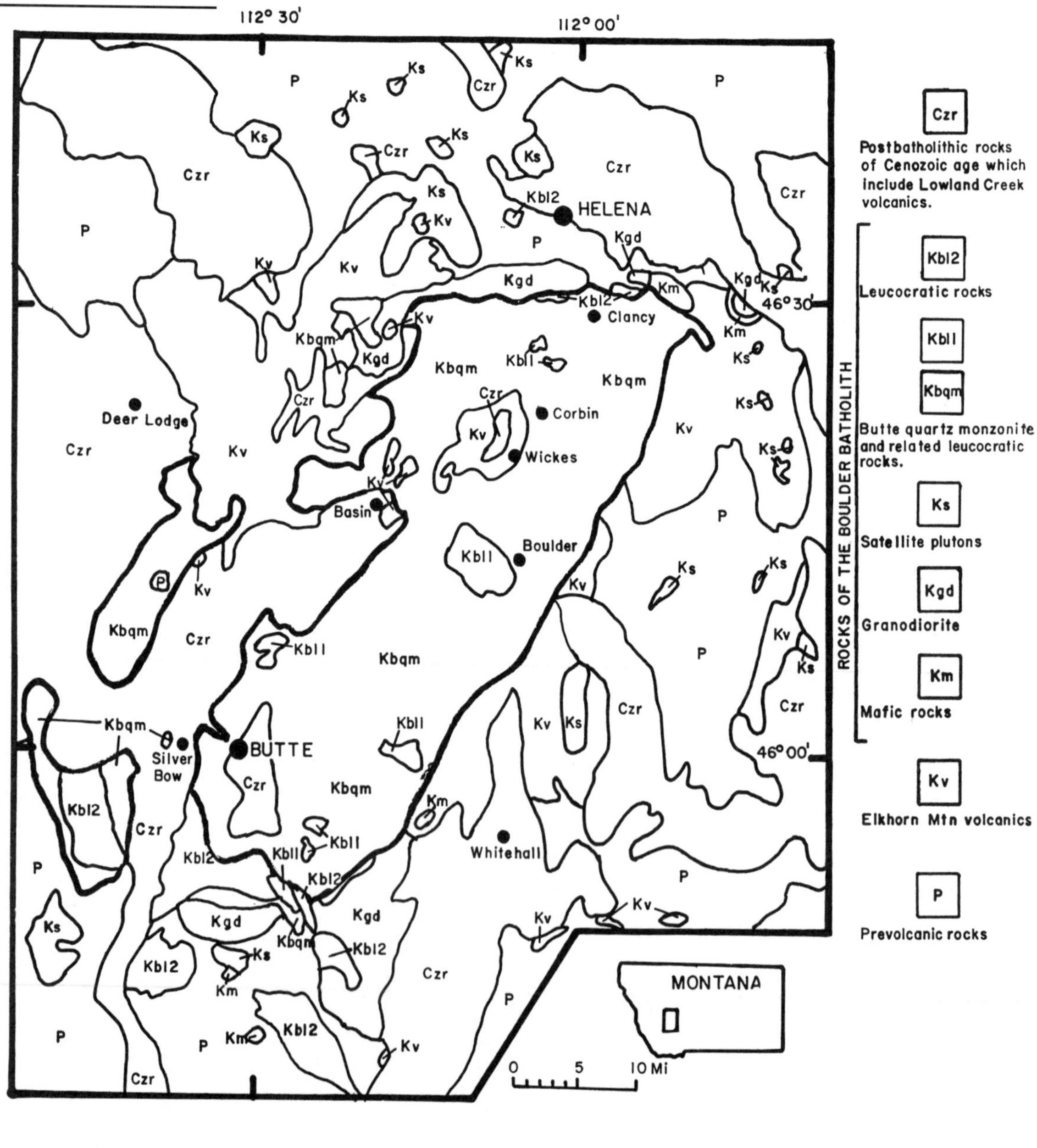

FIGURE 33 - GENERALIZED GEOLOGIC MAP OF THE BOULDER BATHOLITH MODIFIED FROM KLEPPER & OTHERS, 1971

The largest outcrop area of the batholith is occupied by Butte quartz monzonite, but there is a series of more basic intrusives (granodiorite to diorite) exposed along the northwest, north, southeast and south margins of the of the batholith. No such basic rocks are exposed along the eastern margin of the Deer Lodge Valley, but gravity and magnetic surveys (figures 3.34 and 3.35) indicate that they may exist below the cover.

Outcropping generally along the long axis of the batholith (Becraft and others, 1963) and extending from southwest of Butte northeast to the old placer camp of Clancy, is a belt of irregular and dyke-like masses of alaskite, aplite and pegmatite that appear to be more concentrated near the batholith axis (Kbl1 and Kbl2 in figure 3.33). Other leucocratic intrusives are found within the batholith, but some of the larger ones are found southwest of Butte.

A gravity survey (figure 3.34; modified from Klepper and others, 1971 who used a map by Burfeind, 1967) shows an interior low over the central part of the batholith. This low is surrounded by a poorly defined belt of gravity highs associated with the exterior basic intrusives mentioned above. Measured densities on rocks of the Boulder batholith given by Klepper and others (1971) are listed on page 212.

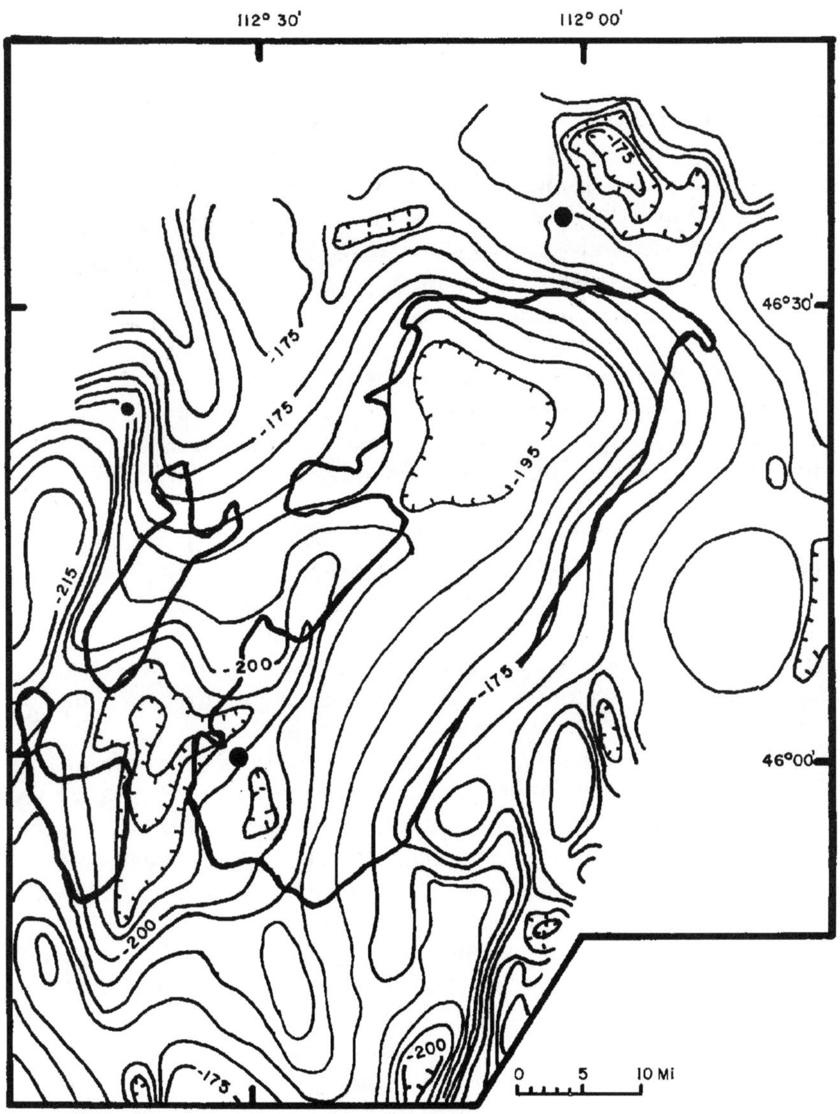

FIGURE 3.34
Bouguer gravity anomaly map of the Boulder Batholith modified from Klepper & Others, 1971

Measured Densities Of Rocks Of The Boulder Batholith

Rock	Type	Density (gm/cm3)
Czr	Postbatholith rocks of Cenozoic age	2.4?
Kbl2	Biotite granite of Knopf (1963)	2.62
	Biotite adamellite of Knopf (1963)	2.65
	Hell Canyon pluton	2.65
	Moose Creek pluton	?
	Climax Gulch pluton	2.67
	Donald pluton	2.64
Kbl1	Alaskite, aplite and pegmatite	2.54-2.65
	Leucocratic variant at Homestake	2.65-2.67
	Leucocratic variant at Pulpit Rock	2.67
Kbqm	Butte quartz monzonite	2.67-2.75
Ks	Satellite plutons	2.65-2.80
Kgd	Unionville granodiorite	2.75-2.82
Ks	Rader Creek pluton	2.72-2.75
	Pluton north of Fish Creek	2.72-2.76
Km	Syenogabbro, syenodiorite & monzonite	2.73-2.90
Kv	Elkhorn Mountain volcanics avg. 2.70	2.60-2.91
P	Prevolcanic sedimentary rocks	2.60-2.78
	Lower Precambrian metamorphic rocks avg.2.8?	2.67-3.10

Airborne magnetics (figure 3.35, modified from Klepper and others, 1971, who used a map by Johnson and others, 1965) reveal a similar pattern: a central magnetic low over the interior and an irregular set of magnetic highs following the basic margin. In general, the susceptibility of rocks in the vicinity of the Boulder batholith is, from high to low, granodiorite - Butte quartz monzonite, Elkhorn volcanics - Lowland Creek volcanics. The susceptibility of the Butte quartz monzonite is about 2000 x 10^{-6} cgs.

This geophysical pattern is similar to that of the Guichon batholith (Ager and others, 1972) which is located in south-central British Columbia. The Guichon batholith is a host for many important porphyry copper deposits. This batholith is older (Late Triassic - Early Jurassic) than the Boulder batholith (Laramide) and has been stripped to a deeper level. The central core of granodiorite/quartz monzonite is surrounded by concentric rings of several varieties of quartz diorites.

These data suggest that the Boulder batholith may contain a deeper central core of rock somewhat more acidic than the present exposed Butte quartz monzonite. This idea is supported by the magnetic low in the core of the batholith, a difficult problem to account for if the Butte quartz monzonite goes to appreciable depth (the Butte quartz monzonite has a fairly high magnetic susceptibility of 2000 x 10^{-6}).

This concept of a more acidic central core affects the accuracy of gravity depth estimates for the batholith bottom. Such estimates have been made by Bonini (1969), Burfeind (1967), and Klepper and others (1971). These authors assume the batholith to be Butte quartz monzonite to the base. When the effect of a more acidic core is considered, the bottom depth would be lessened, but this can be offset by the possibility of more basic rocks near the bottom. Also hard to visualize is the effect of large masses of lower density, Lowland Creek-age-rocks at depth.

The Boulder batholith has very widespread mineralization, located both on the margin and also within the interior. Two ages of mineralization are present: one occurring late in the cooling history of the batholith, and one associated with the Lowland Creek event. Examples of both ages of mineralization will be given. For more information on ages of various parts of the batholith, see Tilling (1968, figure G-1).

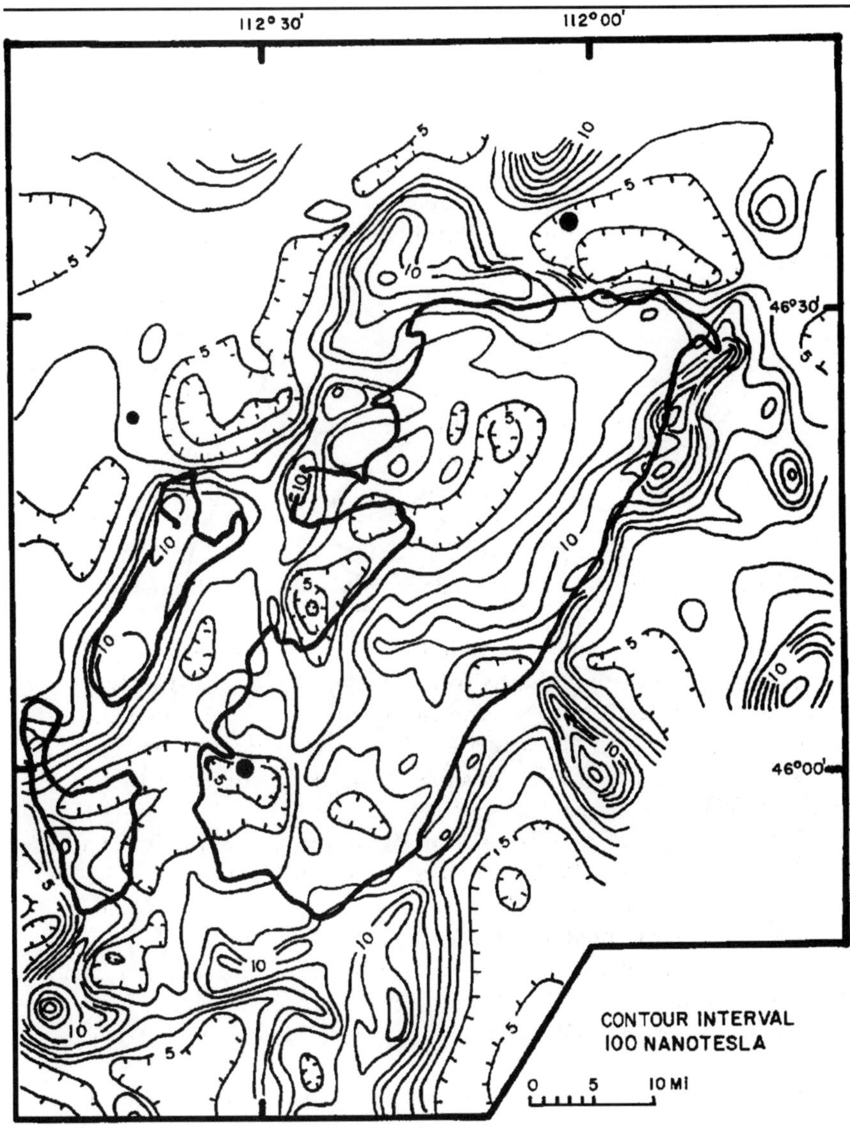

FIGURE 3.35
Aeromagnetic map of the Boulder Batholith modified from Klepper & others, 1971

Mineralization Of Late Boulder Batholith Age

The Butte district (Sales, 1914; Meyer and others, 1968; Miller, 1973) is a major producer of copper, zinc, molybdenum, silver, gold and manganese. The upper part of the district (figure 3.36) is a complex system of veins occupying an area about 5 mi east-west by 2 1/2 mi north-south. The east end of the system has been stepped up and ultimately terminated by late-age faulting on the Continental fault system. On the west end, the vein swarm extends several miles west of Big Butte where it becomes less and less well-mineralized, ultimately degenerating into chalcedony veins of no economic significance. The west end has mostly silver-manganese with some gold, while the east end beyond the Continental fault has disseminated mineralization with a high copper-to-iron ratio and appreciable molybdenite. It is believed to represent a deeper-seated portion of the interior copper/molybdenum center which underlies the unusually well-developed, near-surface vein swarm.

FIGURE 3.36
Generalized map of the Butte District, modified from Proffett, 1973

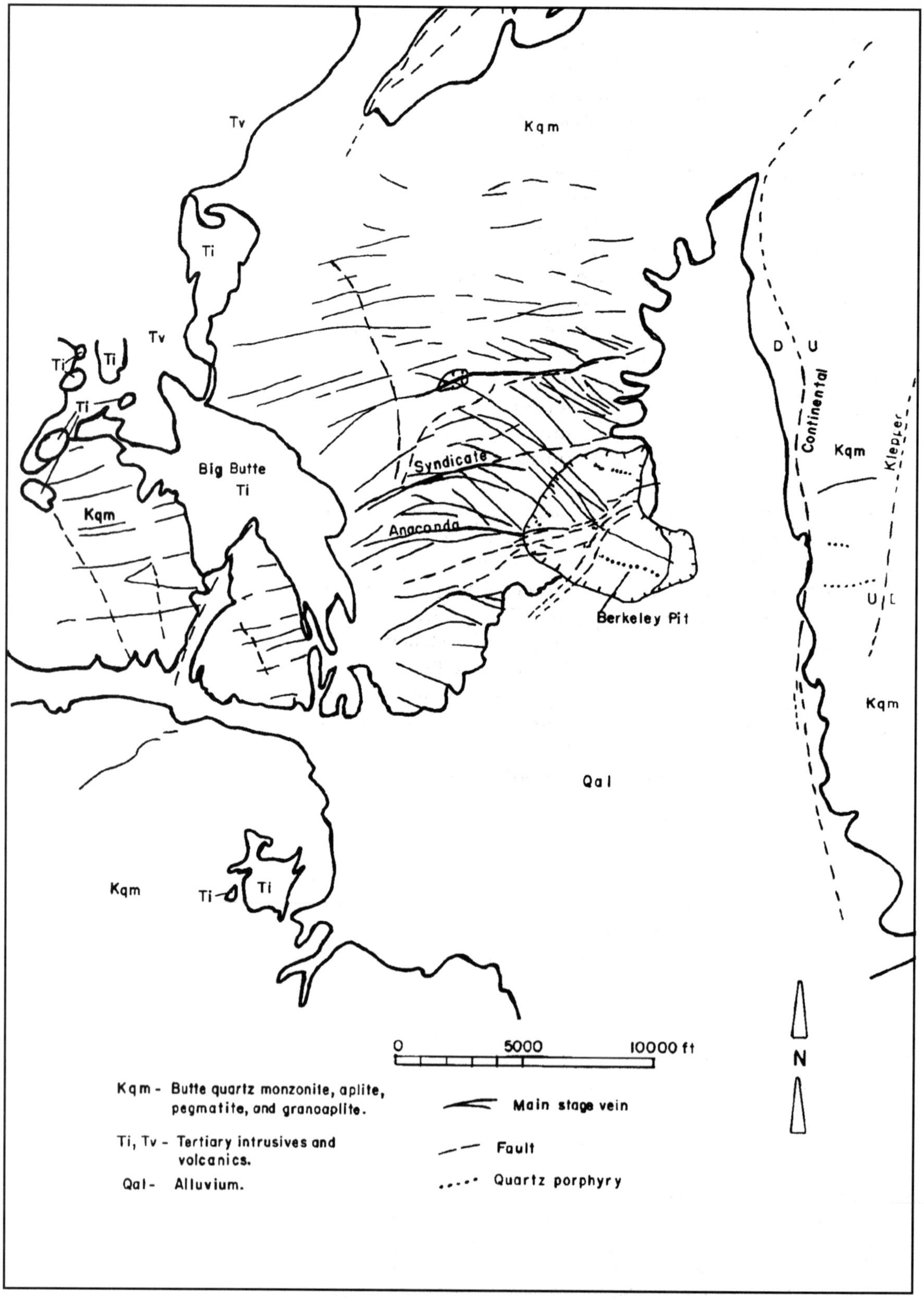

As was early recognized by Sales (1914), three-dimensional metal zoning is well developed. The system zones outward from molybdenum to copper to zinc to silver. Deep post-mineral erosion removed the silver/zinc zone from the center of the district and developed high secondary copper grades below about 300 ft. Secondary silver was also developed. Large tonnages of stockwork copper molybdenum ore occur at depth. This forms a great metal reserve but is essentially below the present economic mining limit. About 100 million tons of ore averaging 4 percent copper were mined from the "horsetail" zones, a part of the complex upper-level vein system. Discovery of more high-grade horsetail zones in Butte is a possibility.

There are scattered quartz porphyry dykes that are closely associated with mineralization. These dykes thicken with depth, but the country rock remains Butte quartz monzonite to the greatest depth explored. There are significant mineralized breccias at intermediate depth (R. N. Miller, personal communication).

It would seem that a localized source of acid intrusive is being approached at depth. It may be a hidden high point on a more acidic core (this idea was early suggested by R. H. Sales).

The massive hydrothermal system which emplaced the veins has developed a thick hood of pyritized Butte quartz monzonite (approximately 10 percent pyrite) which occupies the upper portion of the district. This pyritization has destroyed (by sulfidization) the original magnetite of the Butte quartz monzonite, and apparently the hydrothermal system introduced additional iron and sulfur in the process. This has resulted in two unusual geophysical features:

- The regional aeromagnetic map of the batholith shows a prominent low in the Butte vicinity. This low is about 8 mi east-west by 3 mi north-south. Its deep center occupies the district center. A detailed low-level aeromagnetic survey covers the main Butte district. The detailed results check well with the large area of magnetite destruction. On the west end of the district (just east of Big Butte) where magnetite destruction lessens, the major east-west veins are marked by magnetic lows, because magnetite destruction is greater along them (e.g., the Anaconda vein, Syndicate vein, etc.). Further east, the effect of separate veins merges into the major magnetic low.

- A large gravity high overlies the central metallized (pyritized) core of the district. (This high is not shown on figure 3.34). Valley fill to the south and east of the district complicates interpretation, but the large features can be easily seen. The gravity high indicates that iron as well as sulfur must have been added by the hydrothermal stage.

The Wickes Corbin district (Becraft and others, 1963) lies near the northern end of the batholith. Mineralization of Laramide age includes large areas of hydrothermal alteration and a multitude of veins. Several of the larger veins are the Gregory, Alta, Minah, Mount Washington and Daily. These strike approximately east-west and are quartz argentiferous galena veins with minor sphalerite and other sulfides. Tourmaline is a common vein mineral.

The host rock is a roof pendant of Elkhorn volcanics which may be a block that was down-dropped in the batholith. The eastern portion of the pendant contains a north-northeast-striking graben filled with Lowland Creek volcanic flows. Lowland Creek dykes traverse these flows as well as the earlier rocks of the area. These late volcanics host the Montana Tunnels diatreme discussed below.

The Alta vein is credited with a production of $32 million prior to 1893. Most of the values were in silver and lead. This is the best production record of any mine in the district.

A large but low-grade molybdenum occurrence is located about 2 mi east of Wickes. A pyritized ring encircles the center, and it appears that the original hood has been removed by erosion.

The Wickes Corbin district is part of a 6-mi-wide belt of mineralization that extends from Bernice Station (on the Boulder River 3 mi west of Basin and 12 mi west of Boulder) past Wickes and Clancy, a distance of 24 mi. The belt is laced with a variety of veins ranging from quartz through quartz/chalcedony to chalcedony. The quartz veins may have important amounts of base and precious metals. This belt also includes many acidic intrusives: alaskites, aplites and pegmatites. Breccia pipes are found in the southern part of the belt. Gravity and magnetic lows trace this belt to and past Butte (between Bernice and Butte there is considerable Lowland Creek volcanic cover, but outcrops along the Boulder River are mineralized).

Detailed gravity data are not available at Wickes, but regional gravity indicates that Wickes is the center of the regional gravity low.

Detailed aeromagnetics (survey flown for Anaconda in the '60s by Lockwood, Kessler and Bartlett, Inc.) defines a low extending from the Alta vein 1 mi north of Wickes south past the Daily vein at Wickes. The molybdenum center 2 mi east of Wickes is defined as a rounded magnetic low. This occurs because the upper part of the system including the high magnetite-bearing hood has been eroded. A series of mineralized areas occurs around the outer margin of the batholith. The age seems to be generally late batholith (70 m.y.), but some may be younger. These are mostly gold/silver occurrences. The better-known districts are Rimini, Porphyry Dyke, Indian Creek, Radersburg, Elkhorn, Golden Sunlight, Silver Star, German Gulch, Champion and Emery. Figure 3.37 shows the relationship of these districts and magnetic features.

Some of these districts are associated with apophyses of the border intrusives; other districts are associated with outlier intrusives of the Boulder system. Examples of the latter are Indian Creek, Radersburg and Golden Sunlight.

The Emery gold/silver district is located about 8 mi east of Deer Lodge. A vein swarm is found in the Elkhorn volcanics which bound the edge of the batholith. Overlapping alteration envelopes of these veins have destroyed the magnetite in the Elkhorn volcanics. This has caused a magnetic low. The low is complicated by a north side low of the nearby batholith boundary, but the mineralization low is far enough north to separate.

Lowland Creek Age Mineralization

There is a mineralized diatreme of Lowland Creek age west of the Alta Mine. The operation here is known as the Montana Tunnels (Clark, 1989). A large tonnage of a low-grade gold, silver, zinc and lead ore is being mined by open pit. The diatreme is a vertical pipe in Lowland Creek volcanics about 800-by-1000 ft across. A magnetic low encompasses a large area around the mine.

Further south [south of the junction of Lowland Creek and the Boulder River (9 mi south, 70° west of Basin)], several gold mines within altered Lowland Creek volcanics were developed. The altered and mineralized area lies near the north margin of an unusual magnetic low developed in the Lowland Creek volcanics. It is possible that this was a throat-like feeder for volcanics in this area.

A gold/silver mine, Golden Sunlight (Lowland Creek age?; Smolik and others, 1984) is a mineralized breccia which is closely correlated with several very small Tertiary latite porphyry plugs. The breccia occurs between two of these and contains fragments of latites mixed with those of the host sediments.

The magnetic setting of the mine follows (Johnson and others, 1965). A magnetic high lobe breaks away from the southeast margin of the batholith about 8 mi south of the town of Boulder. This lobe extends 12 mi southeast more or less along the axis of the Bull Mountains. Near Red Mountain, 7 mi northeast of Whitehall, this strong magnetic feature dies, but a weak magnetic nose extends southwest to the Jefferson River. This nose, when detailed by low-level aeromagnetics, divides into several highs representing small exposed and buried intrusives. It is on this line that the Golden Sunlight breccia occurs.

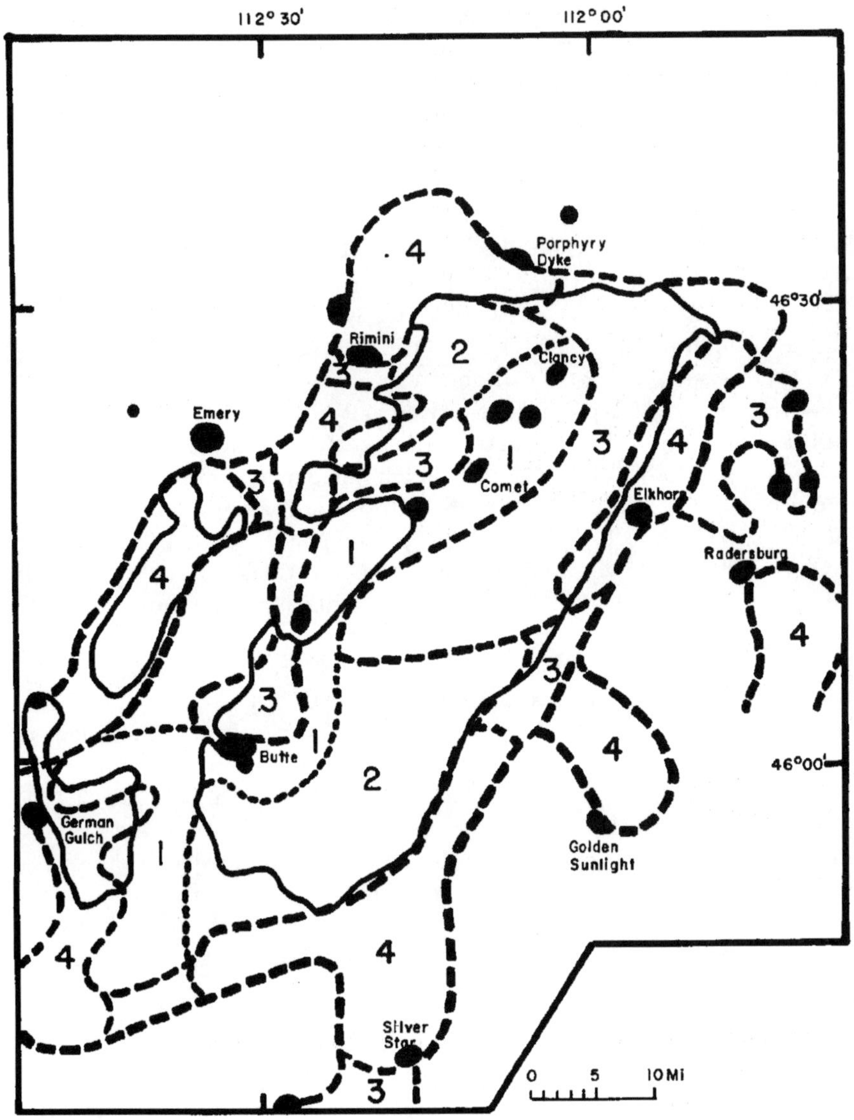

FIGURE 3.37
Distribution of magnetic features and precious metals deposits of the Boulder Batholith. 4 indicates high susceptibility and 1 indicates low susceptibility source bodies. Dashed lines are the boundaries of the source bodies

Summary

To summarize the use of gravity and magnetics in the Boulder batholith setting:

- They can be used to outline the batholith margin where it is covered.
- In the interior, the belt of mineralization (veins, acidic dyking, etc.) is indicated by elongate gravity and magnetic lows. Separate ore deposits are indicated by magnetic lows caused by magnetite destruction. There are also magnetic lows that have other causes.
- At Butte, the intensely mineralized part of the district is indicated by a strong gravity high resulting from sulfides and strong magnetic lows from magnetite destruction.
- On the outer margin of the batholith, the mineralized districts are most often found on magnetic noses that are caused by apophyses of the intrusive.

The Gravity Expression Of Salt Domes

This discussion was written by Leroy E. Brow, B&W Exploration, Inc., 12118 Penthshire, Houston, TX, 713/468-8112.

Salt domes are well-recognized sources of Bouguer gravity anomalies. Because of the economic importance of salt features as a trap for hydrocarbons (and also sulfur in the cap rock of some salt domes), many articles are available concerning the geology and gravity effect of these features. Here we will consider three types of salt uplifts, in terms of the depth-to-top and vertical extent of the salt, and their associated gravity anomalies.

First, we will look at a deep-seated salt pillow where the salt uplift arched but did not penetrate the overlying sedimentary section. Next, we will look at a salt diapir which has penetrated several thousand feet of sediments but whose top is still a few thousand feet deep, say 5000 ft to 10,000 ft to the top, as a general class of features. This type of dome is quite common in the Gulf Coast. Finally, we will consider a diapir which has penetrated to within a few hundred feet of the surface.

To begin, we will consider the basis for the reputation that salt domes have for causing gravity anomalies. For any geological feature to produce a gravity anomaly, the feature must result in either a mass excess (yielding a positive anomaly) or a mass deficiency (yielding a negative anomaly) in the subsurface. The density of salt is about 2.2 gm/cm^3, and the value will be independent of the depth of the salt. (The density may vary somewhat depending on the purity of the salt.) Consequently, if a salt feature is embedded in, say, a carbonate section, the density contrast will be about -0.5 gm/cm^3 (density of salt, 2.2 gm/cm^3, minus density of background, dense limestone, 2.7 gm/cm^3), and this large contrast can produce recognizable gravity anomalies even when the structural relief is moderate and/or the source is quite deep. In general, the density of the Tertiary sand/shale sequence of the Gulf Coast is a function of depth. Figure 3.38 shows the general density versus depth relationship. In addition, the figure shows a constant 2.2 gm/cm^3 salt density. Note that at about 3500 ft (the crossover depth), the sediment density is equal to the salt density. If a salt diapir extends from, say, 20,000 ft to the surface, then that part of the diapir between 3500 ft and 20,000 ft will produce a gravity minimum, and that part of the diapir above 3500 ft will produce a gravity maximum.

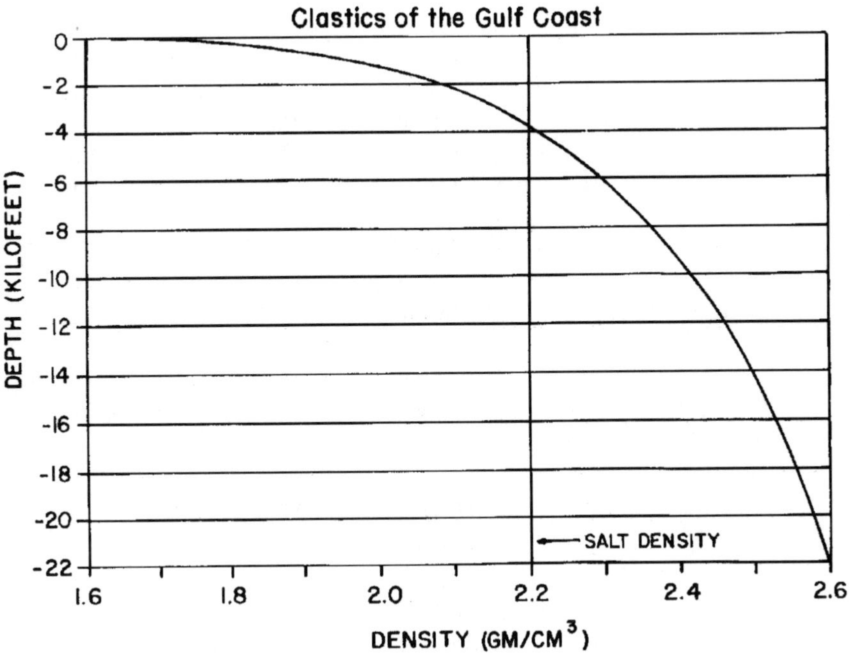

FIGURE 3.38
Density Variations with Depth

Salt Uplifts

Our first example will be a deep salt uplift. Figure 3.39a shows an observed Bouguer gravity anomaly map from an area where salt is known in the subsurface and from easily seen anomalies. The obvious feature on the map is a gravity minimum embedded in a gravity gradient.

Figure 3.39b is one possible contouring of the separate gravity anomalies, and here we see the sought-for anomaly ("residual") to be a roundish gravity minimum which we assume to be caused by a local salt uplift. If we assume the source to be a compact volume (hence a sphere, analytically), we can estimate the depth to the center and relief of the assumed spherical source. Interpretation Chart #1 (figure 3.42) outlines the procedure for making the estimates.

Figure 3.39c shows a cross section through the center of the residual anomaly and compares the observed gravity data with theoretical values for the interpreted sphere. That the fit is quite good is not proof that it is correct but only that it is possible; however, the fit is a comfort to the interpreter. (A borehole located near the center of this feature encountered salt at about 15,000 ft). Aesthetically, we picture the geologic source as relief on a surface of density contrast (salt) where the surface is generally at the deduced depth to center of the sphere and is structured by an amount equal to the calculated radius of the sphere. We account for the mass of the bottom half of the sphere by making the source broader than the sphere's diameter such that the total volume of the structure will be equal to the volume of the sphere.

Salt Diapirs Below The Crossover Depth

The second example is a more typical salt diapir. Figure 3.40a is a Bouguer gravity map from an offshore area. This map shows, in addition to a contouring of the observed data, the actual locations of the observation points. Data acquisition is a subject in itself, but it is worth noting that to sample anomalies of interest in exploration the survey must have an appropriate grid of observation points. The example shown is a square grid with about 3500 ft between samples, and this is quite a good survey in terms of the anomalies which are anticipated for the area. The Bouguer anomaly pattern in figure 3.40a consists of a nearly uniform gradient (gravity increasing to the south) interrupted by a local gravity minimum near the center of the map.

FIGURE 3.39
a) Bouguer gravity map, b) Regional/Residual gravity map, c) Bouguer gravity anomaly for a salt pillow

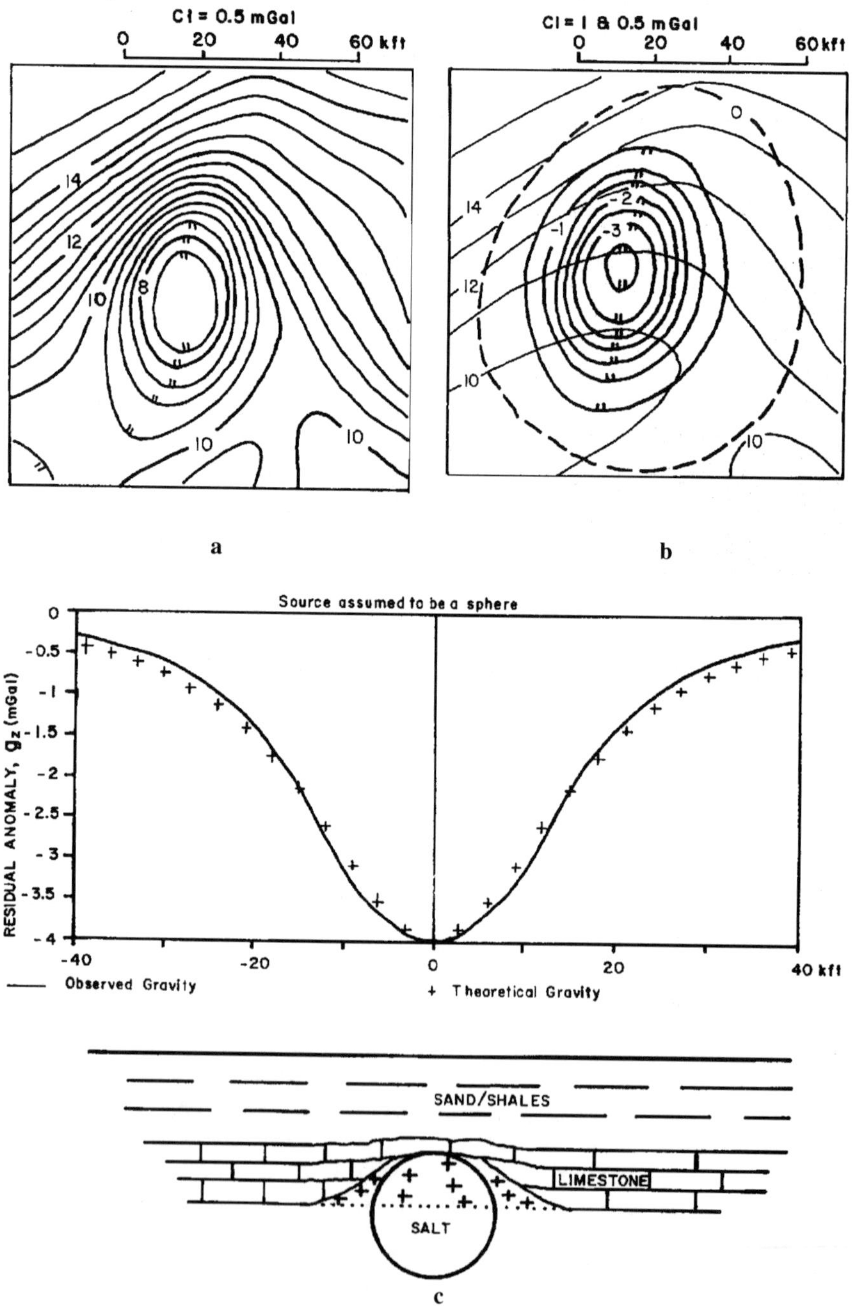

Depth to top = 13.5 Kft
Depth to center = 20.9 Kft
Radius = 7.4 Kft
Density contrast = $-0.5\,gm/cm^3$

Figure 3.40b shows one possible contour disentanglement of the observed pattern into the two features; that is, a uniform gradient (a "regional") and a local circular gravity minimum caused by a salt diapir located at the center of the residual gravity anomaly.

Confidence in the interpretation to this point is quite high, because the control is adequate and the observed anomaly is reasonably large in amplitude. In fact, knowing the existence and position, in plan view, of a geological feature (salt dome in the example) is important, and the interpretation could stop here with the next step being a seismic survey to further define the potential prospect. It is best to attempt some quantitative estimates even if you attach less confidence to the results, because you might rule out some "silly" interpretations. For example, in figure 3.39, if the estimated depth to the assumed sphere happened to be 50,000 ft (a depth which is below the likely base of the sediments in this area), the source might be within the basement rather than within the sediments. On the

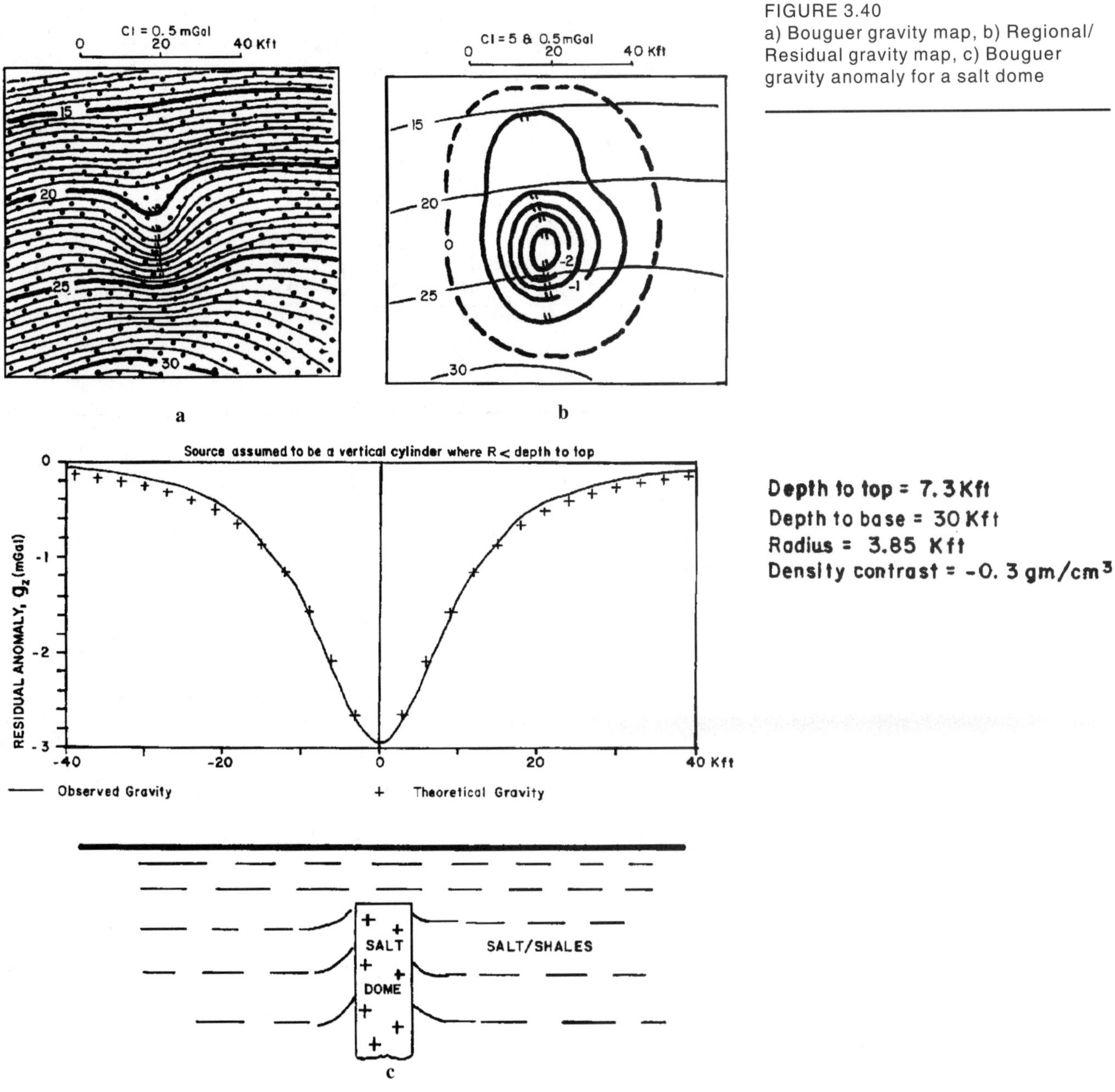

FIGURE 3.40
a) Bouguer gravity map, b) Regional/Residual gravity map, c) Bouguer gravity anomaly for a salt dome

other hand, if the amplitude of the anomaly were large, the resulting calculated radius could produce "silly" consequences (the source might rise above the surface, for example). Hence the purpose of the quantitative work is not to produce values which will be used for drilling estimates, but to check on the reasonableness of the volume and depth which will satisfy a particular anomaly.

Assuming that the source of the residual anomaly in figure 3.40b is a vertical cylinder with a radius which is small in terms of the depth to the top of the cylinder, you can use Interpretation Chart #2 (figure 3.43) to estimate depth to top (and base) and radius of a vertical cylinder which has a theoretical anomaly similar to the observed feature. Figure 3.40c is a cross section through the center of the observed anomaly and also shows the theoretical values for the vertical cylinder deduced by using chart 2. Since the top of the source is below the crossover depth (see figure 3.38) and the radius of the cylinder is probably less than the depth to its top, this type of dome is easiest to analyze both qualitatively and quantitatively. The Gulf Coast basin contains many anomalies of this type.

Salt Diapirs Above The Crossover Depth

The final example, figure 3.41a, has a gravity pattern similar to that shown on figure 3.40a in that a gravity gradient (gravity increasing to the northwest) is interrupted by a large, and complex, gravity minimum.

When the anomaly pattern is disentangled (figure 3.41b) the residual is a large gravity minimum with a central maximum (see figure 3.41c also).

This residual anomaly pattern is probably caused by a salt diapir whose top is quite near the surface such that the minimum is caused by the volume of salt below the crossover depth (see figure 3.38) and the maximum is caused by the salt volume above the crossover depth (with perhaps a part of the maximum being caused by cap rock, if present). Figure 3.41c also shows one possible further separation, in profile form, of the residual anomaly into a discrete minimum and a discrete maximum. Confidence in the qualitative conclusions about the observed anomaly and its interpreted source is reasonably high. To quantify the picture is more difficult, however, because the source does not fit any of the simple analytical shapes used for quantitative analyses. If a computer is available, along with the necessary software and time, the interpreter can model the three-dimensional salt diapir; however, knowing the position of a shallow diapir is probably more important in an exploration sense than trying to determine more precisely the shape of the causative volume by modeling procedures.

We have shown examples of salt features for three depth ranges; that is, deep (12,000 ft or greater), intermediate (diapirs whose tops are below the crossover depth), and shallow (diapirs with considerable volume above the crossover depth). For any interpretation of gravity data one must isolate, either conceptually or actually, the anomalies of interest. In the examples shown, we have effected a contour separation to highlight the discrete anomalies. This anomaly disentanglement is an important part of gravity interpretation, since the discrete anomalies show the locations, in plan, of subsurface masses which may have exploration potential. The quantitative analysis can be as sophisticated as the computer (and programs) available; however, the simple recipes and techniques shown here can be quite useful and the gravity interpretation can often stop at this point, since the next step will be to acquire other data, probably seismic, to further define the potential prospect. All three types of salt structures discussed can, and do, create traps for hydrocarbons and as such are of considerable interest to the explorationist.

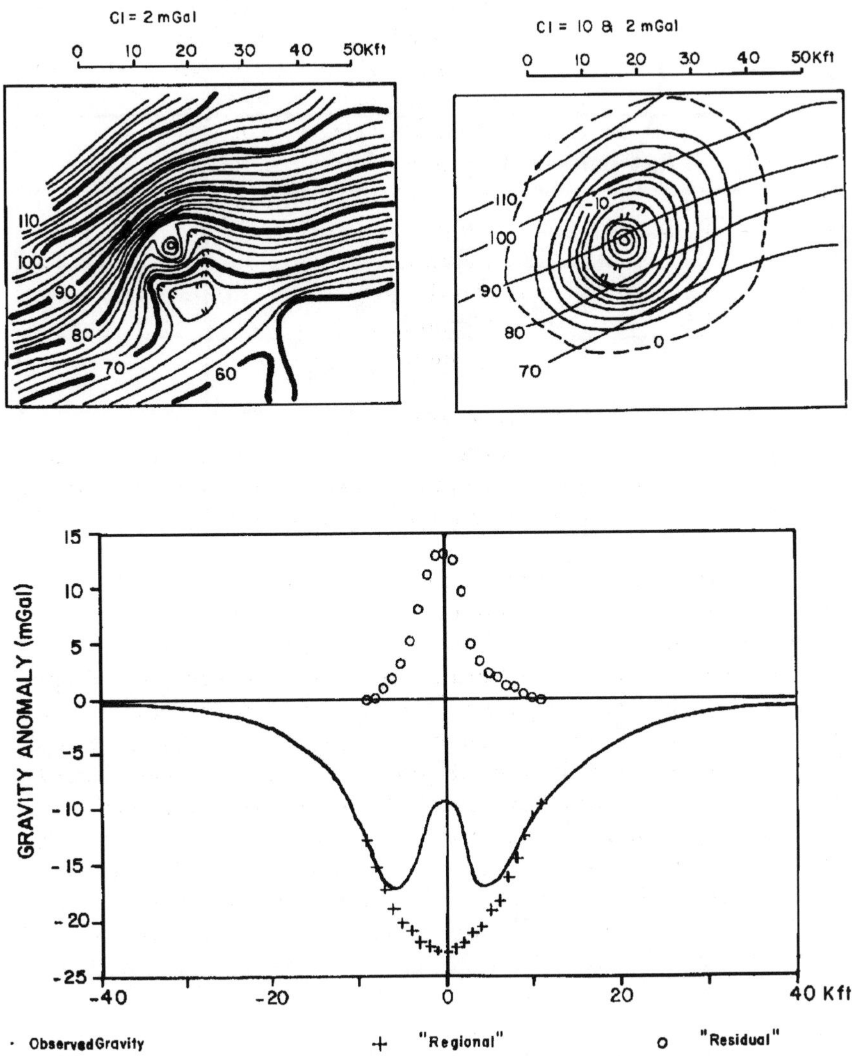

FIGURE 3.41
a) Bouguer gravity map, b) Regional/Residual gravity map, c) Bouguer gravity anomaly shallow salt dome

FIGURE 3.42
Sphere Interpretation Chart

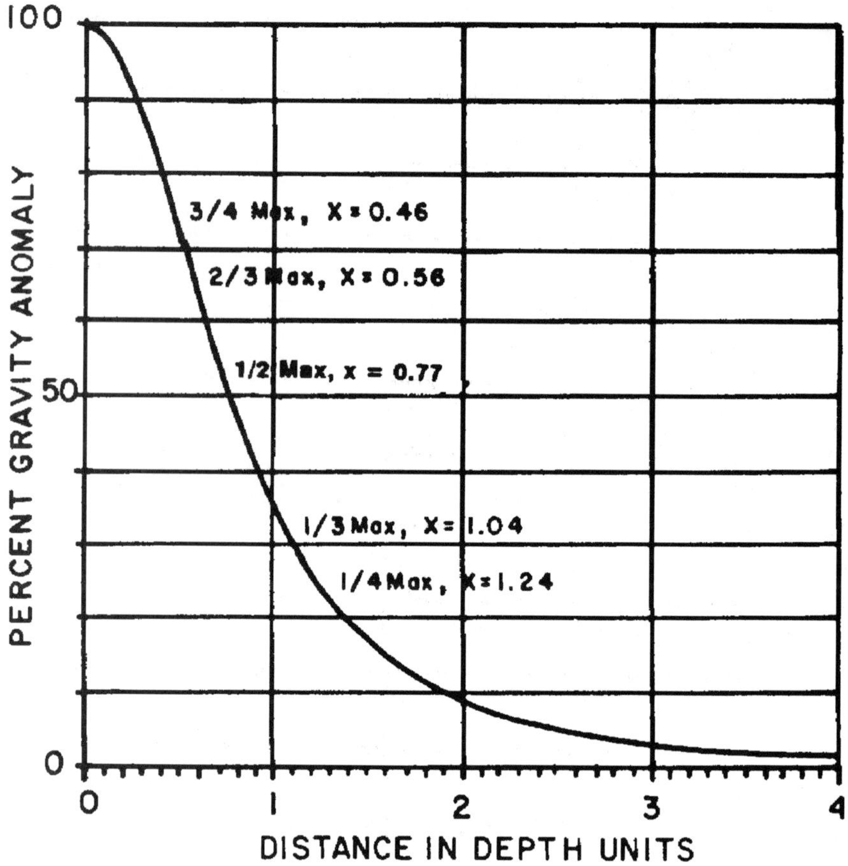

DEPTH DETERMINATION
Shown to the left is the theoretical gravity profile due to a SPHERE, homogeneous in density (or density contrast) and having a depth to its center of 1 unit. Along the right flank of the profile are indicated the positions where the anomaly has reached 3/4, 2/3, 1/2, 1/3, and 1/4 of its maximum value. These positions are, respectively, 0.46, 0.56, 0.77, 1.04, and 1.24 depth-units distant horizontally from the position of the maximum. We shall call these positions 'diagnostic" since from them we may determine a depth to the center of the sphere. The choice of these positions is arbitrary, but the five positions chosen cover that part of the profile which should be well defined by many stations in the field. At any one of these "diagnostic" positions, we determine the depth to the center of the sphere by multiplying the horizonal distance from the position to the maximum by a 'depth index', which is the reciprocal of the appropriate value given above. For example, the 'depth index' for the 1/2 max position is 1/0.77 = 1.305. The 'depth indices' for the five positions shown on the profile are:

'Diagnostic' Position	Depth Index Multiplier
3/4 max	1/0.46 = 2.17
2/3 max	1/0.56 = 1.79
1/2 max	1/0.77 = 1.305
1/3 max	1/1.04 = 0.96
1/4 max	1/1.24 = 0.81

The consistency of the five separate depths obtained is a measure of the fit between the shape of the field profile and the theoretical profile for a sphere. If the five depth determinations are nearly the same, the interpreter is more confident that the anomaly source is a chunky body.

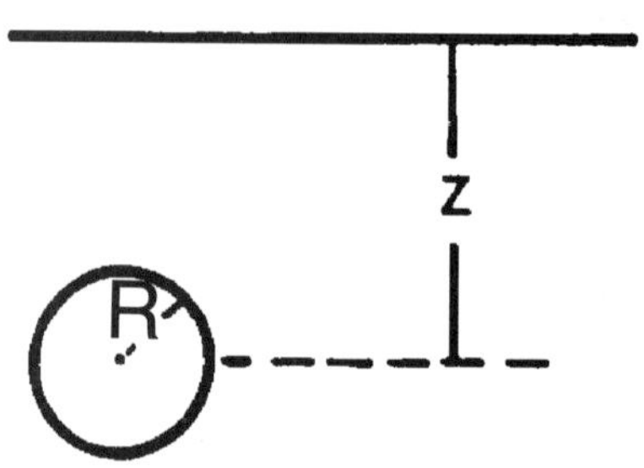

RADIUS CALCULATION
The general expression for the gravity anomaly due to a sphere of radius R, depth to center Z and density contrast DP is:

$$g_z = G\ DP\ (4/3\ \pi R^3)\ Z\ /\ (X^2 + Z^2)^{3/2}$$

Directly above the sphere at X = 0 we have:

$$g_z = G\ DP\ (4/3\ \pi R^3)/Z^2$$

If we take the universal gravity constant as 6.67×10^{-8}, and express Z and R in centimeters, g_z will be given in Gals. We find it convenient however, to adjust the constant so we may express Z and R in kilofeet or kilometers and get the answer in milligals. If Z and R are in kilometers and we want the value of gravity (g_z) in milligals (mGal) use:

$$g_z = 27.93\ DP\ R^3/Z^2 \quad (Z = \text{depth to center})$$

If Z and R are in kilofeet and we want g_z in milligals use:

$$g_z = 8.52\ DP\ R^3/Z^2$$

From either expression above one can solve for a range of R's using a range of density contrasts.

To determine the DEPTH TO THE TOP of the sphere we subtract the value of the radius from the depth to center. We want a radius which together with a reasonable value of the density contrast will make the calculated maximum anomaly match the observed maximum anomaly. This involves trying several combinations of radius and density contrast since NO UNIQUE SOLUTION IS POSSIBLE WITHOUT OTHER INDEPENDENT INFORMATION ABOUT THE DENSITY CONTRAST. The range of likely depths is thus set by the range of density contrasts which are likely for the particular geologic situation. In spite of this ambiguity, the interpretation may have considerable value. For example, if the range of depths to the top should lie within the known depth range of the useful sediments, the interpretation would indicate that the source is present within the sediments and not within the deeper, and more indurated or even metamorphosed, basement rocks.

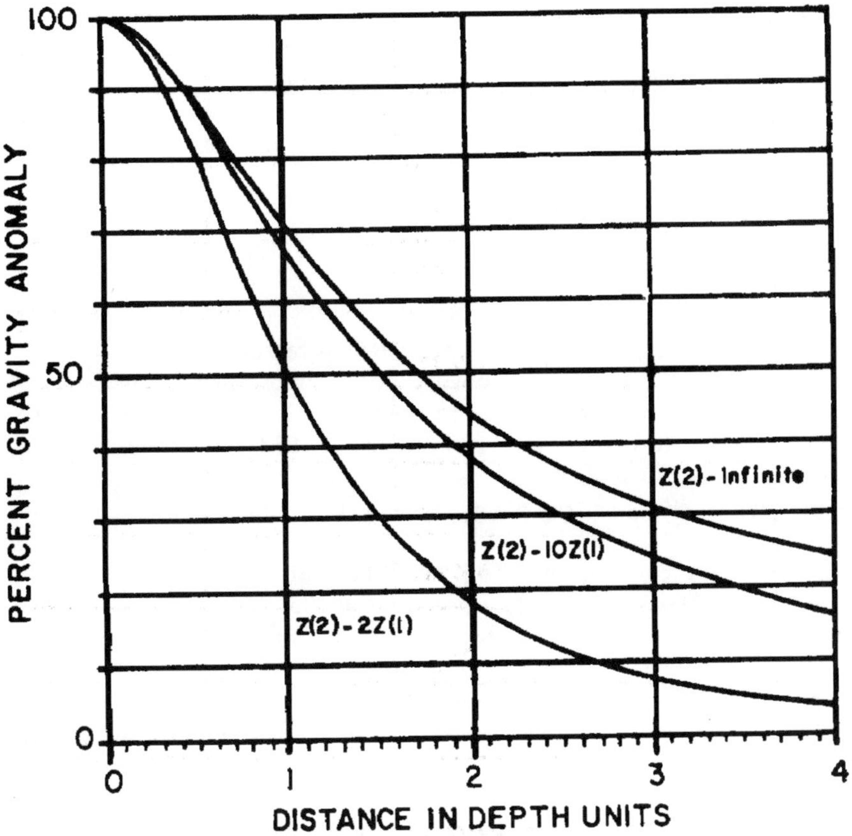

FIGURE 3.43
Vertical Cylinder Interpretation Chart

DEPTH DETERMINATION
To determine the DEPTH TO THE TOP of the anomaly source, assumed to have the shape of a VERTICAL CYLINDER and whose radius is small relative to its depth, refer to the page showing the depth determination procedure when the anomaly source is a sphere. Follow the same procedure using an appropriate five depth indices from the table below. The table lists indices for 9 different cylinders whose DEPTH TO THE BOTTOM ranges from 1.1 times the depth to the top to infinity. The consistency of the five separate depths for a particular choice of cylinder is a measure of the fit between the shape of the field profile and the theoretical profile of the vertical cylinder.

Position Z(2)	3/4	2/3	1/2	1/3	1/4
1.1	2.09	1.69	1.25	0.92	0.78
1.2	2.00	1.64	1.19	0.88	0.74
1.5	1.82	1.49	1.09	0.79	0.67
2.0	1.61	1.33	0.96	0.70	0.58
3.0	1.45	1.19	0.83	0.60	0.50
4.0	1.35	1.11	0.78	0.54	0.44
5.0	1.32	1.08	0.74	0.51	0.41
10.0	1.22	0.98	0.66	0.43	0.34
Inf.	1.16	0.91	0.58	0.36	0.27

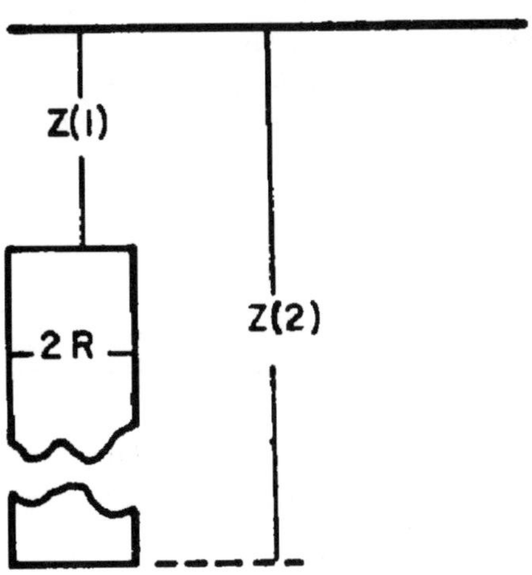

RADIUS CALCULATION
Following Nettleton, the gravity anomaly due to a vertical cylinder of finite depth extent, where the depth to the bottom, Z(2), is N times the depth to the top, Z(1), may be written as given below provided the radius of the cylinder is small compared to Z(1):

$$g_z = G\,\pi\,R^2\,DP\,((X^2 + Z(1)^2)^{-1/2} - (X^2 + Z(2)^2)^{-1/2})$$

At X=O, the position of the maximum anomaly, and for Z(2) = NZ(1) we have :

$$g_z = G\,\pi\,R^2\,DP\,(N-1)/NZ(1)$$

For Z(1) and R expressed in kilofeet and g_z in milligals we have:

$$g_z = 6.39\,DP\,R^2\,(N-1)/NZ(1)$$

For Z(1) and R expressed in kilometers and g_z in milligals we have:

$$g_z = 20.95\,DP\,R^2\,(N-1)/NZ(1)$$

In general we have: $g_z = A\,DP\,R^2/Z(1)$, where: A = constant (N-1)/N.

Solving for R: $R = ((g_z\,Z(1))/A\,DP))^{1/2}$

The chart to the right gives values of A for various Z(2)'s and for distances in kilofeet or kilometers.

Z(2)	A (R & Z in Kft)	A (R & Z in Km)
1.1	0.58	1.90
1.2	1.06	3.49
1.5	2.13	6.98
2.0	3.19	10.48
3.0	4.26	13.97
4.0	4.79	15.72
5.0	5.11	16.76
10.0	5.75	18.86
Inf.	6.39	20.95

OTHER CASE HISTORIES

The following section summarizes the results of a few published case histories. They are included to indicate some of the wide-ranging uses of gravity surveys in exploration. One case, the investigation of the possibility that a fifth force exists, has no exploration use at present.

Massive Sulfide Deposits

The discovery of the Flambeau massive sulfide deposit is discussed by Schwenk (1977). (This article was brought to my attention by Rodney J. Ikola, geophysicist, 2305 East Highway 37, Hibbing, Minnesota 55746, [218] 262-3537). The deposit is located in Rusk County, Wisconsin, on the southern exposure of the Precambrian shield. Bedrock is steeply-dipping metamorphosed volcanic and sedimentary rocks reported to be of early Proterozoic age. Bedrock is overlain by thin remnants of Cambrian sandstone and a thick layer of glacial drift. The thickness of the glacial till and the sandstone is as great as 150 ft.

Geologic reconnaissance by J. S. Phillips of Bear Creek Mining Company in 1966 indicated the potential of the area for massive sulfide deposits based on the examination of three exposures of quartz-sericite schist. Airborne electromagnetic surveys located a number of high conductivity zones. One of these zones was the Flambeau deposit. Drilling and numerous geophysical methods were used to delineate the orebody.

Figure 3.44 shows the Bouguer gravity anomaly along one line over the orebody and g_z for the model shown beneath the gravity profile. The amplitude of the Bouguer gravity anomaly is 0.80 mGal. The dip and depth of the model are based on the interpretation of a ground electromagnetic survey. This interpretation also gave a maximum width for the conductor (100 ft). Densities for bedrock (P_1) and the overburden (P_2) were assumed and the density of the dike (P_3) (and the width of the dike) was adjusted to fit the Bouguer anomalies. The assumed densities were P_1 = 2.7 gm/cm^3 and P_2 = 1.8 gm/cm^3. The density of the dike was determined to be 4.3 gm/cm^3. Density measurements on the host rock averaged 2.6 gm/cm^3. The average density of the near-surface ore averaged 4.3 gm/cm^3.

FIGURE 3.44
Gravity line 1, Flambeau massive sulfide deposit, Wisconsin

MODEL DIMENSIONS
OVERBURDEN THICKNESS - 32 Ft
DIP - 80° N
WIDTH - 60 Ft
LENGTH - INFINITE
VERTICAL EXTENT - 500 Ft
P_1 - 2.7 GM/CM^3
P_2 - 1.8 GM/CM^3
P_3 - 4.3 GM/CM^3

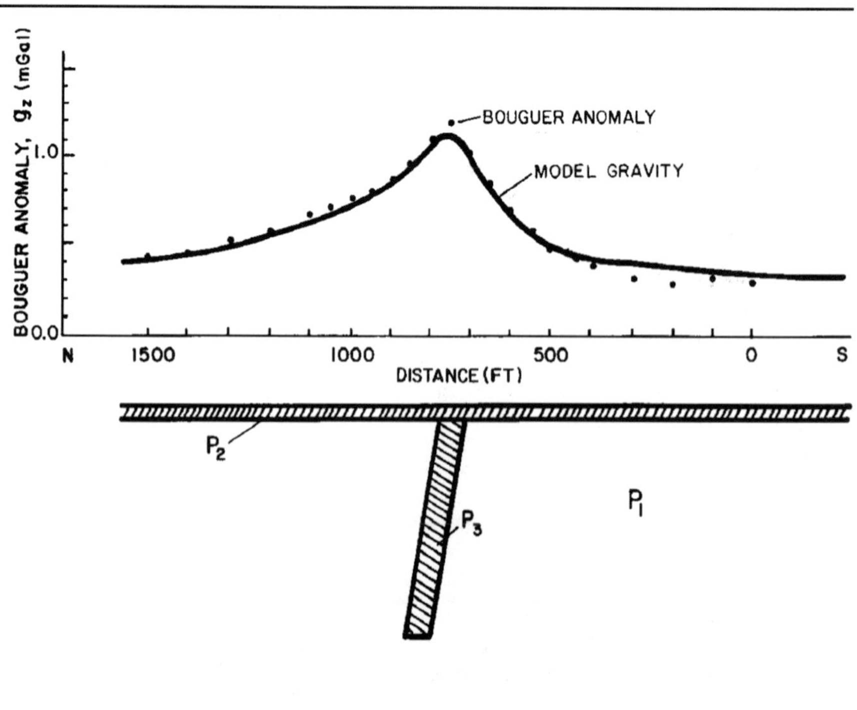

A detailed gravity survey of the entire ore deposit was carried out to determine the anomalous mass of the orebody (Hammer, 1945). Station elevations were surveyed. The maximum residual gravity anomaly that was detected was 1.2 mGal. A total anomalous mass of 6.33 million tons was determined using Gauss' theorem.

The gravity survey probably increased the confidence of the geophysicists that a massive sulfide body existed beneath the covered area. Accurate gravity data are required for massive sulfide exploration, since the amplitude of the gravity anomalies is small.

Layered Mafic Intrusions

Base and precious metals associated with layered mafic intrusions such as the Bushveld complex of South Africa and the Stillwater complex of Montana have large gravity and magnetic anomalies that are caused by thick sections of mafic and ultramafic rocks that have intruded rocks of lower density and susceptibility. For example, in the Bushveld complex, 6 km of mafic rocks overlie 2 km of ultramafic rocks (Guilbert and Park, 1986). These rocks have been intruded into low-density metamorphic rocks.

Gravity data can be used in covered areas to outline the horizontal boundaries of the intrusions, make estimates of their thickness, and define possible structure and/or density variations that occur within the complex. For example, gravity data indicate the Bushveld complex consists of four separate lobes and that the southeastern lobe is largely covered by younger sediments. Gravity surveys over the Stillwater complex and the Glen Mountain complex of southwestern Oklahoma indicate that the exposed portion of these layered mafic intrusions are only a small part of the total mass of the intrusion.

Replacement Orebodies

Underground gravity surveying at Bisbee, Arizona, has been described by Sumner and Schnepfe (1966). Replacement massive sulfide copper orebodies (density up to 4.0 gm/cm^3) occur in less dense Paleozoic limestones (density 2.65 to 2.0 gm/cm^3). Gravity surveys were carried out for many years to aid in detecting the sulfide bodies. The surveys provided three-dimensional residual gravity maps for interpretation. Orebodies that were detected ranged in size from 10,000 tons to 25,000 tons. Pyrite masses 10 times the size of orebodies increased the mass that was detected. Recommended drill holes were successful 80 percent of the time.

The Fifth Force

Assume that a body of mass m is acted upon by the earth's gravitational field $\mathbf{g}_e$ and that no other forces act on the body. The force is given by

$$F = m\, g_e$$

Newton's second law of motion states that

$$F = m\, a$$

and the gravitational acceleration of the body is equal to the gravitational field. This is true if m in the first equation (the gravitational mass) and m in the second equation (the inertial mass) are equivalent. The inertial mass is a kinematic quantity. The gravitational mass is related to the inverse square law of gravitation. There is no reason that the two masses should be equivalent. An analogous situation is that of the electric field where the electric charge of a particle would be required to be equal to the mass of the particle.

Many experiments have verified that, in fact, the inertial and gravitational masses are equivalent. This is referred to as the principle of equivalence. Another aspect of the principle is that the gravitational force that acts on different kinds of matter is independent of the material. This was the experiment that Galileo performed using an inclined plane.

Recently (Goldman and others, 1988), the principle of equivalence has been questioned from both a theoretical and experimental point of view. Theoretically, both a fifth and sixth gravitational force that are different from Newton's inverse square law have been predicted. The range of these forces is predicted to be a few hundred meters to a few hundred kilometers. For ordinary matter in the earth's gravitational field, the deviations from the inverse square law are expected to be small, although both forces depend upon the composition of matter. A significant difference in the two new forces would occur for antimatter such that the gravitational acceleration of an antiproton would be 14 percent greater than the gravitational acceleration of a proton.

A number of new experiments have been carried out to test the principal of equivalence with conflicting results. Gravity measurements in mine shafts in Australia supported a single new repulsive gravitational force that is 1 percent of Newtonian gravity. Another possibility was that two forces exist that cancel to within 1 percent of Newtonian gravity and have ranges of several hundred meters. Experiments with a borehole gravity meter in Greenland suggested a single attractive force with a strength of 1.7 percent to 3.9 percent of Newtonian gravity and a range between 10 m and 1 km. Experiments on a tower in North Carolina suggested an attractive force of 2 percent of Newtonian gravity and a range of 300 m. These experiments require the calculation of the gravity effect of the surroundings which, as we have seen, may be difficult to do.

Other investigations that include re-analysis of old experimental data that originally verified the principle of equivalence have also given mixed results. The results of an experiment that will compare the gravitational acceleration of protons and antiprotons will, hopefully, give more definitive results.

ACKNOWLEDGMENTS

Steve Owens and Diane Wright, geologists, reviewed this chapter. They made many helpful comments and corrections. Thanks for the help.

REFERENCES FOR THE BASICS

References listed here include text books that the reader may find useful for further study of the gravity method as well as references that are given in the text.

Dobrin, M. B., 1976, Introduction to geophysical prospecting: New York, McGraw-Hill Book Co., Inc., 630 p.

Ervin, C. Patrick, 1977, Short note on the theory of the Bouguer anomaly: Geophysics, v. 42, p. 1468.

Garland, G. D., 1965, The earth's shape and gravity: New York, Pergamon Press, 183 p.

Grant, F. S., and West, G. F., 1965, Interpretation theory in applied geophysics: New York, McGraw-Hill Book Co., Inc., 583 p.

Heiskanen, W. A., and Moritz, Helmut, 1967, Physical geodesy: San Francisco, W. H. Freeman and Co., 364 p.

Heiskanen, W. A., and Vening Meinesz, F. A., 1958, The earth and its gravity field: New York, McGraw-Hill Book Co., Inc., 470 p.

LaCoste, Lucien, 1988, The zero-length spring gravity meter: Geophysics: the leading edge of exploration, v. 7, n. 7, pp. 20-24.

Lysonski, Joseph C., and Sumner, John S., 1980, The IGSN 71 residual Bouguer anomaly map of Arizona: Abstracts from the 1980 Society of Exploration Geophysicists 50th annual meeting, Houston, Texas, pp. 179 - 254.

Stacey,F. D., 1969, Physics of the earth: New York, John Wiley Sons, Inc., 324 p.

Talwani, Manik, Worzel, Lamar J., and Landisman, Mark, 1959, Rapid gravity computations for two-dimensional bodies with application to the Mendocino submarine fracture zone: Journal of Geophysical Research, v. 64, pp. 49 - 59.

Watson, Robert J. and Brow, Leroy E., 1966, Gravity and magnetic methods of geophysics: Esso Production Research Co. report.

West, Robert E., 1972, A regional Bouguer gravity anomaly map of Arizona: unpublished PhD dissertation, University of Arizona, 186 p.

REFERENCES FOR CASE HISTORIES

Ager, C. A., McMillen, W. J., and Ulrych, T. J., 1972, Gravity, magnetics, and geology of the Guichon Creek batholith: British Columbia Dept. of Mines Pet. Res. Bull. 62, 124 p.

Becraft, G. E., Pinckney, D. M., and Rosenblum, S., 1963, Geology and mineral deposits of the Jefferson City quadrangle, Jefferson and Lewis and Clark Counties, Montana: U. S. Geological Survey professional paper 428, 101 p.

Bonini,W. E., 1969, Use of gravity anomalies in prediction of the shape of intrusive bodies: S. M. E. - A. I. M. E. preprint 69-L-19, 27 p.

Biehler, Shawn, and Bonini, W. E., 1969, A regional gravity study of the Boulder batholith, Montana: G. S. A. Memoir 115, pp. 401-422.

Burfeind, W. J., 1967, A gravity investigation of the Tobacco Root Mountains, Jefferson Basin, Boulder batholith, and adjacent areas of southwestern Montana: PhD thesis, Indiana University, Bloomington, 146 p.

Bott, M. H. P. and Smith, R. A., The estimation of the limiting depth of gravitating bodies: Geophysical Prospecting, Vol. 6, pp.1-10, 1958.

Clark, M. L., Weitz, T. J., Banning, W. R., Turk, D. J., and Paterson, J., 1989, Montana Tunnels - Startup efficiencies improved, expansion considered: Mining Eng., Vol. 41, No. 2, pp. 87-91.

Ferguson, John F., Felch, Roger N., Aiken, Carlos L. V., Oldow, John S., and Dockery, Holly, 1988, Models of the Bouguer gravity and geologic structure at Yucca Flat, Nevada: Geophysics, Vol. 53, pp. 231-244.

Hammer,Sigmund I., 1945, Estimating ore masses in gravity prospecting: Geophysics, Vol. 10, pp. 50-62.

Grant, F. S., and West, G. F., 1965, Interpretation theory in applied geophysics: New York, McGraw-Hill Book Co., Inc., 583 p.

Guilbert, John M. and Park, Charles F., 1986, The geology of ore deposits: W. H. Freeman and Co., New York, 985 p.

Goldman, Terry, Hughes, R. J., and Nieto, M. M., 1988, Gravity and antimatter: Scientific American, Vol. 258, No. 3, pp. 48-56.

Jachens, Robert C., and Griscom, Andrew, An isostatic residual gravity map of California - A residual map for interpretation of anomalies from intracrustal sources, in The utility of regional gravity and magnetic anomaly maps, William J. Hinze, editor: Soc. of Exp. Geophysicists, P.O. Box 702740, Tulsa, OK 74170-2740, pp. 347-360.

Johnson, H. W., Jr., Henderson, J. R., and Tyson, N. S., 1965, Aeromagnetic map of the Boulder batholith area, southwestern Montana: U.S. Geological Survey, geophysical investigation map GP-538.

Klepper, M. R., Robinson, G. D., and Smedes, H. W., 1971, On the nature of the Boulder batholith, Montana: G. S. A. Bulletin, Vol. 82, pp. 1563-1580.

Knopf, A., 1963, Geology of the northern part of the Boulder batholith and adjacent area: U. S. Geological Survey miscellaneous geological investigations map I 381.

Lyons, Paul L., and O'Hara, Norbert W., Gravity anomaly map of the United States: Society of Exploration Geophysicists, P.O. Box 702740, Tulsa, OK 74170-2740.

Meyer, C., Shea, E. P., Goddard, C. C., Jr., and staff, 1968, Ore deposits at Butte, Montana, in Ore deposits of the United States: A. I. M. E., New York, pp. 1373-1416.

Miller, R. N., 1973, Guidebook for the Butte field meeting of the Society of Economic Geologists: The Anaconda Company, Butte.

Peterson, D. L., 1968, Bouguer gravity map of parts of Maricopa, Pima, Pinal, and Yuma Counties, Arizona: U.S. Geological Survey, Geophysical Investigation Map GP-615, 1 p.

Proffett, John M., 1973, Structure of the Butte district, Montana, in Guidebook for the Butte field meeting of Society of Economic Geologists (Richard N. Miller, ed.): The Anaconda Company, Butte.

Plouff, Donald, 1961, Gravity Survey near Tucson, Arizona: U.S. Geol. Survey professional paper 424 D, pp. 258-259.

Robinson, G. D., Klepper, M. R. , and Obradovich, J. D., 1968, Overlapping plutonism, volcanism, and tectonism in the Boulder batholith region, western Montana, in Studies in volcanology - a memoir in honor of Howel Williams (Robert R. Coats, Richard L. Hay, Charles A. Anderson, eds.): G. S. A. Memoir 116, pp. 557-576.

Sales, R. H., 1914, Ore Deposits at Butte, Montana: Trans, A. I. M. E., Vol. XLXI, pp. 3-109.

Smith, Ronald H., 1985, Isostatic residual gravity map of the United States: National Geophysical Data Center, NOAA, E/GC1, 325 Broadway, Boulder, CO 80303.

Smolik,T. J., Ziesing, G. F., and Loros, E. F., 1984, Golden Sunlight - a new gold mining operation: Mining engineering, Vol. 36, No. 11, pp. 1557-1561.

Schwenk, C. G., 1977, Discovery of the Flambeau deposit, Rusk County, Wisconsin, A geophysical case history: Wisconsin Geological and Natural History Survey.

Sumner,John S., and Schnepfe, Robert N., 1966, Underground gravity surveying at Bisbee, Arizona: Mining Geophysics, volume 1, case histories, pp. 243-251, Society of Exploration Geophysicists, Tulsa.

Swick, C. H., 1942, Pendulum gravity measurements and isostatic reductions: U.S. Coast and Geodetic Survey Special Publication 232, 78 p.

Tilling, R. L., Gottfried, D., 1969, Distribution of thorium, uranium, and potassium in igneous rocks of the Boulder batholith region, Montana and its bearing on radiogenic heat production and heat flow: U. S. Geological Survey professional paper 614 E.

Tilling, R. L., Klepper, M. R., and Obradovich, J. D., 1968, K-Ar ages and time span of emplacement of the Boulder batholith, Montana: Amer. J. Sci., Vol. 266, No. 8, pp. 671-689.

Warren,D. H., 1969, A seismic-refraction survey of crustal structure in central Arizona: Geological Society of America Bulletin, Vol. 80, pp. 257-282.

West, Robert E., 1972, A regional Bouguer gravity anomaly map of Arizona: PhD dissertation, University of Arizona, Tucson, 186 p.

Woollard, G. P., 1969, Regional variations in gravity, in The earth's crust and upper mantle, P. J. Hart, editor: American Geophysical Union Monograph 13, pp. 320-340.

Symbols

Bold-face symbols in the text and symbols with an arrow over them in the figures indicate vector quantities (e.g., **g**). Other symbols indicate scalar quantities or the magnitude of vector quantities (e.g., g). The subscript n indicates a variable subscript where n can be f, 1, or 2. The subscript f indicates a field station. The subscripts 1 and 2 indicate the beginning and ending base stations.

a	Earth's equatorial radius.
a_p	Distance from the earth's rotational axis to a point on the earth's surface (perpendicular to the rotation axis).
$\mathbf{a}_c$	Centrifugal acceleration at the earth's surface.
Bc	Bouguer correction.
c	Earth's polar radius.
CBA	Complete Bouguer anomaly.
Cc	Curvature correction.
dc_n	Gravity meter drift correction.
DG	Anomalous gravity field.
Dr	Gravity meter drift rate.
$Dg_{e\,f}$	The change in the magnitude of gravity. It is approximately equal to DGz. It is the quantity that is compared with gz in gravity interpretation.
DG_z	Vertical component of the anomalous gravity field.
DP	Density contrast.
dr_n	Gravity meter dial reading.
F	Gravitational force
f	Flattening of the earth.
FA	Free air anomaly.
Fc	Free air correction.
$\mathbf{g}$	Gravitational field.
$\mathbf{g}_e$	Gravity.

$\mathbf{g_{en}}$ Gravity.

g_{en} Magnitude of ge. The quantity that is measured with a gravity measuring instrument.

g_o Earth's normal gravity.

g_{oe} Earth's normal gravity at the equator.

g_{oh} Corrected normal gravity at elevation h and latitude L.

g_{ohf} Free air corrected normal gravity.

g_{ohcb} Complete Bouguer and free air corrected normal gravity.

g_{ohi} Isostatic, complete Bouguer, and free air corrected normal gravity.

g_{ohsb} Simple Bouguer and free air corrected normal gravity.

g_{o30} Earth's normal gravity based on the 1930 International Gravity Equation.

g_{o67} Earth's normal gravity based on the 1967 Geodetic Reference System Equation.

g_z Gravity effect. The vertical component of g calculated for a model. It is the quantity compared with Dgef in gravity interpretation.

h Elevation above sea level.

IA Isostatic anomaly.

Ic Isostatic correction.

k_n Gravity meter dial "constant".

L Latitude.

m, M "Point masses".

m′ "Test mass".

P Density.

r Position vector from the test mass m′ to the point mass m or the differential mass dm.

SBA Simple Bouguer anomaly.

Tc Terrain correction.

tc_n Tide correction.

t_n Time.

v Volume.

W Earth's angular velocity.

w Weight.

CHAPTER 4
SEISMIC

APPLICATIONS FOR SHALLOW EXPLORATION SEISMOGRAPHS

Written as an introduction for non-geophysicists, this chapter is intended to be a guide to users of shallow exploration seismographs. (Shallow is used to imply the first few hundred meters below the surface.) Principal applications, most of which are less than 100 m deep, are in engineering surveys, groundwater exploration, geologic mapping, and mining. Exploration for petroleum, covered more thoroughly in other literature, will receive only brief mention here.

Because the methods of shallow reflection have not yet been universally applied to these types of study, the reader will find a strong concentration on the refraction technique. As such, it is highly developmental, both from a procedural and an equipment standpoint.

Fundamental principles and applications

Exploration seismology is a branch of geophysics that examines the subsurface characteristics of the earth. The goal is to produce a geologic cross section representative of the subsurface materials and their geometric shape. The physical principal behind seismology is the difference in velocities of various mechanical vibrations through components of the earth's crust. These velocities are largely dependent on the compressive and shear strengths of the materials and, to a lesser extent, the dimensions and shapes of the structures. Seismic exploration does not identify the material, but measures its engineering properties.

Douglas B. Crice is marketing manager for EG & G Geometrics in Sunnyvale, California. Crice founded Nimbus Instruments in 1971 in Sacramento, California, to manufacture seismic instrumentation for use in shallow subsurface exploration and earthquake studies. It was purchased by EG & G Geometrics in 1977. Prior to founding Nimbus Instruments, Crice was employed at the Earthquake Engineering Office of the California Department of Water Resources for nine years doing work in the geodimeter fault movement investigations and in seismological instrumentation. He holds a B.S. degree in electrical engineering from California State University at Sacramento.

Douglas B. Crice, B.S.
Nimbus Division of GeoMetrics, Inc.

Those engineering properties are the desired information. In other cases, in which dissimilar bodies may have identical properties but no information can be generated to differentiate those properties, the geophysicist may combine seismic with other techniques for differentiation.

If a person had ready access to a structure, it would be simple to measure the velocities. He could put a vibration sensor on one side of the object and strike the other side with a hammer. By measuring the time interval between the hammer blow and the arrival of the shock wave at the other end, he could calculate the velocity from the distance traveled. Seldom is the solution so easy, although this method is occasionally used on rock outcrops and in the laboratory on core samples. In the field, the sensor and energy source are located on the surface, or occasionally down drill holes. The user can measure the travel time between several points by relocating the sensor, using a group of sensors, relocating the energy source, or combinations of these. Each sensor/source location allows the measurement of travel times through different paths. Since each path commonly represents passage through different structures, the data must be compiled and presented in such a manner that the shape and material can be deduced and calculated. This may be a simple task in some applications, and in others may be so complex as to defy analysis. Adding to the problem, each sensor/source combination may record several different arrivals. This occurs to a greater degree in reflection and shear wave studies, less commonly in refraction surveys.

A refraction survey to calculate depth to bedrock overlain by a gravel deposit is a sample application of this technique. The purpose of the survey might be to estimate the cost of cutting a roadbed down to bedrock for highway excavation; or, the purpose could be to find a source of gravel and evaluate it for available quantity. A model of this problem might resemble figure 4.1. If the model doesn't match the results, it will be apparent in the data. Since this survey is neither complicated nor ex-

FIGURE 4.1
Surface refraction survey model

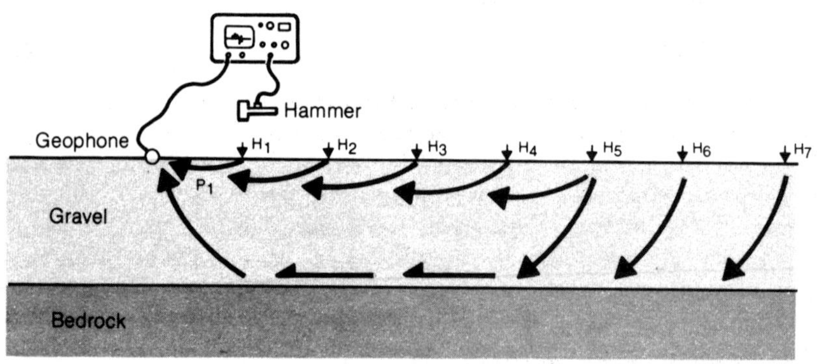

FIGURE 4.2
Time-distance plot surface refraction survey

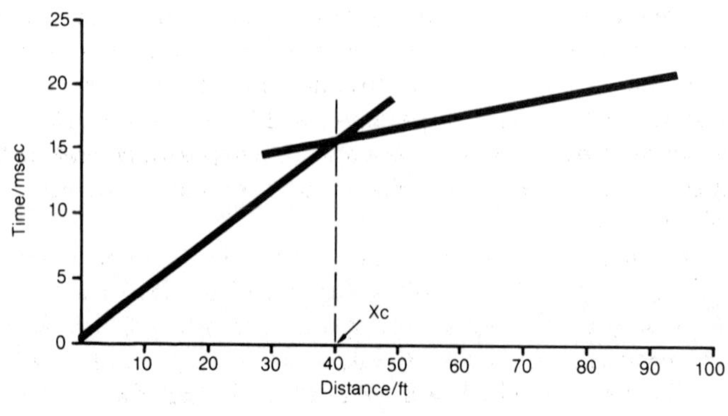

tensive, the field work can be done with a single-channel seismograph, one that uses only a single sensor and a sledgehammer energy source. A line is laid out on the surface, and points at 10-ft intervals are measured. The instrument, sensor, and hammer are connected with the appropriate cables, the sensor is placed at the first point, and the ground struck at the second point. The seismograph indicates how much time elapses between the impact and the arrival at the sensor, 4 milliseconds (msec) in the example. The experiment is repeated; the ground is struck at each 10-ft interval from the sensor, generating the values shown in table 4.1. This table is not diagnostic, except that the time increments become smaller past 40 ft. These data are plotted as a time-distance (t-d) graph in figure 4.2. It is apparent from the graph that the points form two straight lines, and these are drawn in with a straight edge, fitting the points as closely as possible. The first line segment, from 0 to 40 ft, represents the surface gravel layer. The second segment, beyond 40 ft, represents the bedrock. The slope of the first line is 0.4 msec/ft; the reciprocal of the slope is 2.5 ft/msec or 2500 ft/sec—the velocity of the seismic wave through the gravel. Similarly, the inverse of the slope of the second portion is 10,000 ft/sec, a reasonable number for bedrock. A vertical line is drawn from the intersection of the two slopes to the X axis, intersecting at 40 ft. This point is called the critical distance, Xc, and represents the distance from the sensor where an equal amount of time is required for the vibrations to travel directly across the surface layer or for them to travel down to the high velocity layer, across, and back up to the sensor. Beyond 40 ft, it is quicker for the seismic wave to travel down, across, and up, taking the path shown in figure 4.1.

TABLE 4.1
Refraction survey values

Distance/ft	Time/msec
10	4
20	8
30	12
40	16
50	17
60	18
70	19
80	20

The depth to bedrock can be calculated from

$$d = \frac{Xc}{2}\sqrt{\frac{V_2 - V_1}{V_2 + V_1}},$$

or

$$d = \frac{40}{2}\sqrt{\frac{10{,}000 - 2500}{10{,}000 + 2500}},$$

$$d = 15.49 \text{ ft.} \tag{4.1}$$

The bedrock is about 15½ ft below the surface. It would be easy to estimate the cost of excavating to 20 ft for a roadbed: the first 15 ft could be excavated easily, but beyond that drilling and blasting would be necessary. An estimate of gravel quantity could be reached if the depth to bedrock in the total target area were calculated.

Although the example may appear simple, its application is practical and valuable.

Two-layer refraction analysis

The two-layer refraction analysis refers to a geologic structure composed of two layers of material. In nature, it is common to find three, four, or more layers, but the two-layer case is convenient for illustrating seismic principles. The principles developed for the two-layer analysis can be expanded to handle a broader range of structures. The assumption is made that the upper layer will have a lower velocity than the lower layer. Because weathering and decomposition tend to reduce the integrity of the material, resulting in lower seismic velocities, this is a reasonable assumption.

Figure 4.3 illustrates the two-layer case. The upper layer has a velocity of V_1, which is less than the velocity of the lower layer, V_2. An impact is generated at S, which produces a shock wave traveling in all directions, including straight through the surface layer and down toward A. For reasons which will be examined later, the quickest path (the one requiring the least time) from S to B is down to A and across the interface to B. The wave refracted along the interface produces a continuing, traveling disturbance which generates new waves that refract back toward the surface, illustrated as lines AX1, BX1, CXc, DX4, and EX5. In practice, a continuous wavefront is refracted up all along the interface, so any number of points could be located. Xn represents sensor positions in this survey. Because this is a hypothetical model, Xc has been located at the critical distance; the points on the surface where the times required to travel straight through the upper layer or to refract down, across, and up are equal. The angle ϕ depends on the two velocities and may be computed from

$$\sin\phi = \frac{V_1}{V_2}. \tag{4.2}$$

The reader with a physics background may recognize this as being identical to Snell's law from optics.

When the hypothetical arrival times are plotted, the graph shown in figure 4.4 results. From the inverse of the slopes of the two lines, the velocities can be found. In the two-layer case (figure 4.3), the travel time for the direct wave through the surface layer is given by

$$Tc = \frac{Xc}{V_1} \tag{4.3}$$

and for the refracted wave along path SACXc

$$Tc = \frac{Z/\cos\phi}{V_1} + \frac{Xc - 2(Z\tan\phi)}{V_2} + \frac{Z/\cos\phi}{V_1}$$

or

$$\frac{Xc}{V_1} = \frac{2Z}{V_1\cos\phi} + \frac{Xc}{V_2} - \frac{2Z\tan\phi}{V_2}$$

$$= \frac{Xc}{V_2} + 2Z\left(\frac{1}{V_1\cos\phi} - \frac{\tan\phi}{V_2}\right). \tag{4.4}$$

FIGURE 4.3
Two-layer refraction analysis model

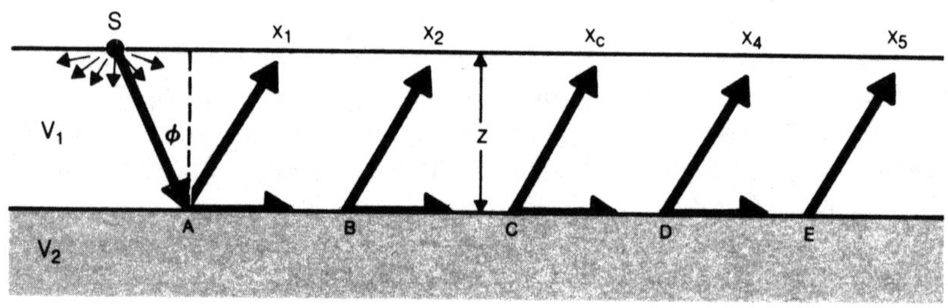

FIGURE 4.4
Time-distance plot, two-layer refraction analysis survey

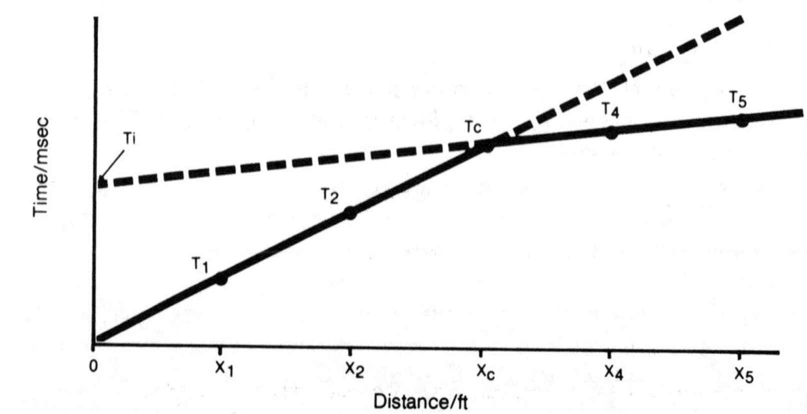

If the relationship $\sin\phi = \frac{V_1}{V_2}$ and substitution formulae from trigonometry are used, it can be shown that

$$\frac{Xc}{V_1} = \frac{Xc}{V_2} + \frac{2Z}{V_1}\cos\phi,$$

$$Z = \left(\frac{Xc}{V_1} - \frac{Xc}{V_2}\right)\frac{V_1}{2\cos\phi} = \frac{Xc}{2}\left(\frac{V_2 - V_1}{V_2}\right)\frac{V_2}{(V_2^2 - V_1^2)^{1/2}},$$

$$Z = \frac{Xc}{2}\sqrt{\frac{V_2 - V_1}{V_2 + V_1}}, \tag{4.5}$$

which was used earlier to find the depth to bedrock.

In addition, an alternate method involving intercept times can be used. In figure 4.4, the line representing the second layer arrivals can be extrapolated to intercept the time axis on the graph. This value is called the intercept time (Ti). Equation 4.4 is applicable for any value of X to compute the transit time through the lower layer, so it can be restated as

$$T = \frac{2Z}{V_1\cos\phi} + \frac{X - 2Z\tan\phi}{V_2}. \tag{4.6}$$

$$T = \frac{X}{V_2} + \frac{2Z}{V_1\cos\phi} - \frac{2Z\tan\phi}{V_2} = \frac{X}{V_2} + \frac{2Z}{V_1\cos\phi} - \frac{2Z\sin\phi}{V_2\cos\phi},$$

$$= \frac{X}{V_2} + \frac{2Z}{\cos\phi}\left(\frac{1}{V_1} - \frac{\sin\phi}{V_2}\right),$$

$$= \frac{X}{V_2} + \frac{2Z}{V_1\cos\phi}\left[\left(1 - \frac{V_1\sin\phi}{V_2}\right) = \left(1 - \frac{V_1^2}{V_2^2}\right) = (\cos^2\phi)\right],$$

$$= \frac{X}{V_2} + \frac{2Z\cos\phi}{V_1}. \tag{4.7}$$

Since this applies to any distance X from the source, the value zero can be substituted to find Ti.

$$Ti = \frac{2Z\cos\phi}{V_1}, \tag{4.8}$$

or

$$Z = \frac{Ti\ V_1}{2\cos\phi}. \tag{4.9}$$

This is another way to compute the depth of the first layer. Either method may be used, but generally Ti can be determined more accurately than Xc from the t-d graph. It requires finding the sine of the angle from the velocities, taking the inverse to find the angle, and then finding the cosine, a simple task with a pocket calculator.

The dipping layer

The case in which the interface is not parallel to the surface is presented in figure 4.5. The resultant t-d plot appears in figure 4.6. The direct wave travel time is given by

$$T = X/vl.$$

The travel time for the refracted wave is

$$td = \frac{X}{V}(\sin\theta + \phi) + \frac{2Zd}{V}\cos\theta. \tag{4.10}$$

These relationships can be used with the example cited earlier, a layer of gravel overlying bedrock, with velocities V = 2500 and V = 10,000.

FIGURE 4.5
Dipping interface model

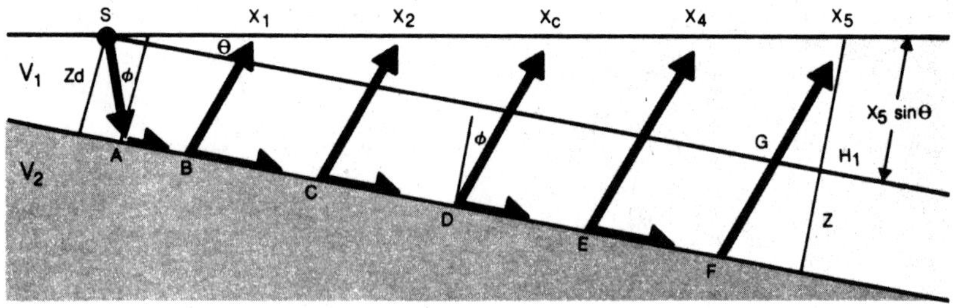

FIGURE 4.6
Time-distance plot, dipping interface

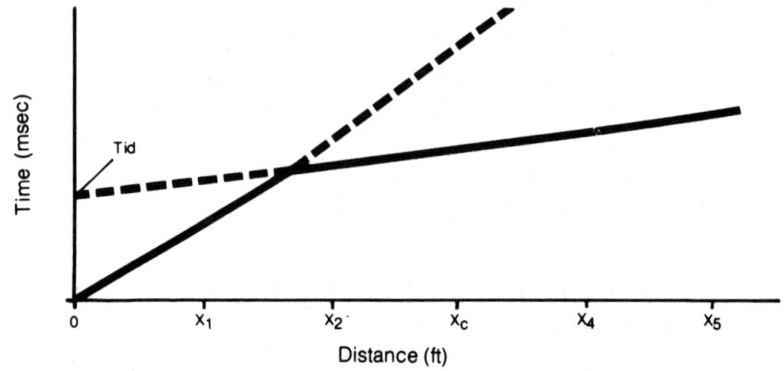

TABLE 4.2
Dipping layer data collected

Distance/ft	Time/msec
10	4
20	8
30	12
40	16
50	19
60	21
70	24
80	26
90	28
100	30

TABLE 4.3
Dipping layer data collected (reverse profile)

Point	Distance/ft	Time/msec
90	10	4
80	20	8
70	30	12
60	40	16
50	50	20
40	60	24
30	70	28
20	80	31
10	90	31
0	100	30

With the source at S and our sensor positions at different distances, the data in table 4.2 are collected. If a standard t-d plot is prepared, a graph similar to figure 4.2 results. The reader should do this as an exercise. The resultant data and graph illustrate an interesting point in seismic exploration: the points do not fall on a perfectly straight line. This is caused by a number of factors including non-uniform velocities in different parts of the material, irregularities in the interface boundary, and errors in measuring the arrival times. The solution requires that the reader locate a line which best seems to fit the points. The actual scatter in a field situation will vary.

The author plotted the data and got the following results:

$V_1 = 2500$ ft/sec
$V_2 = 4500$ ft/sec
$Xc = 45$ ft
$Ti = 8$ msec

If this is treated as the normal two-layer case with parallel layers, a depth of 12 ft for the gravels results. How accurate this number is remains to be seen, but 4500 ft/sec is grossly in error, and certainly does not qualify as a very competent bedrock. A better interpretation would be gravel in clay or highly weathered rock.

The solution to the problem requires the use of a reverse profile. The survey is done again, but this time the source is placed at the last sensor location. The seismograph measures the travel times back to each of the same points, terminating at the previous source location. The numbers in table 4.3 result. A t-d plot containing both sets of numbers as shown in figure 4.7 can be prepared. The reverse profile has the familiar two intersecting lines, except that the slope of the line representing the lower layer is different—it slopes downward. Difficult to accurately measure and plot, the answer for V_2 is $-25{,}000$ ft/sec, the negative velocity resulting from the down slope. Equation 4.10 results in the travel time down the slope. In a similar manner, the travel time for the reverse profile can be found by

$$tu = \frac{X}{V_1}(\sin\theta - \phi) + \frac{2Zu}{V_1}\cos\theta. \qquad (4.11)$$

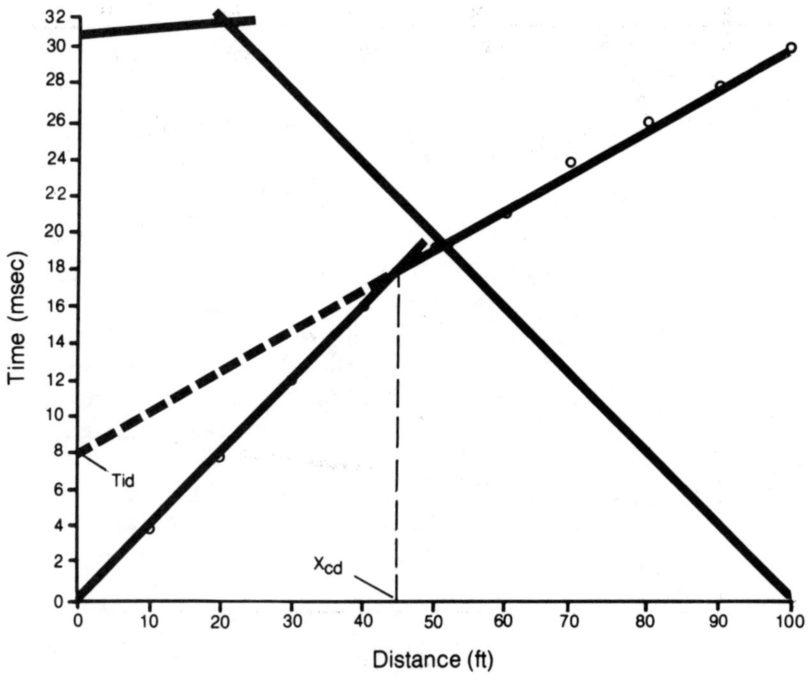

FIGURE 4.7
Two-layer, dipping interface and t-d plot

By differentiating equation 4.10 or 4.11, one can find the apparent velocity as plotted on the t-d chart.

$$\frac{d\,Td}{d\,X} = \frac{\sin(\theta+\phi)}{V_1} = \frac{1}{Vd} \tag{4.12}$$

and

$$\frac{d\,Tu}{d\,X} = \frac{\sin(\theta-\phi)}{V_1} = \frac{1}{Vu}, \tag{4.13}$$

where Vd and Vu are the up and down slope velocities on the t-d plot which is made from the field measurements. Since V_1 is also known from the plot, the angles can be calculated from equations 4.12 and 4.13.

$$\theta = \frac{1}{2}\left[\sin^{-1}\left(\frac{V_1}{Vd}\right) + \sin^{-1}\left(\frac{V_1}{Vu}\right)\right]$$

and

$$\phi = \frac{1}{2}\left[\sin^{-1}\left(\frac{V_1}{Vd}\right) - \sin^{-1}\left(\frac{V_1}{Vu}\right)\right].$$

The values from the t-d plot for T and X (at the endpoints) and the angles from the calculations listed above can be used with equations 4.10 and 4.11 to find Zd and Zu. Applied to the example problem, it can be found that

$$\theta = \frac{1}{2}\left[\sin^{-1}\left(\frac{2500}{4500}\right) + \sin^{-1}\left(\frac{2500}{-25000}\right)\right] = 14^\circ$$

$$\phi = \frac{1}{2}\left[\sin^{-1}\left(\frac{2500}{4500}\right) - \sin^{-1}\left(\frac{2500}{-25000}\right)\right] = 19.7^\circ$$

Since

$$Td = Tu = 30 \times 10^{-3} \text{ sec}$$

$$.030 = \frac{100}{2500}(\sin 14 + 19.7) + \frac{2\,Zd}{2500}\cos 14^\circ$$

$$.030 = \frac{100}{2500}(\sin 14 - 19.7) + \frac{2\,Zu}{2500}\cos 14^\circ$$

Solving:

$$Zd = 10.1 \text{ ft}$$

$$Zu = 43.8 \text{ ft}$$

Since

$$\sin\theta = V_1/V_2$$

$$V_2 = 10{,}330 \text{ ft/sec.}$$

That yields enough information for a cross section of the test area to be drawn. It should be remembered that different velocities for V_2 on the forward and reverse profile imply dipping layers. If the dip were a little less, the V_2 slope would rotate back toward the horizontal and then upward. It can be a perfectly flat line with a slope of zero, which implies a velocity of infinity. If the slope had been much steeper, an angle could be reached in which the second layer would never show up on this limited scale survey. That particular formation will be discussed later, and it would have been apparent were the line also surveyed over a wider area. It should be noted that the single-ended survey gave no clue by itself that there was a dipping layer. For that reason, the double-ended survey is performed routinely unless the user has enough knowledge about the local geology and expected velocities to judge the information from a single-ended study. That the end times were equal in both directions of the survey should also be noted. This must always be true, and serious discrepancies cast doubt on the measurements.

The exact equations used for solving dipping layers can be approximated where the dip is slight or when accuracy requirements are relaxed. For small angles

$$\cos\phi \approx 1 \text{ and } \cos\theta \approx 1,$$

$$\frac{V_1}{V_D} = \sin(\theta + \phi)$$

$$= \sin\theta\cos\phi + \sin\phi\cos\theta \approx \sin\theta + \sin\phi, \qquad \text{(from 4.12)}$$

$$\frac{V_1}{V_u} = \sin(\theta - \phi)$$

$$= \sin\theta\cos\phi - \sin\phi\cos \approx \sin\theta + \sin\phi, \qquad \text{(from 4.13)}$$

so that

$$\frac{V_1}{V_D} + \frac{V_1}{V_1} \approx Z\sin\theta = Z\frac{V_1}{V_2},$$

and

$$\frac{1}{V_2} \approx \frac{1}{2}\left(\frac{1}{V_D} + \frac{1}{V_u}\right). \tag{4.14}$$

This relationship used in the example results in

$$V_2 = \frac{1}{\frac{1}{2}\left(\frac{1}{V_D} + \frac{1}{V_u}\right)} = \frac{1}{\frac{1}{2}\left(\frac{1}{4500} + \frac{1}{-25000}\right)} = 10{,}975 \text{ ft/sec}$$

which is reasonably close to the exact value of 10,330 ft/sec. Use of the average value can result in substantial errors unless the dip angle is very small.

Irregular subsurface layer

The two models have had straight interfaces. Another case in which an old stream channel has cut into the bedrock and a protrusion lies adjacent can be studied. A hypothetical cross section with the refracted wave travel paths is shown in figure 4.8. The resultant t-d plot is shown in figure 4.9. The second layer line in the t-d plot is quite irregular; in fact, it would be difficult to draw a "best fit" line with only the raw data. This type of formation illustrates another aspect of refraction seismic exploration: it is important that the operator have some knowledge of the technique. Only in very simple geologic conditions can the measurements be performed by a non-skilled operator. In this case, for example, the operator might add some additional points off the end in an attempt to collect more data to reach an accurate line fit. In addition, the operator probably would have done some nearby surveys to get a good feel for the actual velocity of the subsurface layer so that grossly erroneous data would stand out.

The subsurface intrusion allows shorter travel times than normal, and the erosion channel produces travel times that are longer than normal. The perturbations in the data will be the opposite of the perturbations in the interface. The graph perturbations will be offset horizontally from their actual position. The amount of offset can be calculated roughly from $\sin \theta = V_1/V_2$ and the depth.

$$\text{Offset} = Z\frac{V_1}{V_2}. \tag{4.15}$$

The reverse profile (figure 4.10) produces a similar result, except that the stream channel does not appear. The points which would have

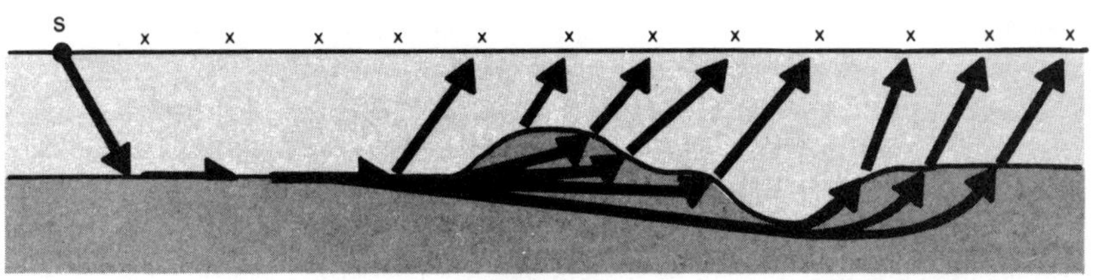

FIGURE 4.8
Two-layer case with irregular interface

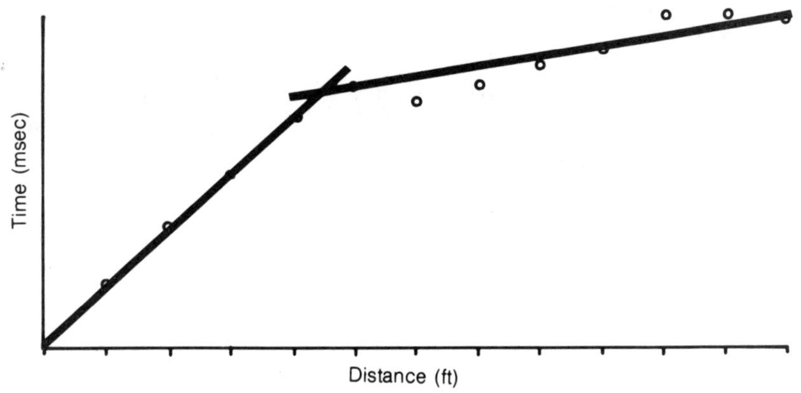

FIGURE 4.9
T-d plot for irregular interface

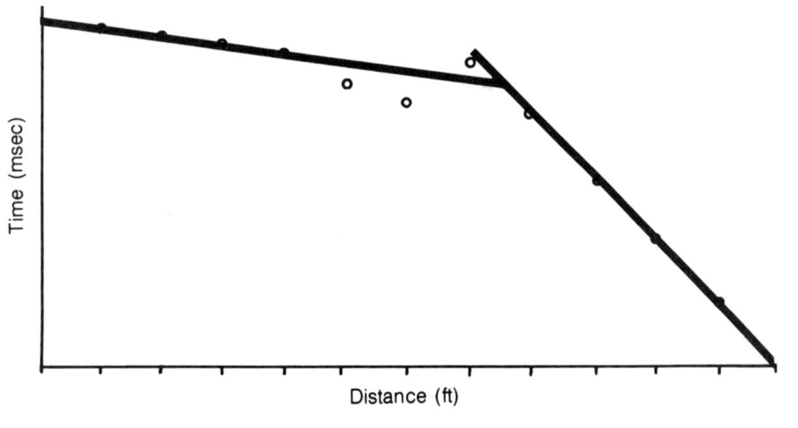

FIGURE 4.10
T-d plot for irregular interface (reverse profile)

reflected the stream channel are short of the critical distance, and the direct wave is the first arrival. The height of the protrusion can be estimated with the time error (measured from the best fit line)

$$h = \frac{\delta t}{\left(\frac{1}{V_1} - \frac{1}{V_2}\right)}. \tag{4.16}$$

Exact solutions of the irregular second layer are not always essential, depending on the object and the relative magnitude of the survey. For example, the excavation estimates and gravel quantities calculated earlier in the chapter are concerned with average amounts of rock. On the other hand, if a person were exploring for placer gold deposits, he would be quite interested in the old stream channel.

Horizontal variations in structure

Numerous structures can be identified by the shape of their t-d plots. A brief examination of several of these and their related plots follows. Each of these examples has a typical cross section and its associated t-d plot for forward and reverse profiles. It should be noted that the shape of the plot might have been different if the surveys were shifted slightly in either direction.

Figure 4.11 shows a horizontal to down-dip structure. In figure 4.12, the forward profile shows the normal two-layer results initially. The slope of V_2 then breaks up (lower apparent velocity) as the refracted wave travels down the incline. In the reverse profile, a portion of the refracted wave travels up-slope giving an apparent velocity higher than V_2, followed by the normal V_2 line.

FIGURE 4.11
Horizontal to down-dip structure

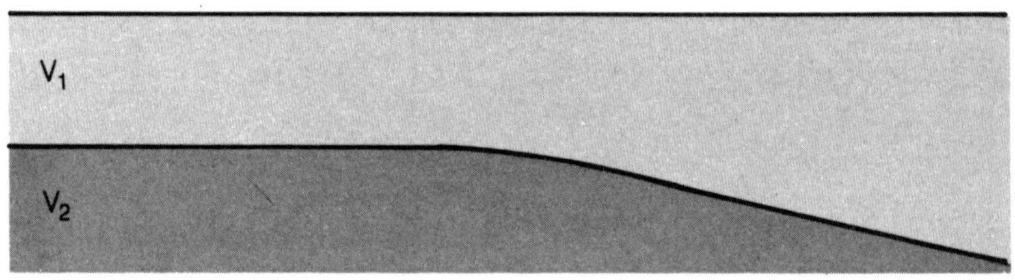

FIGURE 4.12
T-d plot, dip in part of interface

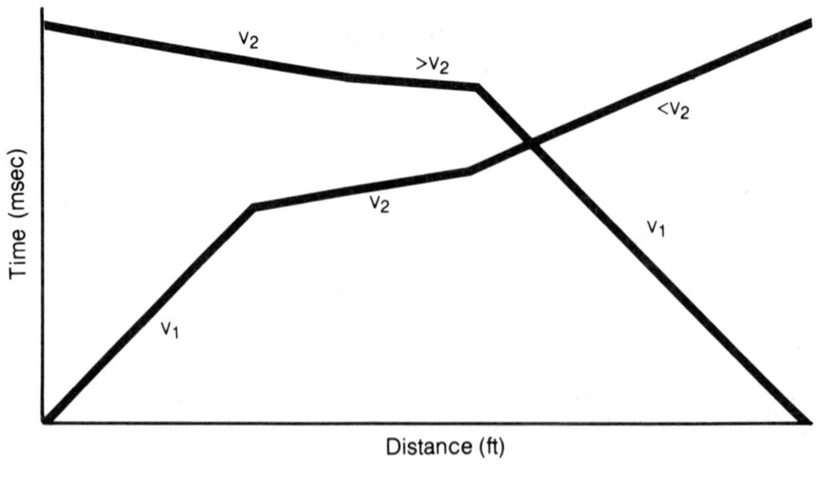

Figure 4.13 shows a horizontal to up-dip structure. Analysis is similar to the previous example. The reverse profile (figure 4.14) is interesting in that its t-d plot has the same appearance as a three-layer profile to be discussed later.

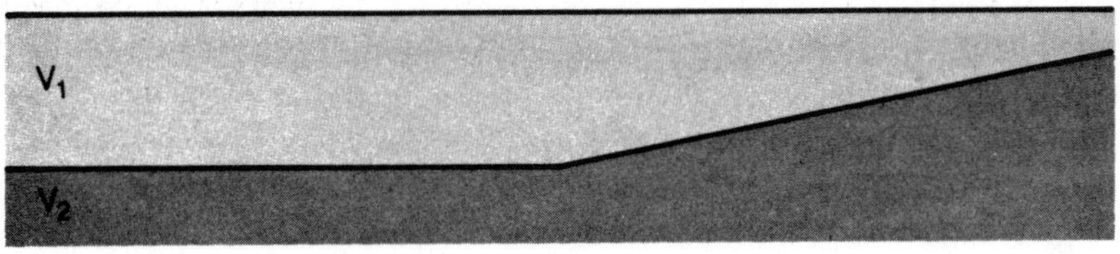

FIGURE 4.13
Horizontal to up-dip structure

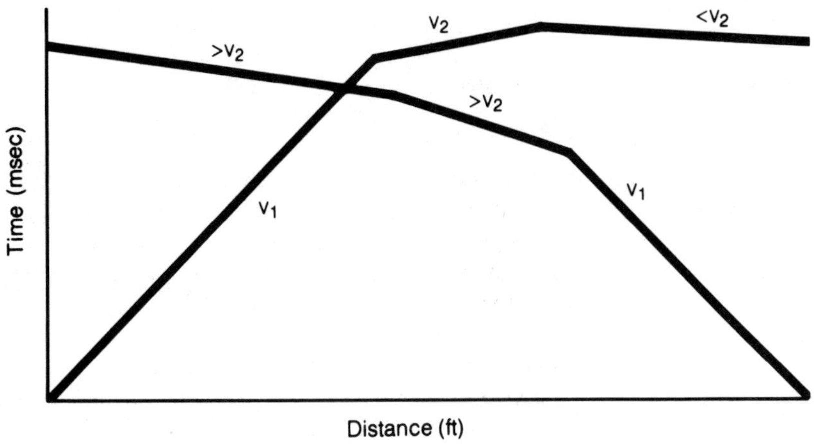

FIGURE 4.14
T-d plot, rise in part of interface

Figure 4.15 shows a subsurface step structure. The V_2 portion of the graph has an offset from the step (figure 4.16). When the survey is toward the down step, the graph is offset up. When the survey is toward the up step, the graph is offset down. The graph often has a discontinuity (an offset with no apparent in between points), because the times are measured at discrete points.

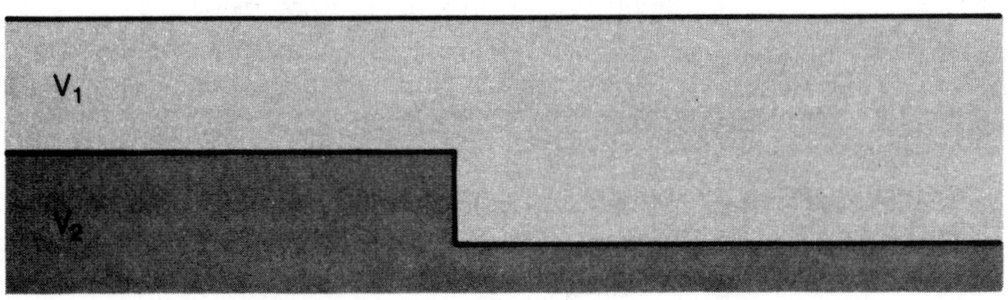

FIGURE 4.15
Subsurface step structure

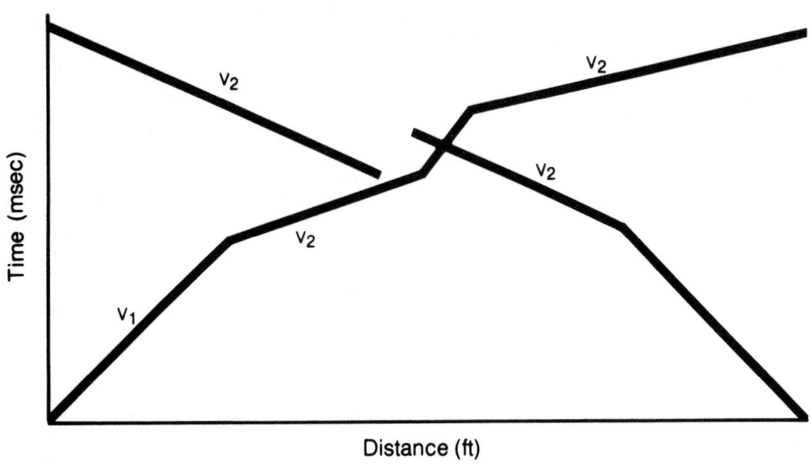

FIGURE 4.16
T-d plot, step in interface

Figure 4.17 shows a low-velocity dike. This structure often indicates an active earthquake fault. The contact area is of lower velocity because of the grinding action of the fault motion. The plot (figure 4.18) shows a short section of low-velocity material, but normally the graph will be discontinuous if the width of the zone is less than the sensor spacing. It should be noted that in this structure, both the forward and reverse V_2 plots were offset upward, where as in the previous example, they were offset in opposite directions.

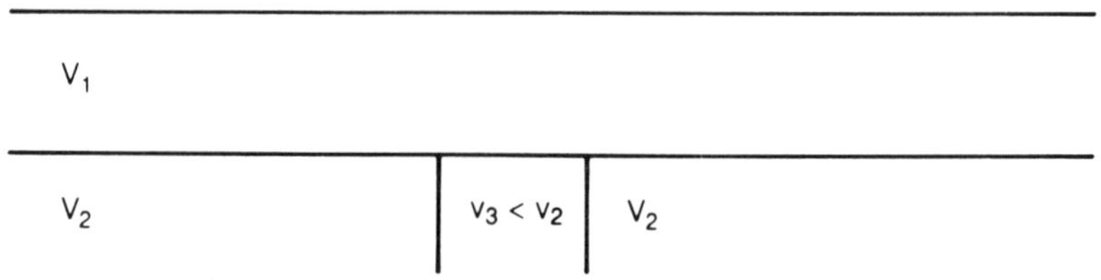

FIGURE 4.17
Low-velocity dike structure

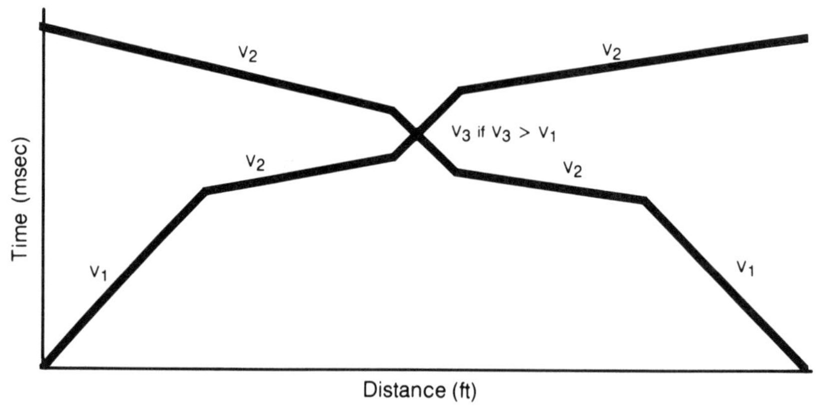

FIGURE 4.18
T-d plot, low-velocity dike

Figure 4.19 shows a high-velocity dike. In this case, both V_2 plots are offset downward. The small V_3 section in the graph (figure 4.20) may not appear if the dike is narrow. In order for the dike to have an effect on the t-d plot, it must have a high enough relative velocity or be thick enough to produce an appreciable change in the travel time. For ex-

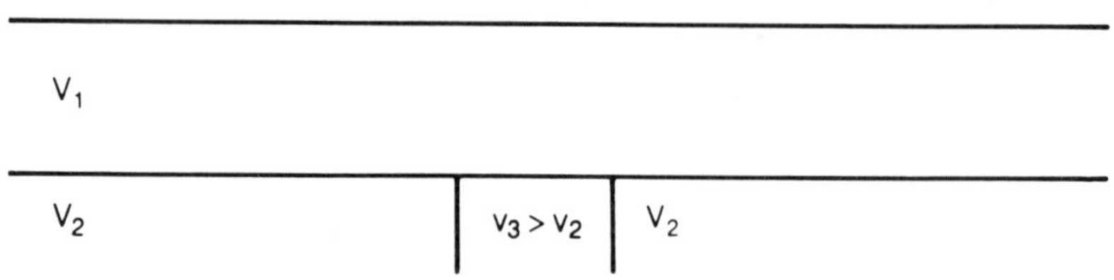

FIGURE 4.19
High-velocity dike structure

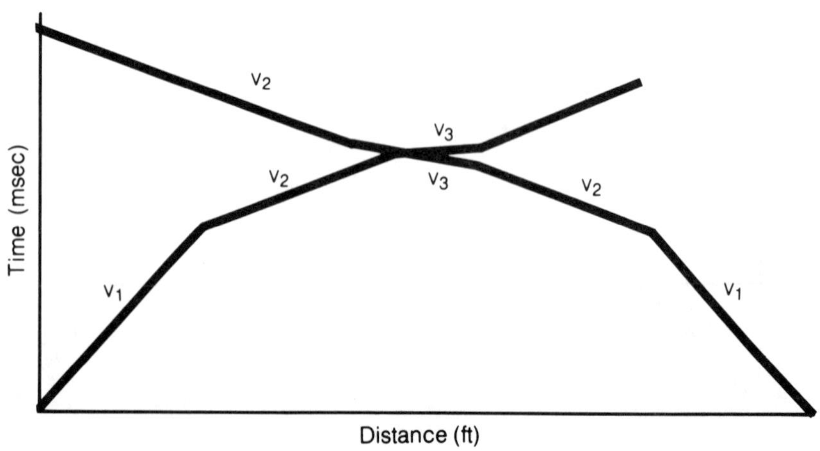

FIGURE 4.20
T-d plot, high-velocity dike

ample, a 10-ft thick dike with a velocity of 20,000 ft/sec in a 10,000 ft/sec bedrock would offset the V_2 graph by only ½ msec. Care would be required to detect this structure, and it would probably go unnoticed in routine surveys. If the base material were slower, about 5000 ft/sec, the time offset would be 1½ msec, a more readily detectable value.

Figure 4.21 shows a vertical fault. This case is similar to figure 4.16, except that the drop is great enough that there are no arrivals from the right side except the direct wave (figure 4.22).

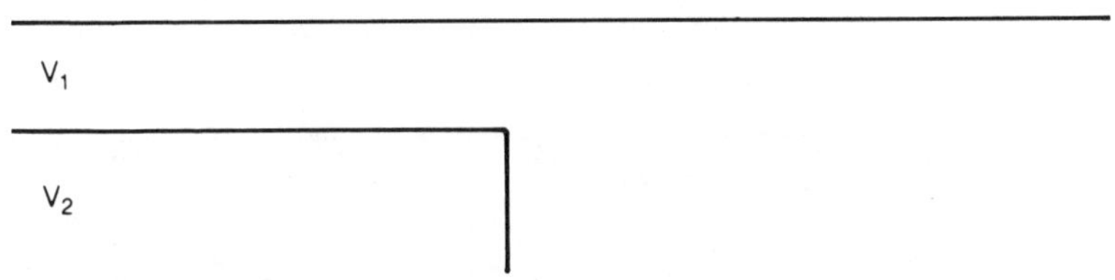

FIGURE 4.21
Vertical fault structure

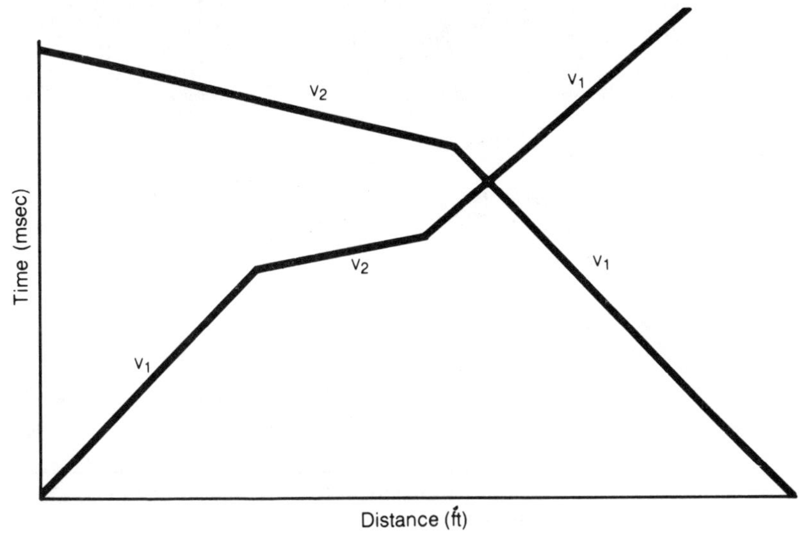

FIGURE 4.22
T-d plot, vertical fault

Figure 4.23 shows a horizontal velocity change. One side of the plot is identical to the three-layer case, once again pointing up the need for reverse profiles (figure 4.24).

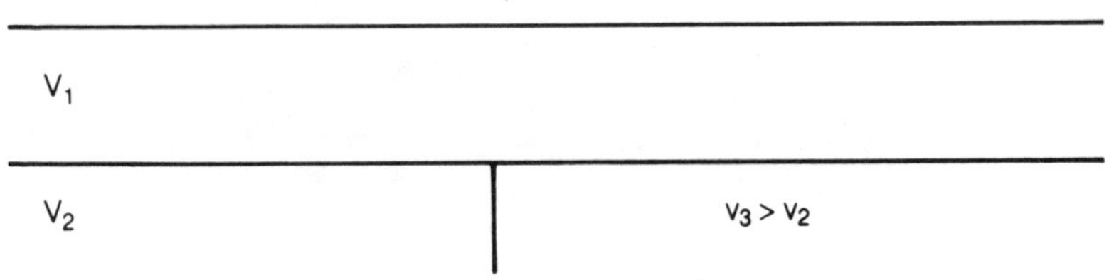

FIGURE 4.23
Horizontal velocity change model

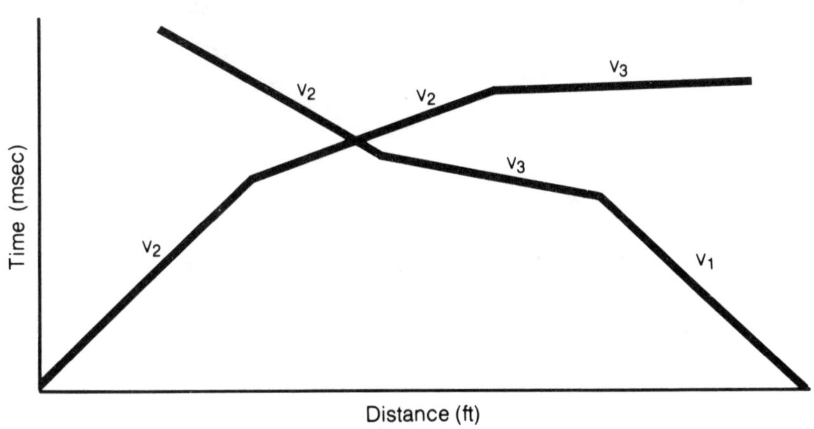

FIGURE 4.24
T-d plot, horizontal velocity change

Figure 4.25 represents an interesting case which most often shows up in a bi-directional survey with indicated velocities of unusually high values such as 25,000 ft/sec or more. It should be noted that there are no intrinsic clues in the t-d plots (figure 4.26); this is not a simple two-layer parallel interface. The selection of reasonable velocities must come from other surveys in the area. Rather than attempt to analyze these data, one might want to repeat the measurement by laying the seismic line perpendicular to the depression and measuring the depth below the center point. Whenever this type of structure is anticipated, such as along riverbeds, the survey should be performed along the channel rather than across it.

FIGURE 4.25
Bowl-shaped depression model

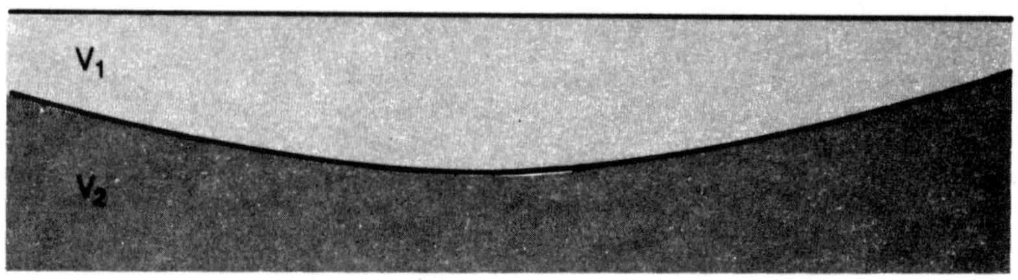

FIGURE 4.26
T-d plot, bowl-shaped depression

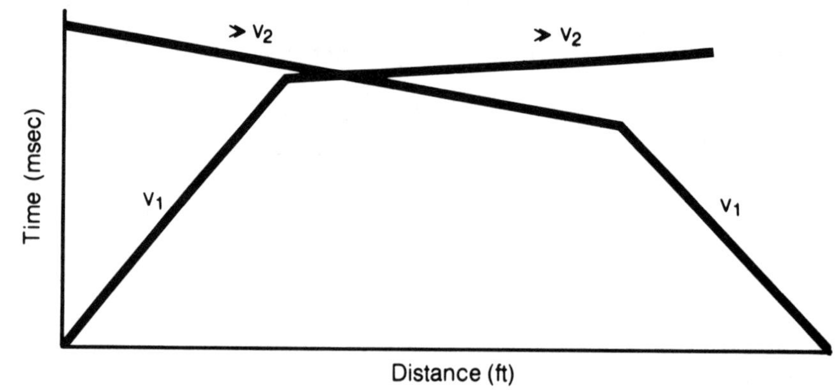

Figure 4.27 is the reverse of figure 4.25, an inverted bowl. In this case, a higher velocity slope often appears in the center of the t-d plot for V_2 (figure 4.28).

FIGURE 4.27
Inverted-bowl model

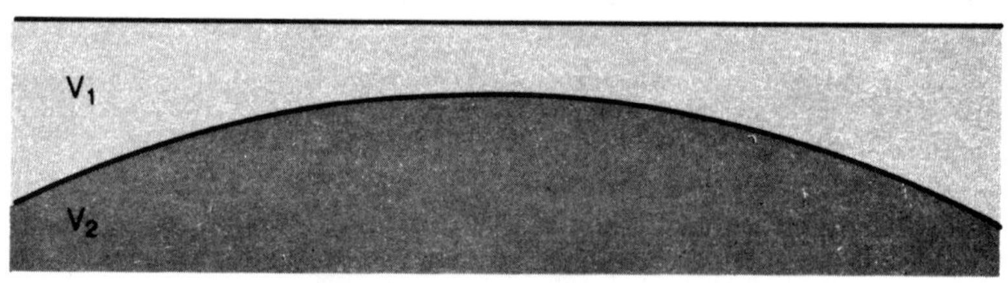

FIGURE 4.28
T-d plot, inverted-bowl

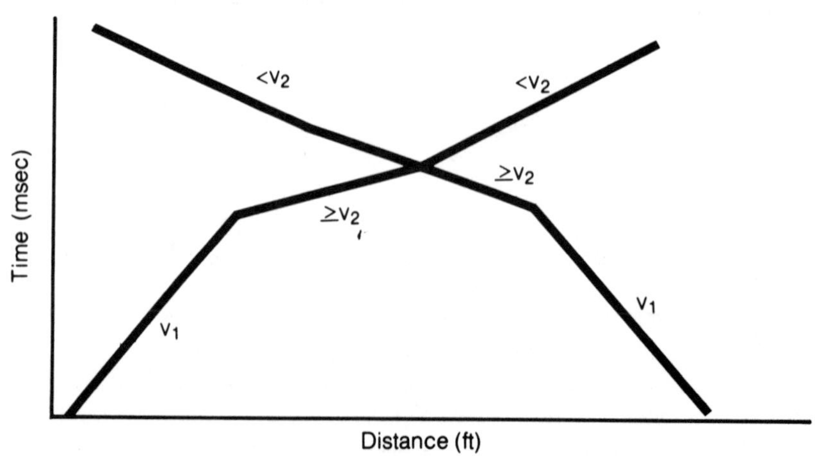

Figure 4.29 shows an irregular surface layer. In this case, the t-d plot for V_2 follows the contour of the ground instead of exhibiting the inverted appearance from the irregular interface (figure 4.30). The best solution to this type of survey is probably reached by taking the variation from the average elevation for each point and adjusting the plotted time by the expected error. An approximation would be h/V_1, and a more accurate solution, $h/(V_1 \cos \theta)$.

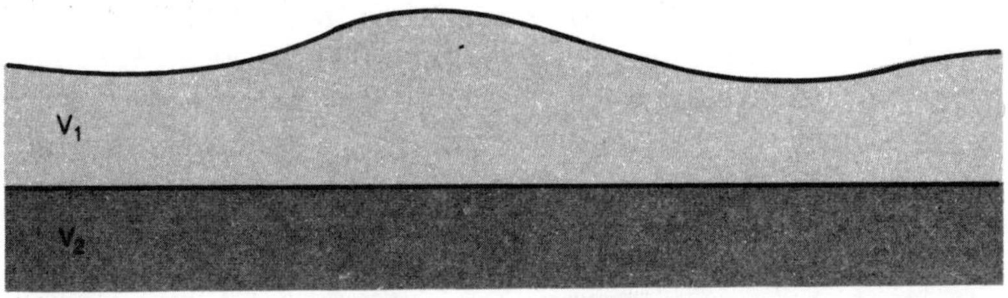

FIGURE 4.29
Irregular surface layer model

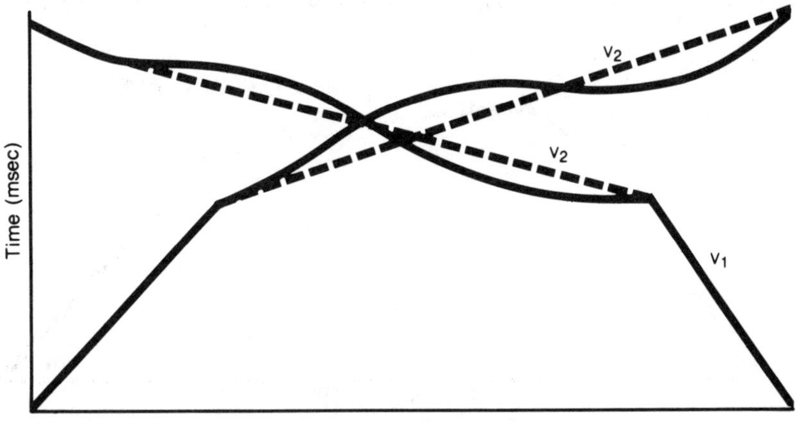

FIGURE 4.30
T-d plot, irregular surface layer

The examples cited here serve as a catalog of t-d plots and geologic structure. Other geologic structures could also produce the identical plot, so the apparent interpretation should always be weighed against other data taken in the vicinity, geologic knowledge based on the operator's experience, and available drilling information.

The occasional appearance of a low-velocity layer has not yet been considered. Because the analysis so far has been based on the fastest travel paths, there is no refractor information from a buried layer with a velocity less than the surficial material. A refracted path down to a lower layer through a low-velocity material can exist, but it will not be indicated on the t-d plot. Calculated depths will be in error, although the surface and buried third-layer velocities will be correct as shown in the t-d plots.

If the low-velocity layer is discontinuous (often the case with sand deposits), assorted t-d plots can be generated, including the examples in figures 4.9, 4.12, 4.16, 4.18, and 4.22. Other hypothetical structures with mixes of high- and low-velocity materials can be postulated to account for the sample t-d plots, but the examples shown represent most common interpretations.

Interpretation requires that a person examine the t-d plots and deduce a geologic structure which would produce the same results. Such deduction should include opinions based on other information, always the case in geologic studies. Once a geologic model is selected which accounts for the shape of the t-d plot, then the physical dimensions can be calculated. The material can be inferred from the velocities, from examination of exposed and borehole samples, and from experience.

Seismic wave motion

At this point, some theory about seismic wave motion is needed. Some of this will be important later in this chapter. Other portions will be purely of theoretical interest, but useful background for future studies the reader may wish to pursue.

As an example, when the end of a long section of some material (see figure 4.31) is struck with a hammer, a zone of compression forms beneath the hammer point and travels down the section. This wave, similar to a sound wave, is called a P-wave, or primary wave, because it is the first arrival. (It may also be called a longitudinal, dilational, or compressional wave.) The velocity of this wave depends on the elastic constants of the material and its density. If E = Young's Modulus, σ = Poisson's Ratio, and ρ = density, then

$$Vp = \sqrt{\frac{E}{\rho}\ 1 + \frac{2\sigma^2}{1 - \sigma - 2\sigma^3}} \tag{4.17}$$

where Vp = the velocity of the P-wave.

FIGURE 4.31
Propagation of a P-wave

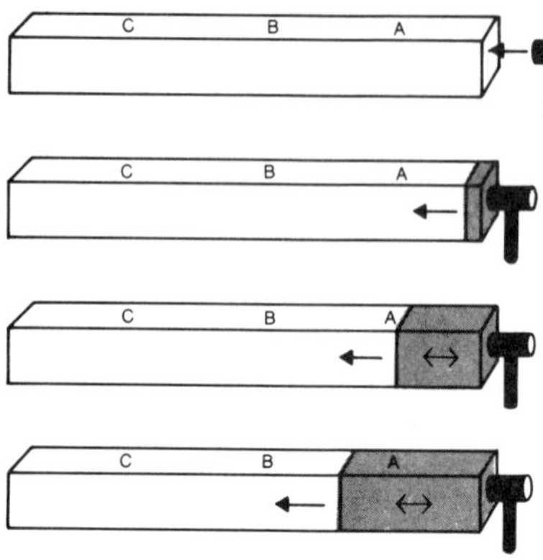

If motion sensors were placed on the beam, oriented so they were sensitive to back-and-forth motion along the long axis, and if the behavior of this motion in time were charted, a graph would resemble figure 4.32 at the three points A, B, and C. This can be done physically with a set of electronic motion sensors and an oscilloscope, generating waveforms very similar to figure 4.32. If a switch on the hammer provided the impact time, it would be a rudimentary, but quite adequate, seismograph.

If the end of the beam is struck on the edge with a hammer (figure 4.33), a portion of the material will deflect in the direction of the hammer blow. This deflection will propagate along the beam as the P-wave did. The same effect can be demonstrated with a rope. If a rope is placed on the ground and the end jerked rapidly upward, a wave appears and travels down the rope. This wave is called a shear wave (also known as a transverse, rotational, secondary, or S-wave). The distinction between particle motion and direction of propagation should be noted. Particle motion (the motion of a small portion of the material) is purely vertical. The distorting influence propagates along the beam, and that is the propagation direction of the wave. The shear wave travels more slowly than the P-wave. With the same elastic constants

FIGURE 4.32
Motion of points on the beam

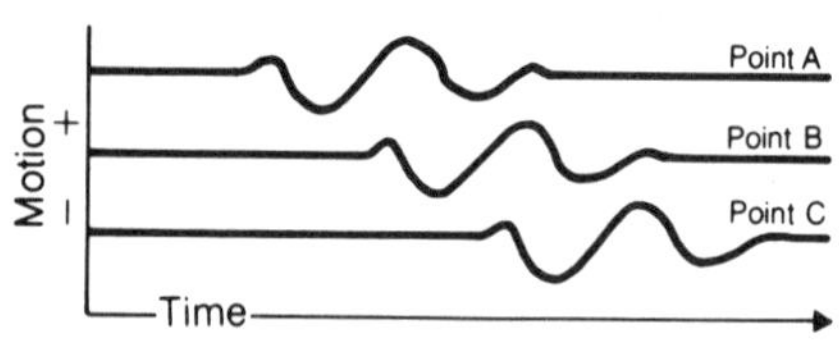

$$Vs = \sqrt{\frac{E}{2\rho(1 + \sigma)}} \tag{4.18}$$

and

$$\frac{Vp}{Vs} = \sqrt{\frac{\frac{1}{2} - \sigma}{1 - \sigma}}.$$

Also,

$$Vs = \sqrt{\frac{\mu}{\rho}}$$

where

Vs = shear wave velocity and
μ = shear modulus.

FIGURE 4.33
Propagation of an S-wave

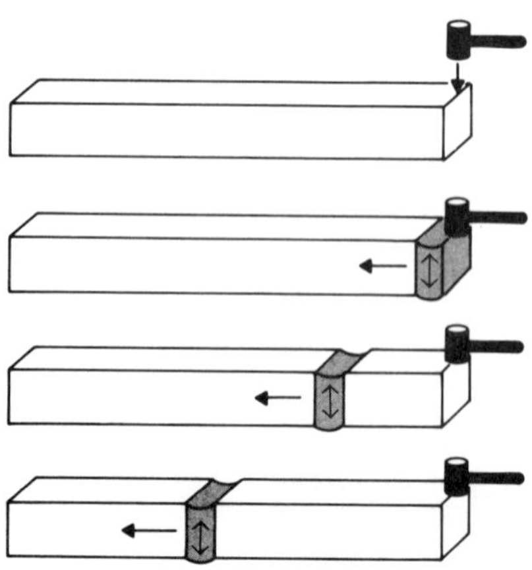

The experiment with the motion sensors could be repeated. This time they could be arranged so that they would be sensitive to up-and-down motion. With the exception of a time change, the results would be identical to figure 4.32.

It should be noted in this experiment that the S-wave particle motion was vertical. This wave is said to be plane polarized and is described as

SV (vertical shear). If the beam had been hit sideways, horizontally polarized waves would have been generated.

The example beam should be considered a small portion of a large block of homogenous and isotropic material, i.e., the material is the same throughout, and its elastic constants are identical in all directions.

If the material is bounded by a surface (such as the ground surface), other types of waves may appear. The principal wave of interest is the Rayleigh wave which has a particle motion that combines the back-and-forth motion of the P-wave with the up-and-down motion of the SV wave. The actual path is elliptical as shown in figure 4.34. Propagation is sideways as in all the other waves and slightly slower than the S-wave. The amplitude of the wave is largest at the ground surface, and the Rayleigh wave particle motion decreases rapidly with depth. The Rayleigh waves tend to be large and often interfere with these other surveys. Ground roll is an expression often used to describe the signals from Rayleigh waves, and numerous procedural and computational methods have been developed in an effort to reduce the problems from them.

FIGURE 4.34
Particle motion for a Rayleigh wave

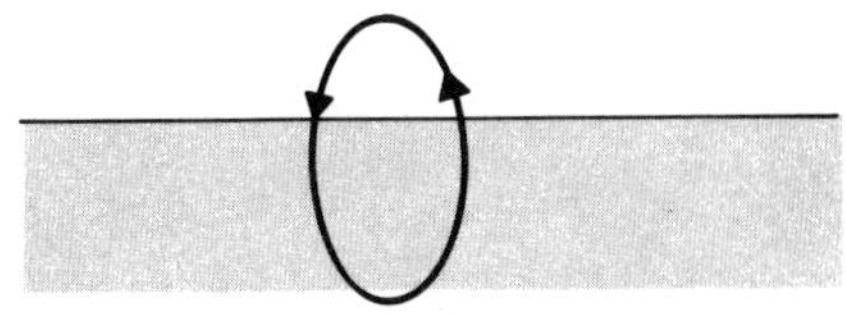

Love waves, horizontal shear waves with a velocity intermediate between S-wave velocities at the surface and in deeper layers, also appear at the surface. They do not normally cause problems in seismic exploration, because most sources are not strong generators of this wave, and most geophones respond only to vertical motion.

Other waves appear in a real, layered earth and in borehole surveys, but these are not significant within the scope of this chapter.

FIGURE 4.35
Dependence of energy upon distance

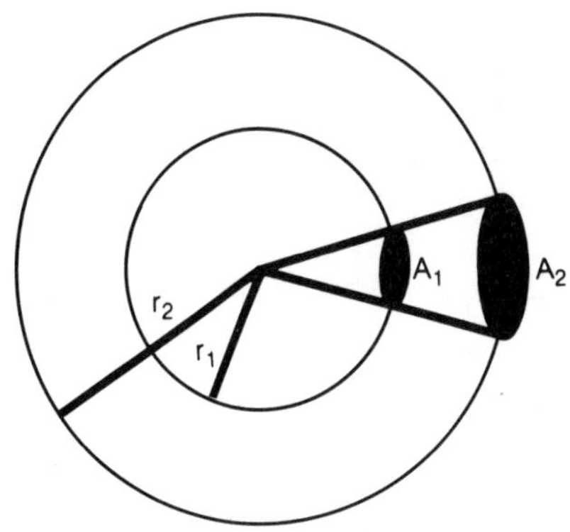

A sledgehammer impact or explosive source generates seismic waves which radiate in all directions. The wavefront has a spherical shape, and as it propagates, the sphere grows larger. The energy of the seismic wave spreads out over a larger and larger area as distance increases from the shotpoint. The area is proportional to the square of the radius, as shown in figure 4.35. Since the area increases, the energy decreases and

$$\frac{E_2}{E_1} = \frac{r_1^2}{r_2^2}. \qquad (4.19)$$

This energy decrease, called spherical divergence, is proportional to the square of the distance.

Energy is also lost through absorption. This phenomenon, associated with energy loss during compression and rarefaction as the wave passes, is exponential with distance. It is possible to write

$$E = E_0 e^{-\alpha x} \qquad (4.20)$$

where α is the absorption coefficient and x is the distance between the points where E and E_0 were determined. α is highly dependent on the frequency of the wave. High-frequency signals are attenuated much more rapidly than low-frequency signals.

It may be relevant to discuss the concept of frequency at this point. In figure 4.32, the graph of the passing seismic signal oscillates to either side of the neutral line. The rate at which this oscillation occurs is called the frequency and is measured in cycles per second or Hertz. The time required to make a full cycle (zero to plus to zero to minus to zero) is called the period (equal to 1/f). Since the wave is moving through the earth, it will cover some distance during a cycle. This distance is called the wavelength and is computed from

$$\lambda = V/f. \qquad (4.21)$$

In shallow exploration, the energy source is usually an explosion, hammer blow, or weight drop. These sources generate a broad band of frequencies, depending on the source and the material. As a rule, small sources generate higher frequencies than large ones, and soft materials

FIGURE 4.36
Reflection and refraction of P-wave at an interface

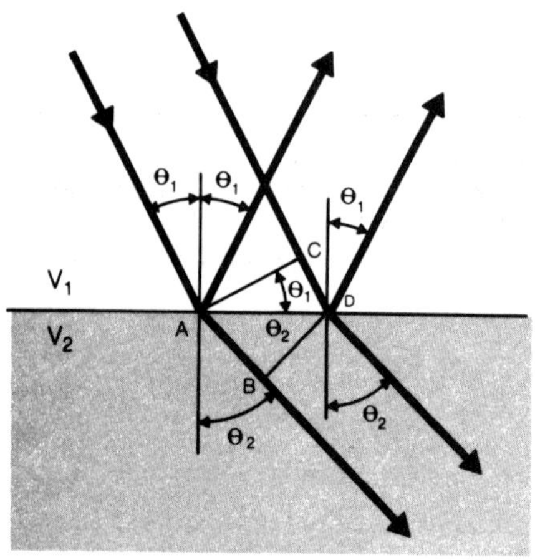

produce lower frequencies than hard ones. Most seismic exploration involves frequencies in the range from 5 to 500 Hz. Since absorption is much more severe at high frequencies, they disappear from the seismic waveform (are filtered out) as the space between the source and sensor increases.

Dispersion is another condition which will change the characteristics of seismic waves. In some materials, the velocity of the seismic wave will vary with frequency. The initial wave will usually contain a complex combination of frequencies. If the material is dispersive, some frequencies will travel faster than others. As a result, the character of the wave will change with distance. Dispersion is not a dominant feature of body waves, but is important in connection with the surface waves.

In addition to absorption and spreading, the partitioning of energy at interfaces is also responsible for the decrease in the energy of seismic waves, the subject of the next section.

Behavior at an interface

The assumption is made that seismic waves have been generated some distance from an interface (the boundary between two materials with different wave velocities). Because of the geometry, it can be assumed that the wavefront is straight, rather than curved, and that it consists of parallel waves. When this wave impinges on an interface as shown in figure 4.36, part of the P-wave is reflected back and part is refracted through, like light waves on a water surface. Each point on the wavefront can be regarded as a new source for waves, a concept described as Huygen's principle in elementary physics. The time for the wave to travel from C to D in the first material is equal to the time, t, required for the wave to travel from A to B in the second material. All the angles labeled θ_1 are equal, as are the angles labeled θ_2.

$$\sin\theta_1 = \frac{CD}{AD}$$

$$\sin\theta_2 = \frac{AB}{AD}$$

$$CD = V_1/t \qquad AB = V_2/t$$

$$\frac{\sin\theta_1}{\sin\theta_2} = \frac{CD/AD}{AB/AD} = \frac{CD}{AB} = \frac{V_1/t}{V_2/t} = \frac{V_1}{V_2} \tag{4.22}$$

The angle θ_2 is called the angle of refraction and equation 4.22 is the law of refraction, known as Snell's law in optics. Figure 4.36, is drawn for $V_2 > V_1$. If V_2 were less than V_1, the ray path would refract toward the vertical, $\theta_2 < \theta_1$. It should be noted that the angle of incidence in either case is the deviation from a line perpendicular to the interface.

Figure 4.36 also shows a ray path reflecting upward. The angle of incidence for this reflected wave equals an angle for the incident wave.

If θ_1 increases, eventually an angle is reached where θ_2 becomes 90 degrees (assuming $V_2 > V_1$). Physically, this means that the wave travels from A to D in the same time that it travels from C to D. Since θ_2 equals 90 degrees, $\sin\theta = 1$, and equation 4.22 becomes

$$\sin\theta_1 = \frac{V_1}{V_2}. \tag{4.23}$$

This is called the critical angle, and this equation was used earlier to solve the two-layer problem. At this point, no energy is transmitted past the interface in an ideal medium. In reality, the disturbance of the materials will produce some signals, but the bulk is refracted along the interface and reflected. The motion along the interface produces a continuous stream of signals back toward the surface.

If the incident wave is inclined at more than the critical angle, there are no refracted signals and all the energy appears in the reflected waves.

The interaction of the particle vibrations at the interface also generates shear waves, so a truer picture is shown in figure 4.37. The incident P-wave, A_0, generates a reflected P-wave, A_1, a refracted P-wave, A_2, a reflected S-wave, B_1, and a refracted S-wave, B_2. Given the velocities of the waves, it can be shown that

$$\frac{\sin\theta_1}{Vp_1} = \frac{\sin\theta_2}{Vp_2} = \frac{\sin\lambda_1}{Vs_1} = \frac{\sin\lambda_2}{Vs_2} \tag{4.24}$$

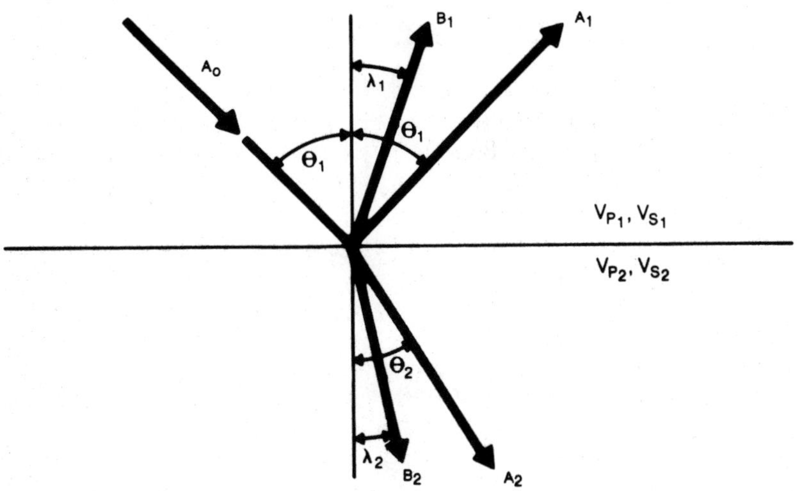

FIGURE 4.37
Waves generated at an interface by an incident P-wave

The partitioning of energy among these waves has not yet been considered. Given the angle of incidence, the densities of the two materials and the velocities, these can be solved for. The equations are complex and lead to intricate relationships (Telford and others, 1976). The case in which the angle of incidence is small is of interest because of its application in reflection seismology. If $\theta_1 \approx$ o, then

$$\frac{A_1}{A_0} = \frac{\rho_2 V_2 - \rho_1 V_1}{\rho_2 V_2 + \rho_1 V_1} \tag{4.25}$$

$$\frac{A_2}{A_0} = \frac{2\sqrt{\rho_1 V_1 \rho_2 V_2}}{\rho_2 V_2 + \rho_1 V_1} \tag{4.26}$$

where ρ = density and Ai is the amplitude of the wave. Amplitude refers to particle motion and is different from energy. In fact, energy is proportional to the square of the amplitude, so

$$\frac{Er}{Eo} = \frac{(\rho_2 V_2 - \rho_1 V_1)^2}{(\rho_2 V_2 + \rho_1 V_1)^2} \tag{4.27}$$

$$\frac{Et}{Eo} = \frac{4\rho_1 V_1 \rho_2 V_2}{(\rho_2 V_2 + \rho_1 V_1)^2} \tag{4.28}$$

$$Et + Er = Eo \tag{4.29}$$

where Er equals the energy of the reflected wave, Et is energy of the transmitted wave, and Eo is energy of the incident wave. The parameter pV is called the acoustic impedance; its symbol is Z. Substituting Z in equations 4.27 and 4.28 simplifies their appearance.

Figure 4.38 shows relative energies for refracted and reflected P- and S-waves as a function of angle of incidence. These graphs are constructed for a specific model interface with given velocities and densities. Different combinations produce quite different figures (Telford and others, 1976). It should be noted that the refracted P-wave has no energy beyond the critical angle.

FIGURE 4.38
Relative energy for refracted and reflected P- and S-waves as a function of angle of incidence

$VP_1 = 12{,}800$
$VS_1 = 8{,}000$
$VP_2 = 21{,}000$
$VS_2 = 12{,}000$
$\rho_1 = 2.4$
$\rho_2 = 2.65$

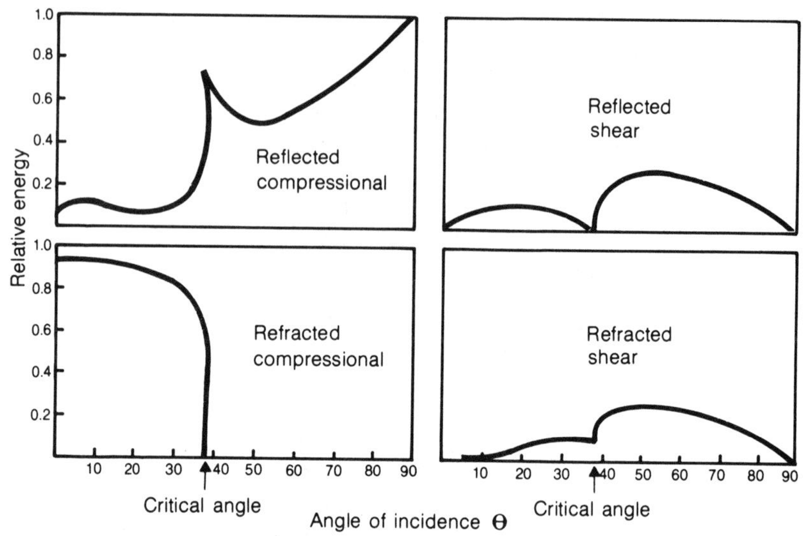

The two-layer case should be considered in light of these other waves (figure 4.39). For simplicity, if only the P-waves are used, figure 4.40 can be constructed. If an arbitrary point X on the surface is established, it can be said that three separate waves pass this point. The direct wave

FIGURE 4.39
Two-layer model, refracted and reflected waves

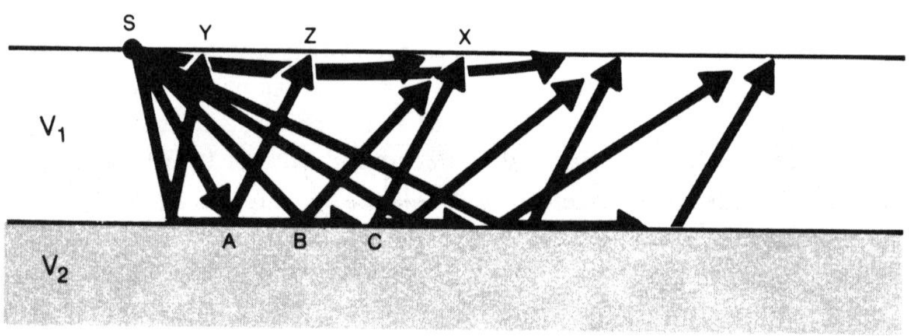

FIGURE 4.40
T-d plot, refracted and reflected waves for the two-layer case

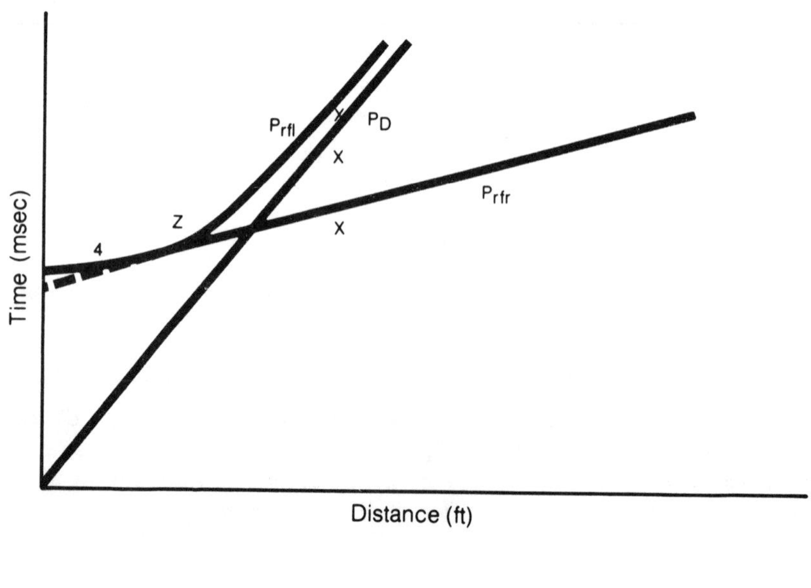

which goes straight across from S to X, the refracted wave with path SACX, and the reflected wave SBX all arrive at different times depending on the location of X. Path SAZ represents a special case in which the refracted and reflected waves are the same. If these arrivals are all plotted on a t-d plot, figure 4.40 results. Because some of these waves are not the first arrivals, they will not be readily detectable on the seismogram, although occasionally they appear quite clearly.

If the reflected shear waves are considered, more arrivals could be added to the cross section and t-d plot. However, normally the sensors used are sensitive on only vertical motion, and the reflected shear waves tend to be longitudinal. Of more consequence is the Rayleigh wave which has a strong vertical motion. It is much slower than the body wave and comes in late, but is usually easily recognized on the seismograph. Incidently, the sound wave from the impact, which travels through the air at 1100 ft/sec, is also commonly seen in the seismogram.

The apparent conclusion is that there are a great many wave arrivals which appear after the first arrival. All of them can be used in seismic interpretation, but are difficult to select in the collage of data. For that reason, refraction surveys almost always use first arrivals.

Multi-layer refraction

The refraction survey for a two-layer earth has been considered in some detail. If there were a third layer, with an even higher velocity, eventually a distance would be reached at which the least time would involve a seismic wave traveling through this layer. Again, this condition is common because the weathering process tends to reduce velocities so that the faster layers are on the bottom. Exceptions will be discussed later.

Figures 4.41 and 4.42 show the typical three-layer model and its corresponding t-d plot. Now there are two critical distances and two intercept times. Before the solution of this problem is considered, the behavior of the refracted wave as it passes through several layers should

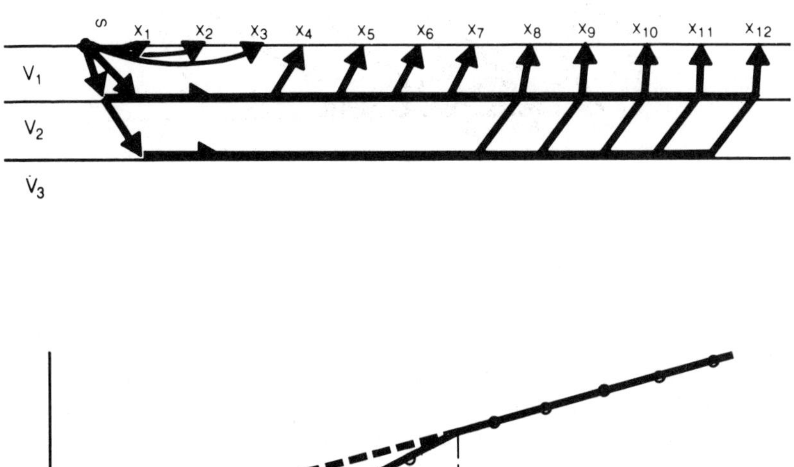

FIGURE 4.41
Three-layer refraction survey model

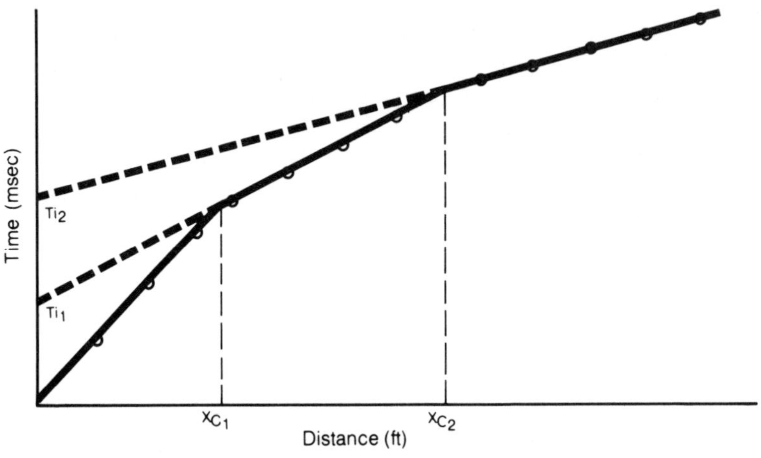

FIGURE 4.42
T-d plot, three-layer refraction survey model

be considered. Such an event is shown in figure 4.43. From the critical angle formula, it is known that

$$\sin\theta_3 = \frac{V_2}{V_4}$$

or

$$\frac{\sin\theta_3}{V_3} = \frac{1}{V_4}. \quad (4.30)$$

FIGURE 4.43
Refraction angles for N-layer case

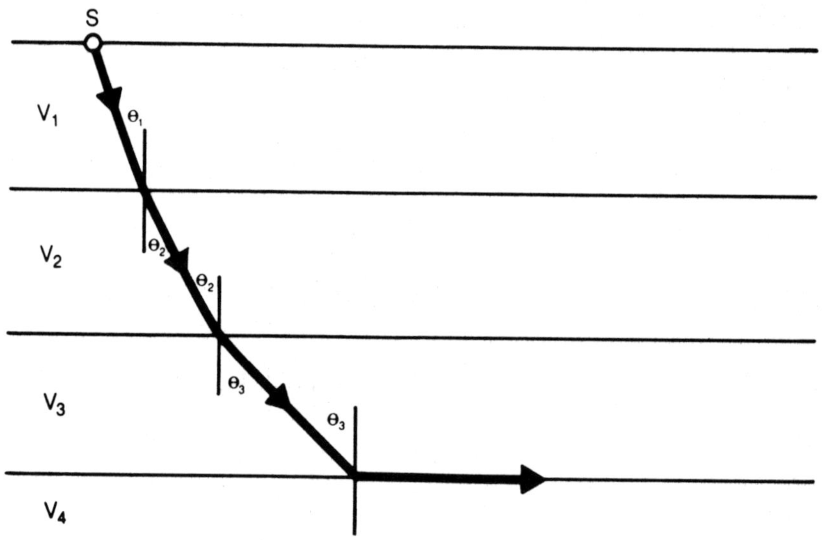

From the study of waves refracted at less than the critical angle (equation 4.24), it is also known that

$$\frac{\sin\theta_1}{V_1} = \frac{\sin\theta_2}{V_2} = \frac{\sin\theta_3}{V_3} = \frac{1}{V_4},$$

so

$$\sin\theta_1 = \frac{V_1}{V_4},$$

$$\sin\theta_2 = \frac{V_2}{V_4},$$

or

$$\sin\theta_i = \frac{V_i}{V_n}. \quad (4.31)$$

The last equation is the general solution for n layers.

In order to solve the three-layer case, one must examine figure 4.44, a simplified reconstruction of figure 4.42. Assume the following: the t-d plot in figure 4.42 has been prepared; the three velocities have been determined; we know X which is the length of the survey; we know T which is the time required for the wave to reach point F at distance X along path ABCDEF. The expression for T is

$$T = \frac{AB}{V_1} + \frac{BC}{V_2} + \frac{CD}{V_3} + \frac{DE}{V_2} + \frac{EF}{V_1},$$

$$AB = EF = \frac{Z_1}{\cos\theta_1} \qquad BC = DE = \frac{Z_2}{\cos\theta_2},$$

$$CD = X - Z_1\tan\theta_1 - Z_2\tan\theta_2 - Z_2\tan\theta_2 - Z_1\tan\theta_1.$$

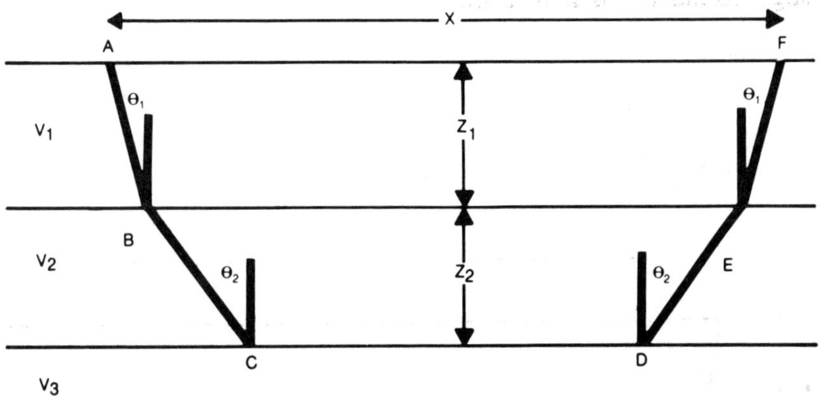

FIGURE 4.44
Three-layer refraction plot

So:

$$T = \frac{2Z_1}{V_1 \cos \theta_1} + \frac{2Z_2}{V_2 \cos \theta_2} + \frac{X - 2Z_1 \tan \theta_1 - 2Z_2 \tan \theta_2}{V_3}$$

$$= \frac{X}{V_3} + \left[\frac{2Z_2}{V_2 \cos \theta_2} - \frac{2Z_2 \tan \theta_2}{V_3}\right] + \left[\frac{2Z_1}{V_1 \cos \theta_1} - \frac{2Z_1 \tan \theta_1}{V_3}\right]$$

$$= \frac{X}{V_3} + \frac{2Z_2}{V_2}\left[\frac{1}{\cos \theta_2} - \frac{V_2 \tan \theta_2}{V_3}\right] + \frac{2Z_1}{V_1}\left[\frac{1}{\cos \theta_1} - \frac{V_1 \tan \theta_1}{V_3}\right]$$

Now

$$\frac{1}{\cos \theta_2} - \frac{V_2 \tan \theta_2}{V_3} = \frac{1}{\cos \theta_2} - \frac{V_2 \sin \theta_2}{V_3 \cos\ _2} - \frac{1}{\cos\ _2}\left[1 - \frac{V_2{}^2}{V_3{}^2}\right]$$

$$= \cos \theta_2 .$$

Likewise:

$$\frac{1}{\cos \theta_1} - \frac{V_1 \tan \theta_1}{V_3} = \cos \theta_1 ,$$

$$T = \frac{X}{V_3} + \frac{2Z_2}{V_2} \cos \theta_2 + \frac{2Z_1}{V_1} \cos \theta_1 . \tag{4.32}$$

This equation can be expanded to an arbitrary number of layers and written in the general form

$$T = \frac{X}{Vn} + \sum \frac{2Zi}{Vi} \cos \theta i. \tag{4.33}$$

If the geometry of the t-d plot is examined, it can be seen that when $X = 0$ the intercept time can be solved for.

$$Ti_n = \sum_{i=1}^{n-1} \frac{2Zi \cos \theta i}{Vi} , \tag{4.34}$$

since

$$\cos \theta i = \sqrt{1 - \sin \theta i} = \sqrt{1 - \frac{Vi}{Vn}^{2}} = \frac{1}{Vn}\sqrt{Vn^2 - Vi^2},$$

$$Ti_n = \sum_{i=1}^{n-1} \frac{2Zi \sqrt{Vn^2 - Vi^2}}{ViVn} , \tag{4.35}$$

where n equals the number of layers.

This equation for the five-layer case can be written

$$Ti_2 = \frac{2Z_1\sqrt{V_2^2 - V_1^2}}{V_1V_2}, \tag{4.36}$$

$$Ti_3 = \frac{2Z_2\sqrt{V_3^2 - V_2^2}}{V_2V_3} + \frac{2Z_1\sqrt{V_3^2 - V_1^2}}{V_1V_3}, \tag{4.37}$$

$$Ti_4 = \frac{2Z_3\sqrt{V_4^2 - V_3^2}}{V_3V_4} + \frac{2Z_2\sqrt{V_4^2 - V_2^2}}{V_2V_4} + \frac{2Z_1\sqrt{V_4^2 - V_1^2}}{V_1V_4}, \tag{4.38}$$

$$Ti_5 = \frac{2Z_4\sqrt{V_5^2 - V_4^2}}{V_4V_5} + \frac{2Z_3\sqrt{V_5^2 - V_3^2}}{V_3V_5} + \frac{2Z_1\sqrt{V_5^2 - V_2^2}}{V_2V_5} + \frac{2Z_1\sqrt{V_5^2 - V_1^2}}{V_1V_5}, \tag{4.39}$$

where Zn equals the thickness of each layer. The depth of each layer is simply the sum of the thicknesses.

An example follows. A hypothetical survey in the field with 16 geophones at 10-ft intervals is run; the times in table 4.4 for first arrivals are logged. The reader should plot this data as a t-d plot exercise. One favorable feature of hypothetical data is that the points fall on four straight lines indicating a probable four-layer case. Normal procedure would involve performing a reverse profile survey to reveal other structures which might produce this same graph. With the t-d plot, the reader should be able to discern the following information:

$V_1 = 2500$ ft/sec
$V_2 = 5000$
$V_3 = 10{,}000$
$V_4 = 20{,}000$
$T_{i2} = 8$ msec
$T_{i3} = 16$ msec
$T_{i4} = 22$ msec.

$$Ti_2 = \frac{2(Z_1)\sqrt{5000^2 - 2500^2}}{(5000)(2500)} = .008$$

$Z_1 = 11.547$ (from 4.36)

$$Ti_3 = \frac{2(Z_2)\sqrt{10000^2 - 5000^2}}{(5000)(10000)} + \frac{2(11.547)\sqrt{10000^2 - 2500^2}}{(10000)(2500)}$$

$$= .016$$

$Z_2 = 20.368$ (from 4.37)

$$Ti_4 = \frac{2(Z_3)\sqrt{20000^2 - 10000^2}}{(20000)(10000)} + \frac{2(20.368)\sqrt{20000^2 - 5000^2}}{(20000)(5000)} + \frac{2(11.547)\sqrt{20000^2 - 2500^2}}{(20000)(2500)} = .022$$

$Z_3 = 28.56$ (from 4.38)

TABLE 4.4
Log of first arrivals

Distance/ft	Time/msec
10	4
20	8
30	12
40	16
50	18
60	20
70	22
80	24
90	25
100	26
110	27
120	28
130	28½
140	29
150	29½
160	30

If the multi-layer structure has dipping interfaces, the solution becomes more complex. A number of analytical methods and computer programs are available to solve this problem. Unfortunately, further discussion is beyond the scope of this chapter.

Borehole seismic surveys

Thus far only surface seismic surveys have been considered. Occasionally, similar surveys are performed in boreholes offering advantages in many instances. One immediate advantage is that borehole surveys are not limited by the usual constraint that each layer have higher velocities. In addition, better resolution can be obtained from borehole surveys. Borehole surveys are often performed for special applications such as shear wave studies, discussed later in this chapter. Occasionally, borehole surveys are performed when the surface area available is insufficient to provide the depth penetration desired.

Borehole surveys are usually described as uphole or downhole. In uphole, a geophone is placed at the surface, and the energy source located down the hole. Normally, the energy source consists of a number of small explosions located at intervals in the hole. (For shallow depths, a single blasting cap is sufficient.) The explosive charges are prepared in advance and lowered down the hole; shot wires are brought up in a bundle. The lowest shot is fired first, then the second lowest, and so on. With that procedure, any damage from the explosion is confined to the portion which has been used.

In a downhole survey, a special string of geophones is lowered into the hole and a single shot or hammer blow used at the surface. This procedure has the disadvantage of requiring special equipment, but preserves the hole for other uses. The geophone array should have a clamping device or be suspended by mud in the hole.

The data from an up/downhole survey can be presented in a sideways t-d plot, usually with a geologic interpretation as shown in figure 4.45.

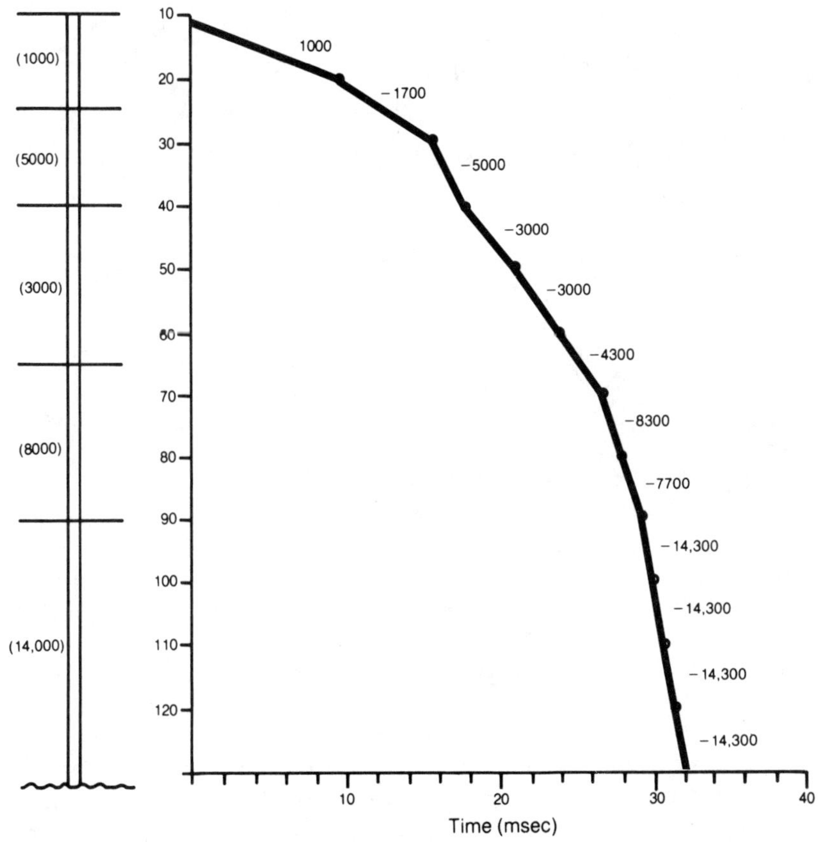

FIGURE 4.45
Example of downhole survey and t-d plot with hypothetical velocities shown in parentheses.

Measured velocities shown on t-d plot.

FIGURE 4.46
Set up for Meissner-type up hole survey

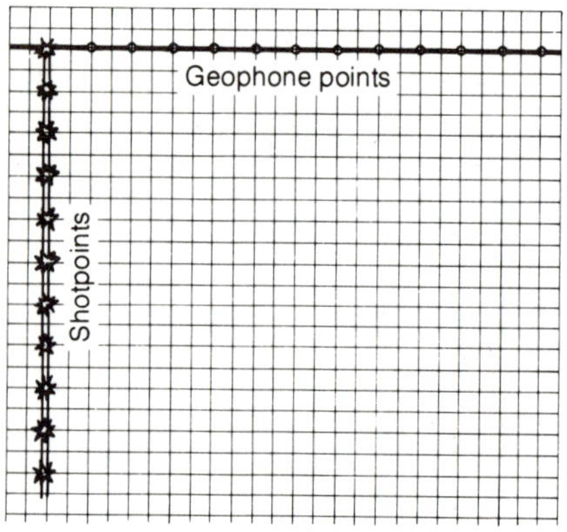

In the example, an arbitrary layered structure with hypothetical velocities shown in parentheses has been established. The example is intended to reflect a real situation in which the shotpoints (or geophones) were located at specific intervals (10 ft). One concession made was the assumption that the arrival times could be measured to 0.1 msec, an unrealistic ideal. In any case, the measured velocities do not correspond exactly to the hypothetical velocities because of the positions of the measuring points. Had the drill hole materials been recorded, it might have been possible to deduce enough information about the layering structure to work back to the true velocities. The shotpoints might also have been located at the material boundaries or at closer intervals. Whether the information from this example would be adequate for this purpose depends on the object of the survey. It is certainly more diagnostic than a simple surface refraction survey would have been, since the low velocity layer would not have been evident.

Once the effort has been expended to perform a borehole survey, it is not much more difficult to lay an array of geophones along the surface if a multichannel instrument is available. This technique (figures 4.46 and 4.47) was described by Meissner (1961).

FIGURE 4.47
Meissner diagram for five-layer example

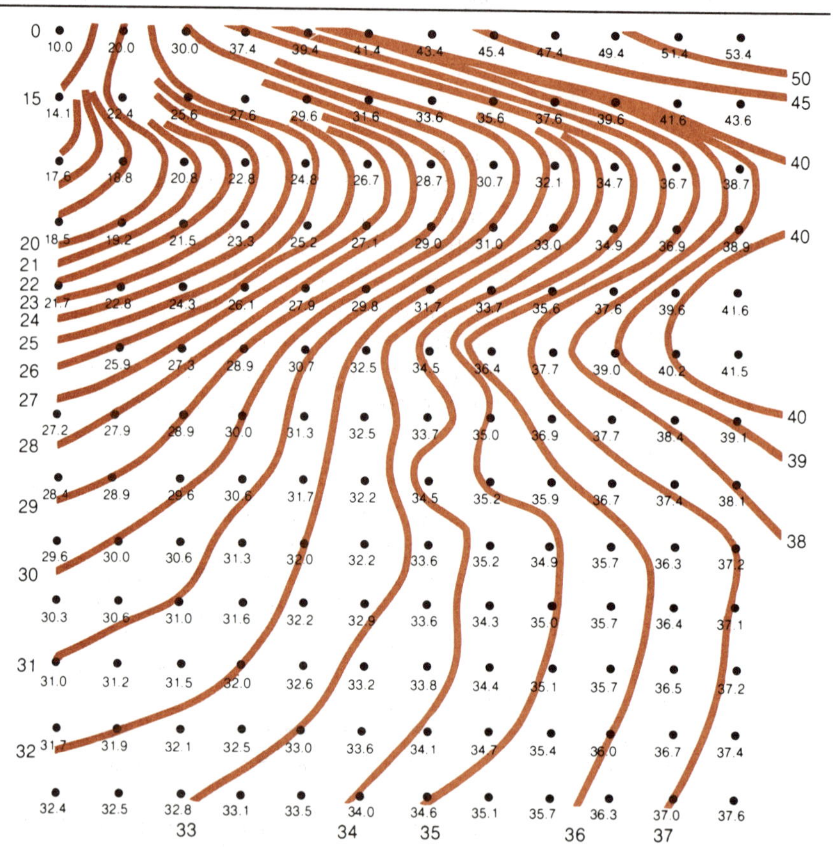

Shear wave techniques and applications

Engineering seismology is a particular branch of geophysics which is of interest to the practicing foundation engineer and engineering geologist. This discipline involves the propagation of vibrations (elastic waves) through the near-surface materials so that dimensions and engineering properties of the various layers can be determined. Relatively broad use has been made of P-waves because of the comparative ease with which these waves can be generated and detected and the simplicity of the calculations to determine the foundation parameters. Increasing interest is developing in the measurement and use of shear wave velocities. These data (1) are more diagnostic than P-waves in terms of shear strength and other engineering features; (2) are used to predict the dynamic behavior of foundation structures; (3) are required for certain major projects such as nuclear power plant sites.

This section is intended to be a primer on shear wave velocity measurements, to provide procedural guidelines, and to describe equipment requirements.

Types of seismic waves

The types of seismic waves of particular interest are P and shear (S). The P-wave is a pressure wave whose wavefront consists of compressions and rarefactions in material. The S-wave is a shearing motion perpendicular to the direction of travel. The velocity of the shear wave is less than that of the P-wave. If the velocities of these waves and the density of the material are known, Poisson's ratio, Shear modulus, Young's modulus, and the Bulk modulus can be calculated. The applicable equations are:

$$\text{Poisson's ratio } \sigma = \frac{1}{2}\frac{(Vp/Vs)^2 - 2}{(Vp/Vs)^2 - 2} \qquad (4.40)$$

$$\text{Shear modulus } G = \rho Vs^2 \qquad (4.41)$$

$$\text{Young's modulus } E = 2G(1 + \sigma) \qquad (4.42)$$

$$\text{Bulk modulus } K = \frac{1}{3}\frac{E}{1 - 2\sigma}$$

$$\rho = \text{density} \qquad (4.43)$$

These are linear, dynamic characteristics and not always applicable to static performance of the foundation material. Nevertheless, one can draw some heuristic conclusions about a material, and some of these conclusions can be supported with empirical data. Opinions based on P-wave velocities are often not reliable. In particular, several low shear strength materials (such as water) have a high confined compressive strength and correspondingly high P-wave velocities. These same materials will have a low shear wave velocity (or none at all). In practice, the shear wave velocity has better correlation with other standard tests for foundation materials.

Measuring shear waves

Shear waves are measured much like P-waves, with similar equipment: a vibration source, a vibration sensor, and an instrument to indicate the travel times of the vibrations. The basic difficulty in measuring shear waves is that they arrive after the P-wave. The operator must select the shear wave in the middle of a complicated signal. This signal will also contain assorted other arrivals from different travel paths, Love and Rayleigh waves, sound waves, and others. For small scale engineering surveys, the practical technique has been to generate a dominant S-wave signal with minimal P-wave energy and then detect the wave with a sensor (typically a geophone) which is most sensitive in the same axis as the shear wave vibrations. A typical system is shown in figure 4.48. It should be noted that strong P-waves are generated along the axis of the energy source; S-waves, along the axis of the survey. There are other techniques, but this one has been popular because of its reliability and simplicity.

Four popular variations on this technique exist: surface surveys, crosshole surveys, downhole surveys, and refracted surveys. The setup for a surface survey resembles figure 4.48. A common system uses a sledgehammer energy source, a wooden plank with a weight superimposed, and a horizontal geophone. The crosshole technique uses two adjacent drill holes, a downhole energy source, and a downhole geophone as shown in figure 4.49. The downhole technique uses a plank on the surface over a single drillhole containing the sensor (figure 4.50). The refracted survey is an extension of the surface survey, except that the

FIGURE 4.48
Basic technique used in S-wave measurements

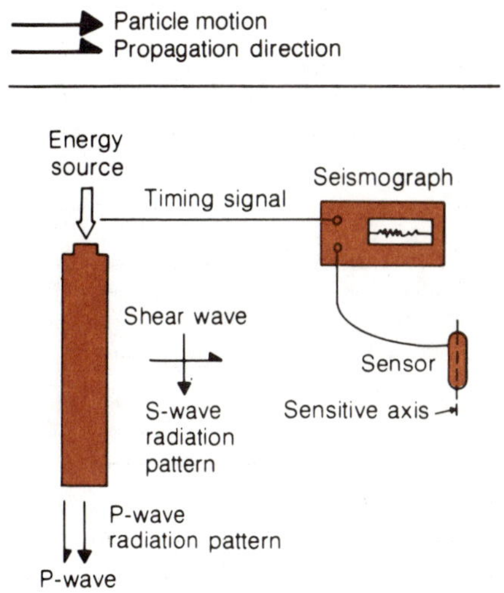

FIGURE 4.49
Crosshole technique

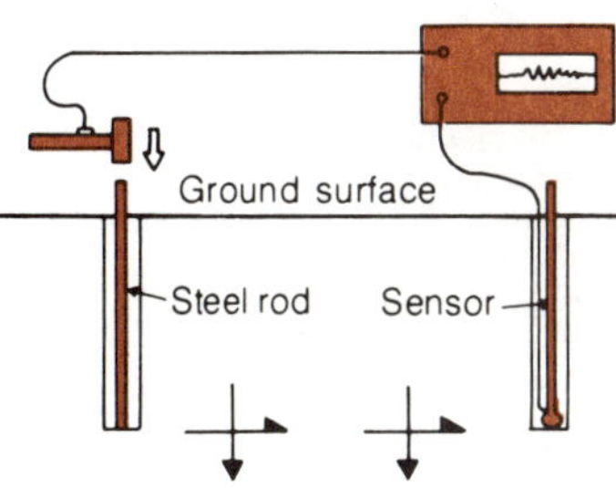

FIGURE 4.50
Downhole technique

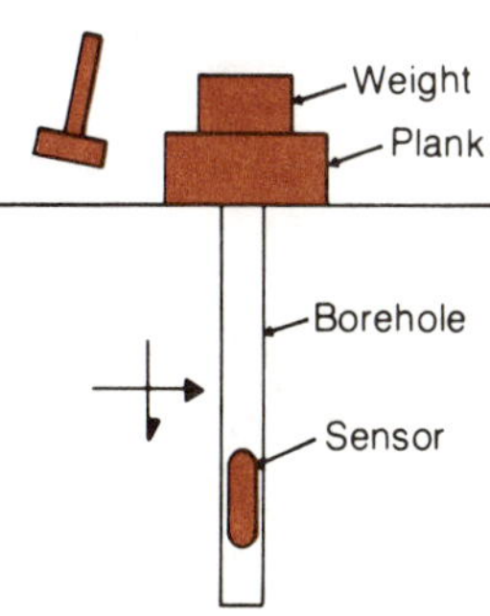

spacing between the source and sensor is extended so that waves will refract through the lower layers of higher velocity as in a conventional refraction survey using P-waves (figure 4.51).

FIGURE 4.51
Refracted S-waves

Distant sensors receive S-waves from multiple directions

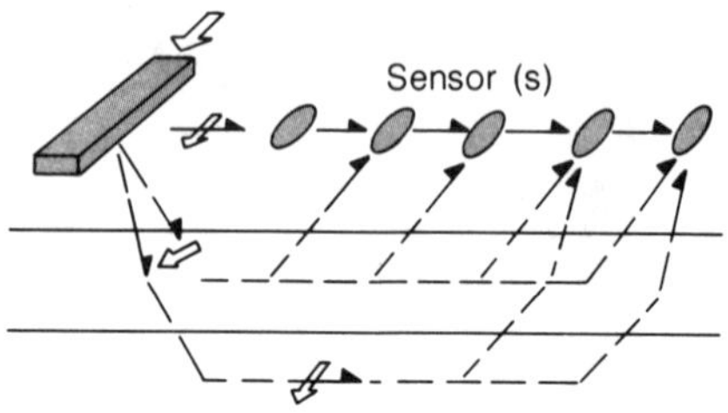

Sensors

The most common sensor used is the geophone: a moving coil and magnet vibration detector. Geophones come in two orientations, horizontal and vertical, and in orthogonal packages, handy in shear wave investigations. Because the waves have different orientations, better decisions can be made if records from both the vertical and from the two horizontal components are visible. This can be done easily on multichannel seismographs and with single channel instruments by electrically switching among the phones and repeating the survey.

Geophones (single and three-component) are available specially designed for borehole use: compact, watertight, and rugged. A few are also available with special clamping devices so the geophones may be held at any elevation in the borehole. These tend to be the most expensive, but are essential for serious shear wave work. A clamping mechanism can be made from the lower cost borehole package with a small innertube attached to one side. The innertube, inflated by a plastic hose and bicycle pump, presses the side of the sensor against the hole wall. This is a low-cost, useful technique. One commercial system uses a water-filled bladder which automatically compensates for the hydraulic pressure in the drill hole. Another uses a spring clamp which releases when the bottom of the hole triggers a release mechanism.

Energy sources

The most popular energy source for shear wave surveys has been the sledgehammer. Highly portable and easy to use, the sledgehammer concentrates nearly all its energy in one axis. Its basic limitation is in the amount of energy available, but for most applications it is sufficient when combined with a signal enhancement seismograph. Timing can be from an inertia switch attached to the hammer or from a geophone placed near the impact point.

Explosives are suitable when used properly. A simple explosion in the ground produces almost no true shear waves. However, the P-waves generated will produce S-waves when they reach an acoustic interface. The resulting combinations of waves are complex and relatively difficult to interpret. Some success has been realized in directing the explosive energy, i.e., the explosive is placed against a metal plate on the side of a small trench. Even with this method, excessive energy will generate elastic deformation which produces waves in all directions. The explosive source should be reserved for large scale surveys in which a hammer is not sufficient and when travel times are long enough for the waves to separate from simple dispersion.

Rotating energy sources and vibrators have also been used successfully. These systems tend to be less convenient and require more sophisticated instrumentation. They are beyond the scope of this paper.

Surface shear waves

The recipe for surface shear waves requires one seismograph, one horizontal geophone, one sledgehammer, and one plank (with automobile). The plank is placed on the undisturbed ground, perpendicular to the axis of the survey, and the car is driven onto the plank. The geophone with its sensitive axis is located perpendicular to the survey axis as shown in figure 4.48. The hammer, seismograph, and geophone are connected. The end of the plank is struck firmly with the sledgehammer, and the waveform can be observed. The shear wave will be recognized as a large, lower frequency displacement arriving sometime after the P-wave. If the system is good, the P-wave may very well not appear at all. The other end of the plank is struck. A new shear wave should appear at about the same time as the previous one; *the direction of motion for the wave will reverse.* Typical records are shown in figure 4.52. The reversal is useful for positive identification of the shear wave and helps the observer have confidence in his work. If the P-wave velocity is to be measured, and it was not readily detectable before, the geophone is rotated 90 degrees and the plank is struck toward its center along the axis of the survey. One limitation of this method is that only the material near the surface is measured, so its practical applications are limited. It might have applications in tunnels, where the material of interest is exposed. However, it is a good training exercise as an introduction to shear waves, requires little preparation, and can be used to evaluate systems prior to purchase.

FIGURE 4.52
Typical records from surface S-wave survey

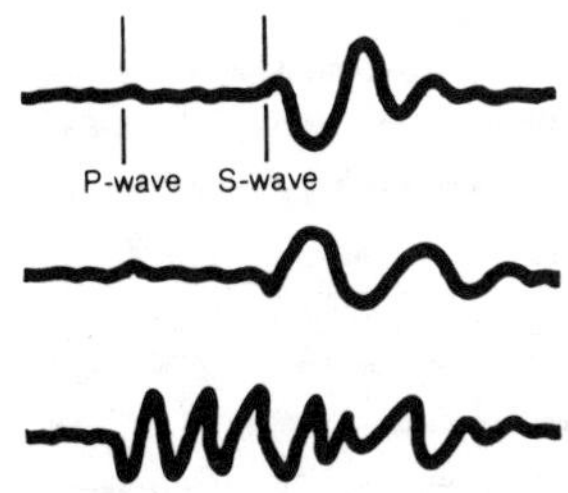

Crosshole surveys

Crosshole surveys use two similar approaches. Popularized by Stokoe and Woods (1972), the first is performed during the drilling operation. Two (or more) holes are drilled side by side to the same depth, and a geophone is placed at the bottom of one hole. A steel rod is placed in the other hole and struck with a hammer (see figure 4.49). The timing can be started from the impact switch on the hammer, or a geophone placed on the bottom of the hole or attached to the rod. If the impact timing method is used, there may be a slight time error due to the switch timing and travel time of the rod. This error can be measured and corrected for if the rod is placed in the same hole as the pickup geophone. After the shear wave travel time is measured at a particular depth, the holes are drilled to a new elevation and the measurements repeated. The process is continued until maximum desired depth is reached. The basic assumption here is that the waves travel directly across the gap. This requires that there be no adjacent layer of higher velocity. In practice, there will be such layers, and they will affect the data if the geophone spacing is large in relation to the depth of the adjacent layer. If there is a higher velocity layer, waves will refract through it and appear in the record. The limited size of the spacing between holes means that an instrument which can measure short time intervals, such as an oscilloscope or signal enhancement seismograph, must be used. It should be noted that there is no provision to reverse the impact direction to record the confidence building wave reversal. This survey can be time consuming because it is tied to the drilling operation.

The other crosshole survey method uses a borehole clamping device attached to the rod and a clamping geophone. One popular clamp resembles an automobile scissor jack. The jack is placed down the hole and expanded by turning the shaft until it presses firmly against the wall. If a bracket is attached to the top of the rod, the impacts can be made in both directions. This method frees the observer from the drilling operation and allows the measurement of reversed polarities. However, there may be a problem in keeping the hole open long enough to perform the survey. The use of steel casing is unsatisfactory because its high velocity will effectively generate waves along its entire length and emphasize the refraction problem. The apparent source point might be displaced to the soil-casing contact point of greatest shear strength. The sensor hole can be cased with plastic pipe, but it is doubtful that this would be suitable for the energy source. The source hole could be cased with plastic pipe that could be extracted just above the jack mechanism. Unavailable commercially, the jack mechanisms are usually homemade.

Downhole surveys

The downhole survey uses a single hole cased with plastic pipe. The gap between the pipe and hole is backfilled with pea gravel or sand. A wooden plank is placed near the hole and weighted down with an

FIGURE 4.53
Typical downhole records

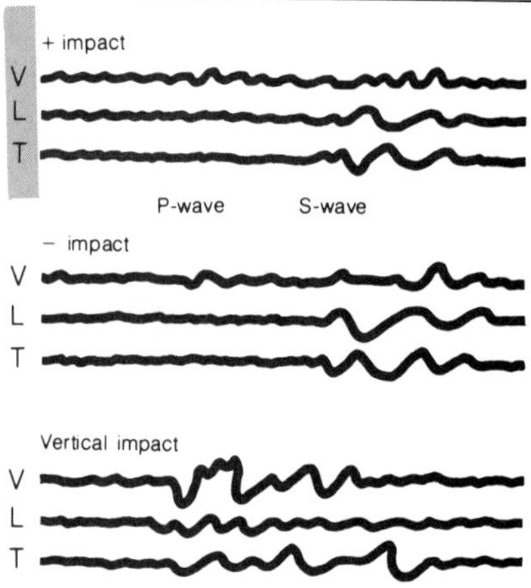

FIGURE 4.54
Composite record from downhole survey

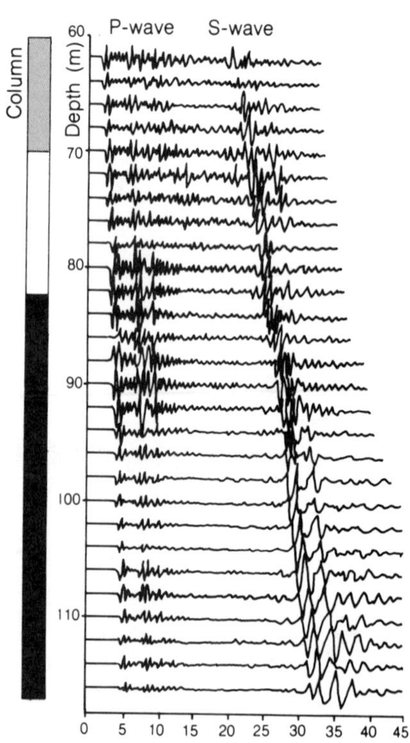

automobile. The sensor, a three-component package with a clamping mechanism, is located down the hole at an appropriate depth. The shear wave is generated when the end of the plank is struck with a sledgehammer, and the vibrations are sensed by one or both of the horizontal geophones. The measurements are repeated on the opposite end of the plank; for the P-wave, the surface of the plank is struck vertically. The records are compared for appropriate reversal of the shear waves and P arrival times. The geophone is moved to a different depth and the survey repeated. The velocities of the various segments are found by noting the differences in the arrival times obtained at different depths. A typical group of records from the three hammer blows is shown in figure 4.53, and a composite survey is shown in figure 4.54. The resolution of the system can be improved if two sets of sensors are used at different elevations and the data recorded simultaneously. With this procedure, differences in arrival times can be picked more precisely. This, of course, requires a multichannel system. The survey can be done with a single channel instrument by repeating the measurements with various components connected to the instrument. The downhole survey is generally a reliable method for routine work. Since the interfaces between layers (which cause refraction ambiguities in crosshole surveys) tend to be parallel to the surface, there are fewer problems from other wave paths.

The downhole method makes it possible to survey to great depths, especially when signal enhancement equipment is used. The equipment, except for the plank, is quite portable, and the user can carry it on commercial airline flights. The one serious limitation in downhole surveys is the difficulty in picking the arrival times with precision. The actual waveforms break fairly smoothly, and it is difficult to pick the exact arrival time. Typical accuracies are ½ to 1 msec. Finding the velocity in a particular layer requires determining the time difference between two of these arrivals. Where thin layers occur, the possible error can be substantial. Signal enhancement is helpful in sharpening up these breaks and averaging out any error in the hammer timing.

Refracted shear waves

The possibility of measuring refracted shear waves is an attractive prospect. If this survey can be performed, a greater area can be covered in a short time without the costly drill hole preparation. The problem is that there are so many arrivals from different waves that it is difficult to select the appropriate shear wave. Vertical impacts and non-directional explosions generate strong refracted P-waves (which generate additional S-waves at each acoustic interface), surface waves, and sound waves, so these sources should be rejected for all but the largest scale surveys. The best results are obtained with the horizontal traction method (plank and hammer). Because the energy will be limited, signal enhancement seismographs provide the best data. Some spacings will not produce any reasonable arrivals, so the recognition will be easier with multichannel instruments (or with a composite record made with a single channel instrument and recorder). The investigator should gain familiarity with shear wave surveys by practicing on the simpler techniques outlined previously. He should start experimenting with refracted shear waves in an area with simple geology (layers parallel with the surface). Success cannot be expected in all cases.

One technique which deserves mention is the reversed impact, reversed polarity method. It requires a signal enhancement seismograph with a polarity switch on the geophone. When the traction method is used, the board is struck a few times on one end, thus stacking the data in the seismograph memory. Then the polarity of the geophone is reversed and the other end of the plank is struck the same number of times. The P-waves and some other extraneous waves have the same polarity regardless of the hammer blow direction and tend to cancel out when the geophone polarity is reversed. The shear waves, which reversed polarity at the same time the geophone polarity was reversed, will enhance.

Pitfalls in shear wave surveys

A number of subtle pitfalls exist in this work. The following items are described and emphasized as a partial list.

1. The existence of other waves should be remembered, and watched for. The shear wave energy should be kept large and the P-wave small to minimize the other signals. If the bedding planes are inclined, this should be considered in planning a survey because they can produce refractions where they would not be expected. If reasonable, holes perpendicular to the bedding planes could be drilled. In the case of vertical bedding planes, the crosshole technique becomes superior to the downhole.
2. Holes should be prepared properly. Thin wall plastic pipe is an excellent casing because the waves which propagate down the pipe attenuate rapidly and do not interfere with the survey. Steel casing, on the other hand, could easily carry its own shear vibrations. Plastic pipe should be backfilled so that it is in good contact with the original material; however, a strong backfill like concrete should not be used because it can carry its own waves. Voids should not be allowed to develop around the pipe, or the voids will reflect the data from some adjacent area where good contact exists. Best results are obtained when the drill hole is small in relation to the plastic pipe and the annular space is backfilled slowly and thoroughly with sand or pea gravel.
3. The geophone should be secured well, which means it should be in firm contact with the edge touching the pipe. If the securing mechanism is at one end of the sensor, there will be an oscillatory mechanical couple between the geophone and wall. This causes the assembly to develop torsional vibrations which generate spurious signals that can mask the shear wave completely.
4. Better results are obtained when the plank is placed on undisturbed ground, rather than on a disturbed or loosely backfilled surface. Locating it on loose granular materials should especially be avoided. If some smoothing must be done, a shearing technique should be used so that the ground below the plank has not been altered. Enough weight and a small enough impact should be used to minimize rolling, bouncing, and sliding of the plank. These motions produce an extended duration signal and some P-waves. Some other energy techniques, such as driving a stake in the ground and hitting it on the side, have been less satisfactory. The resonance effects and the torsional couple produce undesirable extraneous signals.
5. The appearance of a hydro P-wave should be noted. If the material is clean sand or gravel (containing little or no fines) and is saturated with ground water, the P-wave in that layer may be that of water, approximately 5000 ft/sec. The shear wave velocity may be more representative of the material since it doesn't travel through the water. The problem of the hydro-P continues when the P- and S-wave velocities are used to calculate Poisson's ratio. The resultant Vp/Vs ratio is in error for the materials. Careful examination of the waveform will sometimes reveal another P arrival after the hydro-P, and the ratio will be more reasonable when this arrival is used to determine Vp. Further research may reveal whether this second P is a correct value, and whether it may be used with confidence in calculation of the dynamic parameters.

Conclusions

Effective measurement of shear wave velocities can be carried on as a routine practice in foundation investigations. As the operator gains experience, the task becomes less formidable, less time consuming, and generally more reliable. As greater use is made of shear wave velocities, the values will become more meaningful to the geotechnical community, and their diagnostic nature will enhance the results of the study when the values are combined with other tests and drill logs. Shear wave velocities have the advantage that they represent the in-place performance of a large bulk of material, as contrasted with small, selected samples.

Learning experiences should start with the simpler techniques, such as crosshole and downhole. Experience will teach one what to expect and help a person know when things are not working right. It is better to make corrections and adjustments in the field than to try to reconcile mediocre data in the office. After a person becomes proficient with the simpler techniques, he may want to try the more difficult surveys such as refracted shear waves and those involving explosives.

Types of exploration seismographs

As previously discussed, the function of a seismograph is to measure the time between some impulse and the arrival of the vibration at some other point. The impulse is usually an explosion, sledgehammer blow, weight drop, or any number of mechanical sources. A number of innovative methods has been used to measure these arrival times, and most variations are described here. In the case of the mechanical sources, the initial timing is usually provided by an impact switch or vibration sensor at the source point. When explosives (and some high speed mechanical sources) are used, timing is provided by a signal from the detonator.

Single channel instruments normally use a sledgehammer, and the impact position is moved around to provide the timing from several points needed to analyze the results. With explosives, a number of geophones located at various distances is normally used to detect the arrivals from a single shot. Explosives can be used with a single-channel instrument, but it requires the cumbersome task of setting off several charges at different points. This method may be selected occasionally if the need exists to survey deeper than can be penetrated with a hammer. No real limit exists on the depth coverage with explosives.

Seismic timers

Seismic timers are the simplest type of instrument. The basic operation is similar to a very fast electronic stop watch. A switch on the sledgehammer is used to start the clock, and the signal from the geophone provides the stop signal. The display is a digital readout. Because of their simplicity, these instruments can be extremely lightweight, compact, and inexpensive.

Their range is limited, generally being confined to 10 m depths except in unusually good conditions. Care must be taken to operate the instrument properly. The operator must adjust the sensitivity to provide the best range, and he must interpret the readings carefully. An experienced user has a good conceptualization of the idealized waveform.

Seismic timers are excellent for many applications that do not involve surveys to great depths. Outstanding among these are rippability surveys for pipelines and swimming pools, as well as geologic reconnaissance.

CRT display seismographs

The cathode ray tube (CRT) display or oscilloscope type seismographs are the simplest units which actually display the vibration waveform on a screen. Because the waveform can be seen and an observer can make a better selection of arrival time than can an electronic circuit (especially in the presence of noise), they are superior to the timers. The CRT seismograph may be a commercial unit, or it may be a modified industrial oscilloscope. Commercial products are "real time" instruments, which means that the waveform is displayed very briefly while the spot traverses the screen. Modified oscilloscopes often make use of a storage CRT which will hold the display for some time, allowing a leisurely examination of the waveform. Almost as expensive, none of these instruments is as satisfactory as an enhancement seismograph. For these reasons, this is not usually recommended.

Enhancement seismographs (CRT display)

Enhancement seismographs resemble the CRT display seismographs physically, but contain extra circuitry to process the vibration signal. The vibration waveforms are stored in a digital memory, and the contents of the memory are continuously displayed on a screen. When the seismic impact is repeated, the new signal is added to the old one, resulting in a larger amplitude signal. Impacts can be repeated several times and each time the signal grows larger. The random vibrations which normally interfere with the seismic signal do not grow at the same rate. In fact they tend to cancel each other out. Consequently, surveys can be done to greater depths and more accurate time measurements can be performed than with conventional (nonenhancement) seismographs.

The signal enhancement seismograph is the easiest to use primarily because the waveform can be seen clearly, and because of the basic simplicity of the system—a single hammer and geophone. It is also relatively inexpensive. It is an excellent choice for most applications because it will handle a majority of the shallow seismic work. Because it displays the waveforms, is safe and inexpensive to operate, it is excellent for teaching.

Oscillographic seismographs

Oscillographic seismographs record the vibration signals on a chart. Normally these are multichannel instruments used with an array of geophones, each with its own trace on the record. This instrument is commonly used with explosives and is preferred by professional geophysicists. The majority of the instruments have 12 channels, although 24 and six channels are common.

These instruments are most effective when lines longer than 100 m are to be surveyed; and when a significant amount of work makes them cost-effective; when comparisons need to be made among several traces such as in shear wave or reflection surveys. A permanent record is available which is necessary in many applications.

Multichannel enhancement seismographs

Multichannel enhancement seismographs combine the advantages of multiple channels with enhancement. As such, they offer the production and analysis advantages of multichannel and the timing accuracy and hammer range of enhancement. They can be used easily with explosives when desired, or with weight drop systems where explosives are prohibited. Usually, this instrument is a better choice than a conventional oscillographic type, since the difference in cost is small and there are so many significant advantages.

Deep reflection seismographs

Petroleum exploration uses deep reflection extensively. A deep reflection seismograph has a large number of channels and records directly onto computer-compatible magnetic tape. Because of the expense in equipment and personnel, it is not generally used for shallow seismic exploration. A deep reflection seismograph is occasionally used in mining exploration.

Applications of shallow exploration seismographs

Seismograph describes an instrument used for subsurface exploration and for earthquake detection. Instruments described here are only those designed specifically for use in shallow exploration, normally 100-250 m deep. The basic task of a seismograph is to accurately measure the time interval between an explosion or impact, and the arrival of the seismic waves produced at some other known location. These time intervals or travel times of the seismic waves are controlled by the elastic properties of subsurface materials, their distribution, and shape. From travel times measured at several different locations, seismic wave velocities or simply seismic velocities can be determined. These can then be related to the type of material present and used to deduce the location and dimensions of subsurface regions of differing elastic properties.

Some of the many applications of seismic exploration seismographs to engineering, ground-water exploration, mining, and archaeology are described in the following pages.

Depth to bedrock

Determination of depth to fresh bedrock concealed beneath overburden or weathered material is the most common application of shallow seismic methods. Based on the normal geological situation in which seismic velocity increases with depth, the seismic refraction method is usually applied to this problem. Very near surface material is commonly graded into soil or unconsolidated overburden, weathered rock, and bedrock, each respective layer having a higher seismic velocity. The most abrupt velocity increase is generally at the bedrock surface which enables the refraction method to map this surface, even if it is irregular. The thicknesses of the overlying layers can usually be determined as can lateral variations in the overburden or bedrock.

Rippability assessment

A rippability survey is a special application of the depth-to-bedrock survey in which the object is to reduce the cost of excavation by providing information on the strength of subsurface materials to a specific depth.

Again, the refraction technique is used to determine seismic velocities to the depth of the proposed cut. From these velocities, it can be estimated whether the material should be dug, ripped, or blasted. Several major tractor companies and ripping tooth manufacturers publish tables and graphs relating seismic P-wave velocities, rock type, and the ripping ability of various combinations of equipment.

As a general rule, rock becomes marginally rippable with a heavy duty tractor at velocities of about 2500 m/s (8000 ft/s), but there are wide variations with rock type; for specific situations reference material should be consulted. Even then, the best estimates of rippability combine the seismic information with other factors. A user should examine the rock and compare other projects in the area where the same types of materials were excavated. Allowances should be made for traction and the relaxation after overburden removal. Seismic exploration is not a cure-all, but one of several methods to aid the user in determining the economics of ripping. However, an inexpensive rippability survey will sharpen a bid advantage and help prevent economically disastrous estimating errors.

Some skill and experience is required to provide accurate rippability estimates in particular areas since care is needed to obtain quality seismic information. Because the final assessments require integration with geological information, specialist teams working on a regular basis ensure reliable results. The amount of skill required to perform this investigation has occasionally been understated, and the survey has been overpromoted as requiring little or no expertise. This is true for many projects, but some amount of experience is beneficial in obtaining quality information.

Foundation conditions

Seismic surveys are commonly used to assess foundation conditions and to test alternative sites. Because of their small cost compared with drilling, they are best employed in the early stages of engineering investigations before final site selection. The information provided by seismic surveys includes depth-to-bedrock, shape of bedrock surface, variations in bedrock and overburden properties, and expected excavation conditions. Following the seismic survey, preliminary drilling provides detailed information at selected points, whereas seismic data extend the information horizontally. Integration of both sets of data with conventional engineering testing methods provides the most complete site investigation possible without considerable expensive drilling. This minimizes the possibility of foundation design errors caused by unexpected subsurface variations.

Approximate definition of the actual or potential slip plane from seismic work is sometimes possible, but is often difficult due to the high variability of loose materials. Periodic surveys to monitor the variation of seismic velocity with varying saturation have proved useful for this type of problem.

Materials survey

A seismic survey can enable the user to estimate the depth and extent of materials, i.e., shallow sand and gravel or clay deposits. From these figures, the quantity in any particular deposit can be predicted and when coupled with direct sampling and laboratory analysis at specific locations, the advisability of development can be established.

Blasting assessment

In many civil engineering and mining projects, blasting operations are carried out to fracture rock prior to excavation. The intense fracturing caused by blasting usually substantially lowers bulk seismic velocities. The effectiveness of blasting operations can be assessed if seismic refraction surveys are completed both before and after blasting. Such a program allows optimum blast-hole spacings, charge sizes, patterns, and delays to be chosen and can lead to substantial reduction of the amount of explosives required. Also areas requiring secondary blasting can be located from the post-blasting seismic survey.

Dynamic elastic constants

The elastic properties of earth materials may be specified in terms of the usual elastic constants: Young's modulus, Shear modulus, Bulk modulus, and Poisson's ratio. Knowledge of these constants aids foundation design associated with engineering construction. Civil engineers have developed a range of direct testing methods for determination of the static and dynamic elastic constants. These constants can be accurately determined at specific locations, but direct testing methods are often too expensive to use over entire sites. As a result they are often subject to errors caused by undetected changes in geological conditions.

Seismic surveys can provide information on the dynamic elastic constants. Empirical relationships between P-wave velocity and Young's modulus have been known for some time. These are generally useful for unweathered rocks with velocities greater than about 10,000 ft/sec (3000 m/sec). In other conditions, values for dynamic elastic constants can be obtained by measuring both P- and S-wave velocities during the seismic survey. Poisson's ratio can be determined directly from a ratio of these velocities whereas the other constants require density information.

In soils, unconsolidated material, and weathered rocks, dynamic elastic constants from seismic surveys are generally somewhat higher than the values obtained from static tests. This is because they are measured at low stress levels and high loading rates. In many areas, empirical relationships between static and dynamic constants can be established so that sites can be adequately tested with the lower cost seismic method.

Vibration monitoring

In addition to determining geological structure, seismographs are used to monitor vibrations from natural or man-made sources. Measuring the intensity of vibrations is often important in assessing the possibility of damage to surrounding structures from blasting, pile driving, and other engineering works. In these cases, the object of the monitoring is usually legal protection. Vibration levels and frequencies are also commonly measured in buildings or on bridges to assess the effects of traffic, wind, or possible earthquakes.

In vibration monitoring, a seismograph capable of producing a permanent requirement is desirable. Detectors may measure the amplitude of ground displacement, particle velocity, or particle acceleration, and require calibration. Generally, three components at right angles are measured. It is an advantage to use a seismograph which allows individual components to be mixed so that particular modes ofvibration can be enhanced.

Earthquake engineering

Earthquake engineering applies shallow seismic methods to two main areas: foundation site response and fault location. Foundation site response is a special application of the determination of dynamic elastic constants of earth materials discussed previously. Both P- and S- wave seismic surveys are required to determine the linear dynamic constants of the site. These can then be used in computer models to calculate the response of a particular foundation. Thus the response of the foundation to an earthquake can be anticipated. Such information allows more efficient structural design, often incorporating less conservative safety margins and reducing building costs.

Related problems in which dynamic characteristics of earth materials and foundations are important include the construction of structures constantly affected by traffic vibrations. As part of an overall study, including vibration monitoring surveys, seismic surveys can assist the user in designing structures which are resistant to vibration.

Second, seismic surveys are often useful for mapping geological faults. The surveys may detect the low-velocity gouge zone associated with active faults or the lateral velocity change which occurs when dissimilar rocks have been faulted into contact. The strike length of faults can also be determined which is often useful in the assessment of earthquake risk to nuclear power stations or dams. In this application, seismic surveys are often combined with magnetic (and occasionally resistivity) surveys.

Groundwater exploration

Groundwater exploration often makes use of seismic refraction methods to determine the depth of the water table, particularly in sandy areas. Saturated layers generally have a higher velocity than upper dry materials. However, actual velocities can be quite variable, and seismic results should be combined with other data from resistivity sounding or drilling.

Seismic surveys are a valuable aid to the sitting of a well in groundwater development and are best used to locate geological conditions conducive to the accumulation and extraction of groundwater. Specifically, these include fractured areas and collection areas such as buried stream channels or bowl-shaped depressions.

Mining applications

Seismic exploration methods have a varied application in mining, but are usually used to locate subsurface structures conducive to mineral deposition rather than to directly detect mineralization. They may also be used to delineate a deposit once it has been located by other techniques.

Placer deposits

Many important minerals occur as placer deposits in buried river channels which rarely have surface expression. Seismic refraction methods can be used to locate and map such channels particularly where they cut into bedrock. The information obtained includes depth to bedrock, location of channels, and necessary depth of excavation. Conventional refraction and fan shooting methods are useful in this application.

Base metals

Massive sulfide deposits can be delineated with the seismic method. Such deposits containing predominantly lead and zinc minerals usually have a lower velocity than the host rocks, which allows the location and size of the deposit to be determined. Seismic surveys have been used to define massive sulfide deposits whose presence is already known or suspected. Such surveys may be carried out from the surface or underground in drives and adits. A variety of seismic techniques is used, including conventional refraction and fan shooting. Extensions to existing orebodies and the shape of sulfide deposits can sometimes be obtained by the use of seismic methods at much lower cost than drilling. When deposits close to the surface are capped by gossanous material, often not easily recognized, the velocity contrast with the host rocks is enhanced. This allows the seismic method to accurately define the upper part of the orebody as well as provide rippability information on the host rocks.

Coal

Coal seams generally have a much lower velocity than the sediments in which they occur. Thus thick coal seams are potentially good seismic targets. In many cases, however, thin seams are difficult to detect, and special techniques including shallow seismic reflection have to be used. This is an area of active development. A variety of new techniques with greater resolution than conventional methods is being investigated.

Accurate mapping of coal seams and locating of faults, splits, and channel fills require surface, down-hole, and cross-hole seismic techniques. A relatively new technique in coal exploration involves the use of seam or channel waves. In this method a down-hole seismic source is activated within the coal seam and a detector is in an adjacent hole or gallery. If the coal seam is continuous between the holes seam, waves are readily transmitted from the source to the detector with the coal acting as a waveguide. If the seam is interrupted by a fault or stream channel, the distinctive waves are not transmitted across the boundary.

Uranium

Uranium accumulation in buried river channels is well known. Seismic methods can be used to locate and delineate possible uranium bearing channels. In related application where search is directed at sedimentary uranium, the seismic method can be used to determine sediment velocities and basement velocity around the margins of the basin. This allows decisions to be made as to whether sufficient contrast is present to allow seismic methods (usually reflection) to map basement structures conducive to uranium accumulation.

Industrial minerals

Seismic methods are often used to define the limits of deposits of industrial minerals. The use of seismic methods in material surveys for sand and gravels has been mentioned previously. In hard rock deposits such as road metal and magnesite, seismic methods may be used to define the size of the deposit and to calculate the thickness or overlying material. Velocity studies are also employed to determine variations in the deposit and optimum quarrying directions. Seismic methods can also be used to judge the effectiveness of blasting operations used as part of the quarrying operations.

Mining problems

In mining operations, seismic methods can be used to predict variations in rocks. They may be used to locate faults, dikes, and shear zones which can cause problems if unrecognized. Many applications exist in which seismic surveys assist surface and underground mining operations and mine planning.

Further applications

Seismic methods have been applied to many problems in which a variation in subsurface elastic properties was thought to exist. Seismic methods have great flexibility in their scale of operation and can be usefully employed in many areas. They are capable of providing detailed subsurface information over an area at low cost. Further applications follow.

Archaeological surveys usually involve locating large buried stone objects close to the surface. One user reported locating ancient Indian tent sites because the ground was compacted, increasing the seismic velocity.

Soil science. Seismic methods have been used to map variations in soil types and thickness. Unusual soil geochemical patterns in mineral exploration can sometimes be explained by mapping hard, screening layers with seismic methods.

Cave detection (abandoned mine workings). Limestone caves and large sink holes can often be detected with seimic methods if these are large or closely approach the surface. Abandoned underground workings can also be mapped provided they are not too small, although special seismic techniques involving P- and S-wave surveys may be required.

Drilling operations. When deciding the type of drill to be used when drilling overburden and weathered material, one often needs to know the depth and hardness of surficial material. Drilling operations can be costed effectively using seismic methods.

Geothermal studies. Vapor dominated regions show considerable contrast in seismic velocity with surrounding host rocks. The extent of potential geothermal fields can be assessed with shallow seismic methods.

Selected References (Chapter 4)

Bailey, A. D., and Van Alstine, C. L. 1975. *Downhole shear wave and soils study.* West Sacramento, CA: Nimbus Instruments.

Ballard, R. F., Jr., and Chang, F. K. 1973. *Rapid subsurface exploration; review of selected geophysical techniques.* Miscellaneous Paper S-73-36, Report 1. U.S. Army Engineer Waterways Experiment Station, CE, Vicksburg, MI.

Ballard, R. F., Jr., and McLean, F. G. 1975. Seismic field methods for in situ moduli. *Proceedings of the Conference on In Situ Measurements of Soil Properties.* American Society of Civil Engineers, New York, NY.

Fry, A. B. 1963. *A procedure for determining elastic moduli of soils by field vibratory techniques.* Miscellaneous Paper S-72-36. U.S. Army Engineer Waterways Experiment Station, CE, Vicksburg, MI.

Imai, T., and Yoshimura, M. 1975. *The relation of mechanical properties of soils to P- and S-wave velocities for soil ground in Japan.* Tokyo: TN-07 OYO Corporation.

Meissner, R. 1961. Wave front diagrams from up-hole shooting. *Geophysical propsecting* 9, No. 4.

Moony, F. M. 1973. Shear waves in engineering seismology. *Handbook of engineering geophysics.* Minneapolis, MI: Bison Instruments.

Redpath, B. B. 1973. *Seismic refraction exploration for engineering site investigations.* Technical Report E-73-4. U.S. Army Engineer Waterways Experiment Station, CE, Vicksburg, MI.

Romig, P. 1974. New applications of seismic investigations to engineering problems. *Subsurface exploration for underground excavation and heavy construction.* American Society of Civil Engineers, New York, NY.

Schwarz, S. D., and Musser, J. M. 1972. Various techniques for making in situ shear wave velocity measurements—a description and evaluation. *Proceedings of the International Conference on Microzonation for Safer Construction, Research and Application* 2: 593-608. Seattle, WA.

Stokoe, K. H., and Woods, R. D. 1972. In situ shear waves velocity by crosshole method. *Journal of the Soil Mechanics and Foundations Division*, Proceedings, American Society of Civil Engineers 98, No. SM5: 443-460.

Telford, W. M., and others. 1976. *Applied geophysics.* Cambridge: Cambridge University Press.

CHAPTER 4 ADDENDUM
SEISMICS

GENERAL DISCUSSION OF SEISMIC METHODS

The seismic properties of a rock are strongly related to the rock's elasticity and density. If one were to observe the seismic waves traversing the shallow subsurface along a line situated over a typical Basin and Range pediment, one might encounter a 500-ft section of Pleistocene lake beds on the western half, abutting a Paleozoic carbonate. The relationship of the velocity, frequency and amplitude of the seismic wave within the two vastly different lithologies would be evident. These differences are depicted in the table below.

TABLE 4.1

SEISMIC PROPERTY	LAKE BEDS	CARBONATE SECTIONS
Velocity	3.5 fpms	18.0 fpms
Frequency	low	high
Amplitude	higher	lower

Notes: 1. Velocity in ft per millisecond.
2. Frequency and amplitude stipulated in a comparative sense.

FPS = Feet per second
FPMS = Feet per milli-second
MPS = meters per second
MPMS = meters per milli-second

James W. Cooksley, Geophysicist, Cooksley Geophysics, Inc.

Mr. Cooksley is the President of Cooksley Geophysics and is registered as a Professional Engineer, Geologist, and Geophysicist, and is also a Certified Engineering Geologist. He has contributed to the fields of seismic computing and interpretation with his work in areas of complex structure and lithology.

He graduated from the Colorado School of Mines in 1954, with a Professional degree in Geological Engineering. He served in the Army Corps of Engineers from 1954 to 1956.

After working as an engineer and geologist for International Minerals and Chemical Corporation, and the Southern Pacific Company, in 1961 he established his own geological and geophysical consulting office, which grew to become Cooksley Geophysics, Inc. Mr. Cooksley is a member of the Society of Exploration Geophysicists, was a member of the Geophysical Activity Committee, and for three years was Chairman of the Engineering Geophysics, Groundwater Subcommittee.

He is a member of the NWMA and the AIME.

A simple example of the relationship between the velocity of seismic waves and lithology and geologic structure is depicted in table 4.1. This relationship enables the geologist to observe, and in most cases measure or delineate, the following features:

- soil layers and slope failures
- weathered layers
- rock units
- metamorphism and metasomatism (skarns)
- hydrothermal alteration
- faults, sheared zones, tectonic brecciation
- vein structures.

This course in seismic exploration will be directed mainly at the use of seismic velocity in making geologic observations useful to the exploration geologist.

FIGURE 4.1
This illustrates seismic waves propagating through soil and rock units in an hypothetical pediment. Please note the relationship of velocity (in ft per millisecond) to lithology and geologic structure

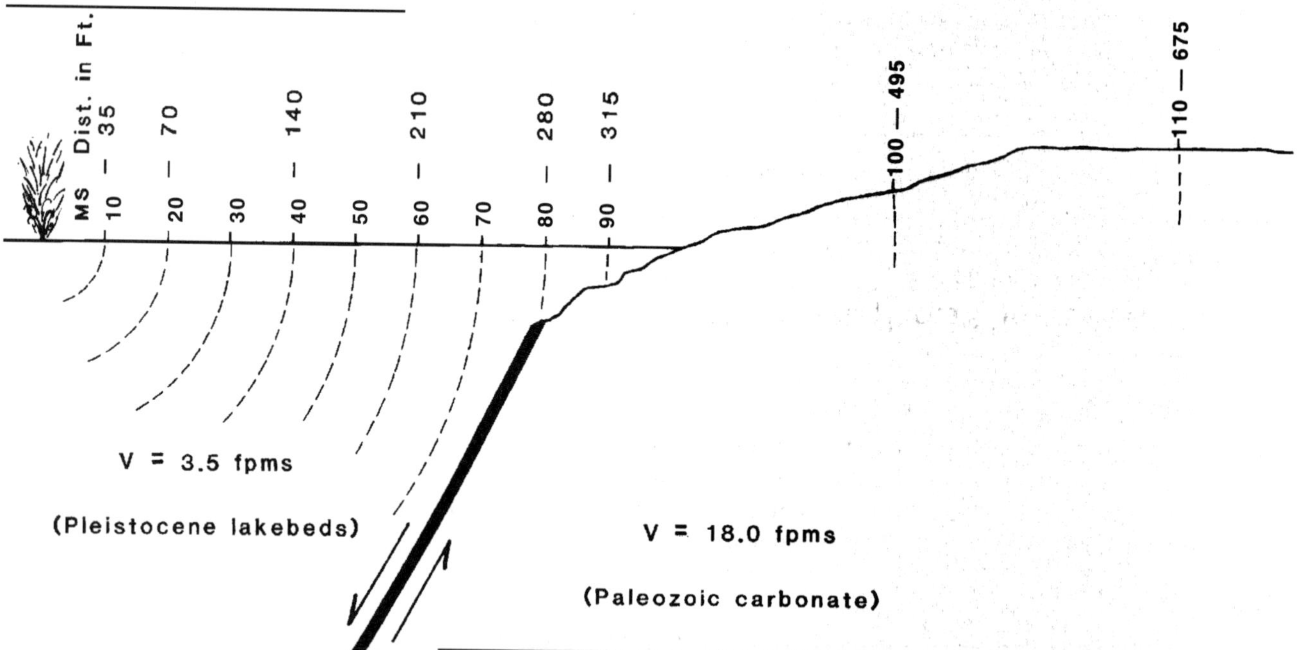

Velocity is the main property used in seismic measurements of depth and rock properties. Currently, little attention is directed toward frequency and amplitude characteristics in shallow refraction seismic investigations. Conversely, this is not true for the reflection seismic method. Although depth is computed as a function of velocity and time, amplitude and frequency along with other parameters such as offset and phase shift are used in the processing and interpretation of reflection data.

Crice (1988), discussed the measurement of seismic velocity from refraction and down hole measurements. Other discussions are presented by Dobrin (1950), Nettleton (1940) and Heiland (1950).

Seismic Techniques

The techniques presently in use in seismic exploration and investigation can be subdivided into three main categories. These are:

- refraction
- reflection
- down-hole and cross-hole.

All three of these techniques find significant use in mining exploration and engineering investigations.

Refraction Method Principles

A refraction seismic discontinuity is at a surface along which the velocities differ, the lower velocity above the discontinuity and the higher velocity below the discontinuity. A velocity inversion is where the higher velocity overlies the lower velocity. Velocity inversions can introduce error in the computing of the data or might even render the refraction method unusable, in which case the reflection method can be a likely alternative to solving the problem. Normally, velocity inversion is the exception rather than the rule.

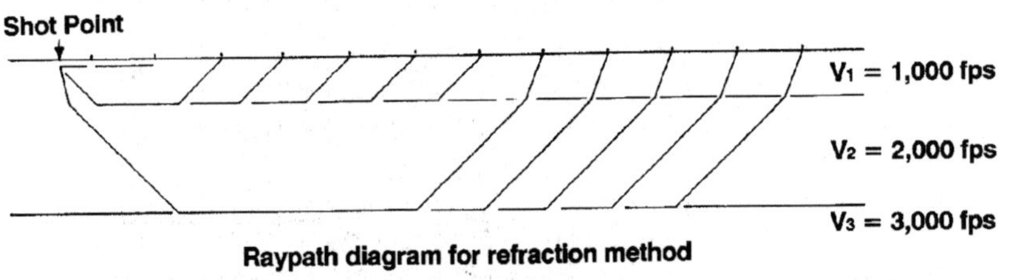

FIGURE 4.2
Section showing ray paths of first arrival seismic waves

In reference to figure 4.2, notice that:

- The shot point or seismic source is to the left.
- The sensors are evenly spaced at 50 ft and are collinear along the surface.
- The velocity increases with depth.
- The differential of the travel time between the sensors becomes smaller as the first arrival of energy is propagated from the deeper discontinuities. Seismic velocity and the depth of the discontinuities are computed from the relationships of these differential travel times.

Applications

- Normally used in shallower surveys — to depths from grass roots down to the 500 to 1,000 -ft depth range.
- Depth to bedrock surveys — pediment surveys are common in the Basin and Range Province.
- Rock rippability measurements for excavation and construction work.
- Mapping bedded mineral deposits.
- Mapping faults.
- Mapping alteration.
- Mapping weathered zones.

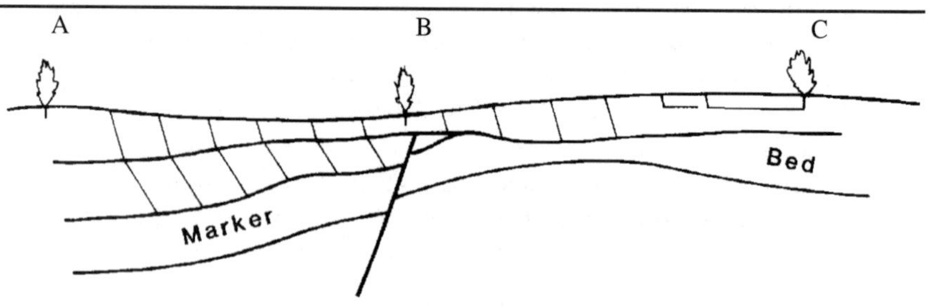

FIGURE 4.3
A typical shooting procedure for obtaining refraction data

SHOT C being shot.
SHOT A will be shot to provide travel-time measurements along the "marker bed" in a direction opposite to those taken from Shot C.
SHOT B shot in center to obtain near-surface velocity data.

Field Procedure Interpretation Methods

In a refraction seismic survey, it is helpful to map the subsurface in terms of velocity units. A velocity unit is a three-dimensional unit of rock or soil that, owing to its elastic properties and density, propagates seismic waves at a characteristic velocity or in a characteristic velocity range.

Anomalously low velocity indicates:

- fracturing......could be associated with faulting
- shearing........could be associated with faulting
- hydrothermal alteration
- weathering
- a combination of two or more of the above.

Apparently, the mere presence of a low-velocity zone does not uniquely designate a particular geologic condition. But with the aid of geologic observation and reasoning, coupled with noting some geometric aspects of the anomalous zones, a unique solution is made possible with the seismic subsurface data. Additionally, the seismic data provides a reasonably accurate measurement of depth and also provides a means to quantify the competence of the rock units.

An important consideration to the interpreter is that the presence of groundwater affects the velocity of rock units. This especially is true in unconsolidated and poorly consolidated sediments which have "dry" velocities substantially less than 5,000 fps which is the approximate velocity of water. In depth to bedrock surveys, this condition is recognized and the necessary adjustments in computing and interpretation can be made. Defining alluvial and other sedimentary contacts in the saturated zone demands more specialized and sophisticated seismic methods and instruments than are used in conventional refraction surveys.

Anomalously high-velocity zones indicate:

- Subsurface rock escarpments where the seismic waves have short-cut through the rock instead of having followed a discontinuity
- Zone of silication or silicification
- Zone of contact metamorphism
- Dike of igneous rock intruding softer rock
- Zones of caliche
- Other.

Faulting is interpreted and delineated from the following refraction seismic criteria:

- Offsetting or total interruption of horizontal and near-horizontal discontinuities
- A zone of lower velocity within more competent, higher velocity units
- General attenuation of seismic waves (otherwise unexplained)
- Local absence of high-frequency energy in areas where high-frequency waves predominate (otherwise unexplained).

Reflection Seismic Measurements

Velocity measurement from reflection data are a little more indirect and complicated than in the refraction method. Given a horizontal ground surface and a horizontal reflection surface, the equation used is:

$$V^2=\frac{x^2}{t^2} \tag{4.1}$$

where: V = velocity (average)
x = distance from shot point to the end of line
t = time measured at end of line minus time measured below the shot point

To compute the depth to a reflection event, the following general equation applies:

$$Z=\frac{Vt}{2} \tag{4.2}$$

where: Z = depth
V = average velocity
t = time of event

Velocities are computed routinely and rather easily from reflection data by employing rather large capacity computers to stack the data on a trial basis at many constant velocities. A velocity function is derived by noting the times at which the various trial velocity plots yield coherent reflection events. Sheriff (1980) discusses this procedure at some length and provides an excellent bibliography.

Reflection of a downward propagating seismic wave takes place where the wave encounters an interface where there is a change in velocity, a change in density, or a change in both velocity and density. The acoustic impedance of a rock is the product of the longitudinal seismic velocity and the density (j = pV).

The reflection coefficient is proportional to the amplitude of the reflected wave and is computed as follows:

$$\textit{reflection coefficient}=\frac{p_2V_2-p_1V_1}{p_1V_1+p_2V_2}=\frac{j_2-j_1}{j_1+j_2} \tag{4.3}$$

An important use of amplitude along a reflected seismic event is that it indicates velocity contrast from one side of the contact to the other. The larger the amplitude, the larger the contrast. It should be noted that offset, the distance from the shot point to the sensor recording the wave, also affects amplitude. If the amplitude changes along the event, a change in velocity contrast is normally indicated. Thus from changes in amplitude along a reflection event, a change in velocity in the horizontal direction in at least one of the units can be inferred. This is an important concept, for if it is true that the velocity of the longitudinal wave propagating through a rock is an indication of the compressibility and rigidity of that rock, then the amplitude of reflected longitudinal seismic waves is an indication of changes in compressibility and rigidity of the rock units at a contact. Further, a change in amplitude along a reflected event indicates a change in the contrast in compressibility and rigidity, such as encountering an increase or decrease of porosity within one or both of the units or encountering the top of a hydrothermal system. Consider, also, encountering a jasperoid mass, a skarn or a silicified mass.

Some General Facts:

Of all the commonly used surface geophysical methods, seismic methods provide the most accurate measurement of depth and the most comprehensive and graphic picture of subsurface soil and rock conditions. Usually an accuracy of 10 percent or better can be attained without drill hole control. With care, a seismic investigation is capable of providing information accurate enough for predicating the design of dams, highrise structures, slope stability programs and other engineering works.

Fundamental to the use of seismic techniques are contrasts in the acoustic impedance of earth materials which permit the reflection and refraction of elastic waves. Seismic exploration is based on the premise that such contrasts exist at contacts between geologic units. Thus, seismic exploration relies on artificially creating a source of seismic energy such that elastic waves will propagate downward through the earth materials, being reflected and/or refracted at discontinuities in the subsurface and being detected at the surface by appropriately emplaced seismic sensors. The theory underlying the application of standard exploration procedures is that individual materials are uniform and homogenous and that paths follow essentially straight ray paths. These theoretical considerations are not always borne out in nature. The technology has been developed sufficiently, however, that the method can be applied effectively. The greater the deviation from the ideal, the more difficult and subjective the interpretation of seismic data becomes.

Seismic properties of rocks, mainly the velocity of propagation, are directly related to the elastic properties and density of the rocks. These properties relate directly to the consolidation and competence of rocks. High seismic velocity, in the order of 10,000 fps or greater, for instance, indicates a fairly competent rock mass. Lower velocity is correlative with weaker rock; the lower the velocity, the weaker the rock. Unconsolidated surficial deposits are about the weakest form of rock material and have a seismic velocity of about 1,000 fps. Both experience and theory indicate that masses of sheared, fractured and weathered rock are detected seismically as zones of lower velocity.

Seismic data are normally acquired by vibrating the ground, using either explosives or mechanical devices, and detecting the seismic waves with sensors appropriately placed along the ground surface.

Computing and interpreting the seismic data might require as much or more effort, in terms of cost, as the field data acquisition phase. The mathematical procedures will be avoided in this presentation, but some emphasis will be directed to interpretation.

Seismic measurements of velocity in the subsurface provide a means to quantify engineering properties for the engineer and to quantify rock competence, alteration and fracturing for the geologist. This fact alone opens a host of approaches to solving geologic problems such as mapping faults, breccia pipes and rock alteration.

Reflection Method Principles

Reflection seismic exploration involves generating seismic waves at or near the surface of the ground, the waves traveling downward to surfaces of geologic features (which geophysicists call discontinuities) where a portion of the total energy of the wave front is reflected back upward toward the surface of the ground, and detecting the reflected waves at the surface of the ground with appropriately placed sensors. The travel time for the wave to complete the trip from the seismic source at or near the ground surface to a discontinuity, and then back up to the surface of the ground, is the quantity that is of primary importance in measuring the depth to the discontinuity. The average velocity of the wave while making its trip into the subsurface must be known or accurately estimated to compute depths accurately. There are several approaches to obtaining the needed velocity information.

Recently, a parameter known as "relative amplitude" has become very important in the petroleum industry in the "direct detection" of hydrocarbons. This parameter appears to have promising applications in the mining exploration field also, e.g., hydrothermal alteration.

Applications of the Reflection Seismic Method

Reflection surveys are normally used in surveys directed to depths in excess of 1,000 ft, but more recently, mining exploration has mandated their use to depths of about 100 ft and less! The success of such shallow, high-resolution surveys depends on:

- physical conditions such as velocity contrasts, air blast and refracted noise, and
- economics.

Depth to pre-mineral rock foundations can be determined and mapped from reflection seismic data. Examples of geologic situations requiring such investigations include areas covered by valley fill, lake beds and volcanic rock. This method is also effective in exploration for mineral-bearing faults, breccia pipes and hydrothermal systems. Certainly, stratiform deposits are amenable to reflection seismic exploration. More recently, the reflection method is gaining use in exploration for ground water.

FIGURE 4.4
Showing ray paths of reflected seismic waves

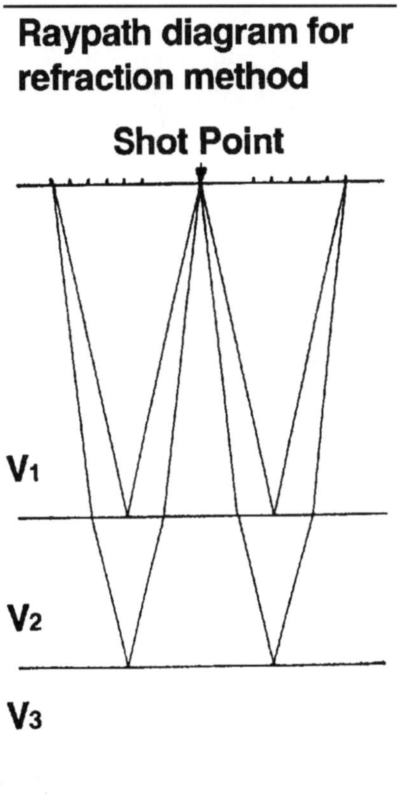

Field Procedures

The normal field setup for reflection work is shown in figure 4.4. Here, the shot point is detonated at the midpoint of a collinear array of seismic sensors which are connected to the recording truck by multi-conductor cable or by telemetry. Other equipment involved include a shot hole drill, a shooting truck and sometimes an auxiliary vehicle or two. For shots away from the recording truck, radio-controlled remote firing systems are commonly used. In areas of shallow bedrock typical of many mining districts, shallow shot holes of only 6 ft to 20 ft in depth can be used. Usually 24 or more sensor stations are placed at a prescribed distance apart, the distance depending on the depth to be explored and some physical considerations. Normally, the station spacing is in the 15 ft to 110 ft range. It is preferred to use at least six sensors placed in an array at each sensor station. In other words, each channel of seismic data (each wavy line) is the response generated from six or more sensing units placed in an array about the sensor station.

Certain late developments in seismic field procedures and shooting techniques in the petroleum exploration field have resulted in being able to complete very large surveys at comparatively low costs while still maintaining a high standard in data quality. At this time, such techniques are applicable to mining exploration only with difficulty because in mining, the areas to be surveyed are smaller and the geological considerations demand a greater amount of flexibility in the field regimen.

Interpretations

Reflection seismic exploration is based on the premise that major contacts will be observable as seismic discontinuities. The presence of a seismic discontinuity is recognizable by the presence of seismic events occurring across the recordings of traces of many sensors at roughly the same time or in a "time-distance trend". A seismic event is an anomalous perturbation in the seismic signal or wave train which is caused by a wavefront impinging on a surface that represents a change in elastic properties and/or density.

With experience, the interpreter can distinguish the events which represent significant "marker beds", intrusives and faults. Familiarity with the geologic section and the types of structural features to be expected in an area are, of course, necessary to making the best possible interpretations of the geophysical data. This is where the input from the resident geologist is most needed.

As in the refraction method, seismic velocity is used as a criterion in assessing the competence and consolidation of the rock units. This observation, of course, can relate to the presence of such features as faults, fracturing, alteration and porosity in the case of lower velocity; and dikes, intrusive plugs, contact metamorphism and silication in the case of higher velocity.

In addition to using velocity as an interpretive parameter, amplitude and frequency may also be used effectively in interpreting areas of fracturing and alteration. Relative amplitude measurement is applicable to detecting zones of altered rock to depths of several thousand feet. Faulting and alteration tend to damp and filter out the higher frequency seismic waves.

Fault interpretation in reflection work employs the same general criteria used in refraction work which is listed at the end of the discussion of the refraction method.

Seismic Velocity Anomalies Caused By Some Types Of Mineral Deposits

Numerous seismic investigations conducted within operating mines and on mineral prospects provide the basic data relating certain mineral deposits to seismic velocity anomalies. The purpose of this chapter is to discuss the theoretical basis, provide a significant but limited amount of experimental data, and present some noteworthy examples illustrating the correlation of ore deposits and related features to seismic velocity anomalies.

The seismic velocity measured within a rock mass is a function of the elastic moduli and the density of that rock mass. Most orebodies possess moduli and densities considerably different than the host rock. Furthermore, rock alteration processes which precede and follow the formation of an ore deposit commonly form an aureole of rock which surrounds the orebody and which possesses moduli and densities significantly different from the host rock. Given that the seismic velocity is a function of moduli and density, it follows that seismic velocity measurements or inferences are applicable to detecting orebodies and associated rock alteration.

For the purpose of discussion, the velocity changes associated with ore deposits have been classified in this text as positive anomalies and negative anomalies to define whether the velocity of the mineral deposit is higher or lower than the adjacent rock mass. Disseminated copper and molybdenum orebodies, breccia pipes, skarn deposits and veins are commonly associated with velocity anomalies.

Seismic velocity measurements taken within large operating open pit copper mines provide much of the experimental basis for concluding that various types of alteration and parts of the ore deposit (e.g. supergene vs. primary) can be defined as zones of characteristic seismic velocities. In the exploration of tungsten-bearing contact pyrometasomatic deposits, positive velocity anomalies represent the predominantly calc-silicate rock mass which is likely to contain disseminated tungsten minerals. Most veins constitute significant velocity anomalies, some positive and some negative. In the exploration of one massive sulfide deposit, the seismic velocity is significantly higher in ore than the adjacent host rock.

Velocity As A Function Of Modulus And Density

The velocity at which elastic waves propagate through matter is a function of the elastic modulus and the density of the matter. The bulk modulus, which is commonly called compressive modulus, is the inverse of compressibility. It measures the elastic property which directly affects the velocity of the pressure wave (p-wave). At this time, the p-wave is the wave used most commonly in seismic exploration. The p-wave deflects forward and back in the direction of propagation and its velocity is usually designated as V_p, or simply v. Only the velocity of the p-wave will be discussed in this chapter.

The equation which mathematically expresses the relationship between the velocity of the p-wave and the bulk modulus and density of a material is:

$$v_p = \sqrt{\frac{k + \frac{4}{3}\mu}{p}} \tag{4.4}$$

p = density
k = bulk modulus
μ = shear modulus

A paradox exists in this equation. According to the equation, if the moduli were held constant, the velocity would decrease as the density increases. But, this IS NOT normally the case in a rock mass. Because the seismic velocity, v_p, normally increases as the density increases, one must conclude that an increase in density is accompanied by an even larger increase in the bulk and shear moduli. Two graphs are shown below; both depict the bulk moduli of naturally occurring mineral materials as functions of their densities. Figure 4.5 depicts the bulk moduli of minerals—naturally occurring aggregates of molecules. Figure 4.6 depicts the bulk moduli of rocks—naturally occurring aggregates of minerals. The data used in compiling these graphs were obtained from Heiland (1940) and Dobrin (1960).

It appears that the modulus, and thus the seismic velocity, is a function of the "packing" of matter within a given volume or space. This statement is obviously a very simplified version of what can be a complicated relationship, e.g., when several types of minerals and pore space are involved. Nevertheless, it does form a necessary part of the foundation for understanding what affects seismic velocity.

FIGURE 4.5
Plot of bulk moduli (k) of minerals vs. their densities (specific gravity), derived from Heiland (1940, p. 78-80, 467)

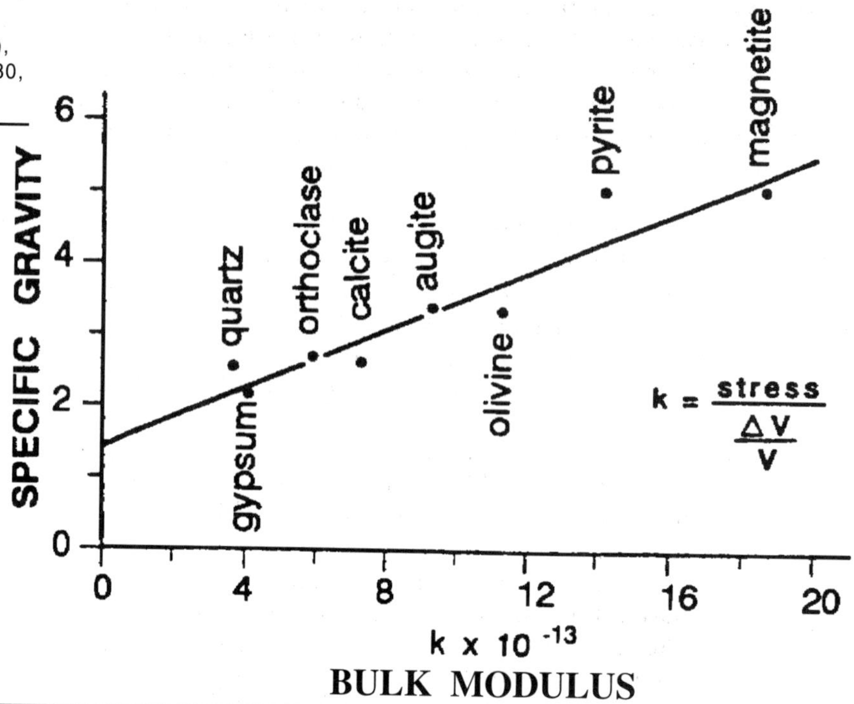

FIGURE 4.6
Plot of bulk moduli (k) of rocks vs. their seismic velocities (V) and densities (specific gravity), derived from Heiland (1940, p.467) and Dobrin (1960, p. 22, 251)

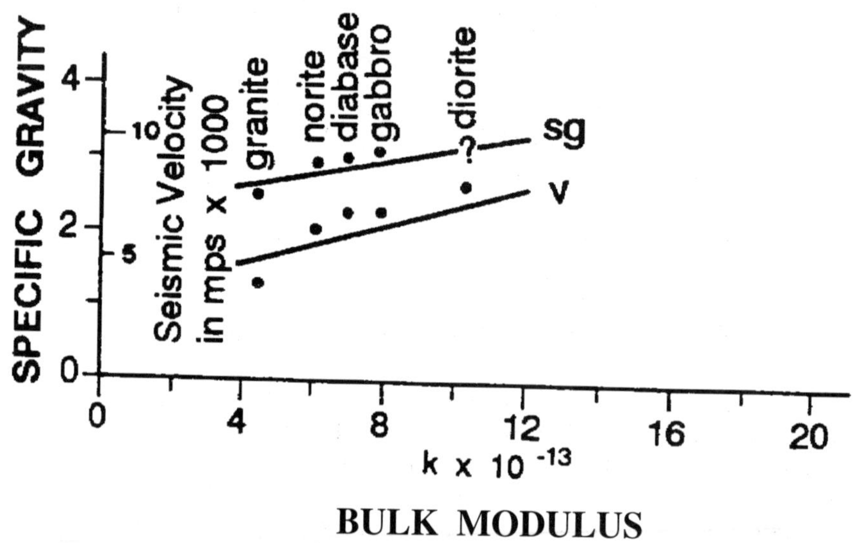

Another necessary part is the fact that although increased density negatively affects velocity in equation 1, apparently there is an unseen effect of density on the modulus which provides a much larger positive contribution to velocity. From the graphs, it is apparent that a small increase in density correlates with a much larger increase in bulk modulus.

Velocity As A Function Of Hydrothermal Alteration
Hydrothermal alteration is defined as those phase changes resulting from the interaction of hydrothermal stage fluids with pre-existing solid phases, such as kaolinization of feldspars. This process also includes the chemical and mineralogical changes in rocks brought about by the addition or removal of materials through the medium of hydrothermal fluids, e.g., silicification. As the harder minerals are

converted to or replaced by softer minerals, the bulk modulus of the rock mass is very significantly diminished. Thus the seismic velocity of the rock mass changes as the feldspars are converted to kaolin or to sericite. Observations taken in several mines indicate that a relatively minor volume of argillization or sericitization causes large decreases in seismic velocity. Conversely, seismic velocity normally increases as a result of the feldspar flooding and garnetization phases of alteration.

FIGURE 4.7
Calculated velocities of sound in air-earth mixtures. From O. C. Lester, 1932

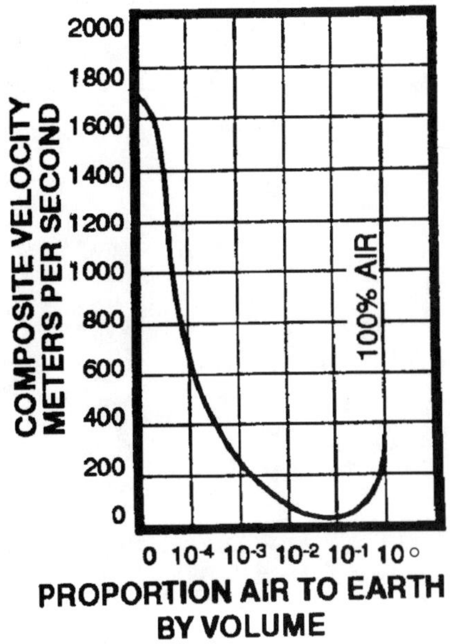

Velocity As A Function Of Porosity

Figure 4.7 depicts the seismic velocity in a porous earth as a function of porosity. The relationship shown in this graph is quite predictable from the relationship demonstrated in figure 4.5 and figure 4.6. Assuming that the pore space would be filled with air or water and knowing that the rock grains are heavier than air or water, it follows that the density decreases as the volume-percent of pore space increases. Thus, a diminished velocity would result as the pore space is increased, because the modulus changes at a greater rate than the density. In this case, both the modulus and density, and thus the velocity, all decrease as the porosity increases.

Another example recognizing that velocity is an inverse function of porosity is discussed in Heiland, C.A., *Geophysical Exploration* (1940, p.474) where he cited using the direct effects of modulus and density. The discussion suggests that the velocity of a porous rock is a function of the velocities and proportions of both the solid component and the material filling the pore space. The equation proposed is as follows:

$$v_1 = \sqrt{\frac{E_1 E_2}{[\,pE_2 + (1-p)E_1][\,p\delta_1 + (1-p)\delta_2]}} \tag{4.5}$$

E_1 = elasticity of air
E_2 = elasticity of earth
p = proportion of air to total by volume
δ_1 = density of air
δ_2 = density of earth
1 - p = proportion of earth to total (by volume)

Although there are some considerations which preclude applying this relationship without at least some modification, it is a way of demonstrating the inverse relationship between porosity on one hand and seismic velocity, modulus and density on the other hand.

The Wyllie relationship (Sheriff, 1973) also expresses the inverse relationship between porosity and velocity. This empirical expression for calculating porosity, from sonic log transit times is:

$$\phi = \frac{\Delta t - \Delta t_m}{\Delta t_f - \Delta t_m} \tag{4.6}$$

where: t = observed transit time
t_f = transit time in pore fluid
and t_m = transit time in the rock matrix.

Knowing that the t's are measured and computed over the same sensor spacing, a, the terms become analogous to the reciprocals of the respective velocities. The reciprocal relations of t with v is shown in the equation:

$$\frac{\Delta t}{a} = \frac{1}{v} \tag{4.7}$$

where a is the sensor interval. Because ϕ and a are constants not related to porosity but only related to the velocity of the rock and the fluid, it is apparent that the porosity increases as velocity decreases.

Void Space

So far in this paper, void space has been discussed as pore space and porosity. This is because the effect of porosity in rock has been extensively studied in petroleum exploration and petroleum engineering and much of these data have been used in compiling the first part of this paper. In mining exploration, mining engineering and civil engineering, other forms and concepts of void space must be considered. The most important of the additional forms of void space is fracturing. Other forms of void space might be argued to be a form of pore space, but they are certainly different in some important respects from the pore space in a sandstone. Some of these forms are as follows:

- voids and "microvoids" resulting from the leaching effect of hydrothermal alteration and weathering
- voids present in breccias
- voids resulting from shear stress.

Not all void space is occupied by air, methane, water, crude oil, and other gasses and liquids. Void space is commonly replaced by calcite, silica, clay, limonite and a host of other epigenetic minerals. In the case of mineral deposits, many of the voids are filled with ore minerals which are normally quite dense and possess a high elastic modulus. Obviously, the contribution of the mineralization to the overall velocity of the rock mass depends to some degree on the volume-percent of the rock which the mineralization constitutes. This, of course, is a function of the void space in the host rock which is available to the mineralizing solutions. The mineralizing process can cause higher velocities by adding rock material, or it can result in lower velocities by removing material.

There are some types of ore deposits in which void space in the host rock is not a factor. In these types of deposits no host rock is involved. Two types which fall into this category are: 1) magmatic segregation, and 2) sedimentary, e.g., evaporates and coal. Although pore space is not a factor in these types of deposits by virtue of their contrasting density and modulus with the surrounding rock units, some can be detected on the basis of seismic velocity.

Mineral Deposits and Void Space

The causative effect of void space with respect to many types of mineral deposits is recognized and accepted by explorationists. Some examples of this premise are:

- the tendency of vein deposits to be emplaced along faults and fractures
- the leaching effect of hydrothermal solutions which, along with fracturing, prepare an otherwise tight, impermeable rock mass for the mineralizing phases of later hydrothermal activity
- the tendency for uranium deposits of secondary origin to be emplaced in or at the contact of porous sedimentary beds or in fractures within igneous and metamorphic as well as sedimentary units.

Velocity Anomaly

A region or zone in the subsurface in which seismic waves propagate at an unusually high or low velocity constitutes a velocity anomaly. An anomaly which has a higher velocity than the surrounding unit(s) is a positive velocity anomaly. A lower velocity than the surrounding rock unit(s) denotes a negative velocity anomaly. The main objective of this chapter is to demonstrate that certain mineral deposits, and certain geologic conditions which relate to mineral deposits, constitute seismic velocity anomalies. Commonly, these velocity anomalies are sufficiently distinctive that they can be recognized on a seismic section and thus constitute a usable criterion for exploration.

Whether the type of anomaly is positive or negative depends on (1) the velocity of the mineral feature and (2) the velocity of the surrounding rock. It is possible for a mineral deposit to constitute a positive velocity anomaly in one geologic environment and a negative velocity anomaly in another geologic environment. For example, a massive sulfide deposit with a velocity of 5,488 mps emplaced in shale and metavolcanic rock of Triassic age possessing a seismic velocity of 3,963 mps would constitute a positive velocity anomaly. Conversely, this same deposit would constitute a negative velocity anomaly if it were emplaced in a typical gneiss of Precambrian age possessing a velocity of 6,707 mps (See figure 4.8). A gold-bearing

FIGURES 4.8
Examples of a massive sulfide deposit constituting a positive velocity anomaly within a host rock of Triassic metavolcanics (above), and a negative velocity anomaly within a host of Precambrian gneiss

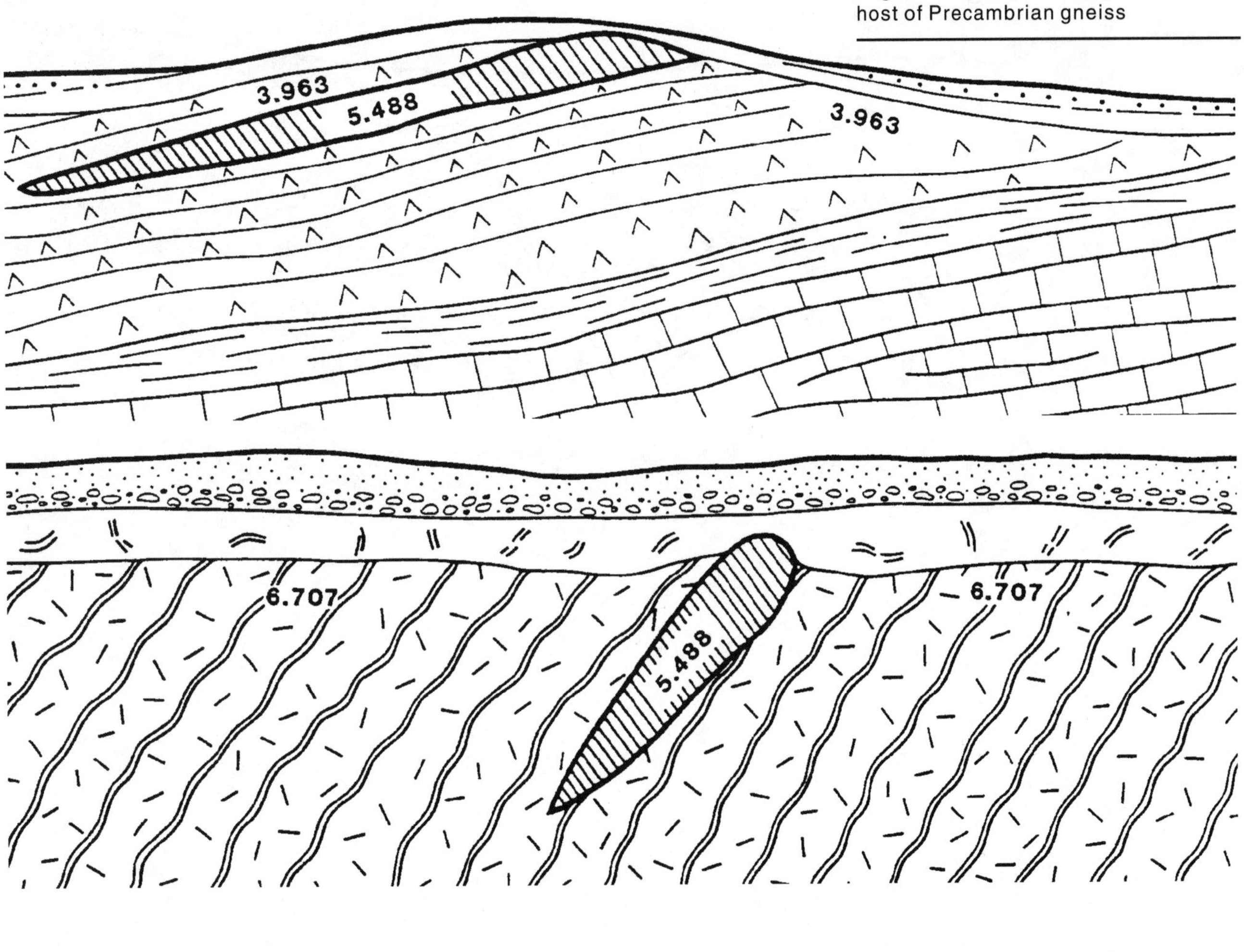

quartz-calcite vein, approximately 25 ft wide and possessing a seismic velocity of about 2,439 mps, would constitute a positive velocity anomaly if it were emplaced in a tuffaceous volcanic unit which has a velocity of only 1,524 mps. Conversely, the same vein would constitute a negative anomaly where it cut an intrusive which had a velocity of 4,268 mps (See figure 4.9). Types and examples of velocity anomalies are discussed further in this paper under the section, EXAMPLES OF SEISMIC VELOCITY ANOMALIES.

FIGURES 4.9
Examples of quartz-calcite veins constituting a positive velocity anomaly within a host rock of tuffaceous volcanics (above) and a negative velocity anomaly within an intrusive host rock

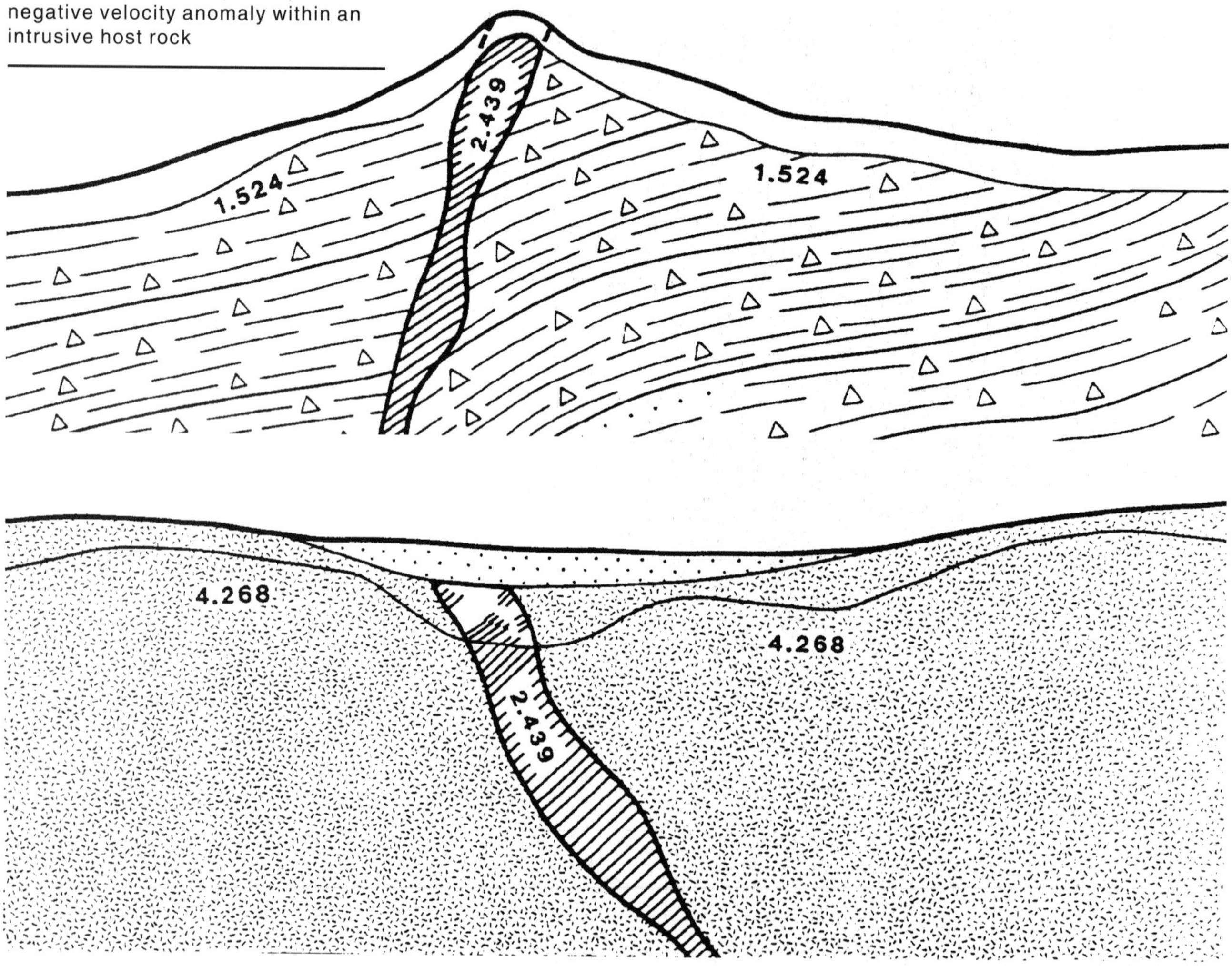

Seismic Velocity Measurements In Mineral Deposits

The following tabulates the location and type of mineral deposit, the velocity measured in the deposit, the velocity measured in the host rock and the type of measurement taken (velocities in meters per millisecond):

TABLE 4.2

LOCATION	MINERAL TYPE	VEL. DEPOSIT	VEL. HOST	TYPE
Yerington, NV	porphyry copper	1.62 oxide, 2.16 mixed ore, 2.67 sulfide ore	3.81	refraction
Berkeley Pit, Butte, MT	porphyry copper	1.86	3.35+	refraction
Twin Buttes, AZ	disseminated copper in a skarn	2.13±	3.05±	refraction
Bloody Mtn., Mammoth, CA	disseminated tungsten in a tactite	3.97 to 10.15	2.70	refraction
Osgood Mtns., NV	disseminated tungsten in a tactite	5.21 to 7.20	3.05 to 3.93	refraction
West Shasta District, Redding, CA	massive sulfide, copper and zinc (in weathered zone)	3.20	1.77	refraction
Calico District, CA	vein deposit, silver	1.42 to 3.05	1.22 to 1.52	refraction
Comstock Lode, NV	gold-silver vein	0.61 to1.80	2.40 to 3.66	refraction
San Juan, CA	stream channel, gold	2.74 blue lead	4.88 slate	refraction
Carrville, CA	stream channel, gold	1.52	2.87 to 5.55	refraction
Galisteo River, NM	stream channel, gold	1.22	3.05	refraction
Southeast, ID	sedimentary bed phosphate	1.22	3.05	refraction direct in hole
Ennis, MT	talc in dolomite and gneiss metamorphic complex	5.55	2.01 to 3.38	refraction
Jack Creek barite prospect, NV	replaced sedimentary	4.57	1.62 to 2.35	refraction

Tactites And Skarns

Numerous ore deposits of copper, molybdenum and tungsten are found within metamorphic zones. The particular metamorphic process involved in these zones is pyrometasomatism, a process in which metasomatic changes take place in rocks, principally in limestone, at or near intrusive contacts, under the influence of magmatic emanations and high temperature and pressure.

Metasomatic generally means replacement. The pyrometasomatic process commonly forms a large mass of clinopyroxenes, garnet, epidote and quartz which possesses extraordinary high seismic velocity. Normally, these masses constitute strikingly obvious, positive velocity anomalies with respect to the adjacent rock masses. They commonly occur at, or within a very short distance of, an intrusive contact.

Significantly large, positive, velocity anomalies were delineated at the Getchel mining district in the Osgood Mountains near Winnemucca, Nev. Figure 4.10 depicts typical positive anomalies in which ore deposits are contained in zones of pyrometasomatic replacement. The zones in this district are characterized by garnet, clinopyroxene, epidote and quartz.

FIGURES 4.10
Seismic section at Granite Creek Mine in the Getchel mining district near Winnemucca, Nevada. Geologic section below depicts a section derived from surface and subsurface geologic mapping and is located nearly parallel to, and approximately 100 meters closer to the intrusive

Examples Of Seismic Velocity Anomalies

SEISMIC SECTION at the GRANITE CREEK MINE

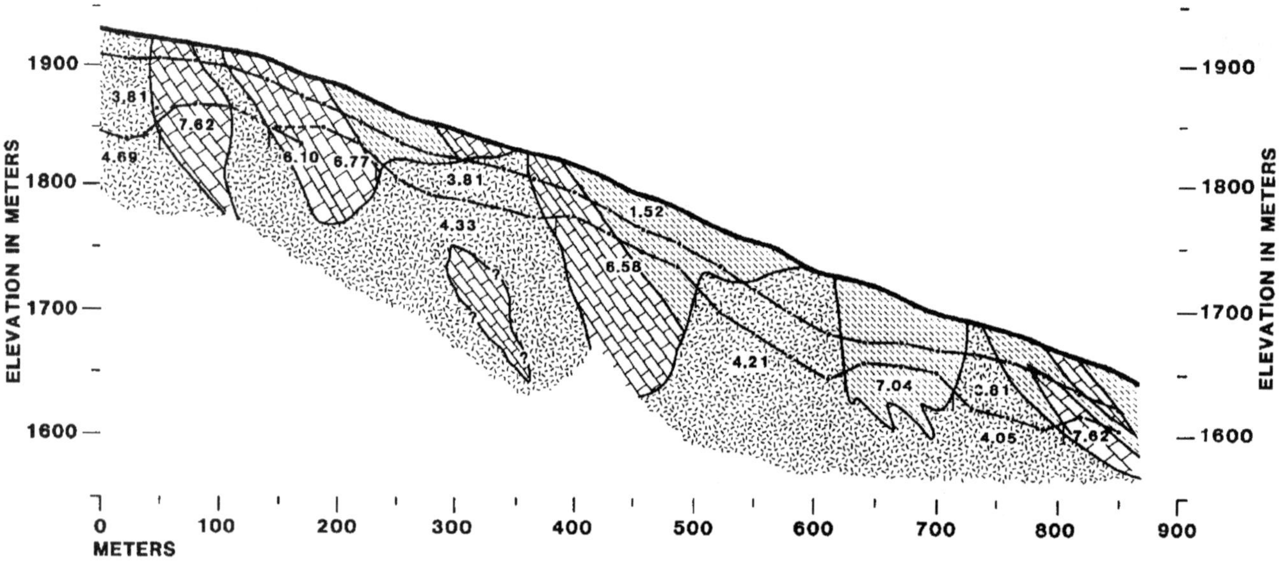

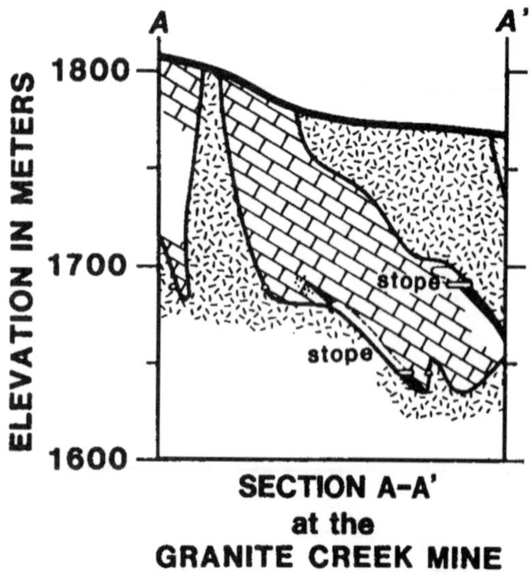

SECTION A-A'
at the
GRANITE CREEK MINE

Seismic Velocity in mpms	Geologic Unit
4.05-4.69	intrusive rock
6.58-7.62	limestone
	hornfels
	tactite

from Hotz, P. E., and Willden, R., 1964.

Velocity measurements inside the pyrometasomatic zone were taken in the Hard Point mine workings on Bloody Mountain near Mammoth, Calif. This deposit differed from those in the Osgood Mountains insofar as the pyrometasomatic zone is composed largely of garnetized rock and is very dense. The extraordinary high velocity of the mineral zone is associated with the zone in which clinopyroxene and garnetization predominates. Figures 4.11 and 4.12 depict the data from the Hard Point prospect.

FIGURE 4.11
Map of underground workings at Hard Point showing contact between orebody and host rock

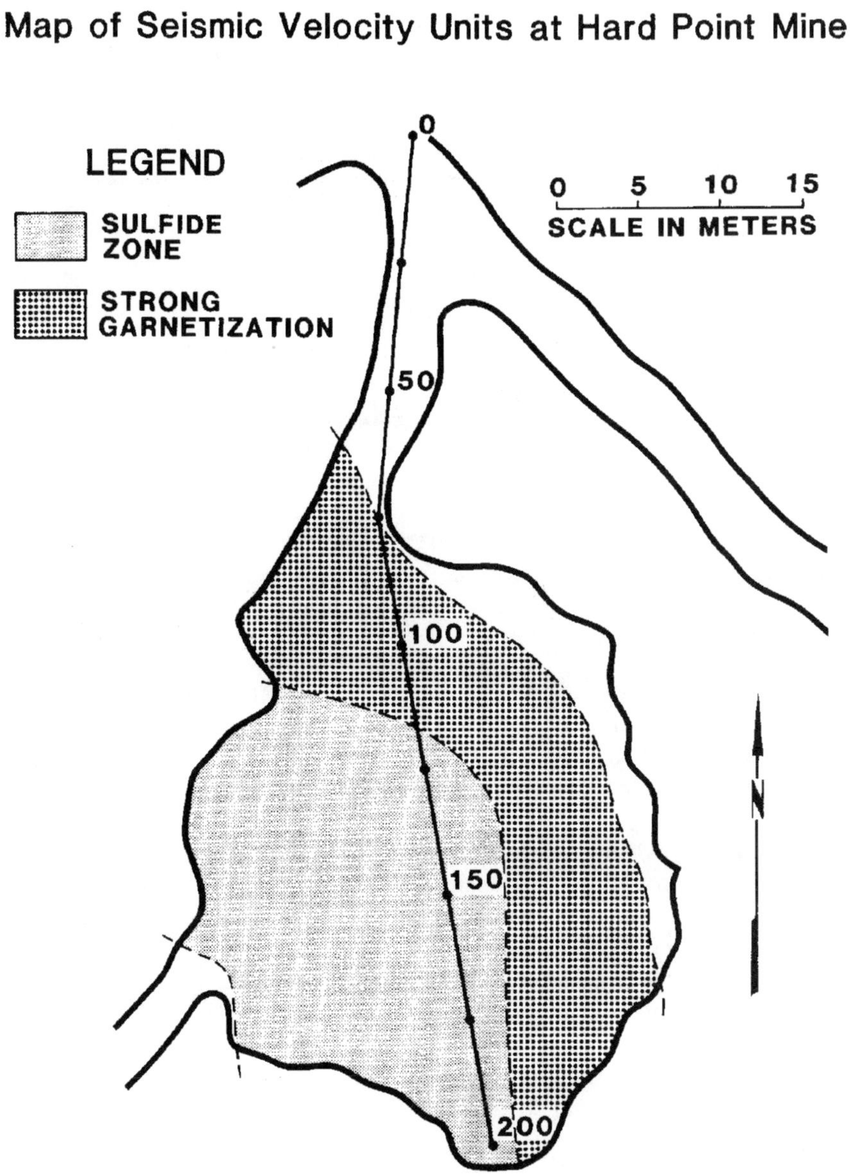

FIGURE 4.12
Derivation of seismic section from the Hard Point prospect, Mono County, California

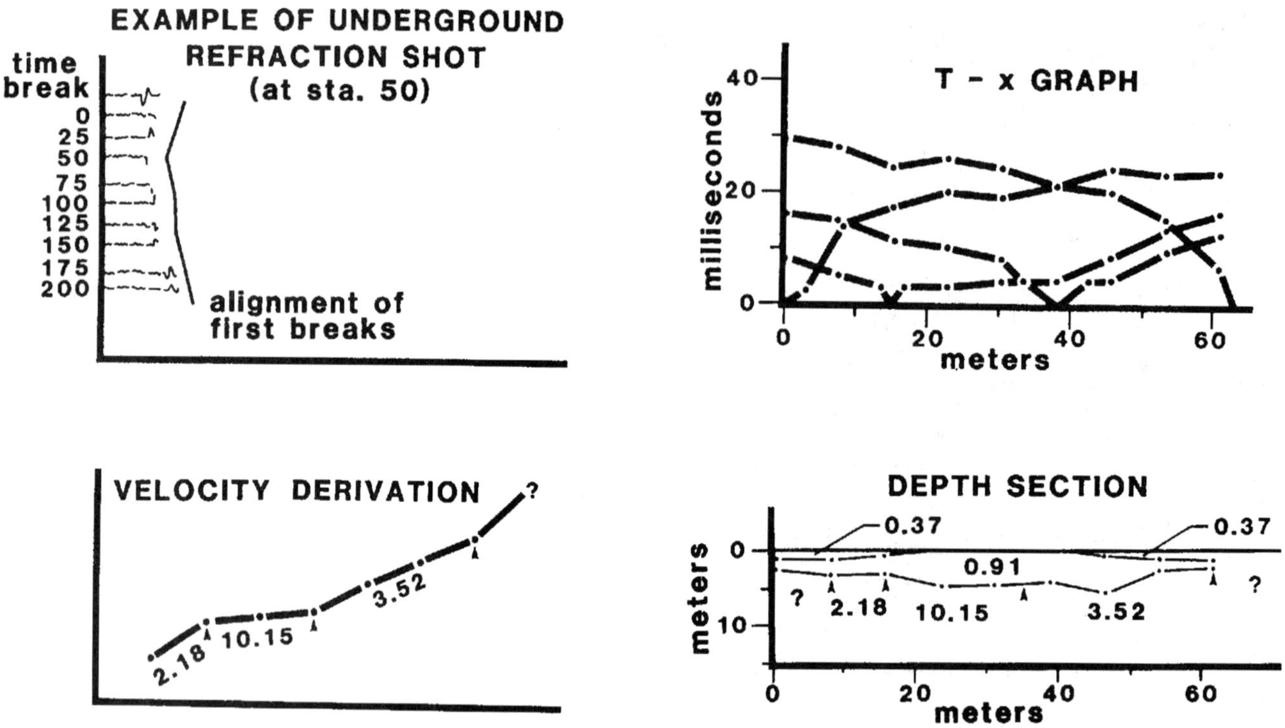

The reflection method is also employed in the exploration of contact pyrometasomatic deposits. The classic application is to find the depth to the intrusive contact underlying a roof pendant or a large area of sedimentary rock which contains significant carbonate units.

In most cases, the depth to the intrusive ranges greatly. Consequently, before drilling, the seismic program should be designed to supply the data necessary to evaluate the configuration of the contact. In some cases, a distinct and anomalously large diffraction of seismic waves originates in a significantly large zone of pyrometasomatic replacement. The presence of a high velocity pyrometasomatic zone tends to "raise" the reflections originating from directly below the high-velocity zone.

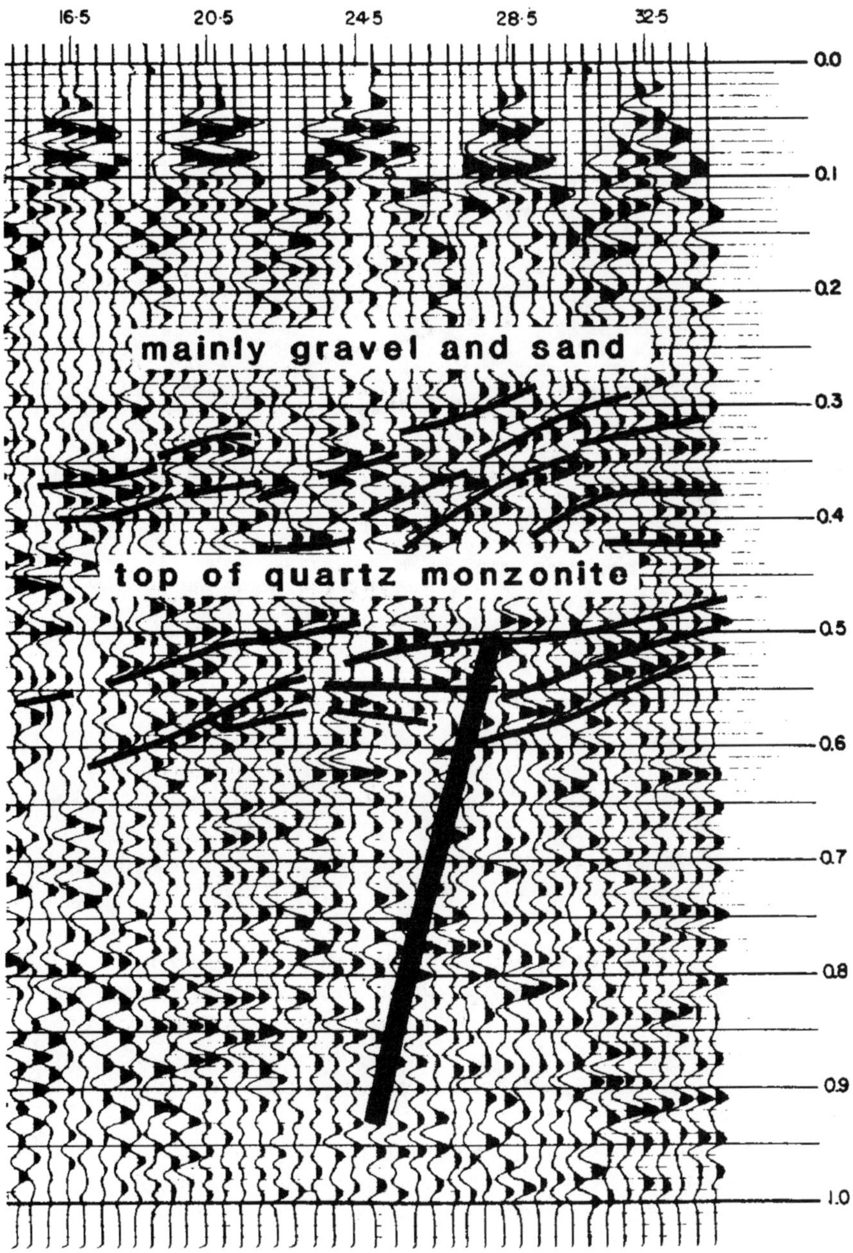

FIGURE 4.13
Reflection seismic section shows contacts ascertained from borehole data (figure 4.14) and downhole velocity measurements from Butte, Montana

Deposits Disseminated In Intrusives

Refraction seismic measurements taken in "porphyry-type" deposits indicate that the various types of alteration affect seismic velocity to different degrees. For example, silication in the form of potassium feldspar flooding results in higher velocities in all areas tested; whereas argillization, sericitization and chloritization result in lower velocities of differing magnitudes, depending on the type and intensity of alteration. Massive silicification in the form of quartz causes higher velocity. Silicification dispersed in the sericite and argillized zones does not normally offset the low velocities associated with the weaker minerals. One set of velocity measurements taken at a major "porphyry" copper mine in the western United States were situated in 1) compacted lacustrine and alluvial deposit, 2) a leached capping, 3) a typical zone of supergene enrichment, 4) relatively barren altered igneous host rock, and 5) unaltered igneous rock. The data are displayed in table 4.3.

FIGURE 4.14
Geologic section derived from seismics and checked with borehole data from (Butte, Montana)

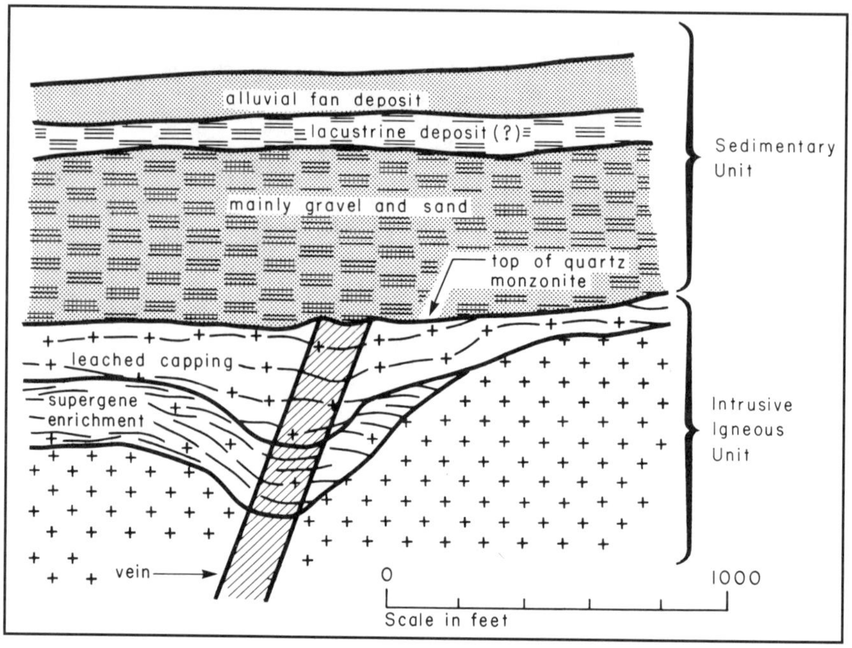

The basic velocity data were used to design and interpret further an exploration program involving the use of the reflection seismic method. The reflection work was directed to an area covered by about 300 m of alluvial deposit. By knowing the velocities of the various units, the reflection coefficients along the various contacts are predicted and the geologic significance of the seismic events are interpreted. In the case of the sections shown in figure 4.9, the main event is at the contact between the alluvium and the bedrock which consists of leached capping. In this case, the alluvium has a significantly lower velocity than the bedrock except where faults and veins emerge. Consequently, the lack of amplitude along this discontinuity is interpreted as a potential exploration target. Later, pervasive events are interpreted as contacts between the leached zone, the supergene zone and the protore.

In another open pit, "porphyry" copper mine, numerous velocity measurements were obtained in the course of a slope stability investigation. The velocities recorded pursuant to the present study were measured in firm, non-moving rock. Two main mineral zones were exposed in the pit: 1) an albite-rich zone in which oxide ore is predominant, and 2) a feldspar-flooded zone in which sulfide ore predominates. Tables 4.4, on next page, presents the velocity data recorded in the two zones.

TABLE 4.3
Seismic velocities measured in lake beds and in four geologic - mineralogic units in a major, open pit, "porphyry" copper mine

UNIT	SEISMIC VELOCITY (IN MPMS)
lacustrine deposit	1.83
leached capping	1.56 - 2.53
supergene mineral zone	1.86
quartz monzonite (altered)	1.91 - 2.16
quartz monzonite (unaltered)	3.05 - 3.51

Zone 1 - Oxide (in sericitic and argillic zone)

VELOCITY	x LENGTH	= VL
1.60	83.84	134.14
1.66	83.84	139.31
1.63	68.60	111.82
1.22	76.22	92.99
1.76	53.35	93.90
1.97	38.11	75.08
1.68	68.60	115.25
	472.56	762.49

Velocity range: 1.22 mpms to 1.97 mpms
Average velocity: VL/L = 1.61 mpms

Zone 2 - Sulfide (in quartz-sericite-feldspar zone)

VELOCITY	x LENGTH	= VL
3.05	83.84	255.71
1.85	38.11	70.50
2.15	68.60	147.49
2.19	83.84	183.61
2.09	83.84	183.61
2.09	83.84	175.23
2.22	83.84	186.12
2.13	22.87	48.71
	464.94	1067.37

Velocity range: 1.85 mpms to 3.05 mpms
Average velocity: VL/L = 2.30 mpms

TABLE 4.4
Seismic velocities measured in two types of mineralized hydrothermal zones in a "porphyry" copper mine. Stipulated in the table are the average velocities measured along the various spreads and the respective lengths along which they were measured. The length-weighted average velocity for each of the zones is stipulated at the base of each tabulation

CASE HISTORIES

Carlin-Type Deposits

The typical Carlin-type deposit is generally defined as disseminated gold mineralization present locally in a quartz-dominated, epithermal phase of a hydrothermal system emplaced in sedimentary host rocks of early-to-middle Paleozoic age. Significant volumes of disseminated pyrite, some of which contain gold, were deposited locally within the hydrothermal system. Host rock lithologies vary from carbonates to chert, siliceous shale and siltstone. Thrust faults and/or detachments have been identified at many of the mines in the district, and they are believed to be related to the formation of the ore deposits. The age of mineralization at the Carlin Mine, Eureka County, Nevada, is thought to be Oligocene by Radtke, A. S. (1985).

The orebodies are contained within a larger mass of hydrothermally altered rock. Commonly, this mass of altered rock is in the form of a stratum or lens which attains a thickness of about 120 ft at the Carlin Mine. In discussing the physical properties of the altered zones, it appears that decalcification and other forms of chemical degradation affect the rock more than the more obvious silicification that seems to pervade in these deposits. These lenses, commonly discordant with the attitudes of the host rock strata, were probably, at least in part, developed along fractures and breccias resulting from detachment or thrust faulting. Seismic reflections of anomalously large amplitude are generated at the contacts of the altered rock masses.

The elastic moduli and densities of the rock within the altered lenses have been significantly changed. The greater the difference in elastic moduli and density between the rock units, the larger the reflection originating at their contact. Consequently, anomalously large seismic reflections emanate from the upper and lower contacts of the hydrothermally altered rock with their unaltered equivalents. Radtke (1985) depicts the oxide orezone at the Carlin mine as being 120 ft thick. In this paper, the oxide zone of Radtke's is being equated with the altered lens. Figure 4.15 shows a portion of one of Radtke's sections. Figure 4.16 is a seismic time section located about 1.5 miles to the northwest and roughly parallel to Radtke's section. Figure 4.17 is an interpreted version of figure 4.16. Given a thickness of the lens equal to 120 ft and a 30-millisecond interval from the top to the base of the lense on the seismic section, the velocity within the lens is 2 X 120 ft/30 milliseconds or 8.0 fpms (ft per millisecond). This compares with velocities ranging from 14.0 fpms to 20.0 fpms measured in unaltered rock of Paleozoic age in this region. From these observations, it may be concluded that indeed one can expect anomalously high amplitude reflections emanating from the surfaces of the altered lens.

The seismic section also depicts high-angle faults and several sections of stratified events. One of the high-angle faults offsets the altered lens and appears to have normal movement. The altered lens is much less affected by the other two faults. Also shown is the contact between the Carlin formation of Pliocene age and the underlying Paleozoic formations.

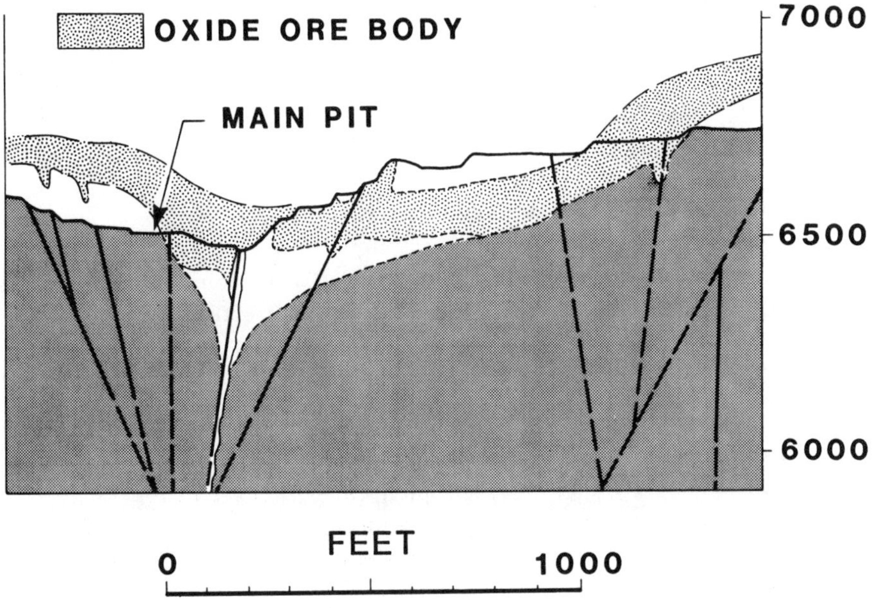

FIGURES 4.15 -4.17
Comparison of seismic sections derived over Carlin-type terrain with a geologic section at the Carlin Mine drawn by Arthur S. Radtke, 1985, in U.S.G.S. Professional Paper 1267. The dashed lines depicting an extrapolation of the pre-mining and/or pre-erosion extent of the oxide orebody were drawn by the author. The horizontal scale of the seismic sections is the same as the geological section, but the vertical scale varies according to the increase of the velocity of the seismic wave with depth. The underlying events indicate the complexity of structural features in the Paleozoic strata

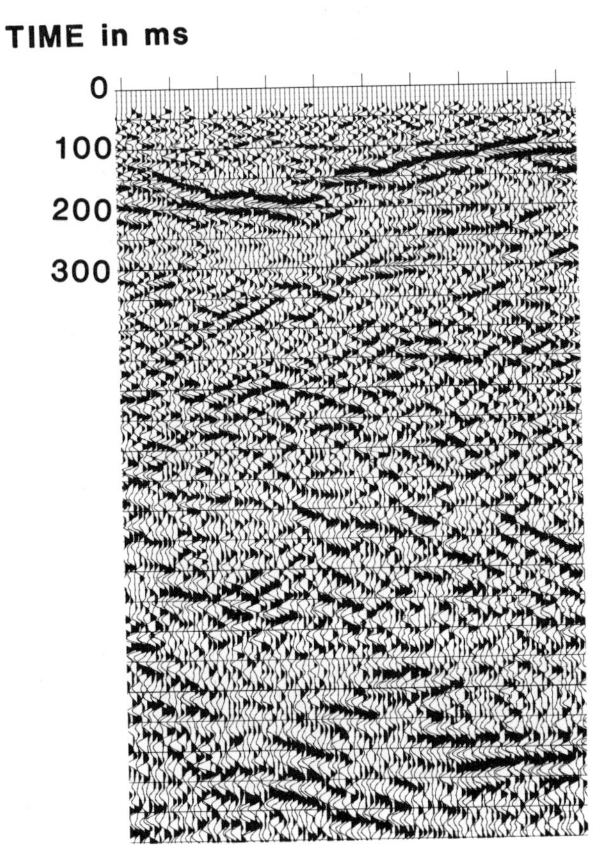

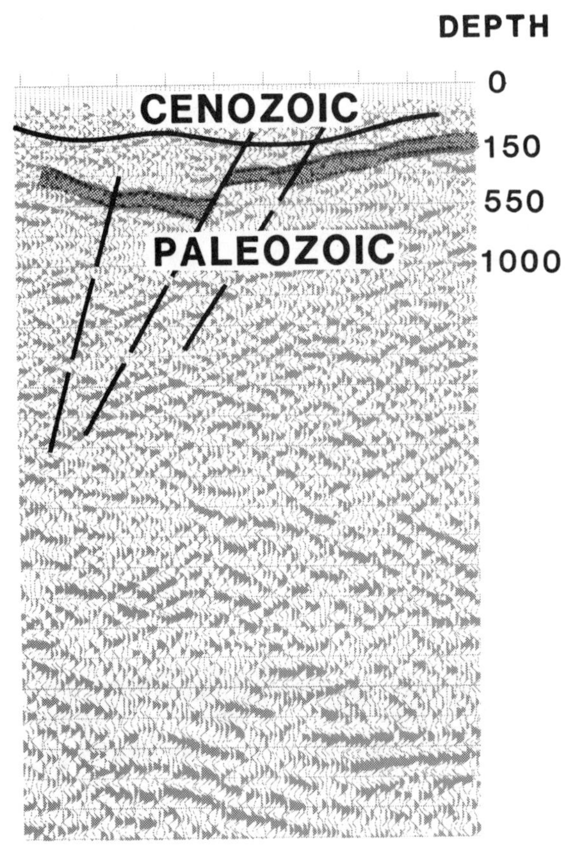

The validity of equating the large amplitude anomalies with the perimeter of the hydrothermally altered rock of Carlin-type deposits is partly substantiated in comparing the seismic sections, approximately one-half mile apart, shown in figure 4.18. The left section depicts the obvious, strong amplitude waves generated at the top and base of the lenticular mass of hydrothermally altered rock. This section also depicts a possible conduit entering the altered lens from below in the form of diminished amplitude and a downward bending of the lower event. This downward bending suggests lower velocity, which equates with less competent rock, suggesting more intense alteration. The area of low-amplitude signal immediately below the large amplitude events is a function of the AGC (automatic gain control) of seismic recording system and is not directly related to the geology. For interpretation purposes, it is useful in comparing this anomaly to others such as the section to the right.

In the case of the left section, the AGC exerted a great deal of suppression to the signals originating from the altered zone, thus the area of low amplitude immediately below the amplitude anomaly. The right section depicts events of much lesser amplitude and coherence in the same region but without the suppressed signal below the events. These lesser events originate along strata in proximity to an unmineralized portion of the thrust fault. No drilling data have been released at the time this paper was written, however, strong circumstantial evidence indicates the above observations to be valid.

FIGURE 4.18
Comparison of seismic section depicting a Carlin-type deposit and a section in which only the thrust fault zone is defined. The sections are about one-half mile apart

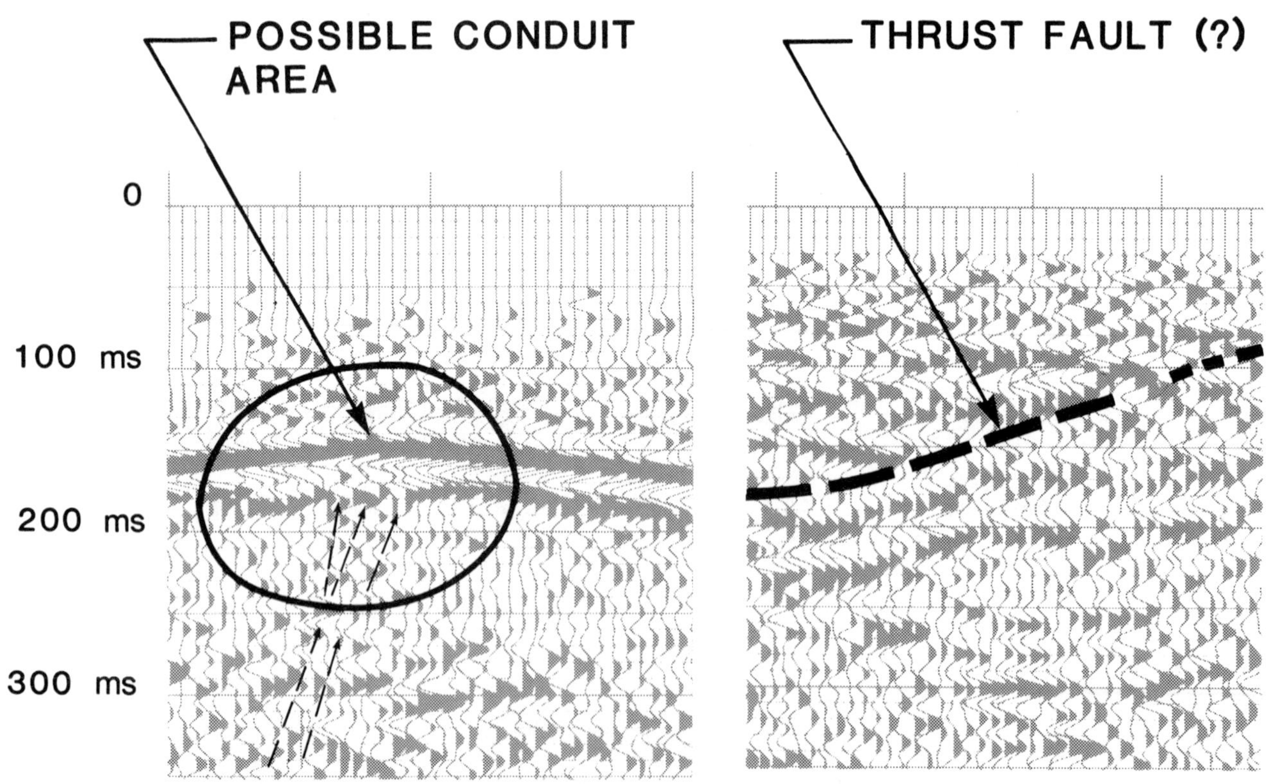

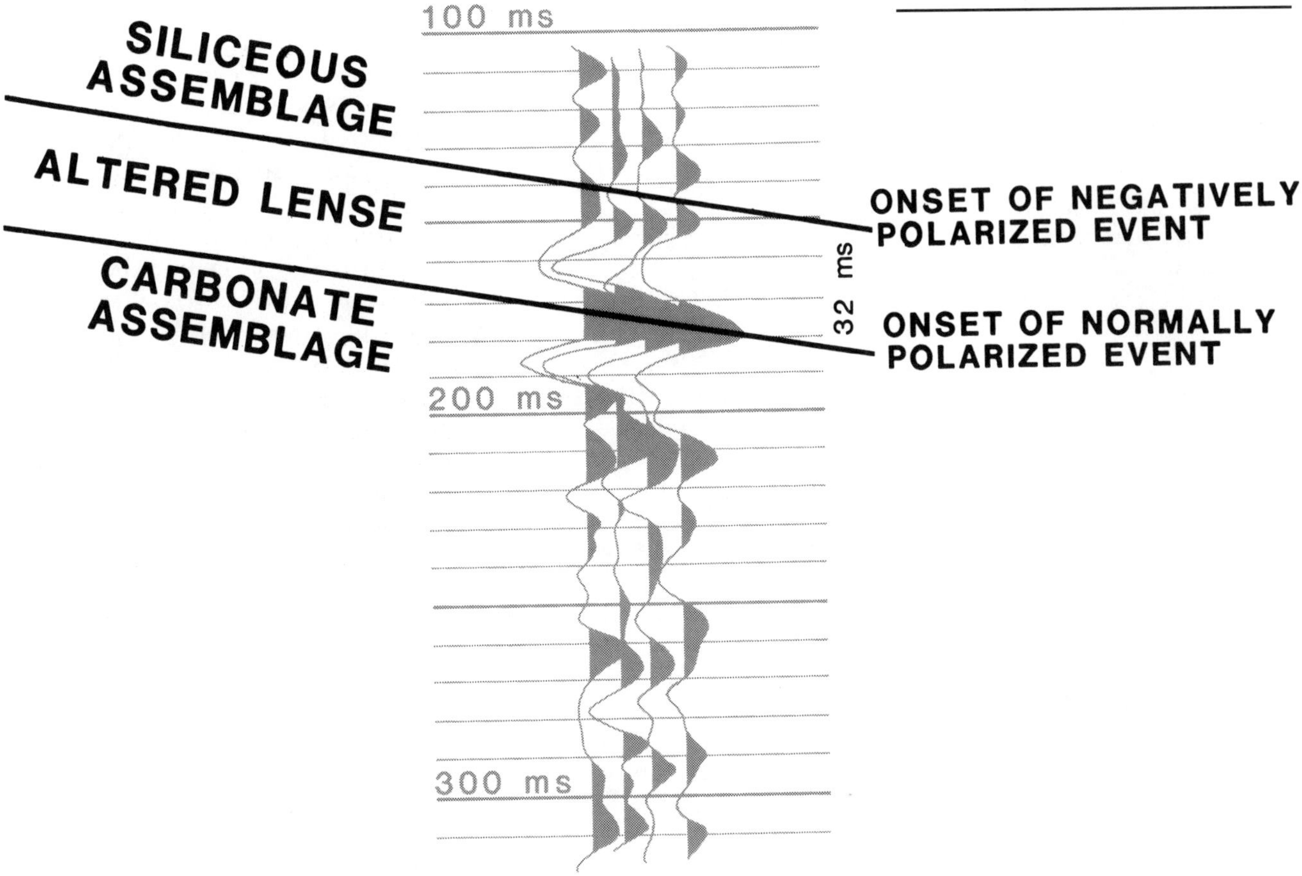

FIGURE 4.19
A typical sequence of four traces on a reflection seismic section. The seismic signature which denotes a Carlin-type deposit is shown

Volcanic-Hosted Epithermal Deposit At Round Mountain, Nevada

In the area of the seismic survey, aside from some locally present thin alluvium, the surficial rock unit consists of a densely welded, rhyolitic tuff. This unit is from 200 to more than 500 ft thick in the survey area and is the host rock for the less extensive "type 1" orebodies. The type 1 orebodies are present mainly along steeply dipping northwest striking fracture fillings and along hydrothermally dilated joints (Tingley and Berger, 1985). Also present in the densely welded tuff are the first prospected veins of the district, some of which dip at less than 35 degrees. Seismic events in the densely welded unit are few and weak.

Below the densely welded tuff, a poorly welded tuff unit has been described as "unwelded tuff" from drill hole samples below the existing mine workings. It is this unit that provides seismic reflection events of anomalously large amplitude and hosts the extensive type 2 orebodies. It is thought that the densely welded unit provided a cap or a trapping condition for the rising hydrothermal solutions and vapors. The poorly welded unit possessed strata with some permeability through which the hydrothermal solutions infiltrated laterally some distance from the source conduits. Thus was the probable "unwelding" of the poorly welded tuff and the formation of the extensive type 2 orebodies (Tingley and Berger, 1985, and Sander, 1988). The type 1 orebodies were formed by what could escape into and through the densely welded trap.

Intuitively, one would predict that the differences in elastic moduli and density between the densely welded and poorly welded tuffs would tend to be amplified by the process of hydrothermal alteration. Because it is doubtful whether any portions of the seismic lines were placed over an unaltered area of the poorly welded tuff, the amount of amplification resulting from the hydrothermal process cannot be assessed. Given the knowledge of the large difference in moduli and density, however, the amplitude of the events originating from the unwelded tuff would be expected to be anomalously large as compared to events from other portions of the stratified volcanic and sedimentary section.

One of the reflection seismic sections acquired at the Round Mountain Mine during 1983 is shown in figure 4.20. Also included is a reproduction of one of Sander's (1988) sections derived from drill hole data in the same area, but the section is not coincident with the seismic section because the two sections are not parallel. Note the similarity of the thickness and configuration of the densely welded and poorly welded (unwelded) units, the correlation of the faults and the location of the paleoridge.

More importantly, the anomalous, strong reflections associated with the unwelded tuff substantiate the prediction that the hydrothermal solutions would attack the poorly welded tuff to a much greater degree than the densely welded unit. This resulted in diminishing the elastic moduli and density in the poorly welded tuff. Consequently, large amplitude reflections are generated at the contact with the altered tuff. The presence of several events instead of only one event at the top and one event at the base might indicate the presence of competent, perhaps siliceous, beds within the unwelded unit.

Another interesting feature appears in the seismic sections. Detailed correlation of the seismic sections with Sander's (1988) sections shows a degradation of signal where Sander denotes the more intense gold mineralization. The main gold mineralization is associated with this late hydrothermal event which perhaps is a later hydrothermal event than the second-stage potassic alteration described by Sander. The signal degradation consists of phase shifting and some diminished amplitude. This phenomenon is shown in figure 4.21 where the seismic section and the geologic section are parallel and only 150 ft apart. This is another example of the change in seismic signal as a function of the change in elastic moduli and density.

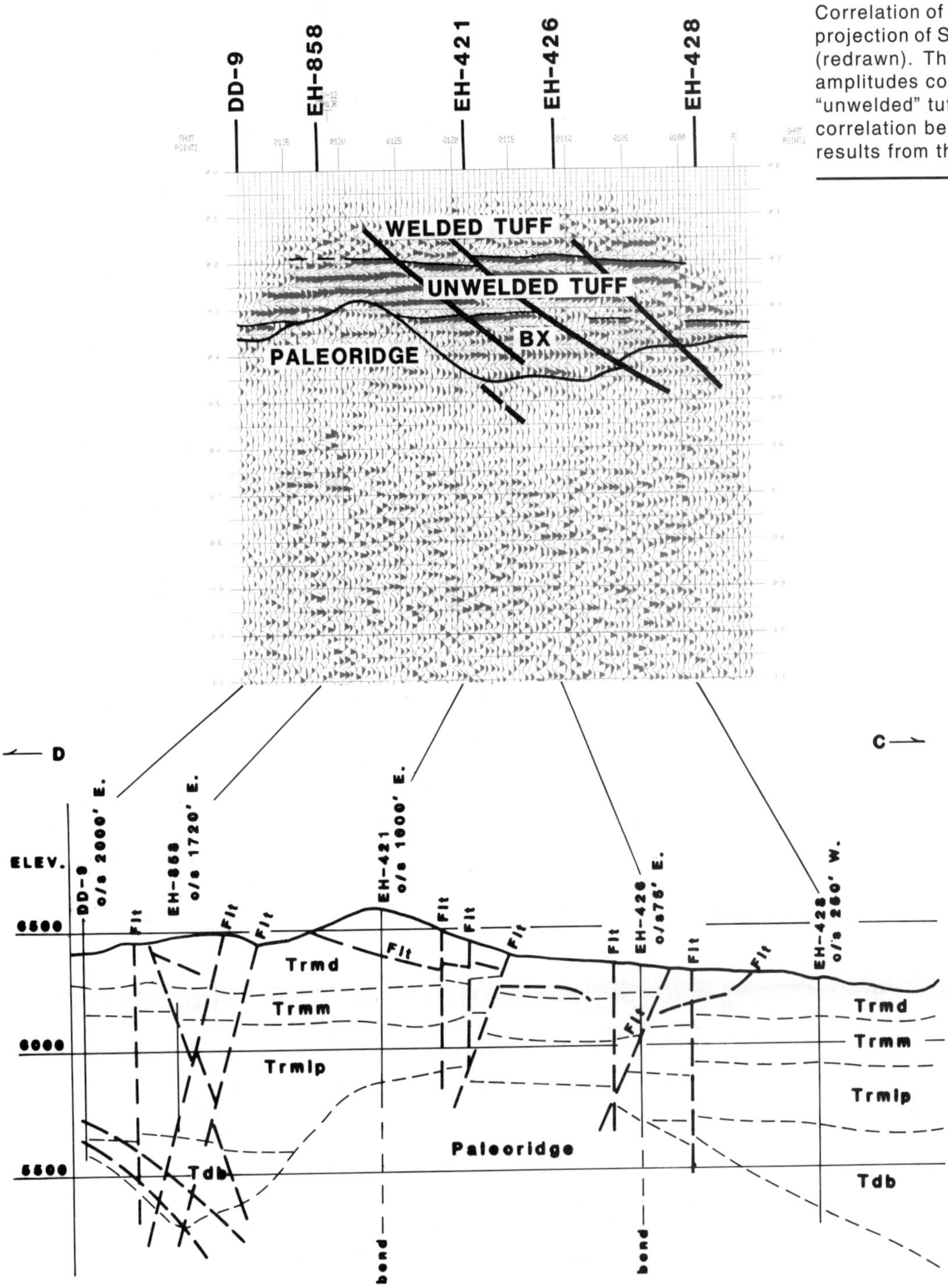

FIGURE 4.20
Correlation of seismic section with projection of Sander's (1988) section (redrawn). The anomalously large amplitudes coincide with the "unwelded" tuff. The lack of exact correlation between the sections results from their not being parallel

FIGURE 4.21
Signal degradation within the "unwelded" tuff unit roughly coincides with orebodies on Sander's (1988) section, which is parallel to, and about 150 ft east of the seismic section

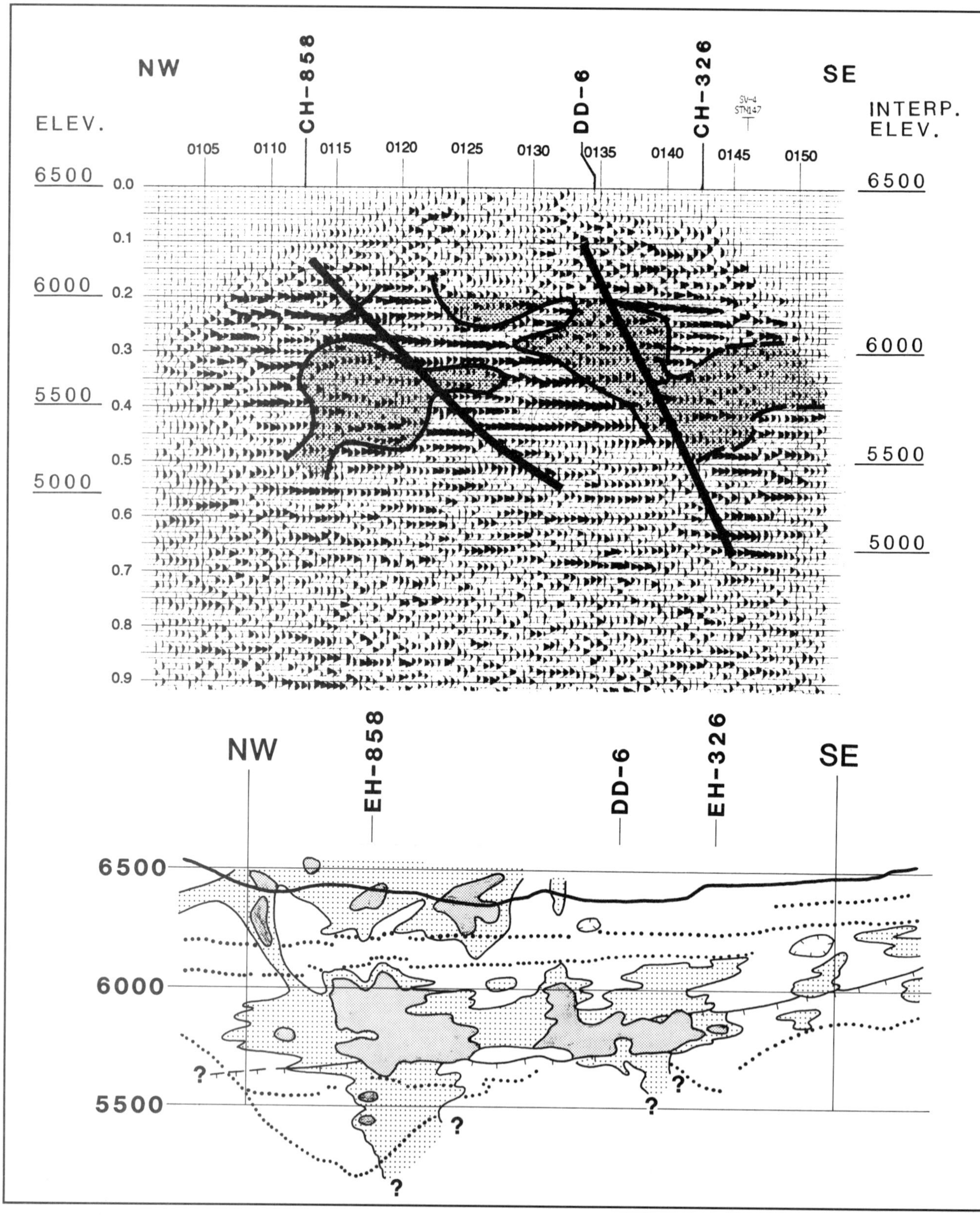

Velocity measurements were made in four drill holes at the mine site. Drill hole 447 yielded measurements on 1) the densely welded tuff, 2) unwelded tuff, 3) breccia, and 4) pre-Tertiary igneous or sedimentary rock. Hole 485 logged 1) alluvium and 2) densely welded tuff. Hole 503 logged only the welded tuff. Hole 558 logged 1) welded tuff and 2) unwelded tuff. Generally, 5.0 fpms velocities predominate in both the densely welded an unwelded tuffs, but some strata at the base of the densely welded tuff possess velocities ranging from 6.3 to 12.5 fpms, averaging about 10.0 fpms.

Figure 4.22 depicts the velocity logs of the holes, using the densely welded tuff/unwelded tuff contact, as interpreted from seismic velocities, as datum.

FIGURE 4.22
Compilation of borehole velocity measurements versus interpretation of lithology from drill cores or cuttings. The datum line is drawn at what the seismic velocity measurements indicate is the contact at the base of the densely welded tuff unit. The competent unit in DD 503 might have been significantly affected by a fault (see map). Note the velocity inversion in the "unwelded" tuff

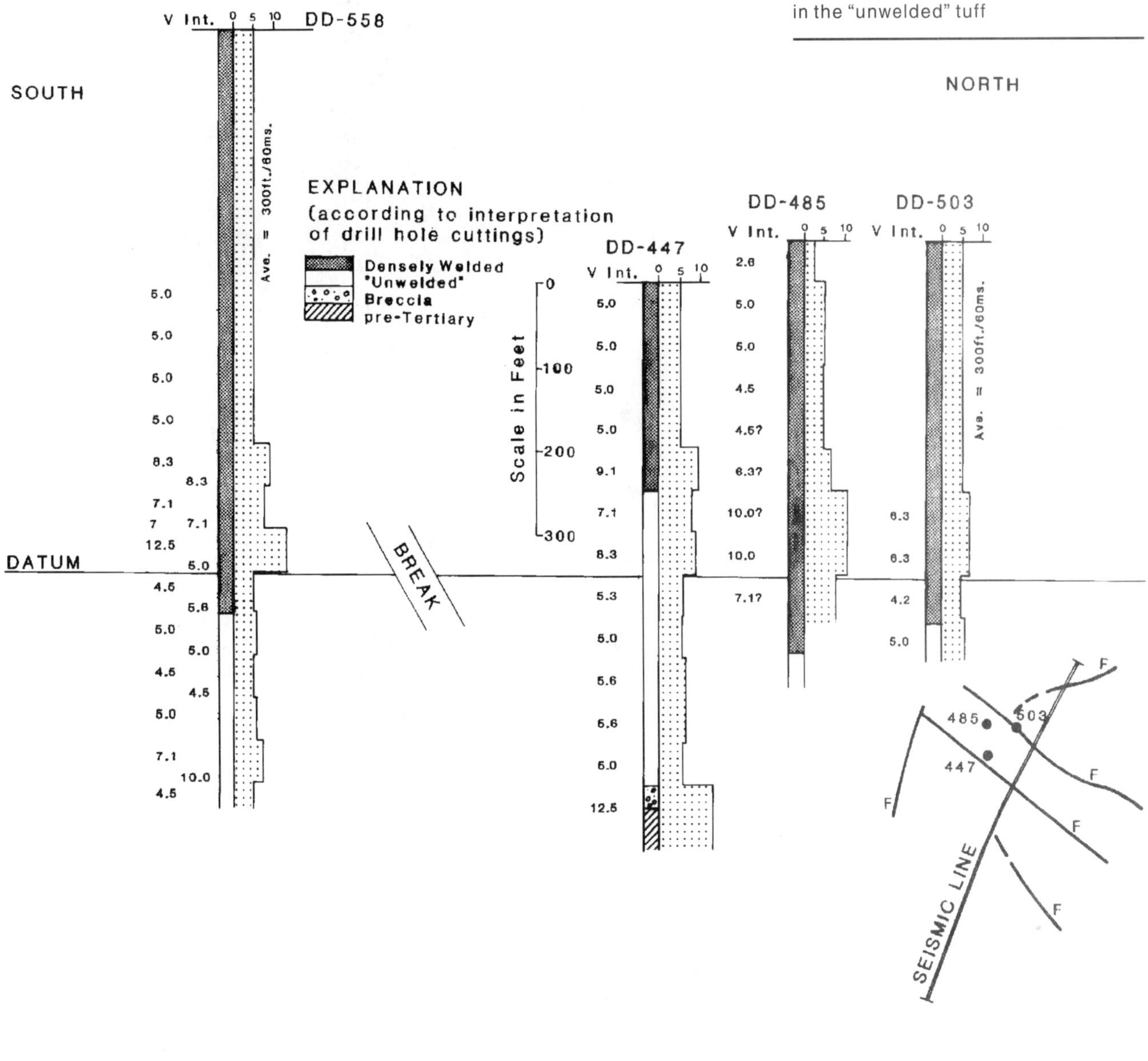

Basin And Range Fault Study

On the seismic section depicted in figure 4.23, strong events are dipping from the range front at 45 degrees. These events are reflections off a typical basin and range fault. Because of the "least time path" principle governing the recording of seismic events, this sloping event must be migrated if one is to observe its true position in both the seismic "time section" and in a true depth, geologic section. The process used in migrating steep events in mining geologic environments is stipulated in the paragraphs below.

Fault Migration Problem

Figure 4.23, seismic section depicting fault migration problem, reference points on fault:

STATION	TIME	VEL/2*	DEPTH(pseudo)
135	325	10.400/2	1690 ft.
125	475	10.765/2	2554 ft.

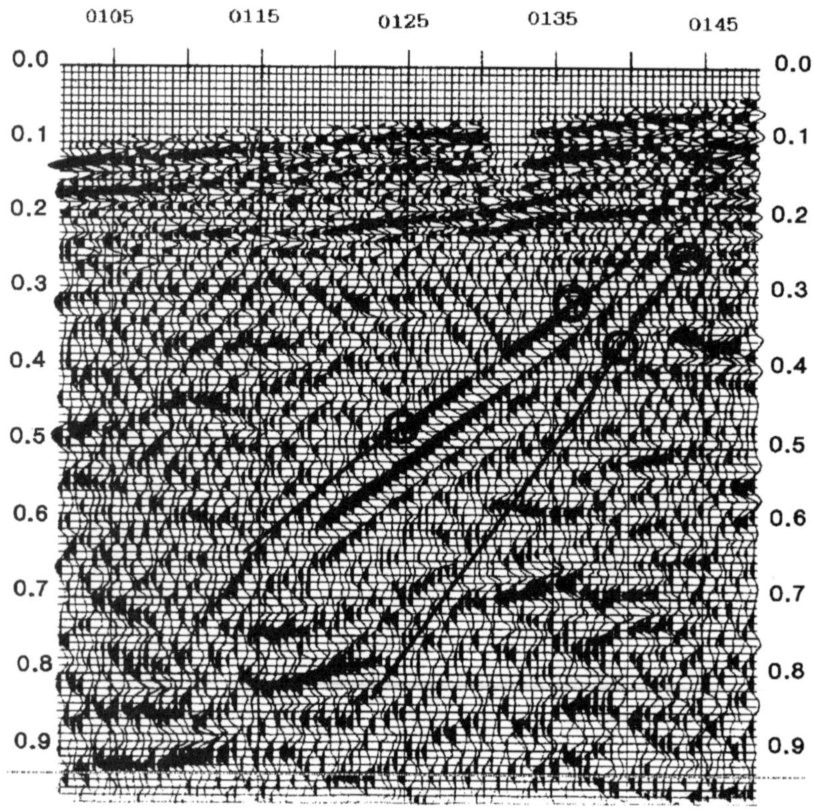

FIGURE 4.23
This is an example of fault migration. This seismic section depicts a fault event which dips 45 degrees and the true position of the same fault. Notice the steeper attitude of the "true fault" and the obvious displacement of events along the true structure

Plot points 1 and 2 vertically below their appropriate stations.

Then plot event by drawing line through points 1 and 2. This is an unmigrated event which has erroneous dip and position.

Draw lines through points 135 and 125 which are perpendicular to the pseudo-dip line which passes through points 1 and 2.

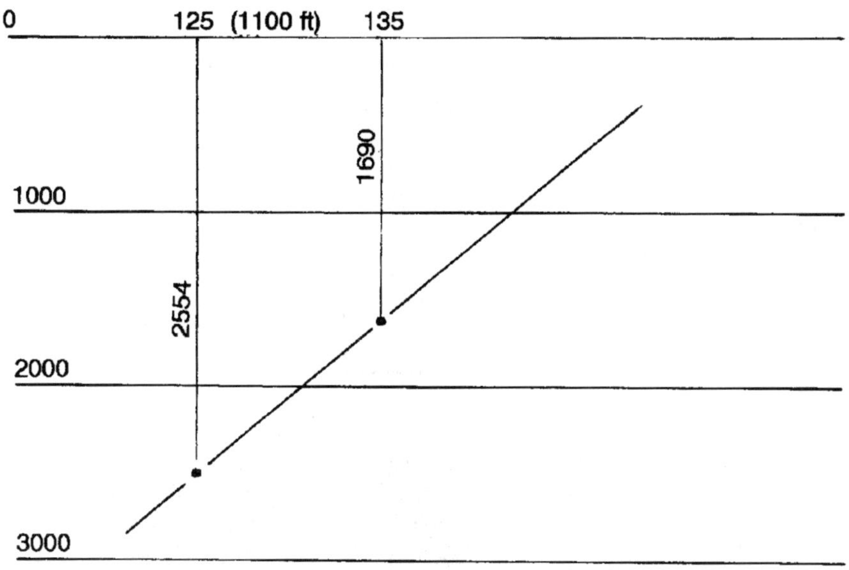

FIGURE 4.24
Illustration for step 1 of the fault migration computations

Plot points 3 and 4 by measuring from points 135 and 125, their respective pseudo-depths along the newly constructed lines which are perpendicular to the pseudo-dip line.

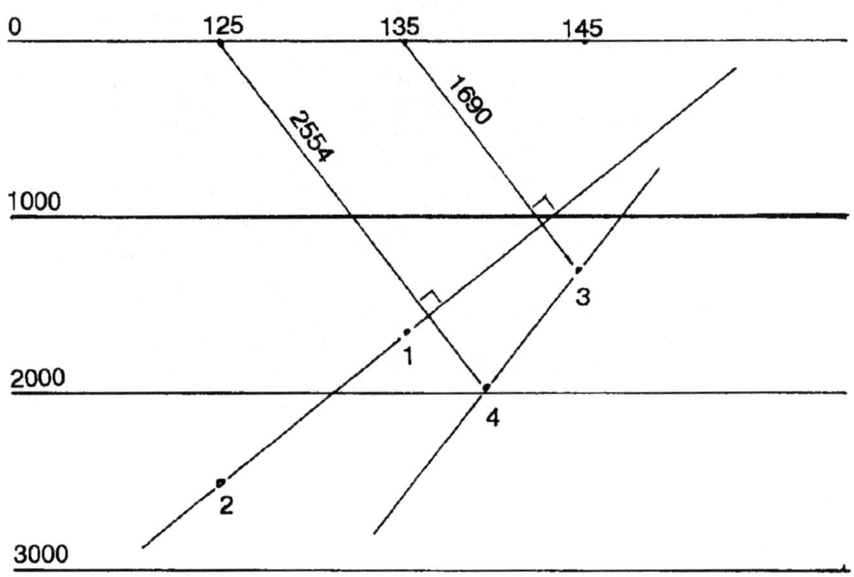

FIGURE 4.25
Illustration for step 2 of the fault migration computations

The line passing through points 3 and 4 represents the true dip of the seismic event which includes points 1 and 2. Points 3 and 4 are the true positions of points 1 and 2.

Carefully scale the depth of points 3 and 4.

Compute times of migrated points 3 and 4.

Point 3 $T = 1280 \times 2 / 10.4 = 246$ ms.

Point 4 $T = 2000 \times 2 / 10.756 = 372$ ms.

Plot points under appropriate stations to see true dip and position in seismic time section. See illustration on page 304.

FIGURE 4.26
Illustration for step 3 of the fault migration computations

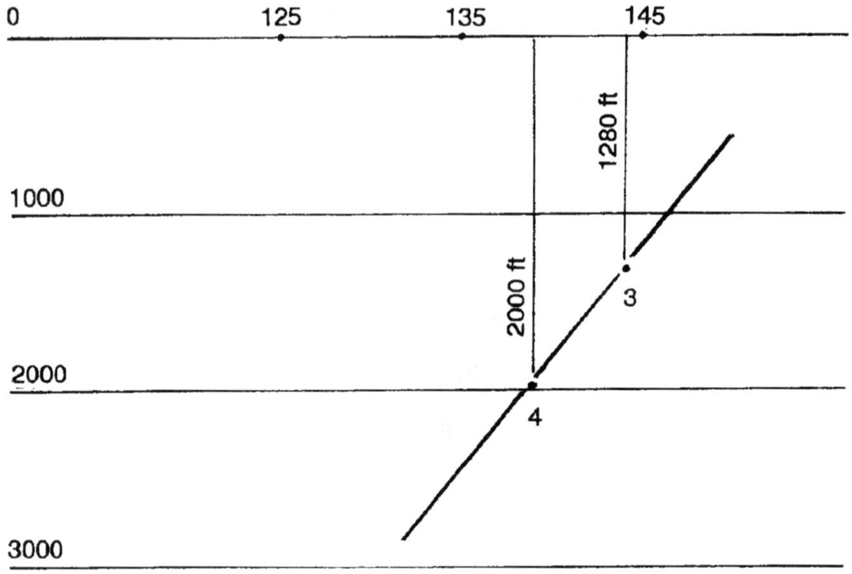

Vein Deposit At Escalante Mine

Escalante typifies a deposit where the orebody possesses higher seismic velocity than the host rock, thus being a positive velocity anomaly. The "bedrock" unit consists of a volcanic sequence of pyroclastic and volcaniclastic beds of siliceous-to-intermediate composition of Tertiary age. Emplaced within this sequence are silica-dominated, mineralized vein structures. The currently mined vein consists of a northeasterly striking vein structure which forms a low ridge. To the north of the mined area, the vein(s) is covered by valley fill to depths in excess of 200 ft in some areas of the survey site.

The volcanic unit is covered by poor-to well-consolidated valley fill. The valley fill constitutes a low-velocity layer providing a large velocity contrast with the underlying higher velocity units. This results in good resolution of the surface of the volcanic unit.

It was judged that a refraction seismic program would define the top of the volcanic host rock and also show the ridge associated with the vein under the alluvial cover. The survey accomplished both and also depicted the quartz vein(s) as zones of significantly higher velocity and post-mineral faults as zones of low velocity.

Structure

Structure in the study area can be divided into pre-mineral and post-mineral faulting. Pre-mineral faulting consists of faulting prior to the formation of vein structures and consists of subparallel northeast striking faults. Post-mineral, northerly striking, fault structures have offset the mineralized zones. These are apparently downthrown to the east, related to basin and range tectonism and produce apparent lateral offsets of the veins. These faults normally constitute low-velocity anomalies and might be differentiated from the higher velocity volcanic unit on the basis of velocity contrast.

Criteria Used In Interpretation of Vein And Fault Structures
Several criteria were considered in the interpretation of the seismic data to provide confirming information in addition to the velocity data depicted on the seismic sections. A list of the criteria follows:

- Velocity has been used to differentiate the higher velocity, volcanic unit from overlying the lower velocity alluvium and to identify irregularities such as veins and faults within the volcanic unit. The veins are represented by higher velocities and the faults are shown as lower velocity anomalies.

- Amplitude variation provided secondary evidence to delineate vein locations. High-or low-velocity anomaly-structures can produce amplitude attenuation of seismic waves as they pass through the accompanying zone of velocity contrast. Amplitude data were plotted on a line location map and were collated to produce zones of both amplitude attenuation and amplitude increase.

- The direction of wave propagation in which amplitude variation takes place might also be used to identify dipping veins or faults. Veins might produce attenuation in one direction of seismic wave propagation, but little or no attenuation, or even an apparent increase, in the other.

- Morphology of the surface of the Tertiary volcanic units was used to help identify veins, since the erosion of a predominantly siliceous vein structure could be expected to produce an erosional remnant of higher relief. This remnant is seen in the depth profile derived from the seismic data as a low ridge in the volcanic unit coincident with a high-velocity zone.

These criteria have been used to delineate possible vein and fault locations. Assignment of vein location confidence was determined on the basis of the above confirming criteria. Areas of high probability for vein location were assigned on the basis of matching very high-velocity zones (greater than 13.00 fpms) and amplitude attenuation related to that event. If the presence of amplitude attenuation was uncertain, horizon morphology was used as the confirming evidence. Concurrence of all three of these should be considered good indication of the existence of vein structures.

Zones of moderate-to-low possibility of vein location were determined by correlating moderately high velocity (10,000 fpms to 12,500 fpms) with confirming data from the amplitude analysis. A pattern of matching attenuation and amplitude increase corresponding to each of the high-velocity (above 13.2 fpms) vein locations was noted in the analysis. This pattern conformed to only one moderately high-velocity zone. In general, zones of moderate probability might consist of scattered horse-tail or other type of subsidiary veins causing a higher average velocity.

Also, this type of zone might correlate with one or more veins of submeasurable size. It has been roughly calculated that a vein, 15 ft in width and a velocity of 13.2 fpms, in a volcanic country rock of

FIGURES 4.27
These sections show geology interpreted separately along each seismic line, and collectively on a map

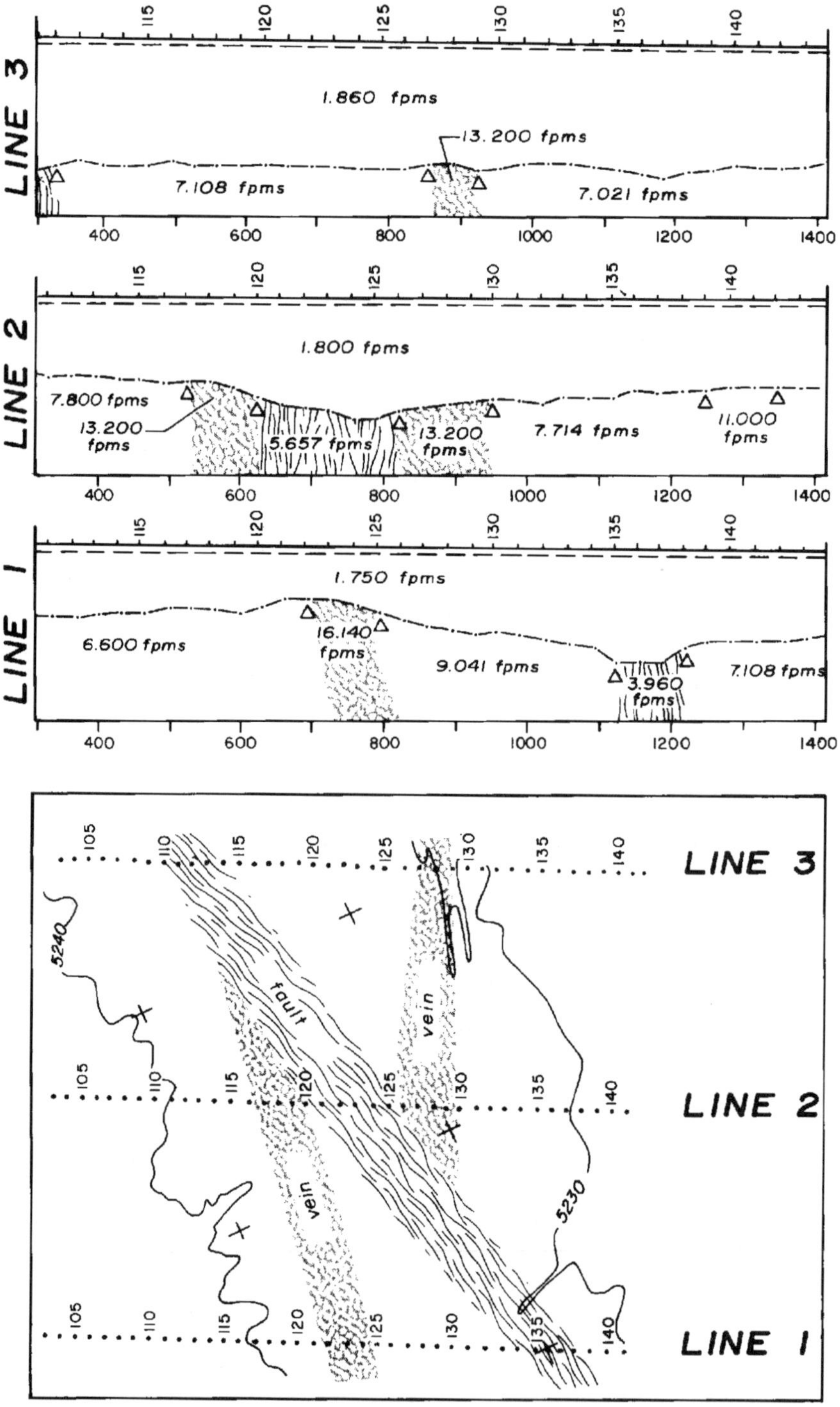

FIGURE 4.28
This reflection section depicts up-warping of events along a known quartz vein at station 370. Also shown are a near vertical zone of incoherent signal at station 410 and one of several faults

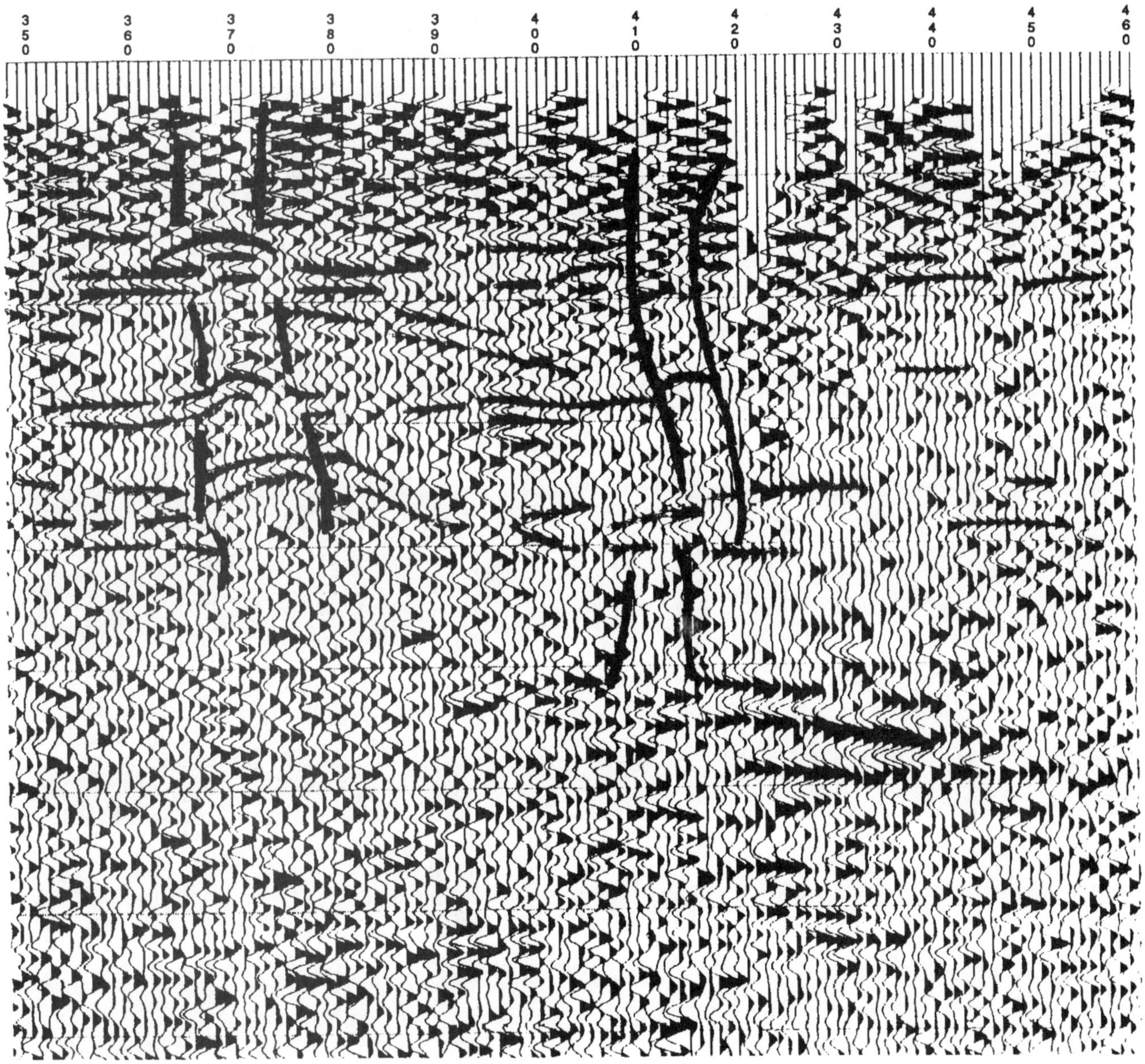

8.0 fpms might not be detectable. Either an increase in vein width or velocity contrast would increase the probability of detection. It should be noted, however, that the zone of silicification which encases the vein adds greatly to the effective detectable width of the vein.

Reflection Seismic Study

The most notable features in the reflection seismic sections derived at Escalante are where up-warping of seismic events appear in proximity to known quartz veins. Given that the higher velocity quartz veins and their zones of associated silica flooding result in higher seismic velocity, the up-warping is quite predictable.

REFERENCES

American Geologic Institute, 1976, Dictionary of geologic terms; Garden City, Anchor Press.

Barnes, H.L., ed,. 1979, Geochemistry of hydrothermal ore deposits; New York, Chichester, Brisbane, Toronto, John Wiley & Sons.

Bateman, A.M., 1951, Economic mineral deposits; New York, John Wiley & Sons.

Crice, D. B., 1988, Practical Geophysics for the Exploration Geologist, Northwest Mining Association, Spokane, Washington.

Dobrin, M.B., 1960, Introduction to geophysical prospecting, New York-Toronto-London; McGraw-Hill.

Heiland, C.A., 1963, Geophysical exploration; New York and London, Hafner Publishing Company.

Holz, P.E., and Willden, R., 1964, Geology of the Osgood Mountains Quadrangle, Humbolt Co., Nevada, U.S. Geologic Survey professional paper 431, 127 p., maps.

Jakoski, J.J., 1961, Exploration geophysics; Newport Beach, California, Trija Publishing Company.

Nettleton, L.L., 1940, Geophysical prospecting for oil; New York and London, McGraw-Hill.

Radtke, A.S., 1985, Geology of the Carlin gold deposit: U.S. Geologic Survey professional paper 1267, 124 pp.

Sheriff, R.E., 1973, Encyclopedia dictionary of exploration geophysics; Tulsa, Oklahoma, S.E.G.

Sheriff, R.E., 1980, Seismic Stratigraphy: International Human Resources Devel., Corp., Boston, 227 p.

Turner, F.J., and Verhoogen, J., 1960, Igneous and metamorphic petrology; New York-Toronto-London, McGraw-Hill.

Weast, R.C., ed., 1973, Handbook of chemistry and physics; Cleveland, Ohio, CRC Press.

CHAPTER 5
MAGNETICS

APPLICATIONS FOR PORTABLE MAGNETOMETERS

Magnetometers

Instrument use

Common portable magnetometers used today are fluxgate, proton precession, Schmidt field balance, dip needle, and other special purpose instruments. Field balances and dip needles are mechanical devices comprised of pivoted magnets measuring vertical or horizontal intensity or field direction; they are being replaced by the more sensitive, less cumbersome fluxgate and proton magnetometers. Portable fluxgate magnetometers employ a saturable core sensor held in a vertical direction to measure vertical intensity with an effective sensitivity on the order of several gammas. Fluxgate magnetometers, too, are slowly being replaced by the proton magnetometer which has greater sensitivity (1 gamma or better), absolute accuracy, no moving parts, and measures total field intensity with freedom from orientation errors. For reasons of its increasing use and because many applications require these features, the proton magnetometer will be the principal instrument under discussion in this chapter. Much of the chapter, nevertheless, applies to vertical component fluxgate magnetometers as well. Anomaly signatures at high latitudes (magnetic dip 70° or greater) are practically identical for the two instruments; at other latitudes they differ significantly.

Sheldon Breiner is president of EG&G GeoMetrics, Inc., of Sunnyvale, California, a company which he founded in 1969 to manufacture geophysical instruments and provide airborne survey services for oil and mineral exploration. Dr. Breiner received his B.S., M.S., and Ph.D. degrees from Stanford University.

Sheldon Breiner, Ph.D.
GeoMetrics, Inc.

Proton magnetometer

The proton precession magnetometer is so named because it uses the precession of spinning protons or nuclei of the hydrogen atom in a sample of hydrocarbon fluid to measure the total magnetic intensity. The spinning protons in a sample of water, kerosene, or alcohol behave as small, spinning magnetic dipoles. These magnets are temporarily aligned or polarized by application of a uniform magnetic field generated by a current in a coil of wire. When the current is removed, the spin of the protons causes them to precess about the direction of the ambient or earth's magnetic field, much as a spinning top precesses about the gravity field. The precessing protons then generate a small signal in the same coil used to polarize them, a signal whose frequency is precisely proportional to the total magnetic field intensity and independent of the orientation of the coil, i.e., sensor of the magnetometer. The proportionality constant which relates frequency to field intensity is a well known atomic constant: the gyromagnetic ratio of the proton. The precession frequency, typically 2,000 Hz, is measured by modern digital counters as the absolute value of the total magnetic field intensity with an accuracy of 1 gamma, and in special cases 0.1 gamma, in the earth's field of approximately 50,000 gammas.

Total field measurement

Total magnetic field intensity, as measured by a proton magnetometer, is a scalar measurement, or simply the magnitude of the earth's field vector independent of its direction. The measurement can be expressed as in figure 5.1 as simply the length of the earth's field vector, F, shown here to be 50,000 gammas. A local perturbation, T, of 10 gammas, as might be measured in any of the applications discussed herein, is shown in figure 5.2 as a vector of arbitrary direction. This disturbance vector adds to the undisturbed field in the usual manner of vector addition as shown in figure 5.2. Notice how the figure would actually appear if both the 50,000 and 10 gamma vectors were drawn to scale. It is clear from the figure, then, that since the proton magnetometer measures only the *magnitude* of the resultant vector whose direction is almost exactly parallel to the undisturbed total field vector, that which is measured is very nearly the component of the disturbance vector in the direction of the original undisturbed total field, or where

FIGURE 5.1
Total field

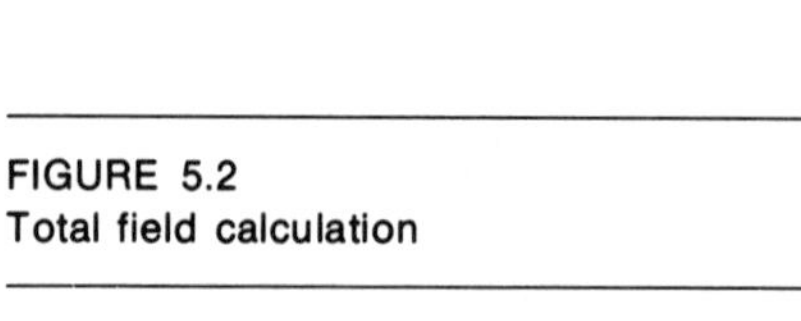

$$|\vec{F} + \vec{T}| \approx F + comp_F T$$

where

$$|\vec{F}| \gg |\vec{T}|.$$

Such conditions are usually valid except in the near field of large steel objects or in the vicinity of iron ore deposits or certain ultrabasic rocks which produce anomalies larger than 10,000 gammas. Thus, the change in total field, $F = comp_F T$, i.e., the component of the anomalous field, T, in the direction of F. (Except where noted, $comp_F T$ will be referred to simply as the anomaly T.) The proton precession magnetometer, for small perturbations, can therefore be considered to be an *earth's-field-determined component magnetometer.*

FIGURE 5.2
Total field calculation

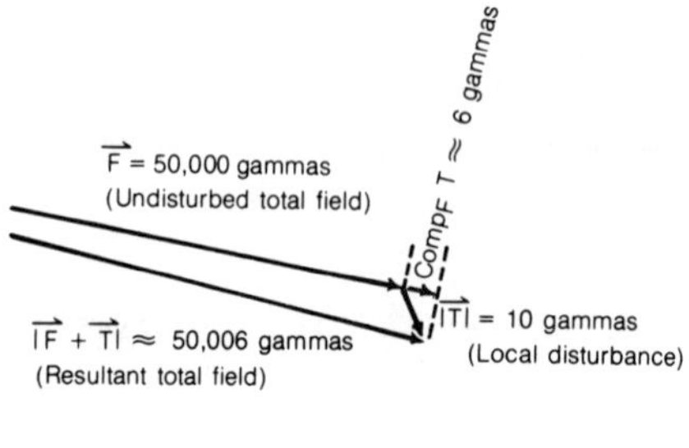

This property of measuring the scalar magnitude of the field, otherwise called total field intensity, is very significant with respect to the asymmetric signatures of anomalies, interpretation of anomalies, and in various special applications. Furthermore, the fact that what is measured is independent of the orientation of the sensor allows the magnetometer to be operated without attention to orientation or leveling which would be necessary with a fluxgate magnetometer on the mobile platform of a person, vehicle, or aircraft. The only limitation of such a scalar measurement, albeit a minor one, is that the component of the anomalous field which is measured is not normally under the control of the observer, but determined by the local direction of the earth's magnetic field.

Limitations of a proton magnetometer

The proton magnetometer has no moving parts, produces an absolute and relatively high resolution measurement of the field, and usually displays the measurement in the form of an unambiguous digital lighted readout. Several operational restrictions exist, however, which may be of concern under special field conditions. First, the proton precession signal is sharply degraded in the presence of a large magnetic field gradient greater than 200 gammas per ft (approximately 600 gammas per m). Also, the signal amplitude from the sensor is on the order of microvolts and must be measured to an accuracy of 0.04 Hz of the precession frequency of several thousand Hz. This small signal can be rendered immeasurable by the effects of nearby alternating current electrical power sources. For these two reasons, a *proton magnetometer cannot usually be operated within the confines of a typical building.* Developments and procedures are presented which minimize these effects for the applications to be described in this chapter.

Earth's field magnetism

The earth's magnetic field resembles the field of a large bar magnet near its center or the field of a uniformly magnetized sphere. Not well understood, the origin of the field is thought to be caused by currents in a fluid conductive core. On the surface of the earth, the pole of this equivalent bar magnet nearest the north geographical pole is actually a 'south' magnetic pole. This paradoxical situation exists since, by convention, a north-seeking end of a compass needle is defined as pointing north yet must point to a pole of opposite sense, or south pole, of the earth's magnetic field. To avoid possible confusion, the magnetic pole near the geographical north pole is referred to as a 'north' pole.

The field, or flux, lines of the earth exhibit the pattern common to a small magnet as shown in figure 5.3. Note that the direction of the field is vertical at the north and south magnetic poles, and horizontal at the magnetic equator. An understanding of this geometry is important with respect to interpretation of magnetic anomalies. The intensity of the field, a function of the density of the 'flux lines' shown in figure 5.3, again behaves as a bar magnet. It is twice as large in the polar region as in the equatorial region, or approximately 60,000 gammas and 30,000 gammas respectively. The inclination from the horizontal and the total intensity are shown in figures 5.4 and 5.5. (NOTE: 1 gamma = 10^{-5} gauss. Gauss is actually a unit of magnetic induction and oersted a unit of magnetic intensity—B and H respectively, in physics nomenclature. By convention in the geophysical community, however, gauss is the unit in cgs of magnetic intensity. Numerically, 1 gamma = 10^{-5} gauss = 10^{-5} oersted = 10^{-9} webers/M^2 = 10^{-9} tesla.)

The earth's total field intensity is not perfectly symmetric about the geographical north pole. First, the north magnetic pole is in northern Canada more than 1,000 miles from the geographical pole (figure 5.3). Also, the earth cannot exactly be represented by a single bar magnet, but has numerous higher order poles and very large-scale anomalous features owing to unknown characteristics of the generating mechanism in the earth's core. In addition, the solar wind, or constant flux of particles and electric currents from the sun, distorts the field lines from what is shown in figure 5.3 to a more or less tear-drop shape with the blunt end towards the sun. The last, but for the purposes of this chapter, most relevant deviation from a symmetric field, is the anomalous set of

FIGURE 5.3
Flux line density

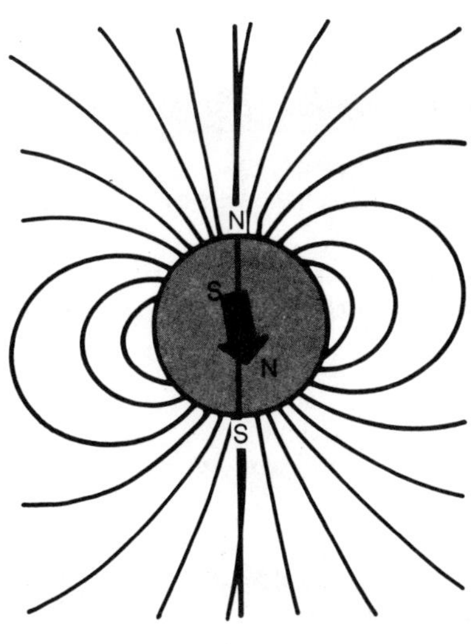

features in the earth's crust caused by local variations in the magnetic minerals or other features of interest which distort the local earth's magnetic field.

FIGURE 5.4
Geomagnetic inclination from the horizontal

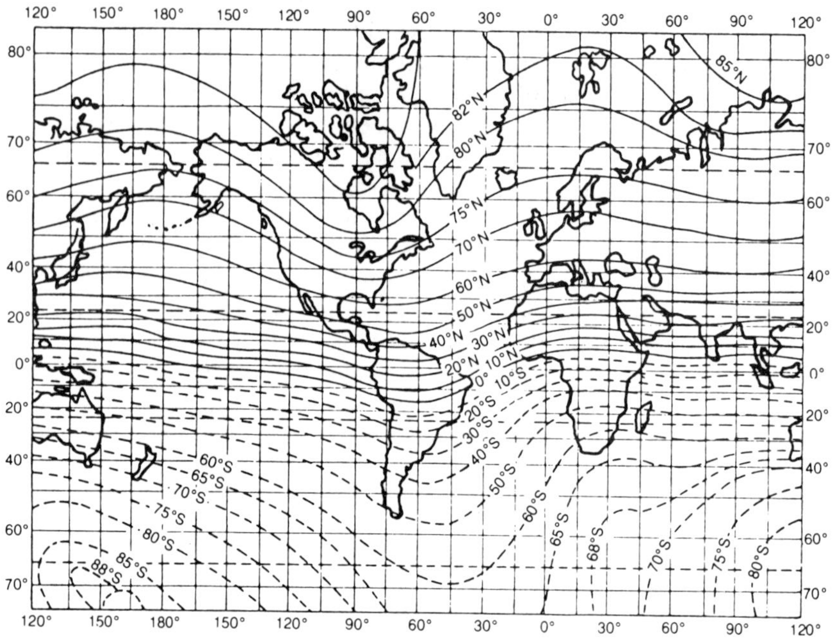

FIGURE 5.5
Total intensity of the Earth's magnetic field

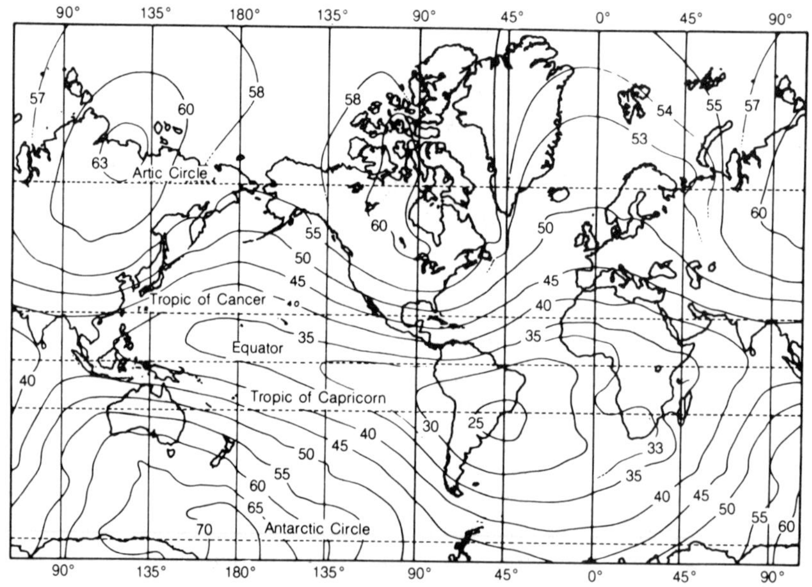

Expressed in kilogammas

FIGURE 5.6
Typical diurnal variations in total field intensity

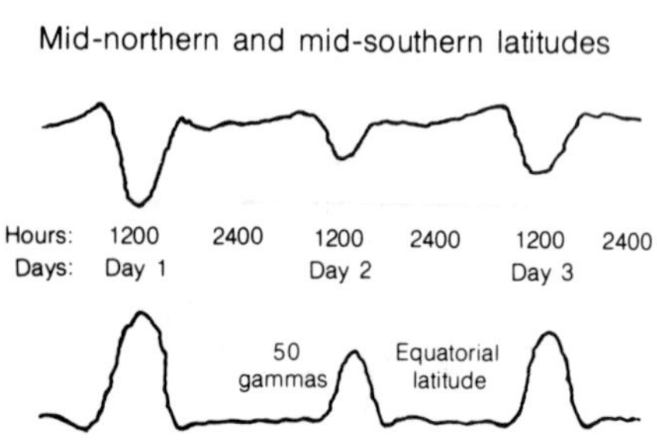

Time variations

The variations described all refer to the spatial variations in the earth's magnetic field, but there are variations in time as well. Significant time variations with periods of seconds, minutes, and hours are the direct or indirect effect of the solar wind referred to earlier as it distorts the magnetosphere or external magnetic field of the earth. Daily or diurnal variations are seen primarily during the local daylight hours shown for typical days in figure 5.6. Diurnal variations are not predictable, may exhibit changes as large as 100 gammas or more, and are often removed from portable magnetometer data by methods described later. Superimposed upon these diurnal variations are more short-period phenomena called micropulsations (figure 5.7) which behave more randomly, have generally smaller amplitude, and may occur at any time of the day or night. Micropulsations occur in a broad range of periods between 0.01 seconds up to several tens of minutes with amplitudes from a thousandth of a gamma to several tens of gammas.

Of still greater concern as a possible source of erroneous data are magnetic storms. Occurring as often as several times per month, their onset is sudden and simultaneous throughout the world. Such storms may vary up to several hundred gammas and last one to several days. (See figure 5.8.) For very important field measurements, particularly for higher resolution measurements, a recording base station or reference monitor is often used. It is examined at the start of each day for an indication of magnetic storm activity and also for subsequent removal of the diurnal variations from field data using time as a correlation.

Over years, the earth's internal or main field also changes slowly through what is termed the secular variation. The inclination, intensity, and even the location of the poles vary slowly at a rate relevant, in the context of this chapter, only to observatory and archival interests. From time to time through geologic history, the main field has even reversed; the consequences of these events are important for a number of portable magnetometer applications covered in the chapter.

Magnetic minerals and iron

The application of portable magnetometers as treated herein has, as its primary objective, the identification and description of spatial changes in the earth's field. The time changes described earlier simply represent noise or interference in the measurements of interest. The spatial variations or anomalies to be mapped for these applications are those which might occur over several feet or several thousands of feet and are usually caused by an anomalous distribution of magnetic minerals or by iron objects or cultural features which may be of interest. The anomalies from naturally occurring rocks and minerals are due chiefly to the presence of the most common magnetic mineral, magnetite ($Fe\ Fe_2O_3$), a dark, heavy, hard and resistant mineral, or its related minerals, ulvospinel, titanomagnetite, and maghemite, which will collectively be referred to as magnetite. The rust-colored, very common forms of iron oxide are not usually magnetic and are seldom related to the source of magnetic anomalies. Other magnetic minerals which occur to a far lesser extent are ilmenite, pyrrhotite (with sulfide mineralization), and others of even lesser consequence.

All rocks contain some magnetite from very small fractions of a percent up to several percent, and even several tens of percent in the case of magnetic iron ore deposits. The distribution of magnetite or certain characteristics of its magnetic properties may be used in exploration or mapped for other purposes. Iron objects in the earth's magnetic field, whether something buried or intentionally planted for subsequent retrieval, would also create a detectable magnetic anomaly. Cultural features associated with man's habitation can frequently be detected through magnetic surveys because of the contrast in magnetite associated with numerous artificial features.

FIGURE 5.7
Typical micropulsations

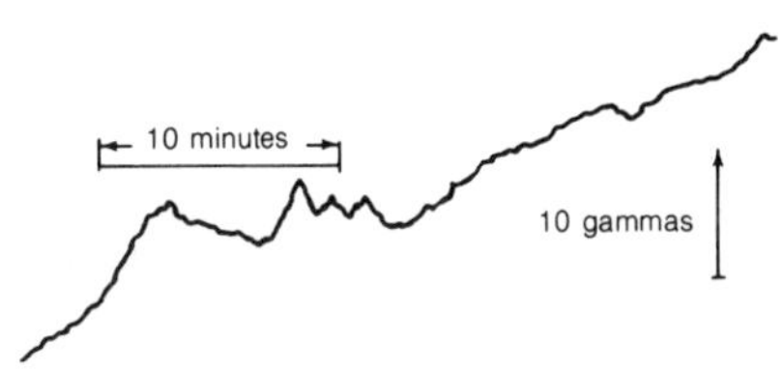

Induced magnetization

Magnetic anomalies in the earth's magnetic field are caused by two different kinds of magnetism: induced and remanent (permanent) magnetization. Induced magnetization refers to the action of the field on the material wherein the ambient field is enhanced and the material itself acts as a magnet. The magnetization of such material is directly proportional to the intensity of the ambient field and to the ability of the material to enhance the local field—a property called magnetic suscep-

FIGURE 5.8
Typical magnetic storm

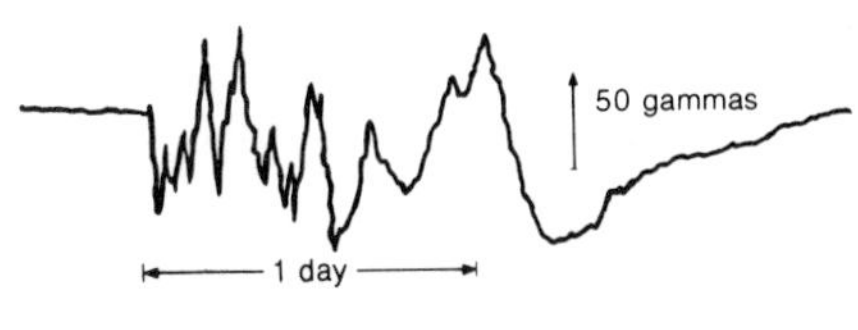

tibility. The induced magnetization is equal to the product of the volume magnetic susceptibility, k, and the earth's, or ambient, field intensity, F, or

$$I_i = kF$$

where I_i is the induced magnetization per unit volume in cgs electromagnetic units, and F is the field intensity in gauss. (Note: in some texts, the specific magnetic susceptibility or susceptibility per unit weight (gram) is used.) For most materials, k is much less than 1 and, in fact, is usually $\pm 10^{-6}$ cgs or smaller. If k is this small and positive, the material is said to be paramagnetic and, when negative, diamagnetic. For magnetite, *k is approximately 0.3 cgs* and is ferrimagnetic; for iron alloys, k may vary between 1 and 1,000,000, and such materials are called ferromagnetic. Both ferrimagnetic and ferromagnetic susceptibility are functions of the field intensity in which they are measured. In all cases in this chapter, the field intensity is assumed to be the ambient earth's field intensity between 0.3 and 0.6 gauss.

A parameter similar to k is the magnetic permeability, μ, which is the ratio of the magnetic induction, B, to the field intensity, F (magnetic induction is the magnetization induced in the material). B includes not only the magnetization of the material, but also the effect of the field itself and is expressed by

$$B = F + 4\pi I$$

where B is in gauss. Therefore as stated above,

$$\mu = \frac{B}{F}$$

and

$$\mu = 1 + 4\pi k.$$

Thus, when k is very small, as in air, $\mu \approx I$ and when k is 0.1 or larger, μ is generally one order of magnitude larger. The susceptibility k can be thought of as the *absolute* ability and μ the *relative* ability of a material to create local magnetization. The measurement of permeability is most often used for materials such as iron, steel and other ferromagnetic alloys, in which μ is much greater than 1.

Inasmuch as magnetite and its distribution are of such great importance for a number of these applications, it is important to understand its relation to common rock types. The susceptibility k of magnetite was given as approximately 0.3 cgs which may actually vary between 0.1 and 1.0 depending upon its grain size and other properties. The magnetic susceptibility of a rock containing magnetite is simply related to the amount of magnetite it contains. For example, *rock containing 1 percent magnetite will have a volume susceptibility of 3×10^{-3} cgs.* Typical susceptibilities of rocks follow, but may vary by an order of magnitude or more in most cases:

altered ultrabasic rocks	-10^{-4} to 10^{-2} cgs
basalt	-10^{-4} to 10^{-3}
gabbro	-10^{-4}
granite	-10^{-5} to 10^{-3}
andesite	-10^{-4}
rhyolite	-10^{-5} to 10^{-4}
shale	-10^{-5} to 10^{-4}
schist and other metamorphic rocks	-10^{-4} to 10^{-6}
most sedimentary rocks	-10^{-6} to 10^{-5}
limestone and chert	-10^{-6}

Typically, dark, more basic igneous rocks have a higher susceptibility than the acid igneous rocks; the latter, in turn, have a higher susceptibility than sedimentary rocks.

Remanent or permanent magnetization

The remanent or permanent magnetization, I_r (the former ascribed to rocks, the latter to metals), is often the predominant magnetization (relative to the induced magnetization) in many igneous rocks and iron alloys. Permanent magnetization depends upon the metallurgical properties and the thermal, mechanical, and magnetic history of the specimen, and is independent of the field in which it is measured. Magnetite may have a remanent magnetization, I_r, of perhaps 0.1 to 1.0 gauss; ordinary iron may have a permanent magnetization between 1 and 10; and a permanent magnet may be between 100 and 1,000 gauss or larger.

The remanent magnetization is important in mapping and interpretation, and in the fields of paleomagnetism; archaeological exploration, and archaeomagnetism. The remanent magnetization of magnetite is, as stated, independent of the present earth's field. By and large, the high values of remanent magnetization are related to the effects of heating, whether naturally heated, as in the case of igneous rocks, or artificially heated, as in the case of baked clay, pottery, and other man-made objects found in archaeological sites. Prior to such heating, small regions, called domains, within each magnetite crystal would be more or less randomly-oriented. During heating, particularly at high temperatures, the domains reorient themselves which, upon cooling, tend to align themselves more or less in the direction of the ambient magnetic field and thus parallel to each other, creating a net magnetization fixed with respect to the object. This remanent magnetization may be as much as 10 or more times greater than the induced magnetization for many rock types. Thus, the net magnetization might be considerably higher than would be indicated merely by consideration of the susceptibilities listed earlier.

The remanent magnetization of a rock or object may or may not be in the same direction, as the present earth's field for the object may have been reoriented and because the earth's field is known to have changed its orientation in geologic and even historic time. Rocks are frequently reversely magnetized so that measurement of this remanent magnetization is a useful aid to interpretation if the rocks which produce an observed anomaly are, indeed, accessible. The fields of paleomagnetism and archaeomagnetism in particular depend upon the precise determination of the orientation of the 'frozen paleofield' as it is measured in a given rock or other specimen.

Field procedures and data reduction

Magnetic cleanliness and sensor positions

Most applications for portable magnetometers require that *the operator be relatively free of magnetic materials on his person.* The importance of checking oneself cannot be over-estimated if measurements on the order of 1 gamma are desired. In field surveys, the magnetic material one may have may include the obvious such as a rock pick, Brunton compass, pocket knife, or instrument console. Not-so-obvious are effects of the pivot in eyeglasses, the pants clip at the top of men's trousers, the light meter in a camera, the magnet in the speaker of a tape recorder, metal in a clipboard, some mechanical pencils, some keychains, and the steel shank in one's shoes or boots. Of course, some of these items cannot be altered or left behind, and some are not significant in any event. The sensor itself should be kept clean to avoid possible contamination by magnetite-bearing dirt on the sensor surface. To check the 'heading effect', i.e., the effect of orientation on the observed field intensity during a field survey, the operator can take readings at each of the four cardinal directions while pivoting about the position of the sensor and note the changes. If the maximum change is typically less than 10 gammas, the average readings on a line will probably not be affected by more than 5 gammas, and individual readings by less than this. Readings along the profile are more or less along a given heading ± perhaps 30° about one orientation. If a sensitivity of 1 gamma is desired, the heading error should be preferably 2 gammas or less. Depending on the desired sensitivity, the operator should make some effort to face in the same direction, if possible, for all readings on a given traverse.

The sensor for a proton magnetometer may be carried on an 8-ft (2.2 m) staff, on a backpack, on an extended staff to 12 ft (4 m) or more as necessary, or by a second person as represented in figure 5.9. Common for field measurements, carrying the sensor on an 8-ft staff removes it sufficiently from the console and the operator so it is not much affected by normal clothing items. The purpose of mounting a sensor on an extended 12-ft or longer staff is to remove the sensor from the locally disturbing effects of highly magnetic surface materials, such as surface laterite, glacial till, or highly magnetic outcropping rocks. The sensor may also be raised in the case of very high magnetic gradients which would otherwise ruin the magnetometer signal and prevent any reading whatsoever (see following sections). Occasional reasons exist for a second person carrying the sensor while the first person carries the console together, perhaps, with magnetic or other materials that must necessarily be on his person such as pick, tape recorder, another instrument, or rock samples.

The sensor may be carried in a backpack pouch for more convenient field operation where 5 or 10 gamma sensitivity is all that is desired, but care should be taken to check the effects of the batteries and console (particularly the very magnetic alkaline batteries). The backpack pouch frees the hands for taking notes and pushing aside the underbrush and, in general, balances the load of the console and decreases fatigue.

FIGURE 5.9
Sensor carrying positions

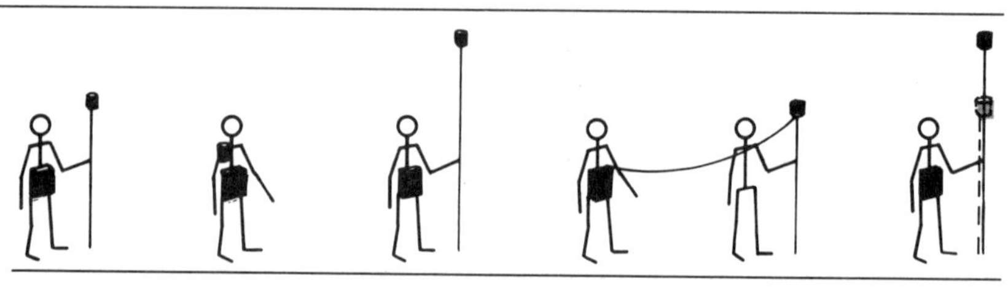

Operational considerations

Valid readings vs. noise

It is important to establish that, in fact, the magnetometer is providing valid readings. The simplest way to confirm that what is being observed is a magnetic field reading and not random, meaningless instrument readings (i.e., noise) is to take several successive readings in one location without moving anything, and note the repeatability. Successive readings should be within ± 1 gamma, ± 0.25 gamma, or ± 1 count for whatever the sensitivity setting. Valid readings should not, under any naturally occurring circumstances including magnetic storms, vary by as much as ± 10 or ± 100 gammas in a few seconds; if such is observed, the readings represent either noise or a degradation of the signal-to-noise ratio with the observed corresponding loss of sensitivity. Under certain circumstances, even successive readings repeating to within several gammas may still represent noise. To confirm that these readings are indeed magnetic field, simply 'kill' the signal by placing, momentarily during the reading, something magnetic adjacent to the sensor such as one's shoe, watch, or certain rocks. Random readings varying by 10 or 100 gammas or more would then be observed; in addition, they would deviate considerably from the readings without the object present. A less certain method is to take readings at intervals of increasing distance from an object or location known to produce a magnetic anomaly.

Typical reasons for a proton magnetometer not producing valid readings may be: electrical noise from AC power lines, transformers, or other radiating sources; high magnetic gradients from underlying rocks, nearby visible or hidden iron objects, fence lines, or improvised iron hardware improperly used near the sensor; improper orientation of the sensor (even when 'omni-directional'); expired batteries, incorrect range setting, or instrument failures; broken or nearly broken sensor cable; and other malfunctions usually described in the instrument operating manual.

Valid but distorted readings may result from several other conditions including the effects of high magnetic gradients, magnetic dirt or other magnetic contamination on the sensor, and any magnetic bias on the operator. Time variations and the effects of direct current in distant power lines and trains can also distort magnetic observations.

Sensor orientation

According to the theory of operation of the proton magnetometer, the total intensity, measured as the *frequency* of precession, is independent of the orientation of the sensor. The *amplitude* of the signal, however, does vary (as $\sin^2\theta$) with the angle between the direction of the applied field within the sensor and the earth's field direction. Variation of signal amplitude does not normally affect the readings unless there is simply insufficient signal to be measured accurately, i.e., a minimum signal amplitude is required above which a variation in amplitude does not affect the readings.

Ideally, the applied field in the sensor should be at right angles to the earth's field direction. The direction of the applied field is governed by the configuration of the polarizing coils in the sensor which are commonly either solenoids (cylindrical) or toroids (ring or doughnut-shaped). The solenoid produces an applied field parallel to its axis, whereas the toroid produces a field which is ring-shaped about the axis of the toroid (consult the instrument operations manual to determine the direction of these axes with respect to the sensor housing). Solenoids are used because they produce a somewhat higher signal than a toroid and are less perturbed by electrical noise, whereas a toroid is inherently omni-directional. For maximum signal amplitude, a solenoid should be held horizontal and in any direction in a vertical field, and should be held vertical in a horizontal (equatorial) field. A toroidal sensor should be held with its axis vertical in a vertical field, and pointing north in an equatorial field for maximum signal. A field which dips more or less than 45° should be treated as though it were a vertical or horizontal field respectively.

Instrument readings

Measurements are normally made at regular intervals along a grid or otherwise selected path whose locations are noted for subsequent plotting. Simple pacing is usually adequate with readings every 6, 10, 50, 100, 500, or even 1,000 ft (2 to 300 m), as anomalies, field, and either geological or search requirements dictate. Traverses may be selected along pathways or other accessible routes and occasional locations noted on an aerial photograph or map using paced distances in between. The density of readings along the traverse should be related to the wavelength of anomalies of interest such that several readings are obtained for any such anomaly. A single trial line with relatively dense stations is usually attempted first to determine the required station density. It is important never to hold the magnetometer sensor within one or two feet of the ground, if possible. Thus, one avoids effects of minor placer magnetite which usually collects on the surface of the ground, as well as the effects of microtopography or outcropping rock surfaces.

Readings may be noted in a field notebook or on a miniature tape recorder, but care must be taken to magnetically compensate the speaker magnet and motor if one is to use a recorder. The convenience of the recorder is that only one hand is needed and the data may be played back for fast, convenient plotting.

Correction for time variations

Some ground magnetic surveys require correction for diurnal and micropulsation time variations. Correction is required if the anomalies of interest are broad (thousands of feet) and typically less than 20 to 50 gammas; if the profile lines are very long; or if the objective of the survey is a good magnetic contour map expressive of deep-seated anomaly sources. Also, if the survey is performed in the high magnetic latitudes in the auroral zone where typical micropulsations are 10 to 100 gammas, correction for such variations would be necessary. On the other hand, if one is merely interested in profile information of anomalies of several hundred gammas or if the anomalies are only 20 gammas but can be traversed completely in less than 5 min, no time variation correction is needed. Perhaps *most surveys* fit the latter criteria and *do not actually require any such correction for time variations.*

The simplest method of correcting for time variations involves repeated readings in the same orientation at the same station at different times during the survey. If a smooth curve is drawn through the readings plotted as a function of time (every hour or so), these values can be subtracted from all other readings provided that each reading also includes the time at which it was observed. To avoid repeated walks to a single reference station, it is also possible to 'double-back' to take a second or third reading on each given traverse to determine at least the time variations for that traverse. Still another technique is to emulate what is done on aeromagnetic surveys; namely, obtain rapidly acquired measurements on tie lines or lines which cross the principal traverse lines at each end and perhaps in the center. The stations common to each traverse and tie lines should be known and occupied while facing the same direction to avoid heading errors. The simplest method for using these tie lines is to make each intersection agree by linearly distributing the error on each traverse line and holding the tie line values fixed—provided the tie line data were acquired rapidly.

A local recording base station, i.e., diurnal station monitor, is the most ideal method and certainly the most accurate for removing time variations. The time variations can readily be removed from each reading, again assuming that the time is noted for each reading on the traverse to within a minute or so of the base station. The base station should not be farther away than 100 mi from the area of the survey for agreement within a few gammas; it should be positioned more than 200 ft away from local traffic and other disturbances. The diurnal base station, if left to continue recording during each evening, can indicate magnetic storms in progress and may be examined at the start of a survey day to determine if any useful measurements can actually be obtained during such conditions. During a magnetic storm, it is best not to obtain field data with the objective of removing the storm variations, as the survey magnetometer and base station may not agree better than 5 or 10 gammas.

High magnetic gradients

In the case where an extremely high magnetic gradient destroys the signal, as evidenced by successive nonrepeating measurements, it may be necessary to raise the sensor up to 10 or 12, sometimes 15 ft in order to move the sensor to a region of lower gradient. This will only happen over outcropping or nearly outcropping large masses of perhaps altered ultrabasic rocks, magnetic iron ore deposits or orebodies containing a large percent of pyrrhotite, and near buried iron objects in the applications for search. Such an event would only occur if the gradient exceeds several hundred gammas per foot. If the span of high gradient is not too wide, it may not actually be necessary to obtain measurements precisely at the highest gradient. Measurements on either side of the anomaly can be extrapolated or used to at least indicate the contacts of such a highly magnetized formation. Furthermore, as the signal disappears and the readings diverge considerably from $\pm$ 1 or 2 counts, it may be worthwhile to note approximate indications of magnetic field gradient which on some instruments are displayed on the front panel as signal amplitude (which is a function of gradient). In areas of highly magnetic surface conditions, as noted previously but where a signal is still obtained, another alternative in acquiring meaningful data other than using an extended staff would be to make two to five measurements per station, for example, at the points of a cross centered at the actual primary station location. The average of these readings would later be used to draw a profile. In this way, some of the surface noise is averaged out.

In the absence of anomalous surface conditions, it may be useful for both geological and search applications to measure the vertical gradient of the total field. The vertical gradient is obtained by making two total field measurements, one over another, taking the difference in the readings and dividing by the distance between them.

Data reduction

The profiles when plotted should be smoothly varying and expressive of the anomalies of interest. (NOTE: The nature of the disturbances or anomalies of interest, their width, character, signature, and amplitude are discussed later.) Should there be an excessive amount of such geologic/magnetic noise at a wavelength much shorter or much longer than is of interest, it is possible to apply simple filtering or smoothing techniques to facilitate interpretation of the profile. As a rule of thumb, *never remove or filter out anomalies whose wavelength is on the order of the depth to sources of interest.* A number of advanced techniques for data enhancement or filtering as employed in airborne surveys or well-gridded ground surveys will not be discussed here, but are listed to acknowledge their existence: vertical derivatives, upward and downward continuation, reduction-to-the-pole, bandpass filtering, trend surface filtering, spectral analysis, trend enhancement, magnetization filtering, and others, most of which are applied to two-dimensional data.

Profile smoothing

Anomalies of very short wavelength (much shorter than the probable depths to sources of interest) may be present and caused by the magnetic effects of the magnetometer operator, or simply by surface magnetization contrasts in the surface or near-surface materials as mentioned earlier. In removing such effects, the eye itself tends to enhance what one is seeking. Another simple and obvious method is to pencil or trace through the noise. A more objective technique is to apply a running average or weighted running average to the data (see figures 5.10 and 5.11). For a 3-point weighted running average, for example, one

FIGURE 5.10
Profile smoothing

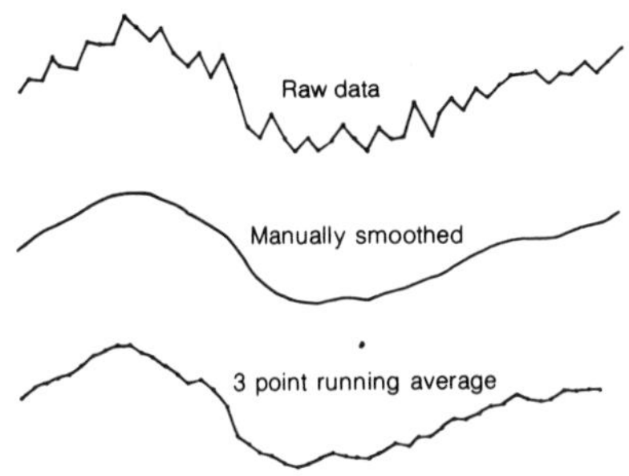

FIGURE 5.11
Running averages

For simple running average

$$\frac{A + B + C}{3} = \overline{B}$$

For 5 point weighted running average

$$\frac{A + 2B + 4C + 2D + E}{10} = \overline{C}$$

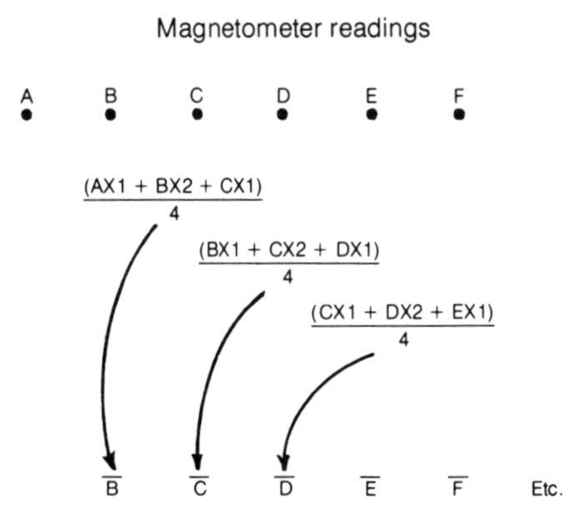

(3 point running average shown in figure 5.10)

would multiply the value at a given station by 2, add the values of the two adjacent stations, divide the sum by 4. This value is then set aside for that station for later recompilation of a new profile while advancing to the next station to perform the same procedure (see figure 5.11). A 5-point running average might use a weighting factor of 4 for the central point, 2 for each adjacent point, and 1 for the outside points, while dividing by 10 to obtain the averaged value. More sophisticated techniques are also possible such as polynomial curve fitting, least squares, and digital bandpass filtering. The number of points or interval over which the averaging or filtering is to be performed for removal of such 'noise' should be much shorter (perhaps 1/5 to 1/10) than the anomalies of interest.

Removal of regional gradients

In most cases, the anomalies of interest usually appear superimposed on a much broader anomaly which is not of interest. Because of the main earth's field or very deep or distance sources, this broader anomaly, or regional gradient, may appear simply as a component of slope in the curve. Although it is subjectively determined, it is often removed from the data in order to better examine the anomaly. This gradient is removed from a single profile as shown by figure 5.12 by drawing a straight line or broadly-curved line through the non-anomalous portions of the curve. The values are then subtracted at each station and replotted to present the 'residual' values, in theory, expressing only the anomalies of interest. In this case, these are the anomalies occurring at the shallower depths. The vertical gradient also serves to remove or largely reduce the regional gradient.

FIGURE 5.12
Removal of regional gradient

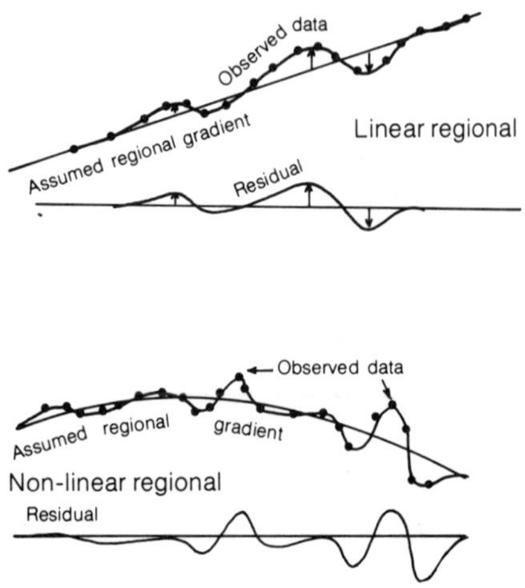

Contour maps

Most survey traverses are not sufficiently close nor well-arranged to allow the compilation of a contour map. Such is usually the case when only mineral exploration data are desired, when broad reconnaissance surveys are done, or when surveys are undertaken in extremely rugged terrain where large areas are otherwise inaccessible. Profiles are, in fact, usually adequate, particularly for anomaly sources with very long, horizontal dimensions. Contour maps are, nevertheless, useful when little is known of the geology or magnetic sources, when the anomaly sources are either small or extremely large, or when ascertaining, on a more objective basis, the distribution of the anomalous masses or very subtle longer wavelength features. Many surveys for search also require broad coverage and perhaps contour map presentation. An alternative to the construction of contour maps for broad two-dimensional coverage is the presentation of 'offset profiles' or profiles plotted on abscissae which also serve to show the location of the traverse lines drawn on a map.

Constructing a contour map requires that large effects from diurnal variations or the heading effects, if any, be removed; that is, that there be a single datum level for all traverses or readings. The process of removing the regional gradients, as described earlier, frequently serves to remove these other sources of errors as well. It is satisfactory as long as one is not interested in the longer wavelength anomalies removed as part of the gradient.

The following guide should be useful to one not familiar with the techniques of constructing a magnetic contour map, or plan view of the anomalous total intensity. A contour or iso-intensity map is analogous to a topographic map. It is a map on which lines (contours) of equal intensity are drawn at convenient and regular intensity intervals, as would be observed were a magnetometer used to occupy every point on the surface of the ground. The contoured values are at best extrapolations and interpolations across areas where measurements are not actually taken. Such a map is drawn with the knowledge that the magnetic field is smoothly varying and on the assumption that one is interested only in broad features expressed by such a map. Features much smaller than the spacing between adjacent traverses should be examined on a profile basis only and should not be sought nor included on a map presentation.

Construction of a contour map

Given a set of readings obtained on a traverse, the time variations, if significant, should be removed, perhaps the regional gradient removed and the profiles smoothed. Values are then selected from these smoothed profiles at widely-spaced intervals not less than about one-half or one-fourth the spacing between adjacent traverses or at similarly spaced but significant points on the profile; namely, maxima, minima, and inflection points. In other words, the values to be contoured should be more or less equally distributed in plain view. Anomalous features which 'obviously' extend across several traverses might be included also. The total intensity values thus selected and representative of the principal features are posted at their proper locations on a base map that can be easily erased and that includes references to location.

Examine the dynamic range of the values and select five or ten intensity levels through this range at convenient values such as every 20, 100, or 1,000 gammas. Draw these contours according to the following instructions and then fill in the intermediate contour lines, i.e., every 10, 50, or 500 gamma contours, depending upon which coarse values were originally selected, until contours appear in all segments of the map. Magnetic intensity values and contours should, in theory, be smoothly varying and should thus be smoothed at the later stages of contouring by removing sharp bends or corners. After such smoothing, other contour lines, as needed to cover the map adequately, are carefully drawn between the fair-drawn contours, and appropriate labels are applied. In areas of steep gradients, only a few coarse contour lines are drawn to avoid numerous and insignificant fine details. Since closed contours (closures) appear the same for maxima and minima, they are differentiated by applying hashure marks or other indications on the inside of the minima.

The position of the various contours is selected by manually (eye and mental calculation or by using proportional dividers, although not really necessary) interpolating linearly between all the neighboring values as shown in figure 5.13. In this case, it was decided to draw contours at 10-gamma intervals. The contour line near data point value 91 would subsequently be smoothed to pass through this data point following the guidelines given above.

Contour lines should never cross nor pass between pairs of data points which are both higher or both lower than the value of the contour. Also, in some regions of zero or near zero gradient such as at a saddle point (region between two adjacent maxima or minima), there exists an ambiguity in the direction of the lines. However, it does not matter under such conditions which of the two possible sets of contours is drawn.

FIGURE 5.13
Interpolation and contouring

•96 Data
■90 Manually interpolated point
—Contours drawn through interpolated points

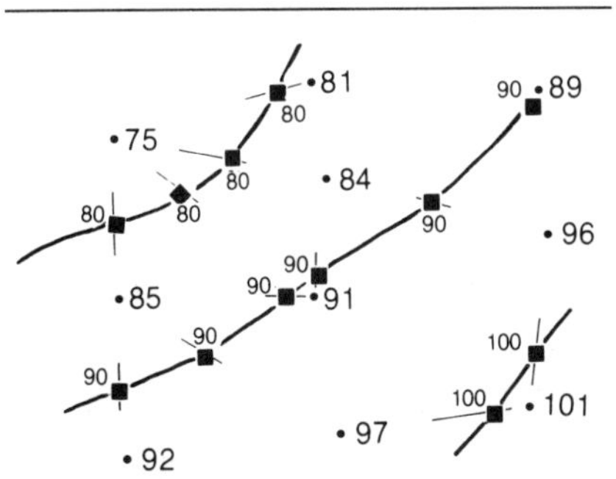

Interpretation

Total magnetic intensity disturbances or anomalies are highly variable in shape and amplitude; they are almost always asymmetrical, sometimes appear complex even from simple sources, and usually portray the combined magnetic effects of several sources. Furthermore, an infinite number of possible sources can produce a given anomaly. The apparent complexity of such anomalies is a consequence of the net effect of several independent but relatively simple functions of magnetic dipole behavior. With an understanding of these individually simple functions and with some reasonable assumptions regarding the geology, buried object, or whatever other source one is seeking to understand, a qualitative but satisfactory interpretation can usually be obtained for most anomaly sources.

For geological reasons as well as for search applications, the interpretation, explanation, and guide presented here are directed primarily toward a qualitative interpretation, i.e., an understanding of what causes the anomaly, its approximate depth, configuration, perhaps magnetite content or mass, and other related factors. But even if qualitative information is derived from the data, it is important to have applied a reasonable amount of care in obtaining precise measurements. Quantitative interpretations are possible, but are applied more often to airborne data, entail relatively complex methods for depth determination, and are the basis for a relatively large body of literature on the subject.

FIGURE 5.14
Total field lines of flux

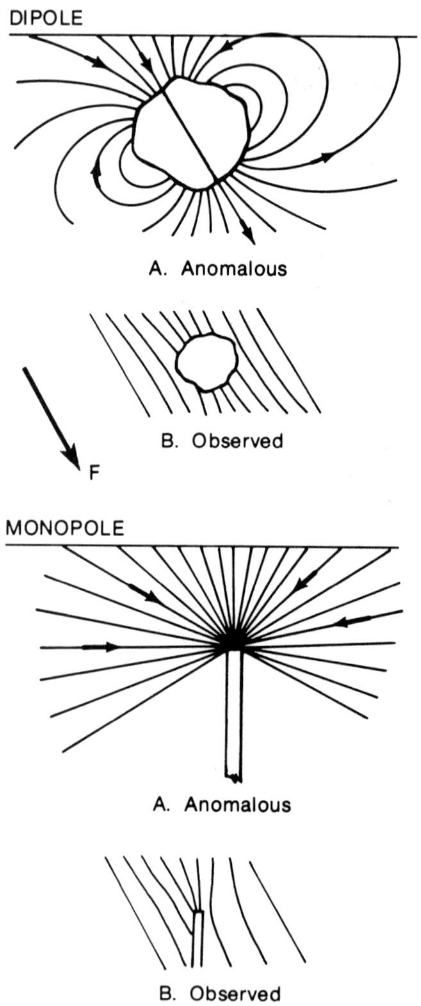

An anomaly represents a local disturbance in the earth's magnetic field which *arises from a local change in magnetization, or magnetization contrast* as it is termed. A profile, for example, over a very broad, uniformly magnetic surface, although magnetic itself, will not exhibit a magnetic anomaly as there is no local change in magnetization. A local increase or decrease, on the other hand, would constitute such a change and produce a locally positive or negative anomaly.

The observed anomaly expresses only the net effect of the induced and remanent magnetizations which usually have different directions and intensities of magnetization. Since the remanent magnetization is so variable and measurements of its properties seldom made, *anomalies are all interpreted* in practice as though *induced magnetization were the total source of the anomalous effects.*

Asymmetry

The asymmetrical nature of total field anomalies is primarily a consequence of the directions of the field lines of the locally created magnet or source and the earth's-field-component-nature of a total field magnetometer in the usually-inclined direction of the earth's magnetic field. Recall that a total field magnetometer measures only the component of any local perturbation which is in the direction of the earth's magnetic field at that point. Anomalies in the earth's field, whether created by induced or permanent magnetization, exist as arrangements of magnetic dipoles, monopoles (effectively), lines of dipoles and monopoles, and sheet-like distributions of such poles. It is important, therefore, to understand the nature of the dipole or monopole field as it will be shown that a summation of such elementary forms will explain the most complex characteristics of anomalies and facilitate their interpretation. Notice, for example, in figure 5.14 the configuration for such fields as they would appear if one were to measure the direction of the anomalous field.

Depth dependence

FIGURE 5.15
Effect of depth on anomaly width

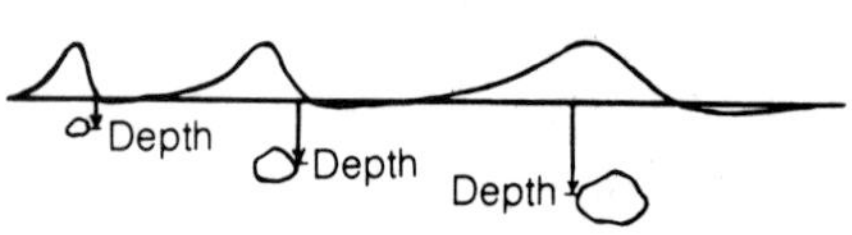

Another significant characteristic of a magnetic anomaly is its variation with the depth between the magnetometer and source; the deeper the source, the broader the anomaly as expressed in figure 5.15. It is this property which enables one to determine the approximate depth to the source independent of any other information concerning the source. *If one familiarizes oneself with only one subject in this discussion on interpretation, it should be the general characteristics of anomaly wavelength, or width, as a function of depth.* A knowledge of this subject allows rapid and easy interpretation of anomalies of interest when numerous anomalies arising from various depths appear in the observed total intensity profile.

Other anomaly shape factors

Other factors which affect the anomaly shape and amplitude are the relative amounts of permanent and induced magnetization, the direction of the former, and the amount of magnetite present in the source compared to the adjacent rocks. The actual configuration of the source, that is, whether it is narrow, broad, or long in one dimension, and its direction in the earth's field also control the anomaly signature.

FIGURE 5.16
Model of common magnetic anomaly sources

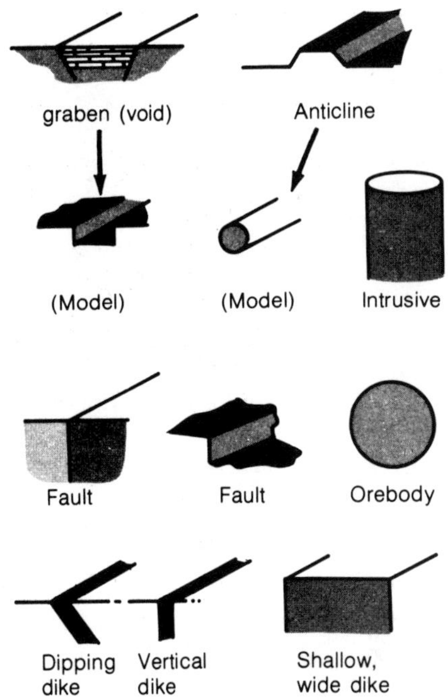

Geological models

Geological anomalies are interpreted in terms of much simplified geological models which facilitate interpretation procedures. The first simplification is the assumption that magnetization is uniform within some elementary prismatic form and that the magnetization is different outside this form, i.e., there is a magnetization contrast. Typical of the kinds of geologic sources that are assumed to cause anomalies are those which are shown in figure 5.16.

As stated, in any potential field method the given magnetic signature can be produced by an infinite combination of sources so that there is no unique interpretation. For example, the same anomaly could be produced by the peculiar distribution of magnetite (unrealistic geologically), and a uniform distribution of magnetite within the prismatic form (realistic), both of which are shown in figure 5.17. It must be emphasized that not only are simplifications required, but a reasonable geologic framework must be used as a guide when considering the various possible sources. A typical set of anomaly signatures of various sources might appear as in figure 5.18.

Elementary dipoles and monopoles

Since anomalies are explained herein in terms of various arrays of dipoles and monopoles, it is important to examine their geometry and in-

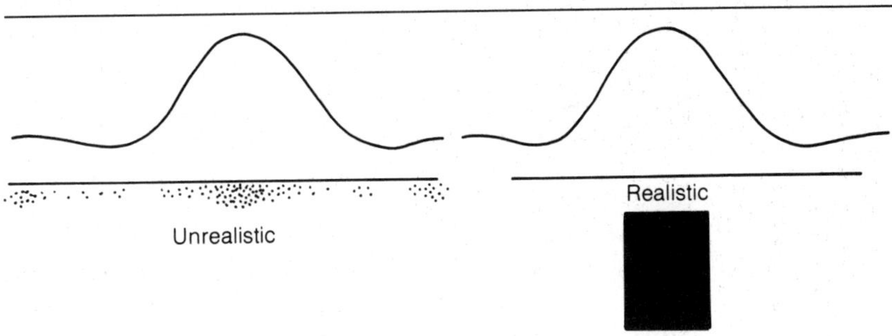

FIGURE 5.17
Possible sources producing same anomaly

FIGURE 5.18
Typical anomalies for simple geologic models

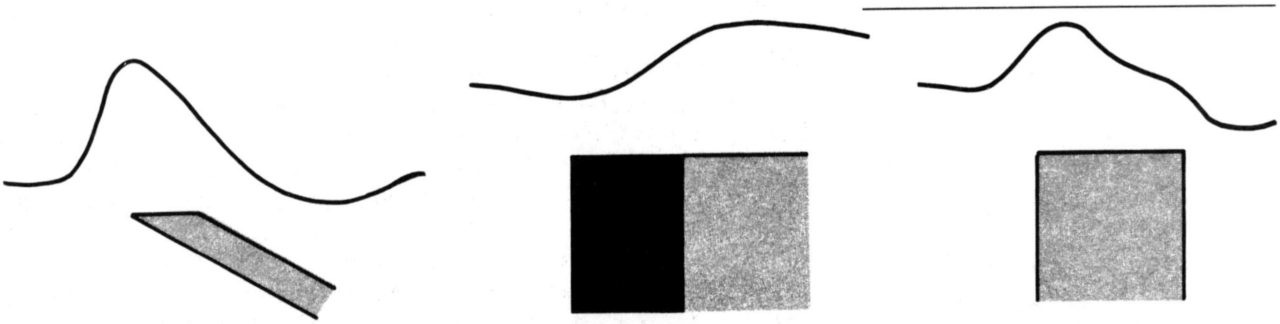

tensity characteristics. A magnetic dipole produces a field with imaginary lines of flux as shown in figure 5.19. The intensity of the field, proportional to the density of the flux lines, is drawn as lines of equal intensity to express this relationship. From figure 5.19, notice that (1) the intensity of the dipole is twice as large off the ends of the dipole as it is at the same distance off the side of the dipole. This explains, for example, why the earth's magnetic field is approximately 30,000 gammas at the magnetic equator and 60,000 gammas at its poles; (2) the direction of the field off the side of the dipole is parallel to the dipole itself, but opposite in sense; (3) the direction of the tangent of the field lines of a dipole are parallel along any radial line from the dipole.

A monopole has field lines which point radially in or out from the positive or negative monopole respectively. The intensity is constant at a given distance and any direction from a monopole. In actual fact, there are no magnetic monopoles, but only dipoles whose ends are far apart. For all practical purposes, however, monopoles exist in terms of the distance to the source and such geological configuration as shown in figure 5.14.

Having outlined the qualitative geometry of the intensity T from a dipole, one can consider the quantitative aspects as follows:

The intensity, T, from a dipole can be expressed as

$$T = \frac{2M}{r^3}$$

along the axis, i.e., off the end of the dipole, and

$$T = \frac{M}{r^3}$$

along a line at right angles to the dipole, i.e., off the side of the dipole, and for a monopole

$$T = \frac{M}{r^2}$$

in any direction from a monopole, where M = magnetic moment and r is the distance to the pole. A more detailed mathematical formulation for the intensity due to a dipole is given later in this section.

Simplified method for total field signature

From the above description of a dipole and monopole and with the knowledge of the earth's-field-component-nature of the total field magnetometer, one can sketch the signature of an anomaly for any given orientation of the dipole (orientation caused by field direction, the direction of remanent magnetization, or by the configuration of the geology). Drawing such signatures at various inclinations of the magnetic field helps one understand the location of the sources with respect to the signature; the dip of the magnetization producing the anomaly, and information related to depth of the source. Remember that all anomalies can be considered as caused by various distributions of dipolar and monopolar sources, and it is possible to produce any anomaly simply by the super-position of such dipole or monopole signatures derived here.

FIGURE 5.19
Lines of flux (—) and lines of equal intensity (——) for a dipole

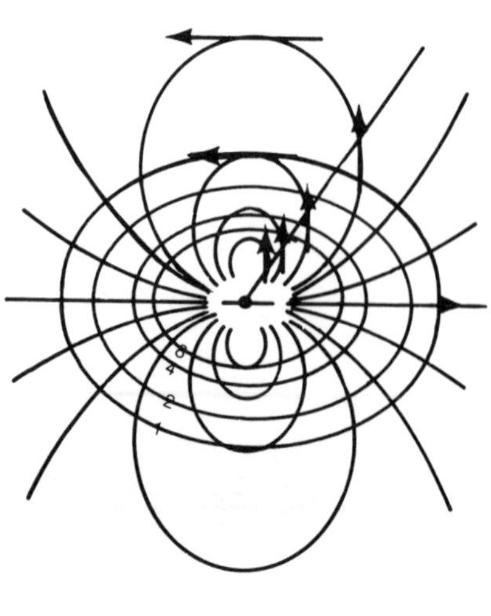

Earth's field component behavior

This method of predicting or drawing the anomaly signature depends upon one property of the field, namely, inclination, and three properties peculiar to the dipole or monopole source, whichever is assumed. The dip of the earth's field is first considered because this is the direction, the only direction, of the components of any local magnetic anomalies which is measured by a total field magnetometer. (If one is using a vertical component magnetometer, this guide still applies except that instead of using the earth's field as the direction of measurement, one uses the vertical.) In other words, the magnetometer will only measure the component of a local perturbation in this direction, i.e., as projected into this direction. See figure 5.20.

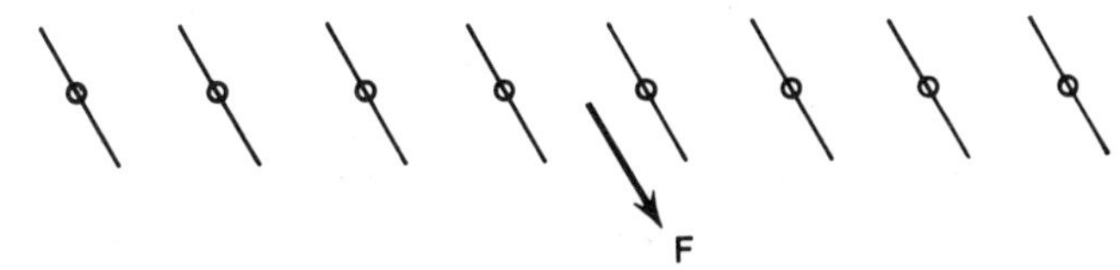

FIGURE 5.20
Direction of measured components

Dipoles vs. monopoles vs. arrays of poles

The decision to use dipoles, monopoles, or other configurations as the model is based upon the manner in which the earth's field induces a local field, and this in turn depends upon the configuration of the geologic body which exhibits the magnetization contrast and the direction of the field. For example, a long body which nearly parallels the earth's field will tend to be magnetized along its long dimension. Furthermore, if the body is sufficiently long with one end near the magnetometer, the anomaly will appear as a monopole seeing only the upper pole with the lower pole removed effectively to infinity. If the same long, thin body were normal to the field, it would then be magnetized through its thinnest dimension producing the sheet-like array of dipoles as shown in figure 5.21.

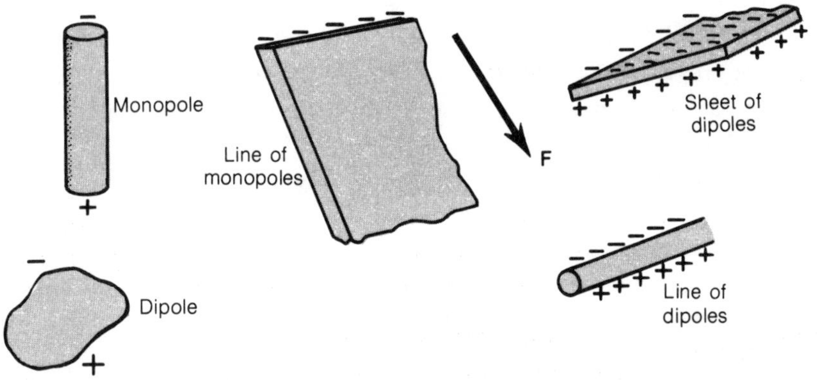

FIGURE 5.21
Sheet-like array of dipoles

One may wish to draw on the typical models depicted in figure 5.16, the array of poles from a uniform earth's field at various inclinations and orientations of the source. Whether the monopoles or the dipoles (and their equivalent line or sheet distributions) are close or far apart determines whether the model is to be considered a dipole or monopole respectively.

Configuration of field lines

The first property of the dipole or monopole to be considered is the configuration of the field lines (see figure 5.14). When superimposed upon the component which is measured by the total field magnetometer, it can be seen that the relative lengths of the disturbance vectors measured are those shown in figure 5.22 for an induced dipole and monopole source. It is the relative length of these disturbance vectors drawn along the total field direction that is measured, each disturbance vector, in turn, weighted by the intensity functions described in the next section.

FIGURE 5.22
Total field components of tangent to field lines of dipole and monopole

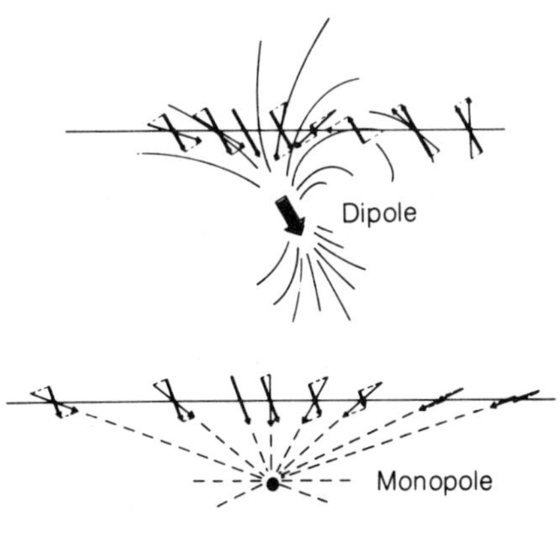

Dipole and monopole fall-off factor

The next factor to be considered is the variation of intensity with distance, i.e., $1/r^3$ and $1/r^2$ factors for the dipole or monopole fields

FIGURE 5.23
Fall-off rate
(Relative intensity or length of vectors in figure 5.22)

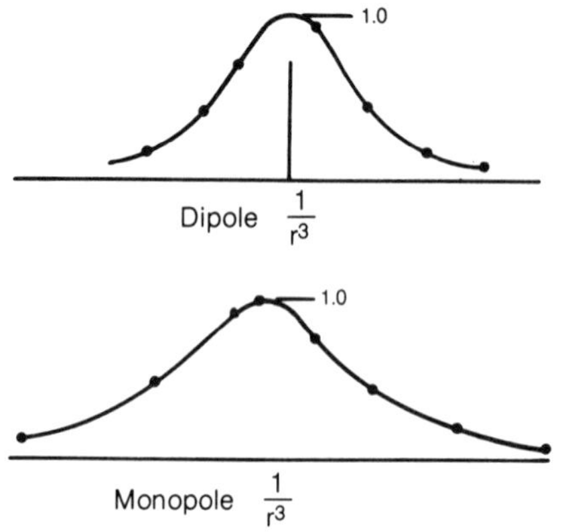

FIGURE 5.24
Aspect factor
(Relative intensity of dipole of figure 5.22 with respect to angle from axis at various points along profile)

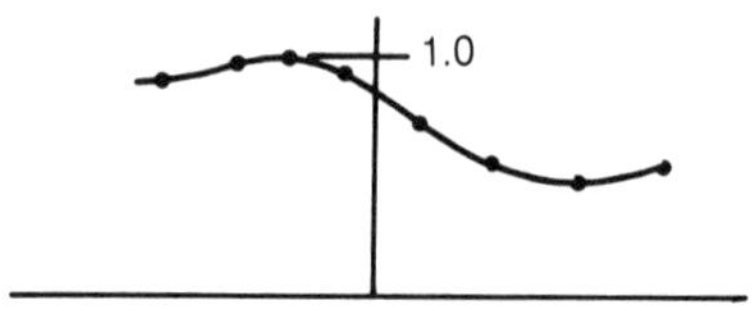

FIGURE 5.25
Dipole and monopole signatures
(from figures 5.20-24)

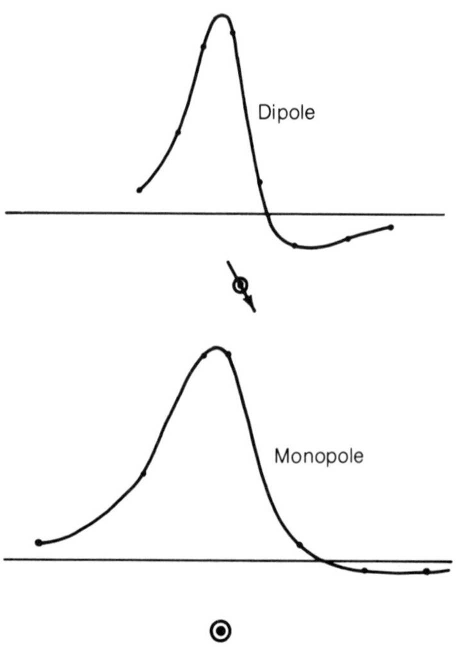

respectively and as expressed in the preceding equations. The relative intensity for dipoles or monopoles as a function of distance to their centers, as would be observed along a traverse, is presented in figure 5.23 and described mathematically under "Anomaly Amplitude" later. This factor multiplies the length of net vectors in figure 5.22.

Dipole factor-of-two

The last consideration really only applies to the dipole and that is a factor of 2 when one is off the end of the dipole compared to a positon off the side. In other words, at a given distance, the intensity varies by a factor of 2 as a function of the angle between the radial line to the dipole and the dipole axis. This function is shown approximately in figure 5.24 for the dipole used in the example. The monopole possesses radial symmetry and therefore requires no such consideration.

Application of method

A dipole and monopole signature is thus constructed in figure 5.25. The amplitude is dimensionless, but can be compared to a real anomaly by multiplying by a single factor derived below from considerations of volume and susceptibility. However, applying these factors even qualitatively should allow one to draw the dipole and monopole signatures for variously inclined fields and geometries. Figure 5.26, for example, is drawn freehand for anomalies in vertical field (90° inclination), magnetic equator, and mid-southern latitudes. By simply sketching in the earth's field direction and the dipole's field lines without consideration of the other last two factors, one can appreciate the basis for:

- a negative anomaly over sources at the magnetic equator;
- absence of anomalies in the central portion of elongate N-S anomalies at the equator;
- both positive and negative fields for almost any anomaly;
- changes in anomaly character for different directions of the dipole;
- assymmetry of anomalies;
- monopole which has only positive sense yet for most inclinations still produces a total intensity anomaly with both positive and negative portions.

The simple exercise of drawing such anomalies may also elucidate other characteristics of signatures, which to those unfamiliar with magnetics or such behavior as shown here, appear complex and difficult to comprehend.

Based upon the above procedures, applied qualitatively, and upon the manner in which lines of flux are induced in various configurations of geologic bodies and ambient field directions and inclinations, it is possible to derive the various signatures shown in figure 5.27 (drawn freehand). By varying the effect of depth as it produces an anomaly of longer wavelength, and by building composite anomalies such as summing the effect of two faults to create a single wide, shallow dike, it is also possible to generate a composite curve demonstrating the effect of different sources and different depths which is the typical observation.

Contour presentation of dipole and prism anomalies

Profiles of total intensity are usually the only form of presentation from ground measurements even when data are taken on a two-dimensional array. If measurements are taken properly, however, it is possible to construct a contour map by the methods described earlier. It is, therefore, useful to examine a few special cases of contour maps that

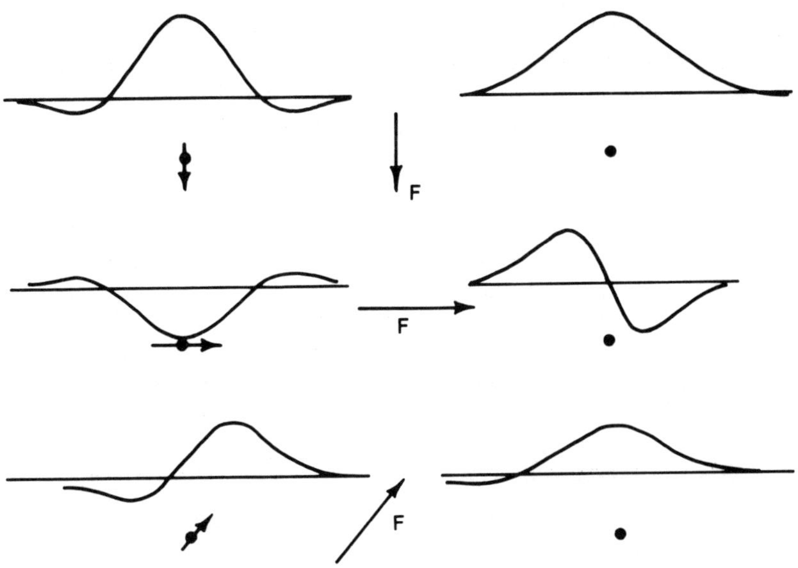

FIGURE 5.26
Freehand sketch of dipole and monopole for various inclinations

● Monopole

↘ Dipole

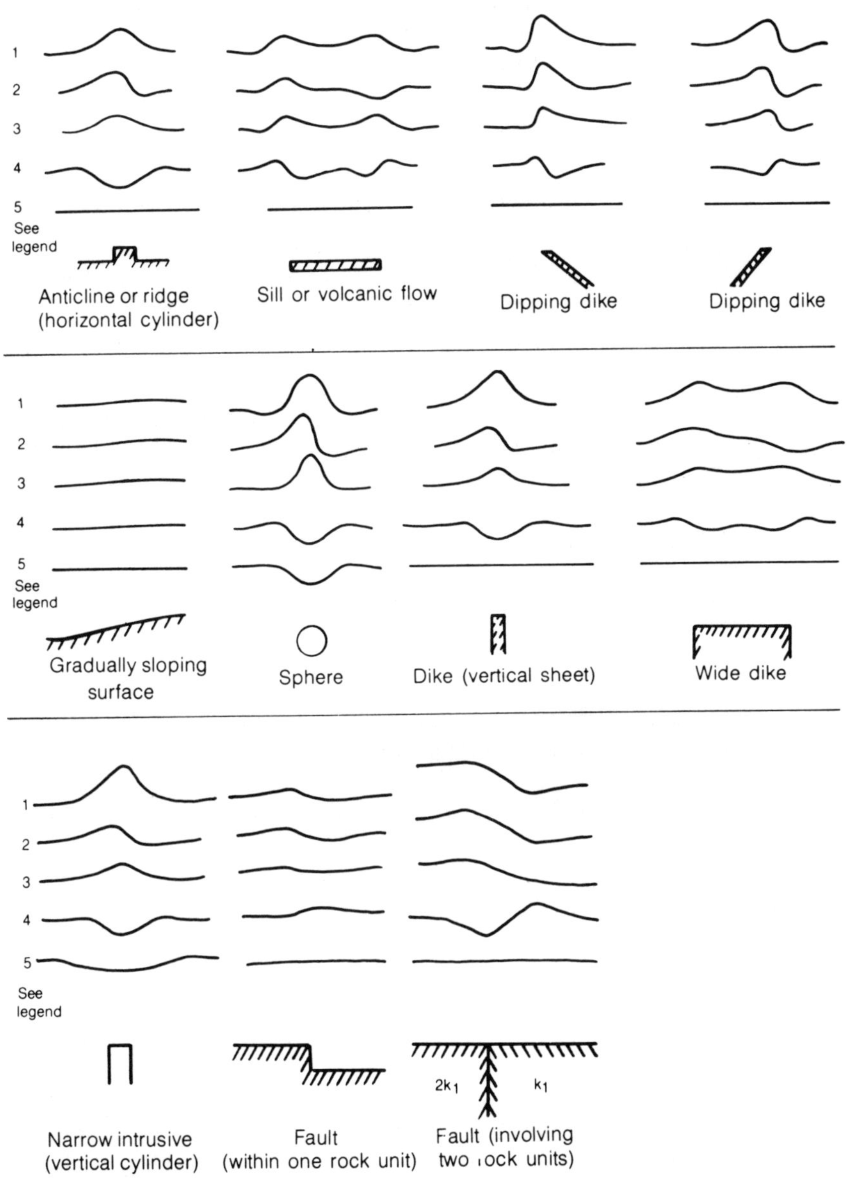

FIGURE 5.27
Anomalies at various orientations and different inclinations

Typical anomalies from various geologic bodies (shown freehand—do not use as quantitative reference).

All anomaly sources except sphere and vertical cylinder are infinitely long normal to the plane of this page.

Legend

1. ↓F
2. ↘F or ↖F traverse N-S (N→) (N←)
3. ↘F or ↖F traverse E-W (i.e., projection of field into plane of this page is vertical)
4. →F traverse N-S (N→)
5. →F traverse E-W (i.e., field is normal to plane of this page)

would be expected over simple sources such as a dipole and a wide, vertical prism in various latitudes. Such a contour map also allows one to extract, even by simple inspection, the appearance of a given profile at various positions over such simple-shaped forms. This is useful information both in search and in geological exploration. Contour maps and selected profiles drawn across the anomaly are sketched in figure 5.28.

FIGURE 5.28
Contour maps of total intensity

Dipole, inclination of F, 60°. (For inclination −60°, i.e., southern hemisphere, N is ↓)

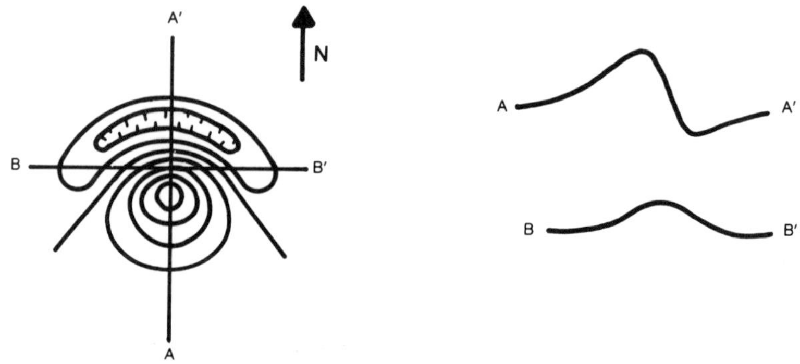

Dipole, horizontal field

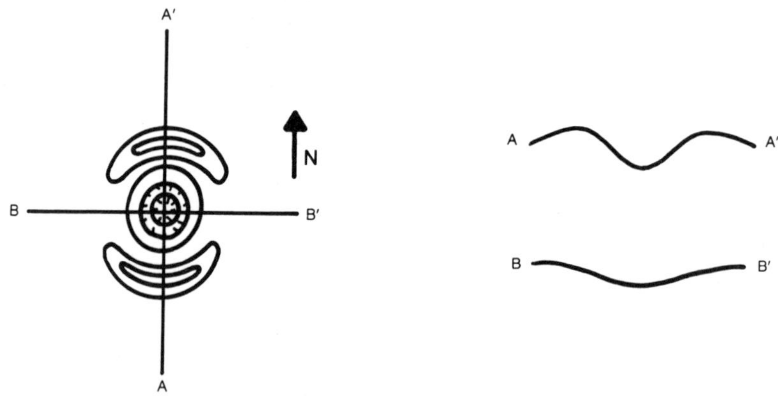

Vertical prism
Vertical field

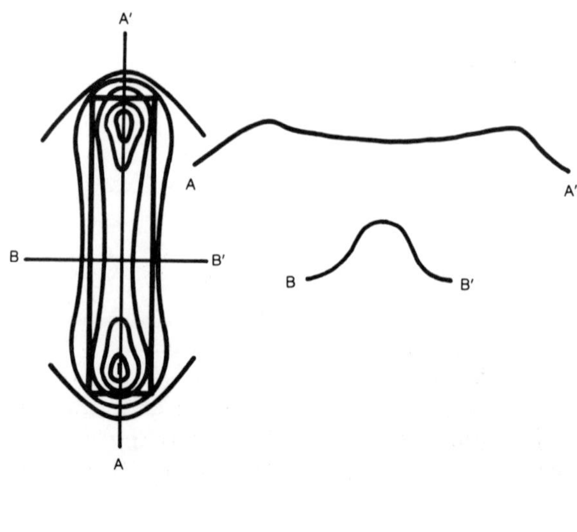

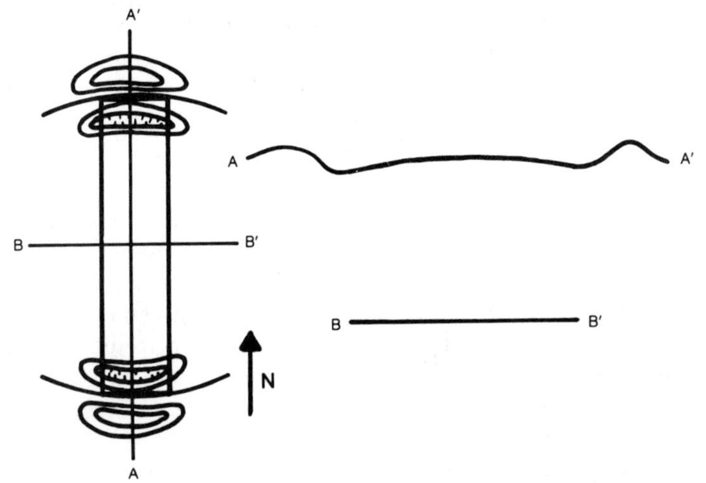

FIGURE 5.28, continued

Vertical prism
Horizontal field

Anomaly amplitude

Amplitude estimates for common sources

The large-amplitude, commonly observed anomalies (several hundred gammas or larger) are almost always the result of a large magnetization contrast, i.e., change in lithology where one igneous rock is in juxtaposition with another or with a sedimentary or metamorphic rock of much lower susceptibility. It must be remembered that magnetization of common rocks varies over six orders of magnitude. Anomalies due to structure alone, i.e., varying configuration of a uniformly magnetized rock, seldom produce anomalies larger than 10 or 100 gammas.

The relative amplitude of a given anomaly (signature) has been shown to be a function of the earth's field direction, the configuration of the source, and the remanent magnetization, if any. The *maximum* amplitude of an anomaly is, on the other hand, largely a function of the depth and the contrast in the mass of magnetite (or iron, for example, in case of search), and to a lesser extent, the configuration of the source. It is of interest to be able to estimate the maximum amplitude for a given source in order to 'model' it for the sake of interpretation. This estimated amplitude can be used with the normalized (i.e., dimensionless) anomaly signatures above and in figure 5.27 to produce the anomaly one wishes for comparison with the observed. Estimation of the maximum anomaly amplitude is also useful in planning a survey or planning the grid and coverage necessary in search applications.

For a few generalized configurations, it is relatively simple to estimate the maximum anomaly amplitude (at a single point above the source) assuming a depth, susceptibility, and much simplified shape of the source. Expressions are given in the literature for calculation of anomalies of more complex figures. Later in this section, the calculation of the complete signature, i.e., the amplitude as a function of distance along the profile for a few simple forms, is given. The methods described herein are merely order-of-magnitude techniques, but are useful for the applications covered by this chapter.

Estimation of the maximum anomaly for comparison with a given source requires first that the signature be studied for the nature of the source; namely, whether the source can be approximated as an isolated dipole, monopole, or line or sheet-like array of such. In the case of the latter two, adjacent traverses or a contour map may be required to determine if it is two-dimensional, i.e., very long normal to the traverse. A depth is then assumed or crudely estimated (according to procedures that follow). In addition, the susceptibility is assumed or, if source rocks are accessible, it can be measured. The following formulas can then be used; they are based upon simplifications and assumptions and are often no better than a factor of two.

The basic expression for estimating the maximum amplitude of any anomaly is

$$T = \frac{M}{r^n}$$

where T is the anomaly, M the magnetic moment, r the distance (depth) to the source, and n a measure of the rate of decay with distance, or fall-off rate (n = 3 for a dipole, n = 2 for a monopole).

Since the magnetic moment M (and k) is usually given in centimeter-gram-second (cgs) units, r must be in centimeters, n in dimensionless, and T is in gauss. To express T in gammas, multiply M by 10^5; if r is in feet, multiply r by 30 and raise the quantity 30r to the exponent n, e.g., if the source is a dipole, then n = 3, and if say, r = 2 ft, M = 1000 cgs, then

$$T = \frac{1000 \times 10^5}{(2 \times 30)^3} = 460 \text{ gammas.}$$

Dipole and monopole signatures in vertical and horizontal fields

The very generalized expression for the maximum anomaly one may expect from a dipole or monopole was presented above in its simplest form. It may be of interest, however, to construct the signature of a dipole or monopole in a vertical or horizontal earth's field as would be observed by a total field magnetometer along a traverse over the source.

Apart from any total field considerations, a dipole has a field with magnitude and direction given by the radial and tangential components, T_r and T_θ, according to the following expression and for the geometry shown.

$$T_r = \frac{2M \cos \theta}{r^3}$$

$$T_\theta = -\frac{M \sin \theta}{r^3}$$

Where the earth's field is vertical or nearly vertical (dip 70° to 90°), the dipole, if induced, would also be vertical, and the total field magnetometer would measure the component, T_z, along this vertical direction, where

$$T_z = T_r \cos \theta + T_\theta \sin \theta$$

$$= \frac{2M \cos^2 \theta - M \sin^2 \theta}{r^3}$$

$$= \frac{M(2z^2 - x^2)}{(x^2 + z^2)^{5/2}}$$

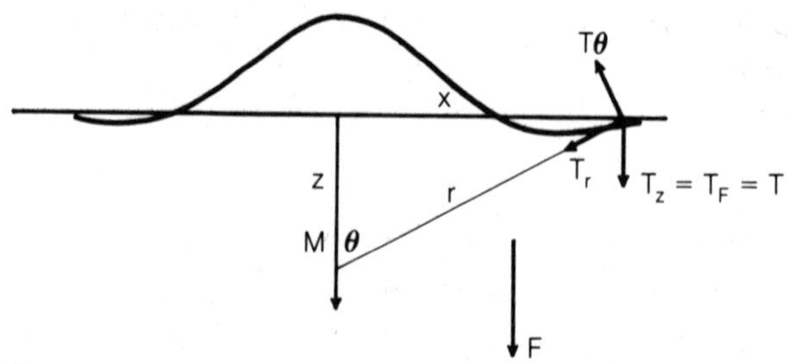

As before $T_z = T_F = T$, the anomaly.

At x = 0,

$$T = \frac{2M}{z^3}$$

at $x = \pm z$,

$$T = \frac{0.175M}{z^3}$$

at $x = \pm \sqrt{2}z$,

$$T = 0$$

at $x = \pm 2z$

$$Tz = \frac{-0.04M}{z^3}.$$

For magnetic equatorial fields, the induced anomaly is horizontal, and the total field magnetometer would measure the components shown and expressed by

$$T_x = T_r \cos\theta + T_\theta \sin\theta$$

$$= \frac{2M\cos^2\theta - M\sin^2\theta}{r^3}$$

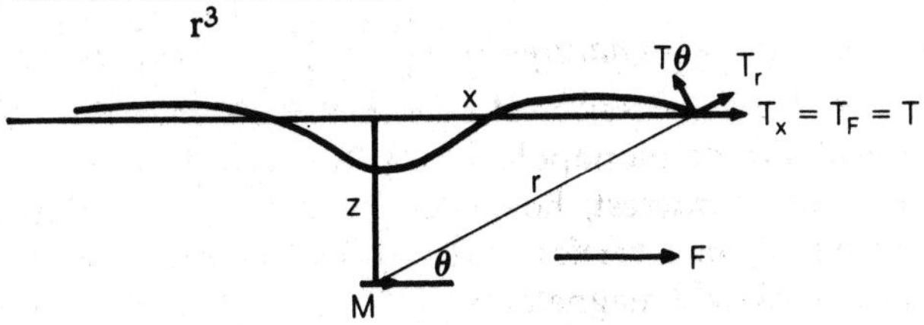

$$= \frac{M(2x^2 - z^2)}{(x^2 + z^2)^{5/2}}$$

as before, $T_x = T_F = T$ the total field anomaly, where

at $x = 0$,

$$T = -\frac{M}{z^3}$$

at $x = \pm \frac{z}{\sqrt{2}}$,

$$T = 0$$

at $x = \pm z$,

$$T = \frac{0.175M}{z^3}$$

at $x = \pm 2z$,

$$T = \frac{0.125M}{z^3}.$$

The monopole shown here has only radial components whose intensity is expressed by

$$T_r = \frac{M}{r^2}.$$

The monopole anomaly in a vertical field as measured by a total field magnetometer would be the component in the z direction (vertical) or

$$T_z = T_r \cos\theta$$

$$= \frac{M\cos\theta}{r^2}$$

$$= \frac{Mz}{(x^2 + z^2)^{3/2}}$$

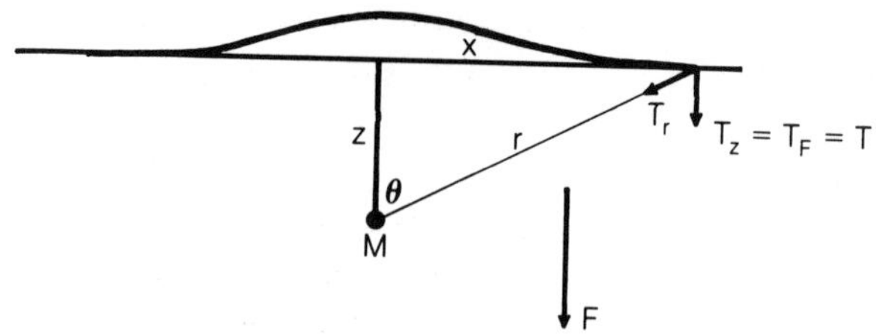

assigning $T_z = T$, the anomaly

at $x = 0$,

$$T = \frac{M}{z^2}$$

at $x = \pm z$,

$$T = \frac{0.35M}{z^2}$$

at $x = \pm 2z$,

$$T = \frac{0.09M}{z^2}.$$

The monopole field in a horizontal field would be measured by a total field magnetometer as the horizontal component, T_x, where

$$T_x = T_r \sin\theta$$

$$= -\frac{M}{r^2}\sin\theta$$

$$= -\frac{Mx}{(x^2 + z^2)^{3/2}}$$

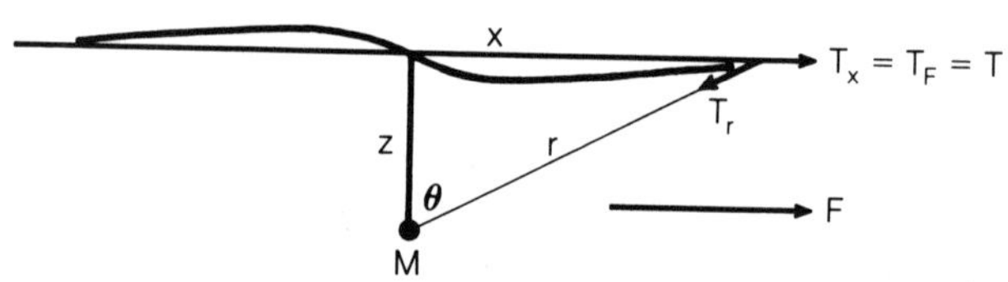

Again, $T_x = T_F = T$, the anomaly, where

at $x = 0$,

$$T = 0$$

at $x = z$,

$$T = -\frac{0.35M}{z^2}$$

at $x = -z$,

$$T = \frac{0.35M}{z^2}$$

at $x = 2z$,

$$T = -\frac{0.18M}{z^2}$$

at $x = -2z$,

$$T = \frac{0.18M}{z^2}.$$

Maximum amplitude given magnetization and generalized form

The magnetic moment M is more usefully expressed as

$$M = IV$$

where I is the magnetization (i.e., magnetization contrast) per unit volume and V the volume. This magnetization is composed of a usually unknown proportion of remanent magnetization, I_r, and induced magnetization I_i. The latter as expressed earlier is

$$I_i = kF$$

where k is the magnetic susceptibility per unit volume and F the earth's field or ambient inducing field. (NOTE: Since I_r is seldom known, an effective magnetization, $I = I_i + I_r$, will always be used. Also it is assumed that $k < 10^{-2}$, i.e., the source under consideration contains less than 10 percent magnetite; then one can ignore what is known as demagnetization effects in the calculation of anomaly amplitude.)

Therefore, for a dipole which can always be assumed for a source all of whose dimensions are small with respect to the distance (less than 1/5 or 1/10) to the magnetometer,

$$T = \frac{M}{r^3} = \frac{IV}{r^3} = \frac{kFV}{r^3}.$$

If the source is approximately spherical, then

$$T = \frac{kF\left(\frac{4}{3}\pi R^3\right)}{r^3}$$

where R is the radius of the source as in figure 5.29.

If the measurement is made along the axis of the dipole (see figure 5.30), then

$$T = \frac{2kF\left(\frac{4}{3}\pi R^3\right)}{r^3}.$$

As an example, consider an orebody 100 ft wide (R = 50), 500 ft deep comprised of 10 percent magnetite (k = 0.3) in a steeply dipping field (60° to 90° latitude in a field of 60,000 gammas:

$$T = 2(0.10 \times 0.3) \times 6 \times 10^4\left(\frac{4\pi}{3}\right)\left(\frac{50}{500}\right)^3 = 14.4 \text{ gammas.}$$

For the same orebody in an equatorial field where F = 30,000 gammas and the induced dipole is now observed at a point on a line normal to the axis (no factor of 2),

$$T = -3.6 \text{ gammas.}$$

Thus, a given dipolar source in an equatorial field will have only one-fourth the maximum anomaly amplitude it would have in a polar region.

The preceding expressions are usually valid only for such sources as a small, distant orebody (containing magnetite), small structure in deep basement, or most objects involved in search applications. The magnetization is expressed in gauss or gammas as desired. Since the anomalies are also expressed in terms of magnetic units, it follows that the units of dimension in the numerator must be of the same order as the denominator since they must cancel. Therefore, for a dipole whose anomaly varies as $1/r^3$ (said to have a fall-off of $1/r^3$), the volume, V, has dimensions of R^3. In the case of a monopole, which varies as $1/r^2$, the magnetic moment, M, is equal to IA where A is surface area and has

FIGURE 5.29
Anomaly of sphere in horizontal field

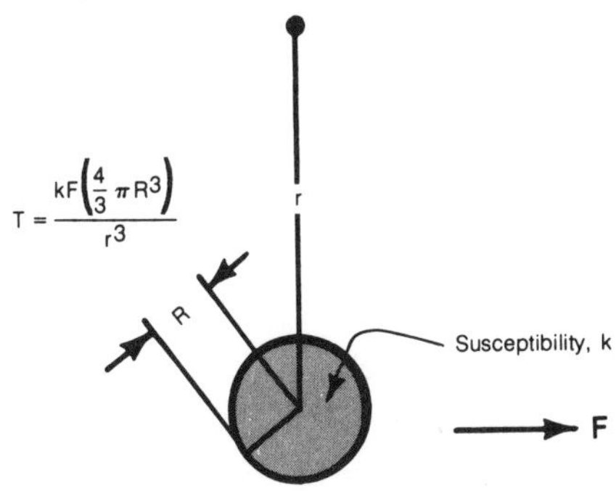

FIGURE 5.30
Anomaly of sphere in vertical field

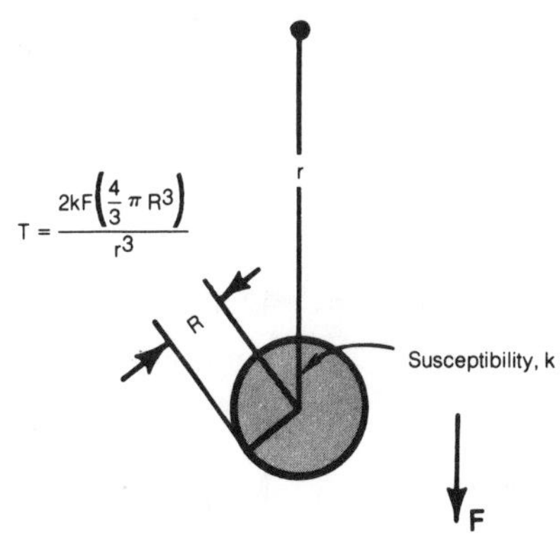

FIGURE 5.31
Anomaly of vertical and horizontal cylinders

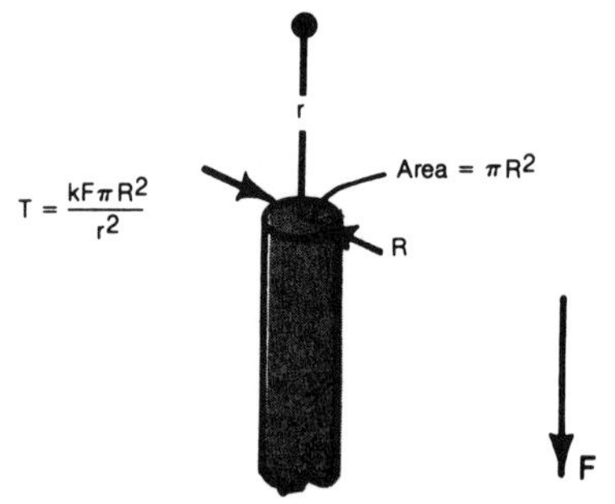

(Note: Also valid for end of N-S horizontal cylinder in horizontal field)

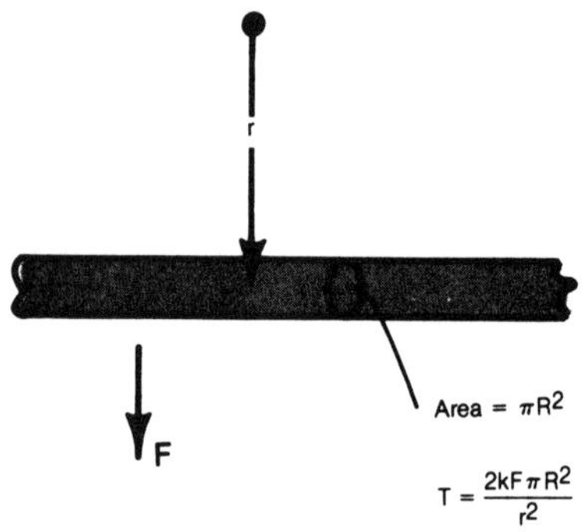

(Note: Also valid for E-W horizontal cylinder in horizontal field)

FIGURE 5.32
Anomaly of narrow vertical dike

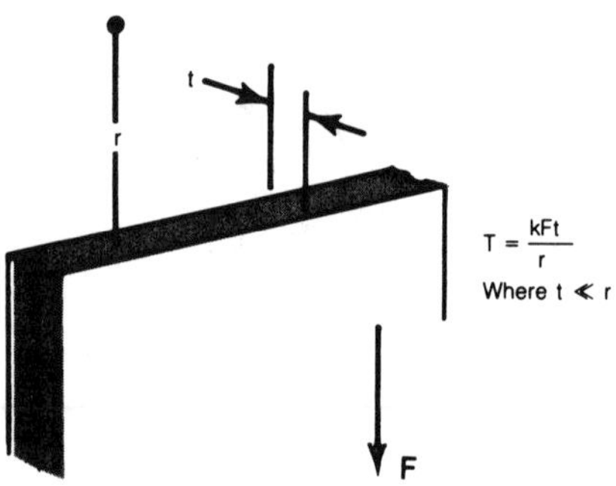

FIGURE 5.33
Anomaly of semi-infinite slab

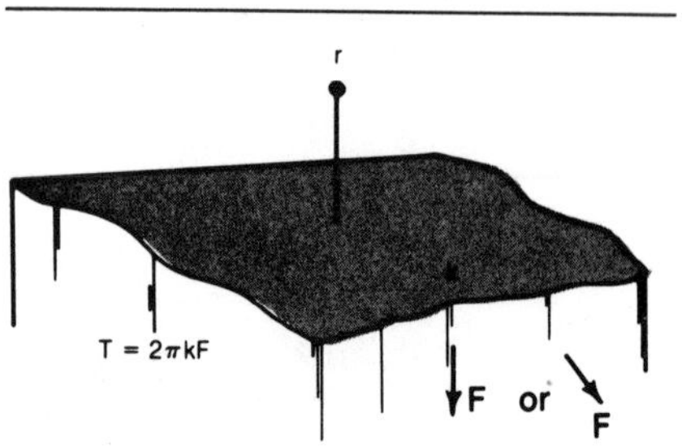

dimensions of R^2. Consider, for example, a vertical basement intrusive in a polar region with an upper surface 1,000 ft in diameter at a depth of 5,000 ft, with a susceptibility contrast of 10^{-2} in a field of 60,000 gammas. Thus,

$$T = \frac{kF\pi R^2}{r^2} = 10^{-2} \times 6 \times 10^4 \times \pi\left(\frac{50}{500}\right)^2 = 18 \text{ gammas.}$$

Horizontal prisms or cylinders also vary as $1/r^2$, with magnetic moment M equal to 2IA (IA for E-W horizontal prisms in equatorial regions) where A is the cross-sectional area of the prism (see figure 5.31). (NOTE: The long horizontal prism varies as $1/R^2$ not because it appears to be comprised of a monopole, but because it is a line of dipoles (in steeply dipping fields) and the effect of adjacent dipoles along an infinitely long line is 'seen' better by the magnetometer at a distant point of measurement than if all the magnetization were concentrated at a point as in an isolated dipole.)

A narrow, vertical dike in steep field or the edge of a horizontal sheet in a horizontal field can be considered as a line of monopoles varying as $1/r$ which is a lower rate of fall-off than a single monopole for the same reasons given for a horizontal cylinder (see figure 5.32). The magnetic moment $M = It$ where t = width of dike. Since the anomaly varies as $1/r$, the dimensions of t are simply length. As an example, a vertical dike might be 100 ft wide, at a depth of 500 ft, with $k = 10^{-3}$ in a field of 50,000 gammas, or

$$T = \frac{kFt}{r} = \frac{10^{-3} \times 5 \times 10^4 \times 10^2}{5 \times 10^2} = 10 \text{ gammas.}$$

A common point of ambiguity arises with such simplified schemes as these in the case of a dike which is nearly as wide as it is deep. In this case, the anomaly is approximated as something between a line of monopoles as above and a sheet of monopoles as shown in the following. Moreover, as the dike is even wider than its depth, it can be approximated simply by two faulted contacts with 'no anomaly' in between.

For a semi-infinite slab of material, such as a rock surface of great thickness and breadth in a non-horizontal field, the flux lines do not vary in direction or density above the slab; therefore, the field does not vary at all with distance to its surface (similar to the limit of the spherical dipole above where R = r) so that

$$T = \frac{M}{r^0} = \frac{2\pi I}{1}, \qquad \text{or} \qquad T = 2\pi kF$$

which is useful in estimating the magnitude of the anomaly at a vertical fault (see figure 5.33). For example, consider two rock types at a vertical contact of $k = 10^{-3}$ and $k = 10^{-5}$ for an effective susceptibility contrast of $k = 10^{-3}$ ($10^{-5} \approx 0$ relative to 10^{-3}) and where F = 50,000 gammas. Thus

$$T = 2\pi \times 10^{-3} \times 5 \times 10^4 = 300 \text{ gammas.}$$

If the rocks had $k = 10^{-4}$ and 10^{-3}, the effective susceptibility contrast would be

$$10^{-3} - 10^{-4} = 10 \times 10^{-4} - 10^{-4} = 9 \times 10^{-4} \text{ and}$$

$$T = 2\pi \times 9 \times 10^{-4} \times 5 \times 10^4 = 270 \text{ gammas.}$$

This simple example of two adjacent rock types is probably applicable in more instances in interpretation than any of the other geometries discussed above.

Anomaly depth characteristics

In a very approximate fashion, the wavelength, or, affective width (or 'half-width' described in the following) of the anomaly and, with more accuracy, the width of certain characteristics of the anomaly such as slope, are measures of the depth to its source. However, recognition of the anomaly, the anomaly 'zero', and certain slopes would not only appear as different values as determined by different interpreters, but they also depend upon what is removed as the regional gradient. More objective criteria are used in some cases such as the nearly straight portions of a slope, and distances and angles between inflection points, peak values, and other anomaly characteristics.

Anomaly width

In general, the anomaly width as shown in figure 5.34 is on the order of 1 to perhaps 3 times the depth. Thus, when an anomaly appears to have a width as such of 100 ft, it is definitely not produced by a source at 1,000 ft or at 10 ft, but more likely by a source between 30 and 100 ft deep (or distant). Such a criterion, approximate as it is, is nevertheless useful for cursory interpretation of profiles and maps.

Anomaly depth estimation

Much is written on the variety and relative merit of methods for estimating the depth to the source of anomalies. Since the magnetometer is primarily a tool for subsurface mapping and detection, it follows that determination of the depth as well as edges of bodies is important in its application to geological exploration and search. The basis for depth determination is presented here in brief. Together with the foregoing background on anomaly behavior, one should at least appreciate how a variation in depth affects an anomaly. In most cases, one need only apply this knowledge qualitatively through visual inspection of a profile. Whatever the requirement, depths may be estimated by visual inspection, several rules of thumb, modeling (i.e., calculation of assumed source and comparison with observed), measured gradient techniques, or various computer-oriented procedures. As was demonstrated earlier, a given anomaly could have an infinite number of possible sources and source depths, but the realistic models that are assumed usually produce maximum depth estimates.

Knowledge of the depth of a particular formation or source may have considerable geological significance as it determines the nature or configuration of a formation, the slope of its surface, and its discontinuities. Knowing the depth to various points on the surface of crystalline rock or magnetic basement allows one to map that surface and its topography and structures to depths exceeding 30,000 ft and to infer thickness of sediments or conformable sedimentary structures above it for exploration of petroleum, sedimentary ores, placer deposits, or groundwater. Areas underlain by pediment or other sedimentary deposits may be ruled economic or noneconomic according to depth. The depth to ore deposits associated with pyrrhotite, magnetite, or ilmenite may be estimated as an aid to a drilling program or even for estimation of total tonnage of magnetic iron ore deposits. Black sand deposits of rutile, zircon, monazite, diamonds, gold, and platinum are often associated with other high density, very resistant, yet magnetic minerals; namely, magnetite or ilmenite. The depth to objects of search, whether buried iron or man-made structures, is invaluable in guiding the subsequent excavation efforts.

Identification of anomaly

The anomaly of interest must be identified and discriminated against the obscuring effects of others. Recognition of the anomaly itself is usually

FIGURE 5.34
Anomaly 'width'

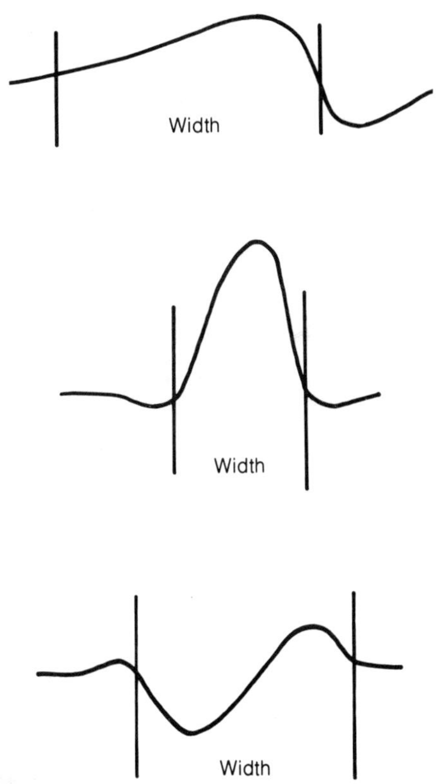

the most difficult aspect of depth determination because of the composite effects of multiple sources and sources at various depths and distances in any direction from the magnetometer. Only the net effect of all anomalies is measured by the magnetometer since it has no inherent discrimination ability at the disposal of the operator. The anomaly should be inspected to ascertain the probable source and, if complex, the possible combination of sources. For example, a wide, shallow dike will appear as two anomalies which may or may not coalesce depending upon the relative width and depth. A very broad anomaly or regional gradient (described earlier) is usually caused by anomalies which are extremely deep or distant or by the normal variation in the earth's magnetic field. If one wishes to remove this gradient, it can be done either by drawing a straight line through the non-anomalous portions of the profile (away from the anomaly of interest) or by drawing a very smooth but broad wavelength curve through the data of much longer wavelength than any anomalies of interest. This regional gradient or background is then subtracted from the anomaly and the remaining, or residual anomaly, replotted. It is this anomaly which is then interpreted for either depth or for amplitude or general configuration of sources as previously described.

Fall-off rate

The variation of anomaly amplitude with distance, or fall-off rate, is important in the interpretation of anomalies. It relates the anomaly to depth; it describes in a general way the configuration of the source; and it assists in determining susceptibility and mass of the causative magnetite. Recall that the anomaly from a dipole varies as $1/r^3$ and that of a monopole as $1/r^2$. The fall-off rate, in actual practice, does not involve precisely such factors or exponents but, in fact, is typically $1/r^{2.5}$, $1/r^{0.6}$, or even $1/r^0$ as described above. In other words, various configurations of dipoles, monopoles, lines, and sheet-like distributions of these poles constitute a continuous series of fall-off rates even in the vicinity of a single anomaly as one is much closer or farther away from the source.

Representing various geologic sources as simple prismatic bodies, one may assume the following fall-off rates: a dipole will be produced by a source all of whose dimensions are small (less than one-tenth compared to the distance between the source and magnetometer). Such a body is rarely seen in nature except as a very confined, usually magnetite-rich orebody. A monopole varying as $1/r^2$ will be produced by a long, thin, vertical prism, such as a narrow vertical intrusive in steeply dipping fields or a horizontal cylinder striking N-S in equatorial fields (e.g., a N-S anticlinal structure on the basement, one end of which is near the magnetometer). A line of dipoles is produced by a long, horizontal cylinder magnetized through its short dimension as in steeply dipping latitudes or striking E-W in equatorial regions. Such a cylinder will also vary as $1/r^2$. A line of monopoles would effectively be observed near one edge of a dike dipping in the direction of the field and would vary approximately as $1/r$. At a point above a horizontal, semi-infinite sheet, the field would vary inversely as $1/r^0 = 1$, which is another way of expressing the fact that the field does not vary at all with distance from a horizontal, semi-infinite sheet of monopoles or dipoles. A wide, vertical dike in a steep field or the edge of a fault might represent combinations between a line of dipoles or sheet-like distribution of monopoles and may thus vary as $1/r^2$ or $1/r^{0.5}$ or less. Figure 5.35 indicates these variations.

Assumptions on maximum amplitude and depth estimates

Unless the remanent magnetization is actually measured, it is generally disregarded, and only the induced magnetization and susceptibility are

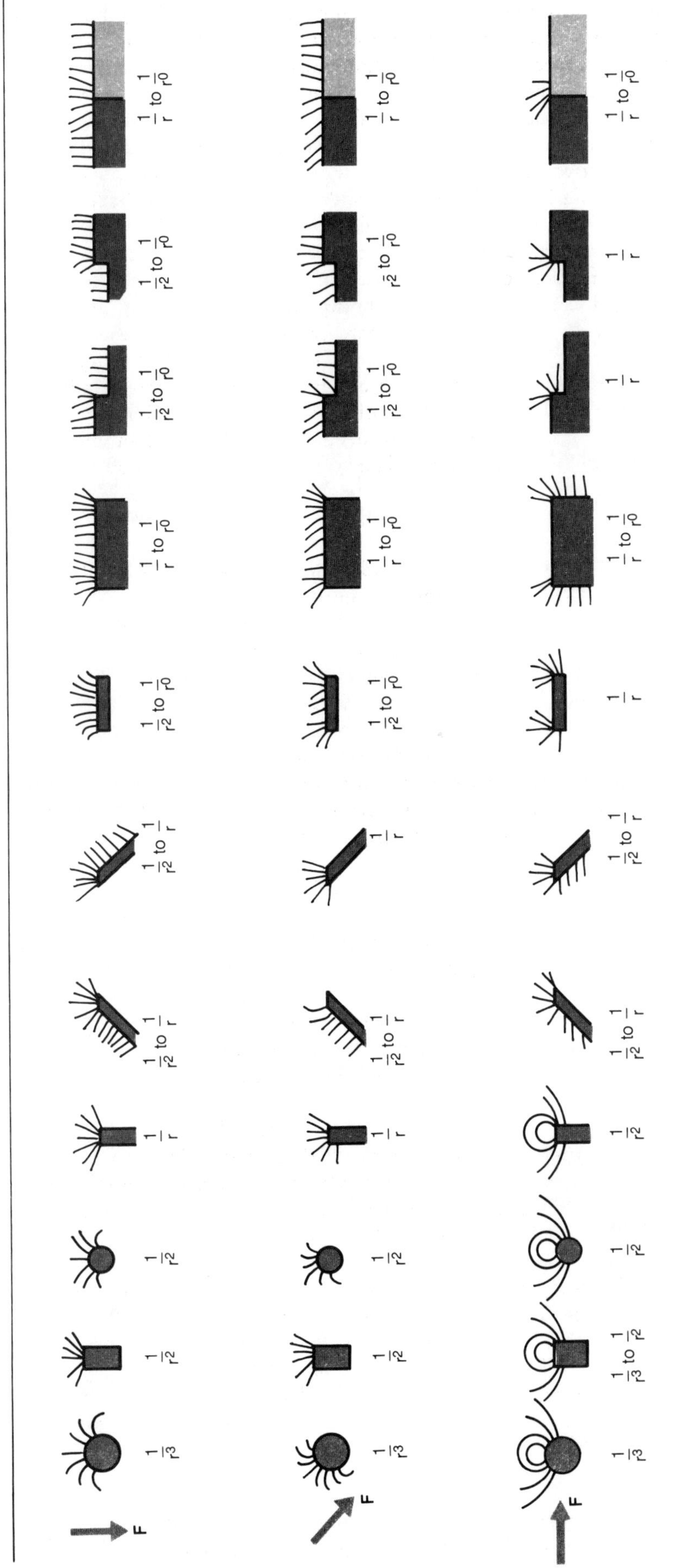

FIGURE 5.35
Field lines and fall-off rates of various geologic models

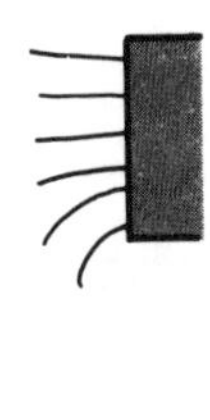

Flux lines superimposed on representative geologic models for various orientations of inducing field. Anomaly amplitude proportional to indicated term of $1/r^n$.

FIGURE 5.36
Half-width rules—vertical field

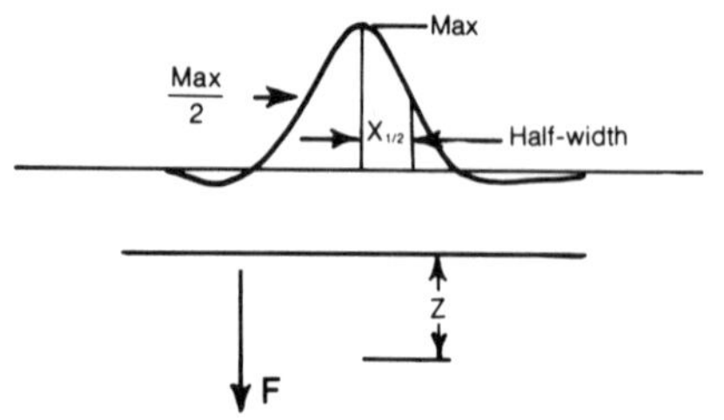

Sphere (dipole)	$Z = 2X_{1/2}$
Vertical cylinder (monopole)	$Z = 1.3X_{1/2}$
Edge of narrow dike (line of monopoles)	$Z = X_{1/2}$
Horizontal cylinder (line of dipoles)	$Z = 2X_{1/2}$

FIGURE 5.37
Half-width rules—horizontal field (equatorial)

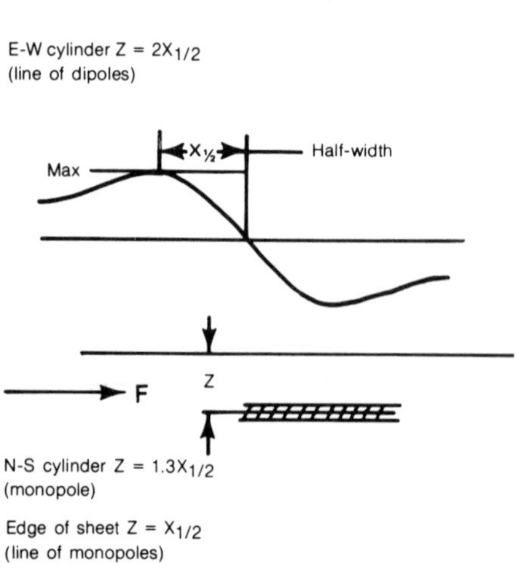

used in these expressions. The magnetic anomaly calculated from these highly simplified expressions represents the maximum amplitude from the local zero, non-anomalous field to the positive peak value in the northern and southern latitudes and to the minimum negative value in equatorial regions. It does not represent the peak-to-peak value which includes both positive and negative portions of the anomaly signature. The depth estimates derived from any of the techniques described are seldom more accurate than 10 percent of the actual depth and sometimes as poor as 50 percent. By theory, most of the estimates are maximum estimates so that the real source will actually be at a shallower depth. Moreover, the 'poles,' or source described frequently throughout this chapter, are within the geologic body or object of search and not simply on the surface; therefore, such depths are again maximum depths.

Half-width rules

In vertical or horizontal fields, it can be shown from the previous expressions for dipoles and monopoles, that for simple forms of anomaly sources, the depth to their centers is related to the half-width of the anomaly. The half-width is the horizontal distance between the principal maximum (or minimum) of the anomaly (assumed to be over the center of the source) and the point where the value is exactly one-half the maximum value (see figure 5.36). This rule is valid only for simple-shaped forms such as a sphere (dipole), vertical cylinder (monopole), and the edge of a narrow, nearly vertical dike (line of monopoles) in the polar regions. At the magnetic equator, the half-width rules are somewhat different with the sphere remaining unchanged; an E-W horizontal cylinder being a line of dipoles, a N-S cylinder being a monopole, and the edge of an E-W striking horizontal sheet representing a line of monopoles. The rules presented in figure 5.37 apply according to the corresponding array of poles. In the case of the latter two, the half-width is the horizontal distance between the point of maximum (or minimum) and zero anomaly. The half-width rules are derived from formulas given earlier in "Dipole and Monopole Signatures in Vertical and Horizontal Fields."

Slope techniques

Perhaps the most commonly used set of methods for estimating depth is that which uses criteria involving the measurement of the horizontal gradient or slope at the inflection points of the anomaly. Based upon empirical observations using computed models, these slopes are measured according to the horizontal extent of the 'straight' portion of the slope (see figure 5.38) or the horizontal extent determined by different combinations of the tangent or slope at the inflection point, and maximum of the anomaly and half slopes. Each of these horizontal distance measurements when multiplied by an empirically-determined factor equals the depth to the top of the anomaly source. (The straight-slope, for example, is multiplied by a factor between 0.5 and 1.5). Detailed explanations of these methods are available in the references cited.

Other depth estimating methods

Modeling techniques require that one examine the observed anomaly for its likely source configuration. A model is assumed, the anomaly calculated, compared with the observed, and repeatedly altered until a satisfactory fit to the observed data is finally achieved. Such work is usually performed on a computer. Other computer-oriented depth estimating methods include programs using Fourier and Hilbert transforms, convolution, and other semi-automated programs which are usually applied to large volumes of data. Gradiometer measurements made with sensors at two points usually vertically arranged can also be used for depth estimates.

Interpretation summary

Interpretation is facilitated if one can thoroughly familiarize oneself with how and why a given source produces an anomaly in the earth's field, the nature of total field measurements, and the general behavior of an anomaly signature with increasing depth. What at first may have appeared complex in the interpretation of field profiles and maps, is more readily understood when the above phenomena are examined one at a time.

The first procedure that should be followed in the interpretation of a given profile is to focus on the anomaly width and shape and attempt to construct at least a mental image of the source (or object in the case of search) in realistic geologic terms and its depth. Use the eye to discriminate against noise and the regional gradient or filter by one of the suggested techniques. Anomalous horizontal gradients should then be used, for lack of any other specific criteria, as an indicator of the edge of subsurface structures producing a magnetization contrast. Most anomalies on any given profile or map represent a simple contrast in magnetization or lithology, i.e., the edge of a body. Attempt to correlate such features on adjacent lines or interpret them as contacts on a total intensity contour map. The cessation, displacement, or interruption of otherwise long or continuous features may also represent significant geologic structural information. However, one must realize also that a magnetic survey is only able to map a contact where there is a magnetization contrast. For example, different lithologies on either side of a long, continuous fault will be mapped only in segments where such contrasts occur.

Changes in the character of the short wavelength anomalies (noise) may also represent mappable information if one is careful to evaluate their typical depth so as not to be mapping irrelevant soil anomalies. Negative anomalies arising from features of locally lower magnetization are as important geologically as the more common positive anomalies. Furthermore, the *most geologically significant anomalies on a given map are probably the more subtle ones and not necessarily the largest, most prominent anomalies.* Lastly, the total intensity profiles and maps are not an end in themselves, but are rendered usable only when expressed in terms of geology (or objects of a search). The more geological information one has (or size, magnetic, or depth information for an object of search), the more valuable the total intensity data becomes.

FIGURE 5.38
Slope method for depth estimation

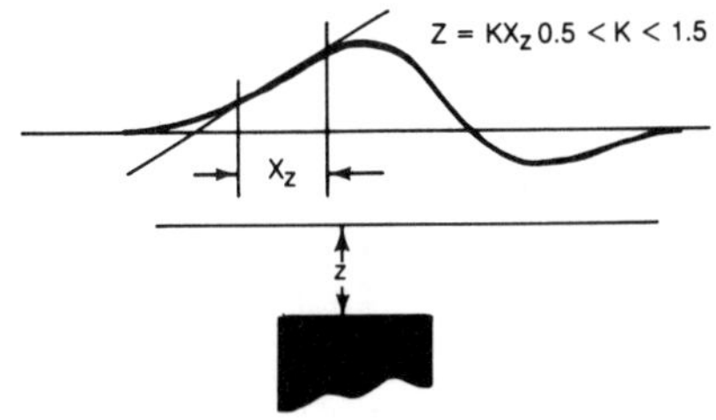

Selected References (Chapter 5)

Aitken, M. J. 1961. *Physics and archaeology.* London: Interscience.

————. *Archaeometry,* a quarterly journal of The Research Laboratory for Archaeology and the History of Art, Oxford University. London: Cambridge Univ. Press.

Bossolasco, M., ed. *Pure and applied geophysics,* a quarterly journal. Basel: Birkhauser Verlag.

Bozorth, R. 1951. *Ferromagnetism.* New York: Van Nostrand.

Chapman, S., and Bartels, J. 1940. *Geomagnetism, Vol. I and II.* London: Oxford Univ. Press.

Dobrin, M. B. 1960. *Introduction to geophysical prospecting.* New York: McGraw-Hill.

————. *Geoexploration,* a quarterly journal. Amsterdam: Elsevier.

————. *Geomagnetism and geoelectricity,* a monthly journal. Univ. of Tokyo.

————.1948, 1956. *Geophysical case histories, Vol. I and II.* Tulsa: Society of Exploration Geophysicists.

————. *Geophysical prospecting,* a quarterly journal. The Hague: European Association of Exploration Geophysicists.

————. *Geophysicists,* The Hague.

————. *Geophysics,* a bi-monthly journal. Tulsa: Society of Exploration Geophysicists.

Grant, F. S., and West, G. F. 1965. *Interpretation theory in applied geophysics.* New York: McGraw-Hill.

Haalck, H. 1934. *Lehrbuch der Angewandten Geophysik.* Berlin.

Heiland, C. A. 1940. *Geophysical exploration.* New York: Hafner.

————. *Izvestia, series geophysics,* a monthly journal. Moscow: Academy of Sciences of the USSR.

Jakosky, J. J. 1957. *Exploration geophysics.* Los Angeles: Trija.

————. 1967. *Mining geophysics, Vol. I and II.* Tulsa: Society of Exploration Geophysicists.

Landsberg, H. E., ed. 1962. *Advances in geophysics,* Vol. 1-15. New York: Academic Press.

Morley, L. W., ed. 1970. *Mining and ground water geophysics.* Economic geology report No. 26. Ottawa: Dept. of Energy, Mines and Resources.

Nagata, T. 1961. *Rock magnetism.* Tokyo: Maruzen.

Nettleton, L. L. 1940. *Geophysical prospecting for oil.* New York: McGraw-Hill.

Parasnis, D. S. 1962. *Principles of applied geophysics.* New York: Wiley.

————. *Prospezioni archaeologiche,* an annual journal. Rome: Lerici Fondazioni.

Runcorn, S. K., ed. 1960. *Methods and techniques in geophysics.* New York: Interscience.

________________. *Transactions, American Institute of Mining and Metallurgical Engineers. Geophysics,* a journal. New York: AIME.

Vacquier, V., Steenland, N. C., and Henderson, R. G. 1951. *Interpretation of aeromagnetic maps.* Geological Society of America, Memoir No. 47. New York: GSA.

CHAPTER 5 ADDENDUM MAGNETICS

THE UTILITY OF MAGNETICS

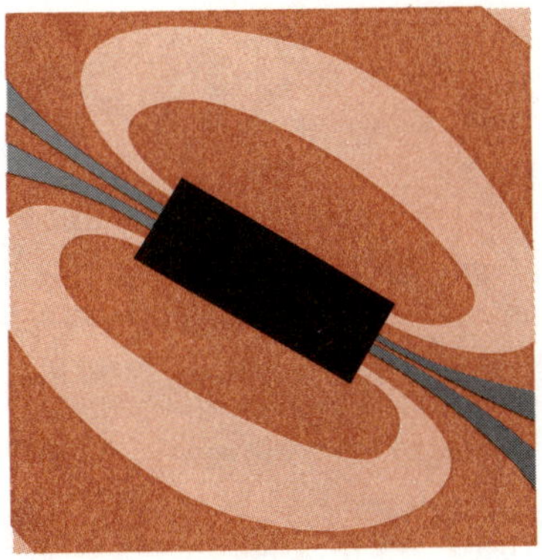

The study of the earth's magnetic field is the oldest branch of geophysics, dating back to the use of the dipping needle and compass in the 18th century in Sweden. Exploration applications surged after World War II, when airborne fluxgate magnetometers were developed as an anti-submarine detection device by the Navy. Widespread use in exploration during the '50s and '60s placed a strong emphasis on regional and national airborne magnetic surveys. Then, in the last 20 years, proton precession, alkali vapor and, more recently, SQUID instrumentation introduced a series of sensitive, portable systems for use both in the study of the earth's geomagnetic field and in exploration for hydrocarbons and minerals.

Perhaps the most versatile of all geophysical prospecting techniques, the magnetic method can be applied on land, in the air, and both on and under water surface. Magnetics is also one of the least expensive methods to apply and has applications that range from micro-surveys in archeology and engineering site investigations to mega-scale studies of crustal tectonic features.

John (Jack) D. Corbett is a consulting geophysicist in mineral exploration, and resides in Denver, Colorado. Formerly Chief Geophysicist of Anaconda Minerals, he has worked for 35 years in precious metal, diamond and industrial mineral exploration in various geologic environments throughout the United States, Alaska, Canada and in Australia and Norway. Graduating from Saint Louis University, he also holds an MS from Colorado School of Mines. During recent years he has been active in introducing geophysics to geologists through short courses, in-house seminars, lectures at several U.S. universities, and as an invited lecturer in the Peoples Republic of China. Involved for many years in the Society of Exploration Geophysicists, he was elected to the position of Vice-President in 1985.

John (Jack) D. Corbett, Consulting Geophysicist

Role of Magnetics

Analysis of local variations in the total magnetic field can yield information on lateral variations of magnetic properties which is directed toward identification of different lithologic units and structure. In an exploration mode, the most common applications of exploration magnetics are to assist geologic mapping of the subsurface through till or shallow soil cover, to locate buried intrusive centers and/or to determine depth of crystalline basement. Few surveys today are oriented toward direct detection of any specific type of mineral deposit; magnetite iron ore, black sands and certain skarn deposits are the occasional exceptions.

The application of magnetics to exploration problems is a multiple-step process. In the data acquisition phase, careful planning will provide a data set of lateral dimensions and precision which will allow interpretation appropriate to the problem. Skillful data processing and reduction will remove extraneous "noise" caused by a wide range of spatial and temporal variations and produce a variety of map and profile products. Interpretation itself is a multi-phased process, from a simple qualitative identification of a geologic anomaly to a quantitative analysis and characterization of the source body.

The interpretation of magnetic anomalies (and other potential field methods) is inherently ambiguous, because any specified anomaly can be caused by an infinite number of possible source distributions. Constraints developed from judicious use of geological and other geophysical data (gravity, electrical depth sounding, etc.) reduces the ambiguity.

The word "anomaly" means different things to different people. To some it is simply any unusual feature of a data set; to others an anomaly is, more specifically, a characteristic perturbation of a normal data field. As used here, a magnetic anomaly embodies data that not only have unusual magnitude (above or below the normal geomagnetic field), but also have a sequence of data points of a character and/or shape which imply a target of interest. Often subtle anomalies of low amplitude, distorted by regional fields or adjacent large anomalies, are as important to the explorationist as large simple closures.

The utility of the magnetic method is enhanced by an understanding and appreciation of the many variables that are involved in the formation of an anomaly caused by a simple source. Variations in the shape of an anomaly are related to the source geometry, dip and depth as well as to external factors such as terrain, regional field, line spacing and height. Because all of these conditions affect an anomaly, an awareness of their influence will aid both in the survey design stage and in interpretation of data.

PRINCIPLES OF MAGNETICS

The earth's geomagnetic field has three principal components: the main field itself, an external field and local perturbations superimposed on the main field. Caused by processes in the interior of the earth, the main field has a large magnitude which varies slowly over time. At present, the earth's field amplitude (T) ranges from a low of about 25,000 nanoTeslas[1] (nT) near the geomagnetic equator to almost 70,000 nT at the geomagnetic poles. The field inclination is horizontal at the equator and vertical at the poles. The external field origi-

[1] formerly gamma (γ); 1 nanoTesla = 1 gamma = 10^{-5} oersted.

nates outside the earth's crust and is associated mainly with electric currents in the ionized layers of the outer atmosphere because of interaction with the solar winds. Travelling along magnetic flux lines, solar winds are ionized plasma or hot-charged particles which transmit energy by wave motion. Local variations in the rock and mineral assemblage of the near-surface crust produce local perturbations that are the anomalies of exploration interest.

The earth's magnetic field (a vector field having both amplitude and direction) is described by an intensity (total field intensity, T), an inclination (I) and a declination (D), figure 5.1. For specific applications, horizontal and vertical field components can be derived from T, I and D.

Rock units in the crust acquire a magnetization in the direction of the earth's field which is referred to as induced magnetization or magnetic polarization. The resulting induced field of a typical finite source body is dipolar: that is, it contains positive and negative elements. In the middle-to-high magnetic latitudes of the Northern Hemisphere, the dipolar nature of an induced anomaly is typified by a positive (high) and related, but subdued, negative (low) on its north side.

Anomalies of interest range in amplitude from a few nanoTeslas (nT) for deep basement or sedimentary anomalies to 1,000's of nT for near-surface mafic rocks or iron formations to 10,000's nT for magnetite iron ore deposits.

Several time-variant or temporal variations occur in the geomagnetic ambient field. A long period, or secular change, occurs slowly over many decades or centuries and modifies inclination, declination and intensity. Such a change can be observed as a change in the magnetic declination as noted on old maps when compared to today's version. A complete reversal of the total field direction, occurring over tens-to-hundreds of thousands of years, is a more dramatic effect of the secular change. Many such reversals have been detected and dated during the Tertiary, figure 5.2.

More important to prospecting are the diurnal variations. A diurnal variation of 10 to 100+ nT occurs regularly on a daily basis, (Breiner, 1973). This diurnal is also related to solar winds, the small effects of which vary with the level of the ionosphere and intensity of solar winds. Micropulsations of 0.001 to 10s nT are random effects lasting from 0.02 to tens of minutes. Magnetic storms also produce a short period random "noise" which may vary up to many 100s nT over periods of a few minutes to hours. Storm effects are unpredictable but are related to and follow solar flaring; commencing abruptly and decreasing slowly over hours or days.

Magnetic Physical Properties

The intensity of magnetization or magnetic polarization is a vector quantity related to the vector sum of two components -- an induced component related to the magnetic field strength and direction, and a remanent component. The induced component, which is lost if the applied field is removed, is always in the direction of the earth's field. Remanence, unlike the induced component, has a direction and amplitude independent of the inducing field at the present time and may have any azimuth and inclination to reinforce or oppose the inducing field. The remanent component is permanent; that is, it remains with the rock if it is moved within, or removed from, the ambient field. In effect, a rock with remanence acts like a bar magnet

FIGURE 5.1
Diagram of the earth's magnetic field components

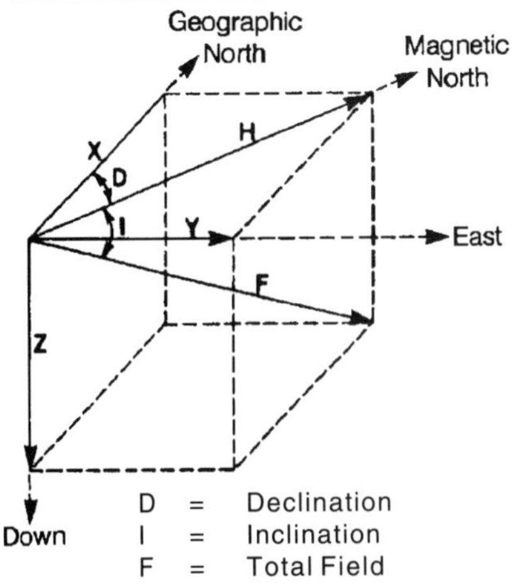

FIGURE 5.2
The magnetic field reversal time-scale over the past 55 Ma (from Harland et al, 1982)

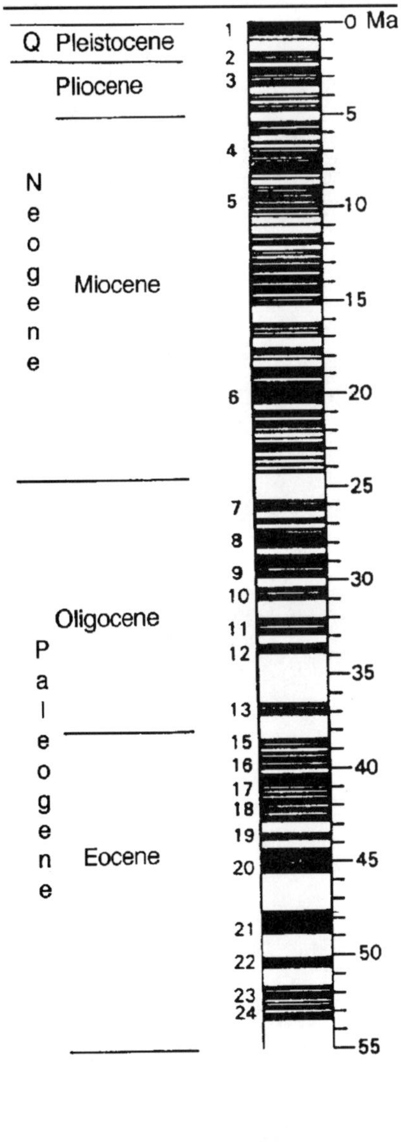

with a separate, independent field which has a strength and orientation of its own. Although considered permanent, remanence does decay slowly over a long period of time.

The induced intensity of magnetization is proportional to the strength of the inducing field. The constant of proportionality between the two, k, is known as the magnetic susceptibility. Stated another way, susceptibility is a measure of the degree to which a substance may be magnetized by induction in the earth's ambient field. Susceptibility is a dimensionless parameter. Noted either as 10^{-6} cgs (i.e., a number like .002) or micro cgs (i.e. 2000), susceptibility may be converted to the occasionally used SI system of units by the relation

$$k \text{ (SI units)} = 4 \ k \text{ (cgs units)}.$$

To a first approximation, susceptibility is related to the magnetite content of a rock. Empirically, the commonly used rule of thumb is that 1 percent magnetite by volume equals 3000 micro cgs (or 4 x 3000 SI units). Magnetite grain size, or the magnetic domains, also has an influence on susceptibility but in practice is a secondary consideration which rarely is evaluated in exploration practice. An approximate range of susceptibilities compiled from several different sources is shown in figure 5.3.

FIGURE 5.3
Magnetic susceptibility chart

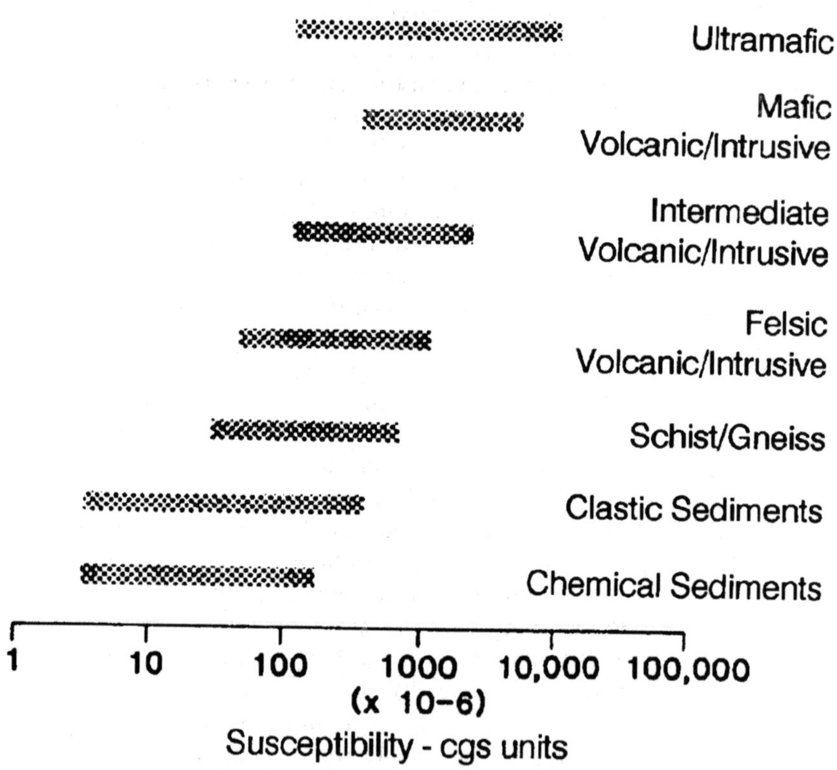

Susceptibility and remanence are highly variable. Charts, such as figure 5.3, must be used with care. Typically mafic igneous rocks have a higher susceptibility than felsic rocks; sedimentary lithologies are usually at the very low end of the scale. Alteration may destroy magnetite and reduce a high susceptibility, but in some types of skarn formation may add magnetite and increase the susceptibility.

The range and variability of susceptibility can easily be in excess of 100 percent or more for one individual, nominally homogeneous, volcanic field. An estimate of susceptibility is required in modelling and interpretation from which a rock type is sometimes inferred. Such inference to a particular rock composition, however, is only a guideline. Specific identification of a particular rock type based only upon magnetic susceptibility is not possible. In practice, it is the susceptibility contrast between two rocks (e.g., a gabbro in a metamorphic terrain, a mafic volcanic in sediments, alteration patterns in an intrusive, etc.) that is of primary interest in the interpretation process. The actual susceptibility of a particular rock unit is secondary in the interpretation process, but it is important in the final evaluation and geologic correlation.

Remanent effects may be small-to-nonexistent or many times larger than the induced effect, may be reversed, or at any angle and azimuth to the induced field vector. A strong remanence at an angle to the induced field will alter the shape and amplitude of an anomaly. Rocks incorporating accessory magnetite in small grains often will have a larger remanent component than those composed of larger grains, because the stability of remanent magnetization is inversely related in part to grain size. Remanence may originate in several different ways and at different times in the history of a rock. The most common, intense and stable component of remanence is thermoremanent magnetization (TRM), which is acquired when a magnetic material is cooled below its Curie temperature (approximately 578° C for magnetite, depending upon impurities) in the presence of an ambient magnetic field. The direction of TRM is parallel to the field at the time of cooling and is quite stable unless reheated. Other types of remanence, related to intrusive activity and heat, metamorphism and/or alteration, can be acquired subsequently. The Curie isotherm occurs at a depth of approximately 35-40 kilometers but may be much shallower in areas of high heat flow.

An igneous rock commonly will acquire a direction of magnetization parallel to that of the earth's field at the time the magma cools below the Curie Point. Time, chemical alteration and other factors can cause an initial remanence to change so that remanence is generally less significant in older rocks and more intense in later volcanics. The strength of remanence is usually quoted as the Koenigsburger Ratio (Q), which is the ratio of remanence to the product of susceptibility and the earth's field. Remanence is often small (low Q) with an orientation subparallel to the present ambient field. If so, the remanent component may be ignored in interpretation.

Magnetite is the primary cause of magnetic anomalies, but pyrrhotite, titanomagnetite and certain maghemite solid solutions are possible contributors. Rock magnetism is a complex topic which incorporates studies of minute grain size and magnetic domains, interaction of particles, solid state solutions, chemistry and changes over time, and is not easily summarized in a few paragraphs. Research and laboratory study of the magnetism of rocks is an ongoing process.

Physical property measurement of surface samples is a definite aid to interpretation of certain anomalies, but the variability and range of sample measurements usually are such that an interpreter has wide discretion and must eventually use his/her judgment and experience.

INSTRUMENTATION

Several types of magnetometers are in use today.

Proton precession magnetometers measure the total absolute intensity, with a minimum orientation requirement, to a precision of 0.1 nT. The intensity is related to the frequency of the precession of protons around a magnetic field. The sensing device is a "bottle" of a liquid rich in hydrogen atoms (e.g., water, kerosene, etc.) surrounded by a coil. An electrical current polarizes the protons in a direction different from the normal alignment parallel to the geomagnetic field. When the current is switched off, the protons precess back to "normal" at a frequency proportional to the strength of the total geomagnetic field. Proton magnetometer readings, therefore, are not continuous. Sensitivity of magnetic field measurement is in part dependent upon the time available for measuring the frequency, decreasing as a higher reading repetition rate is used. As a result, discrete measurements at a precision of 0.1 nT can be collected only once every few seconds, presenting a difficulty for some airborne surveys. Another limitation is that in the presence of strong magnetic gradients (>200 nT/ft), the proton magnetometer signal is sharply degraded and the reading becomes unreliable (Keary, 1984).

Optically pumped alkali vapor (cesium, rubidium or helium) sensors monitor the influence of the magnetic field on the interaction of atoms and light (the Zeeman effect). These magnetometers, which measure the absolute total field intensity, have a rapid response time (i.e., read continuously) and a high sensitivity (better than 0.005 nT in the lab and 0.05 nT in actual field acquisition). The cesium magnetometer is commonly available for use in airborne surveys; ground instruments are still under development. One limitation of cesium instruments, however, is that they are quite insensitive when the axis of the sensor is aligned parallel or perpendicular to the magnetic field. Recent innovations reduce the problem by using multi-celled sensors (personal communication, Scintrex Ltd. publications).

Cryogenic devices, using low temperature physics and known as SQUID (Superconducting QUantum Interference Device) magnetometers, are now available for use in certain limited circumstances although not commonly employed in exploration programs. These instruments, which are maintained at the temperature of liquid helium (4.2°K), have field sensitivities of 0.001 nT and an order of two magnitudes better in the lab (personal communication, R. J. Wold, TerraSense Inc.). One use is the measurement of the magnetic field gradient in three orthogonal directions. Another practical use is the measurement of susceptibility and remanence on small chip samples (e.g., reverse circulation drill chips). Research continues to make a more practical and portable field device for field use.

Fluxgate magnetometers measure only a relative change in the earth's magnetic field and are dependent upon orientation of the sensor coil. Until the last decade, fluxgates were the most commonly used magnetometers for ground and airborne surveys. A sensitivity of 1 nT is typical, although a sensitivity of 0.1 has been achieved in airborne systems. The sensor consists of two parallel solenoids wound in opposite directions on high-permeability cores. A component of magnetic field along the axis of the cores will upset the flux balance and a measure of the strength of that component can be made. Orientation of the core is essential. With a single sensor, usually the vertical component of the magnetic field is measured (from a vertically aligned core solenoid). With three orthogonal cores, the changes in the total intensity of the field can be measured, usually as a scalar function, although a vector representation is determinable. Some supersensitive, three-component fluxgate magnetometers are flown for study of the regional geomagnetic field.

Gradiometer systems are a true field system for measurement of the magnetic gradient (vertical or horizontal) and consist of a differential magnetometer which requires that the distance between sensors is small with respect to the distance to sources being measured. Any of the more sensitive sensors described above may be used. Configurations with a fixed separation of 1 or 2 m are currently in use. Differential magnetometers can be used to remove the effects of time variations, regional anomalies from very distant or deep sources or other coherent noise by common mode rejection. When used as a gradiometer, the measurement made is the ratio of the difference in field intensity to the distance between sensors, which is expressed in nT/ft (or nT/meter).

In North America, proton magnetometers have largely supplanted fluxgates for ground surveys. Analog or digital recording is available. A recent innovation is an "integrated system" consisting of a proton magnetometer and one (or two) coils for measurement of VLF electromagnetic fields. Data are recorded digitally and stored in solid state memory for subsequent transfer to and processing in a personal computer.

In current practice, most airborne surveys incorporate cesium or proton magnetometers connected directly to a digital acquisition system (DAS) which records the magnetic field data, time, location and/or navigation information on tape. The systems measure the scalar total field intensity from which an anomalous or residual intensity is usually derived by removing the earth's main field. For large airborne surveys, the International Geophysical Reference Field (IGRF) is commonly applied to remove the regional field.

Susceptibility and remanence are measured by other types of equipment. Coils and small hand-held AC bridges are used for in situ field determinations of rock susceptibility. In the laboratory, larger coils provide uniform magnetic fields for determination of susceptibility; spinner, astatic magnetometers or SQUID systems provide a remanent measurement. Because remanence includes both an intensity and direction, laboratory determination requires samples that are carefully oriented in the field. Usually large oriented samples are collected in the field from which small cubes are cut for actual measurement. Rock property measurements can be extremely useful in interpretation to facilitate correlation of anomalies to actual rock types and to understand the variability of magnetic properties over a survey area.

INTERPRETATION

The amplitude and form of an individual magnetic anomaly are a complex function of:

- The magnitude and orientation of the ambient magnetic field (the earth's main field),
- The geometry of the crustal (geologic) source,
- The amount of ferrimagnetic material (predominantly magnetite) in the geologic source and its susceptibility contrast to adjacent material, and
- The distance from source to sensor.

FIGURE 5.4a
Effect of varying magnetic field inclination over east-west striking body (north-south profile azimuth)

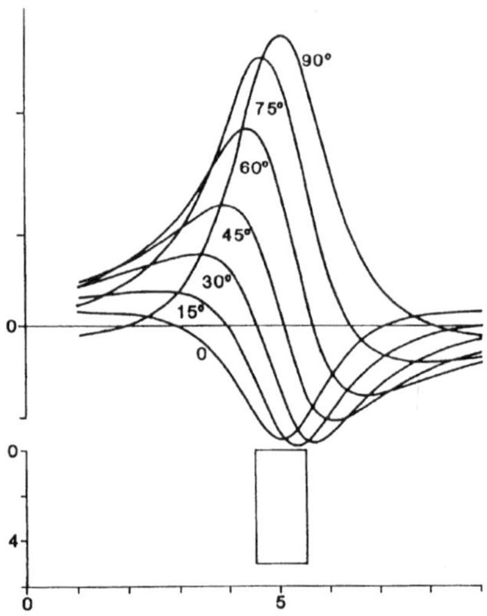

FIGURE 5.4b
Effect of varying dip of east-west striking body (east-west profile azimuth)

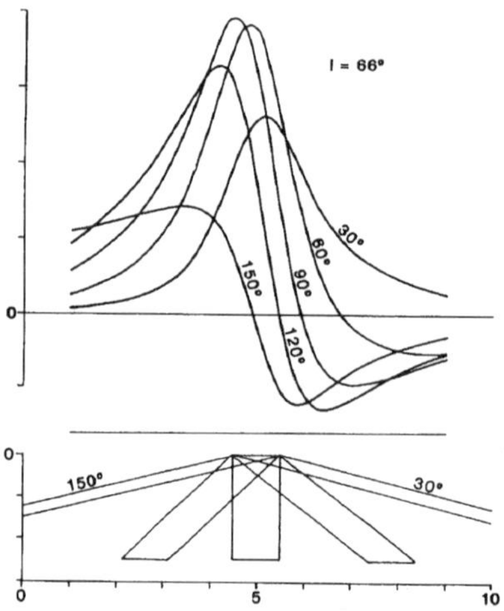

FIGURE 5.4c
Effect of varying dip of north-south body (east-west profile azimuth)

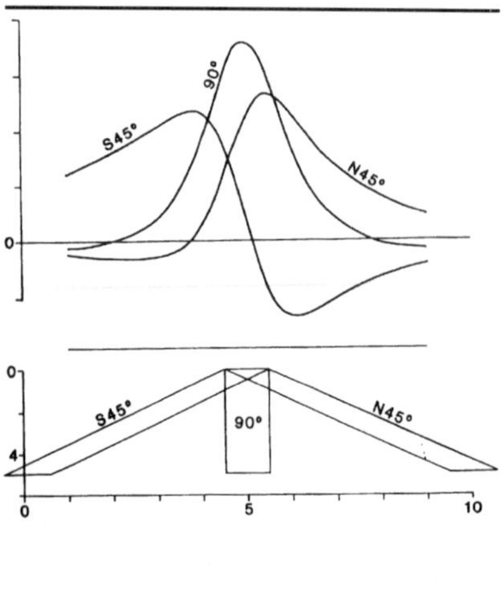

To be effective, an interpreter requires knowledge of the fundamental characteristics and variability of the local magnetic field. The amplitude and characteristic shape of a magnetic anomaly are dependent upon key variables of the source; namely, geometric shape (and dip), volume (cross section area and strike length), depth, strike azimuth and susceptibility contrast. The magnetic field variables of inclination, declination and, to a less obvious extent, the total field strength of the geomagnetic field further modify the anomalous response.

The amplitude of an anomaly, for an induced magnetization component only, may be expressed as

$$A = C\ F\ (k_2 - k_1)/r^n$$

where

- A is the anomaly amplitude,
- C is a constant related to body shape and size,
- F is the ambient field strength at the local inclination angle,
- k is the susceptibility; $(k_2 - k_1)$, therefore, is the susceptibility contrast,
- r is the distance between sensor and source,
- n = 0, 1, 2 or 3 depending upon body geometry.

The amplitude of an anomaly is directly proportional to the size or volume of a body and inversely proportional to the separation distance between source and sensor. Resolution increases as the separation, r, decreases (e.g., as observation altitude decreases). In addition, individual anomaly amplitude depends directly upon the susceptibility contrast between the anomaly source (k_2) and its surrounding environment (k_1), rather than the actual intrinsic susceptibility of the particular anomalous source. Note that the susceptibility contrast may be negative, in which case the positive anomaly is reversed and occurs as a low rather than a high.

The local magnetic field amplitude of an anomaly varies inversely with distance between source and sensor at different rates for different source geometries. Small sources such as thin dikes or volcanic plugs produce a field that decreases approximately as 1/distance cubed ($1/r^3$). Thin volcanic flows fall off at a rate between a $1/r^2$ and $1/r^3$ function. Sources where the size and depth extent (top to bottom) of a source body are much greater than the source-sensor distance to top, such as large igneous intrusions, produce an anomalous magnetic field that decreases approximately as $1/r^1$ to $1/r^2$. In practice in any given survey, anomalous bodies will show a continuum of fall-off rates.

To illustrate the effects of the key variables of depth, width, dip, strike and depth extent, suites of models have been calculated. Each figure consists of a family of curves demonstrating the change in response for one particular variable. The "base model" is a vertical (dip = 90°) tabular body, striking east-west, at a depth (D) of 1000 ft (1000 ft = 1 k ft), a width (W) of 1000 ft, and a depth extent (E) of 5000 ft. All dimensions are normalized to kilofeet (kft) integers. The vertical scale is not specified, as it is directly proportional to the susceptibility contrast at a particular field intensity and inclination.

Figure 5.4a shows the effect of varying magnetic inclination over an east-west striking vertical dike (north-south profile).

The vertical scale is relative. Note that the anomalies are positive symmetric at 90° and negative symmetric at 0°. Figures 5.4b and 5.4c show the effect of dip on the anomaly shape for east-west and north-south striking bodies respectively. The east-west profile over a north-striking, vertical body also is a symmetric anomaly profile.

Figure 5.4d shows the variation with different strike of a tabular body. The profiles for 0° and 90° azimuths are identical to the corresponding curves in 5.4b and 5.4c respectively.

Figure 5.4e shows the effect of a layer termination, where A and C are the east-extending and west-extending, higher susceptibility bodies terminating at 5, and B is the anomaly of a tabular body plotted for comparison.

Figures 5.4f, 5.4g, 5.4h, and 5.4i are calculated to show effects of varying width (W), depth (D), depth extent (E) and strike length respectively, which demonstrate the increase in amplitude resulting from volume and decrease resulting from increased source-sensor distance.

FIGURE 5.4d
Effect of varying strike of the body (profile azimuths are 90° to the strike of the body)

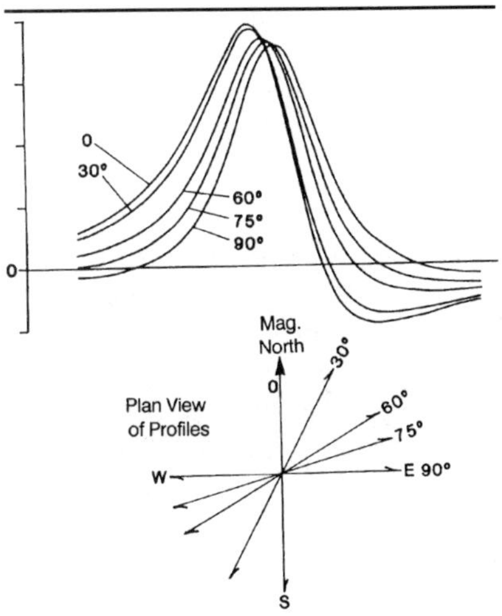

FIGURE 5.4e
Effect of termination edge of east-west body (north-south profile azimuth). Tabular body B is plotted for comparison

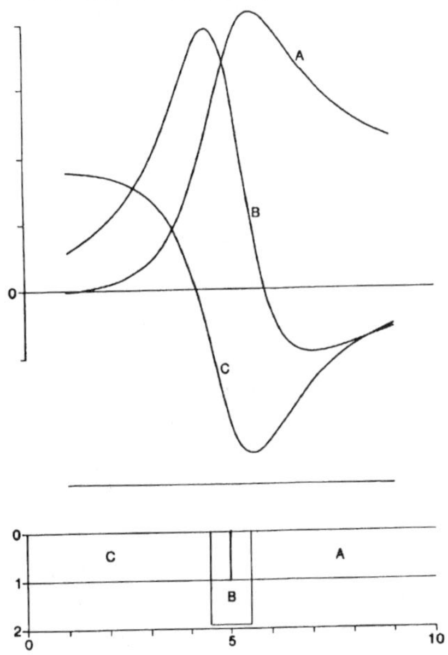

FIGURE 5.4f
Effect of varying width (W) of east-west body (north-south profile azimuth)

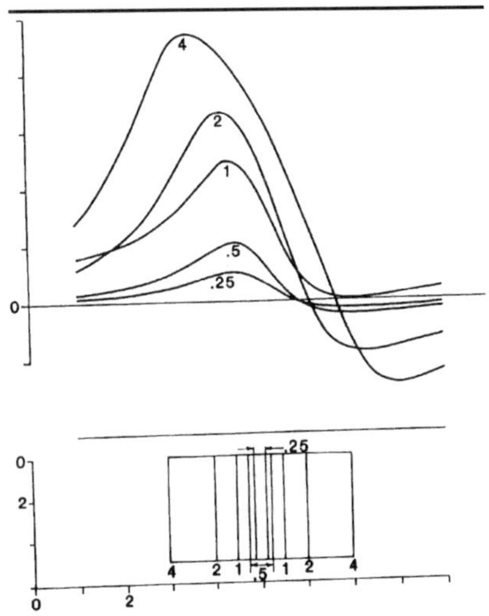

FIGURE 5.4g
Effect of varying depth (D), by changing elevation above ground, of east-west body

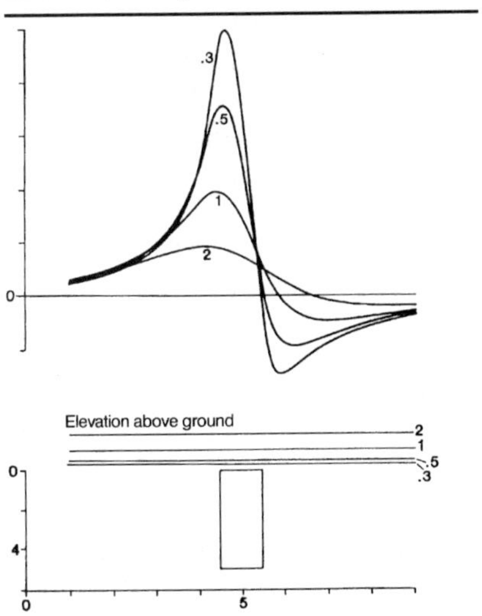

FIGURE 5.4h
Effect of varing depth extent (E) of east-west body

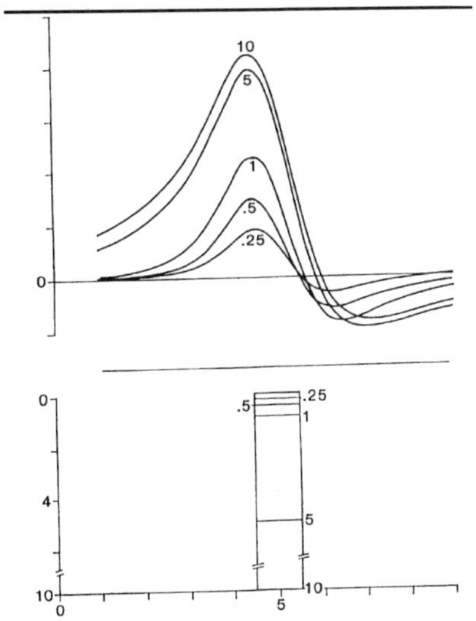

FIGURE 5.4i
Effect of varying strike length of east-west body

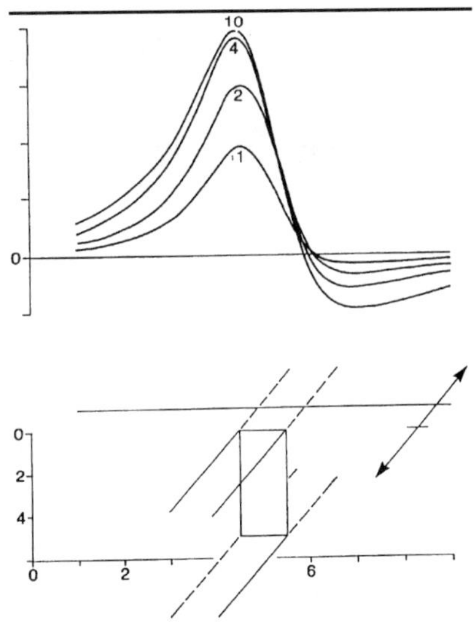

FIGURE 5.5
Effect of remanent magnetization on high-low anomaly pattern

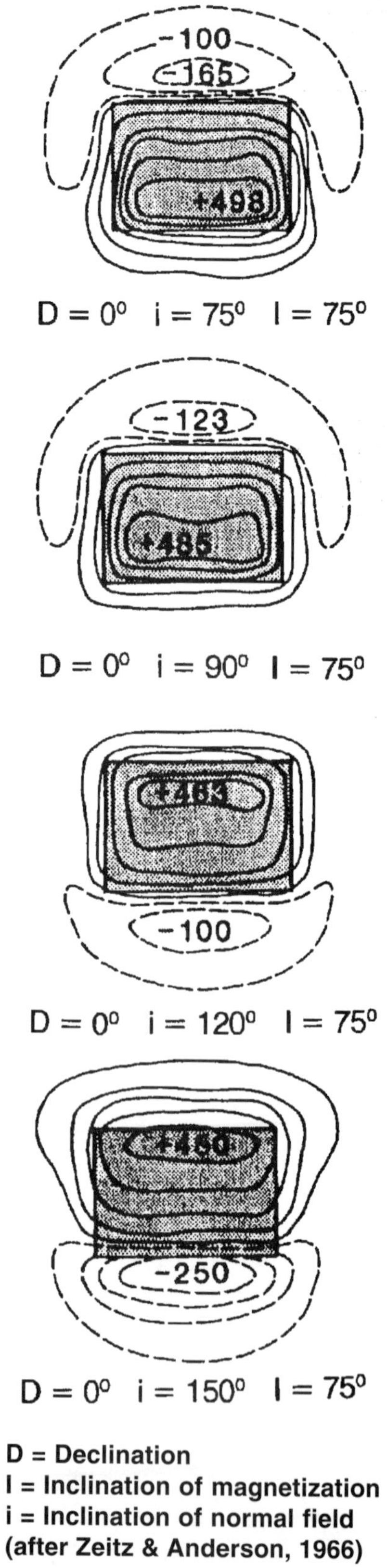

Most magnetic anomalies at the inclination of the west-central United States consist of a high (or positive) and a low (or negative) to the northside. The highs and lows are referenced to the "background" or average magnetic of the area and usually are plotted as a residual anomaly after the main field has been removed. For all but vertical inclinations in the Northern Hemisphere, the low is much reduced in intensity compared to the high and is characteristically located on the north side of the high. The relationship of the high-to-low is a function of dip, depth extent and orientation of the source body and, when sufficiently dominant, the remanence.

Remanence adds a vector of independent amplitude and direction to the induced field vector. If sufficiently strong and at a direction other than the induced field, remanence will alter the anomalous induced field vector and void the "normal" high and northside-low relationship of the induced field, figure 5.5.

If this "northside low" alignment is not positioned close to the line of declination, some remanent effect may be suspected. A strong, fully reversed remanence will produce a negative anomaly with a characteristic "northside" high, which is the reverse of the normal induced-field-only situation. Remanent effects definitely complicate interpretation, but fortunately they often are too small to be significant and may be ignored. In some cases remanence, where suspected from the shape and relative orientation of the low to the high, can be factored into the interpretation process but generally minimizes any quantitative analysis. In general, even when moderate remanent effects are suspected, complete definition is unlikely without actual sample measurements for substantiation.

In areas of outcrop or shallow subcrop, high-susceptibility volcanic terrains give anomaly profiles which are often highly variable showing large variations from station to station. Rarely are such shallow anomalies well defined. Figure 5.6 shows a portion of a line, from Nevada, which crossed a contact from very low susceptibility sediments on the right to a "typical basalt" signature on the center and left.

From the character of the response, two different basalts can be interpreted. At the far left, the more extreme negative variations indicate a remanent component not present in the central part of the line segment. Two distinct units of different age and/or composition are indicated. If an area such as this were surveyed by air, the high and variable amplitude responses from ground data would be attenuated as a result of the increased source-sensor distance. An example of the smoothing effect of an airborne profile (r = 1000 ft) over intense and variable ground data is shown in figure 5.7.

At the typical magnetic latitude of the western United States (60-65°) the source of a magnetic anomaly, unlike gravity, is not located directly under the high but is offset northward because of the angle of inclination of the earth's field. Proper location of the source body is important for geologic comparison or correlation with other geophysical data sets. At 60-65° inclination, the south and north edges of a simple, east-west striking, wide tabular body lie "slightly" to the south of the maximum and to the south of the zero amplitude respectively.

Figures 5.8 and 5.8a show the anomalies, location and rule-of-thumb width estimates for three wide bodies located at 66° inclination for north-south and east-west profiles. Compared to the vertical-sided tabular body (i), the dipping body (ii) and intrusive stock (iii) show similar location and first-approximation width estimates.

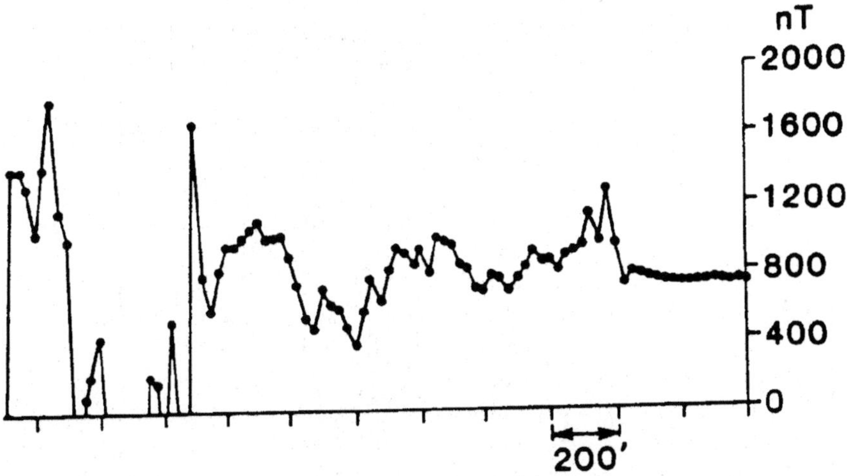

FIGURE 5.6
Ground magnetic profile data showing flat response of sediments (to the right end), and variable signature of two basalt flows, one with an interpreted strong remanence (to the left end)

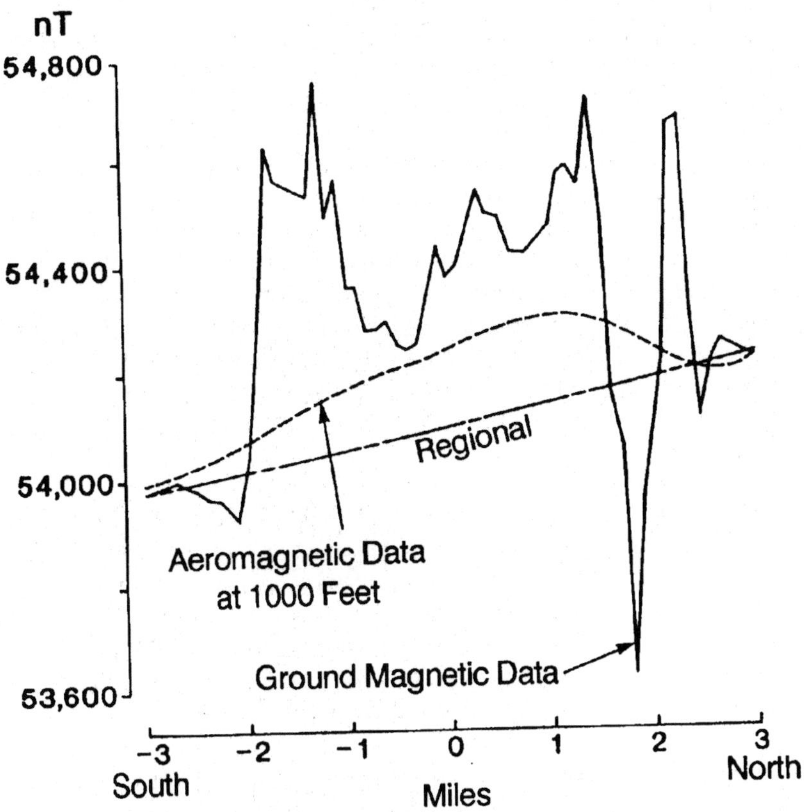

FIGURE 5.7
Comparison of ground and airborne data showing the smooth effect of increased height

A reduction-to-pole calculation, which will be discussed later, would mimic a 90° inclination and center the source under the anomaly. These estimates are crude, but do provide a starting point.

Styles of Interpretation

In interpretation, a magnetic anomaly is defined by its amplitude, width (sometimes half-width is used), shape, strike elongation and azimuth. Isolation and uniqueness enhance geologic importance.

Interpretation can be divided into five overlapping and increasingly complex and costly styles. Interpretation depends in part upon

FIGURE 5.8
Location of source bodies beneath north-south profiles at a magnetic field inclination of 66^0

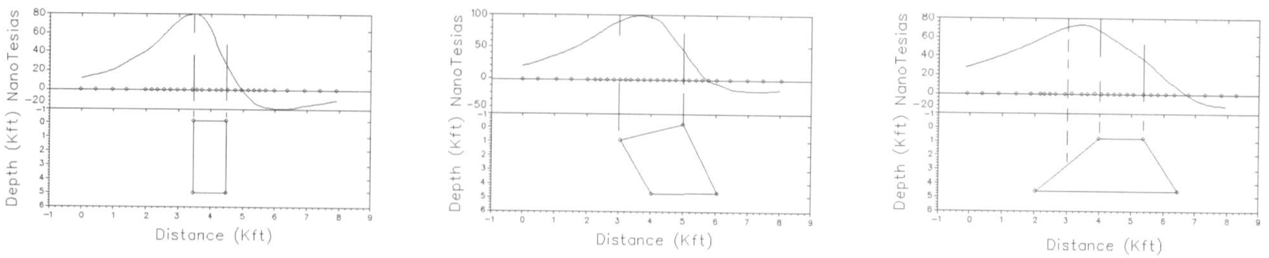

FIGURE 5.8a
Location of source bodies beneath east-west profiles at a magnetic field inclination of 66^0

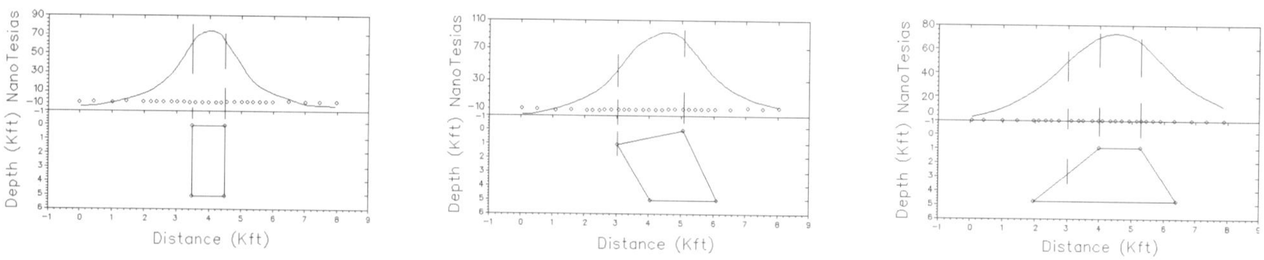

the quality and form of the data set, the knowledge and experience of the interpreter, the time available, and the purpose and requirements (qualitative or quantitative) of the results of the process.

- Inspection: anomalies are identified and evaluated using simple regional removal and rule-of-thumb depth estimates on profiles and/or maps. Inspection concludes with only an identification and location of an anomalous source.

- Curve matching: many suites of model curves and techniques for characteristic curve-matching were developed prior to 1980. These techniques have been supplanted by digital techniques and are primarily used where computer capability is not available.

- Forward modeling: from a preliminary estimate based upon simple theoretical assumptions and/or the interpreter's experience, an anomaly is computed for a specific model. After comparison with the observed anomaly, the model is modified and the anomaly re-computed. Such an iterative process is followed until a satisfactory fit is obtained. This procedure will provide more definitive information on depth, width and/or geometric shape, and susceptibility contrast than simple curve matching. Where available, geologic control should be used to constrain the model and limit the inherent ambiguity of the interpretation.

- Digital filtering: with a digital data base (or analog data converted to digital) isolation and/or enhancement can be improved by application of various types of filters or other sophisticated processing. Color maps are one important output of a digital data set.

- Inverse modeling: a geologic model is calculated directly from profile data, usually first as a 2- or $2\frac{1}{2}$-dimensional body. Inversion does not yield a unique model and must also be constrained to produce a geologically feasible structure. This process is often used in conjunction with digital filtering and production of contour maps of various calculated functions. Many interpreters have concluded that a length-to-width ratio greater than three can be represented by a 2-D feature without introducing appreciable error.

Shaded relief and color maps can and should be used with all styles of interpretation. The interpretation process ranges from a simple style of identification and location to a complex process, which when integrated with geological information, leads to a complete characterization of anomaly sources.

Depth Estimates

Depth estimates can be based on the relatively straightforward half-slope or straight-slope distance (Peters, 1949, and Rao, 1984), forward or inverse modeling of individual anomalies, or complex mathematical computations such as the Werner deconvolution or Compudepth. Figure 5.9 outlines the straight-slope and Peters' half-slope distance measurements on a computed example. For a depth estimate, the Peters' half-slope distance is divided by a factor that varies between 1.2 and 2.0 depending upon whether the source is a thin or thick dike. Significant depth extent of the body is necessary to a reasonable graphical solution. Not infrequently, anomalies of interest are

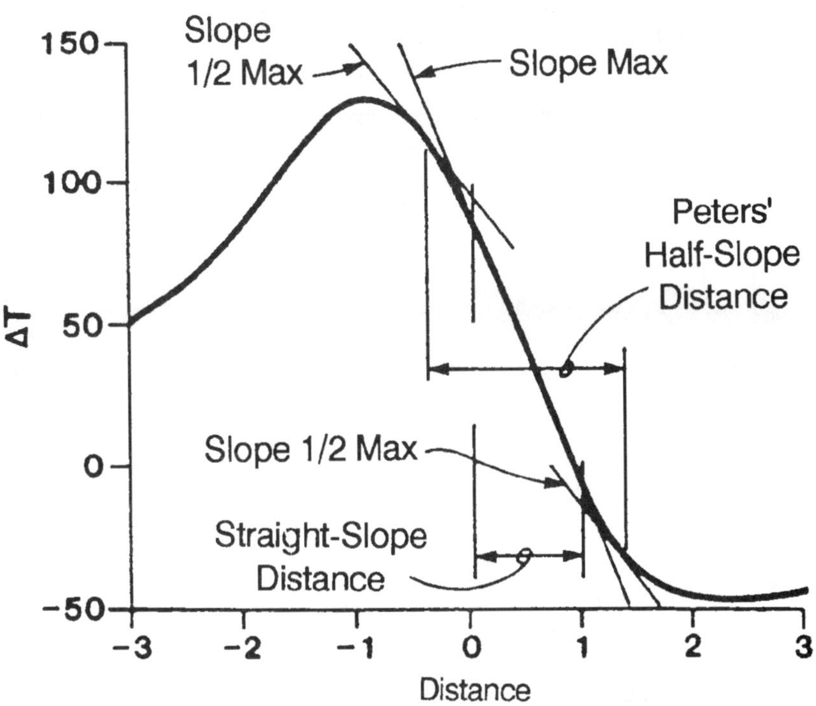

FIGURE 5.9
Measurement of straight slope and Peters' half-slope distances

large relative to the flight height (plus depth) and division by two provides a "better" overall solution. The straight-slope method is a "direct" estimate that is roughly proportional to depth. Depth estimates over thin dikes, however, will be in serious error. For either estimate, slope segments should be picked from profile plots, if possible, because of the averaging and uniform gradient produced by either hand or computer contouring.

Figure 5.10 shows the calculated anomaly of an irregular four-sided polygon. Width can be estimated from the peak-south and zero background locations of the anomaly; depth is derived from the straight-slope estimates.

FIGURE 5.10
Example of slope distance measurements

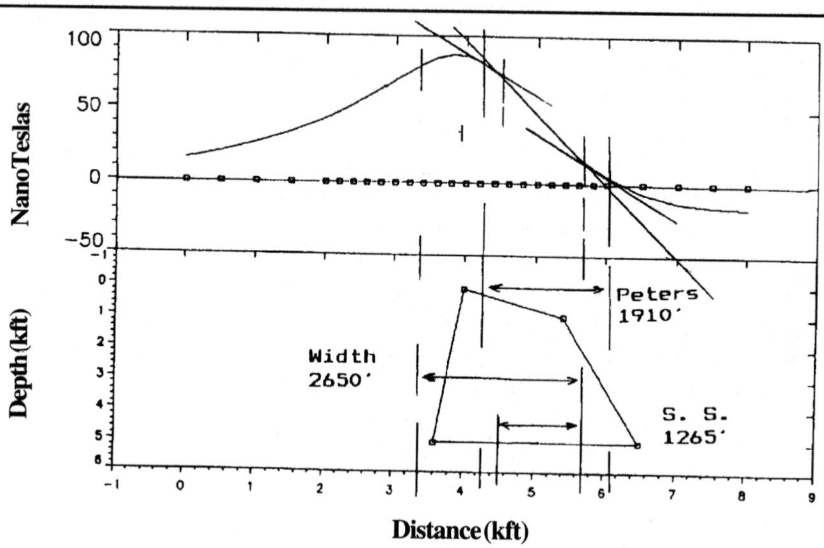

Figure 5.10a compares the original irregular-body anomaly profile (solid line) to the calculated anomaly (square dots) of the rectangular body interpreted from the location and straight-slope depth estimates. The differences can be related to the approximations of the rule-of-thumb estimates and the irregular cross section of the original source body.

FIGURE 5.10a
Interpreted model (square dots) developed from width and slope distance depth estimates, and calculated anomaly of model (solid line)

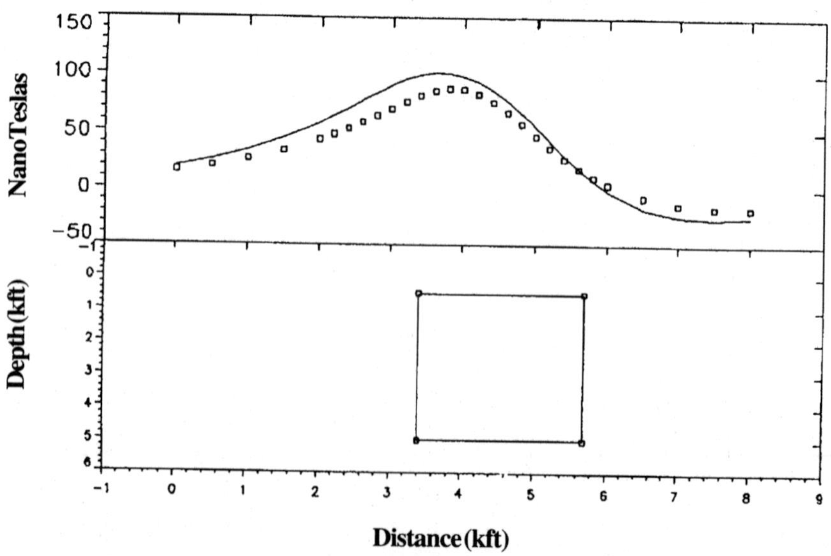

Terrain Effects

Terrain does affect magnetic survey data where it is composed of relatively magnetic rocks. The terrain effect is present whether the data are acquired at constant barometric elevation, draped survey (constant elevation above ground) or ground traverse. Long recognized as producing an anomalous effect, terrain response often is ignored because adequate data are not available for evaluation, analysis and correction. Simply described, the effect arises from the changes in source-sensor distance resulting from the geometry of the surface. The effect is further enhanced by draped or ground lines when the sensor is below the top of an adjacent magnetic body, e.g., against a cliff face, or in a draped survey where the aircraft in a wide valley is below the surrounding mountain peaks. An example of the effect of terrain on both constant barometric and draped surveys is shown in figure 5. 11.

The body indicated has a susceptibility (500 micro-cgs) which is less than "background" (833 micro-cgs). This produces a negative susceptibility contrast and an anomalous low (e.g., magnetite destructive alteration in a granitic host rock). The topographic effect of a high-susceptibility mass (e.g., volcanic or intrusive peak) can lead to seriously erroneous conclusions about the size, depth and/or structural features of any other anomalous source in the vicinity.

Figure 5.12 is the anomaly of an east-west striking terrain feature calculated on a horizontal plane, e.g., a constant barometric survey.

Figure 5.12a is the terrain anomaly added to the characteristic anomaly of a tabular body, which will be evaluated further with a regional in the following sections.

FIGURE 5.11
Calculated model showing comparison of constant barometric and draped flight elevations over a terrain feature

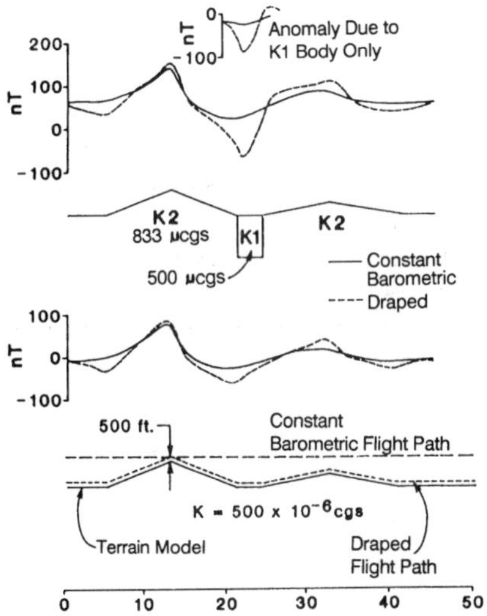

FIGURE 5.12
Calculated anomaly of east-west striking terrain feature (k = 500 micro cgs)

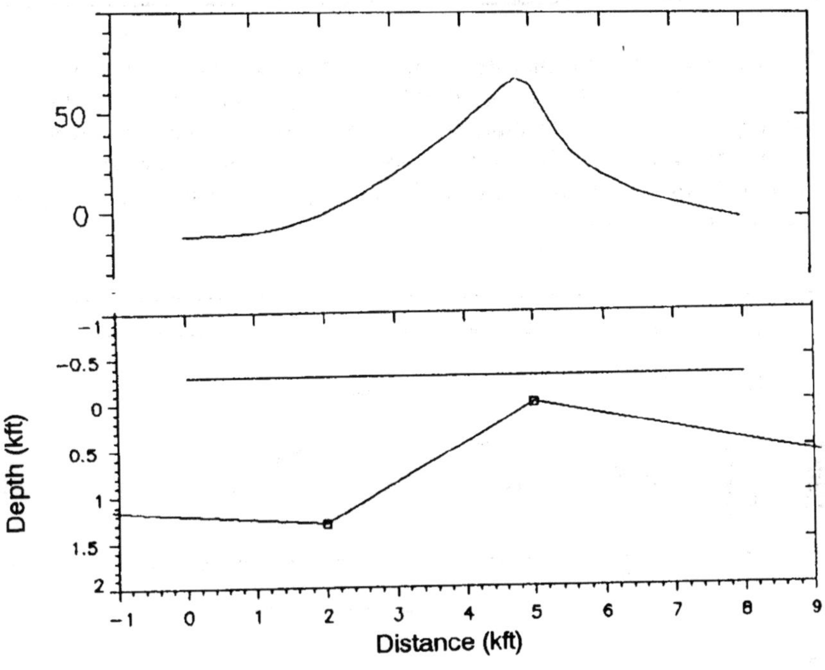

FIGURE 5.12a
Combined anomaly of terrain and rectangular body of susceptibility contrast of 1000 micro cgs

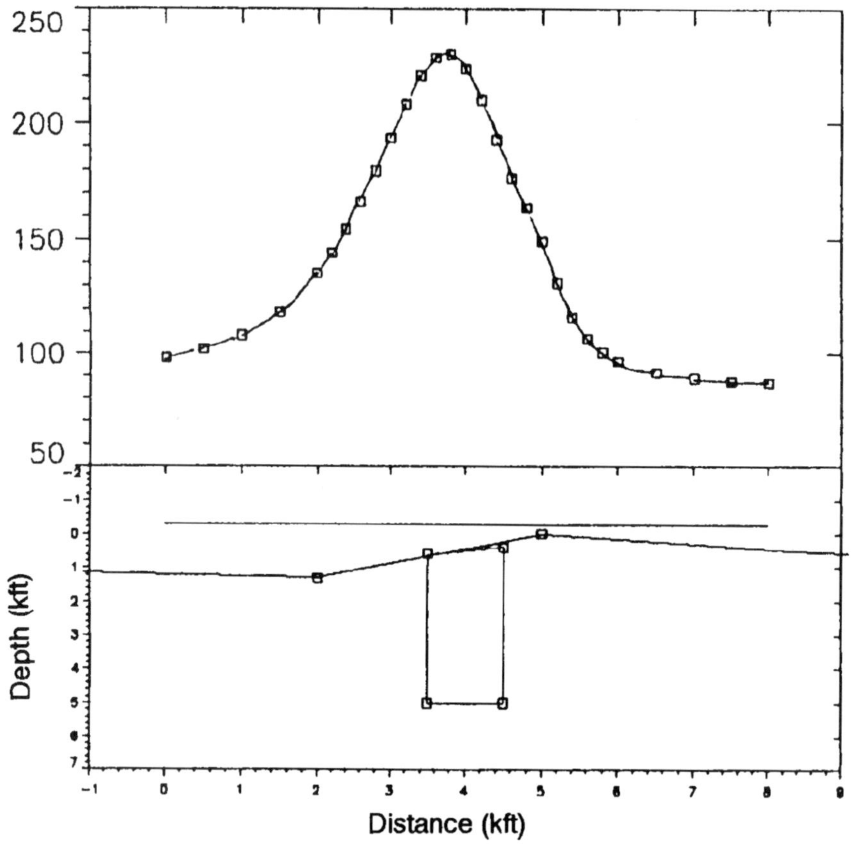

FIGURE 5.13
Schematic showing additive effect of multiple anomalies and regional

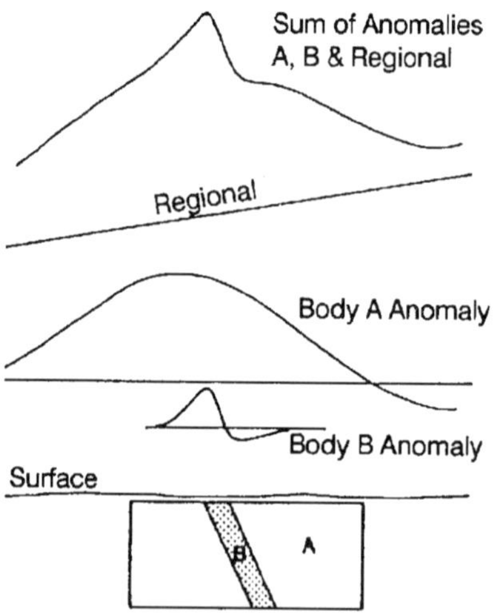

FIGURE 5.14
Anomaly calculated with regional removed

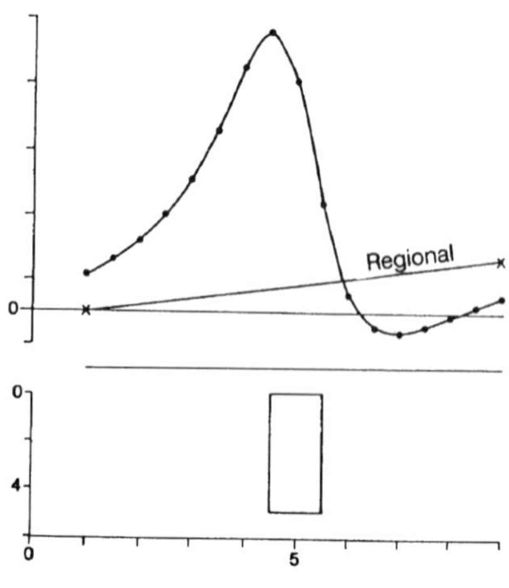

Regional Effect

The intensity of the earth's magnetic field increases to the north (in the Northern Hemisphere). This change in the main field (about 5 nT per mile increase to the north at 60°N) is a regional effect in its broadest sense. For large airborne surveys, application of International Geomagnetic Reference Field (IGRF) formula removes the effect. In local surveys, where the "regional" variation is also related to large geologic features (e.g., changes in crustal lithology, batholiths or other basement intrusives) adjacent to the survey area, the combined regional effect is removed by subtracting a linear or polynomial function. Reduction on either profile or contoured data brings "background" to a consistent horizontal relationship.

Figure 5.13 shows a simple schematic of the regional anomaly of a major geologic feature, block A, and a local anomaly, B, superimposed on a sloping, linear, main field regional. Note how the anomaly of B is distorted by superposition on A, which is further distorted by addition of a linear regional effect.

The anomaly of the tabular body added to a known regional is shown in figure 5.14.

If matched by a modeling inversion routine at the same susceptibility contrast without recognition of the regional, the dipping body of figure 5.14a is derived.

Note the difficulty in fitting the northside low, which is not good but acceptable.

Figure 5.15 shows an interpreted anomaly of figure 5.12a where neither terrain nor regional is assumed. Note the significant depth difference between the original shallow terrain body and the deep source of figure 5.15.

Figure 5.16 shows an inverted model of the same anomaly of figure 5.12a, where no terrain is acknowledged, but a regional is added to achieve a reasonable fit. Again, comparison of the original terrain model to the interpreted model shows significant differences. These examples - of several model sets that might be obtained (the ambiguity problem) - indicate the need to identify both the terrain and the regional and to include their effects, if any, in analysis, model calculations and interpretation.

Both terrain and regional effects, which are important handicaps that limit interpretation accuracy, have been emphasized to demonstrate the need for adequate data on elevation and/or terrain and sufficient data for determination of the regional. It also demonstrates the ambiguity of potential field interpretation. Satisfactory interpretations, however, can be developed with constraints from known geology, terrain and elevation parameters.

FIGURE 5.14a
Anomaly interpreted without recognizing regional slope

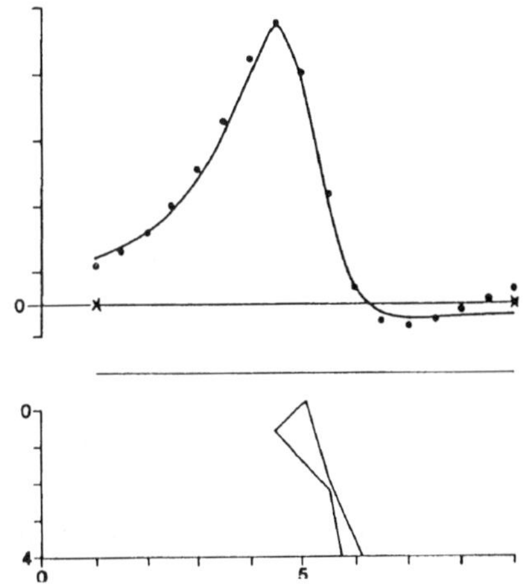

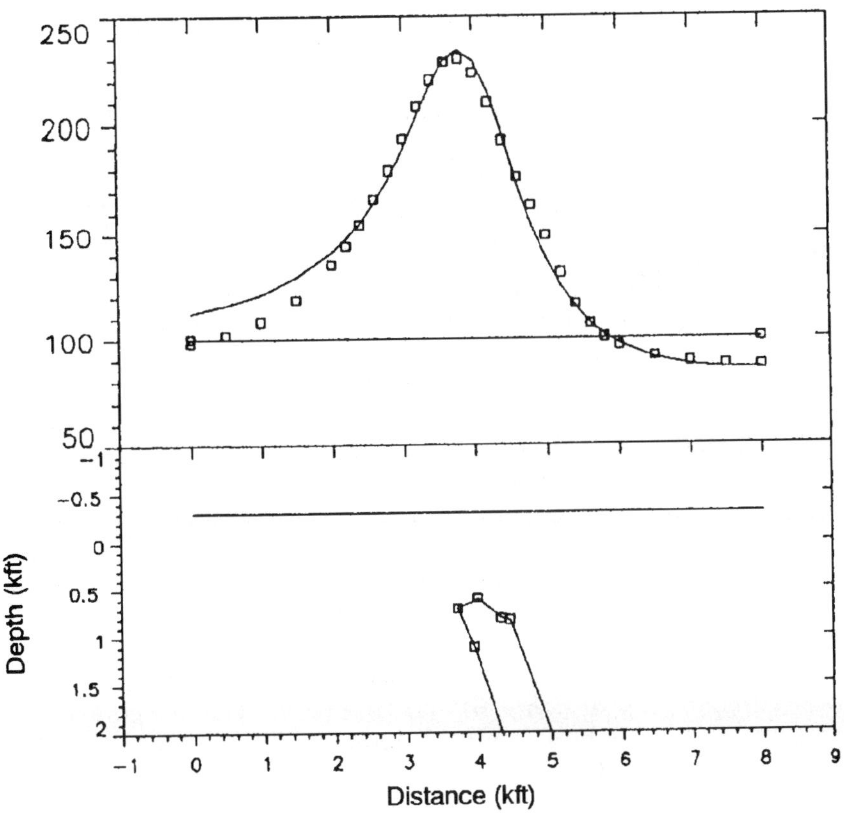

FIGURE 5.15
Anomaly of figure 5.12a interpreted by inversion assuming no terrain effect and no regional

FIGURE 5.16
Anomaly of figure 5.12a interpreted by inversion assuming no terrain effect but adding a regional to achieve a model "fit"

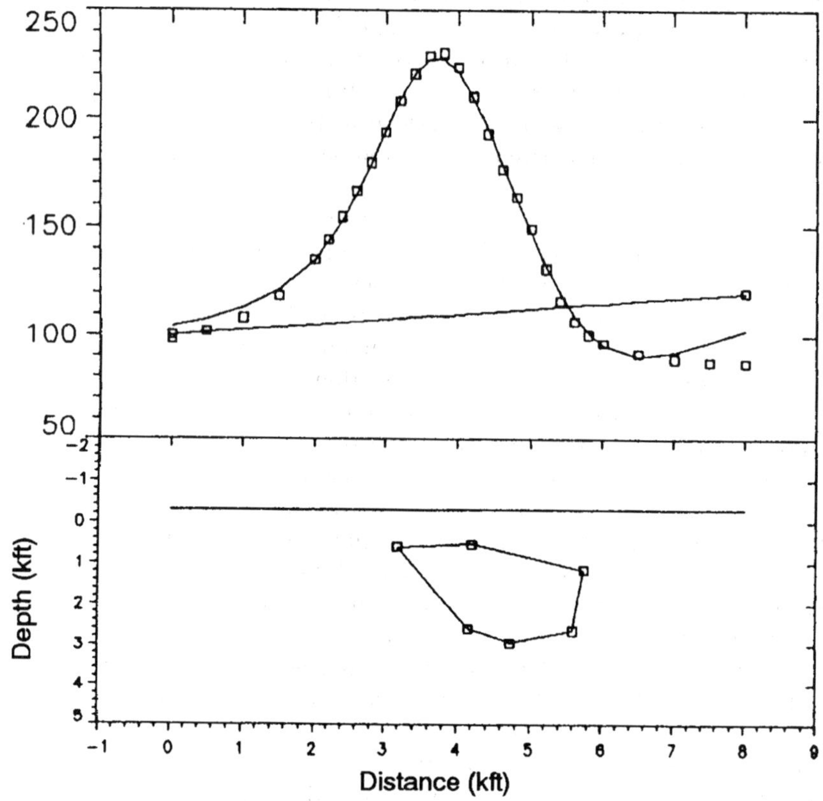

FIGURE 5.17
Resolution of multiple bodies with decreasing flight height

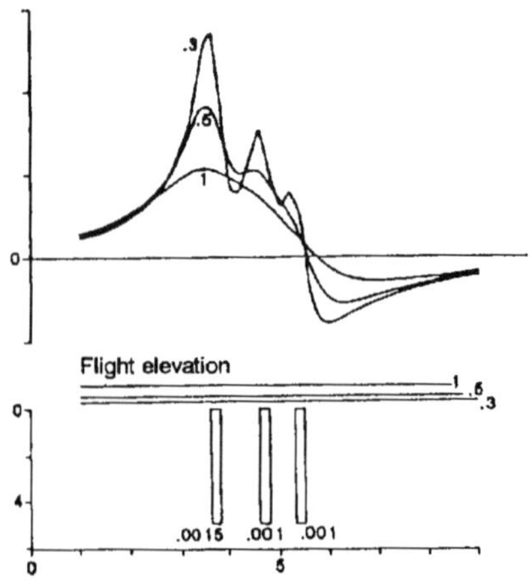

FIGURE 5.18
Interpreted anomaly of unrecognized multiple bodies of high-level anomaly of figure 5.17

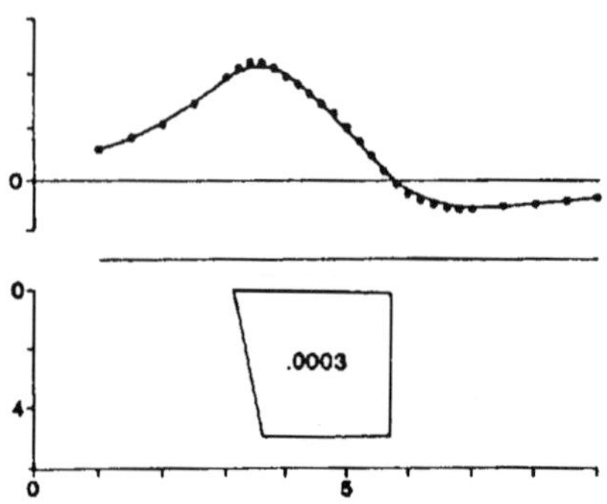

Resolution

Resolution is defined as the identification of small features that make up larger composite anomalies. Resolution of small bodies is significantly affected by flight height. Multiple small, shallow bodies often cannot be discerned in surveys where height plus depth of burial exceeds the separation between bodies. Figure 5.17 demonstrates this problem on a calculated model.

Depths and distances are in kilofeet, susceptibility in cgs units. Using this relationship in the reverse sense, "noisy" ground data can be smoothed by increasing the separation distance, e.g., use of an 8-ft staff for ground surveys over volcanic terrain, or increasing the altitude of an airborne survey. Figure 5.17 shows the smoothing produced by flying at 1000 ft but also shows the lack of resolution. Without any knowledge of the multiple body geology, interpretation of the airborne data of figure 5.17 (or the ground data of figure 5.7) would be vastly different (see figure 5.18).

Smoothing, whether by increased altitude or application of some filter, will alter the character of the anomaly and subsequent interpretation unless the altitude data are known and factored into the interpretation process.

Isolation and Enhancement

An exploration survey detects and records the effects of all variations in magnetic polarization within the earth. An anomaly of interest, therefore, will be distorted by adjacent and/or deeper sources. Broader, low-frequency/long-wavelength[2] anomalies are denoted as

[2] Frequency is the reciprocal of wavelength, and may be described as anomaly cycles per unit distance. Wavenumber is the spatial frequency nomenclature, or the number of wave cycles per unit distance.

regional features; smaller, high-frequency, often intense anomalies, if not the direct target of the survey, are often considered undesirable "noise." Separation of the unwanted effects from the desired anomaly signature produces a residual anomaly. The separation, which may be a subjective determination, is a critical problem area that controls the accuracy of the interpretation process (Hinze, 1989). The contour maps produced after removal of the IGRF are called "Anomaly Maps," but are actually only one step in the separation process.

Separation techniques can be split into two general categories -- isolation and enhancement. Isolation techniques eliminate all anomalies which do not meet a specific set of characteristics; anomalies are essentially unaltered so that quantitative analysis and modeling may be applied directly. By contrast, enhancement techniques accentuate a defined set of characteristics to increase their perceptibility, but in so-doing, usually distort the individual anomaly so that only a qualitative visual interpretation approach is feasible (Hinze, 1989).

All anomalies, even one from a single, isolated source, consist of spectrum of frequencies. The small, shallow, short-wavelength anomalies consist of a higher band of frequencies than broad anomalies from deep sources which are dominated by lower frequency, longer wavelength components. Each of these endpoint types has a characteristic, but overlapping, spectrum.

Filtering is a process of removing a specified range of frequencies from a data set, e.g., the low-frequency regional, or alternatively, the high-frequency noise. Filtering procedures are based upon the frequency components (or more properly, wavenumber), or the reciprocal, wavelength, of an anomaly. Usually applied to large airborne data sets, digital filters enhance individual or multiple anomalies of similar characteristics. Filtering procedures can be accomplished in either the frequency or spatial domain. The specifications of any filter operator must be designed carefully to match the character of the anomalous data and purpose of the process. The data are filtered in either profile form (one-dimensional) or map form (two-dimensional). Often a power spectrum, either one-or-two dimensional, is generated for study and examination of frequency content to determine proper filter coefficients and optimal transfer function.

Reduction-to-pole filters remove the distorting asymmetric shape, assuming induced effects only, from magnetic anomalies caused by the inclination of the induced geomagnetic field. The data are transformed to vertical inclination which removes the "north-side low" of dipolar anomalies, centers the anomaly over the source, and enhances the spatial correlation between the magnetic anomaly and its geologic source.

Frequency (wavenumber) or wavelength filters can be designed to suppress low- or high-frequency components of an anomaly spectrum. High-frequency "noise" resulting from small, near-surface sources is removed by a low-pass filter; the longer wavelength of regional anomalies by a high-pass (low-cut) filter. Bandpass filters are also feasible. The technique enhances rather than isolates particular anomaly characteristics but attenuates amplitude. Because some frequency ranges are eliminated in the process, the anomaly shapes are altered. Use in subsequent quantitative analysis, therefore, is not feasible.

Derivative filters enhance the higher frequency components of anomalies associated with small, shallow sources. The first vertical derivative, or gradient, is used particularly in profile analysis to sharpen small, subtle, flexure-type anomalies superimposed on large regional features. Derivatives do have several particular disadvantages, however, in that they are particularly responsive to noisy data, generating additional noise, and destroy the anomaly slope, eliminating most quantitative work. Depth of a pole source can be determined, however, from the gradient field in combination with the total field. In high magnetic latitudes, the zero value of the second vertical derivative can approximate the outline of the anomaly source. As typical shape and amplitude are lost in any derivative computation, interpreting resulting data in a quantitative fashion is not feasible.

Continuation filters project the observed anomaly field to higher elevations (upward continuation) or lower elevations (downward continuation). Upward continuation attenuates the high-frequency anomalies relative to the low-frequency anomalies and emphasizes the lower frequency, longer wavelength anomalies that may be masked by surficial magnetic effects, in effect isolating larger anomalies from "near-surface noise." Minimizing the effect of higher frequency components is, therefore, another type of low-pass filter which is useful for smoothing and compositing. Downward continuation amplifies the higher frequencies, but often badly distorts the results. One critical advantage of continuation filters is that the geopotential field character is retained and quantitative calculations can be performed on the transformed results.

Strike-sensitive filters selectively pass or reject frequency components which trend over a selected range of azimuth. Trend filtering is similar to band-pass filtering except that the selection of pass-reject is based upon orientation of anomalies and not frequency components. The filter particularly enhances trends when linear patterns are present. Anomaly distortion reduces interpretation effectiveness.

FIGURE 5.19
Filtering effect of 5-point running average

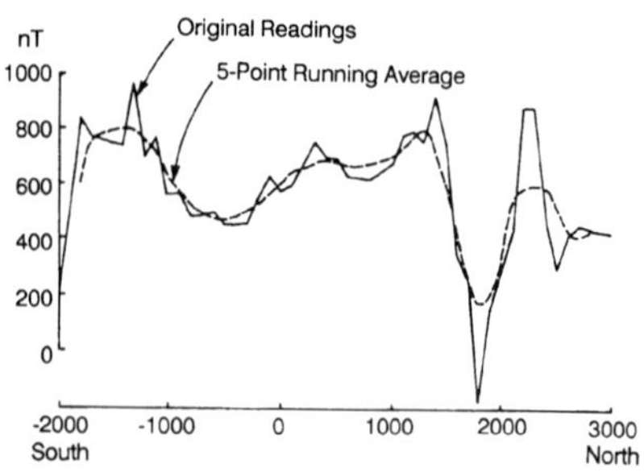

Running average filters are one of the simplest forms of filtering in which a multiple-point average is passed across the data. The moving window of values may be weighted before averaging to retain desired characteristic features of the data set. In figure 5.19, the typical sharp variations of ground data over volcanics are "smoothed" by application of a 5-point running average.

Shaded relief maps are an alternative to strike filtering. For these maps, the anomalous intensity is proportional to reflected light from topography, illuminated from a specific source. The effect is to enhance or delineate low-intensity features of the original map which strike at right angles to the direction of illumination (Dods, et at, 1985). Linear structural patterns are enhanced by shaded relief analysis but are a qualitative representation only.

Geologic Correlation and Integration

Geological constraints are an integral part of geophysical interpretation. Because any specific interpretative solution of magnetic data alone is inherently ambiguous, input of known geological data may provide significant limits. Such information may include:

- Principal lithologies and alteration
- Directions and throw of faulting
- Extent, thickness and variability of volcanics, particularly near surface flows
- Rock susceptibilities of the local, or regional, area,
- Structural history
- Known intrusive phases and differentiation

- Other geophysical survey data, particularly of a scale and extent equivalent to the magnetics (e.g., VLF, radiometrics or air EM).

Geology may be used either to constrain a specific anomaly or, in a broader sense, to delineate a pattern or repetition of a characteristic anomaly throughout an area. Magnetic data may be correlated directly to intrusive contacts, lithology changes, shallow volcanic fields, structural zones, varying depth of basement and/or changes in metamorphic grade. For instance, volcanic centers may be distinguished by a pattern of high-frequency anomalies sometimes with an obvious remanent effect; anomalies may change significantly in amplitude across a major fault or contact; dike swarms may parallel major shear zones and be offset by faulting. On a more regional scale, certain granitic intrusions may be very magnetic while others may have a strong remanent component, suggesting a different age of emplacement. Early work in Arizona suggested that many of the mineralized Laramide intrusives had weak magnetic signatures, often with associated characteristic lows resulting from extensive alteration; conversely, the Precambrian granites and associated schists and mafic volcanics had a higher susceptibility (Brant, 1966). In Canada, many greenstone belts have a significantly different magnetic signature than gneissic terrains.

Susceptibility can be estimated from simple formulae (Vacquier, 1951), sample measurements or, more generally, empirical knowledge of typical rock types.

The style and level of interpretation is variable depending upon survey objectives, data quality, time and resources available, and knowledge and skill of the interpreter. Since magnetic interpretation is neither unique nor static, re-interpretation should be considered whenever additional pertinent geological information becomes available, additional geophysical data are acquired, or new concepts are developed.

SURVEYS AND PROCEDURES

Interpretation depends, in part, upon the proper design of surveys and reduction procedures and, in part, on the quality of the data acquired. For instance, inadequate line or station spacing produce ill-defined anomalies that are not representative of the actual geological sources.

Corrections

The two basic corrections of magnetic survey data are related to temporal (the diurnal variation) and spatial (latitude or regional variation) changes in the geomagnetic field. Micropulsations and magnetic storms are separate problems which, by their random nature, are not correctable. During periods of intense activity, surveys are suspended and data are not collected. In the early years, the diurnal variation was removed by re-occupying a base station every several hours (or flying orthogonal tie lines) and distributing the difference in readings on a proportional basis over the time period. In more recent practice with digital systems available, a base station is set up and diurnal fluctuations of field intensity are recorded automatically every one to three minutes. Any change is algebraically subtracted from the field readings on a station-by-station basis.

The spatial or latitude correction accounts for the change in the geomagnetic field intensity from equator to pole. Nominally in the mid-latitudes, the effect is about 5 nT per mile in a north-south direction (parallel to the declination). If it becomes necessary to remove the effect from extensive ground surveys, the correction is usually removed as a linear function. The latitude effect is intimately associated with regional anomalies of long wavelength related to deep-seated geologic features and cannot be distinguished on small surveys.

For most airborne surveys, the spatial effect is removed by means of a correction for the global anomaly field, defined by an internationally accepted equation called the International Geomagnetic Reference Field (IGRF). The coefficients of this field are modified every five years as new information is received from satellite and broad regional surveys. For most local surveys, the IGRF gradients can be assumed to be linear over the survey area.

Airborne Surveys

The size of an area of interest, inaccessible terrain, economics and, to some extent, timing, predicate the use of aircraft to provide magnetic survey data. Fixed-wing aircraft, or higher cost helicopters, with high-sensitivity proton or cesium magnetometers are available for data acquisition on large blocks of ground. Often a magnetic survey incorporates one or two different directions of VLF-EM field measurements. Occasionally, a radiometric system is combined with magnetic surveys, but one or the other must take precedence because of the different requirements of the two methods.

Survey Scale: regional and detail

Regional surveys cover large areas, hundreds to thousands of square miles, at a relatively wide line spacing and high flight altitude. Detail surveys, on the other hand, are designed to give tighter control at close line spacing and low flight altitude of a relatively small area. In effect, the designation "regional" is a relative term only as a regional mineral-oriented survey might be detailed-size for a large petroleum exploration project.

Regional surveys not only cover larger areas but typically are flown with wide line spacing and high flight altitudes. Line spacings of 0.5 to 2-mi (1000m to 5000 m) or more, and flight altitudes of 1000 ft to 3000 ft (300m to 1000 m) are not uncommon. Airborne surveys are flown either at a constant altitude above terrain (Mean Terrain Clearance - MTC), also called a draped survey, or at a constant barometric elevation above sea level (ASL). The latter are more common in petroleum surveys where a true grid of lines at 1- to 6-mi spacings at 2000 ft to 4000 ft above ground level are typical.

Regional mineral surveys are typically flown at 0.5 line to 1.5 mi grids at 1000 ft MTC or at 2 ft to 3000 ft (ASL) above ground. Detail surveys with today's technology may be as tight as 300 ft (100 m) line spacing and 100 ft MTC using a helicopter platform. Helicopter surveys may be small (less than 500 line mi) but the per-line-mile cost is high.

Instruments and Sensitivity

Most airborne surveys for mineral exploration use either a proton precession or optically pumped cesium magnetometer for total magnetic field measurements. Three-component fluxgate, or other high-sensitivity scalar or vector magnetometers, are flown for special purposes but are not common in mineral exploration.

In airborne surveys, the sensor may be mounted in a rigid tail "stinger" or wing tip mount on fixed-wing aircraft or towed in a bird below either fixed-wing or helicopter aircraft. For any rigid mount, it is mandatory to minimize the effects of the aircraft's magnetic field on the magnetometer sensor(s). Compensation can be either passive or dynamic where heading and other flying variation differences are incorporated directly into the processing routines. Towed sensors are not so affected because of the distance between sensor and aircraft itself.

Proton sensors can be adjusted for high sensitivity (>0.1 nT) but are then limited in repetition rate because of the time required for readings. Cesium sensors have a similar high sensitivity without the repetition rate limitation but do have "dead spots" at polar and equatorial field inclinations.

High-sensitivity data are particularly critical for high-level surveys over sedimentary basins where the total range of anomalous variation is small. Discrete anomalies of several nanoTeslas and contour intervals of several tenths of a nanoTesla are feasible. In most mineral-oriented surveys, however, anomalous range is much higher because of the closer proximity to high-susceptibility source rocks at or near the surface. As result of the greater anomalous range, very high-sensitivity instrumentation is neither as important nor critical.

Repetition Rate

Repetition rate or the interval between samples, however, may be more critical than sensitivity in low-level, mineral-oriented surveys to adequately define small, short-wavelength anomalies. Aliasing occurs at the Nyquist frequency, which is defined as that equal to half the sampling frequency. This means that when the wavelength of an anomaly is less than half the repetition rate, the anomaly is distorted. Anomalies with a wavelength shorter than half of the sampling rate interval alias to longer wavelength anomalies from which they are practically indistinguishable. For example, an aircraft cruising at 120 mi per hour will have an equivalent ground speed of 176 ft per second. A repetition rate of 1.0 second is equivalent to a reading every 176 ft on the ground. In a low-level survey at 300 ft flight height, a thin dike-like body will be detected on only two readings, which can be processed (aliased) into an anomaly of entirely different wavelength, shape and character.

A change in aircraft speed affects the ground distance between readings. For this reason, some airborne systems are Doppler controlled; that is, the reading rate is controlled by a doppler system that provides a reading at a specified ground distance without regard to flying speed.

Aliasing is particularly noticeable on ground surveys where sources are shallow and the station spacing is greater than 100 ft. Legitimate small surface features become single point spikes which become arbitrarily designated as noise. A tighter station spacing will provide better definition of the anomalous character and may enable correlation of the "noise" to specific surface features, e.g., terrain features or volcanic outcrop. Figure 5.20 shows an example of aliasing.

On profile A, data are acquired at the rate of 4 points per unit distance (e.g., a ground line with 25-ft stations paced between 100-ft

FIGURE 5.20
Effect of aliasing

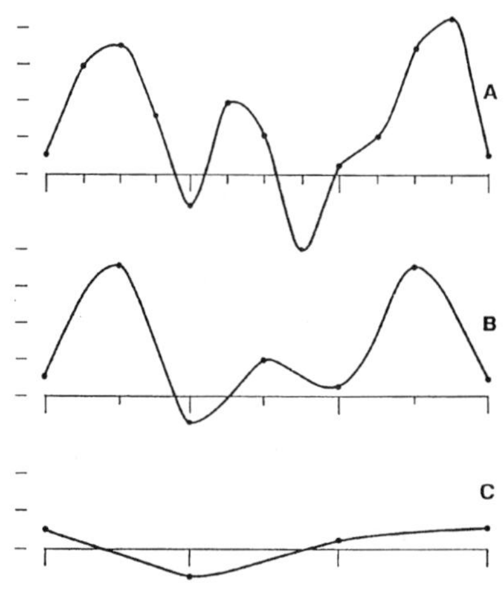

Figure: Example of aliasing
A - 4 points per unit
B - 2 points per unit
C - 1 point per unit

survey points). At two points per unit distance (50-ft readings), the anomalies are still reasonably well defined; however, on C, at one data point per unit distance, the anomalies are completely eliminated.

Line Spacing and Flight Height

Line spacing should be predicated by depth to source and strike length of the anomaly sought, but too often the dollars available dictate the survey specifications. For breccia pipe detection, a line spacing of 0.1 mi (500 ft) with a direction at right angles to the regional geologic fabric may be necessary. For large intrusive targets, a line spacing of 0.5 mi, or even 1 mi, may be satisfactory. Normally, line orientation is established at right angles to regional strike. If an area exhibits strike variability, a square or rectangular grid of lines may be advantageously used. Orthogonal tie lines to aid the reduction process are flown at a spacing of 5 to 10 times the flight line separation.

Figure 5.21 shows a graphic example from northwest Saskatchewan of the line spacing problem. Data from a 1/4-mi grid are contoured and then recontoured using only the lines 1 mi apart. The definition of the bull's eye closure is obvious on the tighter line spacing.

FIGURE 5.21
Anomaly resolution with flight line spacing

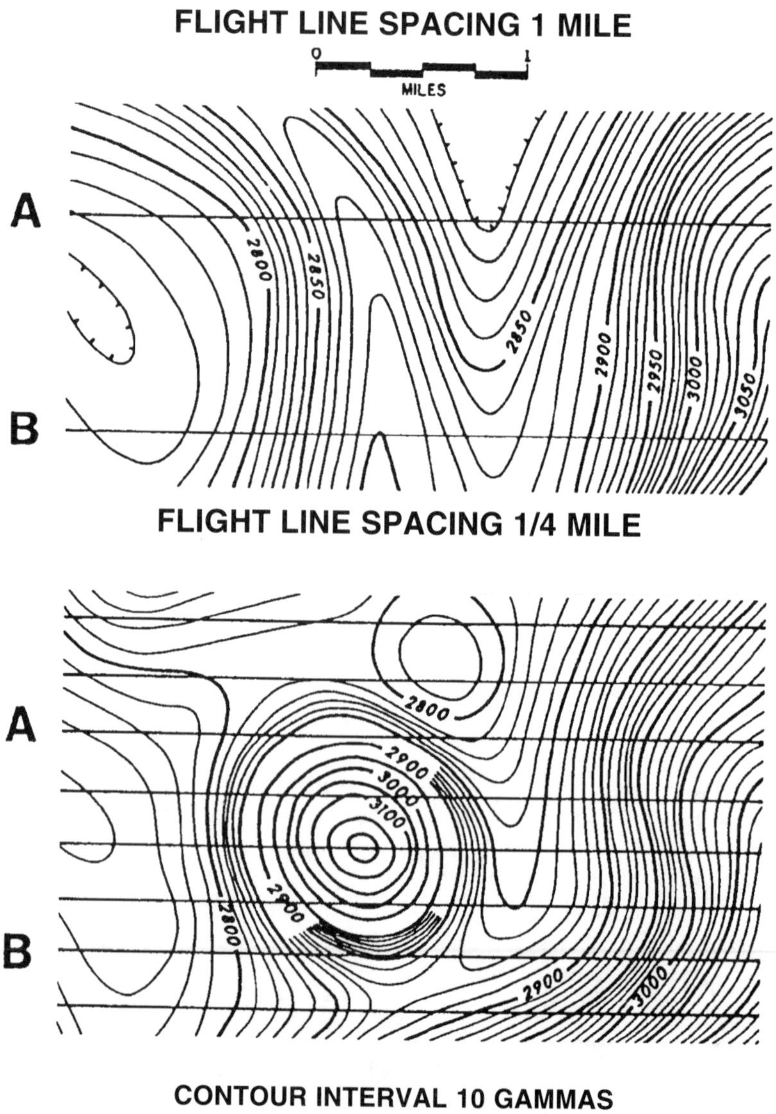

Because anomaly width is a function of distance between sensor and source, definition of small anomalies requires not only a low-level survey with a rapid repetition rate but also a line spacing adequate for along-strike delineation. Ultimately flight height must take into account terrain, depth to expected sources and resolution desired. Standard practice would have at least two lines cross any feature of interest, which is easy to accomplish for long two-dimensional features such as dikes but may be difficult where the principal targets are small, equi-dimensional breccias, volcanic plugs or skarns. As expected, anomaly intensity decreases off the end of an anomaly, figure 5.22.

FIGURE 5.22
Effect of the location of a flight line relative to a finite body on anomaly amplitude

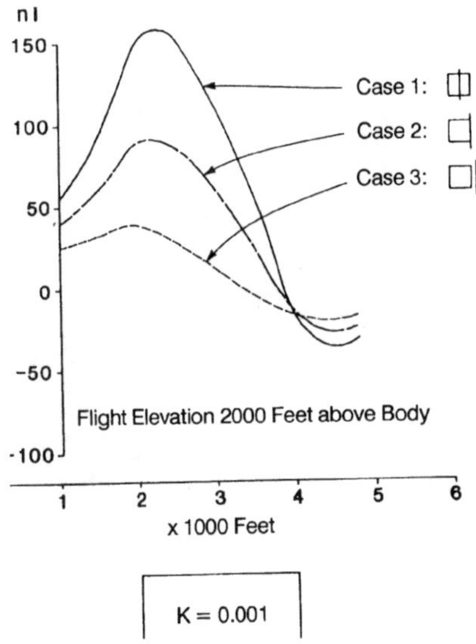

Anomaly wavelength is defined as the distance across the full high-low character of an anomaly. In practice, however, it is imprecisely used as the more dominant and obvious positive segment of an anomaly. For small, near-surface features and Depth/Width ratio > 1, anomaly width is a function of flight height. When the width exceeds the flight height and Depth/Width ratio < 1, the anomaly width relates more directly to the source width.

Flight height is a critical factor in anomaly resolution and definition, previously demonstrated in figure 5.17. At whatever altitude specified, variations commensurate with safe operation of the aircraft must be acknowledged. As with line spacing, cost becomes a factor, particularly when a helicopter is required to meet specifications for a low-level draped survey. There is no simple answer to the question which is better for surveying -- low-altitude, draped surveys or high-level, constant barometric surveys. The answer lies in what is the size and intensity of the anomaly sought, what is the purpose of the survey and what is expected from interpretation. All of these questions are modified by the cost factors of the different survey types.

The contoured data of low-level, draped data and an equivalent high-level, constant barometric survey are shown in figures 5.23 and 5.24. Lines for the low-level survey in the Carlin area were flown with a fixed-wing aircraft at one-quarter mile spacing, north-south, at a nominal 300 ft above ground surface. For comparison to a constant barometric elevation survey, the low-level data were upward continued to 1000 ft and give a reasonable representation of the proprietary high-level data of a different survey. Comparison indicates obvious differences in anomaly patterns resulting from the smoothing effect of the higher level survey. Resolution of the smaller, near-surface features has changed; however, a similarity of the broad features and structures is observed.

Several volcanic units, indicated by the "noisy" tight closure patterns, cover the central and west portions of the survey block; the area to the east and northeast with the bland magnetic signature consists predominantly of older sediments. A deep intrusive anomaly is evident in the magnetic gradient in the southeast corner.

Several magnetic linears indicating fault structures or contacts are identified for comparison. Local linears may be more easily seen on the low-level data, whereas more regional linear features would dominate if a larger area were available for study. The small closure, A, above the title block is a cultural effect of the Carlin operation (only shown on the low-level data). West of this feature, a major north-south linear, L, indicated by the change in anomalous character, separates Paleozoic carbonates from volcanics. Correlated to a known major fault, which is mapped in the south, it can be projected across the map area. The linear, M, is the fault and topographic break, bounding the west side of Boulder Valley. The parallel linear,

FIGURE 5.23
Magnetic contour map of the Carlin area, flown at 300 feet above surface at a north-south line spacing 1/4 mile

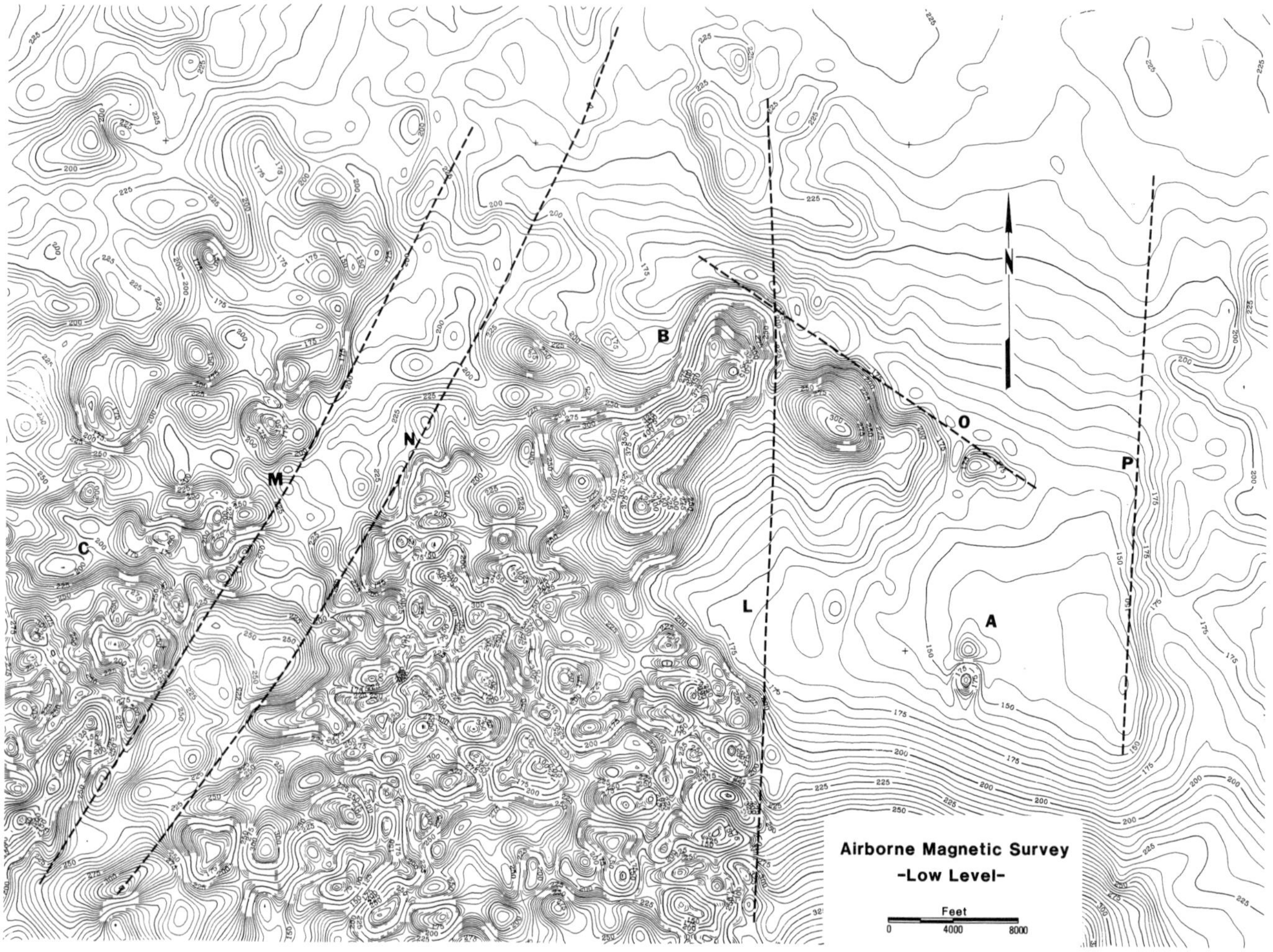

N, is another fault defining a graben (?) along Boulder Valley. The Sheep Creek rhyolite, C, is mapped to the west of this structure. Inspection shows two different magnetic patterns west of this fault indicating different surface rocks and/or lower topography. Closer analysis would show different magnetic "background" levels and high-low relationships, pointing toward a remanent effect in the northern part and suggesting two distinct volcanic units (either time or composition). Both, however, are mapped as the same Sheep Creek rhyolite unit.

Anomaly B is interpreted to be caused by a small diorite intrusive which outcrops along the northeastern margin of the anomaly and is associated with the mineralization at Post-Goldstrike. Several other gold deposits are clustered around the margin of the magnetic anomaly, but no intrusive has been observed. Based upon the magnetics, the intrusive plunges to the southwest. Two other major linears, O and P, are identified to aid comparison of the two data sets. The effect of surface volcanics is obvious. In the high-level

FIGURE 5.24
Magnetic contour map of Figure 5.23 upward continued 1000 ft

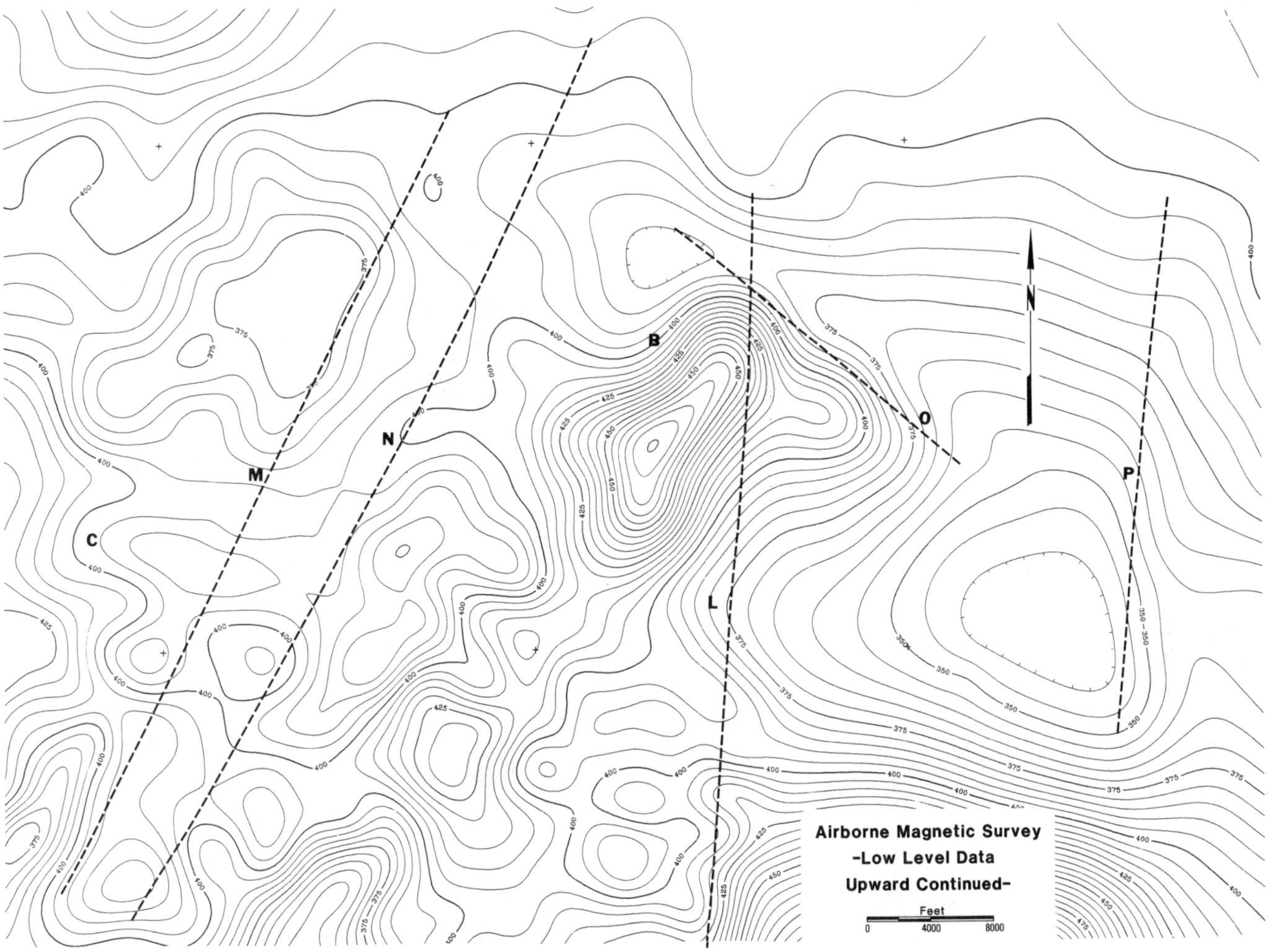

data, larger anomalous features, related to deeper sources, are obscured less by the effects of surface volcanics and are perhaps more immediately evident. Unfortunately, a complete interpretation of the magnetic survey, integrated with geology, is beyond the scope of this paper.

To avoid aliasing errors, Reid (1980) showed that survey lines should have a spacing of no more than twice the sum of flight altitude and estimated depth to sources of particular interest. In addition, a tighter grid gives more data permitting optimum use of the more sophisticated filter computations.

Constant Barometric and Draped Surveys

Both constant and draped surveys have advantages and limitations. Fixed-wing, draped surveys are flown at a low level (300 ft to 500 ft, or even 100 ft in parts of Australia). Rugged terrain or very low-level surveys require a helicopter to maintain adequate elevation control. Draped surveys do not eliminate the topographic effect but do provide the best high-resolution data and anomaly definition of small, near-surface features. Constant barometric surveys, usually flown with fixed-wing aircraft, are less expensive on a per-mile basis. Because of the altitude required to clear the peaks, however, small features in low-lying valleys or basins may not be detected. With increased height, a wider line spacing is usual. The alternatives are either to have a uniform data base for digital processing and interpretation at less cost or to have increased resolution of low-level data at higher cost. The question is whether low-level data continued to an equivalent high-level datum are equivalent to consistent high-level, directly recorded data and whether the low-level anomaly character is important to interpretation, e.g., the merging of multiple small-surface features into a large, single anomalous unit. Strong arguments can be advanced for each type of survey. The answer lies in survey objectives, local geology and topography.

Resolution of multiple complex anomalies and evaluation of aeromagnetic surveys is limited by the line spacing, distance between source and sensor, and the angle between flight line and strike of the major axis of the source body. A structure small enough to fit between two lines will not normally be recognized for what it is, if it is recognized at all.

All magnetic surveys, airborne and ground, reflect terrain features which are composed of higher susceptibility rocks. Draped airborne surveys do not reduce the problem of terrain anomalies; on the contrary, when compared to level (constant barometric) surveys, draping intensifies terrain anomaly amplitude. Therefore, minimizing the terrain effect by choosing a constant barometric elevation level over a low-level draped survey is not the only question to be addressed in choosing which type of survey to fly. Draped surveys have greater resolution for small bodies, especially those occurring in low topography and valleys, and may provide a significant and critical advantage in some areas. On the other hand, this surface noise may obscure important but subtle magnetic features of interest. The principal disadvantages of low-level surveying, however, are the extra cost and greater logistical difficulties of flying close to the ground surface.

Interpretation of data from the two kinds of surveys poses different problems. Low-level, draped data have greater resolution of magnetic sources. Furthermore, the character of anomalies from either topographic lows or high mountainous regions are observed from the same elevation above ground surface, eliminating one variable in comparative interpretation. Therefore, similar sources will cause anomalies of similar amplitude and wavelength. Comparison of anomaly gradients (slopes) is much easier. The main disadvantages are the higher "noise" component and the fact that most interpretation techniques (e.g., depth estimates from slope distances and digital filtering techniques) have been developed for data acquired on a level surface. If filtering techniques developed for level surveys are used without modification (or at least some understanding of the problem), misleading conclusions might result. Constant barometric surveys generally are flown at a higher level, which smoothes anomalies and provides a better data set for quantitative analysis of large, regional-type features but may be handicapped by loss of resolution. Therefore, target size and expected depth are critical parameters in the planning process.

The flight height or aircraft altitude is recorded simultaneously with the magnetic reading. Both the barometric elevation and azimuth altitude above ground as measured by a radar altimeter are recorded, if available. The altimeter record gives the vertical position of the aircraft above topographic features and provides a check of the survey specifications for flight elevation. Some care and judgment must be exercised, because the radar signal may reflect or bounce off tree tops and not ground surface. The altimeter record becomes important in determining depth to sources because the aircraft elevation above ground must always be subtracted from the absolute depth (source to sensor) distance calculation. The altimeter records are plotted with the magnetic data on profiles but are not usually produced in contour form.

Navigation and Positioning

Aircraft positioning, or navigation, is accomplished in several different ways. Traditionally, flight lines were drawn to scale on a map at a scale of 1:24,000 or 1:62,500 depending upon the flight line spacing. A camera recorded aircraft location for reconstruction of the flight path. Specific identifiable points were noted visually for correlation. Photo mosaics provide better control, particularly of small areas. Black-and-white photos of good contrast are often used, but color can add significantly to easy identification of features. In the last decade, color video cameras have been introduced to record the aircraft position.

A common fiducial number is recorded on the magnetometer, altimeter and film records. Later on the ground, points were recovered, line locations determined, errors eliminated and the corrected flight line position plotted. The other data are then referenced spatially to this base for correction and contouring.

In current practice, electronic navigation systems have replaced the old visually aided camera positioning. Doppler systems provide an on-track record between two known points and can be used to trigger a measurement at specific distances on the ground rather than on a uniform time base. The latter may provide readings at considerably different ground separations because of varying ground speed resulting from wind or climbing over terrain. Other navigation systems provide real time location by triangulation between known high-frequency transmitter stations. Several industrial systems are readily available and routinely used, where 2-to 4-or more transponders are put out. When the systems are properly calibrated, the pilot can fly precise lines and know his aircraft position within a few meters. One limitation is that most of these systems are line of sight, but proper locations on mountain tops or high ground can provide excellent position control. When corrected, the X- and Y- coordinates are fed directly into magnetic reduction programs.

Other passive navigation systems (Loran or Shoran) may be used in certain areas and/or conditions where the transmitters are fixed and more-or-less permanent. Most recently, Global Positioning Systems (GPS) are becoming more practical as more satellites are orbited and cost declines to a competitive level. These will provide an X-Y position in digital format for direct use in both air and ground applications.

Data Reduction and Compilation

Most airborne magnetic systems record data in digital form. Five general steps are required to process the data, calculate a unified coherent data set, and produce clean contour maps and/or profiles.

- Locate field data by traditional means on topographic base maps or photographs, by aided electronic navigation systems [Doppler, mini-Ranger, Syledis or GPS (Global Positioning System)], or some combination of both.

- Remove the magnetic diurnal by subtracting the digitally recorded and edited base station monitor readings from the field data, sample by sample.

- Level or tie the magnetic data to minimize magnetic field differences at intersections. Mis-ties are caused by elevation differences, poor locations at intersections, residual effects of the aircraft (poor compensation) and time variations of the earth's geomagnetic field. Poor leveling results in linear steps along single lines and other incursions in contoured data, sometimes called "herringbone." The corrected data are recorded and archived on an "edited data tape."

- Interpolate the leveled data into a regularly spaced grid using minimum curvature, bicubic spline or other interpolation routines. Remove the International Geomagnetic Reference Field (IGRF) to determine the anomalous field resulting from local geology and produce a residual magnetic data set. The grid data thus produced are used for contouring of residual magnetic data and the calculation of any of the two-dimensional digital filters. The data are stored in this second form as a "grid tape."

- Produce and plot any or all of the various profiles and maps required.

Contour Maps

Ordinarily, the primary product of an airborne survey is a contour map of the magnetic intensity. The map scale should be adjusted to a convenient scale to overlay geologic maps, which may be 1:24,000 to 1:62,500 or equivalent metric scales for regional surveys or 1:12,000 or 1:24,000 for smaller detail surveys. More importantly, the contour interval should be adjusted to outline the features of interest. If anomalous range is high and a coarse contour interval is established, subtle features may be lost within contours, particularly in the near-background range.

Color contour maps enhance perception of trends and features. Usually, the hot red colors denote highs, the cold blues are lows, and the pastel yellows and greens designate near background ranges. Some use equal contour intervals and color areas; others prefer a scale with variable intervals which can be adjusted to give smaller intervals in the background range to emphasize small changes and trends, and a larger interval in the high and low ends where small variations are less important. When applied in this latter manner, color becomes a highly useful optical filter.

Profiles are a very useful product for analysis. Usually profiles are produced at a working scale equivalent to available geology. Analog profiles should include two scales of magnetic data, both radar altimeter barometric elevation data and a calculated derivative or other filter, if available. The altimeter records are particularly valuable in interpretation. Stacked profiles (profiles plotted at correct scales on a plan map) can be very useful in recognizing trends and subtle features that may not be as easily recognized on contour maps.

Filtered contour maps or profiles are usually requested later in the interpretation sequence to aid analysis of a particular problem.

One of the newest products of the digital revolution is the shaded relief map. Shaded relief or shadow maps are similar to shaded topographic maps or photos. Maps are computed for a particular sun angle and azimuth illumination which can enhance linear anomalous features striking at right angles to the illumination direction. While qualitative, they provide very interesting overviews of structural orientations, linear patterns of anomalies and local geologic character, as represented in the magnetic data.

Sources of Error

Several sources of error in data acquisition and map preparation must be considered in interpretation. Herringbone, the alternate cyclical pattern of contour lines produced by inadequate flight line location or poor aircraft or heading compensation, is distinctive. Noise spikes cause spot, single line closures, generally of high intensity and short wavelength. Small, single line closures on adjacent lines result from an inadequate contouring algorithm or flight line direction crossing a linear feature at an acute angle. Improper levelling and mis-ties along lines will cause linear features oriented along the flight line.

All of the above should be evaluated and removed, if necessary, in the processing process. Occasionally, however, something slips through into a final product. Alternatively, too much arbitrary smoothing of the data can remove anomalies and/or character from the contour map. Because any contouring process smoothes data between contour intervals, estimates of straight slope distances may, therefore, be in error. In effect, the straight slope portion has become an artifact of the contouring. For this reason, many interpreters prefer to work with profiles rather than only with contour maps.

Ground Surveys

Ground surveys, ideally, should follow the same guidelines for line spacing and station density as airborne surveys. For archeological or engineering site studies, grids of 1 m to 2 m are not unusual. Typical mineral exploration surveys at line spacing of 400 ft to 800 ft and stations at 100 ft are too widely spaced to delineate many exploration targets and may be relatively useless for any type of evaluation other than simple anomaly recognition. Station spacing, in particular, should be no more than 25 ft to 50 ft to define surface features. Paired lines, no more than 200 ft to 400 ft apart, are more useful than a widely spaced grid because line-to-line redundancy can enhance anomaly evaluation and enable better definition than a single line. Inadequately spaced data often are difficult both to evaluate and interpret.

Ground observations are affected by local magnetic sources originating from cultural effects (scrap metal, pipes, etc.), magnetite in sands and soils, and bedrock outcrops, particularly mafic volcanics. For some purposes, the sensor can be placed on a staff 6 ft to 8 ft above surface to minimize the problem of surface noise. Valid surface anomalies are not well defined by stations every 100 ft, which can result in an aliasing problem.

Portable magnetometers for use in ground surveys are generally of the proton precession type. Many now have solid state memory that records all station locations and magnetic readings for later processing. Diurnal variations are monitored and removed by use of a base station recording magnetometer, usually of the same type. The base station should be set up within or close to the survey area, but it must be placed in an easily accessible location that is a magnetically quiet location away from culture effects (moving autos and equipment, buildings and sources of transient DC currents). In addition, care must be taken to place the instrument in a flat magnetic field, not on a steep magnetic gradient, so that if re-established from time to time, a significant change in "background" is not inadvertently input into the survey records. Base station readings, taken every several minutes, are also stored and later dumped into personal computer files with the field data. After corrections are made, the data can be printed as a listing and/or in profile or contour form.

State-of-the-art magnetometer systems may now include two orthogonal coils to measure sequentially the VLF field at one or two separate frequencies.

As with the older magnetometers, care must be exercised to ensure that the operator is magnetically "clean" and that the base station is properly located in a magnetically quiet location away from cultural effects.

Methods of interpretation, processing, filtering and enhancement are much the same as for airborne surveys. Data usually are more limited in scope, however, and evaluated by profile analysis. Location of source bodies, width and depth are the primary variables of interpretation. Rules of thumb for depth, and simple averaging of profile data, are used to initiate 2-D modeling.

Survey Cost Factors

Broken down for an airborne survey, the cost of a ground survey would have similar component parts. An airborne survey would consists of the following cost factors:

- Mobilization and demobilization -- the transportation of aircraft and crew from home base to survey area and return. The charge can be either a fixed or per-mile cost basis.

- Data acquisition -- the daily cost of flying. Often this is fixed on a line-mile (kilometer) basis, but it may be broken down as a fixed daily crew and aircraft expense and a line-mile production charge.

- Data compilation -- the ordinary reduction and processing to a contour map. This often is charged on a line-mile basis but may be included in data acquisition.

- Extra processing or map production -- the base charge covers a predetermined schedule of maps to be produced. Other maps, including filters, extra color and/or shaded relief usually are on an extra charge basis at specified costs, when requested by the client.

- Extras -- surcharges for fuel transport on a helicopter job, standby charges for bad weather or magnetic storms, or other unusual survey costs.

- Interpretation -- interpretation is seldom included in the survey costs, except for a logistics and production report, because the time and effort of interpretation are often an unknown until the data are plotted and reviewed. Interpretation charges may be estimated on a line-mile basis, where this is apropos, or may be a negotiated item dependent upon the complexity of the data set and what is deemed essential and necessary.

Cost effectiveness of a survey begins with a carefully planned and designed survey. Survey specifications start with and are based upon the geology of the area. Interpretation is integrated with geology to provide a timely and useful analysis of anomalies of exploration interest.

SUMMARY

The magnetic field of the earth changes across the surface of the earth. Anomalies, caused by geologic variations in the near-surface crust, are many and varied. The basic relationships of anomalous response to field inclination and source geometry, however, can be defined. Knowledge and appreciation of the differences in anomalous response to the key variables of the source enable an interpreter to achieve a sensible solution in terms of reasonable geology. Interpretation can follow one or several lines of increasing complexity depending upon the goals of the program.

Awareness of the effects that influence anomaly size, shape, depth and the related factors of terrain and regional caused by large adjacent anomalies, can lead to cost-effective use of magnetics as a basic tool in the exploration process. Patterns of anomalies as well as individual anomalies are important, the former more to a mapping program, the latter to location of specific targets of geologic interest. The inherent ambiguity of magnetics can be reduced to an acceptable level by use of geologic constraints moderated by knowledge of the "noise" effects produced by terrain and variable flight elevation.

Proper design of surveys, with input on the size and depth of the expected geological features of exploration interest, is necessary to provide data of a quantity and quality commensurate with the scale and scope of the program. Design foresight also adds significantly to the eventual interpretation process. In a complex minerals exploration environment, in-depth interpretation has many technical facets and relies heavily on experience, skills and integration with geology.

SELECTED REFERENCES

Brant, Arthur A., 1966, Geophysics in the exploration for arizona porphyry coppers, in Geology of the Porphyry copper deposits, Southwestern North America, Ed. by Titley, Spencer R. and Hicks, Carol L., U of Arizona Press, Tucson, AZ.

Breiner, S., 1973, Applications manual for Portable magnetometers, pub. by Geometrics, Sunnyvale, CA.

Clement, W.G., 1973, Basic principles of two-dimensional digital filtering, Geophysical Prospecting, Vol. 21, pp. 125-145.

Fuller, Brent D., 1967, Two-dimensional frequency analysis and design of grid operators, in Mining Geophysics, Vol. II, Soc. Explor. Geophysicists, Tulsa, pp. 658-708.

Graf, R. J., 1986, Compilation and Interpretation of Aeromagnetic Data, internal document, TerraSense, Inc, Sunnyvale, CA.

______ Magnetic Filtering, pub. by Geophysical Service Division, EG&G Geometrics, Sunnyvale, CA.

Harland, W.B., et al, 1982, A geologic time scale, Cambridge Univ. Press, 131 pp.

Dods, S.D., Teskey, D.J. and Hood, P.J., 1985, The new series of 1:1,000,000-scale magnetic maps of the Geol. Surv. of Canada; Compilation techniques and interpretation, in Hinze, W.J., Ed., The utility of regional gravity and magnetic anomaly maps: Soc. Explor. Geophysicists, Tulsa, pp. 69-87.

Grauch, V.J.S., and Campbell, David L., 1984, Does draping aeromagnetic data reduce terrain-induced effects?, Geophysics, Vol. 49, No. 1, pp. 75-80.

Hinze, William J., ed., 1985, The utility of regional gravity and magnetic anomaly maps, Soc. of Explor. Geophysicists, Tulsa, OK.

Hinze, William J. and von Frese, Ralph R.B., 1989, Magnetics in geoexploration, in Geomagnetic methods and structures of the crust and upper mantle beneath India, ed. by B.P. Singh, A special volume of the Proceedings of the Indian Academy of Science, in publication.

de Stadelhofen, C. Meyer, and Juillard, T., 1987, The magnetic anomaly of Lausanne, Switzerland; and interpretation, The Leading Edge, pub. by Soc. of Explor. Geophysicists, Tulsa, Feb., 1987, p 38.

Kearey, P. and Brooks, M., 1984, An introduction to Geophysical Exploration, Geoscience Texts Vol. 4, Blackwell Scientific Pubs., Boston.

Marsh, Bruce D., 1971, Aeromagnetic terrain effects, M.S. Thesis, U. of Arizona, Tucson.

Parasnis, D.S., Principles of Applied Geophysics, 3rd ed., London, Chapman and Hall.

Peters, Leo J., 1949, The direct approach to magnetic interpretation and its practical application, Geophysics, Soc. of Explor. Geophysicists, Tulsa, OK, Vol. 14, No. 3, 1949.

Rao, D.A. and Babu, H.V.R., 1984, On the half-slope and straight-slope methods of basement depth determination, (Short Note), Geophysics, Soc. of Explor. Geophysicists, Vol. 49, No. 8, Aug., 1984, pp. 1365-1368.

Reeves, Colin V., 1985, Airborne geophysics for geological mapping and regional exploration, ITC Journal, 1985-3, pp. 147-161.

Reeves, Colin V., 1989, Aeromagnetic interpretation and rock magnetism, in First Break, pub. by European Assn of Explor Geophysicists, Vol. 7, No. 7, July, 1989, pp. 275-286.

Reid, A.B., 1980, Aeromagnetic survey design, Geophysics, Vol. 45, pp. 973-976.

Reford, M. S., 1964, Magnetic anomalies over thin sheets, Geophysics, Vol. 29, No. 4, pp. 532-536.

_____________, 1980, Magnetic Method, Geophysics, No. 11, pp. 1640-1658.

Vacquier, V. et al, 1951, Interpretation of aeromagnetic maps, GSA Memoir 47, Waverly Press, Baltimore.

Xinzhu, Liu, and Hinze, W. J., 1983, An overview of the utility of isolation and enhancement techniques in gravity and magnetic anomalies, unpublished thesis, Dept. of Geophysics, Purdue University, West Lafayette, IN.

Wright, Phillip M., 1981, Gravity and magnetic methods in mineral exploration, Economic Geology, 75th Anniversary Vol., pp. 829-839.

Zeitz, Isidore and Andreasen, G.E., 1967, Remanent magnetization and aeromagnetic interpretation, in Mining Geophysics, Vol. II, Theory, Soc. Explor. Geophysicists, Tulsa, OK.

CHAPTER 6
ELECTROMAGNETICS

ELECTROMAGNETIC PROSPECTING FOR MINERALS

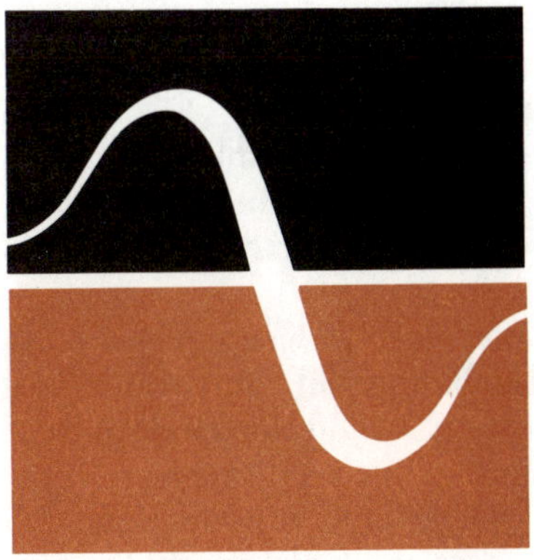

Fifty years ago electromagnetic (EM) techniques began to play a significant role in mineral exploration in Scandanavia and Canada. Early equipment used in those Precambrian terrains consisted of simple devices operating at only one frequency. Interpretation was generally unreliable. Now, a new array of EM techniques is available. In addition to other geophysical techniques, EM is now integrated with geology and geochemistry into most modern exploration programs. Depth penetration of EM methods has greatly improved; this is important in remote areas, in areas with extensive cover, and in areas where mining has gone on for decades. Improvements in EM technology resulted from developments in several areas:

Jan Klein joined Cominco Ltd. in 1974 and has since become chief geophysicist. He was employed by Scintrex Ltd. for nine years, and developed worldwide experience in all aspects and types of ground and airborne mining geophysics during that period. Klein holds a degree in mining engineering (exploration option) from the Technological University of Delft, Netherlands.

Jan Klein
Cominco Ltd.

Jules J. Lajoie has been employed by Cominco Ltd. as research geophysicist since 1973. His main interest and work has been in electromagnetic and induced polarization exploration methods. Lajoie obtained an honors B.Sc. degree in physics from the University of Ottawa, a M.Sc. degree in geophysics from the University of British Columbia, and a Ph.D. in geophysics from the University of Toronto. His doctoral thesis was on computer modeling of conductive host rock effects in electromagnetic exploration.

Jules J. Lajoie, Ph.D.
Cominco Ltd.

- A better understanding of the physics and mathematics of EM fields
- Modelling of EM geologic targets with the aid of metal/graphite plates and computers
- Development of small electronic components
- Digital recording systems (both for airborne and ground equipment)
- Provision for rapid plotting, filtering, and interpretation of EM data with aid of computers.

Improvements in EM equipment were also guided by feedback from drilling geophysical anomalies, conductivity measurements on core and rock specimens, and a better understanding of geology.

Primarily used for base metal exploration, EM techniques enjoy a wide application, including the search for uranium, coal, oil and gas, gravels, and fresh water. EM has also been used in civil engineering and geologic mapping applications.

Only the most common and readily available EM techniques used in mining exploration will be reviewed. The chapter will not repeat several excellent textbooks and articles that are available on the application of EM techniques. Attached references list only those publications relevant to the objective of the text: to make the geologist more comfortable with those EM methods currently used in exploration. Books referred to give in-depth treatment of the theory and application of particular topics. Referenced articles normally treat a specific aspect of the EM technique, e.g., interpretation or a good case history.

Theoretical considerations

This theoretical section is intended to provide the reader with an intuitive understanding of electromagnetic methods. Mathematics are kept to a minimum.

Resistivities of rocks

Resistivity of a rock or individual mineral is a physical property, defined and presented in detail in the chapter by Dr. P. Hallof, "Grounded Electrical Methods" (see chapter 2). The unit of resistivity (ρ) is ohmmeter. Conductivity (σ) is simply the inverse of resistivity, and the unit is "mho per meter."

Histograms in figure 6.1 summarize the resistivity ranges for common rock types. Note that the resistivity scale is logarithmic. Resistivities of most common rocks usually range from 100 to 100,000 ohmmeters (Ω-m); resistivities of most metallic luster minerals* range from 10^2 to 10^5 Ω-m, a difference of about six orders of magnitude. Electromagnetic methods exploit this large variation in resistivity. The most common conductive metallic minerals found with EM methods are pyrite, pyrrhotite, galena, and chalcopyrite. Unfortunately, graphite is highly conductive and commonly responds to EM systems. Water-saturated and clay-rich surficial materials can be good conductors. In some parts of the world, such as in Australia, areas of evaporative groundwater can also be highly conductive.

FIGURE 6.1
Histograms of the resistivities of common rocks

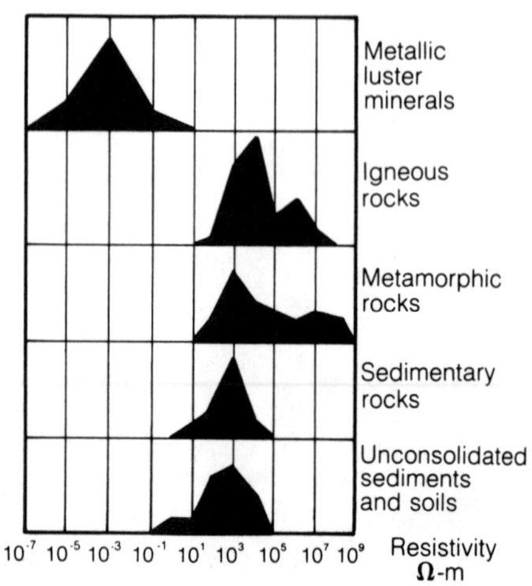

In a discussion about the use of electromagnetic methods in the search for metallic minerals, one critical point must be emphasized: within the target zone, the conducting minerals *must be electrically interconnected* so that the *whole body* is an electrical conductor. If the conducting minerals are not well interconnected, the target may not respond to EM methods. For example, massive sulfide deposits are usually excellent electromagnetic exploration targets, whereas porphyry copper deposits are not because the conducting minerals are not well interconnected.

* Sphalerite is a non-conductor, even in the high iron variety which can have a metallic luster. Graphite is a highly conducting mineral.

Electrical methods vs. electromagnetic methods

The main differences between electrical and electromagnetic methods are outlined in this section. In "Grounded Electrical Methods," electrical methods have been thoroughly described.

With electrical methods, the *source* consists of electrical current injected into the ground through two electrodes. The transmitted current waveform may be dc, low frequency sinusoidal (up to about 10 Hz), or rectangular, as in induced polarization surveys with a frequency of about 0.1 Hz. The energizer, therefore, is the electrical current injected into the ground through current electrodes.

With electromagnetic methods, the source most commonly consists of a closed loop of wire in which ac current flows. It can be a small, portable transmitter coil of 2-3 feet in diameter, in which case there are many turns of wire. Alternately, the source can be a large transmitter loop on the ground, as large as 1 mile in diameter. The frequency of the transmitter current can range from about 0.1 to about 10,000 Hz; it is more commonly in the 100-2500 Hz range.

Electric current in the transmitter loop generates a magnetic field. The magnetic field is the energizer in electromagnetic methods as compared with electric current in electrical methods.

In terms of *response*, with electrical methods, anomalies result from resistivity contrasts. For example, if a target with a resistivity of 10 Ω-m is in a host rock with a resistivity of 100 Ω-m, the same anomaly results as if the target had a resistivity of 100 Ω-m in a host rock with a resistivity of 1000 Ω-m. In both cases, there is a resistivity contrast of 10. This example holds as long as the transmitter frequency is low enough that there is no appreciable electromagnetic induction in the rocks.

With electromagnetic methods, anomalies are due more to *absolute* resistivity rather than resistivity *contrasts*. The two examples mentioned previously for electrical methods would *not* produce the same anomalies with electromagnetic methods.

The electromagnetic induction process

The electromagnetic induction process is conceptually summarized in figure 6.2. A power supply powers an EM transmitter which outputs a time-varying electric current into a transmitter coil. The current in the transmitter coil generates a magnetic field of the same frequency and phase. Lines of force of this magnetic field penetrate the earth and may penetrate a conductive body. When this occurs, an electromotive force or voltage is set up within the conductor, according to Faraday's Law:

$$EMF_c = M_{TC} \frac{di_T}{dt}$$

where

EMF_c is the electromotive force or voltage in the conductor,

M_{TC} is the mutual inductance between the transmitter and the conductive body in the ground and is a complex number,

and

di_T/dt is the time rate of change (derivative) of the current (i_T) in the transmitter loop. (6.1)

Current will flow in the conductor in response to the induced electromotive force. These currents will usually flow through the conductor in planes perpendicular to lines of magnetic field of force from the transmitter, unless restricted by the conductor's geometry. Current flow within the conductor generates a secondary magnetic field whose lines of force, at the conductor, are such that they oppose those of the primary magnetic field. The receiver coil, at some distance from the transmitter coil, is therefore energized by two fields: from the transmitter and from the induced currents in the ground.

FIGURE 6.2
Generalized picture of electromagnetic induction prospecting

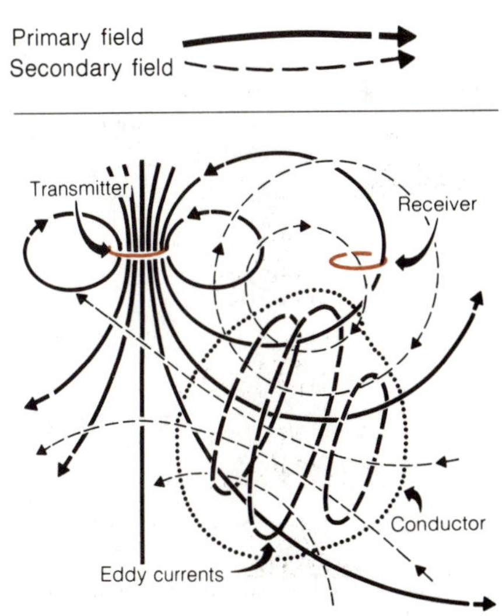

From Faraday's Law, the EMF induced in the receiver may be expressed as

$$EMF_R = M_{RT}\frac{di_T}{dt} + M_{RC}\frac{di_C}{dt}.$$

where

di_T/dt was defined previously and M_{RT} is the mutual inductance between the receiver (R) and transmitter (T),
M_{RC} is the mutual inductance between the receiver (R) and conductor (C) in the ground, and is a complex number,

and

di_c/dt is the time derivative of the current induced in the conductor (C).

Frequency domain electromagnetic method

In the frequency domain method, the transmitter emits a sinusoidally varying current at a specific frequency. For example, at a frequency of 100 Hz the magnetic field amplitude at the receiver will be that shown in the top part of figure 6.3. Because the mutual inductance between the transmitter and conductor is a complex quantity, the electromagnetic force induced in the conductor will be shifted in-phase with respect to the primary field, similar to the illustration in the lower part of figure 6.3. At the receiver, the secondary field generated by the currents in the conductor will also be shifted in-phase by the same amount.

We now have three methods of measuring and describing the secondary field:

▪ *Amplitude and phase* The amplitude of the secondary field can be measured and is usually expressed as a percentage of the theoretical primary field at the receiver. Phase shift, the time delay in the received field by a fraction of the period, can also be measured and displayed.
▪ *In-phase and out-of-phase components.* The second method of presentation is to electronically separate the received field into two components as shown in the lower part of figure 6.3. The first component is in-phase with the transmitted field while the second component is exactly 90° out-of-phase with the transmitted field. The in-phase component is sometimes called the real component, and the out-of-phase component is sometimes called the "quadrature" or "imaginary" component.

Both measurements require some kind of a phase link between transmitter and receiver to establish a time or phase reference. This is commonly done with a direct wire link, sometimes with a radio link, or through the use of highly accurate, synchronized crystal clocks in both transmitter and receiver.
▪ *Tilt angle systems.* The simpler systems are tilt angle systems which have no reference link between the transmitter and receiver coils. The receiver simply measures the total field irrespective of phase, and the receiver coil is tilted to find the direction of maximum or minimum magnetic field strength. As shown conceptually in figure 6.2, at any point the secondary magnetic field may be in a direction different from the primary field. With tilt angle systems, therefore, the objective is to measure deviations from the normal in field direction and to interpret these in terms of geological conductors.

FIGURE 6.3
Generalized picture of the frequency domain electromagnetic method

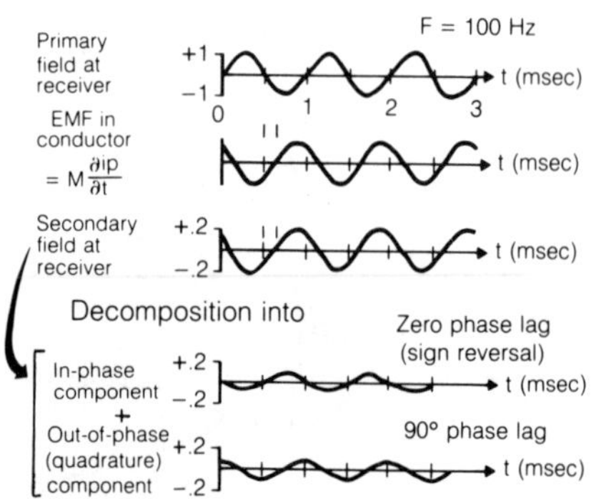

The response function

Now that the in-phase and out-of-phase components have been defined, we can investigate their response to different conductors. First, the response parameter of the conductor is defined as the product of conductivity-thickness (σt), permeability (μ), angular frequency ($w = 2\pi f$), and the square of some mean dimension of the target (a^2). The response parameter is a dimensionless quantity. In MKS units, a poor conductor will have a response parameter of less than about 1, whereas an excellent conductor will have a response value greater than 1000. The relative amplitudes of in-phase and quadrature components as a function of response parameter are given in figure 6.4 for the particular case of the sphere model in a uniform alternating magnetic field. One notes that for low values of the response parameter (<1) the sphere will generally produce a low amplitude out-of-phase anomaly; at moderate values of the response parameter (~10-100), the response will be a moderate amplitude in-phase and out-of-phase anomaly, whereas for high values of the response parameter (>1000), the response will usually be in the in-phase component. Although figure 6.4 shows the response only for the particular case of a sphere in a uniform field, the response functions for other models are similar.

In frequency domain EM, the depth and size of the conductor primarily affect the amplitude of the secondary field, whereas the quality of the conductor mainly affects the ratio of in-phase to out-of-phase amplitudes, a good conductor having a higher ratio and a poorer conductor having a lower ratio.

Time domain electromagnetic method

In time domain EM, the transmitter transmits nearly square current pulses of alternating sign, as shown in the top part of figure 6.5. The primary magnetic field at the receiver, therefore, will have the same form. Now, because the electromotive force (EMF) induced in a conductor is proportional to the time derivative of the primary field, an EMF is induced in the conductor only when the primary field is changing as shown in the second graph of figure 6.5. The induced EMF in the conductor will produce a current whose amplitude and decay rate (after the EMF stops) are dependent on the quality of the conductor. A poor conductor will have a relatively higher amplitude current flow at first, but it will decay quickly. Of course, a very poor conductor will have no currents induced at all. A good conductor will have a relatively lower amplitude of current flow at first, but it will decay more slowly. The secondary field at the receiver will have the same form as the current flow in the conductor, and this is shown in the lower two plots of figure 6.5.

The time domain EM receiver is synchronized to the transmitter and measures the received magnetic field in between the transmitter pulses at as many as 10 channel locations as shown in the lower plot of figure 6.5.

In time domain EM, the depth and size of the conductor are reflected primarily in the amplitude of the secondary field, whereas the quality of the conductor is reflected mainly in the rate of decay of the secondary field, a good conductor having a long decay and a poor conductor having a short decay.

Interpretation procedures

The interpretation of electromagnetic surveys follows basic steps. Most of the common EM systems have nomograms associated with them. (Nomograms are diagrams on which the measured parameters, e.g., in-phase and out-of-phase components, are plotted for varying model conductivity and one or more geometrical factors, e.g., depth to top, thickness, etc. The model responses available for most electromagnetic methods are those of the long, thin dike (to model thin tabular massive sulfide bodies), a homogeneous earth, and a layer for simulating conductive overburden.

The first step is to attempt to determine from the shape of the anomaly, a simple model geometry which can be thought to approximate the cause of the anomaly. The second step is to measure characteristics

FIGURE 6.4
In-phase and out-of-phase response of a sphere in a uniform alternating magnetic field, as a function of response parameter

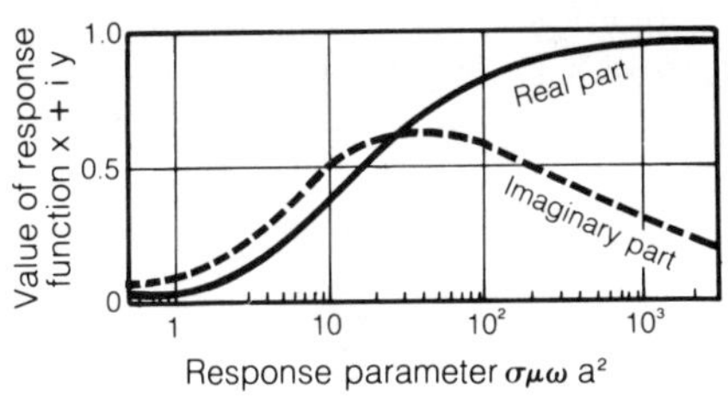

FIGURE 6.5
Generalized picture of the time domain electromagnetic method

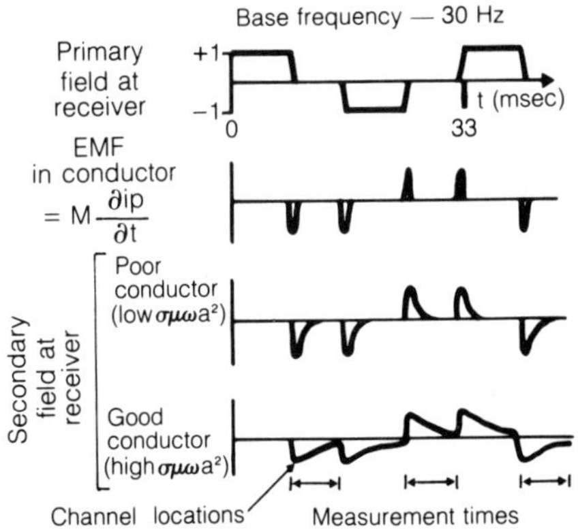

of the anomaly such as in-phase and out-of-phase amplitudes, and to plot these on the appropriate nomograms. From the nomogram and the shape of the anomaly, one can generally arrive at estimates for:

- Quality of the conductor
- Depth to top of the conductor
- Conductor thickness
- Dip
- Strike and strike length

Airborne EM systems

The earliest airborne EM (AEM) system was designed by Hans Lundberg shortly after World War II. The first truly operational system was that of Inco and McPhar in 1950. Geological mapping with AEM really began in 1965 with the development of the Barringer Input System.

The first AEM systems were simple, low frequency quadrature or in-phase/out-of-phase instruments with relatively high noise levels and long time constants. Today, multifrequency or multichannel EM systems are available with low noise levels and short time constants which result in higher resolution between closely spaced conductors; deeper penetration; better discrimination between good and poor conductors; and distinction between bedrock and overburden sources.

Some brief comments will be made regarding airborne EM surveying, followed by a detailed description of the different systems. Figure 6.6 shows examples of current airborne EM configurations.

FIGURE 6.6
Examples of airborne EM installations and configurations

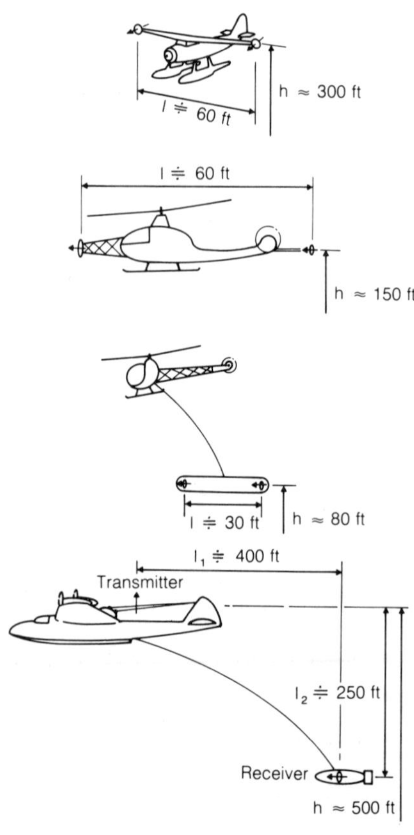

General description

A brief description of types of aircraft, AEM crew, flight records, line spacing, auxiliary equipment, penetration and resolution, as well as other parameters, follows. Several *types of fixed-wing aircraft and helicopters* are used for AEM surveying. The most common fixed-wing aircraft are the following:

- Canso — PBY
- Islander and Trislander
- DC-3
- Otter (single and twin)
- Skyvan

Fixed-wing EM systems are preferable in flat terrain; for survey areas near an airbase; and for large-scale reconnaissance surveys.

The most common helicopters used for AEM surveying are the following:

- Bell 206
- Gazelle-Lama-Alouette
- Hughes 500
- Sikorsky 55 and 58

Helicopter EM systems are preferable in remote areas; in mountainous terrain; and for accurate detail flying because helicopter EM systems generally have high resolution (the ability to record separate anomalies caused by two closely spaced conductors). In general, helicopter EM systems have less depth penetration (maximum depth at which a conductor is detectable in a particular environment) than fixed-wing mounted systems.

The *spacing between flight lines* usually ranges from ½ to ⅛ mile depending on the survey objectives.

Production rates for fixed-wing EM surveys and helicopter-borne surveys compare as follows. Fixed-wing EM survey production is great and can be as much as 60-80 mph. Minimum line length can be as much as 5 miles. Total survey cost is presently between $30 and $50/line mile (including interpretation and report).

Helicopter-borne surveys are often along short lines (contracts are usually for 3-4 mile minimum line length) resulting in an overall effective survey production rate of 30-40 mph. Thus, the helicopter cost alone can be some $10-20/line mile. Total costs might range from $40-70/line mile (including interpretation and report).

An *airborne EM crew* might consist of pilot (engineer); navigator/co-pilot; operator/electronicist; dataman; and geophysicist for "redball mapping" (rough plotting of the most obvious anomalies in the field).

Auxiliary equipment in the survey aircraft usually includes power supplies; recorders (analog and digital); altimeter; camera (strip or frame); navigation devices; magnetometer; radiometric equipment.

In terms of *flying speed*, fixed-wing aircraft are faster and fly up to 150 mph; helicopter speed ranges to 60 mph.

Flight records will show several EM parameters depending on the system used, e.g., in-phase and out-of-phase channels, six input channels, quadrature data, and field strength for VLF. Other parameters commonly recorded with the EM data are magnetometer (and spectrometer) data, altimeter and camera frame number (= fiducial), and other time or positioning information (figure 6.7). Normally, one camera frame per second is taken resulting in an accurate position fix once every 85-225 ft depending on the aircraft speed. The strip camera gives a continuous (often less clear) picture of the ground but a time or fiducial mark only once every second or 10 seconds.

Recorded in figure 6.8 are six channels of input, altimeter, magnetics (fine and course), and fiducials. These figures show the anomalies recorded on two adjacent flight lines in opposite directions over a dike-like sulfide body which dips at approximately 40° to the north. The down-dip record above shows a single response, whereas the up-dip record below shows a characteristic double-peaked anomaly. A measurement of the amplitude ratio of these two peaks is used to determine the amount of dip. When responses such as these are plotted, and provided that the flight lines alternate in direction, recognition of dipping conductors becomes a simple matter.

It is imperative that a proper *flight log* be kept by the flight crew to enable the dataman to edit all analogue tapes and camera film so that anomalies and other data can be accurately positioned on mosaics or maps. This flight log should show date, hour, flight number, line number and direction, start and end fiducials, equipment calibrations, weather, and general comments.

Films will be developed immediately after a survey flight and the flightpath traced on maps or mosaics to verify that the survey area was properly covered. In practice, only one film frame or fiducial is recovered per half mile or mile of traverse depending on the map scale and accuracy required. The flightpath between fiducials is then assumed straight and flown at a constant speed.

Reflights or detail flights can be planned in the field after the data checking and flightpath recovery. Initial anomaly picking or "redball" mapping is often done in the field to enable staking or to permit an immediate follow-up ground verification. Final plotting of anomalies and interpretation are done later in the home office.

Results of AEM surveys are presented on topographic maps, photo mosaics, or overlays (figure 6.9). The scale is often chosen so that flightlines are ½ in. apart. These maps show topographic features; flightlines and flightline numbers; fiducials and fiducial numbers; anomalies, e.g., circles or bars with labels indicating conductor strength or value, interpreted depth below surface, and magnetic coincidence in gammas; and anomaly zones. Other maps normally resulting from an AEM survey are magnetic maps; EM profiles (stacked along flightlines); filtered data, and other computer output.

FIGURE 6.7
Example of a dual frequency in-phase and out-of-phase helicopter EM recording. Other traces show the altimeter, magnetometer (fine and coarse), power line noise monitor, and fiducial information. (See also figure 6.8.)

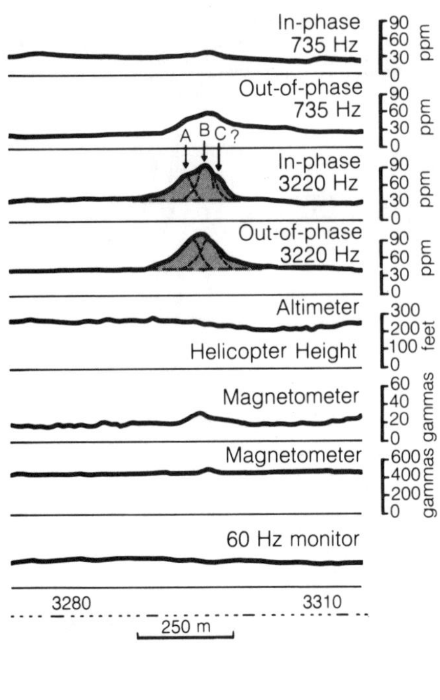

Compatibility with radiometric equipment is usually no problem with on-board EM or helicopter towed bird systems. Fixed-wing towed bird systems usually fly too high for a radiometric system to be effective.

Many factors dictate the *effective depth penetration and resolution* of an EM system. The complexity of this topic has caused considerable disagreement among EM experts for years. In general, it can be said that short, fixed, coupled systems such as those towed by a helicopter have a lesser depth penetration (approximately 150 ft below surface) but greater resolution (better than 100 ft) than fixed-wing towed bird systems. The penetration for the input system may be as much as 400-500 ft, but its resolution is only 150-200 ft. One of the first reviews on this topic was by Pemberton (1962). Penetration is a function of coil configuration; distance between Tx and Rc; frequency(ies) used; time constant of the system and flying speed; sensitivity (anomaly recognition) over background response; aircraft flying (height) characteristics; and target σt versus host rock and overburden σt.

The description of different EM systems has been based on a combination of two related factors, *coil configuration* and the *EM field parameters measured*. The division is as follows:

- Rigid boom — In-phase and out-of-phase
- Towed bird — Quadrature (or out-of-phase), transient, amplitude
- Fixed source — Field strength ratio and phase difference
- VLF — Dip angle of the field
- AFMAG — Field strength

Table 6.1, taken from P. Hood's well-known yearly review, "Mineral exploration trends and developments" (1979), lists some of the technical specifications of most AEM systems available today. Table 6.2 shows a somewhat similar compilation and grouping (without technical details) as compiled by the authors. This grouping is used to discuss some of these systems in more detail later. These descriptions are framed around unorthodox case histories that often refer to pitfalls not normally found in textbooks and articles.

FIGURE 6.8
Examples of input data

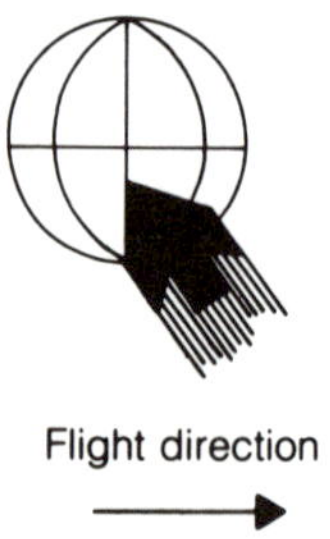

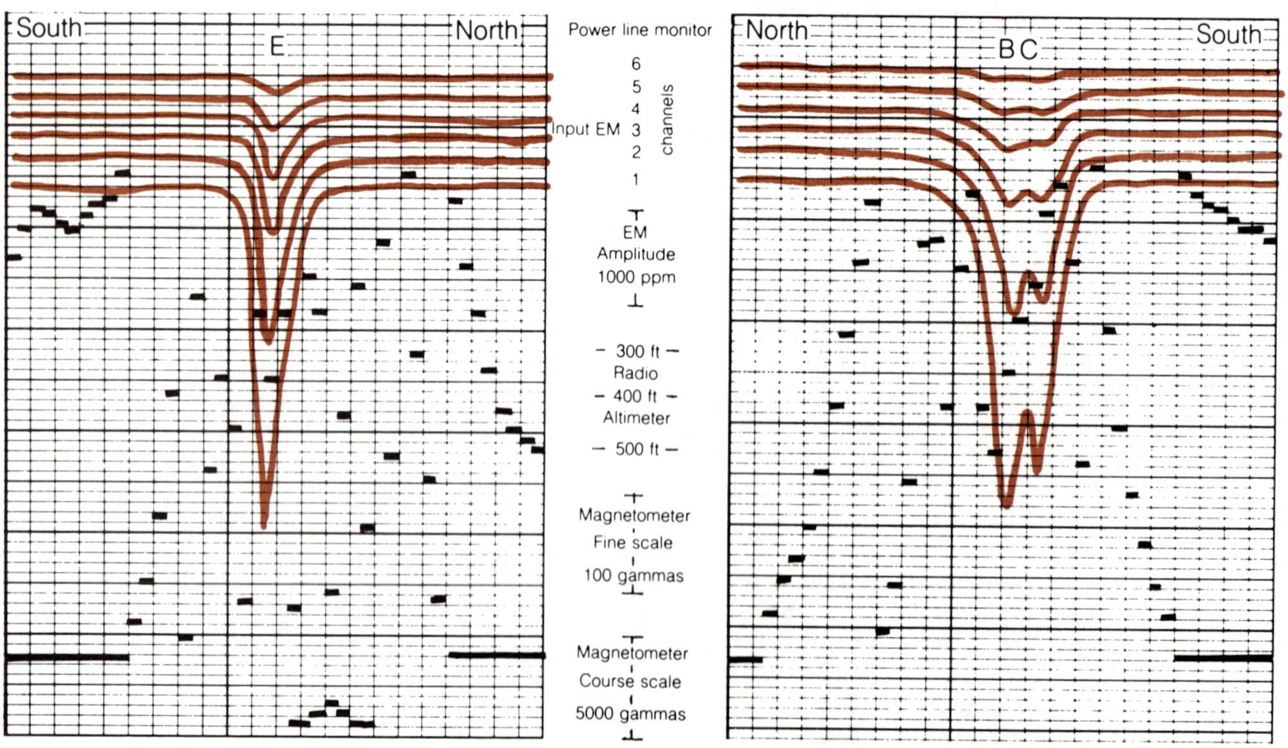

The figure on this page is referred to in the preceding section *General description.*

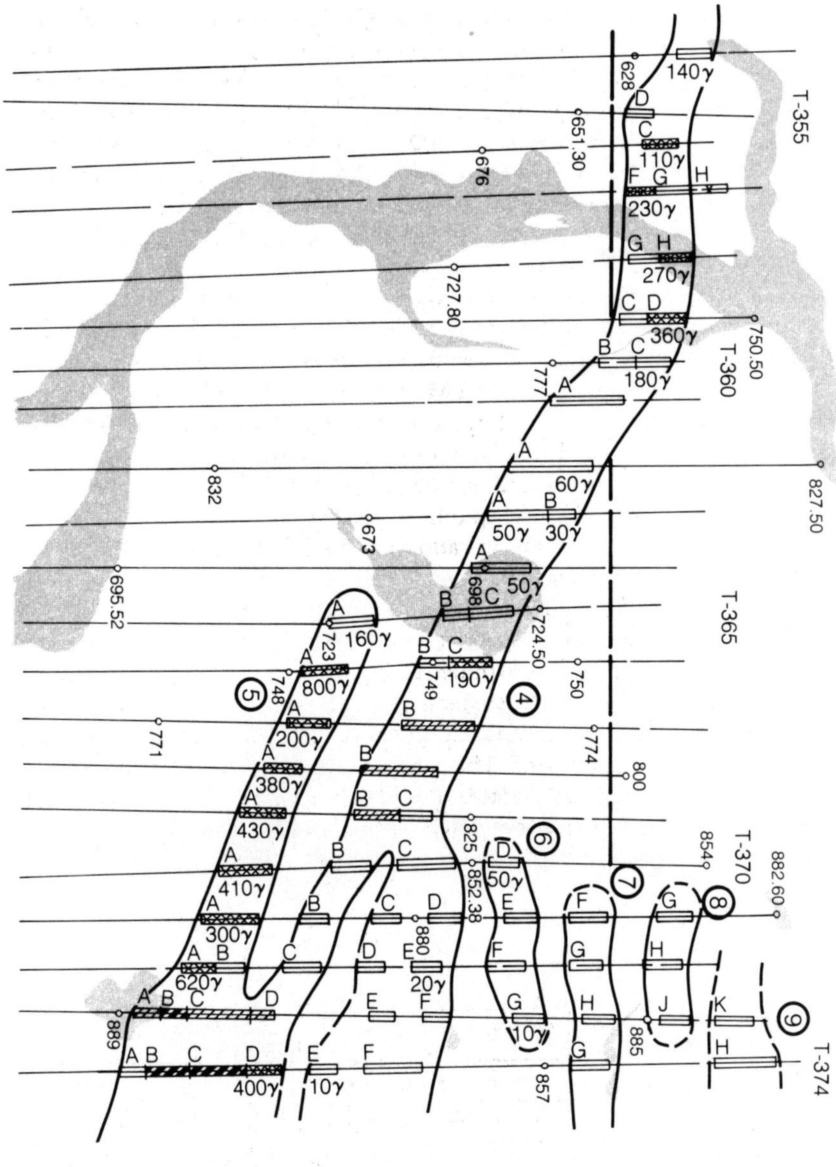

FIGURE 6.9
Typical airborne EM presentation. Flight lines, fiducials, line numbers, conductors, and their interpretation

Mean terrain clearance — 150-200 ft
Traverse interval — ⅛ mi
Horizontal control — based on photo laydown
Scale 1 in. = 1,320 ft

1A anomaly
1B anomaly
2A anomaly
2B anomaly
3 anomaly
X type anomaly

The table on this page is referred to in the preceding section *General description.*

TABLE 6.1
Airborne electromagnetic survey systems available for purchase or as a contract service

EM System Trans. Rec. Tx Rx Flight Dirn	Manufacturer/ contractor — Survey platform H=Helicopter FW=Fixed Wing	Normal Survey height in feet Trans. Tx	 Rec. Rx	Tx-Rx separation in feet	Freq. (Hz)
	Kenting-Scintrex Tridem — Canso FW	164	164	83	500 2000 8000
	Barringer — Aerodat H	98	98	29.5	900
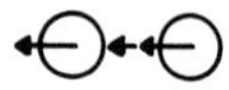	Geonics — EM33 H	98	98	20	400- 4000 (normally 736)
Vertical coaxial	Northway — LHEM 200 — LHEM 250 H	98 98	98 98	29.5 29.5	4000 1000
Rigid coupled	Sander H	98	98	23	500- 4000
	Scintrex — HEM-802 H	98	98	27	735 & 3220
Multicoil boom Rigid coupled 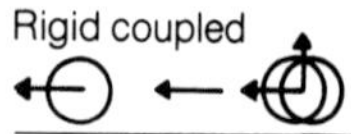	Dighem II H	98	98	27.8	900
Max coupled Towed bird 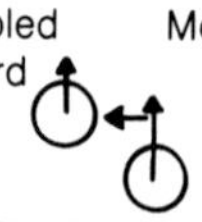	McPhar—Quadrem FW	459	164	885; 295 behind and 295 below	95, 855, 2565, & 7695
Max coupled Towed bird 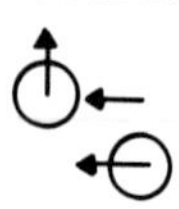	Barringer — Mark 6 Input Geoterrex — Canso Questor — Skyvan & Trislander FW	394	164	246 behind and 246 below	286 pulsed ½ sine waves per sec
 Wing mounted Max coupled	Geoterrex — Otter So quem Emal 19 — Cessna L19 FW	164 164	164 164	691 361	320 625

Measure in phase out of phase	Noise level ppm	Response from overburden	Power requirements
in phase & out of phase	20	Small but increases with freq. to moderate	28V dc 500W
in phase & out of phase	2	Small	28V dc 340W
in phase & out of phase at 2 time constants	<0.5	Small at low freqs.	28V dc 225W
in phase & out of phase	5 5	Moderate Small	28V dc 25W 28V dc 60W
in phase & out of phase	1 in phase 2 out of phase	Small at low freqs.	28V dc 75W
in phase & out of phase	1	Small at lower freq.	28V dc 35W
in phase & out of phase	1	Removed by differencing measured signals	28V dc 370W
out of phase	500	Medium	28V dc 2000W
Transient decay at 6 intervals	50	Large on channel 1, decreasing rapidly with higher channels	56V dc 3000W
in phase & out of phase in phase & out of phase	20 35	Small Small	28V dc 500W 28V dc

The table on this page is referred to in the preceding section *General description.*

TABLE 6.1, continued
Airborne electromagnetic survey systems available for purchase or as a contract service

EM System **Trans. Rec.** **Tx Rx** **Flight** **Dirn**	**Manufacturer/ contractor — Survey platform** **H=Helicopter** **FW=Fixed Wing**	**Normal Survey height in feet** **Trans.** **Tx**	 **Rec.** **Rx**	**Tx-Rx separation in feet**	**Freq. (Hz)**
Ground loop 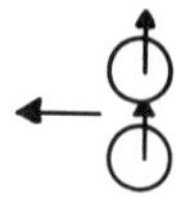	Scintrex — Turair II H	Ground Loop	98	7.9 for Rx coils	200, 400 & 800
	Barringer — E-Phase	N/A	295	infinity	3 simult. freq.
	— Radiophase H or FW	N/A	295	infinity	14-28 KHz
	Geonics — EM 18 H or FW	N/A	98 984	infinity	12.25 KHz
Very low frequency (VLF)	Herz — Totem 1A H or FW	N/A	98 984	infinity	12-25 KHz
	McPhar — Kem H or FW	N/A	98 984	infinity	12-25 KHz
	Scintrex — SE 90 H or FW	N/A	98 984	infinity	12-25 KHz

Measure in phase out of phase	Noise level ppm	Response from overburden	Power requirements
Field strength ratio and phase difference	0.1% FSR 0.1° PD	Varies with frequency	28V dc 15W mg for ground coil
out of phase out of phase	0.01% 0.01%	Variable at different freqs.	28V dc 42W
in phase & out of phase	0.5%	Large	28V dc 50W
Total field & vert. quad.	Depends on distance of Tx	Large	28V dc 14W
Dip angle & amplitude	0.5 μV at sensor coils.	Large	28V dc 10W
Vertical field vector in in phase & out of phase wrt hori. vector	<1%	Large	28V dc 30W

The table on this page is referred to in the preceding section *General description.*

TABLE 6.2
Airborne EM Systems

Common AEM Systems Commercially Available in North America During The Late 1970's

	System	Manufacturer/Contractor
Group 1 (Towed Receiver)	Inco Input Mk-6 Quadrem	Inco Geoterrex, Questor McPhar
Group 2 (Rigid Boom)	Dighem I, II EM-33 HEM 801, 802	Dighem Geonics; Aerodat Geophysical Surveys Geoterrex Scintrex
	Tridem	Kenting, Scintrex
Group 3 (Ground Source)	Turair	Scintrex
Group 4 (Distant Field)	EM-18 Kem SE-90 Totem AFMAG } — VLF	Geonics (Several Contr.) McPhar (Several Contr.) Scintrex Herz (Several Contr.) McPhar

Older systems still available: F-400; Barringer and Northway HEM; E-Phase, Radio-Phase; Rio Mullard; EM-30, etc.
Future systems under development: Unicoil; Cotran.

Towed receiver systems

Towed receiver systems are clearly distinct from the rigid boom systems because the receiver (Rx) coil is towed behind the aircraft, and the transmitter (Tx) coil is attached to the aircraft. The separation between them is, therefore, variable. The advantage of this configuration is that the larger separation between Tx and Rx as compared to that in rigid boom systems makes it much easier to achieve greater depth penetrations of up to 500 ft. The disadvantages, on the other hand, are twofold: first, the large Tx-Rx separation induces currents in surrounding rocks and if they or the overburden are appreciably conductive, many spurious anomalies can result; second, because the towed bird plunges, yaws, and rotates in turbulent air, measurements of relative amplitude of the in-phase components are impossible; therefore, only the out-of-phase component can be measured in the frequency domain mode (FD). In the time domain (TD) mode, measurements can only be made when the transmitter current is turned off. To counteract this limitation and to arrive at estimates of conductor quality, these systems take measurements at more than one frequency in the FD mode or at more than one time delay in the TD mode.

McPhar Quadrem system is a multifrequency airborne electromagnetic system. The transmitter can be mounted in either a helicopter or fixed-wing aircraft. The magnetic dipole antenna radiates at five frequencies: 95 Hz, 285 Hz, 855 Hz, 2565 Hz, 7695 Hz. The transmitter dipole axis (line passing through the center of the coil and perpendicular to the plane of the coil) is horizontal and in line with the flight path. The receiver sensor is towed in a bird attached to a 400-ft cable. The receiver dipole axis is vertical for maximum coupling with the magnetic field produced by the transmitter. The receiver measures out-of-phase components of the secondary magnetic field.

Figure 6.10 shows a flight record with the five channels of EM information in parts per thousand, an altimeter trace, a power line interference monitor, and a fiducial marker. The short vertical lines near the top denoted as D4 to M3 are anomalies which have been picked from the record.

FIGURE 6.10
Example of a Quadrem flight record

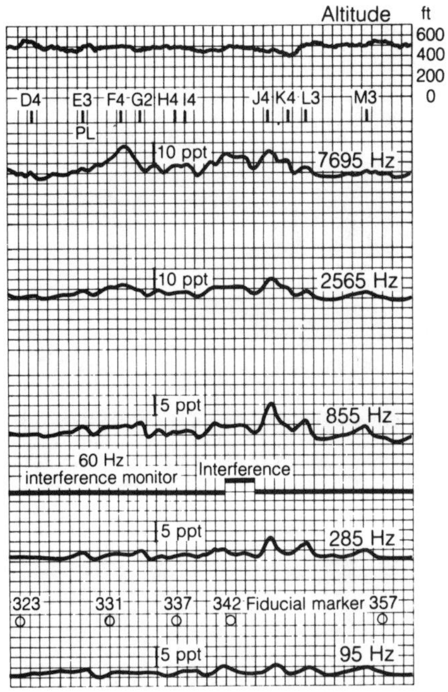

Interpretation of conductor quality and depth is done by comparing anomaly amplitudes with model curves. Figure 6.11 shows such an example. The anomalies at the five frequencies are reproduced on the left side. The five anomaly amplitudes have been measured and plotted against frequency on log-log paper. The model curves for a thin conductor, however, have anomaly amplitudes plotted against σtf (frequency x conductivity-thickness (σt)). Since σt is the quantity to be determined, the model curves (on transparent paper) are overlaid on the plot of anomaly amplitudes and shifted laterally until a good fit is obtained. The two model curves in figure 6.11 have been thus shifted. A frequency of 1000 Hz on the flight data corresponds to a σtf of 2000 on the shifted model curves.

Therefore, the σt of the target causing this anomaly is estimated at

$$\frac{\sigma tf}{f} = \frac{2000}{1000} = 2 \text{ mhos.} \tag{6.2}$$

The two model curves in figure 6.11 are for elevations of 500 and 600 ft. The flight height, therefore, is estimated at 560 ft.

Subtracting the flying altitude of 500 ft, the depth of the conductor is estimated at 60 ft. This is only an approximation because the position of the bird is somewhat uncertain.

The *Barringer input system* is a towed bird time domain electromagnetic system. The transmitter loop is formed by a horizontal loop wound around the aircraft. To make its area as large as possible, it is fixed at the wingtips, nose, and rudder. The receiver coil is towed behind the aircraft about 200 ft below and 350 ft behind. The transmitter loop is energized by half sine wave pulses of alternately opposite polarity. As shown in figure 6.12, the on and off times are 1.1 and 2.37 msec, respectively. The fundamental frequency is, therefore, 1444 Hz. Voltage in the receiver coil is proportional to the time derivative of the magnetic field. In the absence of any anomalous fields, this is as shown in the second plot on the left side of figure 6.12. When conductive material is present, however, currents will be induced in them by the transmitter pulses. During the transmitter off period, they will decay almost exponentially. Induced currents will produce secondary magnetic fields, and the net result as far as induced voltage in the receiver is concerned is as shown in the third plot on the left side of figure 6.12. During the transmitter off time, the voltage decay in the receiver is integrated in six-gate intervals and reproduced on a strip chart recorder as shown on the lower right of figure 6.12.

The overall signal amplitude is controlled by the depth of the conductor, whereas the relative amplitudes of successive channels are controlled mainly by the quality of the conductor. The latter is illustrated in figure 6.13 where the six channel responses are shown for five conductors of identical geometry but differing conductivity. The poorest conductor has a stronger anomaly on channel 1, but the anomaly decays quickly and can barely be detected on channels 5 and 6. On the other hand, the best conductor has only a moderate amplitude anomaly on channel 1, but the decay is much slower so that there is still a recognizable anomaly on channels 5 and 6.

The shape of the anomaly is sensitive to conductor dip. This is illustrated in figure 6.14. A dip estimate can be made by measuring the relative amplitudes of the smaller precursor peak and the main peak.

Interpretation of conductor depth and σt is done by comparing anomaly amplitudes with a nomogram (figure 6.15). The six anomaly amplitudes are plotted on an overlay as points on the bar scale shown on the right side of figure 6.15. The bar scale is then moved laterally and vertically (but not rotated) until a good fit of the six anomalous amplitudes is obtained with the theoretical curves. Conductor depth and σt are then interpreted on the top graph according to where the 1000-ft mark of the bar scale falls. In the example of figure 6.15, the depth is 50 ft and the σt is 60 mhos.

The Barringer input system has been used to detect more than a dozen massive sulfide orebodies in Canada. Figures 6.16 and 6.17 show examples of flight records over the Thierry and Magusi River deposits in the province of Ontario. Both are very good conductors because all six channels are anomalous.

FIGURE 6.11
Example showing Quadrem interpretation of conductor quality (σ t) and depth to top (d), Cavendish Township, Ontario, Canada

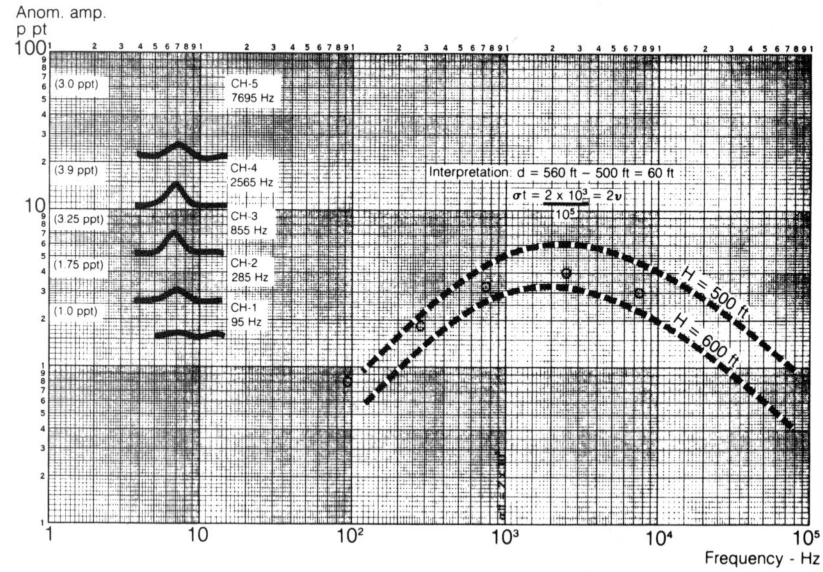

FIGURE 6.12
Principle of operation of the input system

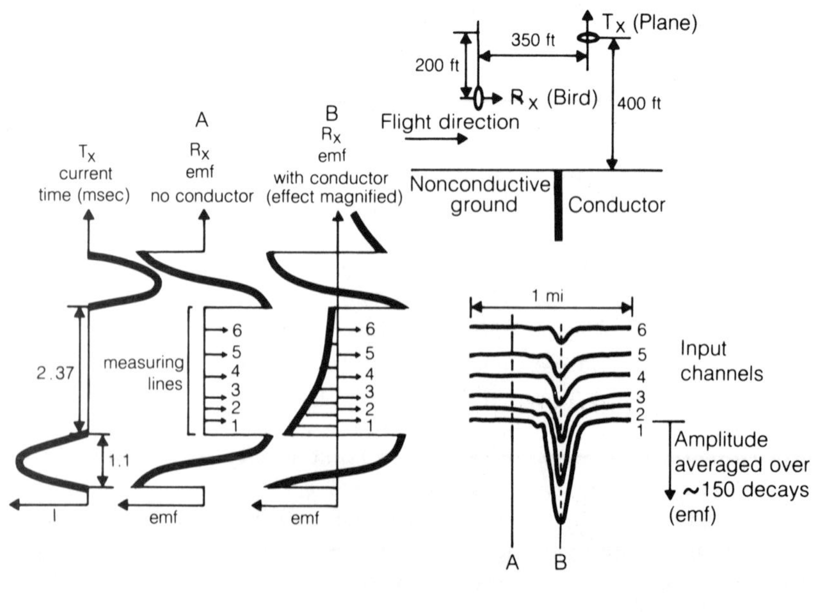

The figures on this page are referred to in the preceding section *Towed receiver systems.*

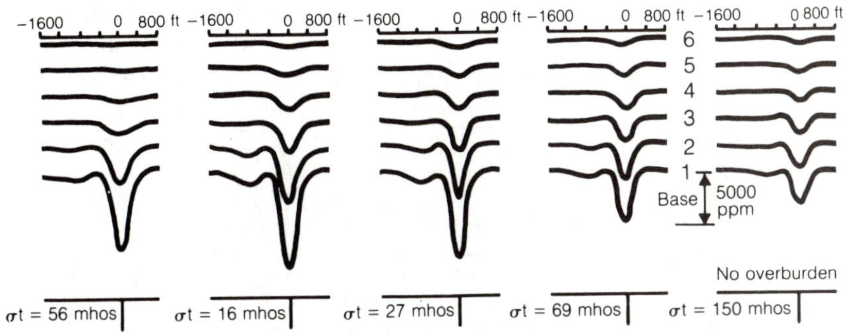

FIGURE 6.13
Model input profiles over vertical half planes of various conductivity thickness (σ t) product

FIGURE 6.14
Effect of conductor dip on an input anomaly

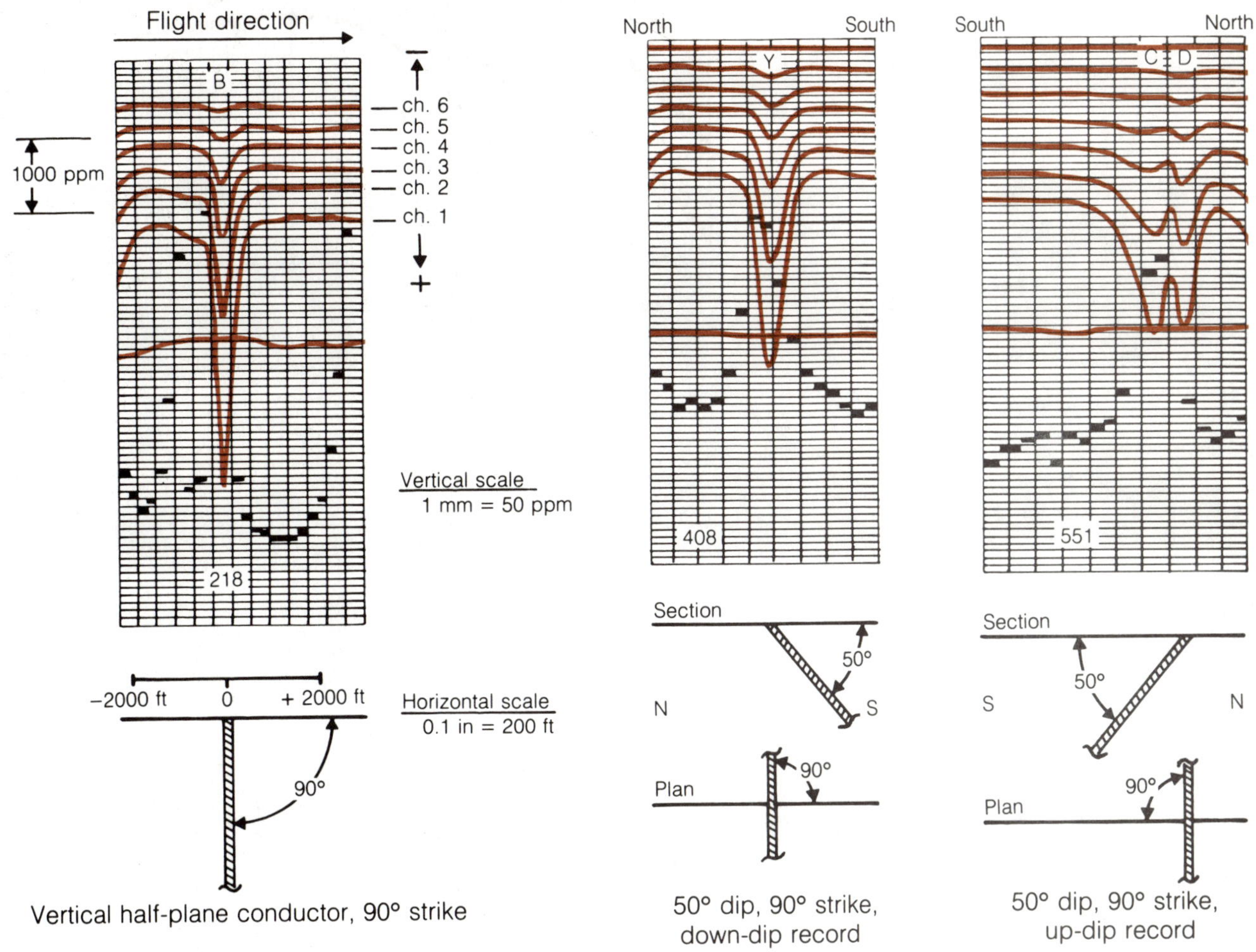

The figure on this page is referred to in the preceding section *Towed receiver systems.*

FIGURE 6.15
Nomogram for interpreting conductor σt and depth to top from input anomalies

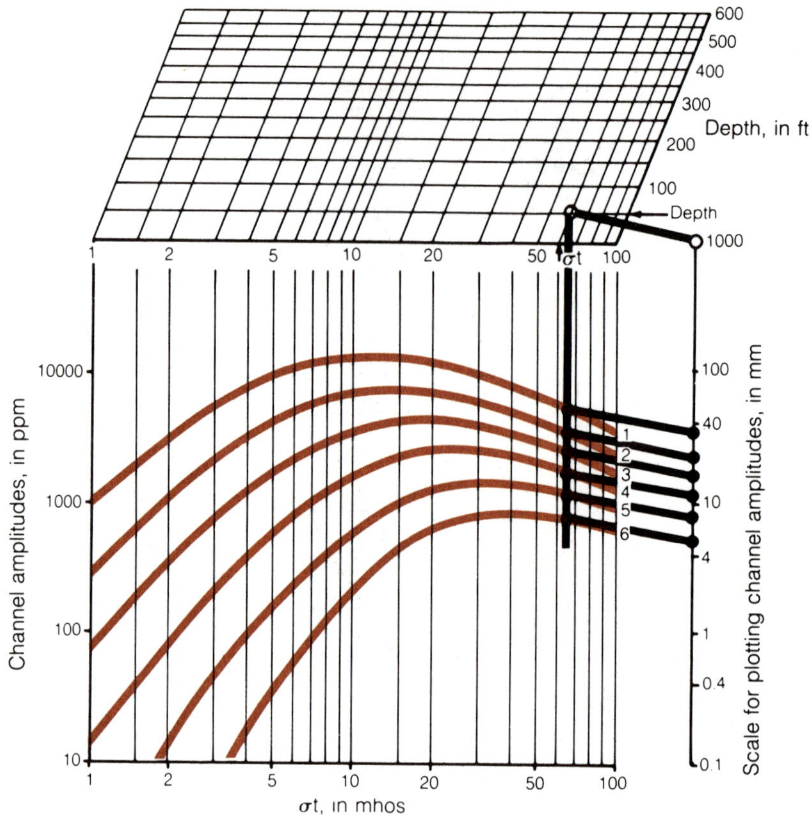

The figures on this page are referred to in the preceding section *Towed receiver systems.*

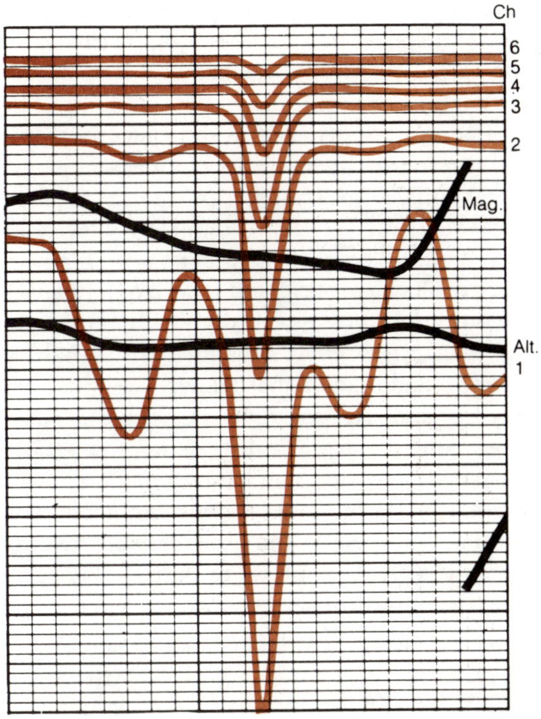

FIGURE 6.16
Example of an input discovery

Magusi River discovery, Iso/Copperfields
Barringer/Questor Mark VI Input

Location: 20 miles northwest of Noranda, Quebec
Tonnage: Unknown
Grade: Hole No. 2 intersected 97.2 ft of 1.63% Cu, 5.06% Zn, 0.97 oz Ag
Geology: Quartz-sericite-ankerite schist; altered gabbro
Mineralogy: Pyrite, chalcopyrite, sphalerite, silver
Apparent conductivity-width: 12 mhos
Discovery: Airborne geophysics

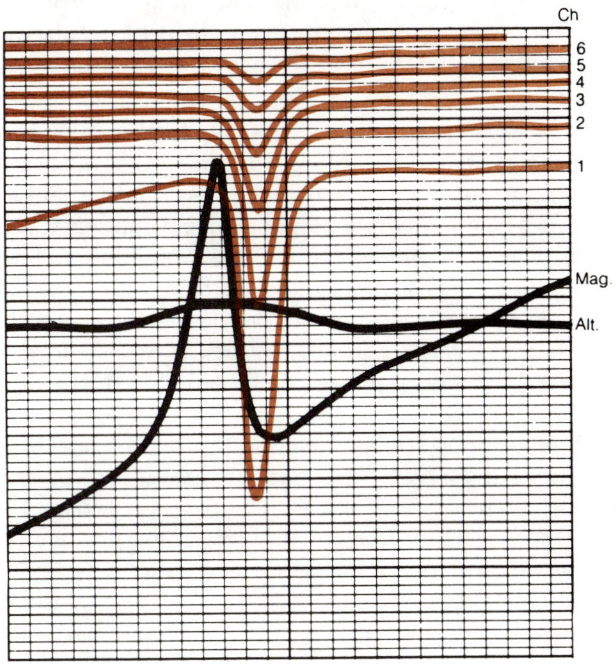

FIGURE 6.17
Example of an input survey over a producing mine

Thierry Deposit, Umex
Barringer/Questor Mark VI Input

Location: 6 miles west of Pickle Lake, Ontario
Tonnage: 10,000,000
Grade: 1.68% Cu, 0.18% Ni
Size: Strike length: 3,200 ft
Depth extent: 1,500 ft
Geology: Metamorphosed basic and ultrabasic intrusives
Mineralogy: Pyrrhotite, chalcopyrite, pyrite and pentlandite
Apparent conductivity-width: 20 mhos
Discovery: Ground magnetic and EM surveys

Rigid boom systems

Two groups of rigid boom AEM systems are used today: systems permanently mounted on the wing tips or fuselage of fixed-wing aircraft and systems installed in a helicopter-bird. The helicopter EM systems have the coil assemblies rigidly installed in a 20-30 ft long, lightweight, torpedo-shaped tube (figure 6.6). This bird is attached to the helicopter by a 100-ft tow cable, which also provides electrical connections. The transmitting and receiving modules are installed in the helicopter. The transmitter and receiver coils are normally mounted in a vertical plane, either coaxial for the helicopter and Canso-Tridem (fuselage) systems, or coplanar for the wingtip system. These vertically mounted coil systems respond strongly to steeply dipping bedrock conductors. The newer Dighem II helicopter EM system has a vertical and a horizontal pair of coils. The horizontal coil configuration couples well with horizontal sheets such as overburden. The interpretive strength of the Dighem system is enhanced by the coil combination that allows the system to distinguish between flat-lying and steeply dipping sources. Some systems have more than one frequency, e.g., Tridem or Scintrex HEM-802. These multifrequency systems have additional interpretation power in that they can distinguish more clearly between poor and good conductors. Overburden recognition is based on the broadness of the out-of-phase response together with a small in-phase response.

It is not possible to say which system is best. There are geological environments and conditions where one system will be more advantageous than others. Most systems perform well in many exploration areas.

High frequencies give strong amplitudes, resulting in easy anomaly recognition. The disadvantage is that the higher frequencies also produce a strong and often varying background response in areas with conductive overburden, which makes target recognition more difficult. The use of multifrequency systems is, therefore, very helpful and allows the interpreter to distinguish between the different types of conductors.

FIGURE 6.18
Example of an HEM-802 flight record

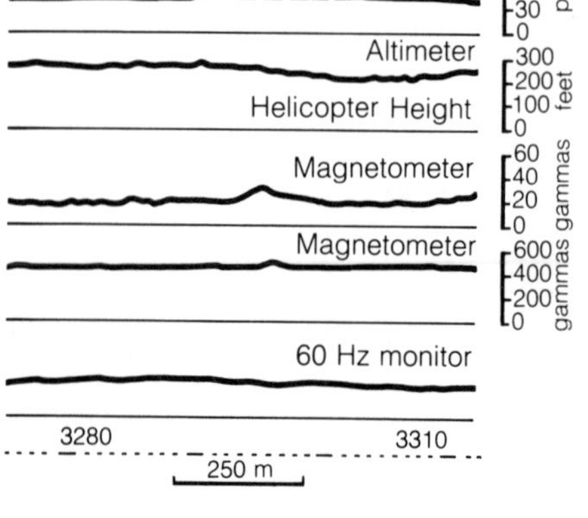

Figure 6.18 shows an idealized analog record obtained with the *Scintrex HEM-802 helicopter system.* Two or possibly three conductors (A, B, and C) are clearly visible in the 735 out-of-phase and 3220 in-phase channels. The 3220 out-of-phase channel shows only a single broad anomaly which is a combination of responses from conductors A and B. Probably both have very similar out-of-phase amplitudes. The 735 in-phase channel shows only a weak response from conductor B. The background on all channels has a very low noise level which suggests that no conductive overburden or host rock is present in the survey area. The character of the responses A, B, and C is typical for parallel bands of steeply dipping formational conductors (e.g., graphite, pyrite, pyrrhotite). A phasor diagram as shown in figure 6.19 can then be used to interpret these conductors. Interpretation diagram shows the response of a vertical half-plane obtained with a vertical coaxial EM system and coil separation of 28 m. The system is flown perpendicular to the vertical half-plane (points A and B refer to conductors A and B in figure 6.18). The 3220 Hz data give 2.5 and 3.5 mhos for conductors A and B respectively, and very shallow depths. Although having a weak IP response, the 735 Hz data also confirm this. A slight shift in the arbitrarily chosen in-phase zero level by 2 ppm will change the calculated σt value and create a slightly wider range from 2 to 4 mhos. Such a range is still narrow and will most likely not alter the priority grading of these conductors.

Figure 6.20 shows a line of digitally recorded and computer plotted airborne data collected with the *Dighem II system* over the Montcalm deposit (Fraser, 1978). The Dighem II system operates at about 900 Hz. The different channels show the magnetic and altimeter records near the top followed by in-phase and out-of-phase channels of the conventional vertical coaxial coils (configuration similar to the Scintrex 802), and the in-phase and out-of-phase channels of the specially installed horizontal coplanar coils. Other channels are computer-derived traces. Flat-lying conductors such as conductive overburden produce anomalies with amplitudes that are twice as strong with the horizontal coil pair as with the vertical coil pair. A simple subtraction of two times the amplitude of the vertical coaxial in-phase response from that of the horizontal coplanar coil in-phase response removes the overburden effect and leaves only anomalous responses caused by steeply dipping bedrock sources as a result. This subtraction is shown in the difference channels. Such filtering of the data tells us directly that most or all of the anomalous behavior in the out-of-phase channels appears to be caused by overburden. A small but sharp in-phase anomaly remains visible in the difference channel and is caused by the Montcalm Township deposit. This conductor also has a magnetic anomaly caused by the pyrrhotite in the deposit. The bottom channel shows, in the form of spikes, the computer derived conductivity-thickness product (conductance) for the conductor, using a vertical half plane model. The other lower channel shows the half space resistivity computed for the horizontal coplanar coil data. No values greater than 1000 Ω-m are plotted. Note the resistivity low of approximately 20 ohms reflecting the Montcalm deposit. Obviously, this kind of overburden response removal can be important in certain areas, whereas a two-frequency helicopter system might be preferred in others.

Magnetite, when present in a concentrated form, may cause in-phase responses that oppose and, therefore, reduce the response of a good EM conductor. In extreme cases the in-phase response may even reverse sign. Figure 6.21 shows a situation in which magnetic susceptibility masks the EM effect from the conductor in the two Dighem II in-phase channels (22 and 24). The difference channels for in-phase and out-of-phase (33 and 34) suggest, however, a weak bedrock conductor. The small but sharp out-of-phase peak on channel 23 might alert an experienced interpreter, but he might not recommend a follow-up on such a response without the availability of the difference channels.

Figures 6.22, 6.23, and 6.24 show some *Scintrex Tridem*-simulated responses over bedrock conductors underneath a varying overburden (see also section on HLEM). Figure 6.22 shows the change in response for a 13.5 mho steeply dipping massive sulfide sheet (e.g., 30-ft-thick body of .7 Ω-m resistivity) with a varying overburden σt using a 2000-Hz vertical coplanar system. The .3 mho overburden may represent a 10-ft-thick clay layer of 10-Ω-m resistivity increasing to a layer approximately 30 ft thick for the .9 mhos case. The in-phase amplitudes vary only slightly. The out-of-phase amplitude changes dramatically with a sign reversal for the thickest overburden. Figure 6.23 shows the in-phase and out-of-phase responses employing three different frequencies (500, 2000, and 8000 Hz) over the two center cases of figure 6.22. The out-of-phase response is reduced to 0 in the .3 mhos (overburden) − 8000 Hz situation and has changed sign when the overburden is thicker or more conductive (.6 mhos). These simulated cases show only the effect of the bedrock conductor (σt = 13.5 mhos). The effect of the overburden itself also causes a change in the background response. This effect will be more pronounced in the higher frequencies than in the lower ones. The reason for this is that overburden is normally a poor conductor producing weak responses at low frequencies, whereas good bedrock conductors produce stronger responses for most frequencies. Figure 6.24 depicts this clearly. The response from a good conductor covered with overburden stands out in the low frequency channels but is hidden in the high frequency responses which are mainly caused by overburden. The medium conductor shows up clearly only in the high frequency channels.

This example illustrates again that in order to choose the right frequency, a certain knowledge about the target and host rock conductivity is required so that the responses of the economic conductors will be optimized and those of shear zones and overburden will be minimized. A better solution is to use a multifrequency system. This improves the interpretation and enhances the chance of selecting the right (economic) targets for follow up. Figure 6.25 illustrates in a phasor diagram that the main effect of overburden is to decrease the quadrature response. This diagram clearly shows that a conductive overburden causes a decrease of the free air out-of-phase response of a bedrock conductor, and that this effect is strongest at high frequencies and for good bedrock conductors and overburden (σtf = 200K and overburden σtf = .3K). The effect is much smaller at lower frequencies or for less conductive overburden and bedrock sources (σtf = 6K and overburden σtf = 1K). The in-phase response decreases quite rapidly when the quadrature anomalies are negative.

Results of the Tridem system over the New Insco deposit in Quebec are shown in figure 6.26. This deposit contains approximately one million tons of massive sulfide ore averaging 2.79 percent Cu. A conformable, tabular, massive sulfide body in Precambrian volcanics, the deposit has a strike length of about 300 ft and dips south at 55°. The dip extent is about 600 ft and the surface thickness is about 60 ft. The strongest in-phase responses along this flight line are caused by the orebody (note the coincident magnetic peak of 275 gammas); weaker responses reflect the overburden, and in this case are not strong enough to cause any obvious distortions of the out-of-phase responses. Figure 6.27 shows these data plotted on a simplified phasor diagram calculated for an infinite vertical sheet. It appears that the depth to the top and the σt values increase for lower frequencies. The depth corrected for the aircraft altitude increases from 40-80 and 120 ft, and the t values from 2-5 and 40 mhos. How is this possible? Two reasons can be given, and these must always be kept in mind when interpreting geophysical data. First, an interpretation diagram based on an infinite vertical sheet was used. This model simplifies theoretical calculations. The drilling indicates, however, a conductor of finite size. To compute phasor diagrams for many different models, especially finite bodies, is very expensive and cumbersome. To a first approximation, the infinite sheet model is quite accurate in most situations. The second reason is related to the irregularities of an orebody. Again, the model assumes a well-defined sulfide sheet without changes in thickness or σt value. In nature we often see a weathered top or a weak conductive zone surrounding the sulfide core. After an orebody is drilled off and cross sections are drawn, it is often recognized that the conductivity indeed increases with depth as suggested from this interpretation.

The final example in this group shows the effect of overburden on the response of a good bedrock conductor using the Tridem system. Figure 6.28 shows the responses over the Magusi orebody, about 1 mile west of the New Insco orebody. Magusi orebody has a strike length of 1500 ft and a dip extent of at least 1350 ft. It strikes E-W and dips 50°S. Near surface, thickness of the body is about 90 ft. Host rocks are conformable Precambrian volcanics. The orebody contains over 4 mT averaging 1.2 percent Cu, 3.55 percent Zn with Au and Ag. Four zones of moderate conductivity (1-15 mhos) are clearly defined by the Tridem survey. The main zone is a sharply peaked, high amplitude response corresponding to the ore zone and yields a σt value of 15 mhos and an interpreted depth of 45 ft below surface. There is a coincident magnetic peak of 35 gammas. The interesting feature is the reversal of the 8000-Hz quadrature peak which is attributed to the phase rotation effect of the conductive overburden. The overburden is interpreted (using a phasor diagram for a horizontal sheet) to vary from .1-.4 mhos. The other three zones may be interpreted in two ways: they may be caused by weak bedrock conductors (1-5 mhos), or they may reflect variations in overburden thickness or conductivity. Based on the weak coincident magnetic responses, the first interpretation is preferred.

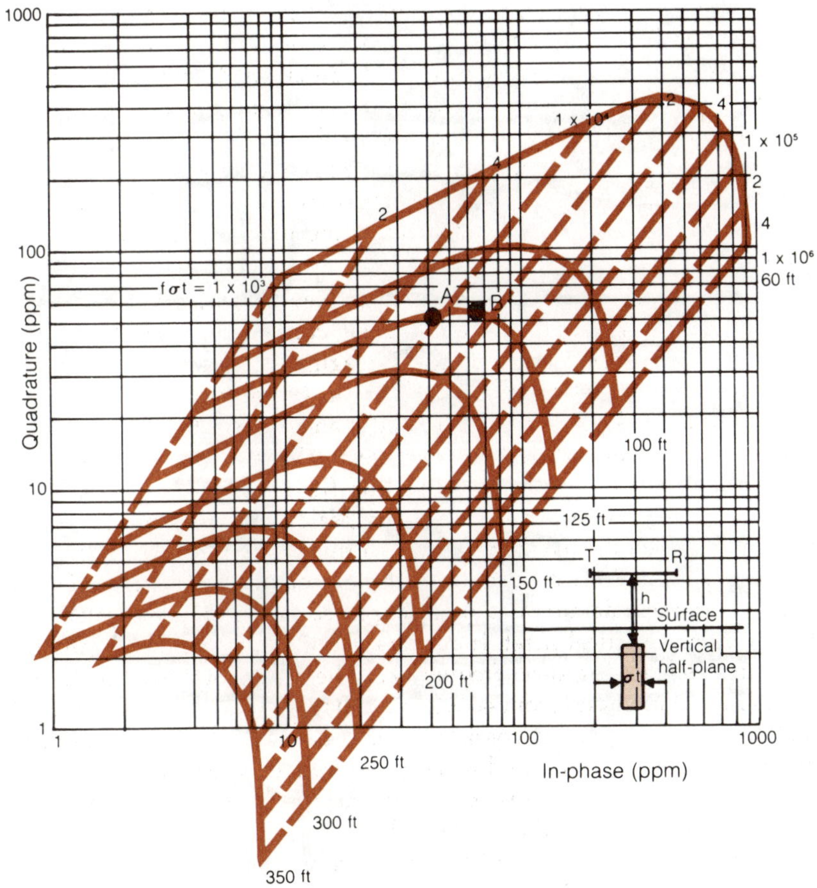

FIGURE 6.19
Scintrex HEM-802 for vertical thin sheets

Vertical coaxial coils
Average separation = 28 ft
Frequencies: 1 — 735 Hz or 520 Hz
2 — 3220 Hz

The figures on this page are referred to in the preceding section *Rigid boom systems.*

FIGURE 6.20
Dighem II recording over the Montcalm Township copper-nickel discovery

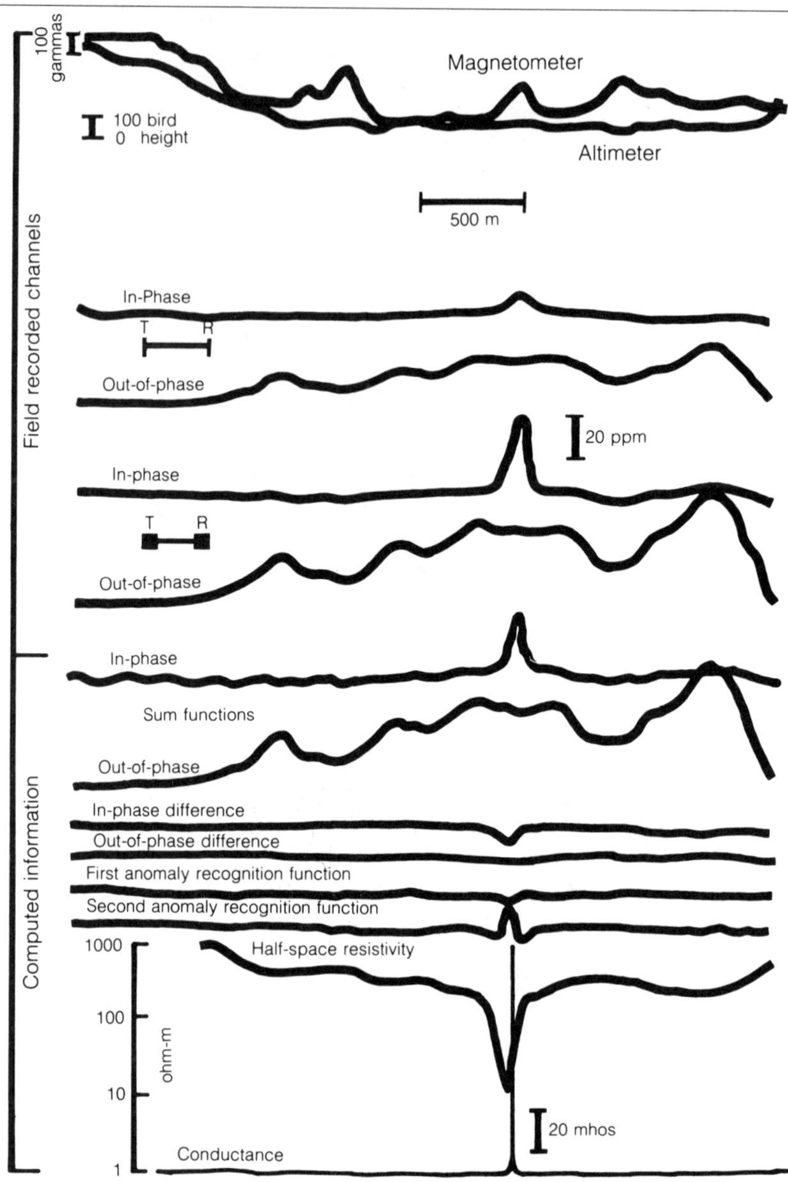

FIGURE 6.21
Dighem II recording over a magnetite-rich bedrock conductor overlain by conductive overburden

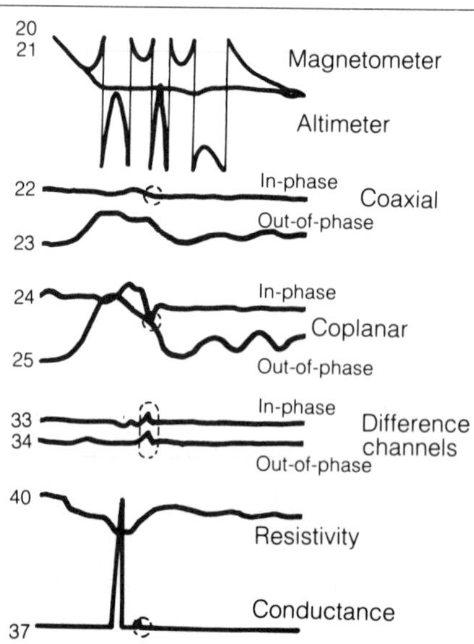

The figures on this page are referred to in the preceding section *Rigid boom systems.*

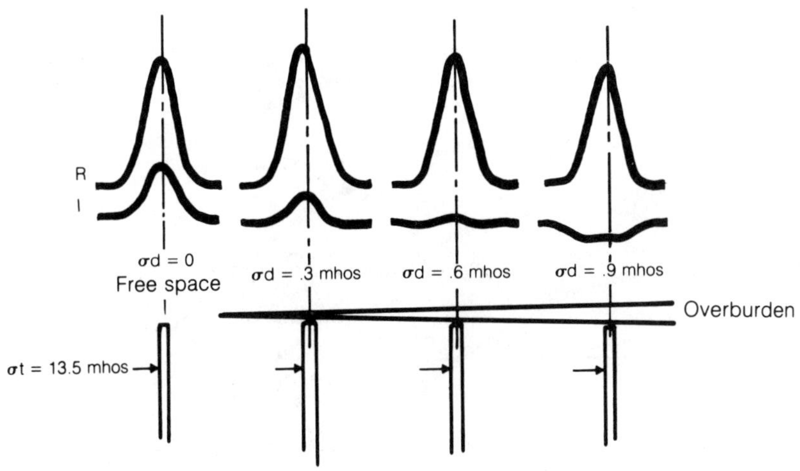

FIGURE 6.22
Change in in-phase and out-of-phase response

A vertical coplanar coil system (2000 Hz) over a steeply-dipping bedrock conductor (σt = 13.5 mhos) covered by an overburden of increasing conductivity (σd = 0–.9 mhos)

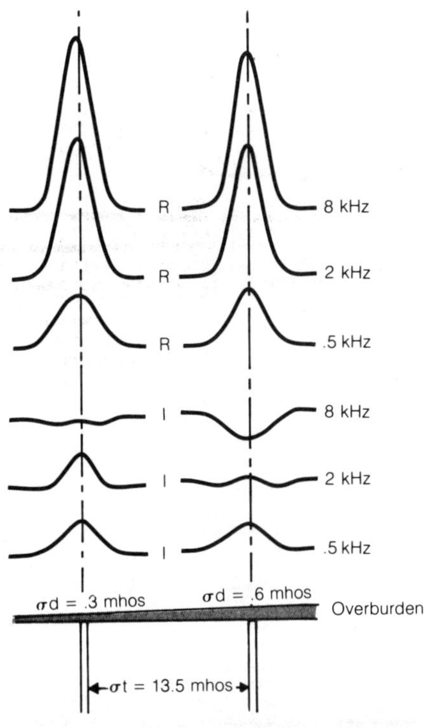

FIGURE 6.23
Change in in-phase and out-of-phase response using a vertical coplanar coil system (2000 Hz)

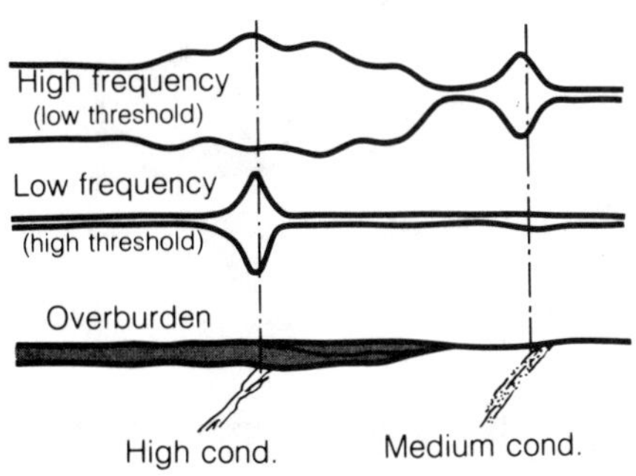

FIGURE 6.24
Simulated responses over a terrain with several conductors

The figures on this page are referred to in the preceding section *Rigid boom systems.*

FIGURE 6.25
Overburden over vertical conductor

Helicopter system
Flying height—120 ft

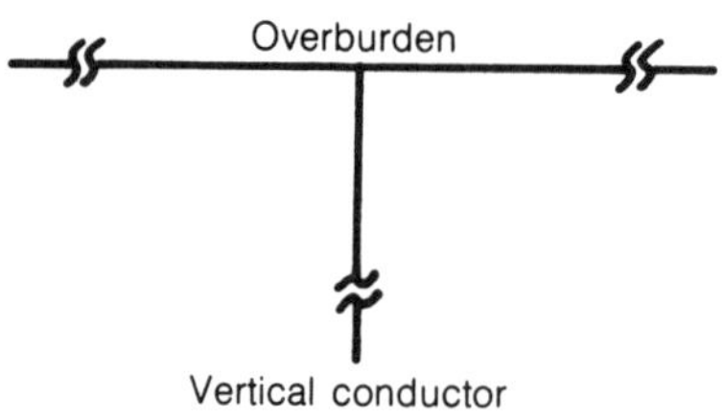

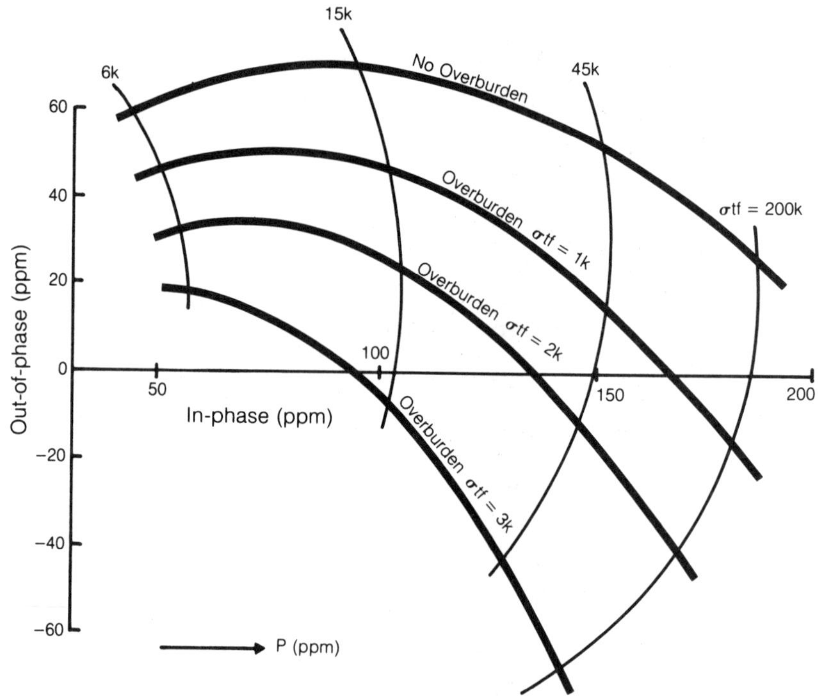

FIGURE 6.26
Tridem traverse over the New Insco deposit, Quebec, Canada

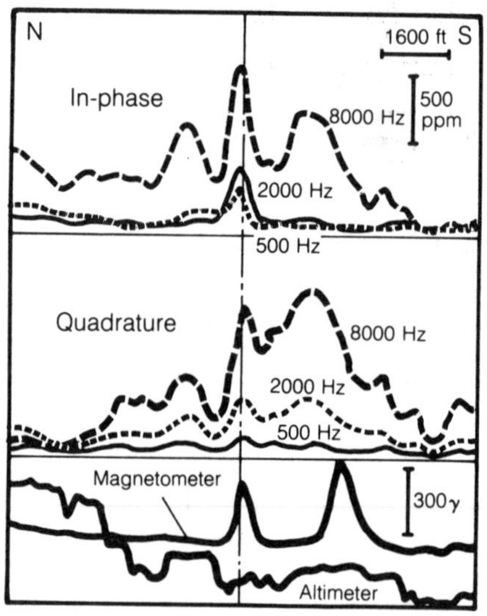

FIGURE 6.27
Simplified Tridem phasor diagram with responses plotted (vertical sheet), New Insco deposit, Quebec, Canada

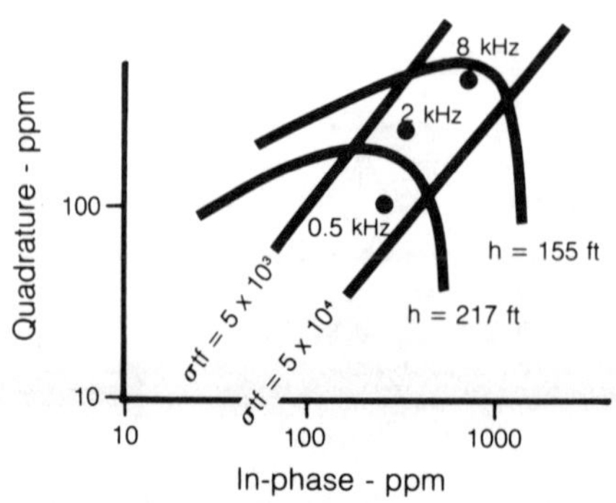

The figure on this page is referred to in the preceding section *Rigid boom systems.*

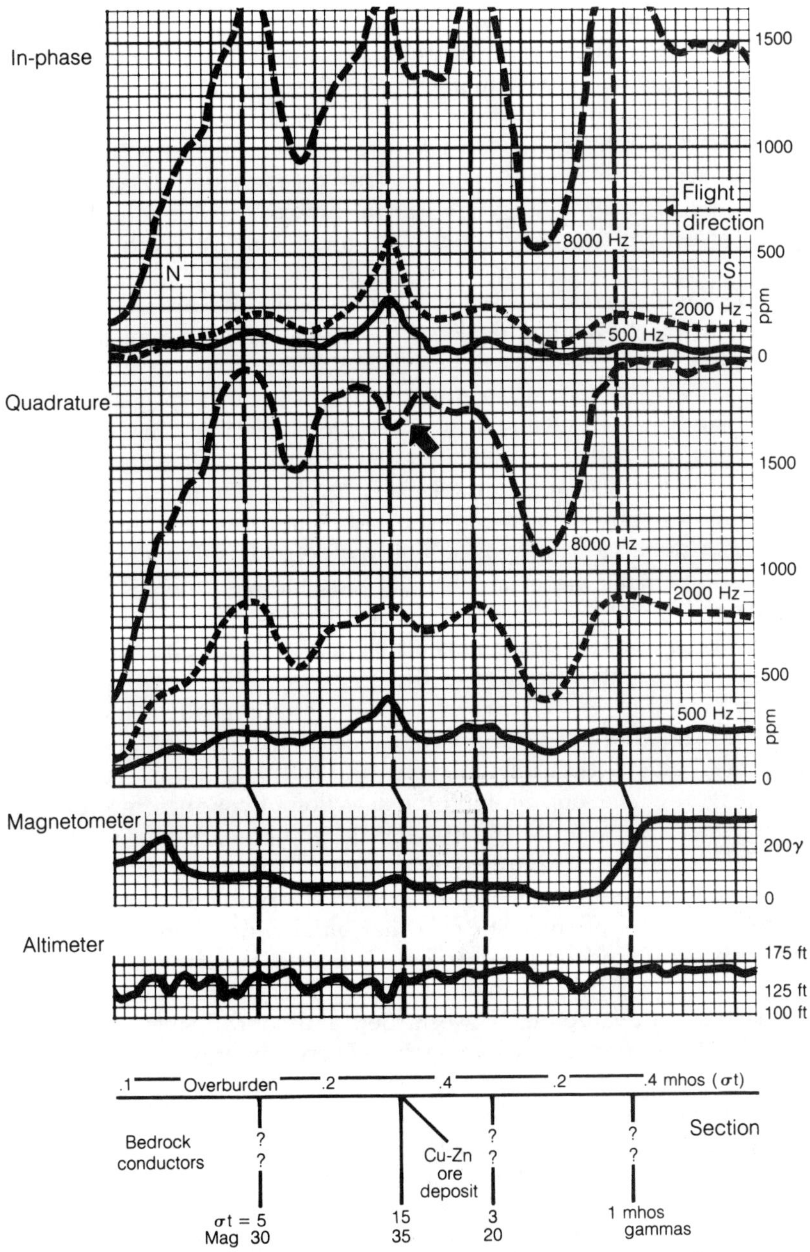

FIGURE 6.28
Tridem traverse over the Magusi deposit, Quebec, Canada

Ground source—Turair

The only system available today employing a ground source (transmitter) and an airborne receiver is Turair, its name derived from "Turam in the air." It was developed primarily to overcome the depth penetration limitations of most conventional systems. As stated before, helicopter systems have a maximum search depth of 200 ft. Most rigid boom fixed-wing systems might be able to see as deep as 300 ft; the input system can see 400 ft and occasionally as much as 500 ft. These maximum search depths can result in severe limitations especially in rugged terrain or in areas with extensive conductive cover. The first Turair surveys were, therefore, flown in areas of Manitoba where the overburden and non-conductive Paleozoic sediments reach over 500 ft thick.

FIGURE 6.29
Turair ground loop source and receiver flight line pattern

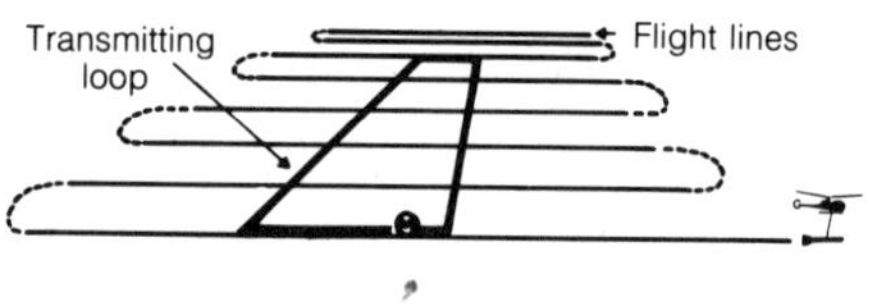

The Turair system consists of a large ground loop source. Frequently 2 x 2 miles or 2 x 4 miles, the loop consists of a single strand of gauge 16 or 18 magnet wire energized by a 15 kw motor generator set (see figure 6.29). At speeds of about 30 mph, a helicopter using a specially designed dispenser can lay out the wire. The motor generator can also be transported by air so that the technique can be used in remote areas. The frequency of the system is variable, but 400 Hz is commonly used. The receiver is a gradiometer consisting of two horizontal coplanar coils mounted some 10 ft apart in a bird towed 100 ft below the helicopter. Two components are measured: the ratio between the field strengths (FSR) and the phase difference (P) of the signals received between the two coils. These two parameters are similar to the in-phase and out-of-phase components discussed in the previous section.

FIGURE 6.30
Variation in Turair responses over a series of closely spaced, near-surface conductors near Timmins, Ontario, Canada

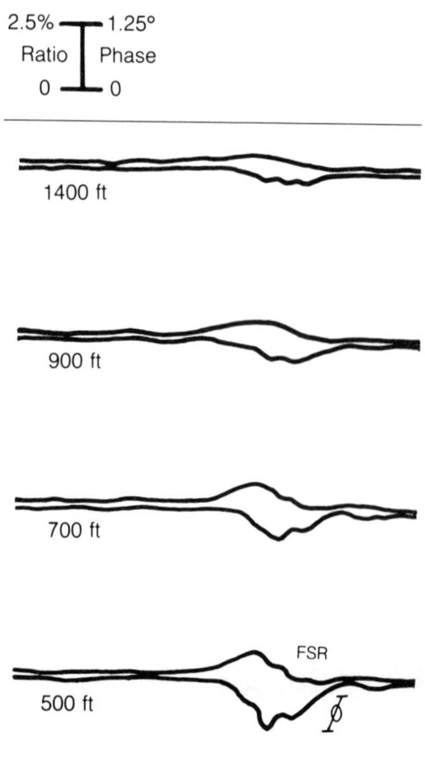

Limitations of the system are two: the requirement of laying out a loop which reduces the effective line mileage that can be flown in a day and the "blind" zone created very near the loop wire. Very strong gradients near the loop prohibit the detection of conductors within this zone which can be up to 1000 ft wide depending on loop size and conductivities of the overburden and bedrock. This problem is normally overcome by planning the survey in such a way that the loop is placed outside the area of interest or so that blind zones from one loop are properly surveyed from another loop. It is also important to lay the loop on the footwall side of shallow-dipping formations. This optimizes electromagnetic coupling between loop and conformable targets, and facilitates the recognition of anomalies. These limitations and pitfalls are well known to companies using this technique. Because of these restrictions, the cost of Turair is about 30-50 percent higher than conventional helicopter-borne in-phase, out-of-phase systems.

Figure 6.30 shows the free air rate of fall-off from a group of closely spaced shallow conductors (Shield area of Ontario). These zones would have been detected at even higher altitude. It is difficult to calculate how much non-conductive cover (limestones, overburden) would make this multiple zone undetectable; a conservative estimate is 1000 ft.

A Turair flight line obtained with the original system (figure 6.31) shows the sharp gradients near the loop, some conductors within the loop area (e.g., marked 1), and a good conductor with a coincident magnetic anomaly located outside the loop (marked 2). This conductor is further outlined in figure 6.32 together with the ground follow-up data. The ground Turam data are noisy because of very local, near surface conditions. After calculations of the secondary electromagnetic field (the field from the conductor with the effect of the primary field fully removed) were made, the Turam data resulted in a curve with a strong inflection point located directly over the conductor. The near absence of a phase difference response (Turair and Turam) suggests that the conductor has a high σt value (180 mhos). Drilling intersected a 10-ft massive pyrrhotite/pyrite zone within a quartz biotite schist beneath

500 ft of Paleozoic limestones. The conductor has a strike length of about one mile. The strength of the Turair anomaly indicates that this zone would still have been detected under a thicker package of limestones (up to 800-1000 ft).

Figure 6.33 shows another example from the same area. This conductor was detected along several lines and shows a reversal in the field strength ratio. This crossover-type anomaly suggests a wide zone with the return current (negative part of the response) flowing in a horizontal plane and in that section of the conductor zone farthest away from the loop. The phase difference and magnetic anomaly (70 gammas) show a single but rather broad anomaly supporting that interpretation. It should be noted here that the location of the return current flow of narrow vertical bodies is normally too deep and the effect is not seen in the overall response of a conductor. The ground follow-up (figure 6.33) again shows a broad response. The Turam data do not show the effect of the return current flow because of the different location and size of the Turam transmitter loop.

Drilling of this anomaly intersected a 350-ft-wide zone consisting of a sequence of a narrow band of sulfides, a 50-ft-wide zone of partly serpentinized amphibolites, another narrow (3 ft) sulfide zone, 120 ft of gabbro and a 9-ft-wide zone of pyrite-graphite mineralization under some 550 ft of Paleozoic sediments. In addition, the upper 90 ft of the basement were locally weathered and leached, bringing the cover to a thickness of 650 ft. This weathering and leaching might at least in part have caused the return current flow. The Turair and Turam results indicate a moderate conductivity of about 15 mhos.

FIGURE 6.31
Original Turair recordings over two areas in Manitoba, Canada (See also figures 6.32 and 6.33.)

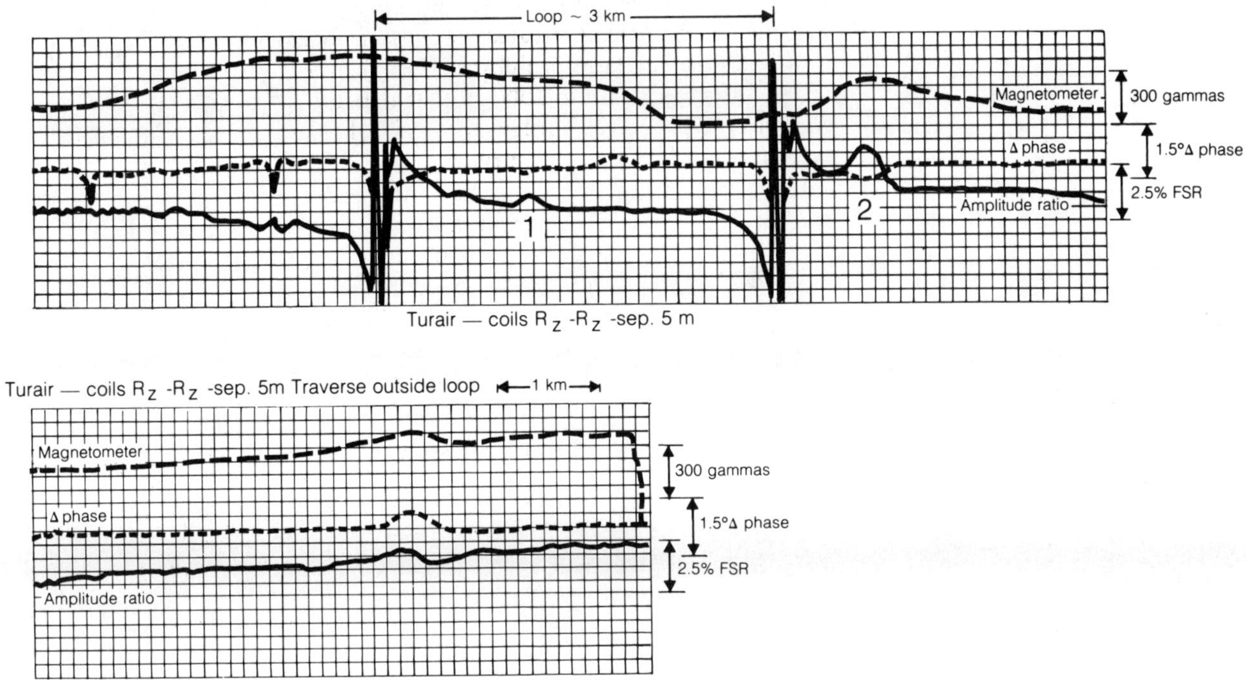

The figure on this page is referred to in the preceding section *Ground source—Turair.*

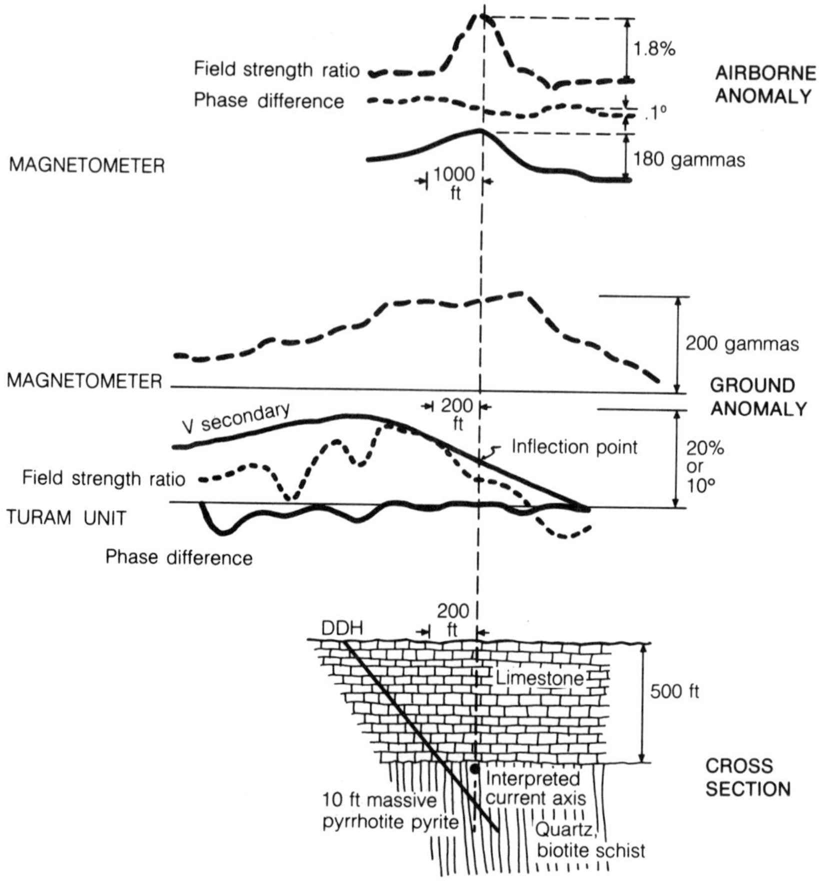

FIGURE 6.32
Turair, ground follow-up and drill information from a massive sulfide conductor covered by 500 feet of Paleozoic limestones in Manitoba, Canada

The figure on this page is referred to in the preceding section *Ground source—Turair.*

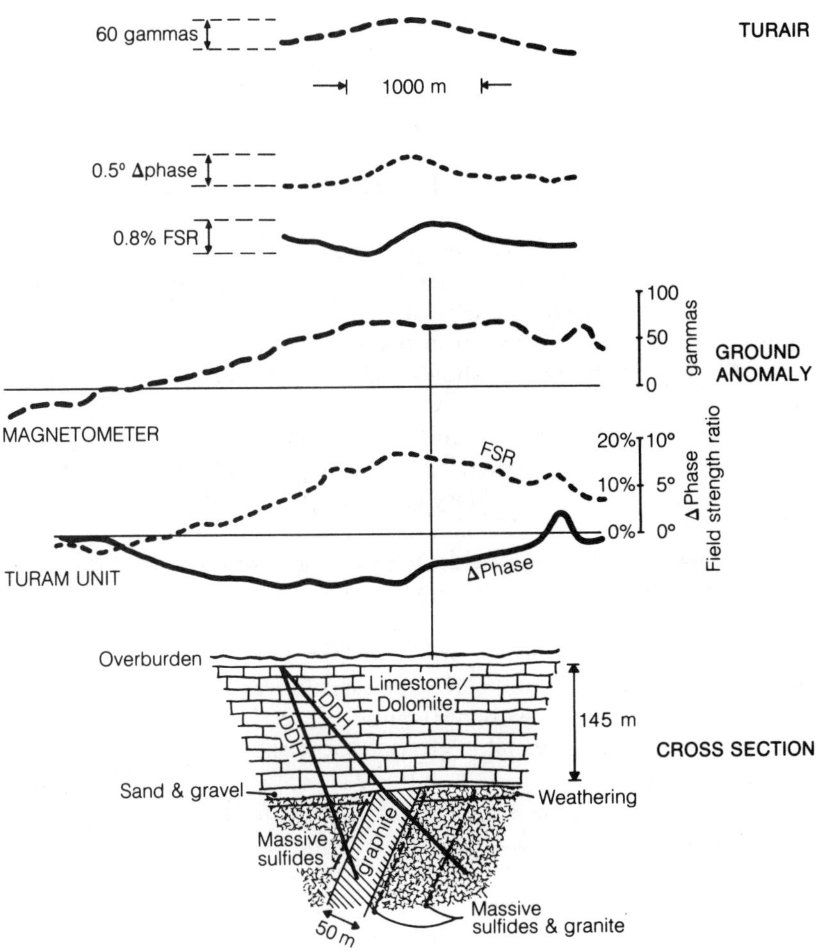

FIGURE 6.33
Turair, ground follow-up, and drill information from a conductive zone covered by 550 ft of Paleozoic limestones in Manitoba, Canada

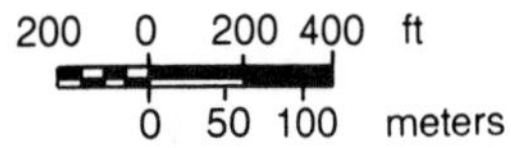

The VLF method

The VLF (very low frequency) method uses powerful radio transmitters set up in different parts of the world for military communications (see figure 6.34). In radio communications terminology, VLF means very low frequency, about 15 to 25 kilocycles/sec. Relative to frequencies generally used in geophysical exploration, this is actually very high.

These powerful radio transmitters induce electric currents in conductive bodies thousands of miles away. Induced currents produce secondary magnetic fields which can be detected at surface through deviations of the normal VLF field. The VLF method is relatively inexpensive and can be a useful prospecting tool.

Successful use of VLF requires that the strike of the conductor be in the direction of the VLF station so that the lines of magnetic field from the VLF transmitter cut the conductor. The upper half of figure 6.35 shows the magnetic field vector in relation to the transmitting antenna. The lower half of figure 6.35 shows that currents will be induced in conductor C1 but not in conductor C2, because the lines of magnetic field cut conductor C1 but not conductor C2.

Figure 6.36 shows schematically how the secondary field from the conductor is added to the primary field vector, so that the resultant field is tilted up on one side of the conductor and down on the other side. A VLF receiver measures the field tilt and hence we have the tilt profile shown in the upper part of figure 6.36.

Some receivers measure other parameters such as the relative amplitude of the total field or any component, and the phase between any two components. Figure 6.37 shows a comparison of the main types of measurements made with different VLF receivers. Table 6.3 summarizes the currently available ground and airborne VLF systems and their manufacturers.

Interpretation is quite simple. The conductor is located at the inflection point marking the crossover from positive tilt to negative tilt, and the maximum in field strength. One cannot make reliable estimates of conductor quality, however. A rule of thumb depth estimate can be made from the distance between the positive and negative peaks in the tilt angle profile. The major disadvantage of the VLF method, however, is that the high frequency results in a multitude of anomalies from unwanted sources such as swamp edges, creeks, and topographic highs. It is sometimes impossible to get a powerful enough VLF station to be near the strike direction of the expected conductor. On the other hand, the tendency for VLF to respond to poor conductors has aided in mapping faults and rock contacts.

Figures 6.38 and 6.39 illustrate dip angle and field strength profiles, respectively, over two sulfide conductors in Cavendish Township, province of Ontario, Canada. The RADEM VLF receiver used the PANAMA VLF station at 24 KHz. The field strength was adjusted at 100 percent at 0+00, line E. Two conductive zones are clearly depicted by both the dip angle and field strength data.

Figure 6.40 shows an instructive comparison of standard helicopter EM data (HEM) and airborne VLF data (EM18). VLF data have many anomalies, and it would be nearly impossible to differentiate between those that are geologically significant and those that are not. By comparison, the HEM clearly distinguishes between strong and poor conductors by using the in-phase/out-of-phase ratio and anomaly amplitude. VLF data pick up more conductors because the frequency is much higher.

The figures on this page are referred to in the preceding section *The VLF method.*

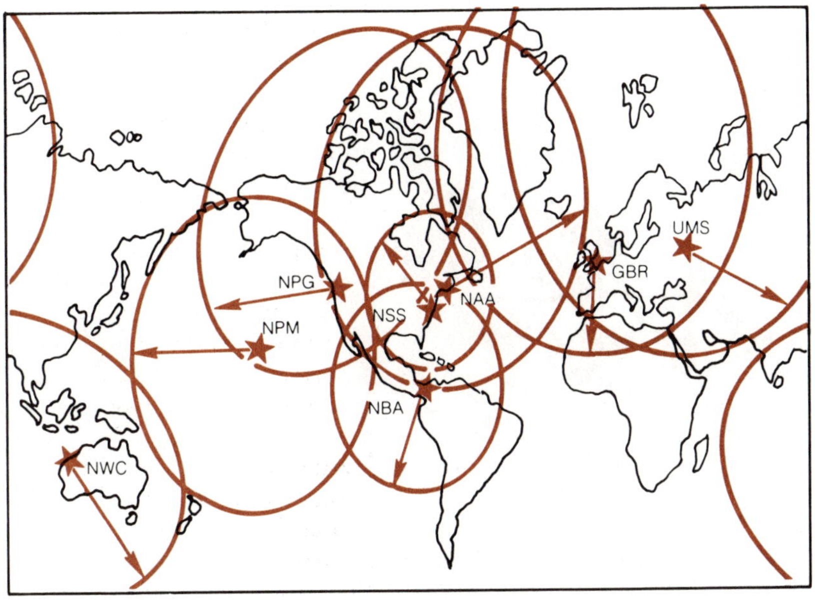

FIGURE 6.34
Locations of well-known VLF transmitter stations

Coverage shown only for well-known stations

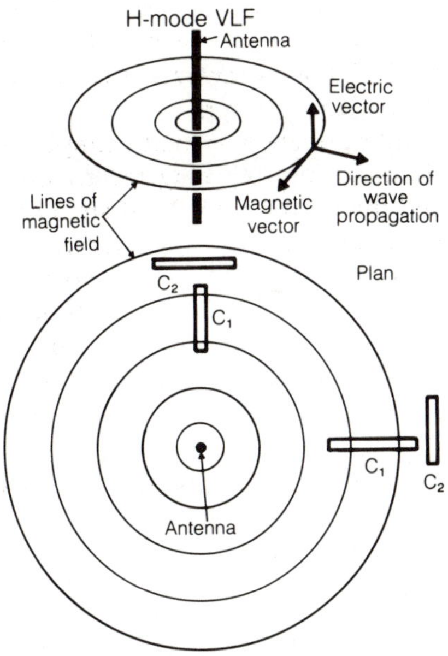

FIGURE 6.35
The VLF field

The figures and table on this page are referred to in the preceding section *The VLF method.*

FIGURE 6.36
Tilt of the VLF field vector over a conductor

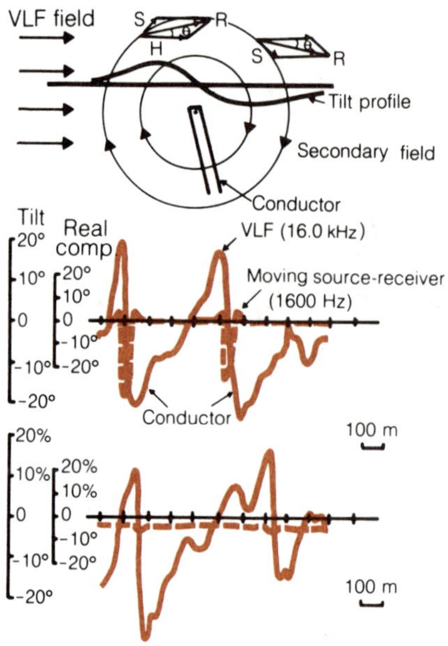

FIGURE 6.37
Comparison of VLF instruments

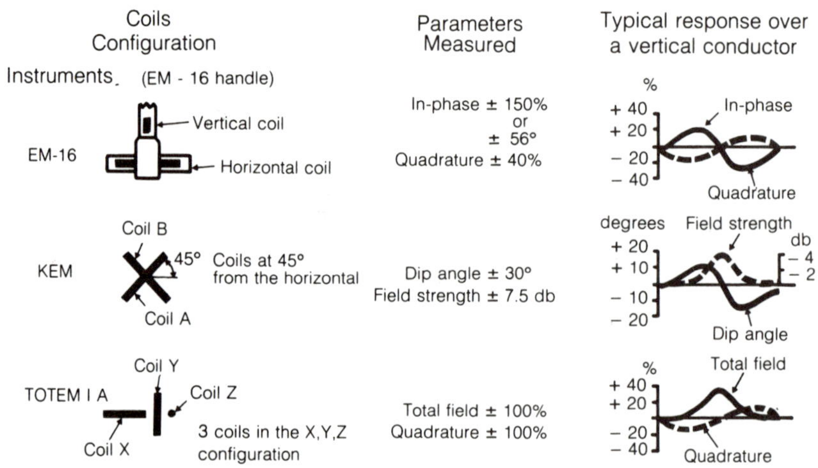

TABLE 6.3
Currently available ground and airborne VLF systems

Airborne	Ground
Total field and vertical quadrature component Herz Industries Ltd., Totem 1A	Orientation and magnitude of the major and minor axes of the ellipse of polarization Crone, Radem Phoenix, VLF-2 Scintrex, SE 81
Horizontal field amplitude & in-phase and quadrature components of vertical field Geonics, EM 18 Scintrex, SE90	Vertical in-phase and quadrature components Geonics, EM 16
Total field strength and dip angle McPhar, Kem	

The figures on this page are referred to in the preceding section *The VLF method.*

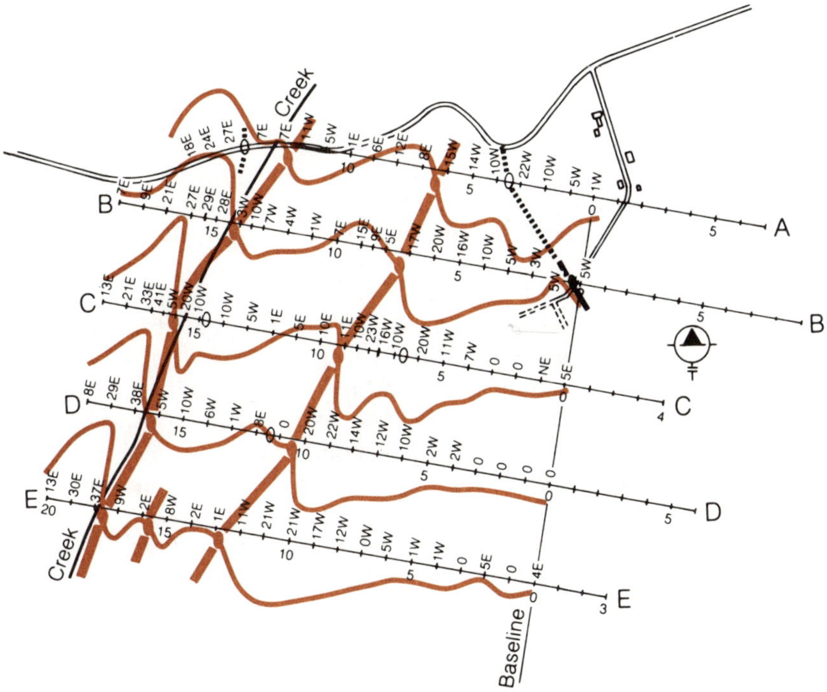

FIGURE 6.38
Example of a VLF tilt angle survey
Geophysical test range, Cavendish Township, south of Gooderham, Ontario, Canada

Crone-radem-VLF-EM survey
Dip angle of the resultant field in degrees
Station — Balboa, Panama 24.0 kHz

Conductor
Indicated conductor
Drill hole

30°
20°
10°
0°
Vertical scale

200 ft
Horizontal scale

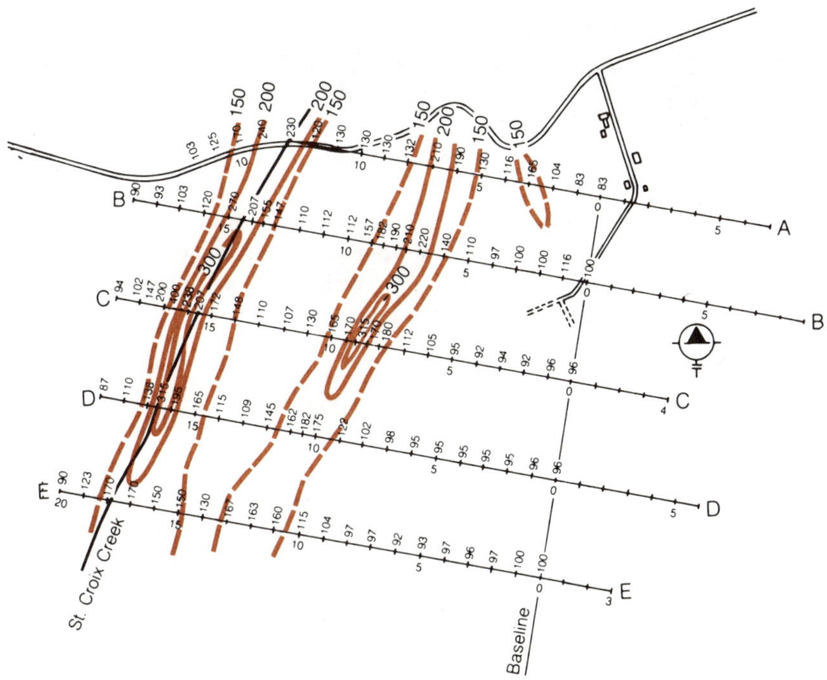

FIGURE 6.39
Example of a VLF field strength survey
Geophysical test range, Cavendish Township, south of Gooderham, Ontario, Canada

Crone-radem-VLF-EM survey
Resultant field strength (% of normal field strength)
Station — Balboa, Panama 24.0 kHz

FIGURE 6.40
Comparison of airborne EM and airborne VLF data, northern Hudson Bay area, Canada

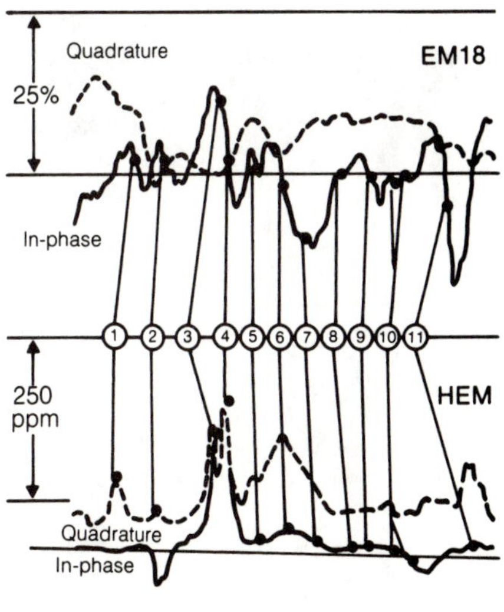

Ground EM systems

Ground EM equipment can be divided into two groups: active systems employing their own transmitter, and passive systems employing natural fields, e.g., AFMAG. The best known passive EM systems are the VLF-EM receivers which use strong signals transmitted by a worldwide network of submarine communication stations. Most of the older active ground systems consisting of a receiver and transmitter combination employ only a single frequency because of weight and design limitations. Modern electronic components permit the construction of relatively lightweight multifrequency or time domain EM equipment. The frequency range for most active systems is from 50-5000 Hz. Passive systems operate at very low or very high frequencies, e.g., AFMAG at 8 Hz, and VLF-EM in the 15-25 KHz range. The power source for the active systems can be either rechargeable batteries or a small semiportable motor generator set.

The magnetic field is transmitted and received through a coil of wire which consists of multi-turn (100-1000 turns) wire loops. The EM transmitter is designed to create a large enough electromagnetic moment to energize the ground (and conductors therein) and to enable the return of a strong enough secondary signal back to the receiver coil. The coils are either air- or iron-cored. The former are flat and as much as a few feet in diameter. The iron-cored coils are normally rod-shaped, a few feet long, and only a few inches in diameter.

The distance between the transmitter and receiver can vary widely. The conventional horizontal loop EM (HLEM) systems may operate with the two coils as close as 50 and as far as 800 ft apart. Coil separation for the vertical loop EM (VLEM) systems varies from 400 to 1200 ft. The Turam technique employs a single turn transmitter loop as large as 2000 by 4000 ft; the receiver consists of a gradiometer, two similar receiver coils 100 to 200 ft apart. The gradiometer can operate as far as 2000 ft from the edge of the transmitter loop. Depending on the signal strength; VLF-EM receivers can operate up to several thousand miles away from the transmitter station.

The weight of these systems ranges from a few pounds for the VLF-EM units to 30 lbs for the battery-operated ($<$25 watts) HLEM and VLEM transmitters. Overall shipping weight of a motor generator-operated TURAM unit (up to 2500 watts) is about 450 lbs.

FIGURE 6.41
Configuration of tilt angle devices. Configuration A is the standard VLEM setup.

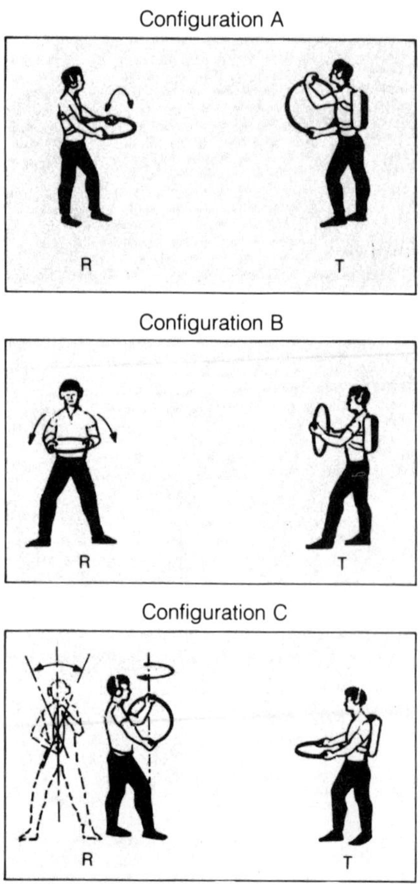

Tilt angle systems

One of the oldest EM systems is the tilt angle system, the most popular of which is vertical loop EM (VLEM). Vertical loop EM refers to the most commonly used transmitting configuration in which the plane of the transmitting coil is directed toward the receiver. Other transmitter setups may be used, but because the name of the system derives from the most common parameter measured, tilt angle device is appropriately named (figure 6.41). The significant difference between the vertical and horizontal loop and other compensation type (e.g., TURAM) measuring systems is that the transmitter and receiver are not interconnected.

The vertical loop system is used primarily in two different modes or field setups: *broadside* and *fixed transmitter* (figures 6.42 and 6.43). In the broadside technique, both coils move together along two roughly parallel lines. The transmitter is oriented as described above, and the receiver is rotated out of the horizontal plane around an axis connecting the transmitter and receiver. The amount of tilt in degrees and the direction of the plane of tilt (e.g., this is E or W along an E-W line) are recorded. The data are plotted and the crossover point determines the position of the conductor. The broadside technique normally uses a grid of uncut pace and compass lines. The line connecting the most well-defined crossover points will be used to establish the base line of a proper grid. Detail work is then done using the fixed transmitter technique. The transmitter is now positioned above the best (broadside) crossover point, and the receiver moves from station to station along cut lines. A good working scheme has to be set up so that the transmitter plane and horizontal axis of the receiver are as nearly coplanar as possible for each measuring station. Normally, several lines are covered from one transmitter setup. The location of the crossover points determines the position of the conductor (figure 6.44).

VLEM is a fast technique that works well in flat terrain. The advantage of the VLEM system is that transmitter orientation and coil separation do not have to be overly accurate. The strike direction, especially in complex geologic situations of folding and variations in strike, can be accurately determined. Two disadvantages of VLEM are 1) its limited use in rugged terrain and 2) its limited capability in quantitative interpretation such as dip, depth, and conductivity-thickness product. Some interpretation diagrams are available, but these are not very precise, and most of the interpretation is restricted to conductor location (figure 6.45). In this illustration, unit h is the distance between the transmitter and the picket line along which the transmitter moves. If the conductor is deeply buried, $h>0.2$, the effects of dip become difficult to distinguish from those of depth in these curves.

Some general qualitative rules for interpreting the VLEM data follow: strong, well-defined, sharp crossovers normally reflect shallow massive sulfides or graphite horizons (figure 6.46); weak and less well-defined crossovers often represent shear or fault zones. Weak, well-defined crossovers with the opposite peak responses more widely separated than for the first case can indicate massive sulfides subcropping at greater depth.

A special adaptation of the tilt angle technique for *rugged terrain* is the Shootback method. Both coils are similar and can be used as transmitter or receiver. The procedure follows (figure 6.47): the lead coil is held horizontal and transmits; a null is then obtained with the trailing receiver by turning this coil around a horizontal axis perpendicular to the traverse line; the dip of the receiver coil is then noted; next, the functions of transmitter and receiver coils are reversed; both dip measurements are added, with proper attention to the sign, and the total value is plotted in a standard profile. The technique was introduced to the exploration scene about 15 years ago by Crone Geophysics. Simple, mainly qualitative interpretation diagrams are available (figure 6.48).

The figures on this page are referred to in the preceding section *Tilt angle systems.*

FIGURE 6.42
Field setup and results over a steeply dipping bedrock conductor using VLEM. Fixed transmitter setup

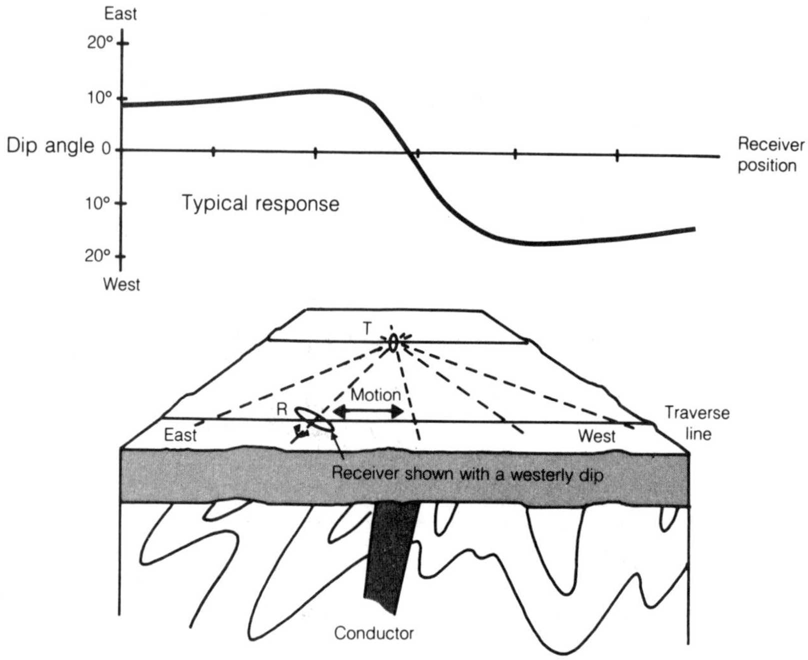

FIGURE 6.43
Field setup and results over a steeply dipping bedrock conductor using VLEM. Parallel line technique

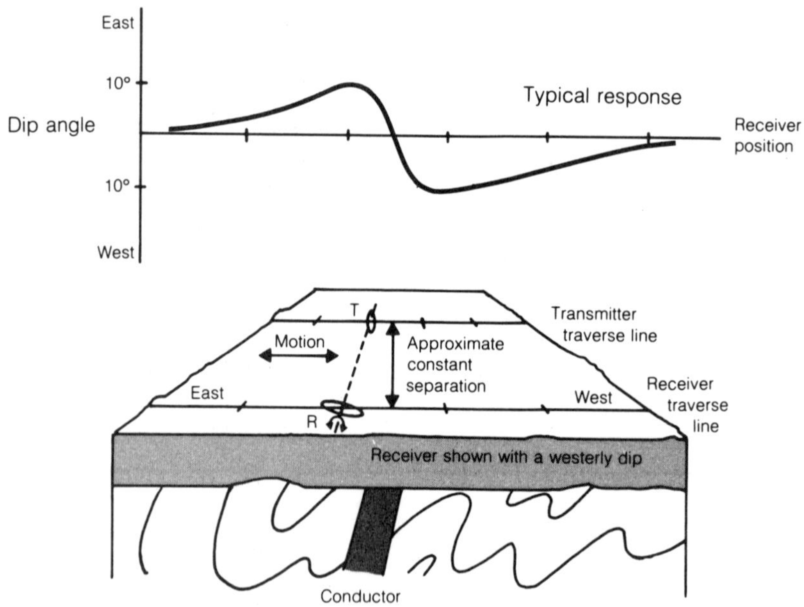

FIGURE 6.44
Fixed transmitter VLEM plan over shallow massive sulfide conductor

T_1 and T_2 denote fixed transmitter locations and the lines surveyed with each.

200 0 200 ft
+20° 0 −20° tilt

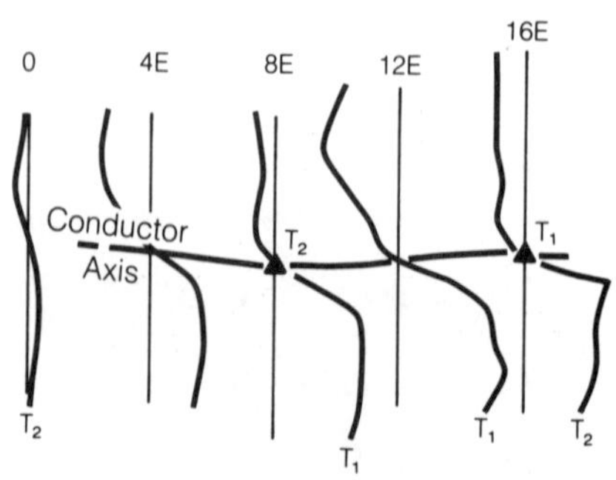

The figures on this page are referred to in the preceding section *Tilt angle systems.*

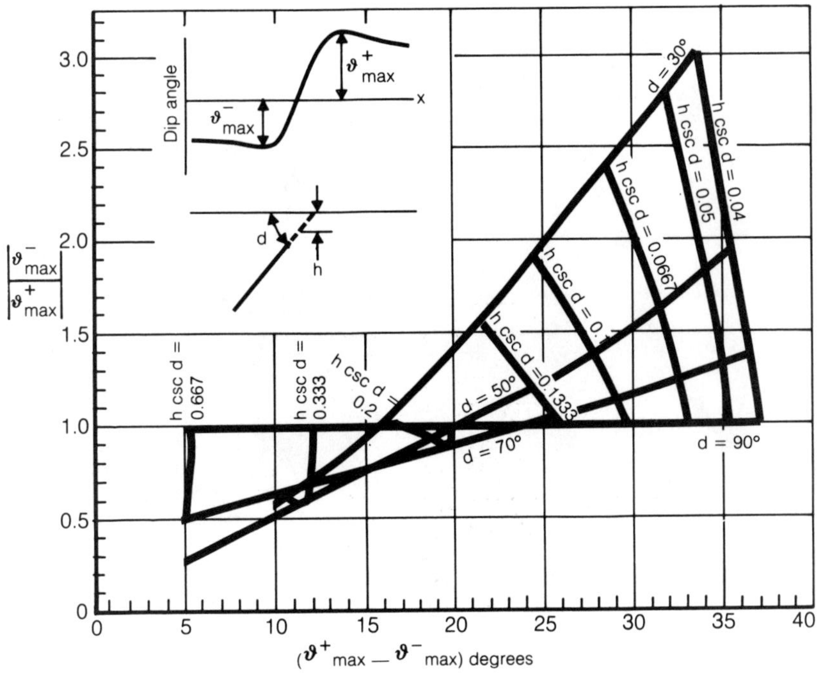

FIGURE 6.45
"Characteristic curves" for the conducting half-plane model using the vertical-loop. Fixed transmitter method.

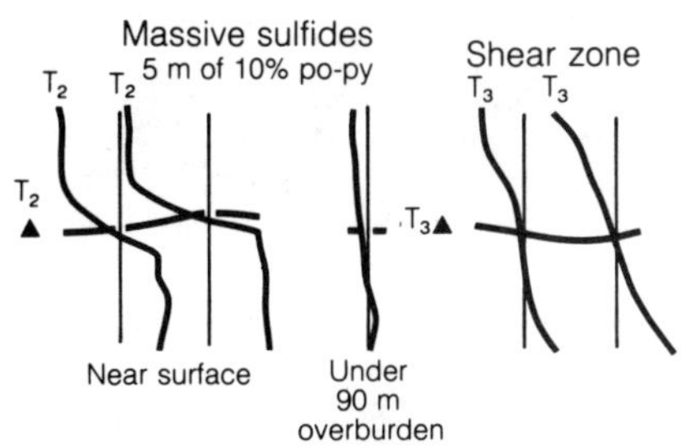

FIGURE 6.46
Examples of VLEM results over three different types of conductors

200 0 200 ft
+20° 0 −20° tilt

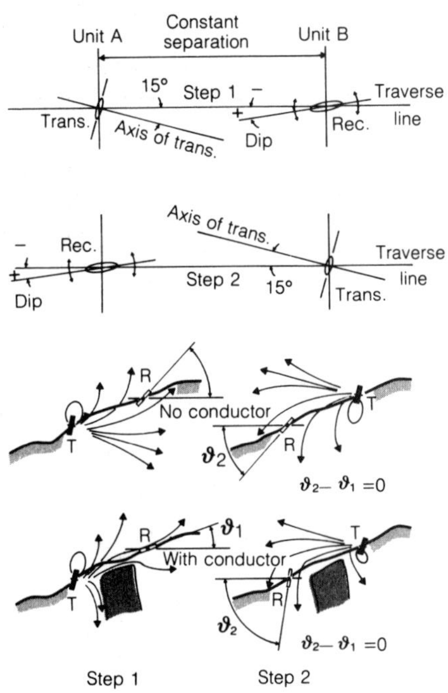

FIGURE 6.47
The Shootback method

The directions of the transmitter coils are shown in the original vertical shootback setup rather than in the more commonly used horizontal mode.

The figure on this page is referred to in the preceding section *Tilt angle systems.*

FIGURE 6.48
Model and field data obtained with the Shootback technique

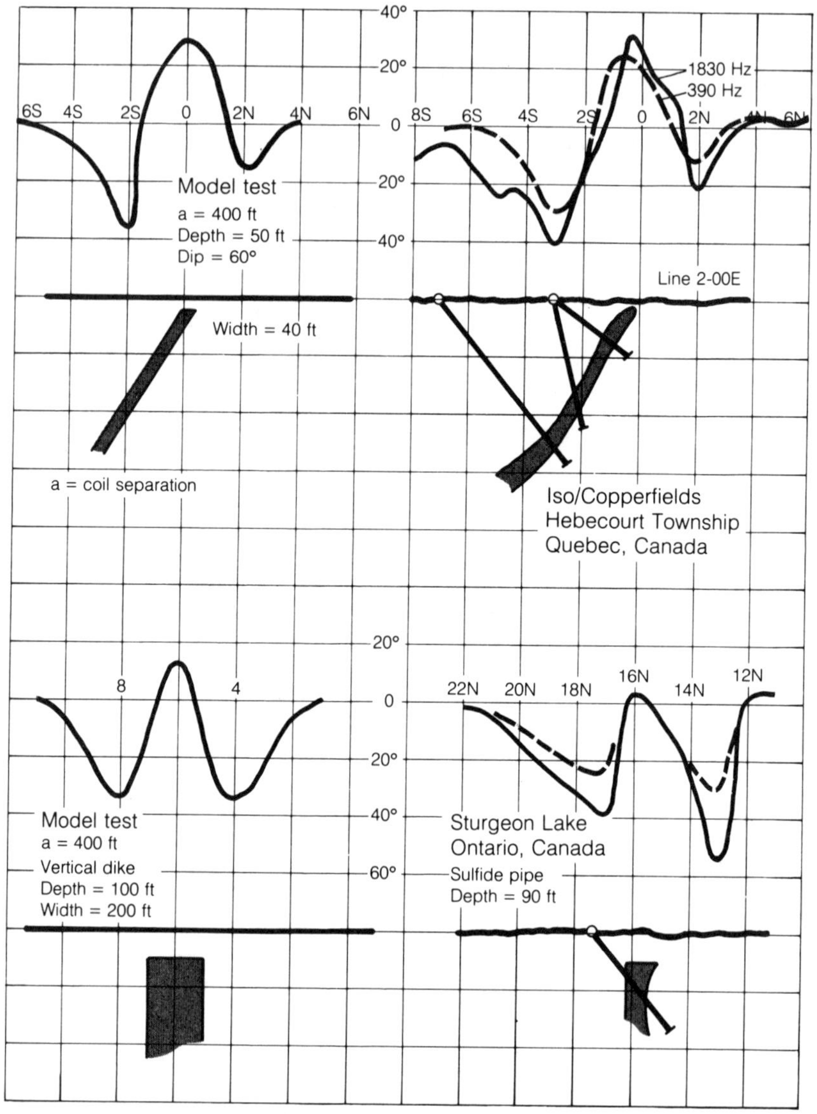

Horizontal loop systems

The horizontal loop electromagnetic system (HLEM) is the most popular EM system in use today. The transmitter and receiver are interconnected by a reference cable permitting the measurement of the EM field (primary plus secondary) at the receiver as a percentage of the theoretical primary field at the receiver (figures 6.49, 6.50, and 6.51). This system allows one to quantitatively interpret dip, depth to top, conductivity thickness (σt), and overall thickness.

The latest versions of HLEM systems have as many as five binary spaced frequencies, e.g., 111, 222, 444 Hz, etc. Interconnecting cables are available in several standard lengths, e.g., 25, 50, 100, 150, and 250 m. The components measured are the in-phase and out-of-phase in percentage of the primary field.

In comparison to the VLEM technique, great care has to be taken to keep the distance between the coils constant from station to station, and to hold the coils in one common plane (horizontal in flat terrain). A discrepancy of 1 percent in the coil separation causes an in-phase error of 3 percent which can be important when looking for very weak conductors in the 5 percent peak-to-peak amplitude range.

Depth penetration of the HLEM systems is somewhat less than that of the VLEM systems. This is due to the length limitations of the interconnecting cable. The largest HLEM cable is 250 m, whereas VLEM systems can operate at up to 500 m separation.

Depth penetration is in most situations not more than 50-60 percent of the coil separation. In situations where the coil separation can be kept to within .5 percent and no conductive overburden is encountered, the depth capability may be increased to 75 or 80 percent of the coil separation. A knowledge of the estimated overburden thickness is important when conducting an HLEM survey.

In complex situations, the strike direction (or continuity) of HLEM anomalies cannot be determined as clearly as with the VLEM technique. On the other hand, the response to short conductors is better for HLEM than for VLEM; the major advantage of HLEM over VLEM is that the results can be interpreted quantitatively. HLEM is, therefore, a more detailed and refined technique and can only be properly used along suitably prepared grid lines.

To ensure sufficient data for proper interpretation, one must make measurements at stations not more than half of the coil separation apart in non-anomalous areas and more frequently over anomalies. The two components measured, in-phase and out-of-phase, are plotted, and simple interpretation schemes are used to calculate dip, depth, and σt (figures 6.52 and 6.53). Extensive documentation on the interpretation of HLEM is available (e.g. Nair and others).

The first step in interpreting HLEM data is to establish background in-phase and out-of-phase values away from the anomaly to be interpreted. The peak amplitudes C_1, C_2, Q, and R and a are then measured with respect to this background (or zero) level. The width of the conductor (or closely spaced parallel bands of conductors and called W) is calculated by subtracting the coil separation (b) from the distance between the zero value points of the in-phase and out-of-phase curves (distance a). The widths calculated using the in-phase or out-of-phase profiles might differ somewhat. Only the value obtained from the cleanest curve should be considered correct. However, a weak conductive halo around a very good conductor often causes a wider out-of-phase than in-phase profile. The next step is to estimate the dip of the conductor using a dip monogram. The final phase is to select the proper phasor diagram (figure 6.54). The example of figure 6.52 requires a 40° dip diagram to estimate depth (d) and σt values. Most phasor diagrams are universal and show these values in terms of coil separation and σtf

FIGURE 6.49
HLEM field lines and responses over a vertical half plane

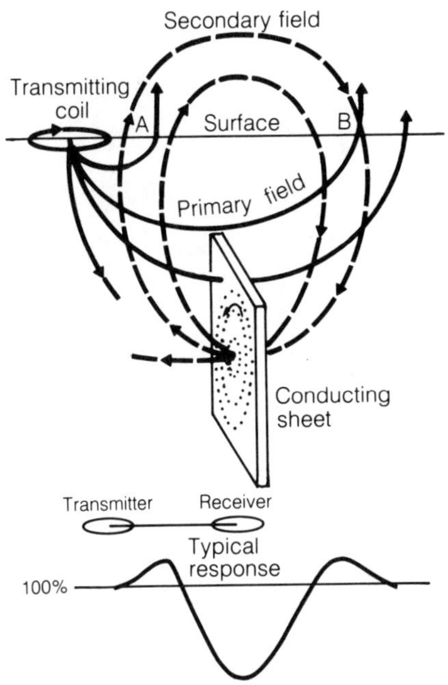

product (f = frequency). Simple calculations convert these values into depth and σt.

In many field situations, the background values are nearly equal to the instrument zero level. Data obtained with a multifrequency HLEM system should show depth and σt values that are very similar for all frequencies. In areas where conductive overburden or host rock is present or where the conductor is electrically saturated at higher frequencies, these calculated values will often differ considerably. In such situations it is imperative to use a multifrequency (three or more) approach to obtain reasonable depth and σt estimates. The most accurate depth and σt estimates will be obtained using data collected with the lowest frequency (that is, the lowest frequency that produced data of sufficient amplitude to be interpreted).

The effect of conductive overburden on a bedrock conductor is to rotate the phase (the inverse tangent of the ratio of the out-of-phase anomaly to the in-phase anomaly) of the anomaly toward the in-phase component. Figures 6.55 and 6.56 show the effect of anomaly *rotation* for a steeply dipping bedrock conductor under conductive overburden. The effect can be strong enough to *reverse* the sign of the out-of-phase component. It is easy to understand that single frequency data will not provide insight into this problem or its effect. Phase rotation causes a conductor to appear a *deeper* and *better* conductor when the data are plotted on a standard phasor diagram (i.e., a diagram for a bedrock conductor under non-conductive overburden or in free air). An example of phase rotation with increasing frequency is given in figures 6.57 and 6.58. The estimated background response levels for both in-phase and quadrature components are shown for each frequency. The background response is assumed flat over the zone of the anomaly. A sounding interpretation using the single thick layer model gives a layer resistivity of about 40 Ω-m and layer thickness of about 28 m (i.e., a σt of 0.7 mhos).

The peak negative in-phase and quadrature anomaly amplitudes (relative to background) are plotted in figure 6.58. The solid arrow indicates the frequency migration that would have occurred on the argand diagram, from 222-3555 Hz, if a half-plane conductor model in free-air were applicable. The actual field results are seen to deviate substantially from the free-air case. This demonstrates the phase rotation of the anomaly of a bedrock conductor, caused by conductive overburden. By assuming the 222 Hz data to approximate the free-air case, the conductor depth may be interpreted as 35 m and the conductivity-thickness product as 30 mhos. It is possible to attempt a better interpretation by extrapolating back in frequency until the effects of the conductive layer may be assumed negligible. Such a procedure would decrease the conductor depth somewhat.

The gravity data show a strong regional component with a possible 0.2 mgl positive residual anomaly centered at about 275N. Although the existence of the residual anomaly may be subject to debate, the important point is that the combination of the gravity data and the frequency response in the electromagnetic data strongly suggests the presence of a bedrock conductor as opposed to an overburden trough. The magnetic results are featureless over the conductor, suggesting little magnetite or pyrrhotite in the conductor.

A more complex (but fortunately not frequent) situation arises when currents induced in overburden and/or host rock are galvanically channeled into a bedrock conductor (current gathering). Calculating the d and σt values for data at different frequencies will again suggest that the standard phasor diagram should not have been used. The lowest frequency will, as in the case of phase rotation, provide the best estimate for d and σt. (See Fraser, 1978.) Details on treatment of data obtained with the HLEM systems in complex environments can be found in several in-house publications by J. E. Betz, Geophysical Consultant, Toronto.

Measurement with different frequencies and coil separations might provide different d and σt estimates even when no conductive overburden or host is present and no electric saturation is apparent. Interpretation diagrams are based on *homogeneous* sheet-like models, whereas orebodies are irregularly shaped, not sheet-like, and have varying σt values. This should always be kept in mind when using interpretation schemes.

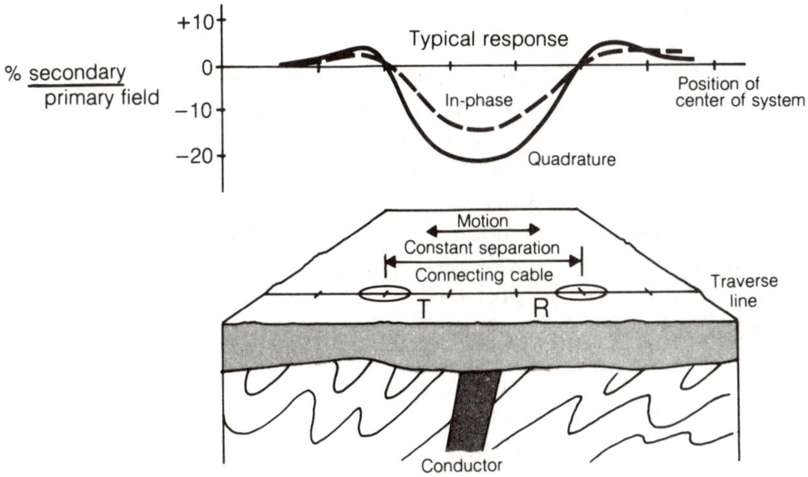

FIGURE 6.50
Field setup and results over a steeply dipping bedrock conductor using HLEM

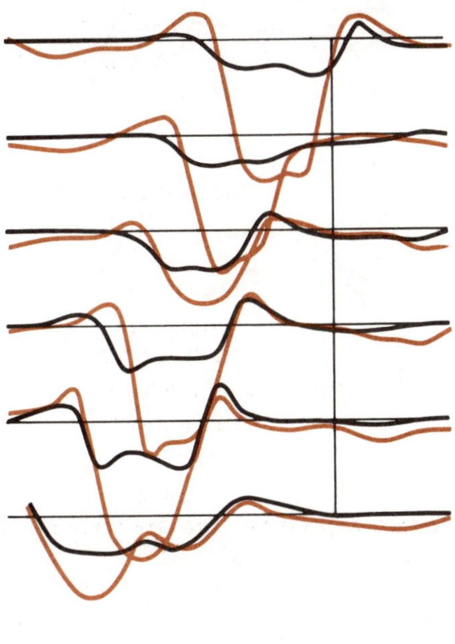

FIGURE 6.51
Typical presentation of HLEM data

Scale 1:2500; solid line in-phase; dashed line out-of-phase; in-phase and out-of-phase scale 1 cm = 10%, coil separation 100m; frequency 444 Hz.

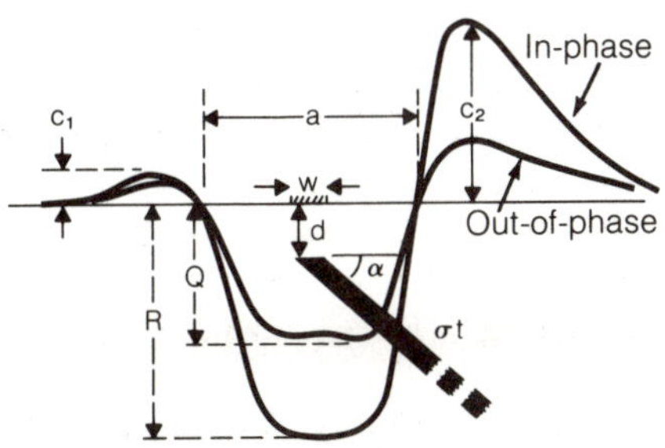

FIGURE 6.52
HLEM interpretation

The figures on this page are referred to in the preceding section *Horizontal loop systems.*

FIGURE 6.53
HLEM interpretation

a-b=w, width of conductor
e.g., 120-100=20 m

C_2:C_1 → α, dip e.g., 40°

R:Q → σt, conductivity x thickness
→ d, depth to top
e.g., 13 mhos, 16 m

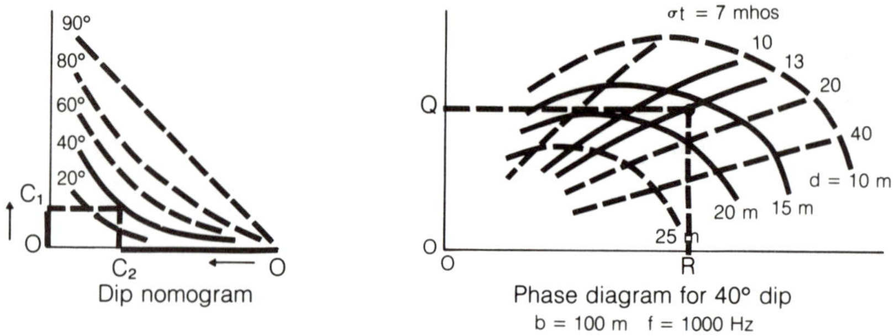

FIGURE 6.54
Examples of HLEM phasor diagrams for bedrock conductors of varying dip

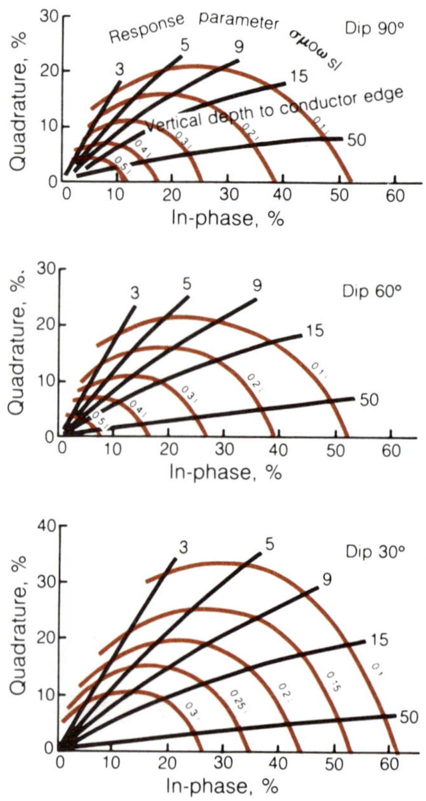

The figures on this page are referred to in the preceding section *Horizontal loop systems.*

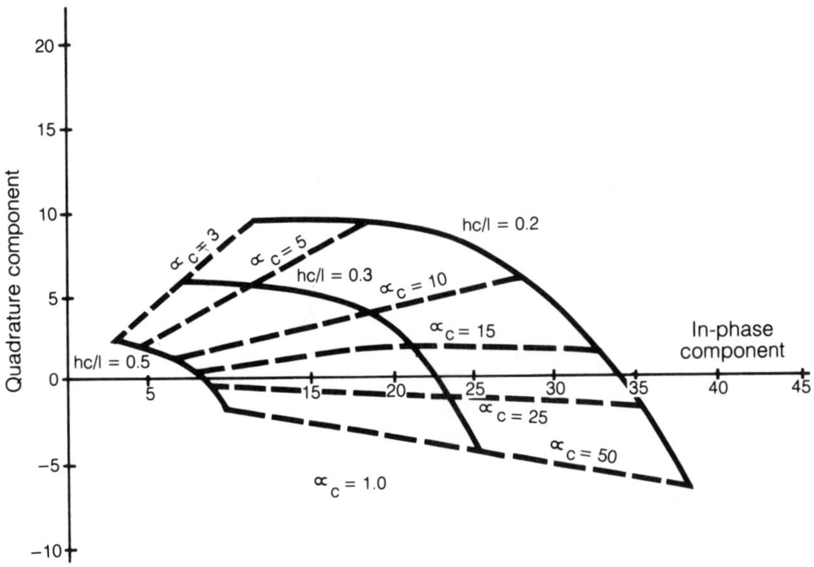

FIGURE 6.55
Phase rotated response diagram

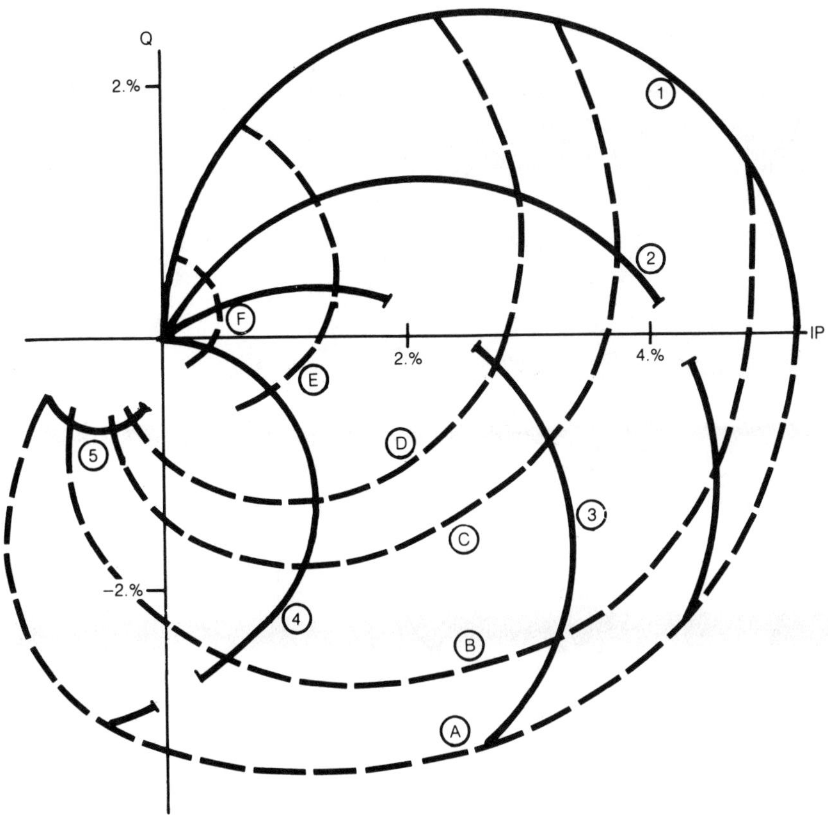

FIGURE 6.56
Phase rotated response diagram.

Effect of different overburden conductivities:
L (1) no conductive overburden
L (5) highly conductive overburden
(A) strong bedrock conductor
(F) poor bedrock conductor

– – σ_p Constant
—— σ_L Constant

σ_L (mho/m)	$t\sigma_p$ (mho)
① 0.	Ⓐ 1000
② 1×10^{-2}	Ⓑ 30
③ 2×10^{-2}	Ⓒ 10.5
④ 4.5×10^{-2}	Ⓓ 7
⑤ $1. \times 10^{-2}$	Ⓔ 3
	Ⓕ 1

The figures on this page are referred to in the preceding section *Horizontal loop systems.*

FIGURE 6.57
Electromagnetic, gravity, and magnetic data over Abitibi clay belt of Ontario, Canada

Horizontal loop EM

Coil spacing = 150 m

—— In-phase

—— Quadrature

+20%
0%

100 m

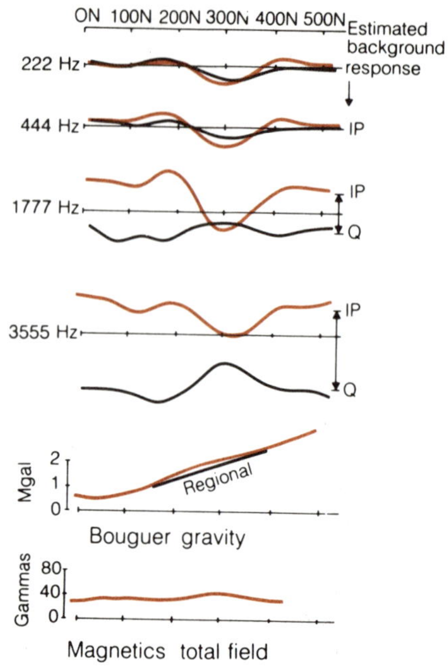

FIGURE 6.58
Diagram of peak negative in-phase and quadrature anomaly amplitudes

Conductor I

Theoretical (free air)

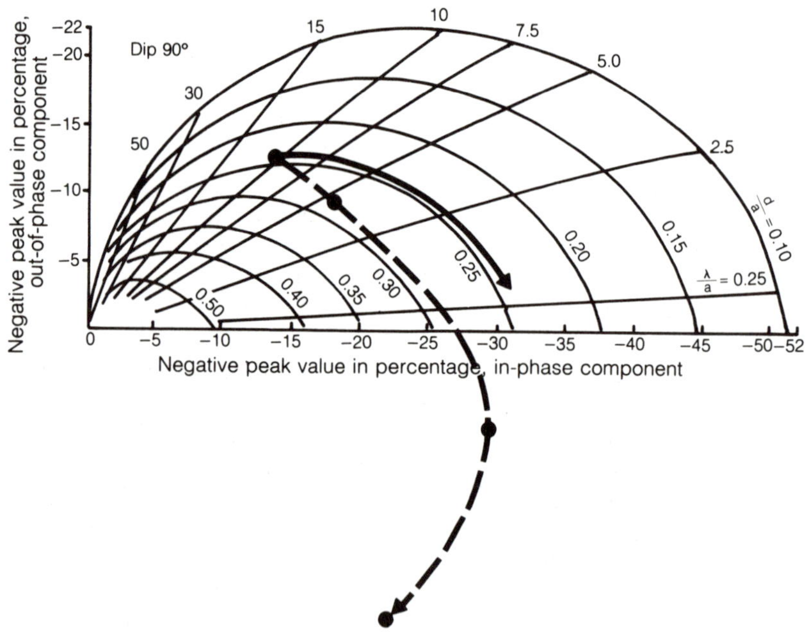

Turam

The Turam method dates back to 1925 when it was first used in Sweden. An excellent doctoral thesis by Bosschart describes the Turam method, and much of the discussion herein is extracted from this work.

In the Turam method, a large loop of wire is laid out on the ground. A common size is 4000 by 2000 ft as shown in figure 6.59. The wire is energized by a transmitter powered by a semi-portable motor generator. The surveying is done along traverse lines perpendicular to the long side of the loop. The Turam receiver consists of two coils 100-200 ft apart, both connected to a compensator on one of the coils. The compensator measures and displays the amplitude ratio and phase difference between the signals measured in the two coils.

FIGURE 6.59
General layout of the Turam method

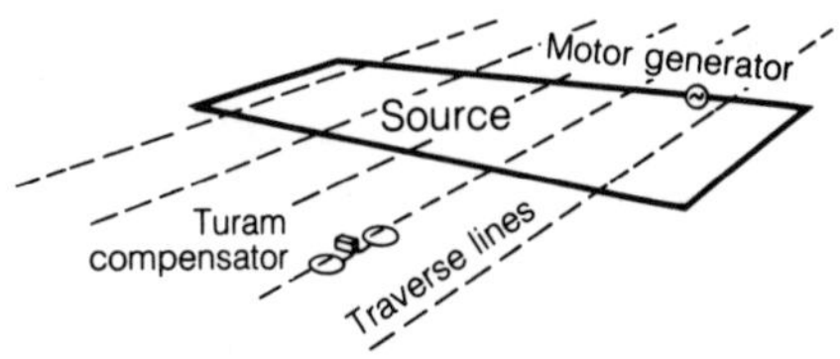

The physics of the Turam method are shown schematically in figure 6.60. The upper diagram shows the two receiver coils connected to the compensator. A null adjustment on the compensator allows the operator to measure the field strength ratio and phase difference between the two coils. The second diagram shows the lines of force of the primary magnetic field from the Turam wire, and the lines of force of the secondary magnetic field from the current induced in the conductor. On the transmitter side of the conductor, the vertical component of the secondary field will add to the primary field while, on the other side of the conductor, the secondary field will subtract from the primary field. This is demonstrated in the lower diagram, where the secondary field is shown superimposed on the primary vertical field strength. The receiver system consisting of two coils measures the ratio of vertical field strength between two stations, 100 or 200 ft apart. The measured ratio is then divided by the theoretical ratio and the result is called the Reduced Field Strength Ratio. This is shown in the lowermost diagram of figure 6.60.

Figure 6.61 shows an example of Turam model responses for different conductor depths. In this case, the conductor is 500 m from the wire and parallel to it, has a strike length of 300 m, and dips 60° away from the wire. Four quantities are plotted on each profile: the reduced field strength ratio (FR), the phase difference between the two coils (P), the secondary field in phase component (R), and the secondary field out-of-phase component (I). As expected, the anomaly amplitude drops off sharply with increased depth.

A similar diagram which shows the effect on the anomaly of conductor dip is shown in figure 6.62. Note that at a particular dip of 110° (or 70° towards the loop), the anomaly becomes extinct. For dips greater than 110°, all the anomalous parameters (FR, P, R, and I) are reversed in sign. The reason the anomaly disappears at the extinction angle of 110° is that the lines of primary magnetic field parallel the conductor and do not cross it; therefore, no induction takes place in it. Planning a Turam survey requires placing the transmitter loop on the up-dip side.

A rough depth interpretation to the current axis may be obtained through the half width of the field strength ratio and phase anomalies. (Half width is simply the width of the anomaly at half amplitude.) Interpretation of conductor quality is made through the use of a nomogram as shown in figure 6.63.

Figure 6.64 shows an example of a Turam survey in Isle Dieu Township in the province of Quebec, along with a geological plan at 150 ft below surface, as determined by drilling.

Comments regarding effects of conductive overburden and hosts in the HLEM section also apply in the Turam case and will not be repeated here.

The advantage of Turam is its greater depth penetration over other systems such as HLEM. This is by virtue of the large loop. The disadvantages are the poor dip interpretation, poor thickness determination, and the necessity of laying out and retrieving a large loop of wire.

FIGURE 6.60
Schematic presentation of primary and secondary fields for the Turam method

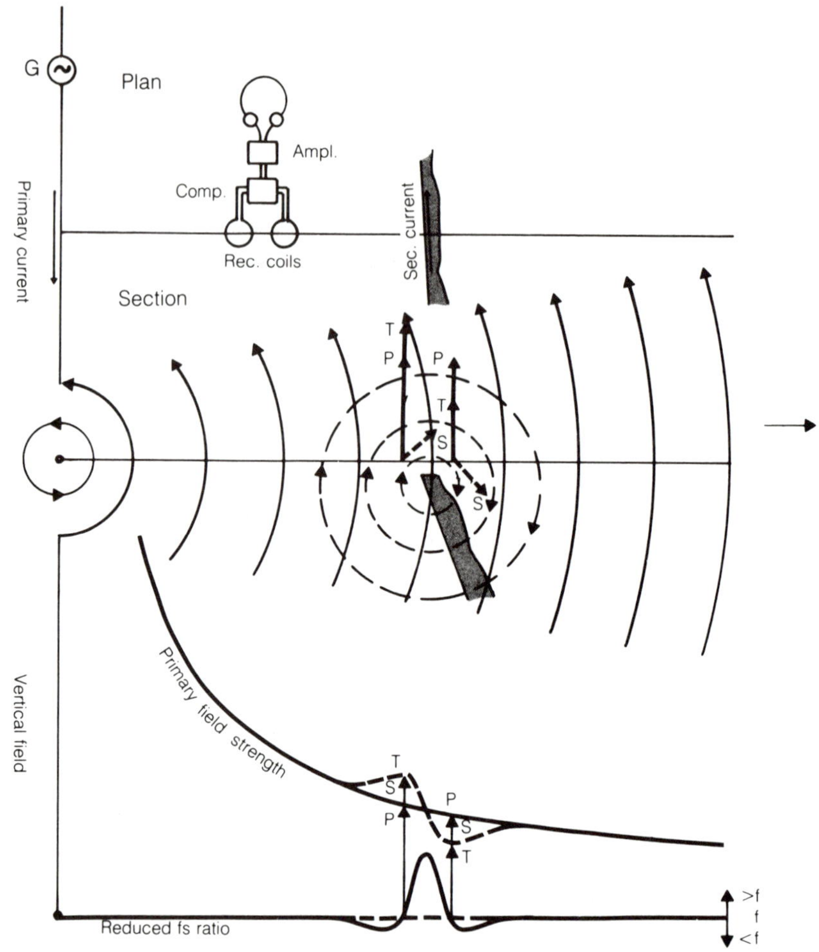

FIGURE 6.61
Profiles showing field strength ratio, phase difference, and complex components for different depths of the conductor

$\lambda = 3.5$
$L = 300m = H$
$\phi = 60°$
$x = 500m$

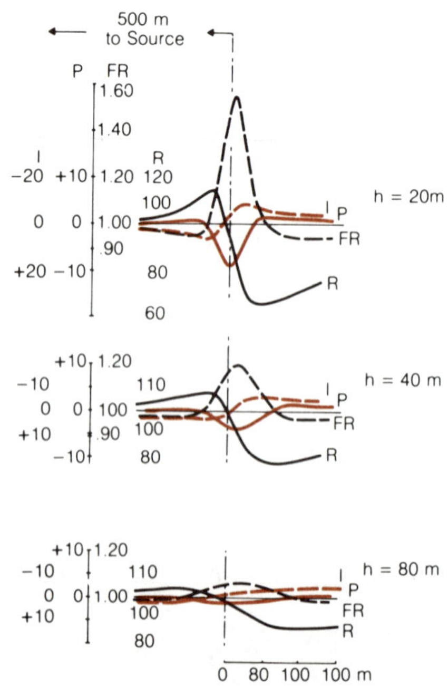

The figures on this page are referred to in the preceding section *Turam*.

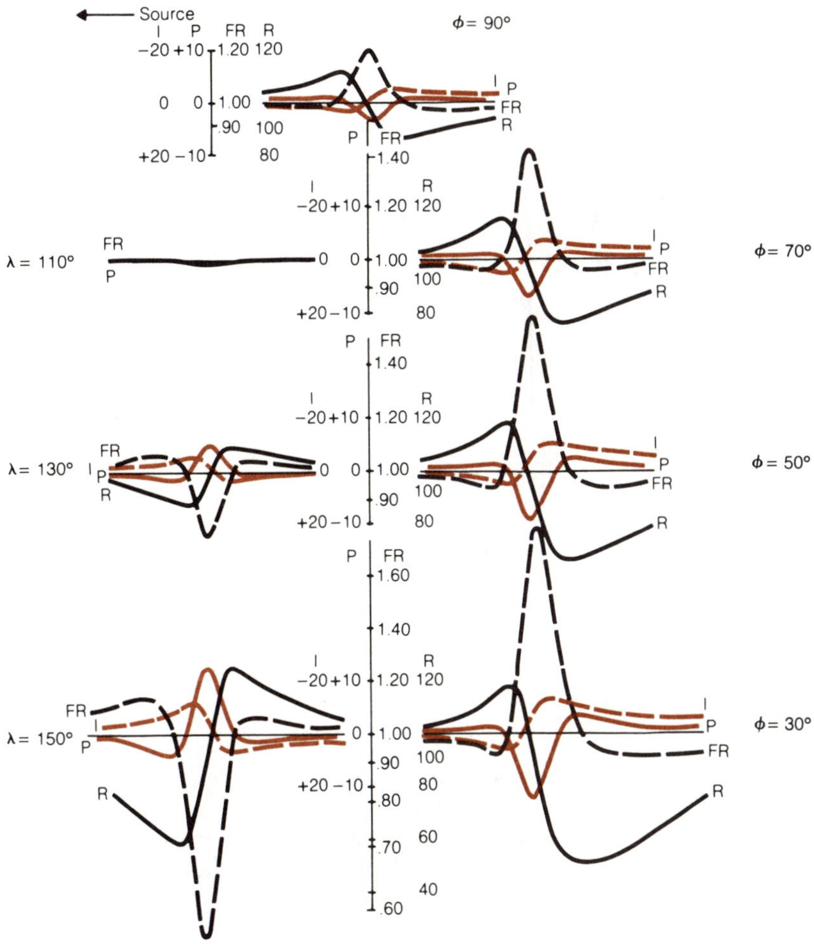

FIGURE 6.62
Profiles showing field strength ratio, phase difference, and complex components at different dip angles

λ= 3.5
h = 20 m
L = 300 m = H
x = 500 m

0 50 100 150 m

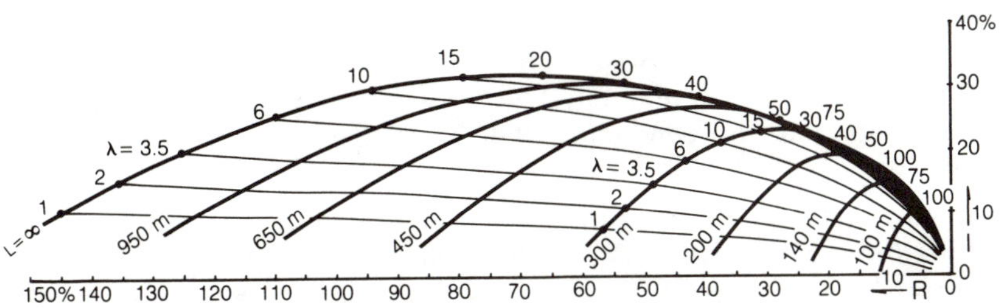

FIGURE 6.63
General response diagram for thin conductors of variable size

h = 20 m
φ= 60°
Tr. loop 600 × 1200 m

 The figure on this page is referred to in the preceding section *Turam.*

FIGURE 6.64
Turam survey in Isle Dieu Township, Quebec, Canada

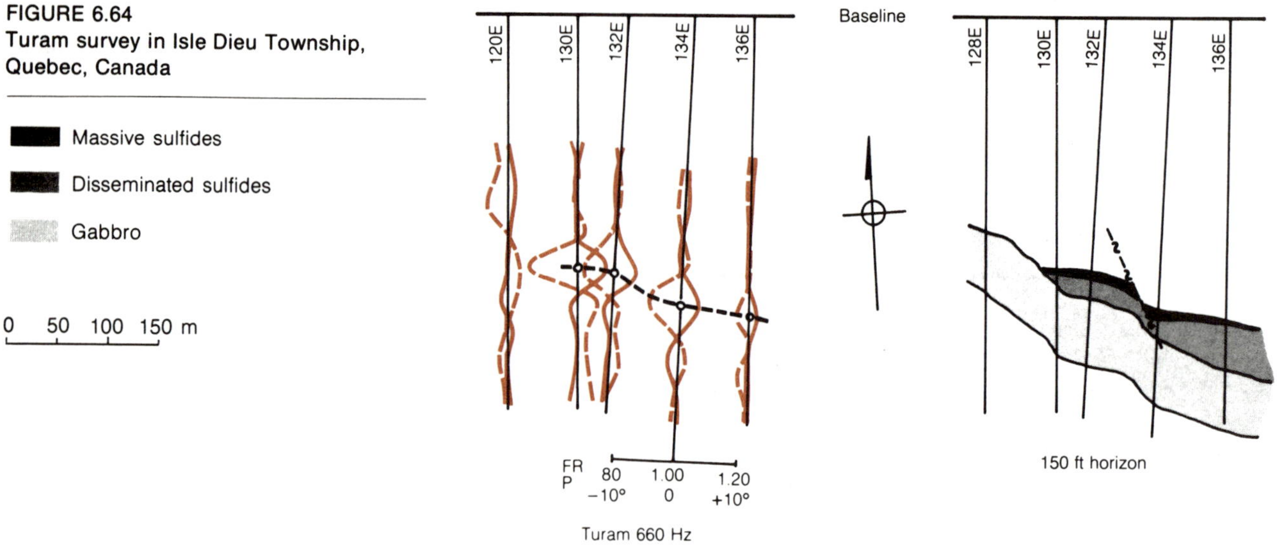

FIGURE 6.65
Examples of three transmitter receiver setups for drill hole EM work

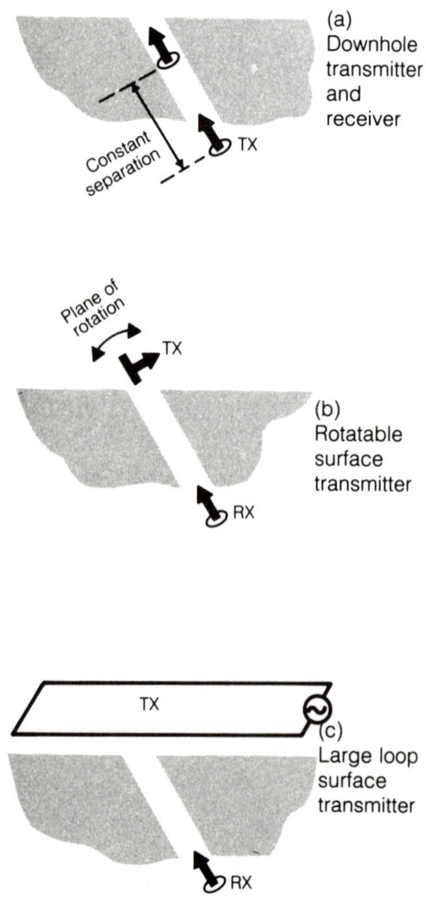

Drill hole methods

Drill hole EM methods are used increasingly in exploration geophysics, primarily to verify whether or not an EM drill target has been hit. Figure 6.65 shows three common transmitter and receiver setups for drill hole EM systems. Only the large loop surface transmitter system will be discussed, as it seems to be enjoying the greatest success at this time.

The principle of operation is shown in figure 6.66. The primary magnetic field from a transmitter loop on surface induces electric current in a buried conductor. A secondary magnetic field will be generated, and it will oppose the primary magnetic field at the conductor. If the drill hole passes through the conductor, a sharp negative anomaly should be obtained. On the other hand, if the drill hole passes outside the conductor, a broader positive anomaly will result because the secondary field, laterally outside the conductor, will be in the same direction as the primary field.

Figure 6.67 shows an example of a downhole EM survey with the Crone Pulse EM system, using a large loop on surface. This is a multichannel time-domain EM system. A strong negative anomaly is obtained on the massive sulfides, and a weaker anomaly coincides with the disseminated sulfides.

New developments

Research and development in EM reflect two main objectives: first, to increase the depth of exploration (e.g., geothermal exploration); second, to develop systems which will have increased capability in discriminating between various conductors in complex geologic environments. The first objective is usually achieved by using a large loop source to generate a large dipole moment. The second objective is reached by using multifrequency systems or preferably multichannel time-domain EM systems.

Table 6.4 summarizes some of the more modern instruments being used today. Because they have special applications, their cost is high and certainly beyond the reach of the average prospector. Nevertheless, it is anticipated that the knowledge and experience gained with such systems will eventually filter down to the simpler systems used in more conventional EM surveys.

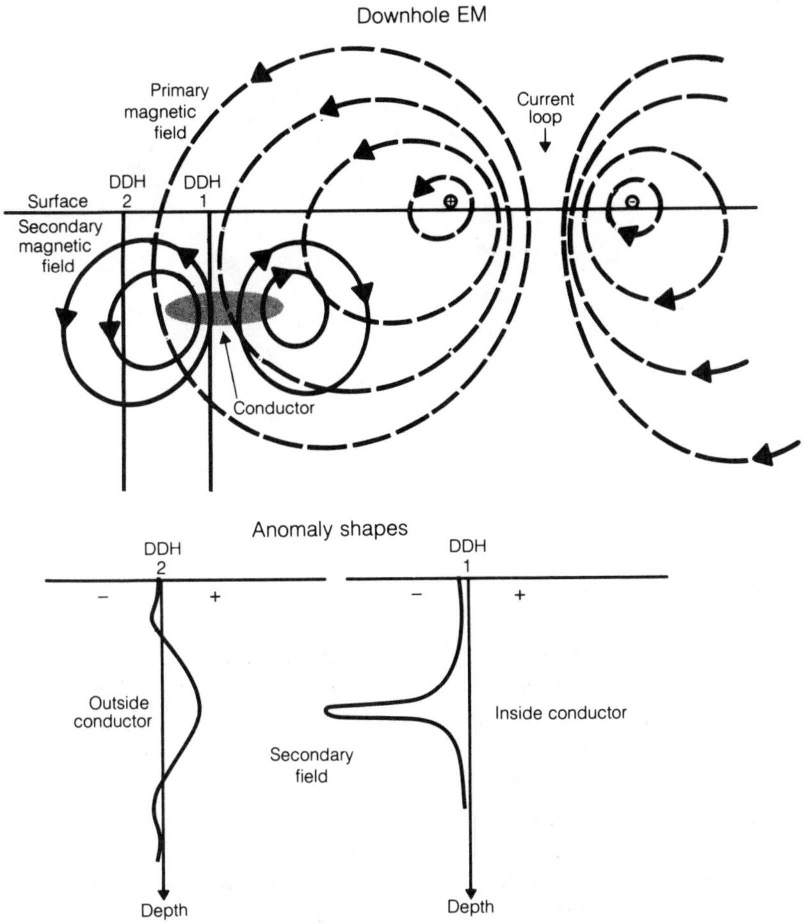

FIGURE 6.66
Principle of operation of downhole EM

TABLE 6.4
Examples of new wideband EM systems

System	Organization		Geometry	Major application	Availability
Maxiprobe	Geoprobe	FD	Coil-coil (large separation)	Layered environments	Contract surveys
EMR 14	U. of Utah	FD	Coil-coil (large separation)	Layered environments and mining	Research and some contract surveys
EM-60	U. of California — Berkeley	FD	Very large TX loop	Geothermal	Research
UTEM	U. of Toronto	TD	'Turam' type	Mining exploration	Research and some contract surveys
EMP	Newmont	TD	'Turam' type	Mining exploration	Proprietary
The following have somewhat less depth penetration than the preceding systems because of lower power and/or transmitter loop size.					
Sirotem	Geox-Phoenix	TD	Superposed Coils	Mining exploration	Sale and contract surveys
PEM	Crone	TD	Coil-coil (small separation)	Mining exploration	Sale and contract surveys

The figure on this page is referred to in the preceding section *Drill hole methods.*

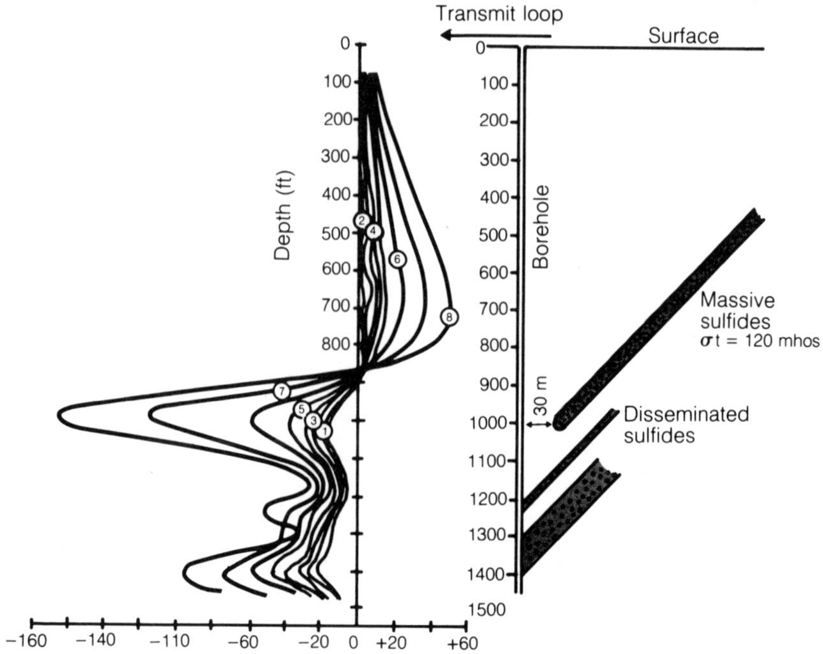

FIGURE 6.67
Example of a multi-channel time domain downhole EM survey. Borehole pulse EM from Gertrude orebody (INCO), Sudbury, Ontario, Canada

Selected References (Chapter 6)

Betz, J. E. 1974. *Test program report with additional comments and posters on the max-min electromagnetic system.* Toronto: Apex Parametrics Ltd.

Bosschart, R. A. 1964. Analytical interpretation of fixed source electromagnetic prospecting data. Doctoral thesis. Holland: U. of Delft.

Caven, R. J. 1976. Geophysical surveying at the Umex Thierry deposit. *Canadian Mining Journal* 97, No. 10: 34-43.

Crone, D. 1977. *Exploration for massive sulfides in desert areas using the ground pulse EM method.* Mississauga, Canada: Crone Geophysics Ltd.

Fraser, D. C. 1978. Geophysics of the Montcalm Township copper-nickel discovery. *Canadian Mining & Metallurgical Bulletin* 71: 99-104.

Ghosh, M. K. 1972. Interpretation of airborne EM measurements based on thin sheet models. Ph.D. thesis. Canada: University of Toronto.

Grant, F. S., and West, G. F. 1965. *Interpretation theory in applied geophysics.* New York: McGraw-Hill.

Hood, P. 1977. Mineral exploration trends and developments in 1977. *Canadian Mining Journal* 99, No. 1: 28-69.

Keller, G. V., and Frischknecht, F. C. 1966. *Electrical methods in geophysical prospecting.* London: Pergamon Press.

Lajoie, J. J., and West, G. F. 1977. Two selected field examples of EM anomalies in a conductive environment. *Geophysics* 42: 655-60.

Lowrie, W., and West, G. F. 1965. The effects of a conducting overburden on electromagnetic prospecting measurements. *Geophysics* 30: 624-32.

Lazenby, P. G. 1973. New developments in the input airborne EM system. *Canadian Mining & Metallurgical Bulletin* 66, No. 732: 96-104.

Marley, L. W., ed. 1967. *Mining and groundwater geophysics.* Economic Geology Report No. 26, Geological Survey of Canada.

Nair, M. R., Bisivas, S. K., and Mazumdar, K. 1968. Experimental studies on the electromagnetic response of tilted conducting half planes to a horizontal loop prospecting system. *Geoexploration* 6: 207-44.

Palacky, G. J. 1975. Interpretation of input AEM measurements in areas of conductive overburden. *Geophysics* 40: 490-502.

Palacky, G. J., and West, G. F. 1973. Quantitative interpretation of input AEM measurements. *Geophysics* 38: 1145-58.

Parasnis, D. S. 1966. Mining geophysics. In *Methods in Geochemistry and Geophysics* Vol. 3. Amsterdam: Elsevier.

Pemberton, R. H. 1962. Airborne electromagnetics in review. *Geophysics* 27: 691-713.

Seigel, H. O., and Pitcher, D. H. 1976. *The Tridem airborne electromagnetic system: a multi-purpose, natural resource mapping tool.* Toronto: Scintrex Limited Application Brief 76-3.

Society of Exploration Geophysicists. 1966. *Mining geophysics.* Vol. I and II.

Telford, W. M., and others. 1976. *Applied geophysics.* Cambridge: Cambridge University Press.

Figure Credits (Chapter 6)

Figures 6.1, 6.2, 6.4 & 6.6
Reprinted with permission of F. S. Grant and G. F. West, *Interpretation theory in applied geophysics*, 1965.

Figure 6.7
Reprinted with permission of Scintrex Ltd., Concord, Ontario, Canada.

Figure 6.8
Reprinted with permission of Questor Survey Ltd., Mississauga, Ontario, Canada.

Figures 6.10 & 6.11
Reprinted with permission of McPhar Geophysics Ltd., Willowdale, Toronto, Canada.

Figure 6.12
Reprinted with permission of G. J. Palacky and G. F. West, Quantitative interpretation of input AEM measurements, 1973.

Figure 6.13
Reprinted with permission of G. J. Palacky, Interpretation of input AEM measurements in areas of conductive overburden, 1975.

Figure 6.14
Reprinted with permission of Questor Survey Ltd., Mississauga, Ontario, Canada.

Figure 6.15
Reprinted with permission of G. J. Palacky and G. F. West, Quantitative interpretation of input AEM measurements, 1973.

Figures 6.16 & 6.17
Reprinted with permission of Questor Survey Ltd., Mississauga, Ontario, Canada.

Figures 6.18 & 6.19
Reprinted with permission of Scintrex Ltd., Concord, Ontario, Canada.

Figures 6.20 & 6.21
Reprinted with permission of D. C. Fraser, Geophysics of the Montcalm Township copper-nickel discovery, *Canadian Mining & Metallurgical Bulletin* 71: 99-104, 1978.

Figure 6.22-6.24
Reprinted with permission of Scintrex Ltd., Concord, Ontario, Canada.

Figure 6.25
Reprinted with permission of M. K. Ghosh, Interpretation of airborne EM measurements based on thin sheet models, Ph.D. Thesis, University of Toronto, 1972.

Figures 6.26-6.28, 6.30-6.33
Reprinted with permission of Scintrex Ltd., Concord, Ontario, Canada.

Figure 6.34
Reprinted with permission of Geonics Ltd., Mississauga, Ontario, Canada.

Figures 6.35 & 6.36
Reprinted with permission of D. S. Parasnis, Mining geophysics, in *Methods in geochemistry and geophysics Vol. 3*, 1966.

Figure 6.38
Reprinted with permission of Crone Geophysics, Mississauga, Ontario, Canada.

Figure 6.39
Reprinted with permission of Geonics Ltd., Mississauga, Ontario, Canada.

Figure 6.40
Reprinted with permission of Scintrex Ltd., Concord, Ontario, Canada.

Figure 6.41, 6.42, 6.43, 6.45, 6.47
Reprinted with permission of F. S. Grant and G. F. West, *Interpretation theory in applied geophysics*, 1965.

Figure 6.48
Reprinted with permission of Crone Geophysics, Mississauga, Ontario, Canada.

Figure 6.49
Reprinted with permission of ABEM, Bromma, Sweden.

Figure 6.50 & 6.54
Reprinted with permission of F. S. Grant and G. F. West, *Interpretation theory in applied geophysics*, 1965.

Fiugre 6.55
Reprinted with permission of W. Lowrie and G. F. West, The effects of a conducting overburden on electromagnetic prospecting measurements, *Geophysics* 30: 624-32, 1965.

Figures 6.59-6.64
Reprinted with permission of R. A. Bosschart, analytical interpretation of fixed source electromagnetic prospecting data. Doctoral thesis, U. of Delft, Holland, 1964.

Figure 6.67
Reprinted with permission of Crone Geophysics, Mississauga, Ontario, Canada.

Table 6.1
Reprinted with permission of P. Hood, Mineral exploration trends and developments in 1978, *Canadian Mining Journal* 100, 1: 28-69, 1979.

CHAPTER 6 ADDENDUM ELECTROMAGNETICS

BROAD BAND ELECTROMAGNETIC SYSTEMS

The information presented in this chapter concentrates on broad-band ground electromagnetic (EM) techniques and is an extension of the previous chapter, by Klein and Lajoie. Refer to the previous chapter for basic background information on time and frequency domain EM systems.

EM systems with one or two frequencies work in simple geology with resistive host rocks. For complex geology and conductive host rocks, however, broad-band systems are needed. Developments during the past 15 years have seen the evolution of new time and frequency domain EM systems that gather large amounts of data in a short amount of time and provide information on depth, dip, and size of both conductive and resistive features.

Controlled source audio-frequency magnetotellurics (CSAMT) is a frequency domain sounding system that was first suggested by Myron Goldstein in his doctoral thesis at the University of Toronto in 1971 (Goldstein, 1971). For this method, depth of exploration is a function of the frequency of the transmitted signal and the resistivity of the ground being explored. It is independent of receiver-transmitter separation. With encouragement from EXXON Minerals, Zonge Engineering built the first commercial digital receiver system for CSAMT in 1977, followed by French (BRGM) and Canadian (Phoenix) systems in the 1980s.

Kenneth L. Zonge,
Zonge Engineering & Research Organization, Inc.

Ken Zonge is president of Zonge Engineering & Research Organization, Inc., of Tucson, Arizona, which he founded in 1972 to manufacture geophysical instrumentation and provide ground surveys for mineral exploration. Since that time, the company has become involved in field surveys and equipment sales worldwide in exploration programs for minerals, hydrocarbons, geothermal resources, and engineering and environmental concerns. Dr. Zonge received his B.S. degree from the University of Alaska, his M.S. and Ph. D. degrees from the University of Arizona, and is presently an Adjunct Professor with the University of Arizona.

The first research into time domain EM (TDEM) or transient EM (TEM) methods was implemented by the Russians in the late 1950s. By 1960, they had developed the MPPO-1, a single-loop TEM system, which used the same loop for both the transmitter and receiver. This model was brought to Australia by Western Mining Corporation in 1970, and the South Australian Bureau of Mineral resources acquired one in 1976. By 1964, George Keller at the Colorado School of Mines had begun development of the CSM Megasource TDEM program, which was the first large offset TEM (LOTEM) system. In 1972, Crone developed their PEM (Pulse EM) system which initially was a loop-loop system. Also during that year, the University of Toronto developed the first UTEM system. A version of this system became commercially available through Lamontagne Geophysics in 1981.

In 1974, Newmont Exploration Limited developed the first high-powered, three-component, digital system for in-house use. By 1977, the Australian government (CSIRO) had developed the SIROTEM single- or coincident-loop system, based on the early Russian design, and began extensive testing and exploration programs. In 1980, Geonics introduced their EM-37 TEM system, and the Russians produced a modified unit called the MPP4.

Today, various types of TEM systems exist, including: Australian SIROTEM MKIII, Lamontagne Geophysics UTEM-3, Geonics EM-41 and PROTEM systems, Crone PEM, Newmont EMP (still proprietary), Keller/Strack/Vozoff LOTEM, Phoenix V5, and Zonge Engineering GDP-16 TEM and NanoTEM.

Depth of exploration for TEM systems can be dependent on transmitter loop size, which determines the received signal level, but independent of array geometry for certain configurations such as central loop.

INTRODUCTION TO CSAMT

The CSAMT electromagnetic sounding technique uses a fixed grounded dipole or horizontal loop as an artificial signal source. CSAMT is similar to the natural source magnetotellurics (MT) and audio-frequency magnetotellurics (AMT) techniques, with the main difference being the use of an artificial signal source for CSAMT. The source provides a stable signal, resulting in higher-precision and faster measurements than are usually obtainable with natural-source measurements in the same spectral band. The controlled source can also complicate interpretation by adding source effects, however, and by placing certain logistical restrictions on the survey. In most practical field situations, these drawbacks are not serious, and the method has proven particularly effective in mapping the earth's crust in the range of 20 m to 2000 m.

CSAMT was initially introduced in the mid-1970s as an EM method for finding conductive orebodies. More recently, the method has been used for mapping narrow, resistive, dike-like silicified zones likely to host gold mineralization. In general, the technique has been successfully used in exploration for massive sulfides, base and precious metals, petroleum and geothermal resources; for mapping structure and lithology; in searching for sources of groundwater and groundwater contamination; and for engineering applications.

A CSAMT source usually consists of a grounded electric dipole 1 km to 2 km in length, located 4 km to 10 km from the area where the measurements are to be made. The frequency band for typical instruments is between 0.125 and 8,000 Hz, with measurements most commonly made in the 16- to 8,000-Hz range. Magnitude and phase are normally measured for one electric (E) and one magnetic (H) field component (for example Ex and Hy, with Ex parallel to the transmitting dipole and Hy perpendicular) See figure 6.1. In reconnaissance mode with an eight-channel receiver, up to seven electric field dipoles can be measured simultaneously with just one magnetic field measurement. For detailing modes, electric and magnetic orthogonal pairs (Ex, Hy) are normally measured. For widely spaced soundings or for research applications, vector and tensor measurements employing two electric field components (Ex, Ey) and three magnetic field components (Hx, Hy, Hz) should be considered.

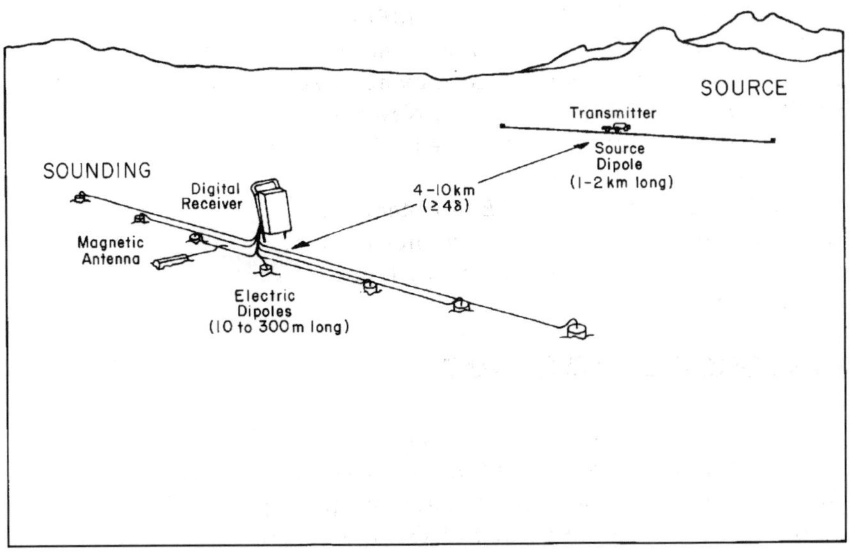

FIGURE 6.1
Field setup for a scalar reconnaissance CSAMT survey using multiple E-fields and one H-field measurement. This is often referred to as a CSAET (controlled source audio-frequency) configuration

Figure 6.2 provides schematic examples of different field configurations. Figure 6.2a shows the setup for a standard, single-station, scalar CSAMT sounding measuring Ex and Hy. This requires a two-channel receiver and provides a sounding at one station at a time.

A good reconnaissance CSAMT configuration is the multiple electric field measurement which is also called the controlled source audio-frequency electrotelluric (CSAET) method. Using an eight-channel receiver, one can measure up to seven electric fields for each magnetic field as shown in figure 6.2b. The one magnetic field measurement is then used to normalize each one of the electric fields for calculation of the Cagniard resistivity and phase difference. The horizontal component of the magnetic field (Hy) usually varies relatively smoothly along a survey line, permitting sparse measurement of this component on a reconnaissance basis. For detail work, it is recommended that one make scalar measurements as in figure 6.2a or at least reduce the number of electric-to-magnetic field measurements to 3:1 or 2:1.

If measurements are to be made on a sparse grid -- for example, making individual soundings every kilometer or every 500 m instead of running continuous stations along a line, one may consider vector

FIGURE 6.2
Field configurations for: (a) a scalar CSAMT survey, (b) a multiple E-field reconnaissance CSAMT survey, or controlled source audio-frequency (CSAET) survey, (c) a vector CSAMT survey, (d) a tensor CSAMT survey. An alternate set-up for the tensor survey would be to have one transmitter at its present location to the west of the receiver station (oriented N-S) and the other to the north (oriented E-W)

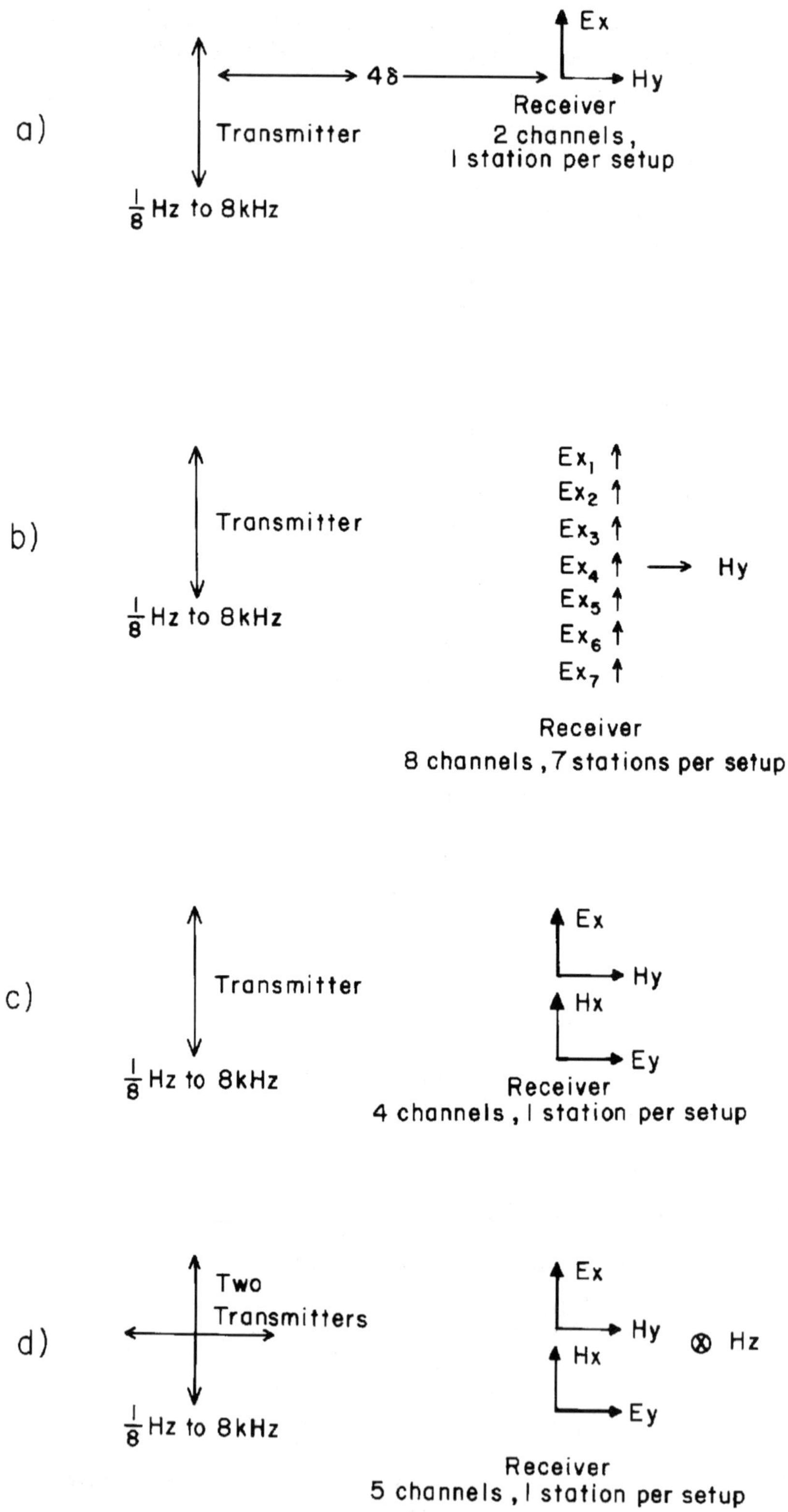

or tensor measurements. (See figures 6.2c and 6.2d). These measurements can provide information on anisotropy (current channeling) and 2-D or 3-D behavior of the subsurface on an individual station basis. Vector or tensor arrays are not necessary for normal continuous line profiling, since variations from station-to-station provide information on geologic dimensionality.

Of the four arrays shown in figure 6.2, the most commonly used systems are scalar CSAMT (figure 6.2a) and the reconnaissance CSAET measurement technique (figure 6.2b).

Grounded dipoles detect the electric field, and magnetic coil antennas sense the magnetic field. The ratio of orthogonal, horizontal electric and magnetic field magnitudes (e.g. Ex and Hy) yields the apparent resistivity. This is usually referred to as the apparent or Cagniard resistivity after the French geophysicist who was instrumental in the development of the magnetotelluric (MT) method in the early 1950s (Cagniard, 1953). The difference between the phase of the electric and magnetic fields yields the impedance phase, which in this text will be called the phase or phase difference.

A parameter used extensively in EM work is skin depth for frequency domain systems (diffusion depth in time domain systems).

Skin depth, is equal to $503\sqrt{\frac{\rho_a}{f}}$ meters; where ρ_a = apparent (measured) resistivity, and f = signal frequency. This is the depth at which the amplitude of a plane wave signal has dropped to 37 percent of its original value.

AMT depth of exploration or depth of investigation, D, is equal to $\delta/\sqrt{2}$ or $356\sqrt{\frac{\rho_a}{f}}$ meters. This equation holds for CSAMT when the separation between the transmitter dipole and the receiver station is greater than three skin depths or 3 δ. Although this equation predicts unlimited depths of investigation under the right circumstances, we have found that in actual practice the maximum depth of investigation is limited to about 3 km.

Lateral resolution is controlled by the electric field dipole length, which normally is between 10- and 200-m. Vertical resolution is 5 percent to 20 percent of the depth of exploration, depending upon resistivity contrasts, geologic complexity and electrical noise.

Table 6.1 lists the basic equations most commonly used for calculating apparent or Cagniard resistivity, skin depth and depth of investigation. These equations hold for a layered earth with the transmitter placed an infinite distance from the receiver. The limitations brought on by a finite distance between transmitter and receiver will be discussed later. Figure 6.3 shows how to determine exploration depth from resistivity and frequency.

If it is impractical to set up a large dipole for a transmitting source, one can also use a large loop of wire for the transmitting antenna. The main drawback of the loop source is its inefficiency in coupling energy into the ground. Given a dipole and square loop of the same size (say a 1 km dipole and a 1 km square loop) it takes eight to ten times more loop current than dipole current to get the same signal level at the receiving site. For this reason, loops are seldom used as transmitting sources. The Cagniard resistivity and phase measurements are the same in the far field for both sources, but differ in the transition and near-field zones. These three "data zones" will be discussed in detail later.

TABLE 6.1
Useful equations for CSAMT

Cagniard resistivity, ρ_a:

$$\rho_a = \frac{1}{5f}\left|\frac{E}{H}\right|^2_{\perp} \quad \text{(ohm-meters)}$$

Phase difference, ϕ:

$$\phi = \phi E - \phi H \quad \text{(milliradians)}$$

Skin depth δ is defined by:

$$\delta = \sqrt{\frac{2}{\mu\sigma\omega}} = 503\sqrt{\frac{\rho_a}{f}} \quad \text{(meters)}$$

Equivalent depth of investigation, D:

$$D = 356\sqrt{\frac{\rho_a}{f}} \quad \text{(D in meters)}$$

Wavelength, λ:

$$\lambda = 2\pi\delta$$

Where: E = electric field in mV/km

H = magnetic field in gammas (γ) or nanoteslas (nT)

ϕ = phase in milliradians

μ = magnetic permeability of air = $4\pi \times 10^{-7}$

σ = conductivity in siemen/meter = $\frac{1}{\rho}$

$\omega = 2\pi f$, f = frequency in Hz

ρ = resistivity in ohm-meters = $\frac{1}{\sigma}$

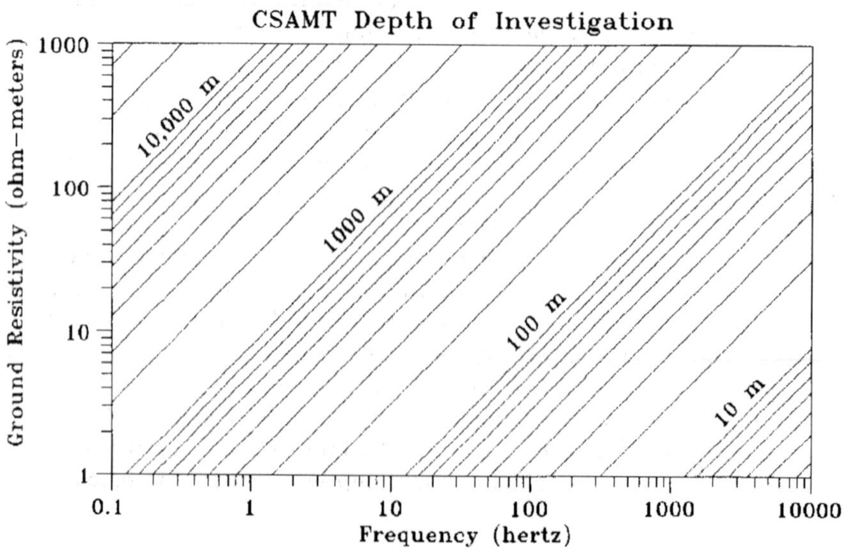

FIGURE 6.3
Effective investigation depth for CSAMT as a function of frequency and ground resistivity

Far-Field or Plane-Wave Zone

The electric and magnetic fields that are generated by driving current into the ground with the large transmitting dipole propagate along the surface of the ground and penetrate nearly vertically into the ground at distances beyond half a wavelength (approximately 3δ) from the transmitter. This zone of vertical penetration is called the far field, and in this area the electric and magnetic fields behave as plane waves, similar to natural source MT and AMT fields. Therefore, one can use the simplified MT and AMT equations for modeling structure in the far field.

Figures 6.4 and 6.5 are 2-D models showing the CSAMT far-field Cagniard resistivity and phase, and the E-field and H-field response when running across a vertical contact with the E-field oriented perpendicular (H-field parallel) to structural strike. Notice the sharp response to this contact of resistivity, phase and the electric field, and the null response of the magnetic field.

Now contrast this picture with figures 6.6 and 6.7 which show the results of running across the same formation with the E-field parallel (H-field perpendicular) to strike. Notice the broad response of the apparent resistivity to the contact for this configuration, as well as the more complex response of the phase and the diffuse response of the E-field. Notice that the H-field responds sharply to the vertical contact with its perpendicular orientation.

One can infer from the layered-earth models which follow and these two simple models that (1) the horizontal magnetic field responds primarily to changes in layering and is nearly oblivious to 2-D and 3-D features when oriented parallel to them, but the electric field responds sharply to lateral changes when oriented perpendicular to them, and (2); the response of the magnetic field when oriented perpendicular to strike, as in figure 6.7, will show a moderate lateral change at the interface and the electric field will vary smoothly across the interface.

The strong reaction of the electric field to contacts, when oriented perpendicular to strike, is the reason why it is usually best to run with the electric field perpendicular to strike when trying to get good lateral resolution. The inability of the magnetic field to delineate 2-D and 3-D features when oriented parallel to strike is the basis of the CSAET measurements mentioned previously.

One can also get some useful information from figure 6.6 on lateral effects in CSAMT measurements. Notice that when running a survey with the electric field oriented subparallel to an off-line fault, the resistivities at depth can be shifted to values either higher or lower than they actually are, depending upon the resistivity of the material on the far side of the fault. This is one of the pitfalls with acquiring data (actually any kind of data) on a single-line basis and then trying to perform meaningful interpretation from that single line without the advantage of additional input.

Before leaving the far-field regime, let us look at several other models and examples of data presentation. Figure 6.8 shows the 2-D model results of a 250- by 500-m prism of 5000 Ωm material buried 250 m below the surface in a uniform 100 m half-space. The E-field is oriented perpendicular to and the H-field parallel to strike for this example. Again, the apparent resistivity accurately resolves the lateral position of the prism. Notice the high-resistivity shadow extending to lower frequencies (lower depths) beneath the block.

FIGURE 6.4
Apparent resistivity (a) and phase (b) for a buried vertical contact with the E-field oriented perpendicular to strike

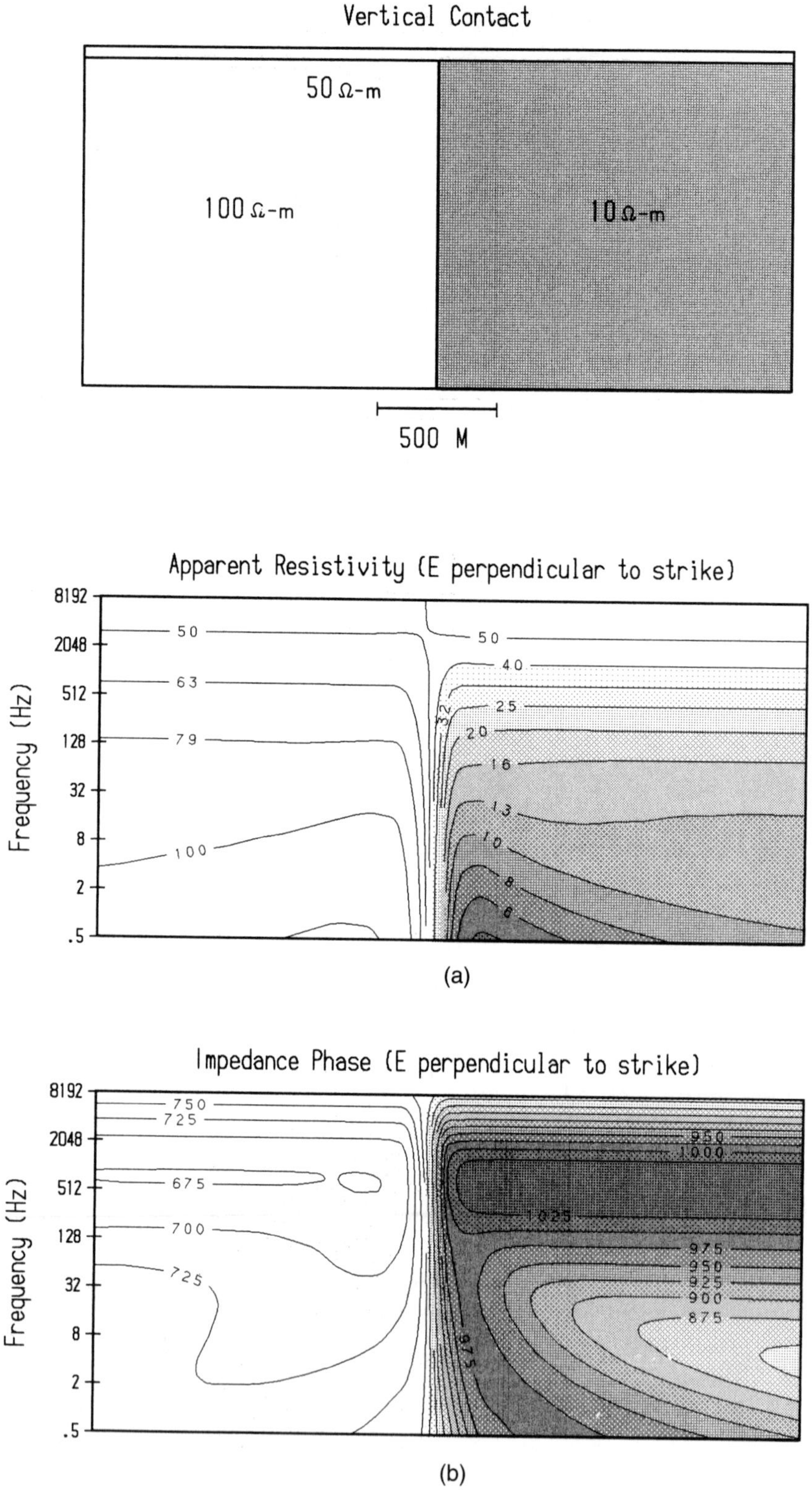

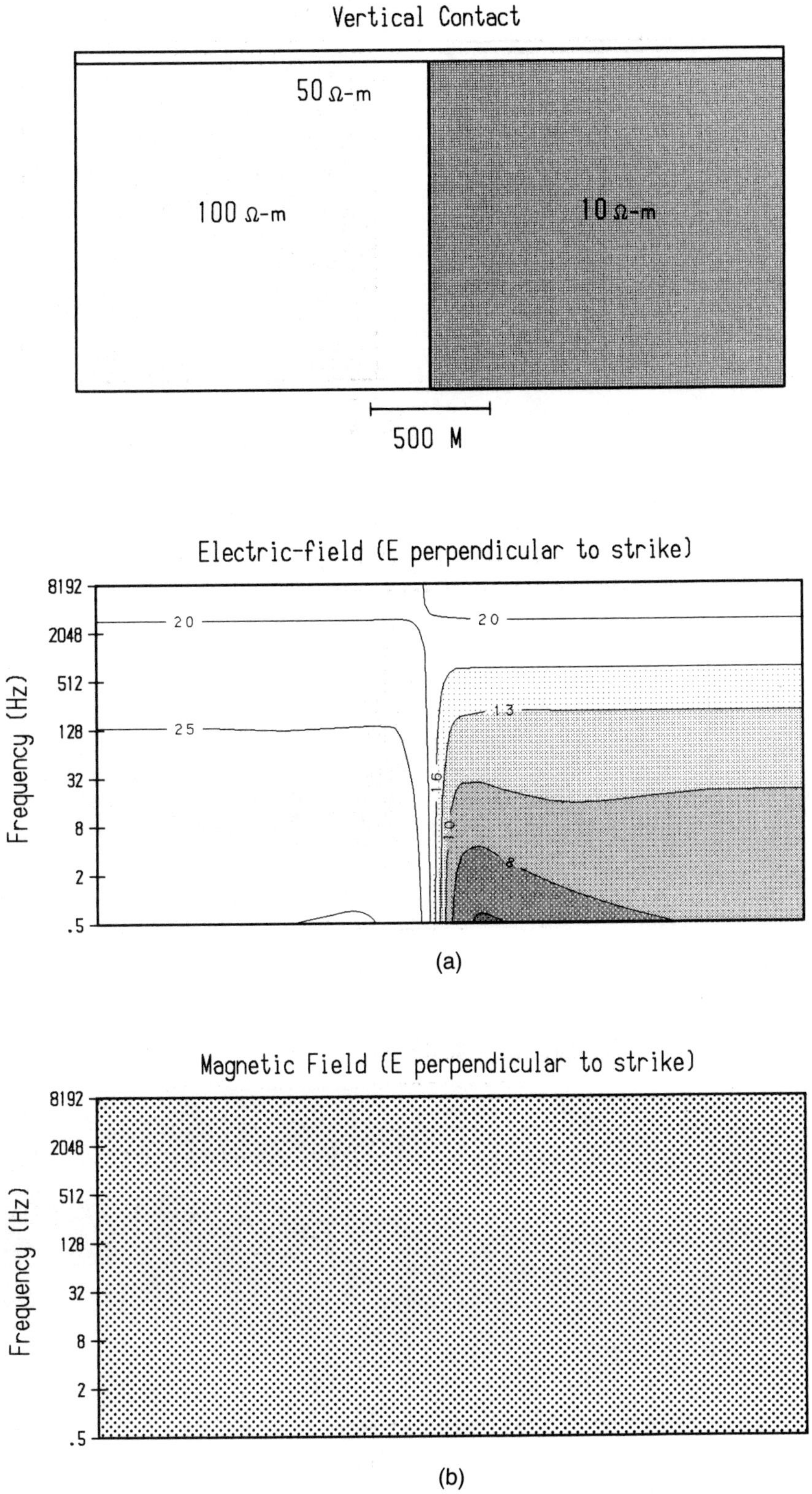

FIGURE 6.5
Electric (a) and magnetic (b) field responses for a buried vertical contact with the E-field oriented perpendicular to strike

FIGURE 6.6
Apparent resistivity (a) and phase (b) for a buried vertical contact with the E-field oriented parallel to strike

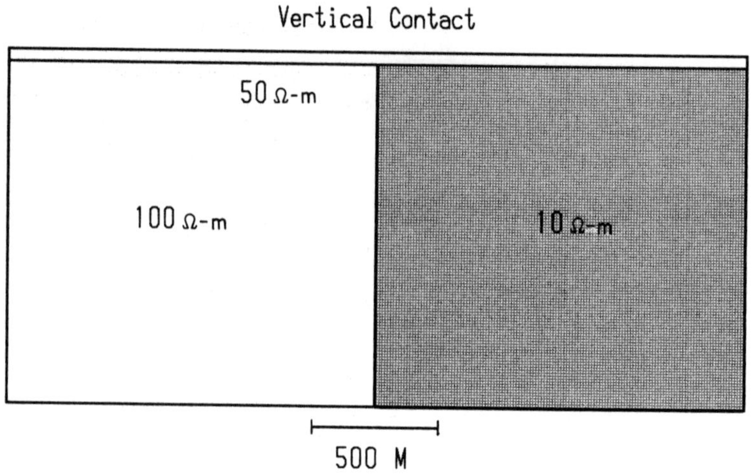

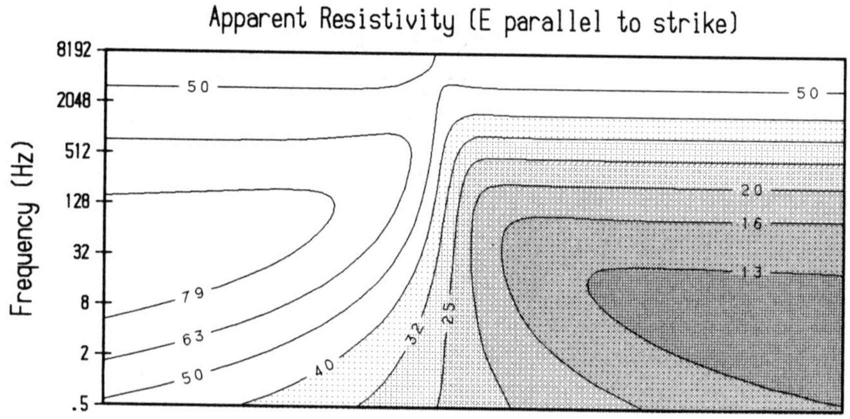

(a)

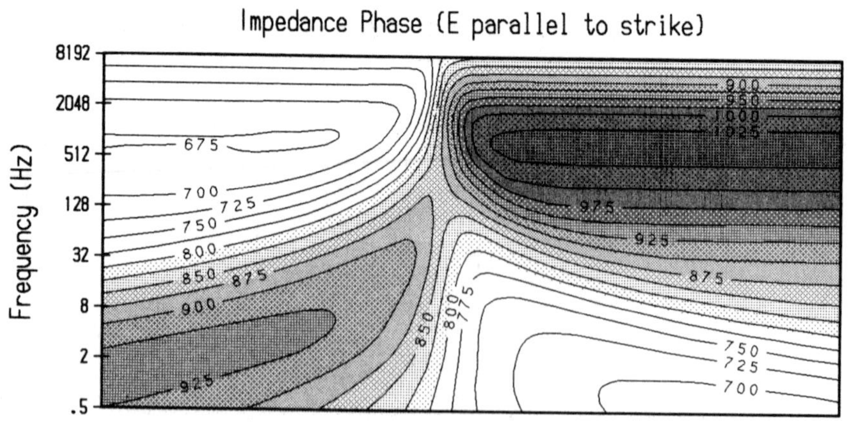

(b)

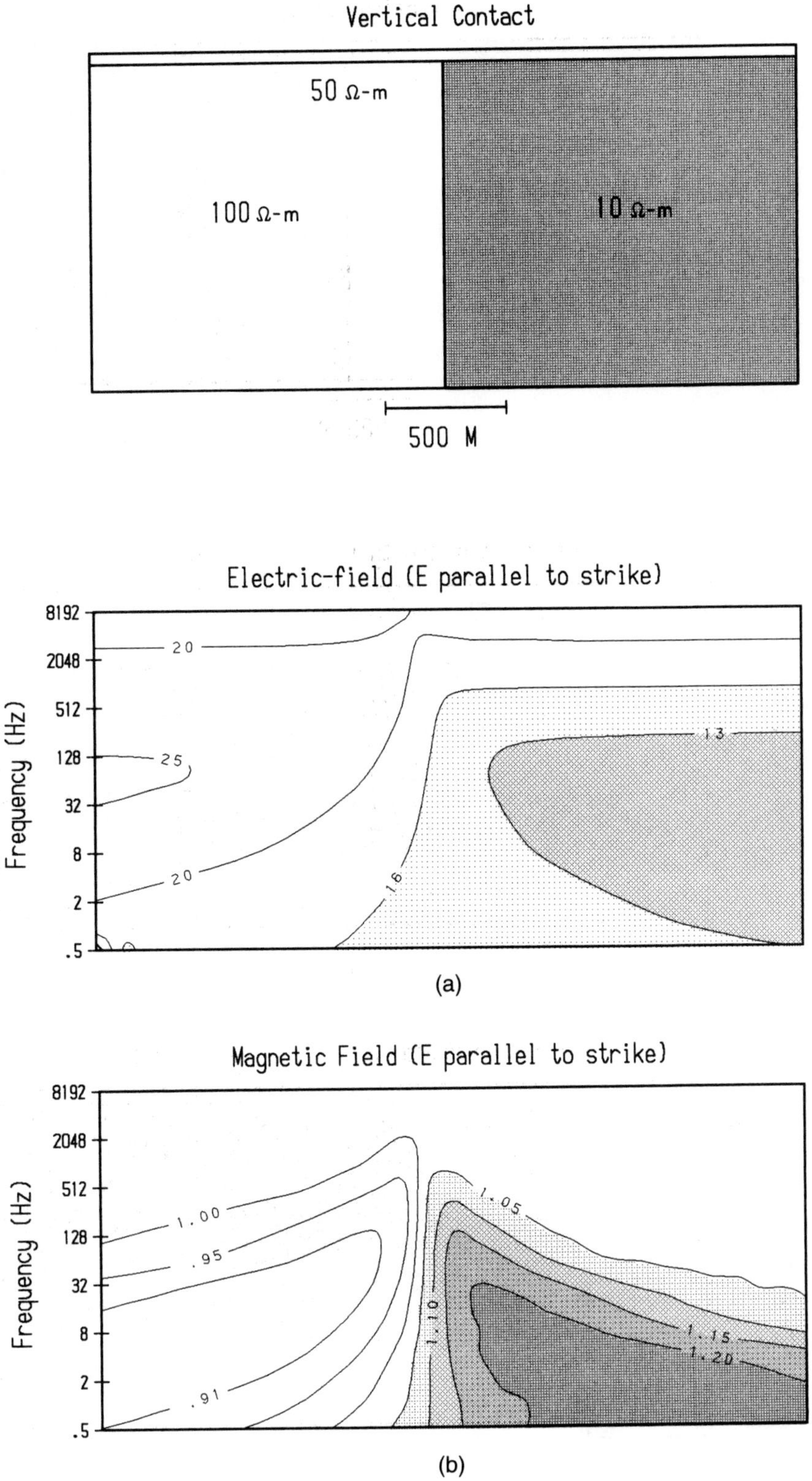

FIGURE 6.7
Electric (a) and magnetic (b) field responses for a buried vertical contact with the E-field oriented parallel to strike

FIGURE 6.8
Apparent resistivity (a) and phase (b) for a 5000Ωm prism buried 250 m in a 100Ωm half-space. E-field oriented perpendicular to strike

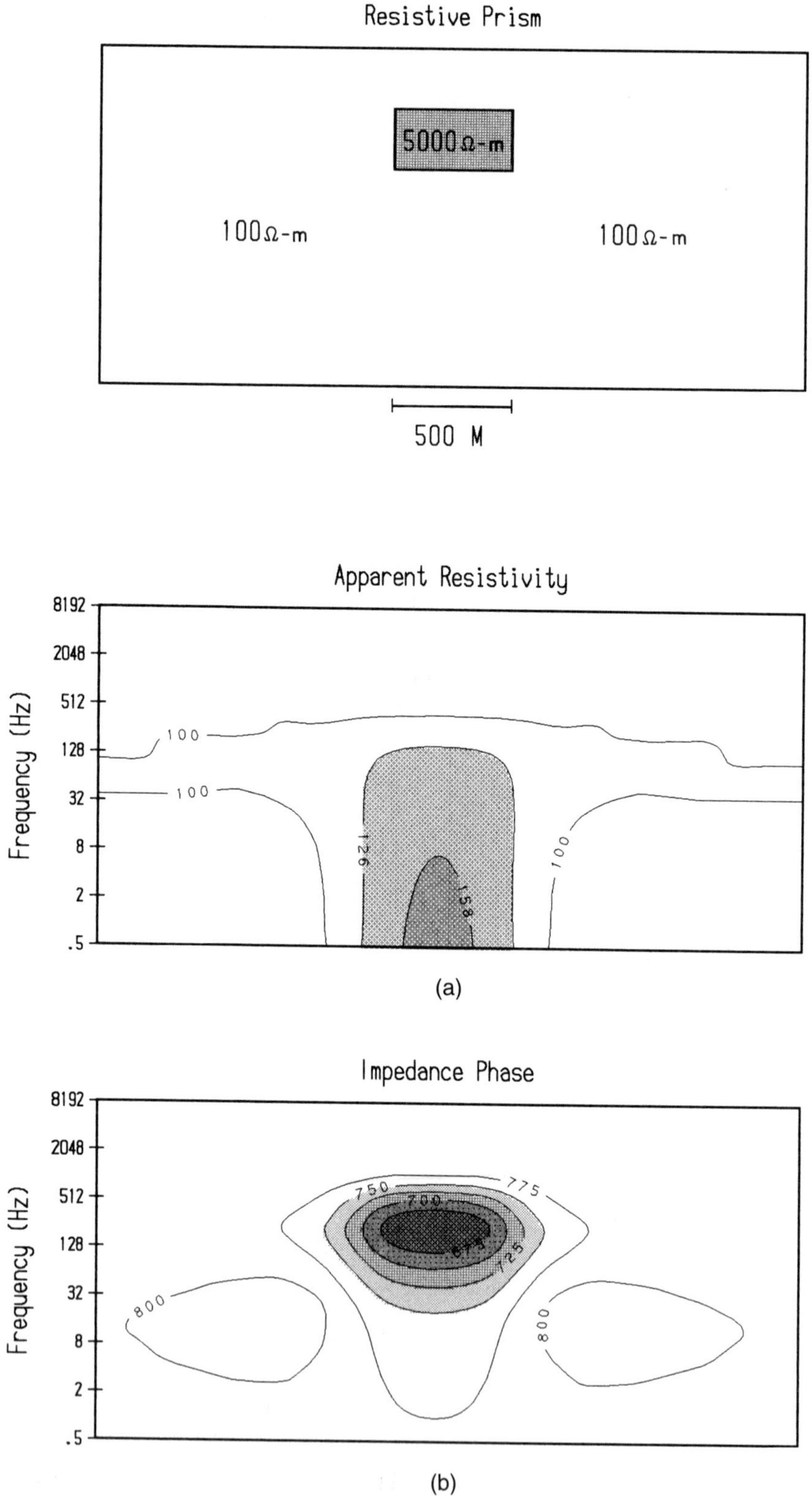

One would naturally ask if the bottom of the block can be resolved and the difference can be recognized between a near-surface prism and a vertical dike-like feature with infinite depth extent. In actual field practice, it appears that one can at least differentiate between a prism with limited depth extent and a dike-like feature. The depth to the top surface of both features can be determined, but measuring the actual depth extent of the bottom of the prism is usually not possible.

When dealing with a 1-D layered earth, the depth to the top and bottom of a conductive or resistive layer can be calculated; but when these features take on 2-D or 3-D characteristics, it becomes more difficult to determine their depth extent.

Various plotting techniques, modeling and inversion can often help in visual interpretation. To determine if this response can be detected as resulting from a prism rather than a dike-like feature by using simple modeling techniques, a smooth-model inversion was performed on the data. The results are displayed in figure 6.9, with depth along the vertical axis. Notice the high-resistivity bull's eye centered around the location of the prism, indicating limited depth extent. This figure was generated using a layered-earth inversion to approximate the 2-D environment.

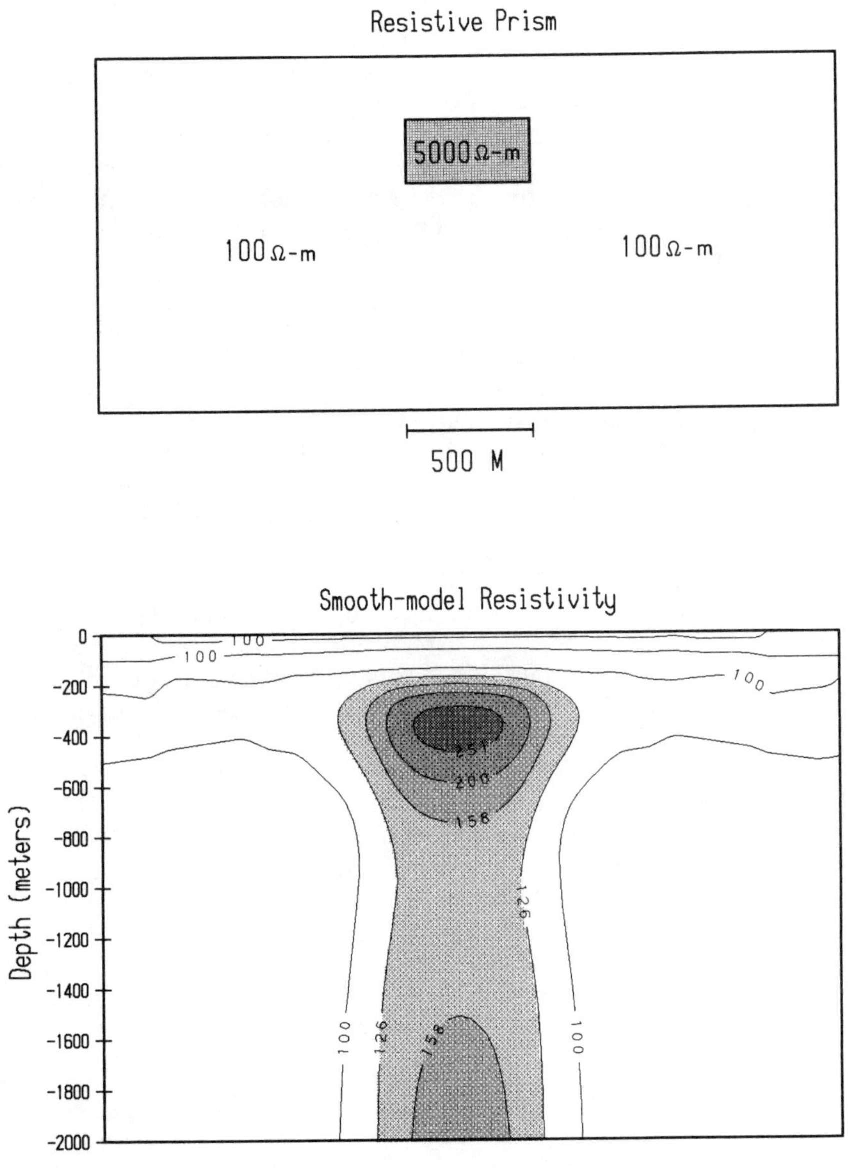

FIGURE 6.9
Smooth-model inversion results for data from figure 6.8, using layered-earth approximations for the 2-D model

Compare this data set with figure 6.10 which is the modeled data set from an infinite dike with the same lateral dimensions and same depth as the resistive prism. The resistivity value is reduced to 200Ω m to give the same modeled resistive intensity as figure 6.8. Notice the different shapes of the two models, giving more hope that differentiation between the two features could be obtained in field data. Figure 6.11 presents the smooth-model inversion of the data. Compare this resistive dike data with the response resulting from the resistive prism of figure 6.9. From these data sets, it appears possible to discriminate between near-surface prisms and buried dikes.

Figure 6.12 displays the response of a conductive prism of 10 Ωm imbedded 250 m beneath the surface in a uniform half-space of 100 Ω m material. Figure 6.13 presents the smooth-model inversion data showing a bull's eye centered about the location of the prism, indicating that the response can be detected as being from a buried prism. Actual field measurements with a controlled source usually details the 2-D or 3-D nature of these features.

Please note that whenever the term "smooth-model" is used in this text, it means approximating a multi-dimensional response with a series of layered-earth responses which are constrained to minimize the vertical variation in resistivity from one layer to the next — that is, to provide the smoothest model possible that will fit the field data.

Near-Field and Transition Zone

Up to this point, we have been discussing the results of modeling using a plane-wave source which is applicable to AMT, MT or far-field CSAMT measurements. The types of fields generated by a controlled source at a fixed distance from the measuring site must also be reviewed. First, review the 3-D modeling comparison between plane-wave and controlled sources. Figure 6.14 depicts a model used to compare MT and CSAMT results (Butterworth, 1988). This is a 3-D version of the model in figure 6.12, with the E-field oriented parallel to the longest dimension of the conductive block. (In normal field practice, one would orient the E-field to be perpendicular to the longest dimension, as in figure 6.12).

Figures 6.15 and 6.16 show the apparent resistivity and phase measurement results over a conductive block using a plane-wave (MT) source. Notice the conductive shadow beneath the block extending to the lowest frequency measured. This response is similar to that observed in figure 6.12.

Figures 6.17 and 6.18 show the results of making measurements with the source placed at station 0 to simulate the CSAMT configuration. Notice the limited depth extent of the conductive response of the block compared to figure 6.15. This is mainly a result of the changing E- and H-field characteristics encountered at the lower frequencies which control the depth of investigation.

When the electrical distance between the transmitter and receiver becomes less than 3 δ (about 1/2 wavelength - see table 6.1) at the frequency being measured, the electric and magnetic fields change gradually from plane-wave to curved, and the Cagniard resistivity formula no longer provides realistic apparent resistivity values. When operating in the far-field or plane-wave zone over a layered earth, both E- and H-fields drop off as $1/r^3$, where r is the separation between the transmitter and receiver, and both fields vary as a

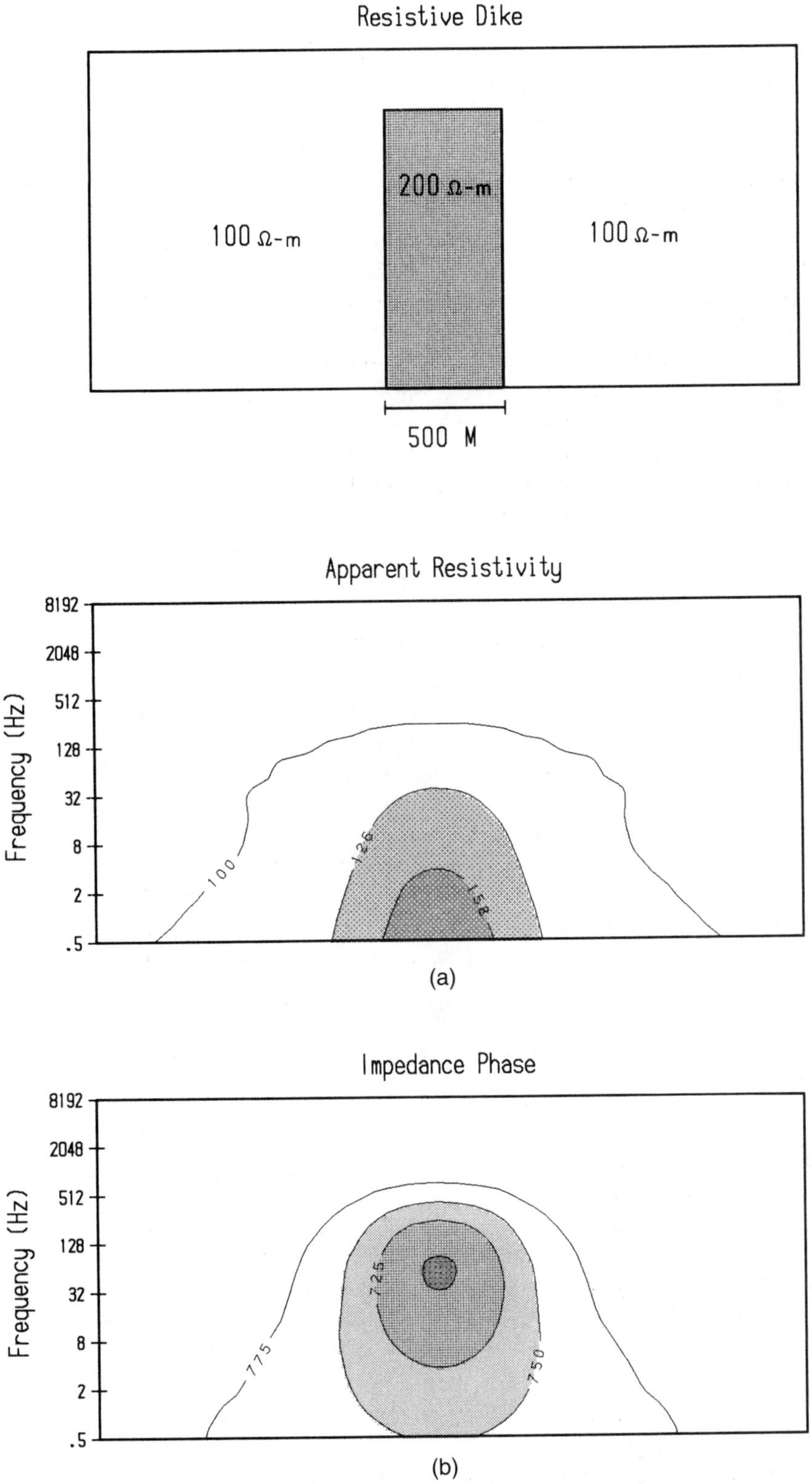

FIGURE 6.10
Apparent resistivity (a) and phase (b) for a 200 Ωm dike buried 250 m in a 100Ωm half-space. E-field oriented perpendicular to strike

FIGURE 6.11
Smooth-model results for data from figure 6.10

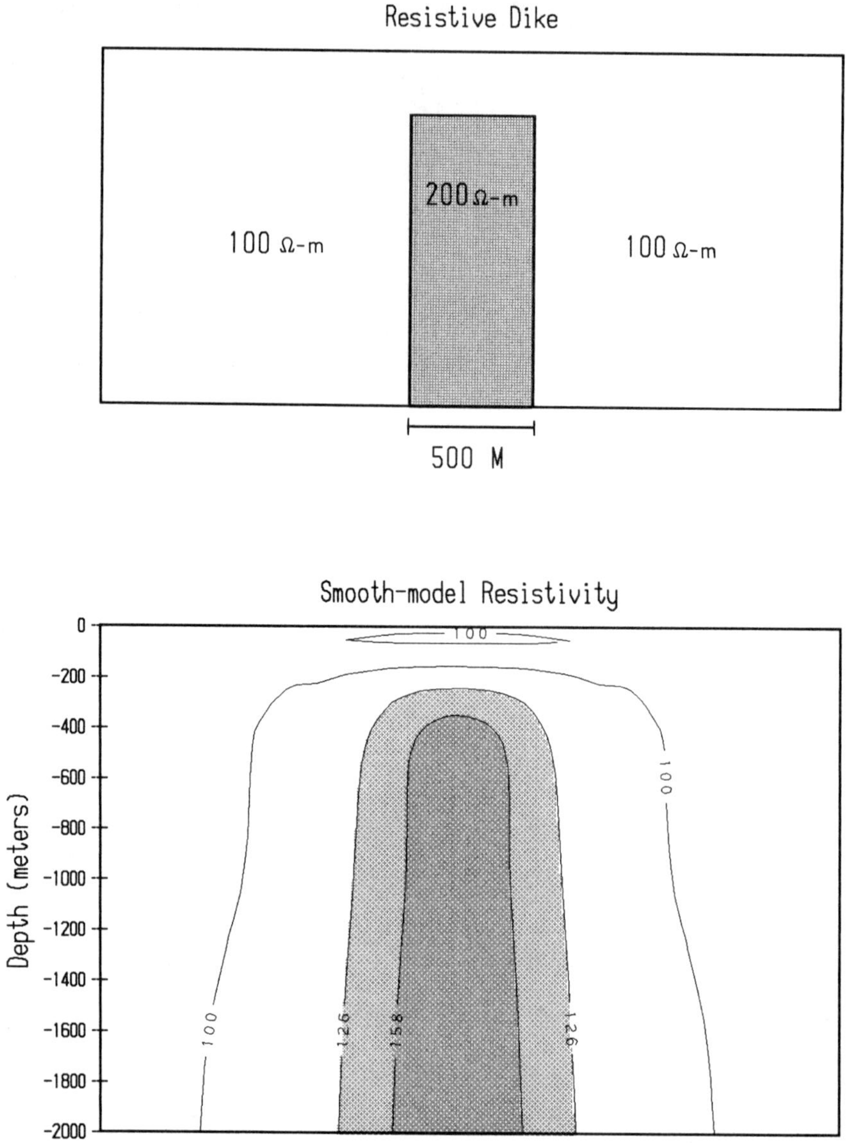

function of frequency and earth resistivity. In the transition zone, the H-field drop off begins to change to $1/r^2$, and the dependency on frequency and earth resistivity begins to change as well. The E-field continues to drop off as $1/r^3$ and retains its function of earth resistivity, but its dependency on frequency also begins to change.

In the near field, the H-field decays at $1/r^2$, becomes saturated and no longer varies as a function of frequency or resistivity. The E-field still remains a function of resistivity, decays at $1/r^3$ but is independent of frequency. It is in this near-field zone that depth of investigation becomes independent of frequency and dependent upon array geometry. This is the condition that actually puts a practical limit on the depth of exploration for CSAMT soundings.

For consistency, the Cagniard resistivity calculation is used through the transition zone and into the near field, although the values calculated in these areas do not reflect the actual resistivity values of the earth.

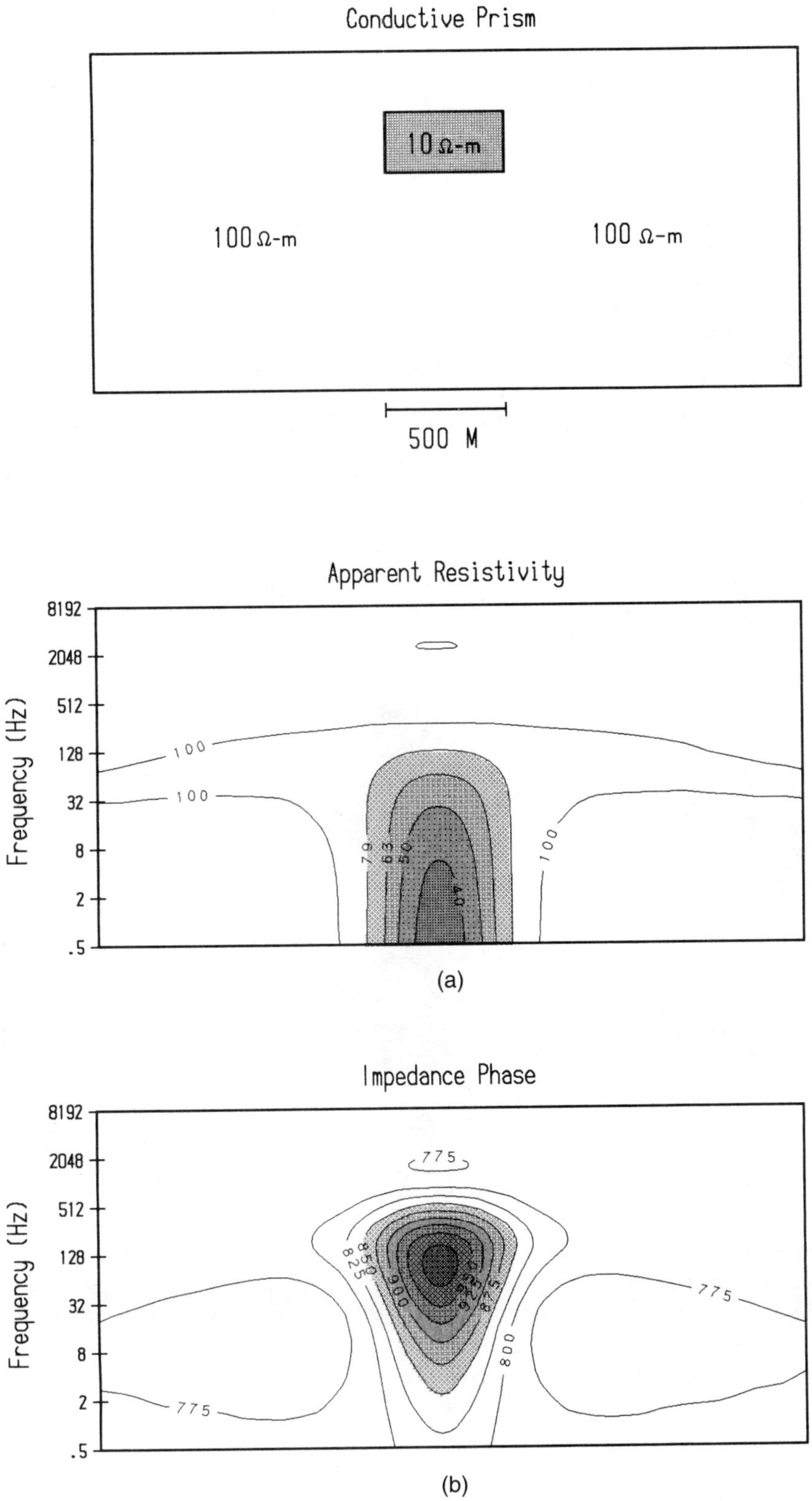

FIGURE 6.12
Apparent resistivity for a 10Ωm prism buried 250 m in a 100Ωm halfspace. E-field oriented perpendicular to strike

FIGURE 6.13
Smooth-model results for data from figure 6.12

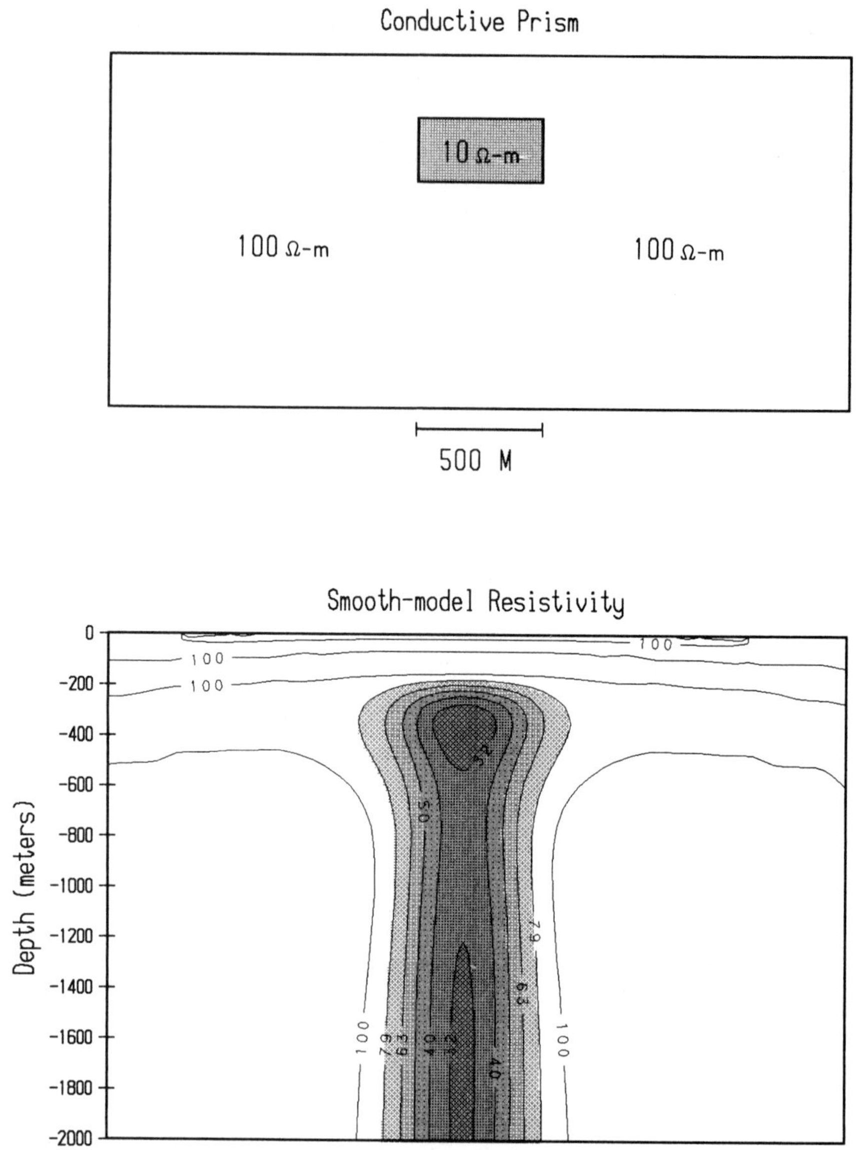

Please refer to figure 6.19. Here, the data are displayed in what is called a "log-log" format -- that is, log resistivity versus log frequency, low frequencies (greater depth) to the left, high frequencies (closer to the surface) to the right. This plot gives a comparison between the responses resulting from a plane-wave, natural source, and a controlled source placed 8 km from the receiver site over a layered earth, showing the CSAMT far field (64 to 4096 Hz), transition zone (8 to 64 Hz) and near-field areas (below 8 Hz). The transition zone can take on various forms, as will be seen in following examples, and the near-field curve will approach a 45-degree slope when in full saturation at the lowest frequencies. This apparent increase of resistivity in the near field always forces high resistivities at depth, but these values are just artifacts of the calculation procedure.

The next six figures will display the four components of electromagnetic measurements — apparent resistivity, phase, E-field and H-field -- and demonstrate how these parameters change between the far field, transition zone and the near-field as transmitter/receiver (Rx/Tx) distance, resistivity and layering are made to vary.

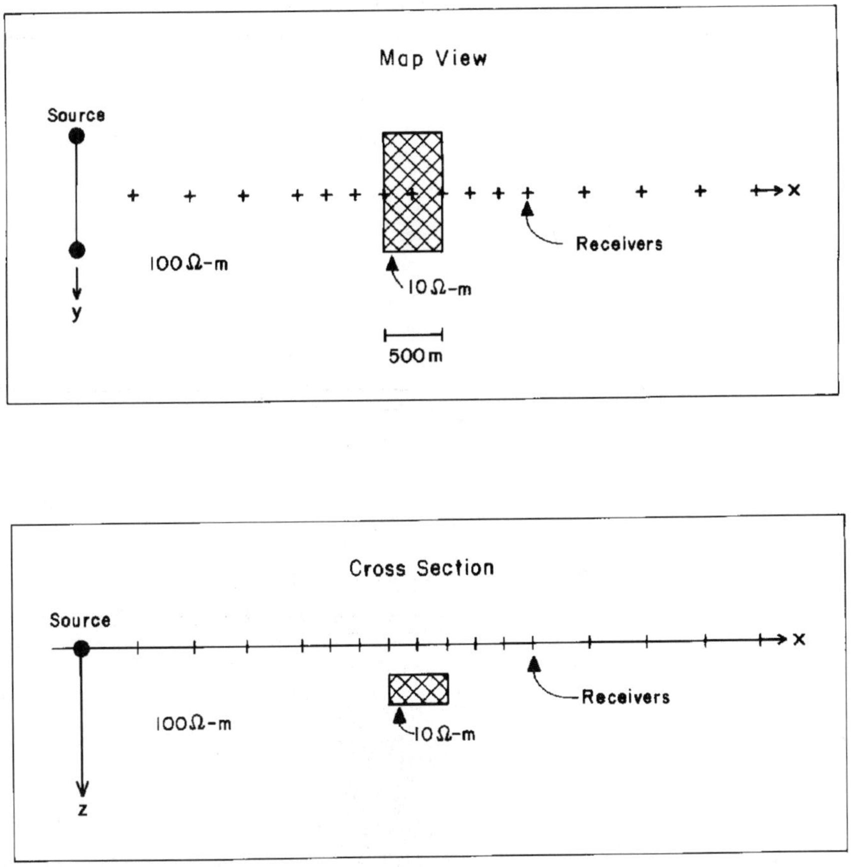

FIGURE 6.14
Model configuration for comparing plane-wave and controlled source measurements over a 10 m block buried 250 m in a 100 m half-space (from Butterworth, 1988)

FIGURE 6.15
Apparent resistivity pseudosection over a conductive block in a uniform background with a normally incident plane-wave. (from Butterworth, 1988)

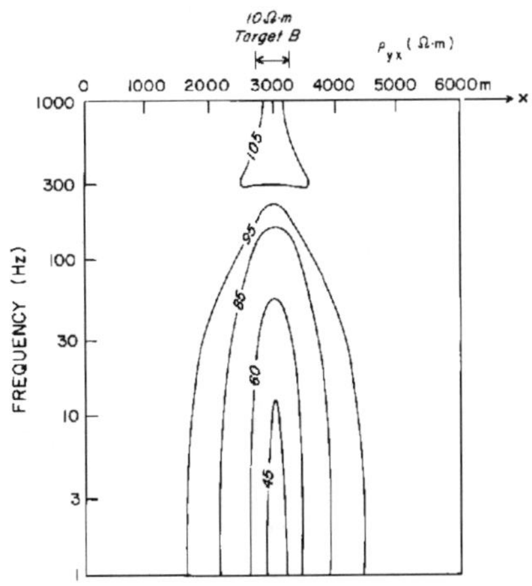

Figure 6.20 shows the behavior of these components over a 100 Ω m homogeneous half-space, for receiver/transmitter separations of 1-, 3- and 10-km. Notice how the apparent resistivity becomes artificially high at low frequencies, even in a homogeneous half-space, when using a finite separation between transmitter and receiver. The apparent resistivity enters a broad transition zone and near field at relatively high frequencies for close Rx/Tx spacing, whereas for a 10 km spacing, the resistivities are constant down to about 4 Hz. Also, notice how the magnetic fields (d) flatten out and saturate at the lower frequencies for the 1- and 3-km separations.

Before going further, examine the values of the phase (or phase difference or impedance phase -- they are all the same) in figure 6.20b. Notice in the far field (high frequencies), the phase response is a constant 785 milliradians (45 degrees). This is the phase response of a homogeneous half-space. We can see here, and in the examples that follow, that the phase difference is proportional to the slope of the resistivity. Actually, it is proportional to the derivative of the log of resistivity versus the log of frequency. The significance of this relationship is that given a single measurement of resistivity and phase, one can tell from that single measurement whether the material at a lower frequency (lower depth) or higher frequency (shallower depth) is increasing or decreasing in resistivity.

FIGURE 6.16
Phase pseudosection over a conductive block in a uniform background with a normally incident plane-wave (from Butterworth, 1988)

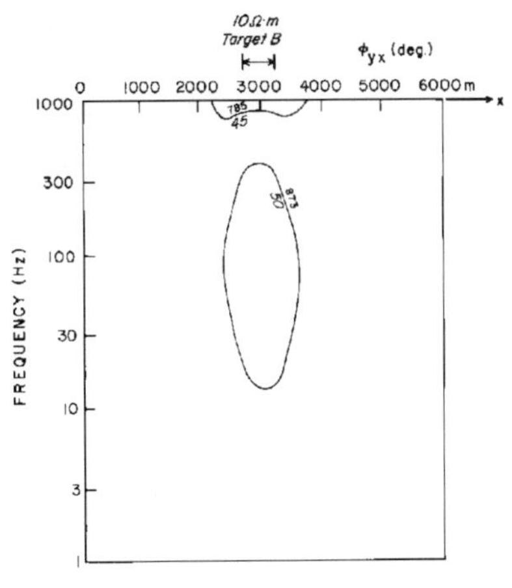

FIGURE 6.17
Apparent resistivity pseudosection over a conductive block in a uniform background with a controlled source located at station 0. (from Butterworth, 1988)

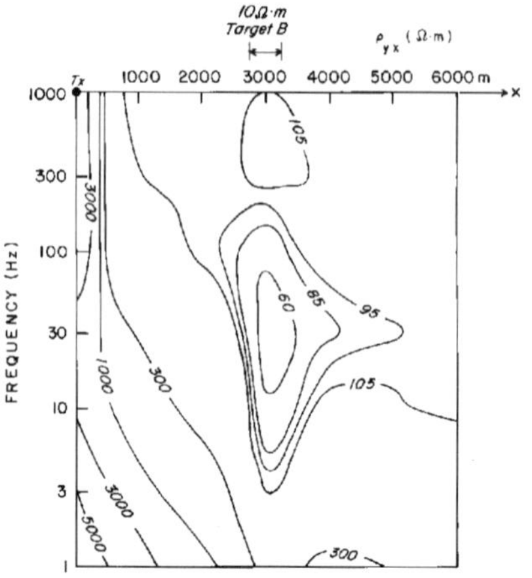

FIGURE 6.18
Phase pseudosection over a conductive block in a uniform background with a controlled source located at station 0. (from Butterworth, 1988)

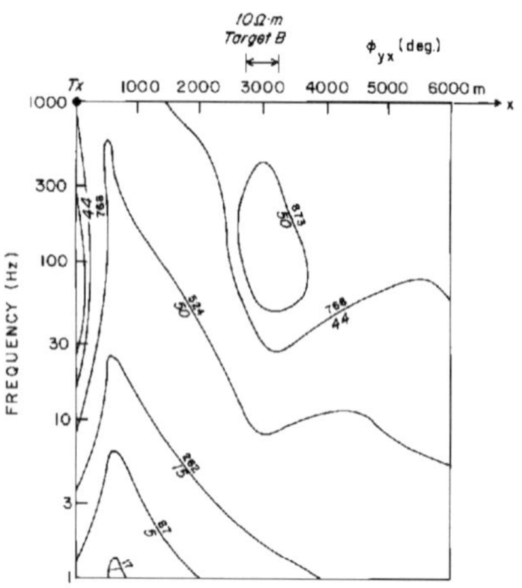

Following are some general rules for phase interpretation:

FAR FIELD PHASE VALUES:

- 785 milliradians - homogeneous half-space response
- greater than 785 milliradians - indicates going from high to low resistivity layering with depth
- less than 785 milliradians - indicates going from low to high resistivity layering with depth.

TRANSITION ZONE AND BEGINNING OF NEAR FIELD:

- greater than 0 mr - indicates basement with a low resistivity contrast (<1:10)
- less than 0 mr - indicates basement with a high resistivity contrast (>1:100)

NEAR FIELD: All values tend toward 0 milliradians

Now, let us return to the modeled E- and H-field plots in figure 6.20. Notice the effect of the transition zone and near field in the field component data. In figure 6.20c, these areas are clearly seen in the 1 km Rx/Tx separation data. The E-field data goes through a level change (transition zone) between 8 and 128 Hz to a low-frequency (near-field) value of one-half of the value at the highest frequencies (far field). The magnetic field, shown in figure 6.20d, goes from a sloping response at the high frequencies to a flat response at low frequencies. The flat area is the near field, the sloped area the far field, and the area of transition between the two is the transition zone. If one compares all four plots, it will be seen that these near-field/far-field/transition characteristics are consistent with frequency between the corresponding curves.

Figure 6.21 shows the effects of varying the ground resistivity while keeping the receiver/transmitter separation fixed at 10 km. As predicted by the skin-depth formula (table 6.1), the higher the resistivity, the quicker the soundings get into the near field. Or, the higher the resistivity, the shorter the electrical distance between the transmitter and receiver. Notice that the form of these curves -- apparent resistivity, phase, E-field and H-field -- are virtually identical to those in figure 6.20, just shifted up or down in magnitude except for phase which is basically the same in both figures. Notice how all of the H-field curves converge at low frequency at the same value, regardless of ground resistivity, indicating that the H-field is not a function of resistivity once it gets well into the near field. The E-field, however, still varies with ground resistivity when in the near field.

Figure 6.22 is a two-layer model, 10 Ωm over 100 Ωm, with the transmitter located 10-km distant from the receiver, configured to show the effect of a conductive surface layer. The top layer has thicknesses of 1, 3, 10, 32 and 100 m. Notice that all curves are tending toward convergence at the lowest frequencies in the 100 Ωm material and begin to enter the near-field/transition zone around 8 Hz. The phase data changes and begins to go toward zero at 8 Hz, also. In the apparent resistivity plots, the influence of the 3 m layer is detected at the highest frequencies, whereas the 32 m layer is seen above 2048 Hz, and all frequencies above 256 Hz are in the 100 m layer. This can be seen in all of the other components, including the phase which shows a constant 785 mr phase shift above 512 Hz for the 100 m layer (the high frequencies cannot see through the low-resistivity layer), and the phase is just beginning to nudge below 785 mr below 4096 Hz in the 32 m layer. The H-field in the far field (above 8 Hz in this model) will follow a 27 degree slope in a homogeneous half-space.

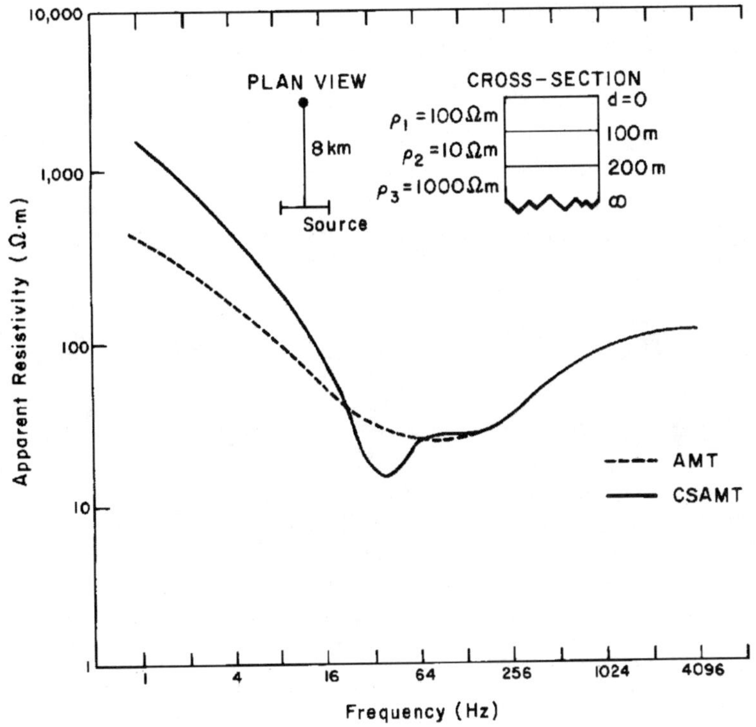

FIGURE 6.19
Comparison between measurements with a plane-wave source and a controlled source over a 3-layer earth. The far-field (plane-wave zone), transition zone and near-field areas are shown. Notice how the curves match in the far-field

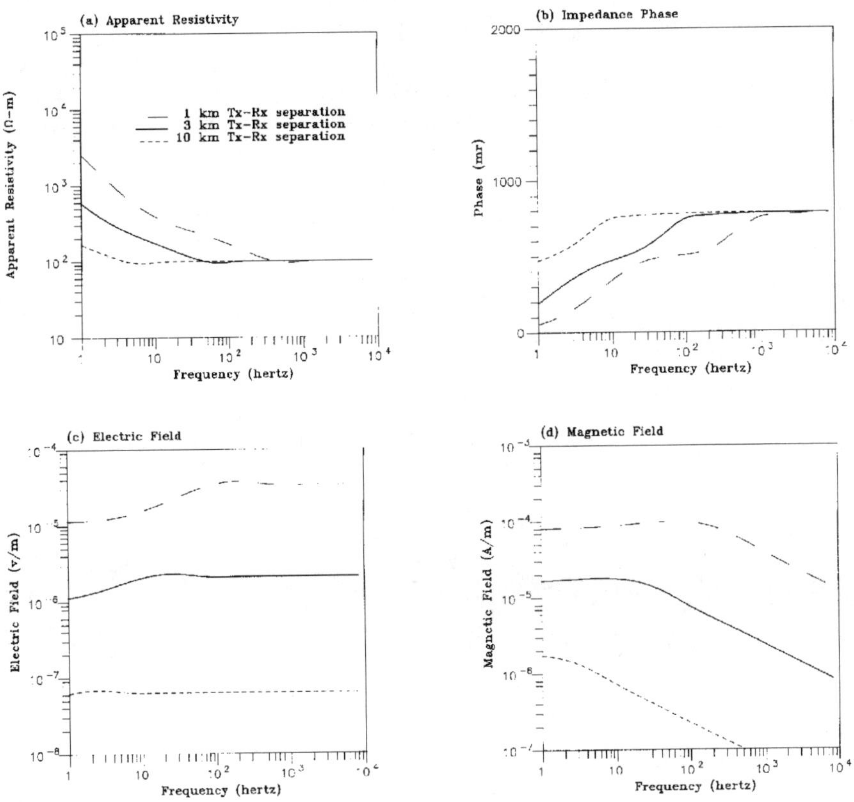

FIGURE 6.20
Effect of separation between transmitter and receiver on (a) apparent resistivity, (b) phase, (c) E-field and (d) H-field over a 100Ω m homogeneous earth

FIGURE 6.21
Effect of ground resistivity on (a) apparent resistivity, (b) phase, (c) E-field and (d) H-field over a homogeneous earth

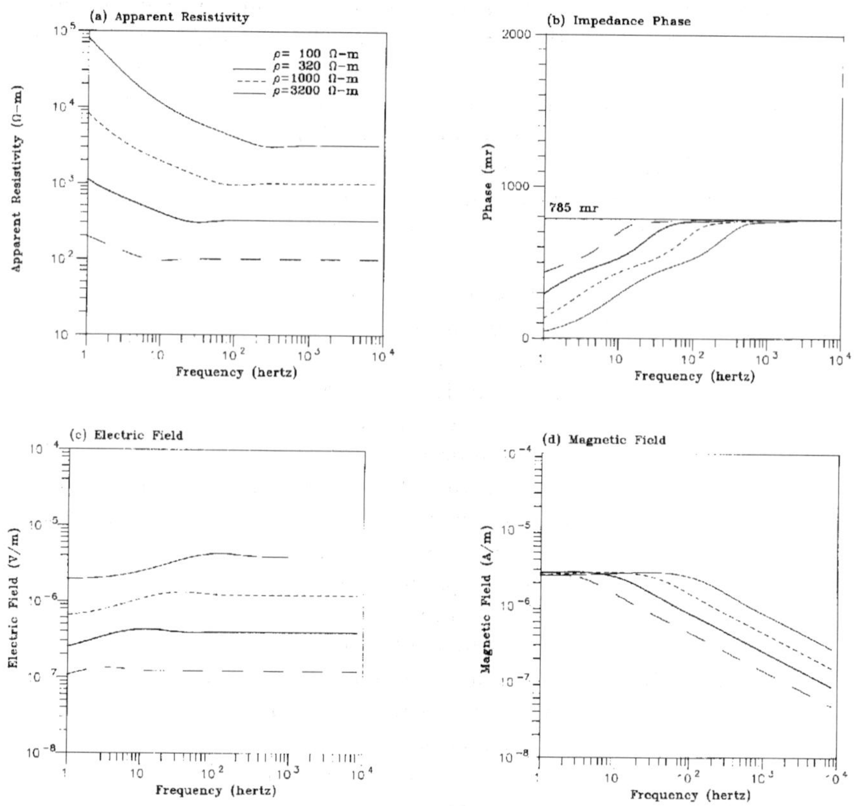

FIGURE 6.22
Effect of the thickness of a conductive surface layer on (a) apparent resistivity, (b) phase, (c) E-field and (d) H-field

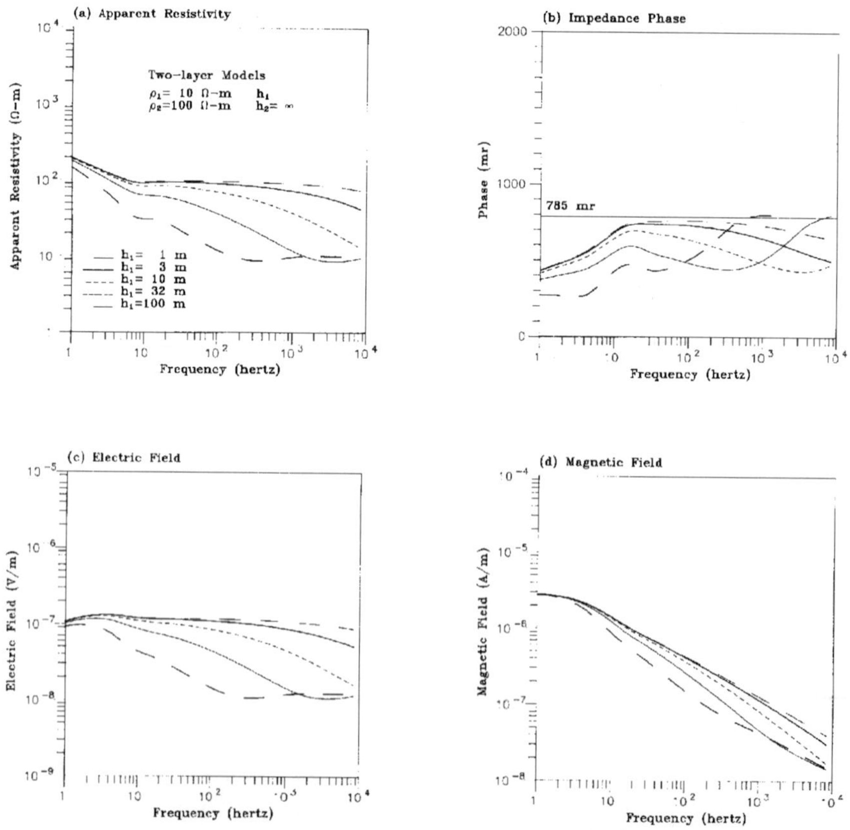

In going from high to low frequencies (near-surface to depth), if the slope decreases, it means the presence of a more conductive layer. If the slope increases, a more resistive layer is being entered. In this example, the curve for the 1 m thick layer can be used as a 27 degree reference line at all but the highest frequencies. The curve for the 100-m layer is subparallel to this line only for the highest frequencies and then begins to increase in slope — indicating contact with a higher resistivity layer at depth. Notice also that the curves for the 3-, 10- and 32-m layers all indicate a transition into a higher resistivity medium at depth with little information about the resistivity of the surface layer.

Figure 6.23 demonstrates the effect of varying surface resistivity while holding the thickness of the surface layer constant at 10 m with the transmitter located 10 km from the receiver. The surface-layer resistivity varies between 1, 10, 100 and 1000 Ω m, with the lower layer held constant at 100 Ω m. The 100 Ω m surface layer is included to permit comparison with a homogeneous earth response, and the 10 Ω m layer is identical to the 10 m thick 10 Ω m layer in figure 6.22. The near-field/transition zone begins around 8 Hz, as can be seen in the apparent resistivity, phase, E-field and H-field curves. All curves are converging at the lower frequencies which are sampling the 100 Ω m lower layer. Notice in the apparent resistivity data that it takes a large frequency range to see through the 1 Ω m layer, although the bottom 100 Ω m layer is sensed as early as 4096 Hz, since the phase response is less than 785 mr at this frequency and decreases steadily as one moves to lower frequencies.

Figure 6.24 displays the effect of a conductive mid-layer in a three-layer model. The top layer is held at 100 Ω m and is 100 m thick, while the middle layer is 10 Ω m and varies in thickness between 1, 10, 100 and 1000 m. The bottom layer is 100 Ω m, and the transmitter is

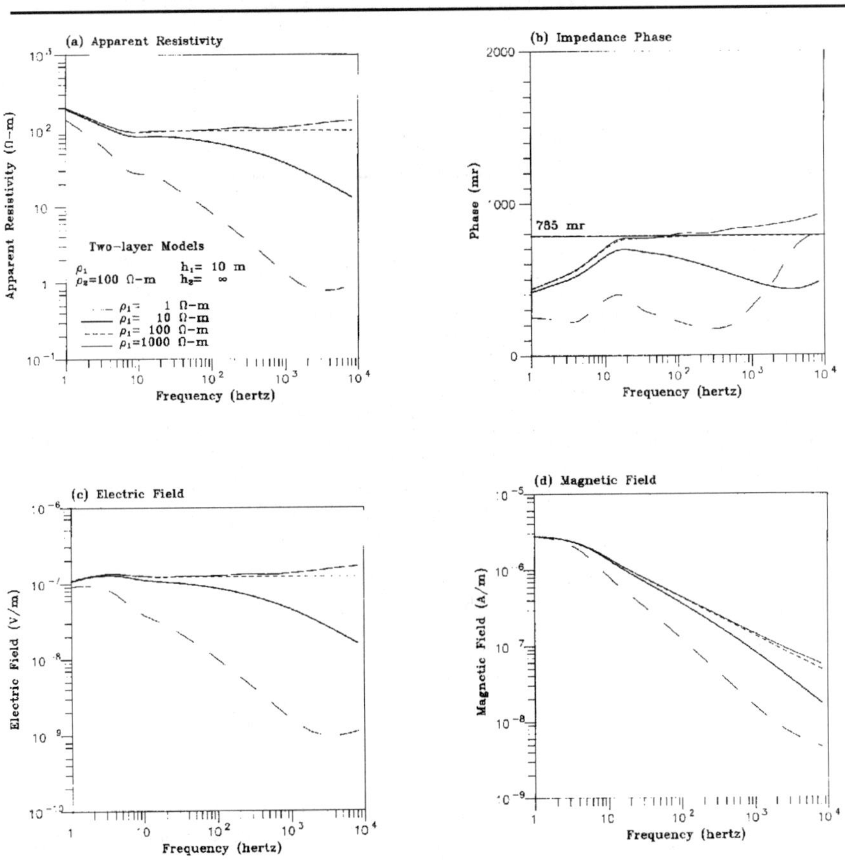

FIGURE 6.23
Effect of varying the resistivity of a 10 meter thick surface layer on (a) apparent resistivity, (b) phase, (c) E-field and (d) H-field

is located 10 km from the receiver. Note that the central 1 m layer is not evidenced in any of the data, the 10 m layer is barely resolved, the 100 m layer is fully resolved, and the 1000 m thick layer influences most of the data. The bottom of the 1000 m layer is evidenced by the sharp dip in the apparent resistivity and E-field curves at 2 Hz. This notch in the transition zone indicates that a high resistivity interface occurs at depth.

FIGURE 6.24
Effect of a conductive mid-layer with variable thickness on (a) apparent resistivity, (b) phase, (c) E-field and (d) H-field

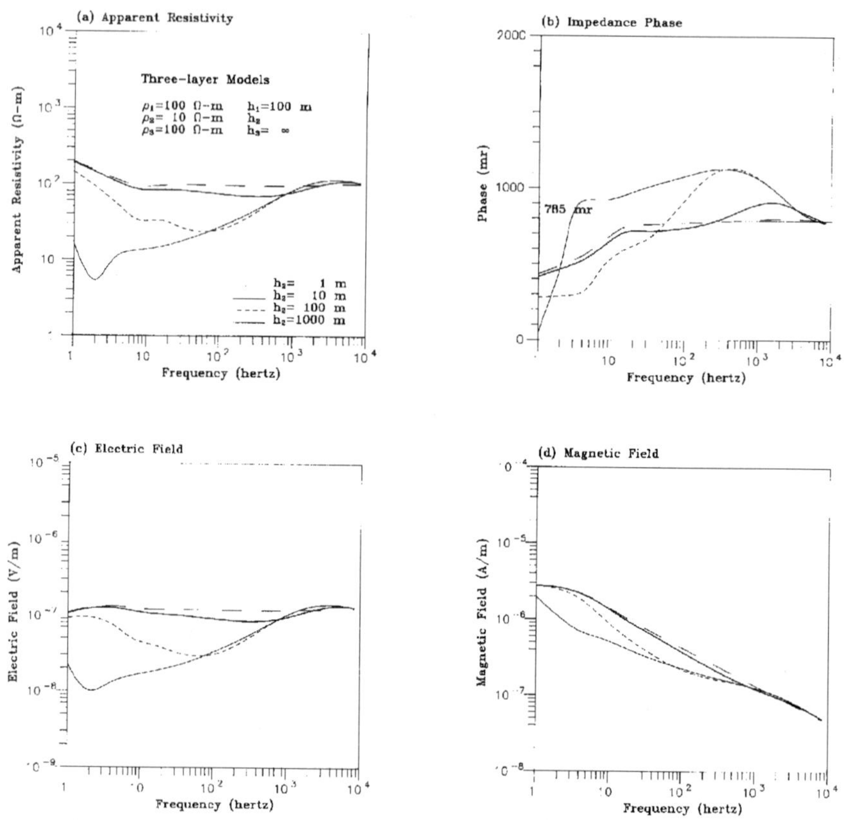

Another way of looking at this in terms of data progression from high-to-low frequencies is: the 10 m layer is weakly observed in all components; the 100 m layer is entered into and passed thru; the 1000-m layer is entered into and exited at the transition zone. A pronounced notch in the transition zone data is always produced when the electrical depth to a resistive layer coincides with the transition zone. Notice that the phase response at 1 Hz for the 1000 m layer is headed very sharply toward zero. Continuing the model to lower frequencies, one would find that the phase would go negative. Negative phases are often observed in data where the basement is highly resistive. This is the only time when negative phases are generated in a truly one-dimensional layered earth.

Figure 6.25 demonstrates the effect of a 100 m thick mid-layer with resistivities that vary from 1 Ω m to 1000 Ω m. As to be expected, the curves are less responsive to resistive layers as compared to conductive layers. Notice that the phase response for the 1 Ω m layer goes negative at the lowest frequency. This is the result of the extreme contrast in resistivities (1:100) at the lower interface.

Compare figures 6.22 through 6.25, and notice that all of the examples break into the near field/transition zone at approximately the same frequency, 4 to 8 Hz. This is because the perturbations have been relatively small and have not affected the depth of exploration to a great extent, and the receiver/ transmitter separation was held constant.

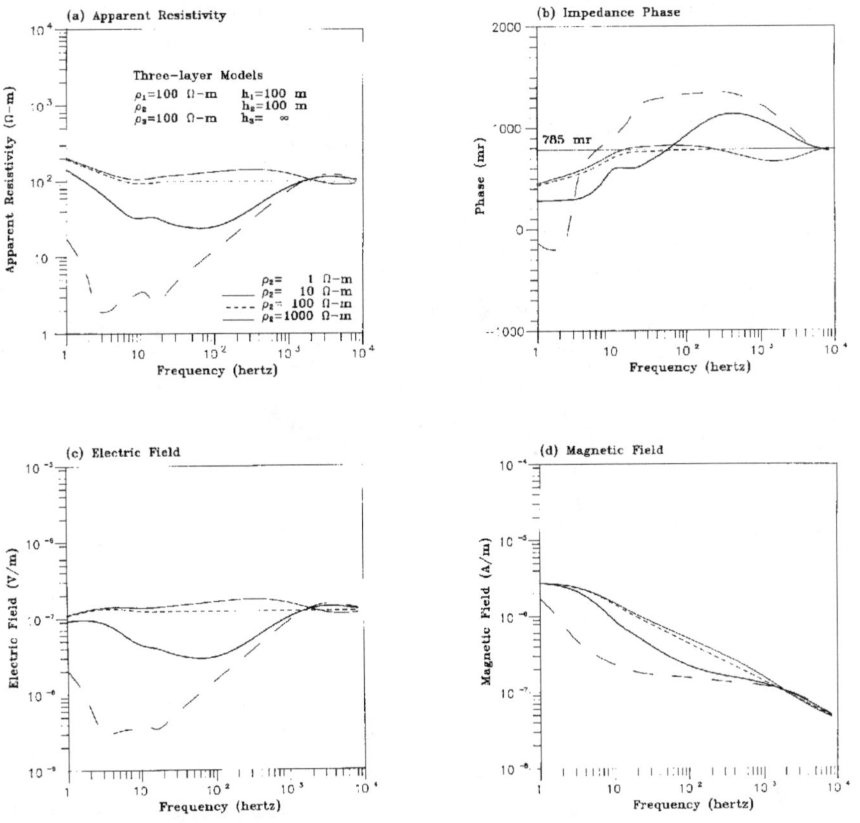

FIGURE 6.25
Effect of a 100 m thick mid-layer with variable resistivity on (a) apparent resistivity, (b) phase, (c) E-field and (d) H-field

Figure 6.26 shows the results of varying the Rx/Tx separation over a three layer earth and the corresponding changes in transition zone activity. Notice how the intensity of the transition zone notch varies as it passes through the effects of the central layer. This effect is what makes interpretation of CSAMT data with MT programs impossible.

So far, only model data have been reviewed to get a feel for the basics of CSAMT. We will now get more involved with field data and data interpretation. Before doing this, review some of the plotting methods that will be used. Figure 6.27 shows several of the basic methods. Figure 6.27(a) is the individual station plot called the log-log plot for modeled data. Figure 6.27(b) is the log-frequency pseudosection used in the 2-D models. Figure 6.27(c) is a pseudo-depth section obtained by plotting the frequency axis as a square root function, since this is proportional to depth of exploration (table 6.1). Notice how all of the high-frequency values are crunched together near the surface. Figure 6.27(d) is a calculated depth section based on the depth of exploration formula using the square root of measured resistivity divided by frequency. For slowly varying resistivities, there is often little difference between (c) and (d). Compare (b) and (d), however, and notice that depth section is much easier to interpret as far as depth to target is concerned. Pseudosections similar to figure 6.27b with the resistivity or phase values plotted at each frequency level are the easiest to use for data quality checking.

FIGURE 6.26
Results of varying the Rx/Tx separation over a 3-layer earth and the corresponding changes in transition zone activity

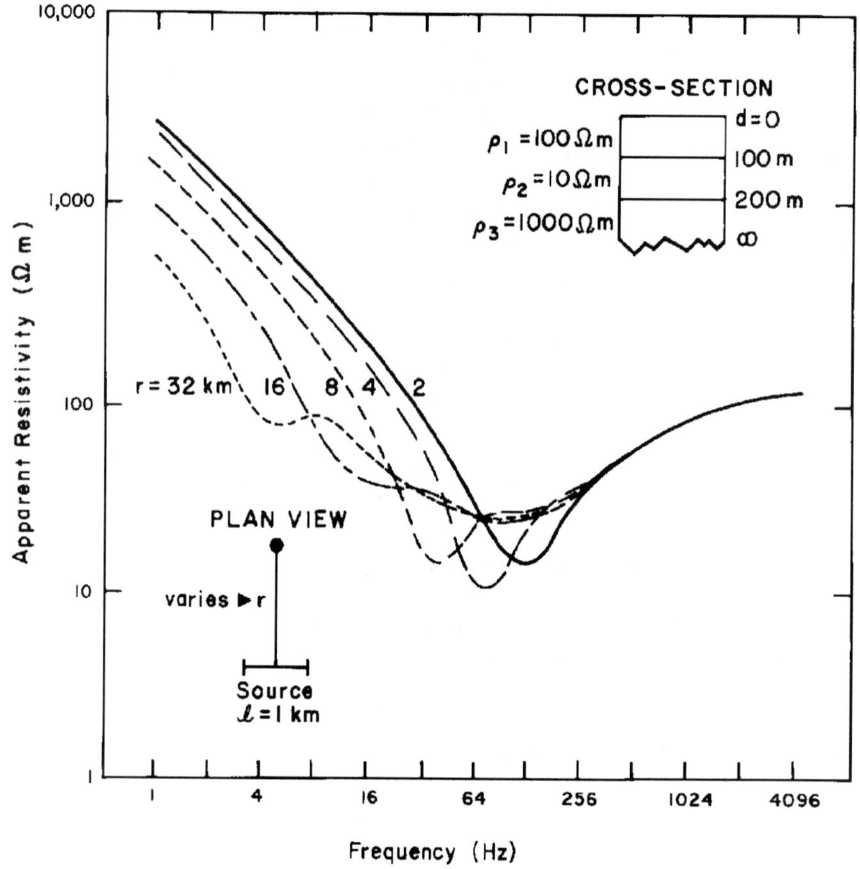

FIGURE 6.27
Some plotting conventions for CSAMT data. See the text for explanation

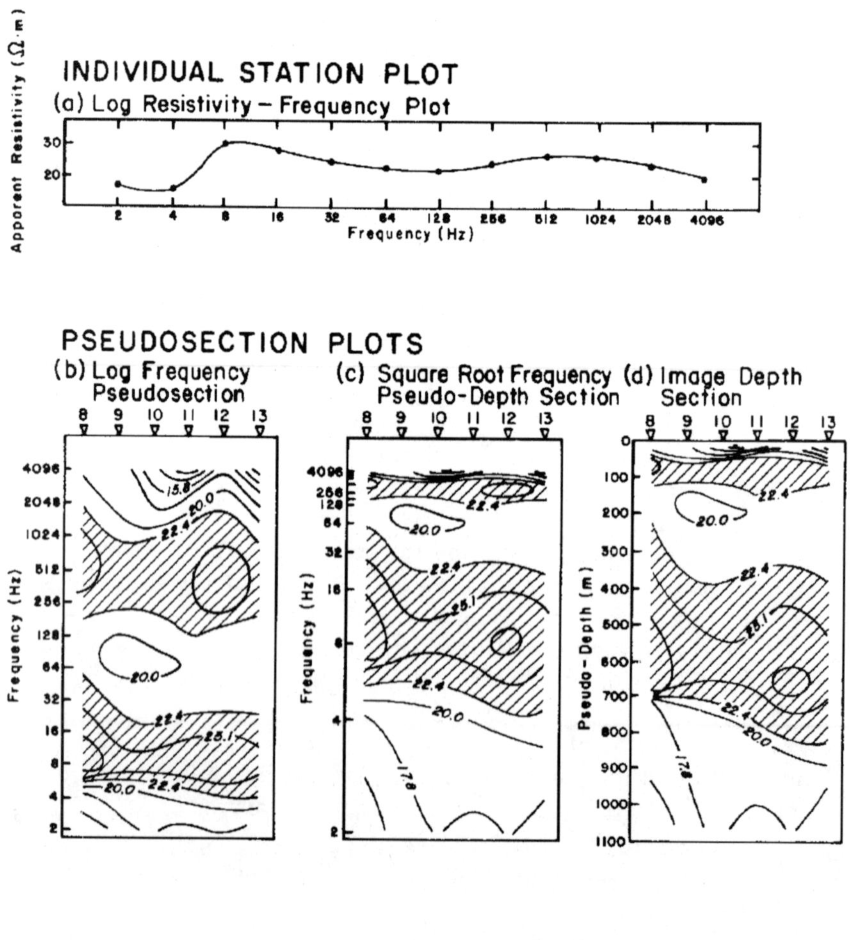

DEPTH OF EXPLORATION

As observed in the graph in figure 6.2 and the equations in table 6.1, the depth of exploration or investigation is related to the square root of ground resistivity and the inverse square root of signal frequency. Although the equations do not provide any limit to the depth of exploration obtainable, the maximum usable depth achieved in practice is usually between 2 and 3 kms. As a general rule, when sounding over a relatively homogeneous half-space, the separation between transmitter and receiver should be about five times the depth of exploration. Therefore, to see down to 1 km, one should have receiver/transmitter separation of about 5 km. If the background resistivity is 1000 Ωm, one will only have to sound down to about 100 Hz to penetrate 1 km; if the background resistivity is 100 Ω m, one will have to sound down to about 10 Hz, etc. Refer to figure 6.2 for more information.

The limiting factor on depth of exploration with all of the data in the far field is usually signal level. Remember that the electric and magnetic fields drop off as the inverse cube of the separation distance ($1/r^3$) between the receiver and transmitter, so the signal level drops off very quickly. It is not unusual to work in environments where the background noise is more than 10 times the signal level. Most surveys are run with receiver/transmitter separations between 5 and 15 km.

LATERAL RESOLUTION

Lateral resolution is mainly determined by the size of the E-field dipole. Theoretically, the dipole can be made as small as one wishes to get the desired lateral resolution, but again signal strength and noise enter the picture. Received signal strength is proportional to the length of the dipole, so if the dipole size is cut in half, signal strength is cut in half. The smallest dipole we have used in normal production is about 25 ft (8 m).

Figure 6.28 is a field example of the effect of decreasing dipole size to increase lateral resolution. This short survey was run over a hydrothermally altered area with a precious metals mine located near station 8. The object was to map the extent of alteration in an effort to locate possible mine extensions. The initial 60 m E-field dipole data mapped out a central resistivity low between stations 4 and 8 with the resistivities about a factor of 2 or 3 lower than the surrounding values. The 30 m dipole data shows more detail within this area, mapping out two conductive zones — one beneath station 8 where the known mineralization existed and the other beneath station 3. Further reduction of the dipole size to 15 m indicated that the western conductive zone is located between stations 2.5 and 3 and is considerably more conductive than the surrounding rock.

Figure 6.29 compares dipole-dipole resistivity data with CSAMT data over the same line. The depth of exploration desired was at least 100 m, so the dipole size (a-spacing) for the resistivity measurements was fixed at 60 m, which provides an effective depth of 100 m to 120 m at N=6. Since the depth of exploration for CSAMT is dependent only on resistivity and frequency, the E-field dipole size was set to 15 m for maximum lateral resolution. The effective depth of exploration at 32 Hz within the conductive zone is over 800 m, while in the surrounding rock it exceeds 3000 m. Although both types of data display the effects of the same conductive dike-like feature, it is much easier to distinguish the lateral extent in the CSAMT data.

FIGURE 6.28
A field example showing increasing resolution with decreasing E-field dipole length. From Zonge and Hughes, 1991

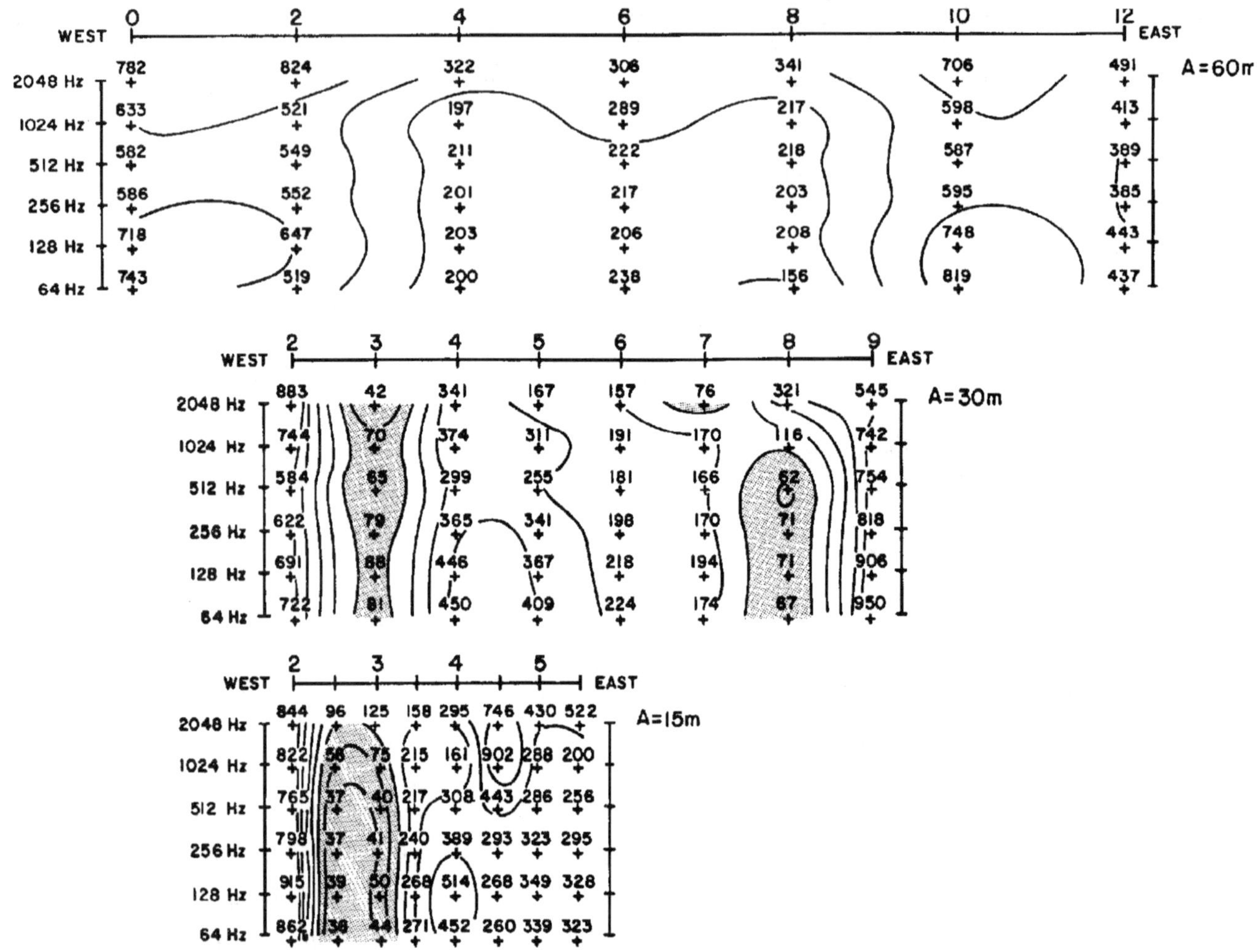

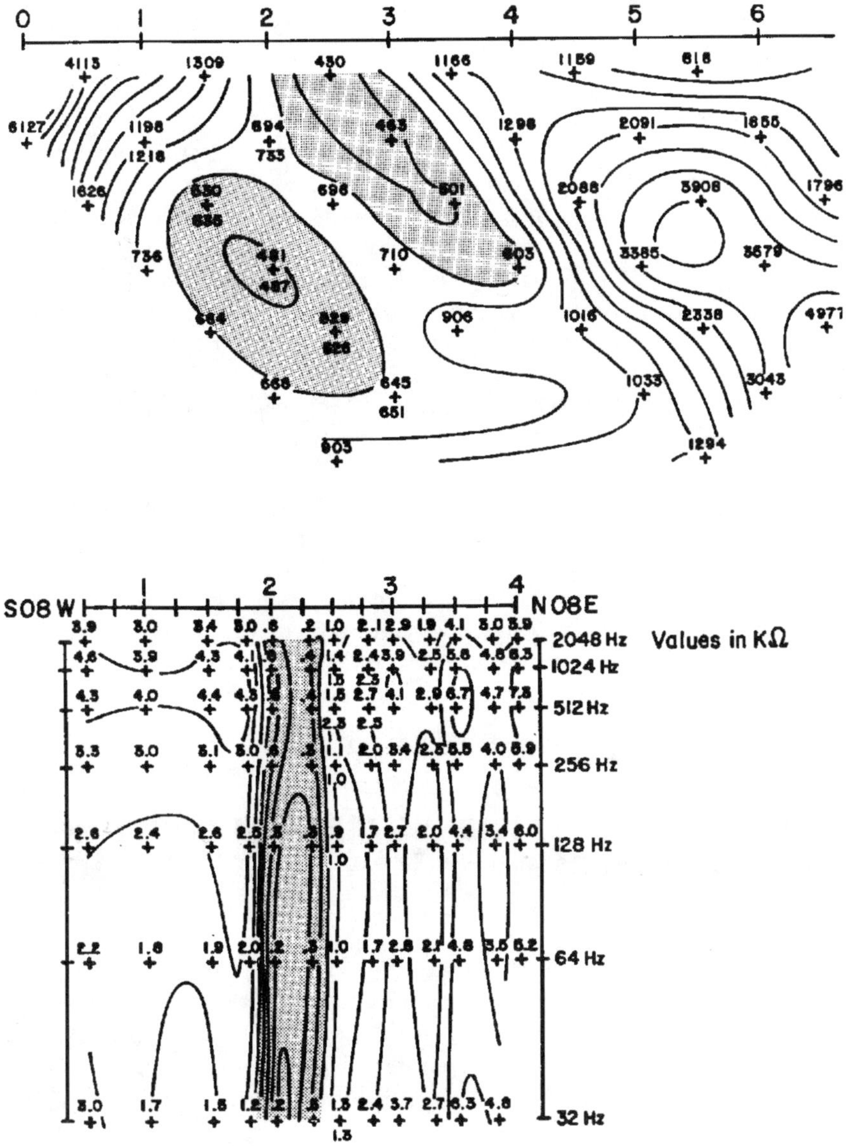

FIGURE 6.29
A field example comparing a dipole-dipole resistivity pseudosection (top) with a CSAMT resistivity pseudosection (bottom) along the same line. A-spacing for the dipole-dipole survey was 60 m; the E-field dipole size for the CSAMT survey was 15 m. From Zonge and Hughes, 1991

INTERPRETATION HINTS

Before continuing further into more field data, a brief discussion of some basic ideas on data interpretation will be presented. Some of the conditions were reviewed affecting CSAMT measurements, along with models of layered-earth configurations and examples of lateral resolution. How does one apply this knowledge to interpretation of field data?

To begin interpretation of CSAMT data, use a minimum of four different types of plots: apparent (Cagniard) resistivity pseudosections, phase pseudosections, log-resistivity versus log-frequency plots, and log-H-field versus log-frequency plots. These are the type of plots shown in figure 6.27a and 6.27b with posted values as well as contours on the pseudosections.

The first step is to check the data quality by scanning the log-resistivity plots looking for sharp variations that may be caused by various sources of noise or grounded culture (pipelines, powerlines, fences, etc.). Then check to see where the near field is coming in, if indeed it is. This can be done by scanning the log-resistivity plots for notches in the low-frequency data. Also, observe the behavior of the H-field in the log-H-field plots. Does the H-field response flatten out at low frequencies as seen in the layered-earth figures? This information is needed to avoid misinterpreting the data at low frequencies.

The next step is to scan the apparent resistivity section noting contacts, high- and low-resistivity features, layering and any other interesting patterns that might come to light. Compare these features with the phase section to see if they have corresponding perturbations in the phase data. Activity in the resistivity section that does not have corresponding movement in the phase data often indicates that the effects in the resistivity section result from near-surface features. If you are not looking for near-surface features, then you may want to consider methods for removing these features in the final interpretation.

Now, correlate all of this with known geology. As an additional aid, model the data or construct a simple inversion by plotting the data as a function of depth by using the depth of investigation equation in table 6.1. Calculate an equivalent depth at each frequency in the far field by using the measured apparent resistivity. Notice that the depth equation does not hold for the transition zone or the near field. Extracting definitive information from these two areas usually necessitates modeling the CSAMT data.

FIELD EXAMPLES

INTERPRETATION INCLUDING CURRENT FOCUSING OR STATIC EFFECTS

So-called static effects are caused by surface or near-surface resistive or conductive features with limited lateral and depth extent, or they can be caused by sharp topographic relief.

How are static effects defined in CSAMT data? The accepted definition is as follows: if the apparent resistivity values between sounding sites are shifted by a constant with no corresponding change in phase, then static effects exist. This can be best observed in the log-log plots of apparent resistivity curves. Static effects will translate the curves vertically, without changing their shape and with no change in phase between the soundings. Static effects are actually a direct current phenomenon resulting from current focusing and can be removed by several different techniques.

Since phase is proportional to the slope of the resistivity, resistivity can be derived from phase. This derived resistivity is free of static offset effects. Also, since static effects are just an offset (multiplier) on the apparent resistivity curves, one can normalize the curves using data from a non-affected area in a brute force correction method. Various types of filtering have been tried with CSAMT and natural source AMT data with varying degrees of success. Whatever method is tried, the phase information needs to be included as part of the correction, since phase in not influenced by static offset.

To demonstrate static effects and their removal, look at an extreme example from northern Arizona. This CSAMT survey was run over several prospect areas searching for breccia pipes containing uranium in association with massive sulfides, using 150 ft electric-field dipoles. CSAMT was used, since the targets were narrow and deep. Figure 6.30 shows the apparent resistivity and phase pseudosections for one line.

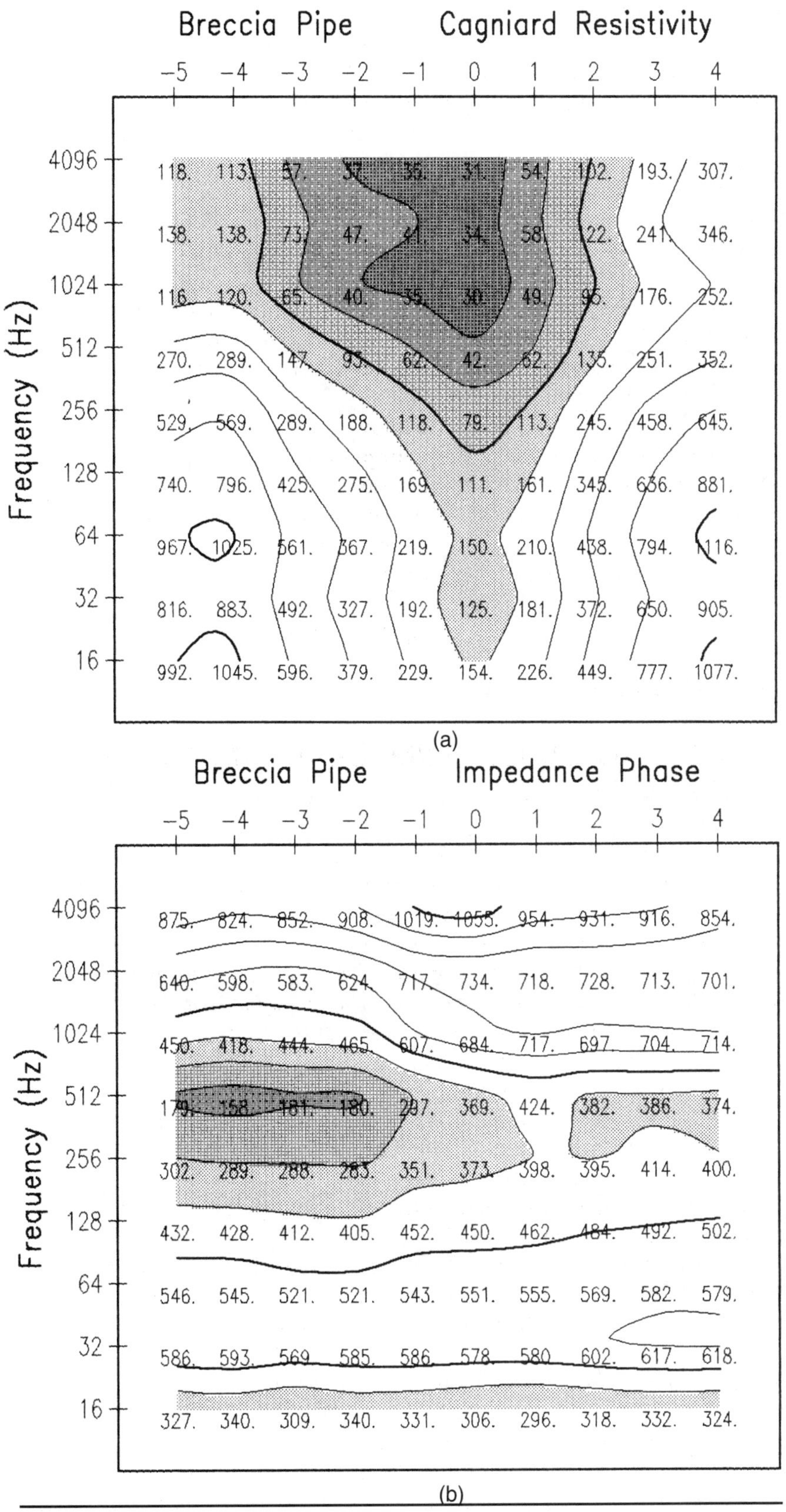

FIGURE 6.30
(a) Apparent resistivity pseudosection over a prospect area possibly hosting a breccia pipe. Dipole spacing is 150 ft (46 m). (b) Phase pseudosection

Figure 6.31 has the same data plotted as in figure 6.3a, in square-root-frequency pseudo-depth format. Figures 6.32 and 6.33 contain the apparent resistivity and H-field log-log plots.

FIGURE 6.31
Same data as in figure 6.30, with the frequencies plotted in equivalent depth format: (a) apparent resistivity, (b) phase

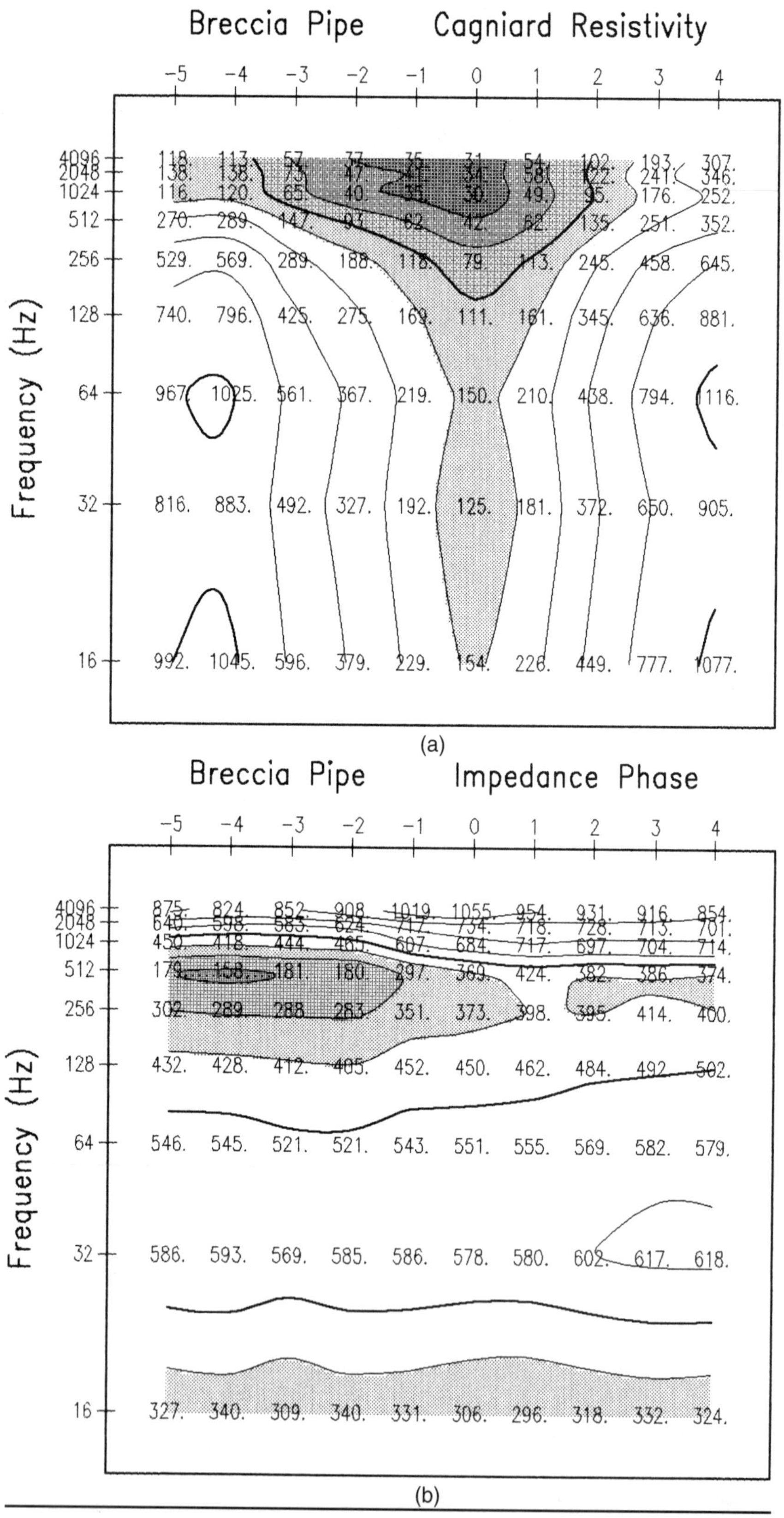

In the log-log resistivity plots of figure 6.32, notice the vertical translation of similar-looking resistivity curves indicating the possibility of static offset — especially on the eastern half of the line. Notice also the notch at 1024 Hz which looks like a transition-zone notch, except that the low-frequency data flatten out rather than continuing on at a 45-degree slope (as one would expect in the near field). The H-field plots in figure 6.33 confirm that the notch at 1024 Hz is in the far field and must be geologically or structurally derived. There is insufficient information to determine if the data approach the transition zone at 16 Hz, but the available information suggests that this data set is mainly in the far field.

Now, observe the apparent resistivity section, figure 6.30, and especially the pseudo-depth section, figure 6.31. This latter figure has the basic appearance of what one might expect from a conductive breccia pipe. Observe the phase in figure 6.30, however, and notice the lack of correlation between the two plots. As an additional interpretational aid, the log-log plots of several phase response

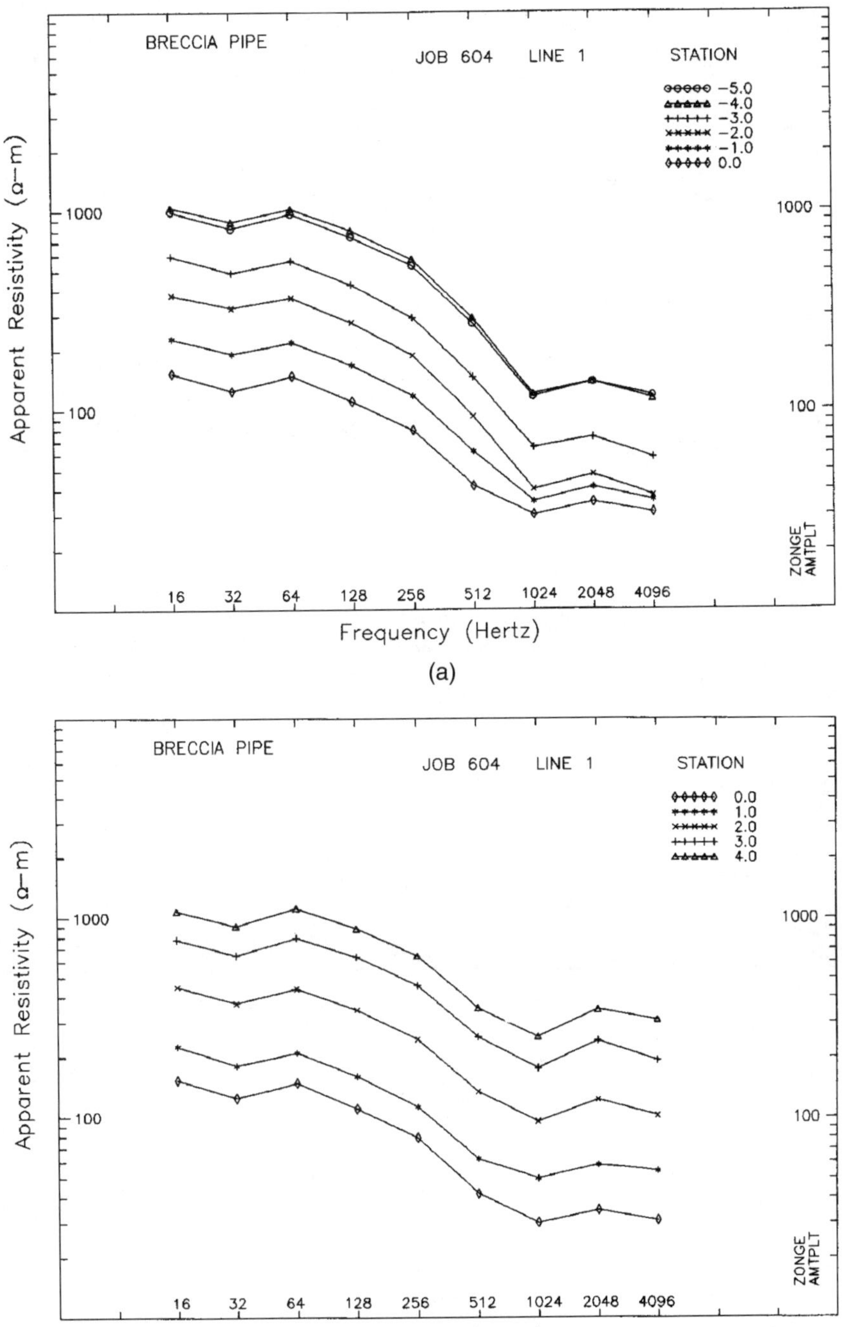

FIGURE 6.32
Apparent resistivity log-log plots: (a) western half, (b) eastern half of the line

FIGURE 6.33
Log-log H-Field plots: (a) western half, (b) eastern half of the line

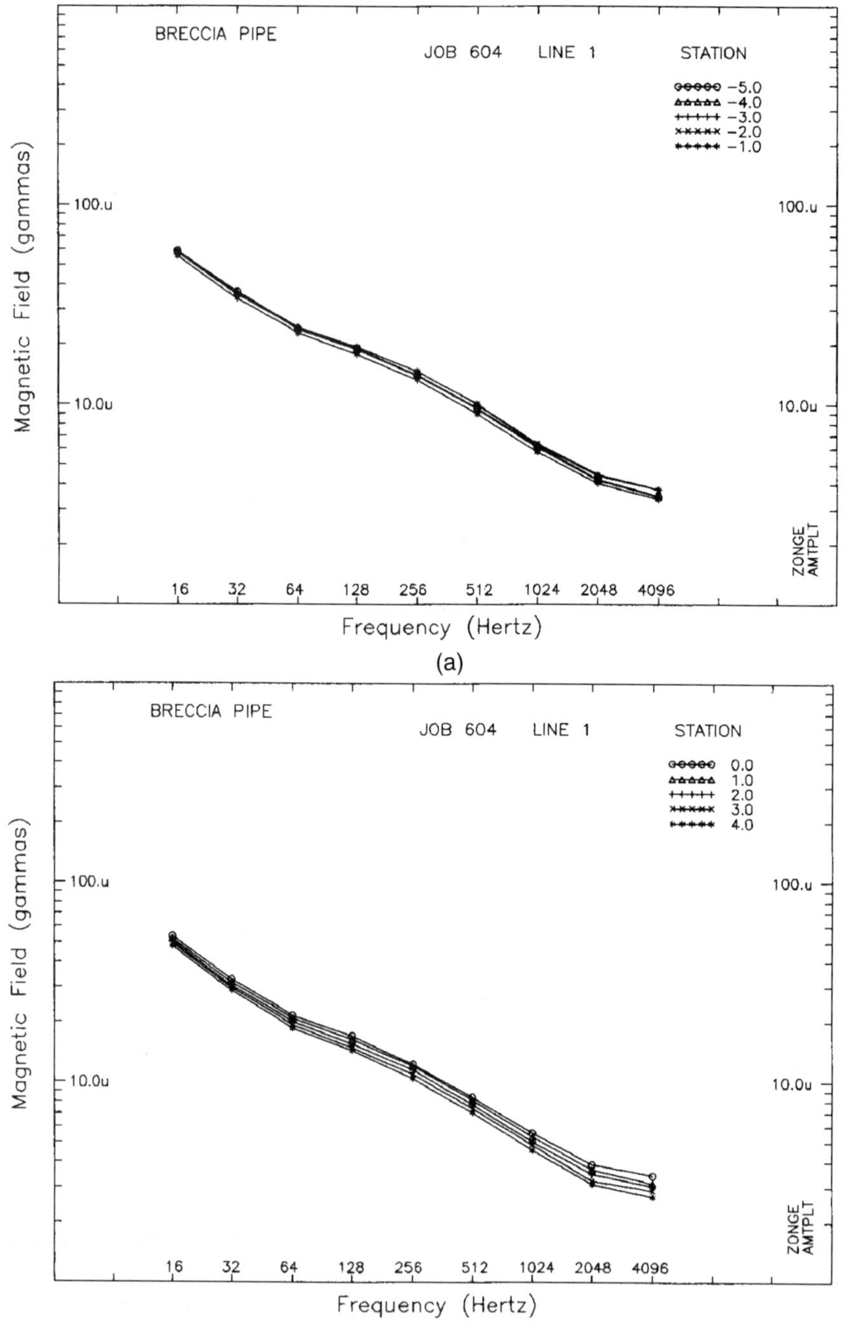

curves are presented in figure 6.34. Plots are shown for stations -3, -1 and 3. Curves for stations -2 through -5 are basically the same as station -3, and curves for stations 0 through 4 are basically the same as 3. Notice the difference in the phase response between the western and eastern halves, with station -1 being a transition between the two.

Considering all of the data plots,some preliminary conclusions can be tabulated without any additional data processing:

- Static offset effects occur (observed in the vertical translation of the log-log resistivity curves).

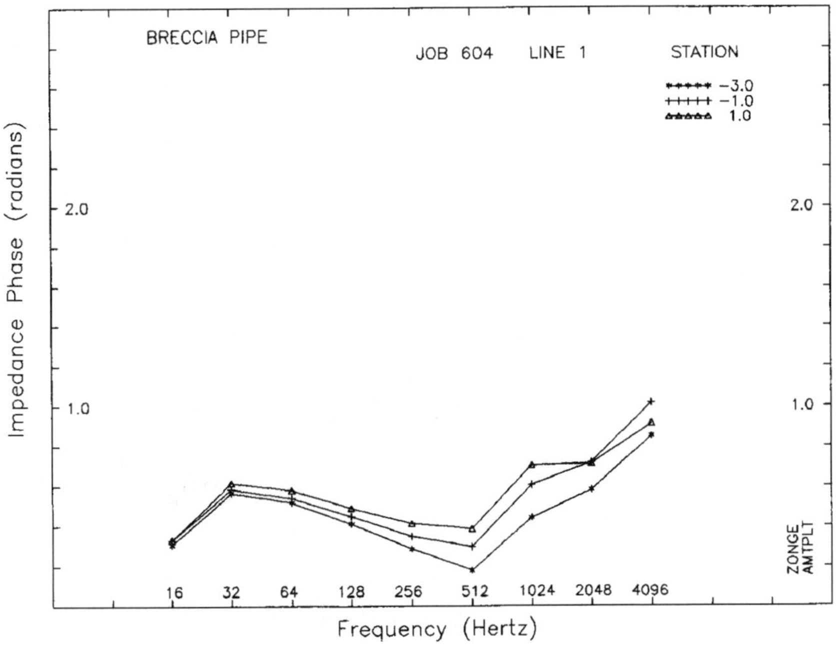

FIGURE 6.34
Log-linear phase plots for stations -3, -1, and 3

- Most, if not all, of the data are in the far field.

- There is a distinct difference in structure between the eastern and western halves of the line (observed mainly in the phase data at this point).

- The area is covered with a conductive layer of varying thickness (observed in both the resistivity and phase data). Recall that phase values greater than 785 mr indicate resistivities are decreasing with depth, and phase values less than 785 mr predict resistivities increasing with depth.

One might obtain the idea from this discussion that phase is an important interpretation parameter. This reasoning is correct, as we will continue to see in future examples.

Since static offset effects exist in the data, these effects must be removed to see how the final resistivity section will appear. Figure 6.35a shows the results of shifting the resistivity values using a brute-force, phase-weighted technique, and figure 6.35b was constructed using resistivity values derived from phase. The plots for both figures were derived using a Bostick-Niblett inversion. For this inversion, the depths are calculated from the static-corrected resistivities and then the resistivity values are modified by the phase values to bring them closer to the actual resistivity of the rocks providing the response. This is a poor-man's inversion technique, but it works quite nicely on CSAMT data gathered in the far field.

Basically identical results are obtained for both static removal techniques. These depth sections show a fault or contact occurring near station -1 with little other variation except in the surface layers. Drilling confirmed that this was not a breccia pipe.

This field example was included to provide an extreme example of static shift and its accurate removal. You will see in the following examples that static shift is not normally a problem, but one should be aware of its effect on data, and how to detect it, to avoid interpretation problems.

SENSITIVITY TO RESISTIVE DIKES AND A COMPARISON BETWEEN DIPOLE-DIPOLE RESISTIVITY AND CSAMT MEASUREMENTS

This example is included to demonstrate the sensitivity of CSAMT measurements to resistive dikes, to provide additional data processing procedures, and to give a comparison of dipole-dipole resistivity and CSAMT data over the same line using the same E-field dipole spacings (300 ft).

Figure 6.36 shows the observed Cagniard resistivity along the line with several prominent features. A deep resistive zone can be seen centered beneath station 6 and a strongly resistive dike-like zone coming down from the surface beneath station 11.5. A deep conductive area appears to flank the dike-like feature. Either of the resistive features could represent silicified structure that would be of interest from a gold exploration standpoint.

Figure 6.37 is a pseudosection of the phase difference between the E- and H-fields. The deep resistive zone is manifested by the low phase numbers in that area, but notice that there is very little

FIGURE 6.35.
Static-corrected depth sections using Bostick inversion on (a) static-corrected Cagniard resistivity and (b) resistivity from phase

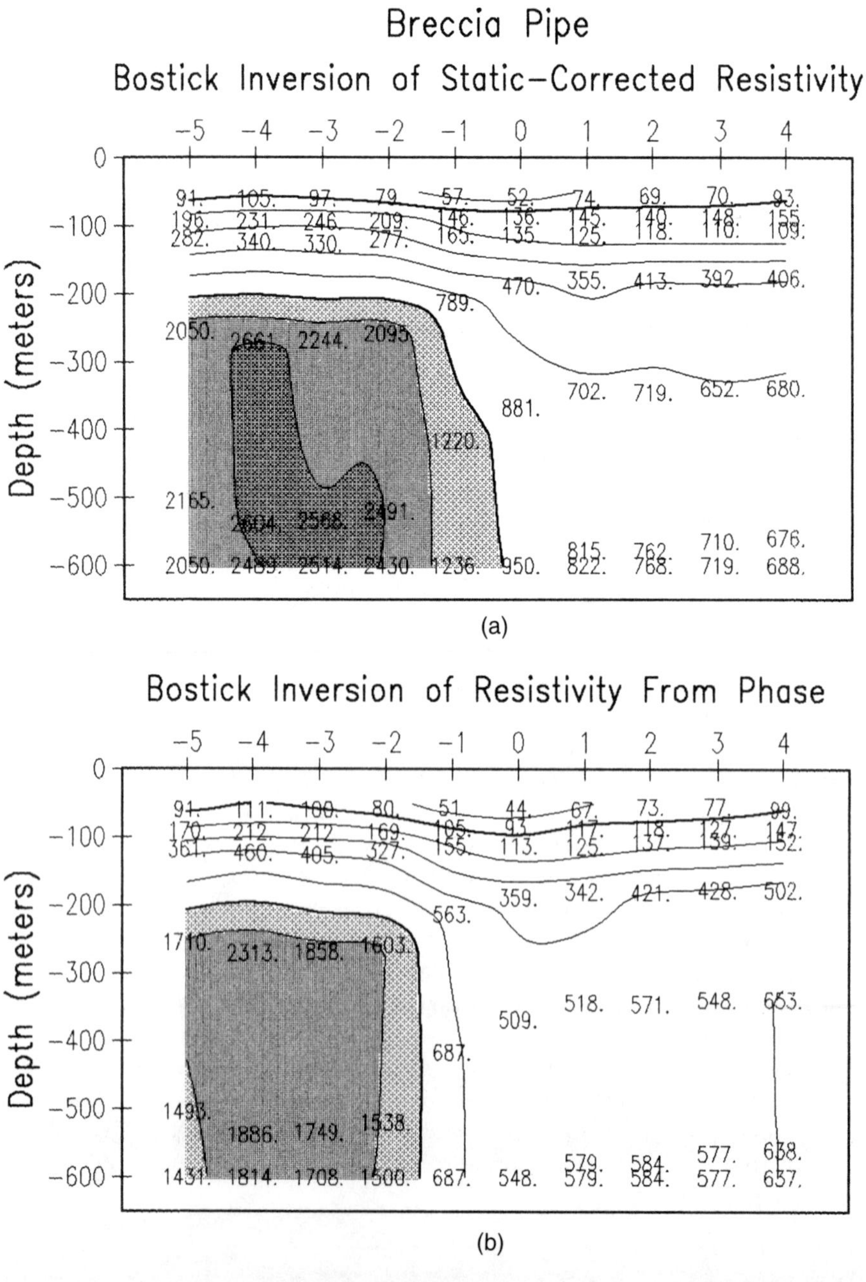

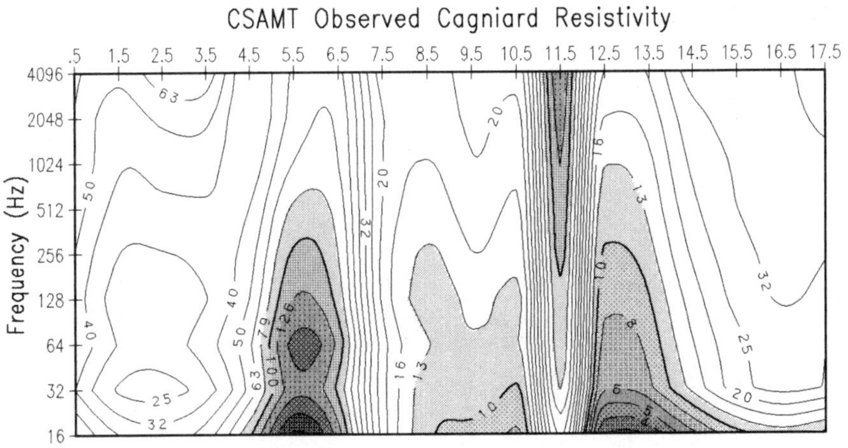

FIGURE 6.36
Observed Cagniard resistivity pseudosection over a prospect in the northwest United States (from Zonge and Hughes, 1991)

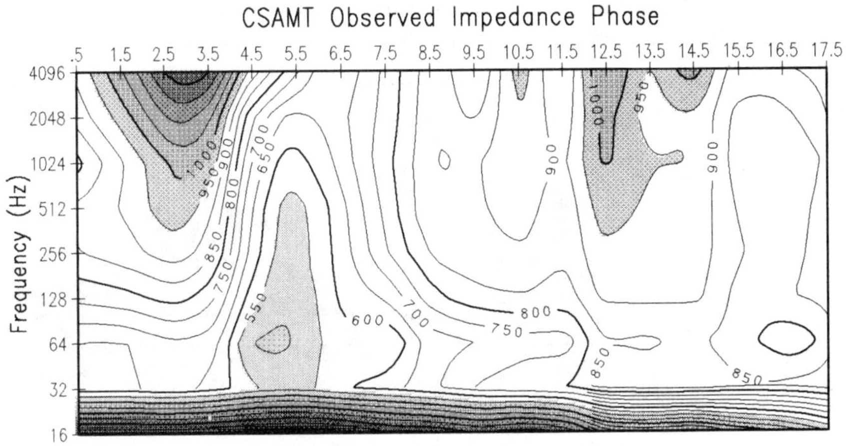

FIGURE 6.37
E-H phase difference (or impedance phase) pseudosection (from Zonge and Hughes, 1991)

FIGURE 6.38
Detail Cagniard resistivity values over the near-surface resistive feature

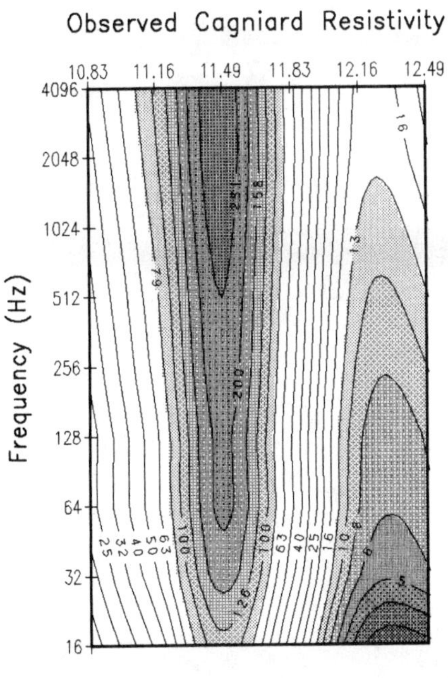

evidence of the resistive feature beneath station 11.5, leading us to believe that this is an artifact of a near-surface resistive block, or static-offset effect. Notice the intense gradient of the phase response below 32 Hz which results from transition/near-field effects.

Figure 6.38 shows the results of a detail survey over the resistive zone centered beneath station 11.5. Notice that the influence of this feature extends across four stations.

Figure 6.39 is a static-corrected resistivity pseudosection. Notice that the high-resistivity feature beneath station 11.5 has disappeared, substantiating our contention that this was a near-surface effect with shallow depth extent and probably diminishing its interest as a gold target.

Figure 6.40 shows the results of a smooth-model inversion of the static-corrected apparent resistivity. Notice the deep resistive target is now centered beneath station 5.5 and that there is a broad, very low resistivity (alteration?) zone stretching from station 10.5 to the end of the line.

In comparison, figure 6.41 shows the dipole-dipole resistivity results over the same line. Notice that the data are mostly affected by near-surface geology. The resistive surface block beneath station 11.5 appears as a pantsleg which masks the conductive zone on that end of the line. The resistive zone beneath station 5.5 is not evident at all, since the dipole-dipole data do not have sufficient depth extent.

FIGURE 6.39
Static-corrected Cagniard resistivity pseudosection (from Zonge and Hughes, 1991)

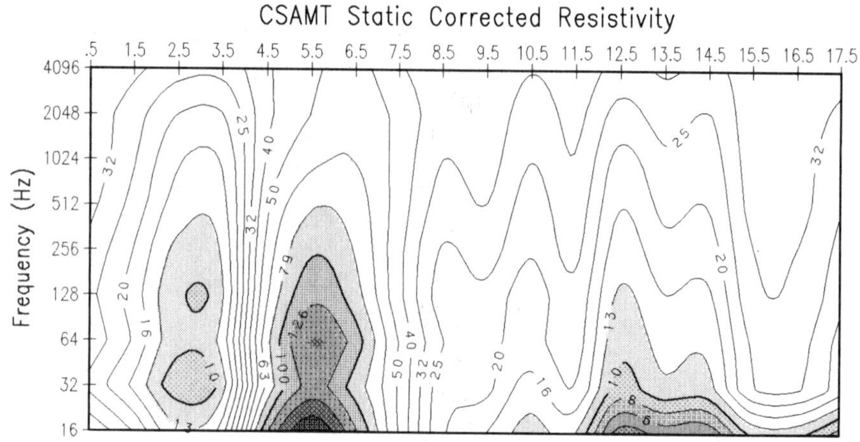

FIGURE 6.40
Smooth-model inversion of static-corrected resistivities.(from Zonge and Hughes, 1991)

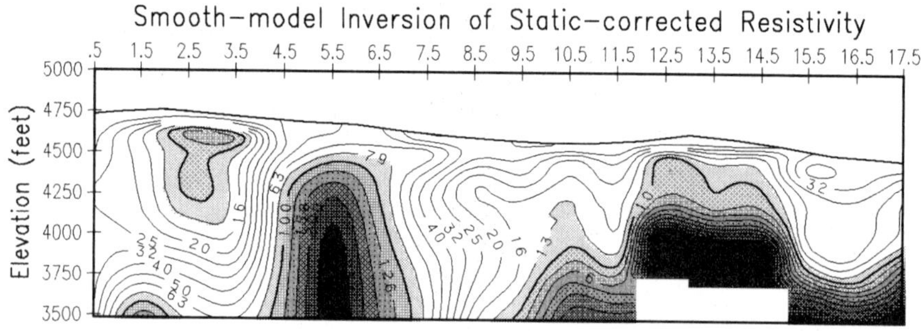

FIGURE 6.41
Apparent resistivity dipole-dipole pseudosection over the same line as figure 36 (from Zonge and Hughes, 1991)

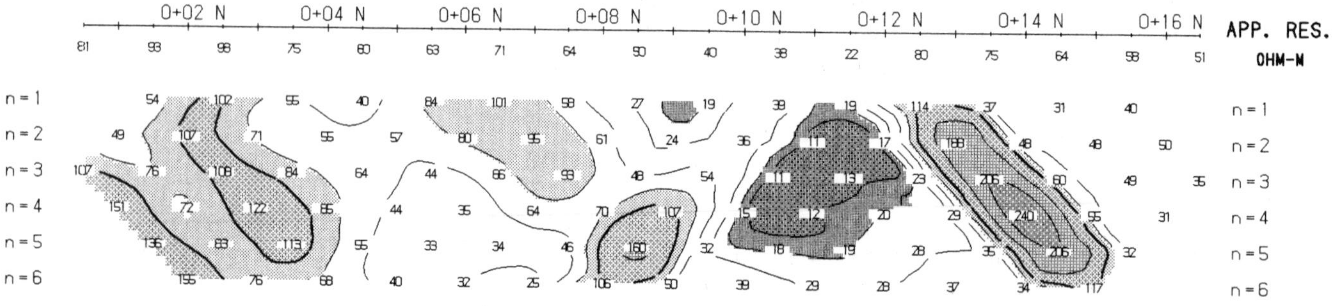

MASSIVE SULFIDE EXPLORATION, CRANDON, WISCONSIN

This field example includes the first CSAMT data set acquired with the first Zonge digital receiver system. It is included to show that one can get meaningful data in the near field when looking for massive conductive features.

The Crandon massive sulfide orebody is located in the Rhinelander-Ladysmith greenstone belt of northern Wisconsin. As described in a review by Schmidt et al. (1978), the body was estimated to have reserves of 62 million metric tons of ore, consisting of 5 percent zinc, 1 percent copper, and low grades of silver, gold and lead. The sulfide content of the ore zone averages about 65 percent, primarily consisting of pyrite, with moderate amounts of sphalerite and chalcopyrite, and small amounts of galena, quartz, chlorite and sericite. The ore zone is an elliptical, lens-like, nearly vertical tabular structure averaging about 100 ft in width and having a length of about 5000 ft. It is covered by 150 ft of glacial overburden and extends to a depth of about 1500 ft. Figure 6.42 shows a geologic cross section based on early drilling data.

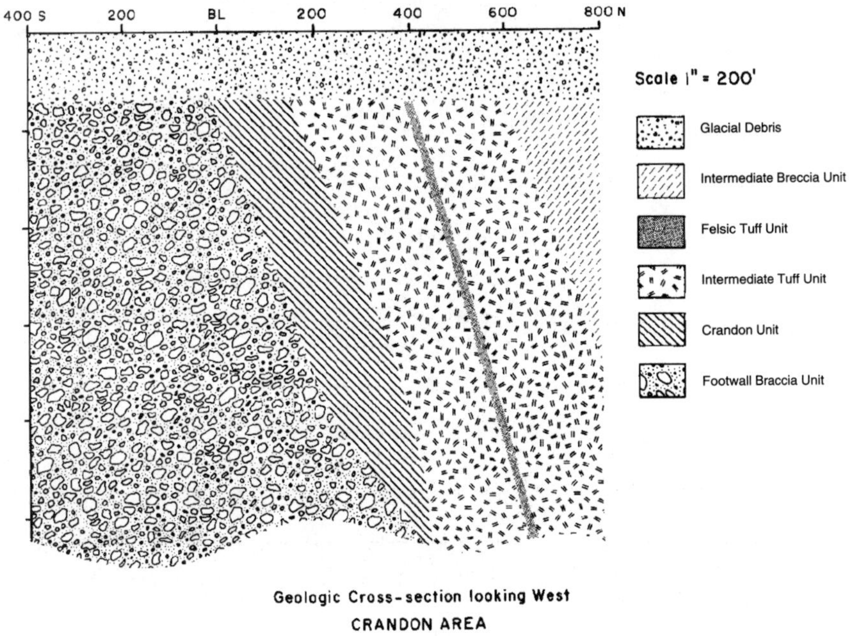

FIGURE 6.42
Geoelectric cross-section of the Crandon orebody. (From Schmidt et al, 1978). The approximate resistivities for the main units are: glacial debris, 200Ωm; footwall breccia, 2000 Ωm; hanging wall tuff, 2000Ωm

Both multi-frequency resistivity and IP data (using 300-ft dipoles) and CSAMT data (using 100-ft dipoles) over this orebody were obtained. The IP apparent resistivity data (figure 6.43) show a symmetric, strongly conductive feature over the orebody. The relatively large 300 ft a-spacing, along with the 150 ft of cover, broadens and weakens the strength of the anomaly, but it is still plainly discernable. A moderate polarization high is also associated with the high sulfide content of the body.

Scalar Ex/Hy CSAMT data were obtained along the IP line using a 100 ft E-field dipole with a 2500-ft transmitter dipole placed 2 mi from the line. The Cagniard resistivity data in figure 6.44a show artificially high near-field resistivities, which is the result of the close receiver/transmitter separation (this was the first CSAMT survey run, and source effects were not as well understood) and the highly resistive host rock. Near field saturation occurs below 512 Hz (see figure 6.45). Superimposed upon the transition zone and near field data is a strongly conductive feature near station 0. The conductor is well correlated with the location of the sulfide body. The data show a sharp southern boundary, indicating a strong resistivity contrast between the sulfide body and the footwall breccia to the south. The northern boundary is more diffuse, caused in part by the northerly dip of the orebody.

The phase data for this survey appear a bit odd, since most of the data are acquired in the near field/transition zone where the back

FIGURE 6.43
Dipole-dipole data collected over the Crandon orebody: (a) apparent resistivity and (b) phase, using 300 ft dipoles

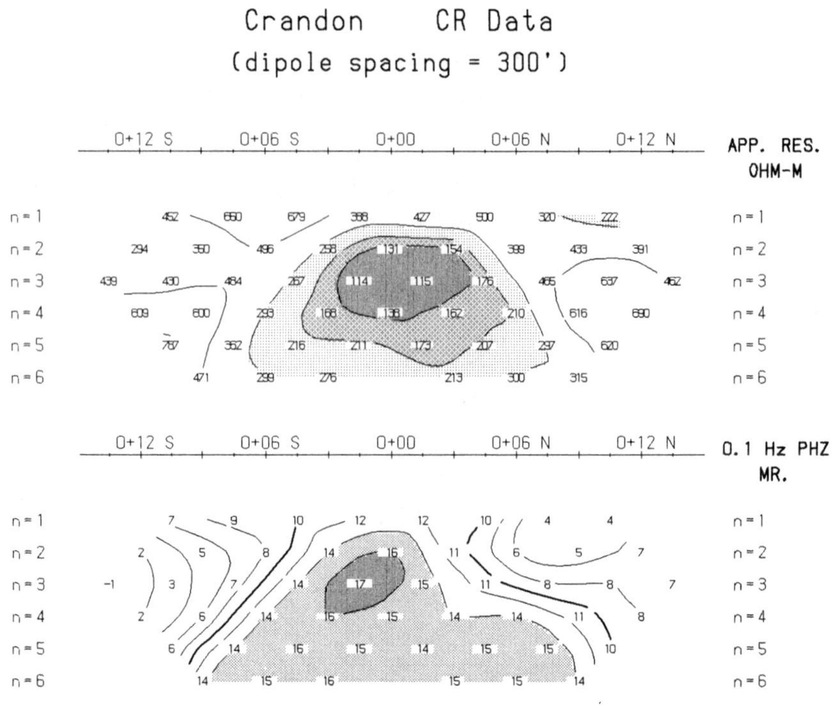

ground phase is close to zero. The relatively high phase values centered over the orebody coincide with the highly conductive Crandon formation.

MASSIVE SULFIDE EXPLORATION, NORTHEASTERN USA

The target in this example was assumed to be a volcanogenic, massive sulfide deposit in an area that appeared slightly anomalous in an airborne survey and strongly anomalous in follow-up dipole-dipole resistivity and IP work. The assumed depth of the target dictated the use of large dipoles, which in turn provided poor lateral resolution. Drilling hit small, very high-grade intersections of ore, but the main body (if one existed) could not be identified. A large number of geophysical techniques, both surface and downhole, were used, but drill hole intersections continued to be small and unpredictable. Surveys that were run included dipole-dipole resistivity and IP, TEM, PEM, magnetics, gravity, complex resistivity, shallow HEM, downhole PEM and downhole MMR. Drilling continued, but after more than 30 drill holes in the general area, upper management decided there was no "main body" and that the intersections in the drill hole were too small to justify further work.

During the latter part of the work, the client's crew was trained in CSAMT and then processed and modeled the data that they acquired. A gridded CSAMT survey was run over the area using 30 m E-field dipoles. With the results, the managing geologist was able to obtain enough budget for one more drill hole. The drill site was chosen based on the modeling of the CSAMT data. At 185 m, the drill went into ore and continued in the ore for over 60 m. The property

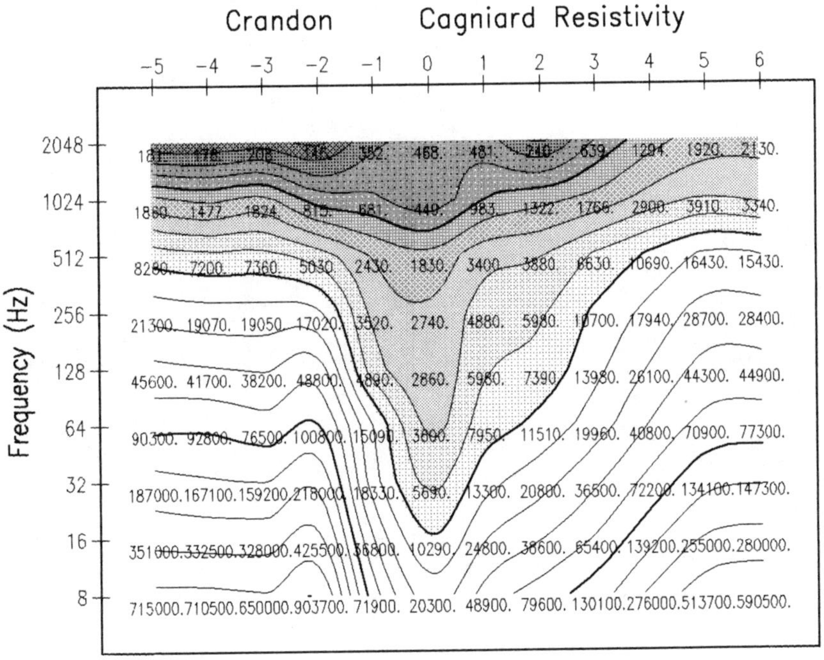

(a)

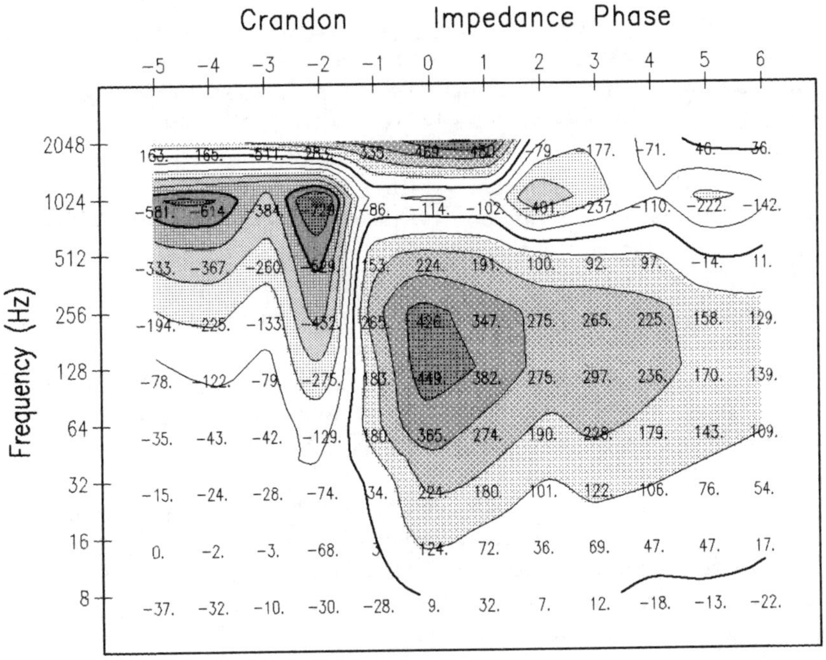

(b)

FIGURE 6.44
CSAMT data collected over the Crandon orebody: (a) Cagniard resistivity, (b) phase, using 100 ft E-field dipoles

has now been transferred to the mine development branch of the company and environmental permitting is underway for a new mine.

The main problem for this project was one of resolution — both lateral resolution and differentiation between shallow and deep targets. Figure 6.29 shows a comparison between data sets gathered in this area: 61 m dipole-dipole data and CSAMT data using 15 m E-field dipoles. Both data sets show the effects of a shallow conductive

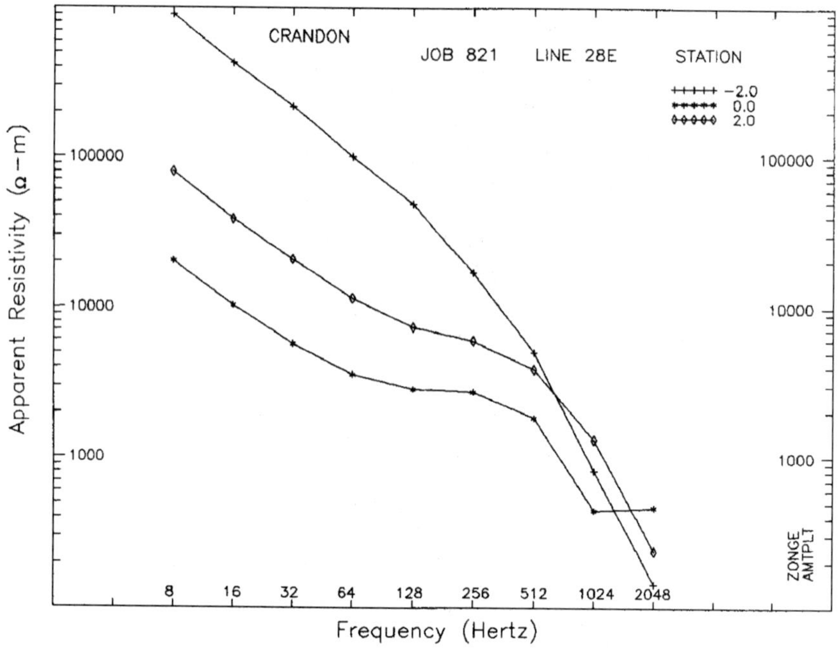

FIGURE 6.45
Log-log apparent resistivity curves for stations -2, 0 and 2. Same data as in figure 6.44 (a)

body, but the exact location and size of the body is not clear in the dipole-dipole data. To sound down to 200 m and below (where the main body was located), the dipoles for the resistivity sounding would have had to be more than 100 m long, further decreasing resolution for the dipole-dipole survey. With CSAMT, the lateral resolution is a function of E-field dipole length, and depth of exploration is dependent upon frequency rather than geometry. Therefore, shallow features such as this one could be identified easily, and the deeper targets resolved with equivalent accuracy.

Figures 6.46 and 6.47 help illustrate the problem created by multiple small targets. Figure 6.46 is a Cagniard resistivity pseudosection along one of the grid lines, and figure 6.47 shows the associated phase difference pseudosection. Notice that there are three possible targets in the resistivity section: one between stations 117 and 118, one between stations 120 and 122, and one beneath stations 128 and 129. Upon examination of the phase difference data (figure 6.47), however, it can be seen that only one area remains anomalous at depth — the area between stations 120 and 123. The other two zones are probably static offset effects resulting from small, shallow features. Once again, the phase data play a large part in accurate interpretation of CSAMT data.

Multiple, small, shallow targets were one of the main problems with the other geophysical techniques used in this area. The geometry necessary for high resolution did not permit adequate depth of exploration. CSAMT was able to provide both high resolution and large depth penetration simultaneously. CSAMT saw the small features the other techniques had been detecting, but it also saw the target they had missed. That literally saved the project.

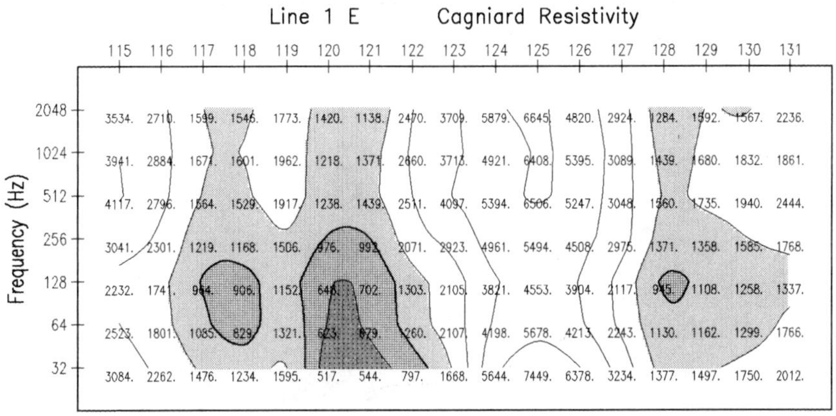

FIGURE 6.46
Cagniard resistivity over a massive sulfide target

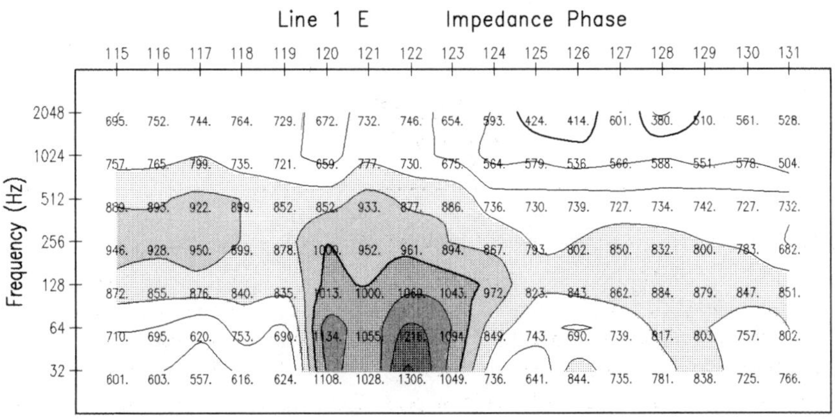

FIGURE 6.47
E-H Phase difference over a massive sulfide target

MAPPING ALTERATION AND SILICIFICATION, GOLDEN CROSS MINE, NEW ZEALAND

Epithermal gold deposits are characterized by strong variations in electrical resistivity. This typically results from alteration of silicate minerals and clays and results in increased conductivities; whereas remobilized silica deposition results in increased resistivities. Contrasts of four orders of magnitude can be observed within hydrothermal alteration zones.

The new Golden Cross orebody is the only major epithermal gold discovery in New Zealand this century. Earlier mining activities south of the new deposit produced a documented 386,000 oz of gold-silver bullion. Drilling at the new site predicts a total estimated yield of more than 600,000 oz. Figure 6.48 shows the location of the Golden Cross mine in the Coromandel Kaimai volcanic zone on the northern island.

Figure 6.49 shows the general geology of the area, and figure 6.50 is a geologic cross section along line 4850N. There are four main groupings of rock types which determine the geophysical signature of the Empire deposit (Hay, 1989):

- Silicified ore zone
- Surrounding alteration zone
- Unaltered host material
- Overlying andesite.

The silicified ore zone is characterized by low porosity, high resistivity (1000 ohm-meter) rocks containing 2 percent to 3 percent sulfides and no magnetite.

FIGURE 6.48
Location of the new Golden Cross Mine in New Zealand

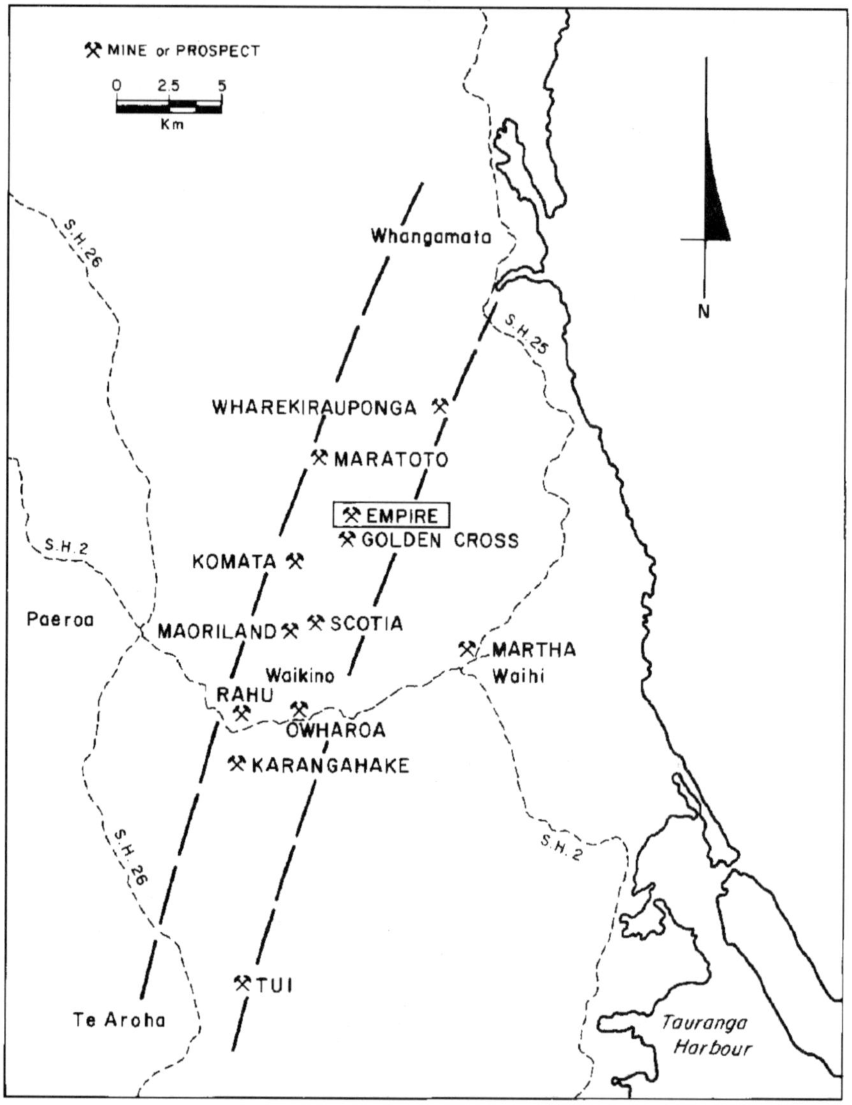

Surrounding argillic altered rock is low resistivity (≈1 Ω m) resulting from clays in the heavily altered zones. The formation contains up to 5 percent sulfides in the stockwork above the ore zone and up to 3 percent sulfides in the clay-altered zones. No magnetic minerals are found in this area.

The propylitic altered zones are characterized by low resistivities (≈10Ω m), up to 1 percent sulfides and a lack of magnetic materials.

Unaltered host rocks and overlying andesites have a resistivity of about 100 Ω m and are variably magnetic.

This new orebody was discovered as a result of ground follow up over an area delineated as low magnetic relief by an airborne magnetometer survey flown in 1978. Ground follow up consisted of geological mapping, rock chip and soil sampling and vertical diamond drilling designed to follow the gently dipping stockwork down dip from its outcrop position.

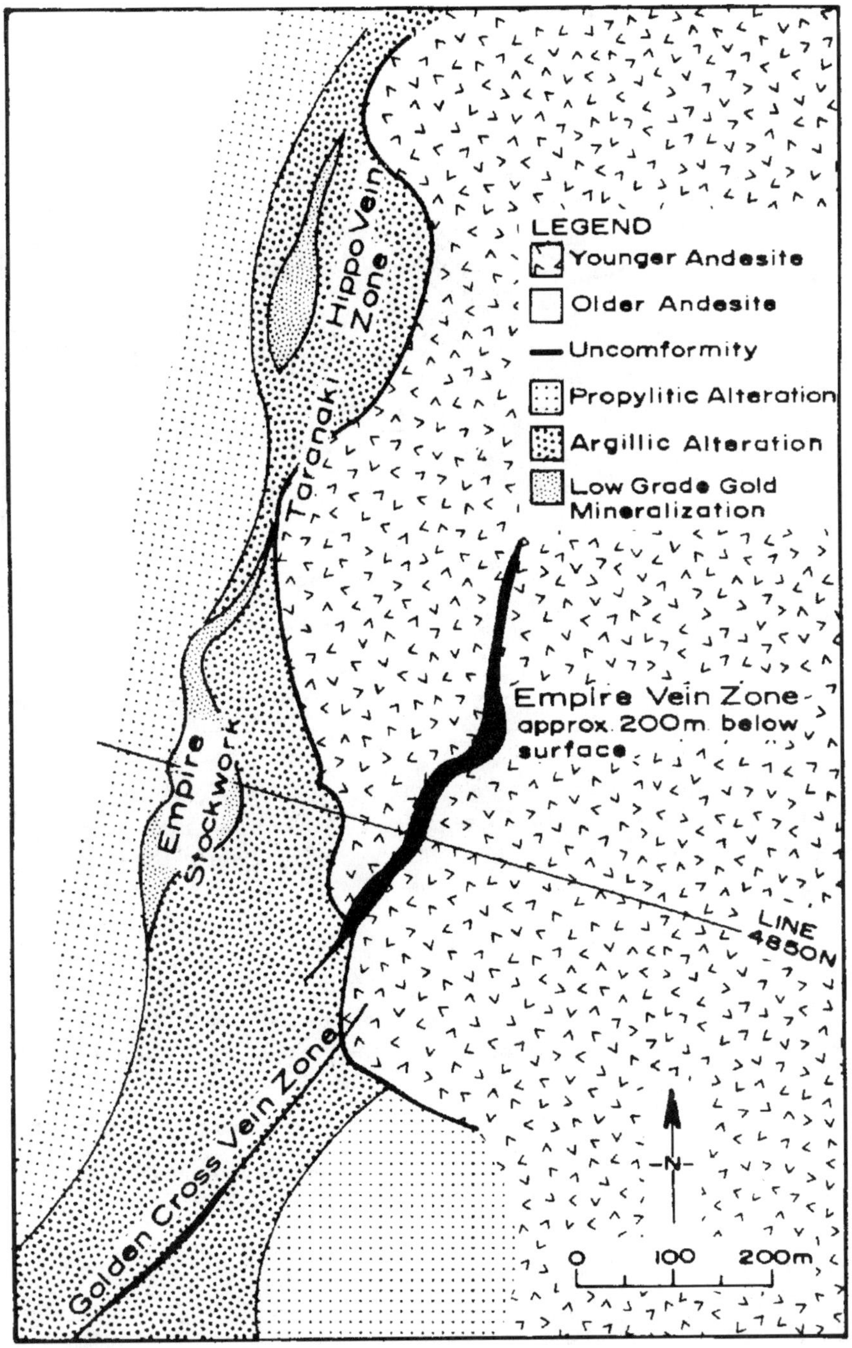

FIGURE 6.49
General geology of the Golden Cross/ Empire vein zones (after Hay, 1989)

After the initial discovery hole was completed, a number of different geophysical techniques were run over this area in an effort to find some method that would detect the orebody. Air and ground magnetics tended to outline the outcropping altered zones. Resistivity-IP (time-domain dipole-dipole and gradient) mapped alteration but did not delineate the high-resistivity ore zones. The IP data were very noisy and inconclusive, and subtle IP highs were drilled without finding gold. Perhaps the highly altered rock overlying the stockwork shielded the IP response. None of the surface techniques used at Golden Cross located the Empire vein zone.

FIGURE 6.50
Geologic cross section along Line 4850N (after Hay, 1989)

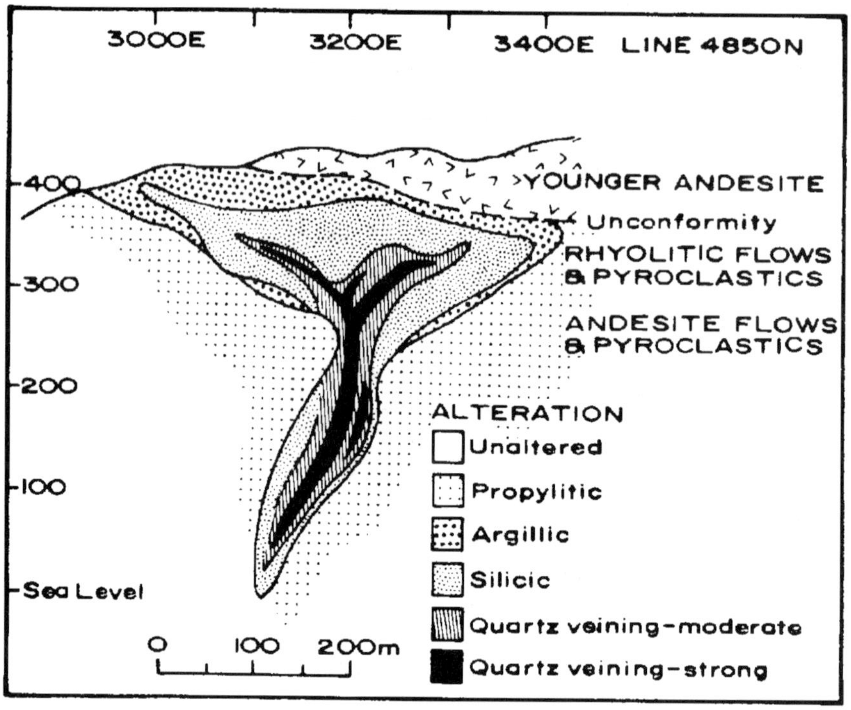

In 1986, it was decided to run a test CSAMT line over the orebody, using 50-m dipole spacing, to see if this technique would provide useful information. Figures 6.51 and 6.52 show the observed Cagniard resistivity and phase difference measured along line 4850N. Notice the strong contact to the east of station 3000E and the resistive zone beneath station 3200E which is flanked by low resistivity material on either side. The phase pseudosection shows similar activity: a contact east of station 3000E, a resistive zone beneath station 3200E with conductive zones coming near to the surface on either side, and a conductive zone extending deeper to the east.

FIGURE 6.51
Cagniard resistivity along Line 4850N (after Zonge and Hughes, 1991)

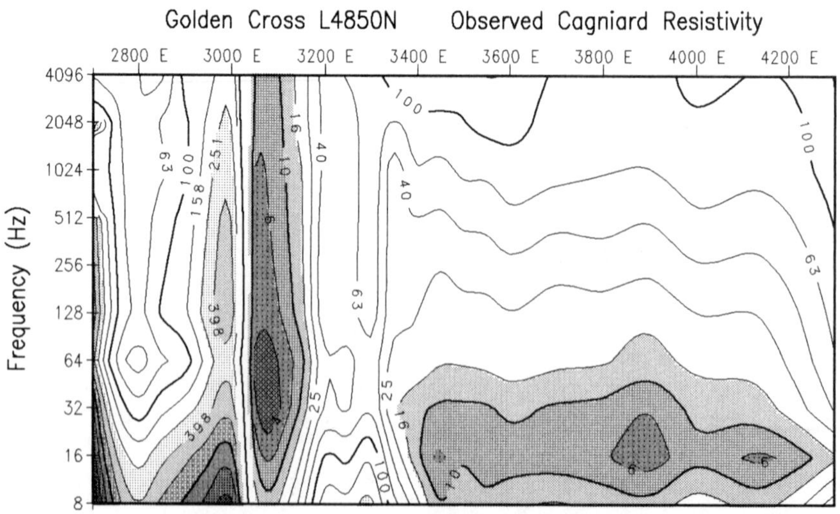

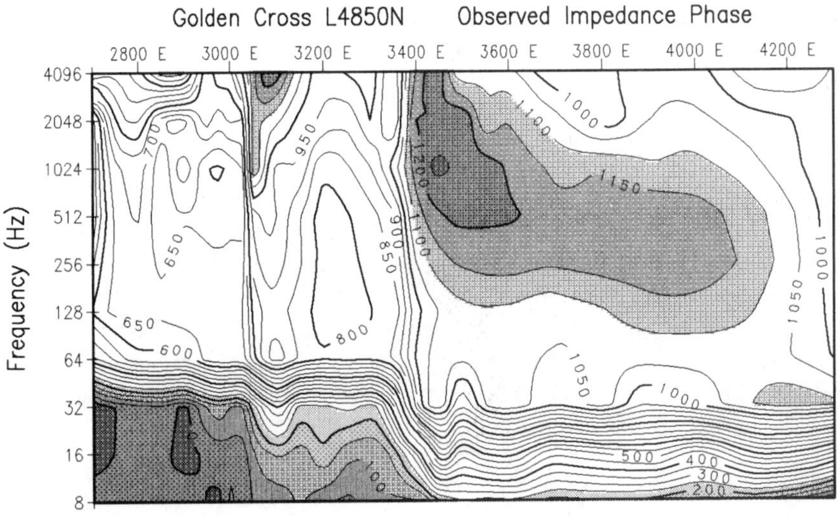

FIGURE 6.52
E-H Phase difference along Line 4850N (after Zonge and Hughes, 1991)

Figure 6.53 gives the results of smooth-model inversion of the Cagniard resistivity. This section shows a break in the conductive layer beneath station 3200E with a depth extent of several hundred meters. This feature coincides exactly with the Golden Cross orebody.

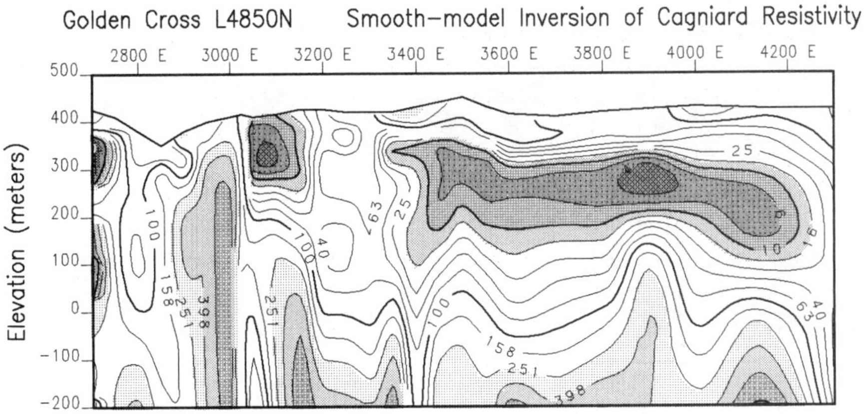

FIGURE 6.53
Smooth-model inversion of the Cagniard resistivity for Line 4850N (after Zonge and Hughes, 1991)

This test survey provided the only surface geophysical information that could be used for delineating the Golden Cross orebody.

Appreciation is given to Cyprus Gold New Zealand Ltd. for permission to use this data set.

NEVADA PEDIMENT SURVEY

This case history is included as an example of high density data plotted in plan map form. A grid of 31 lines was run with a CSAET setup using six 400 ft E-field dipoles and one H-field measurement at the center of each six dipole spread. The transmitter was placed 6,800 ft to the south of the southernmost line (Line 00N). Spacing between lines is 400 ft except for two detail lines that were run with 100-ft dipoles and centered at 200 ft between the reconnaissance lines. The target was silicified resistive features hosting precious metals.

Figure 6.54 is a comparison of 2048 and 32 Hz data. Dark shades are conductive, and light shades are resistive. Notice the general decrease in resistivity as frequency decreases (depth increases). Also notice the four features which are outlined to demonstrate some of the interpretational characteristics of this data set. "A" points out a shift in position of a conductive area which implies a dipping conductive layer. This is much better shown in the depth plots in figure 6.55. "B" points out a conductive feature that is present at all frequencies, indicating a static effect. "C" is not a static effect, but is a dramatic change in resistivity with depth — going from a conductor at high frequencies to a resistive feature at depth, making this a possible gold target (silicified zone in a highly altered environment). "D" is another static effect. This time it is a resistive feature.

Figure 6.55 shows the modeled, static-corrected data. Notice the near-surface noise (near-surface changes in resistivity) in the 100-ft depth section, and also notice that they disappear in the 400-ft depth section as might be expected. Shallow features do not dominate the results. Static effects "B" and "D" in figure 6.54 have been removed. "A" shows up much clearer as a dipping conductive layer, and "C" comes through as a strong resistive feature at depth.

FIGURE 6.54
Plan view maps of high density Cagniard resistivity at 2048 and 32 Hz. Gathered over a Nevada pediment prospect

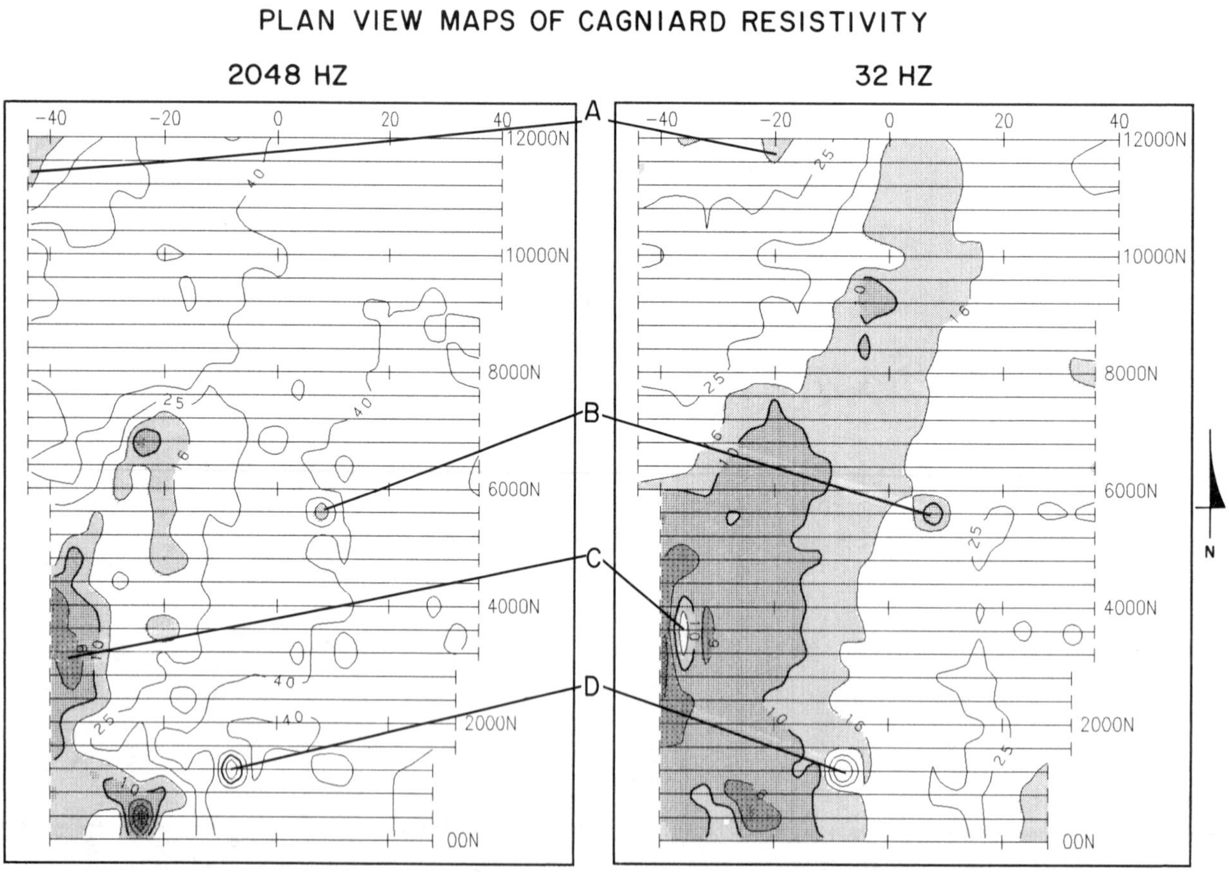

FIGURE 6.55
Plan view maps of static-corrected and smooth-modeled Cagniard resistivity for the 100 ft and 400 ft levels

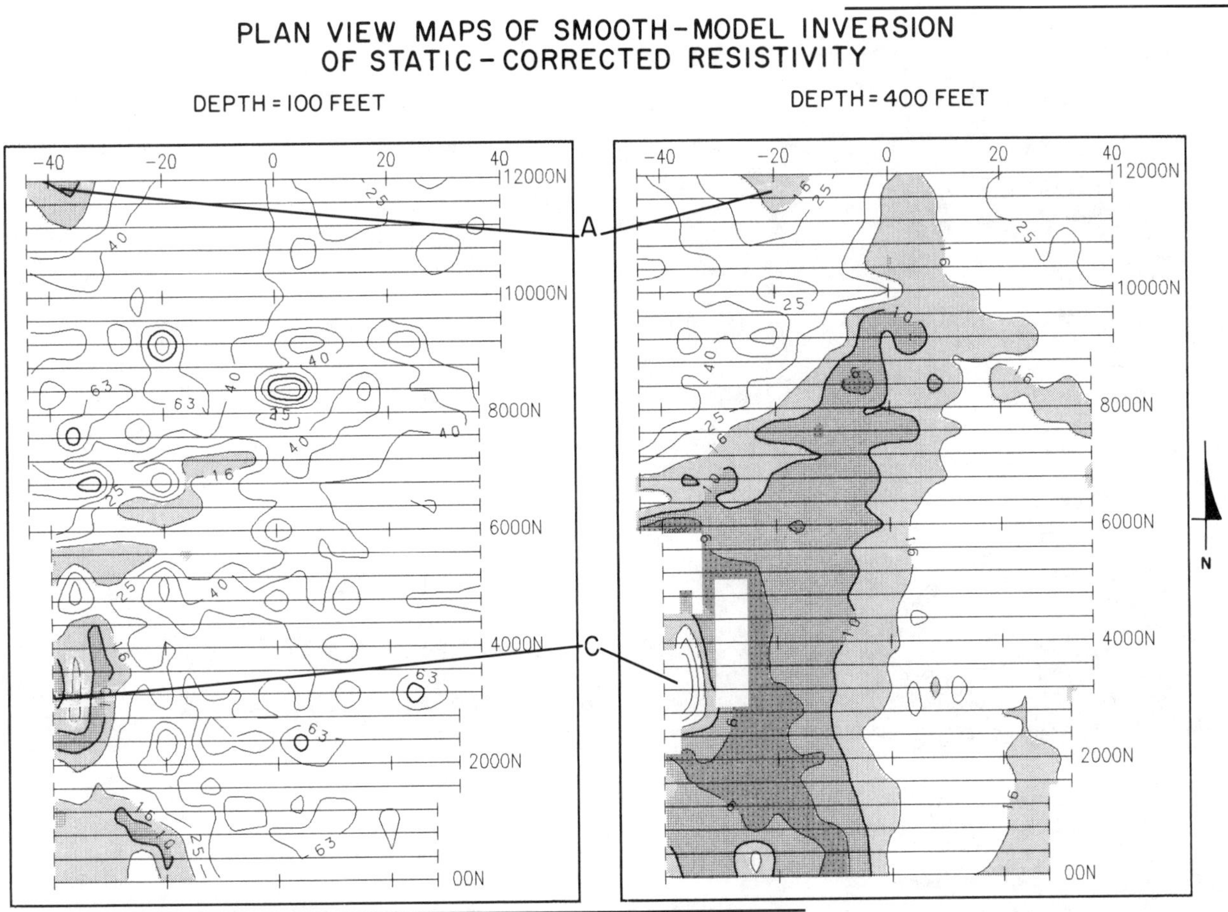

Feature "C" was sufficiently interesting to go back and place two detail lines straddling Line 3600 using 100 ft electric field dipoles to check out the validity of this resistive pinnacle. The results are shown in figures 6.56 and 6.57. Figure 6.56 compares the 400 ft and 100 ft data showing the detail of resistor "C". Notice the excellent correlation between the 100 ft and 400 ft data. Figure 6.57 compares the two high resolution lines placed 400 ft apart and straddling line 3600, showing consistency and repeatability between the two lines.

The lowest frequency used on this survey was 32 Hz which was not low enough to acquire data at depth in the very conductive zones. Hence the missing depth data on portions of the plots. Recall that depth of exploration is proportional to the square root of resistivity divided by frequency. The lack of data in conductive zones was not considered to be a detriment in this survey, since the object was to find resistive targets.

FIGURE 6.56
Model results of Line 3600N using 400 ft E-field dipoles and high resolution. Line 3400N using 100 ft E-field dipoles

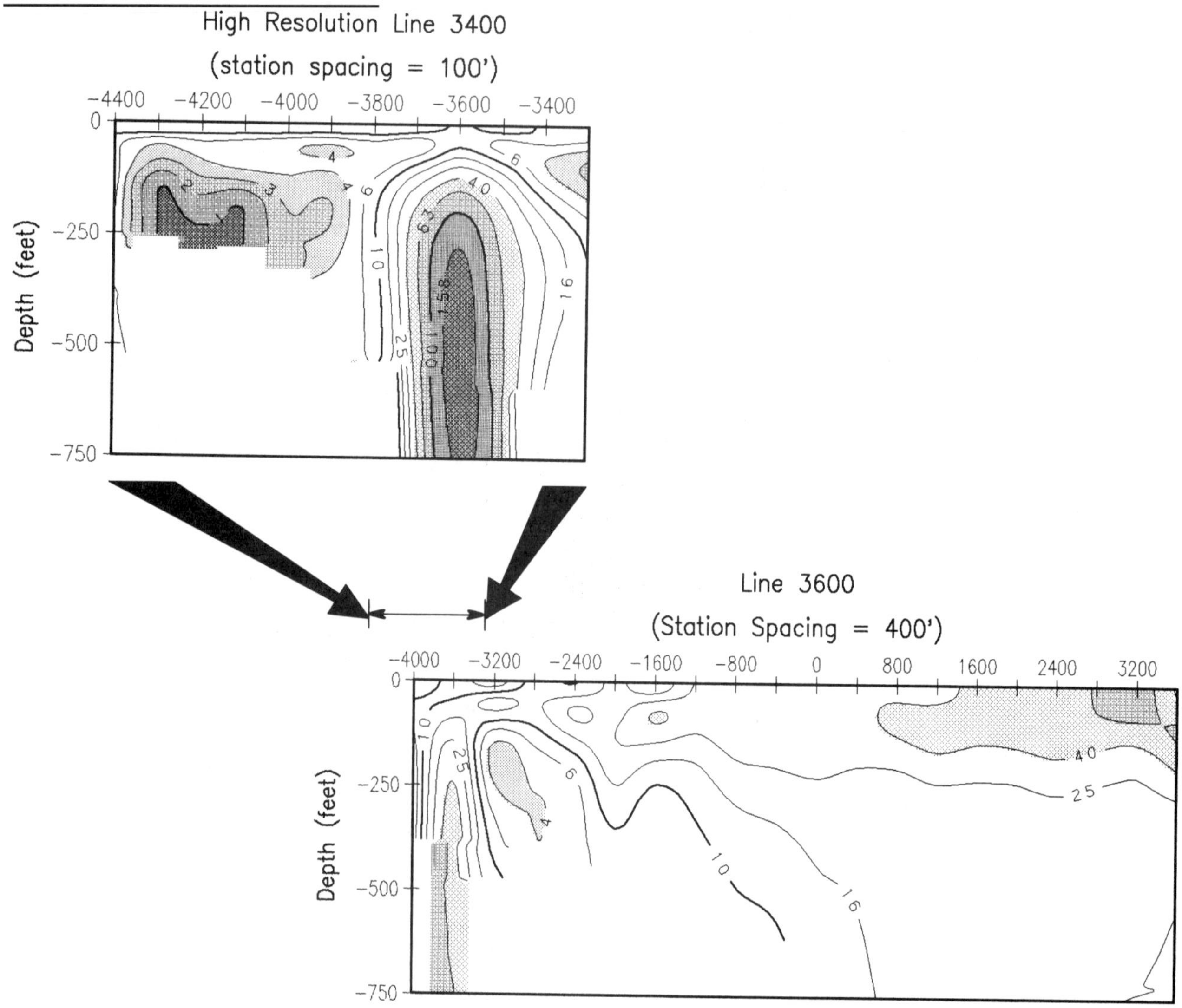

DETECTING SUBSURFACE WATER AND STRUCTURE

CSAMT has been used extensively for structure mapping in mineral and hydrocarbon exploration, civil engineering and in searching for potable water. This example is a survey that was conducted north of Phoenix, Arizona, for a potential commercial development of a piece of desert. The object was to help determine whether sufficient water was available on the property to support development.

The only geophysical data for the area was regional gravity. Figure 6.58 shows the surface geology. Granitic outcrops trend approximately N50°W. The outcrop is discontinuous along Skunk Creek, suggesting structural control in this area.

Two drill holes were placed in the southwestern part of the property based on surface geology. The holes were drilled to a depth of 500 ft. Water was encountered near 300 ft in hole B and at 140 ft in hole A, but saturated layers were thin and there was insufficient flow.

FIGURE 6.57
Comparison between smooth-modeled inversions of the two high resolution lines

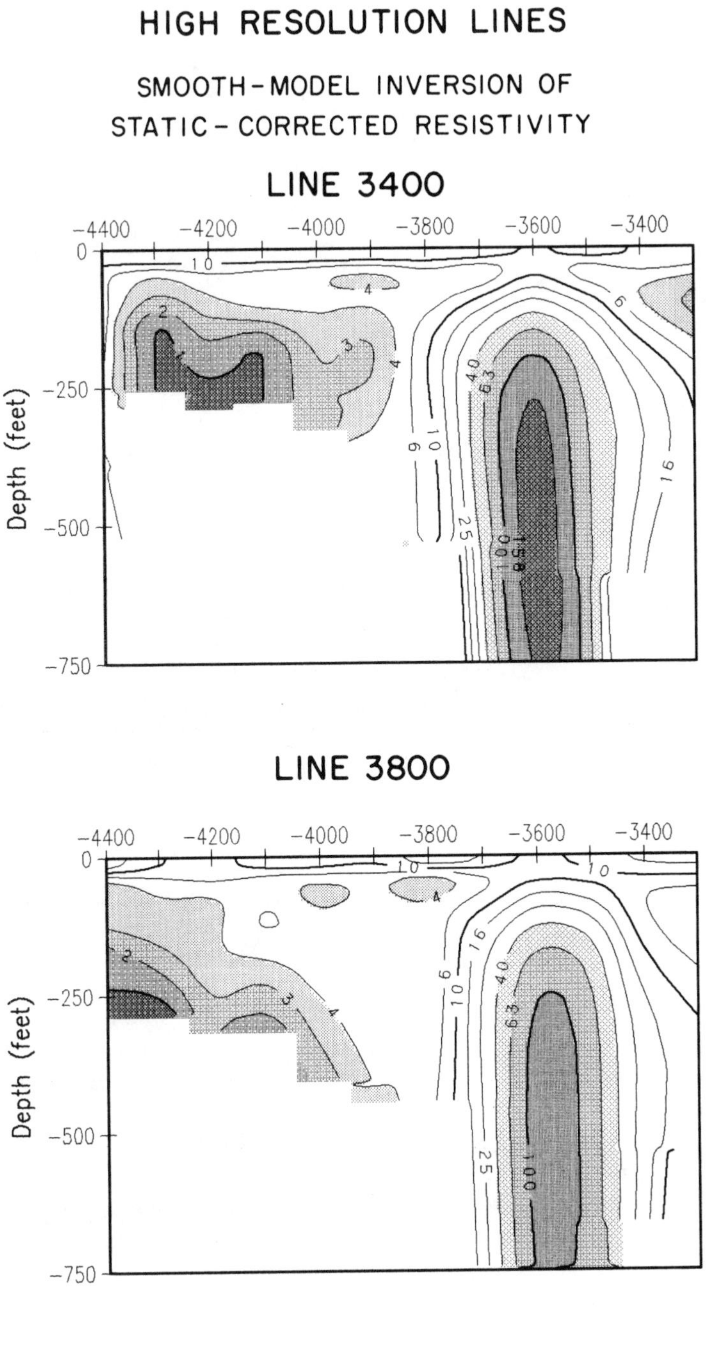

Three lines of CSAMT data were then obtained to provide information about subsurface structure. The apparent resistivity data shown in figure 6.59 indicate a fault at station 8, which lies in Skunk Creek. All three lines show the fault as a NNE-trending feature.

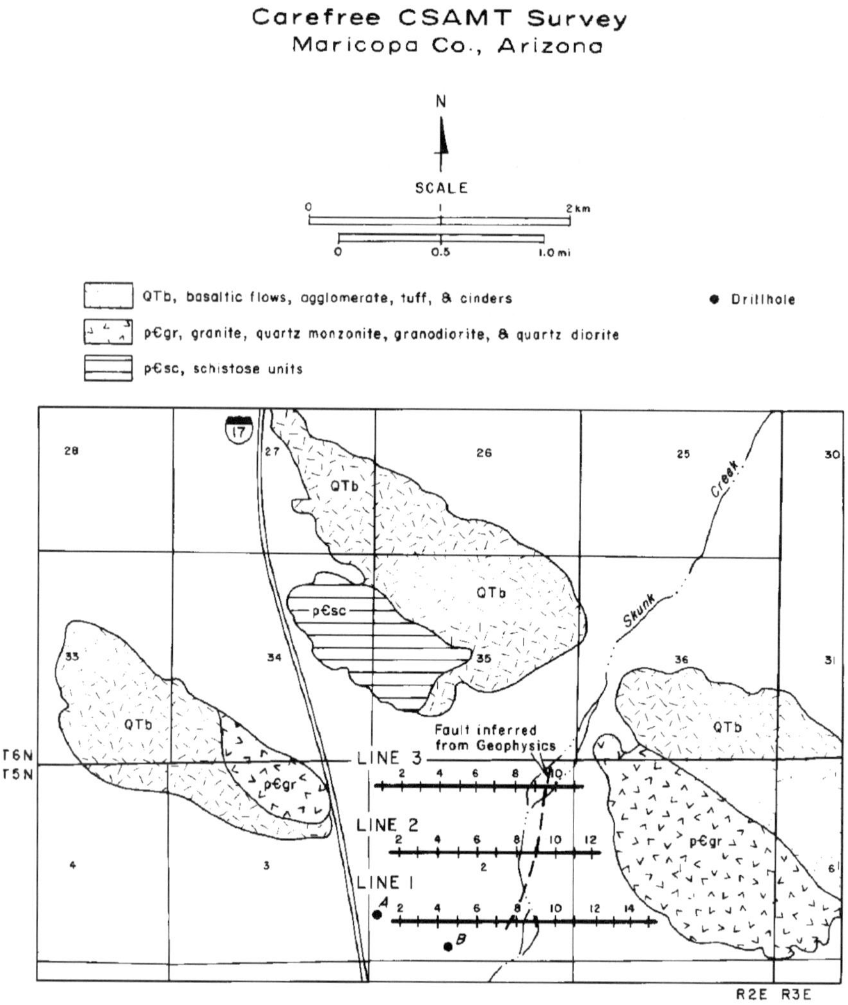

FIGURE 6.58
Surface geology of the groundwater survey area (after Zonge and Hughes, 1991)

The data exhibit two types of behavior. As shown in log-log plots in figure 6.60, the stations well away from the fault show type A behavior, with the shallow notch near 64 Hz. This notch is the standard transition-zone notch that is a function of transmitter/ receiver separation and geology (resistivity contrasts). Stations near the fault show type B behavior, with a sharper notch. Computer modeling indicated that a conductive layer near 1000 ft was needed to provide this type of behavior. This suggested that an unsuspected aquifer could occur deeper than the testing done to date. The decision was made to deepen hole B to investigate the conductor near 1000 ft.

Drilling encountered a thick water saturated zone at 1040 ft (figure 6.61). The saturated layer was more than 200 ft thick and showed a pressure head of 360 psi on a 6 in borehole. It is estimated that with a 17 in borehole, the well will produce from 600 to 1000 gpm.

The CSAMT survey was significantly less expensive than the drilling program and provided information which was crucial to the success of the project. Drilling to 1000 ft would have been risky in this area without some geophysical basis. Based upon the geophysics, however, continued drilling located water which will support commercial development of the property.

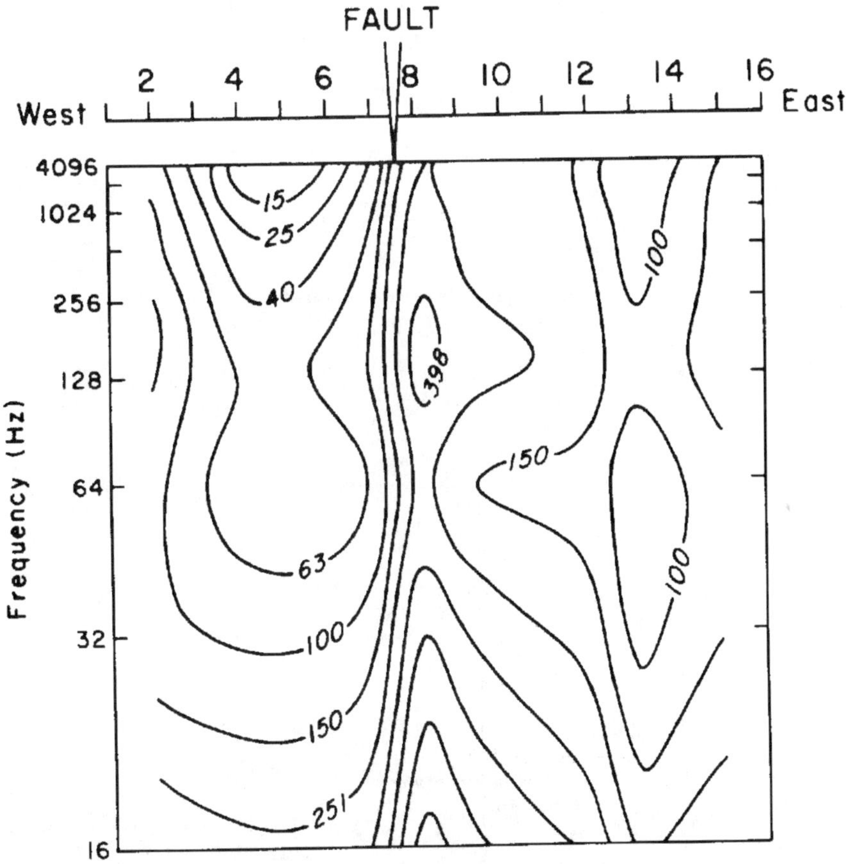

FIGURE 6.59
Cagniard resistivity for Line 1 (after Zonge and Hughes, 1991)

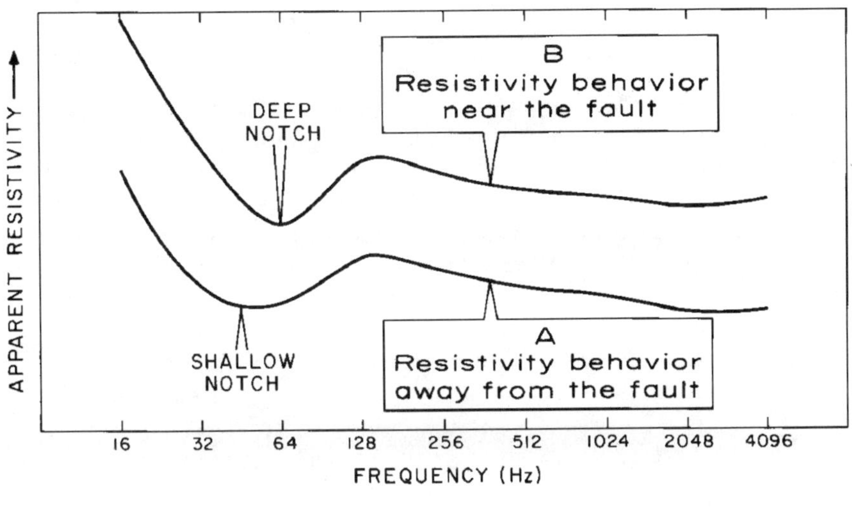

FIGURE 6.60
Comparison between the two types of Cagniard resistivity curves measured along Line 1 (after Zonge and Hughes, 1991)

TRACING INJECTION FLUIDS

CSAMT has been successfully used to trace the migration of conductive fluids for in-situ mining projects and injection tests. In this example, an in-situ leaching project was recovering only 10 percent of the injected fluid, and it was necessary to determine where the rest of the fluid was going. Figure 6.62 displays the modeled CSAMT data which was processed it to give resistivities at fixed levels. The

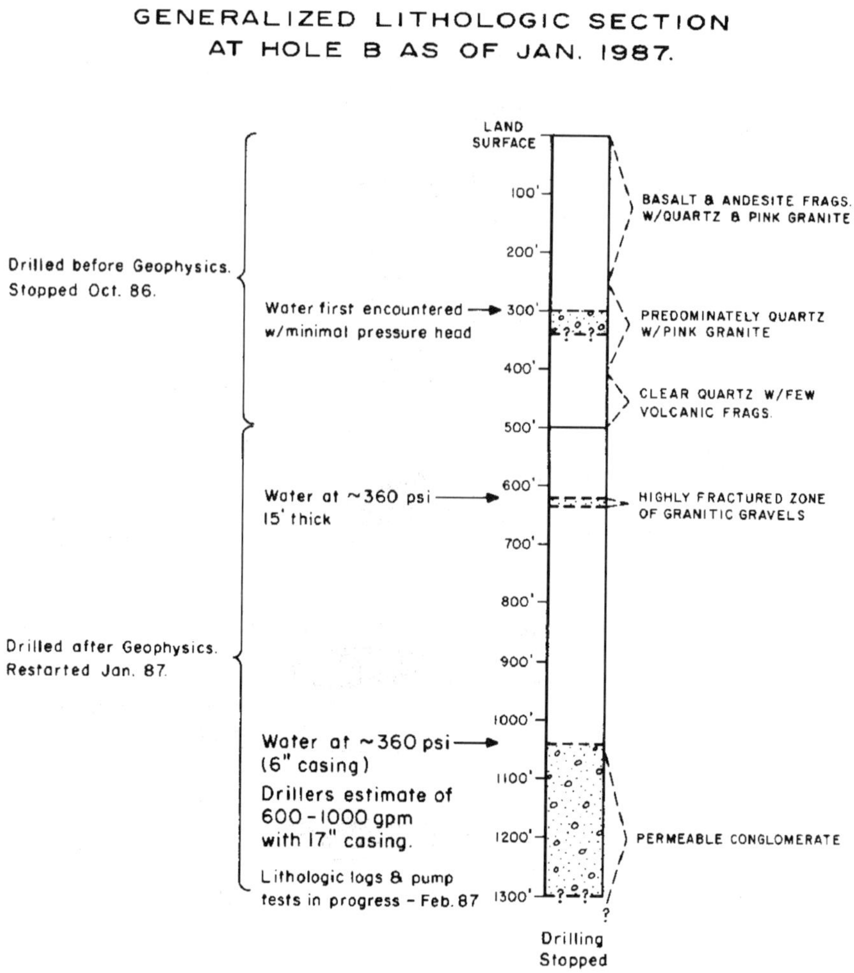

FIGURE 6.61
Generalized lithologic section of drill hole B (after Zonge and Hughes, 1991)

heavy shading indicates conductive zones, with lighter shades being more resistive. The conductive fluid is associated with the dark feature which is centered at the 250 m level and migrates toward the bottom right-hand corner of the individual plots as depth increases.

Prior to the CSAMT survey, a well was placed at the 750 m level to intercept the fluid. Apparently the fluid was migrating more slowly than anticipated by the project hydrologists. A well placed at the 450 m level intercepted fluid and enhanced recovery began.

These results show that you can measure through conductive features and can get a good indication of dip for such structures.

Two surveys were run over this area one month apart to see if the fluid motion could be traced. Results of the survey demonstrated that in some areas, the apparent resistivity at depth was changing in excess of 1 percent per day. In figure 6.63, the darkest shades indicate a 30 percent change between surveys. Notice that the major changes at depth are in the bottom right hand portion of the plot, indicating that the fluid is penetrating to the lower levels as predicted by the hydrologists.

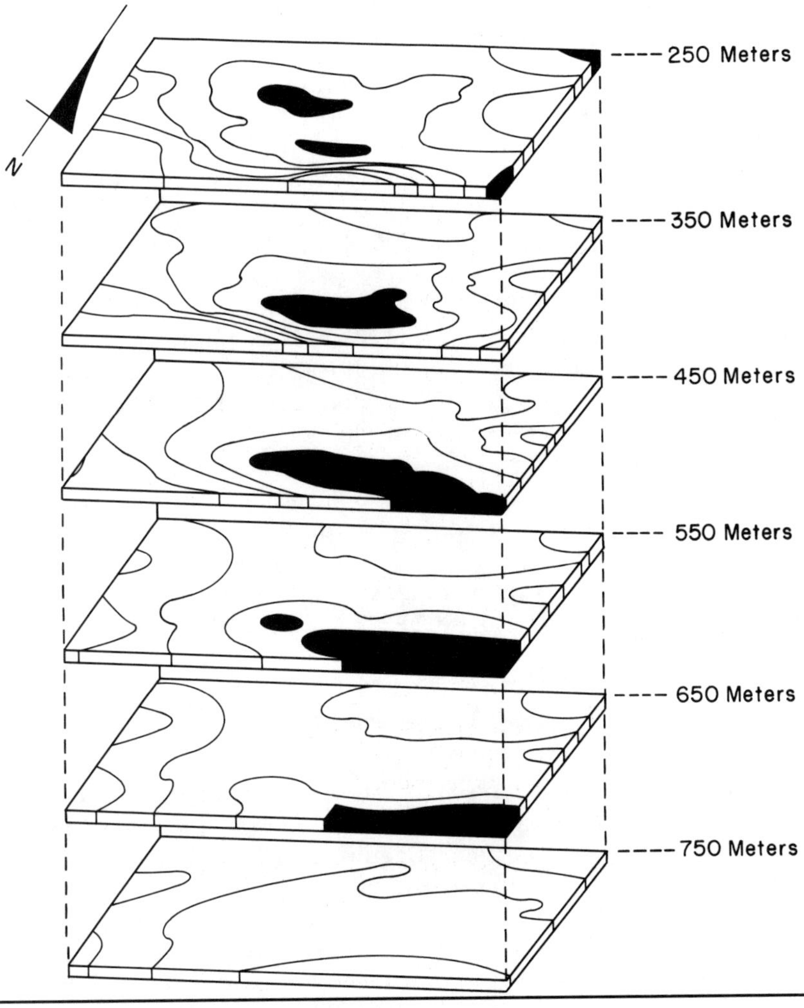

FIGURE 6.62
CSAMT data inversion results: plan views at constant depth (adapted from Zonge and Hughes, 1991)

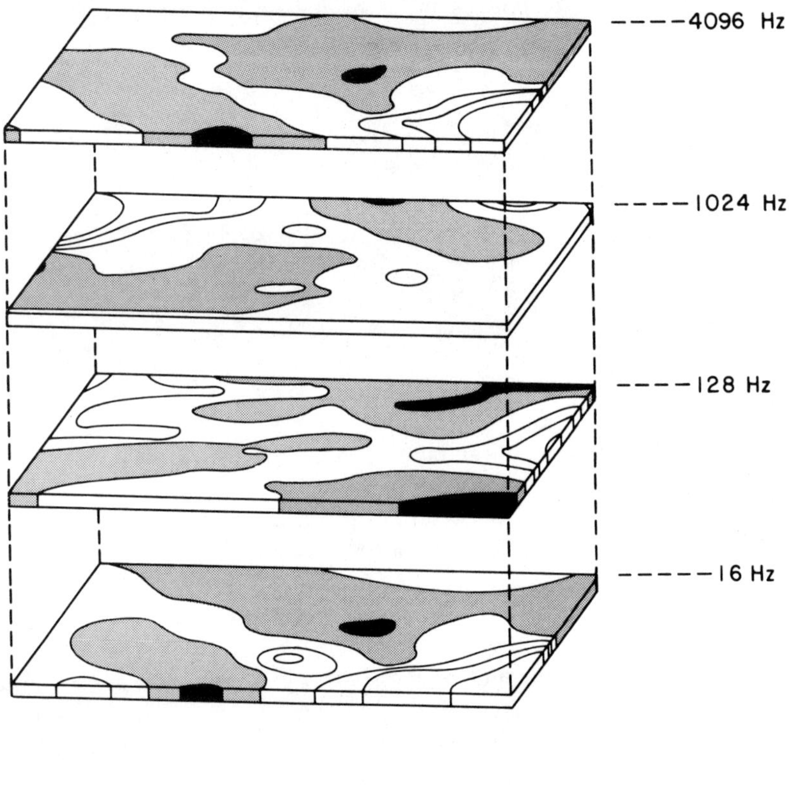

FIGURE 6.63
Percentage change per day for selected frequencies

INTRODUCTION TO TEM

The transient electromagnetic or TEM technique, which is sometimes called time-domain EM (TDEM) or pulse EM (PEM, EMP), can be used for vertical depth sounding or profiling. The equipment consists of a transmitter and receiver which can be contained in a single box (with an external power source for the transmitter) or used with separate transmitter and receiver enclosures. The depth of exploration attained in a vertical sounding configuration can vary from tens of meters to more than 1000 m, depending upon transmitter loop size, available power from the transmitter and ambient noise levels.

TEM was first used in Russia and Australia to sound through conductive overburden in exploring for base metal deposits. The technique has found extensive use in Russia and China for hydrocarbon exploration and is used for base metal exploration worldwide. Today, in addition to base metal and hydrocarbon exploration, TEM is used in exploration for geothermal sources, mapping structure and lithology, searching for sources of groundwater and groundwater contamination, and for engineering applications.

The TEM vertical sounding arrays produce essentially the same type of data as the CSAMT method, but the effect of lateral changes in resistivity are smoothed with TEM, and the logistics are completely different.

ARRAYS AND SENSORS

The TEM source consists of a fast turn-off, high-current transmitter driving a large loop placed on the ground. The actual size of the loop depends upon the depth of investigation desired and the amount of area to be covered. A battery-powered transmitter will produce currents of up to 5 amperes in a typical loop, and high-powered generator-driven transmitters can produce 50 amperes or more. The current is cycled on and off in pulses of alternating polarity. Receiver measurements are made during the off time when only the secondary fields are present. A pulse of energy is put into the ground during the current turnoff time, and the voltage generated in the receiver loop or magnetic coil antenna is measured as the induced energy diffuses down into the earth.

Typical time intervals or delays monitored on the decay waveform vary from 50 microseconds to hundreds of milliseconds for mining applications. For shallow engineering applications, time windows can vary from microseconds to perhaps 10 milliseconds. Each decay waveform is measured with logarithmically spaced data points (time windows) which give a relatively uniform depth measurement. Unlike reconnaissance CSAMT, TEM measurements are made one station at a time, but the measurements are made quickly since each transient is equivalent to a wide spectrum of individual frequency measurements.

Figure 6.64 shows the three basic TEM configurations: a) the vertical sounding mode, in this case an in-loop configuration; b) the profiling mode using a fixed-loop configuration; and c) downhole measurements using a surface transmitter loop and a drill-hole receiver probe.

To obtain the precise timing accuracy needed to measure responses in the microsecond range, it is necessary to have the transmitter and receiver synchronized — that is, both must run at the same frequency and switch at exactly the same time. This can be done by using a direct connection between the receiver and transmitter or by using a very accurate, low-drift crystal clock in each unit. The two methods for controlling the timing are shown schematically in figure 6.65c for the in-loop sounding array.

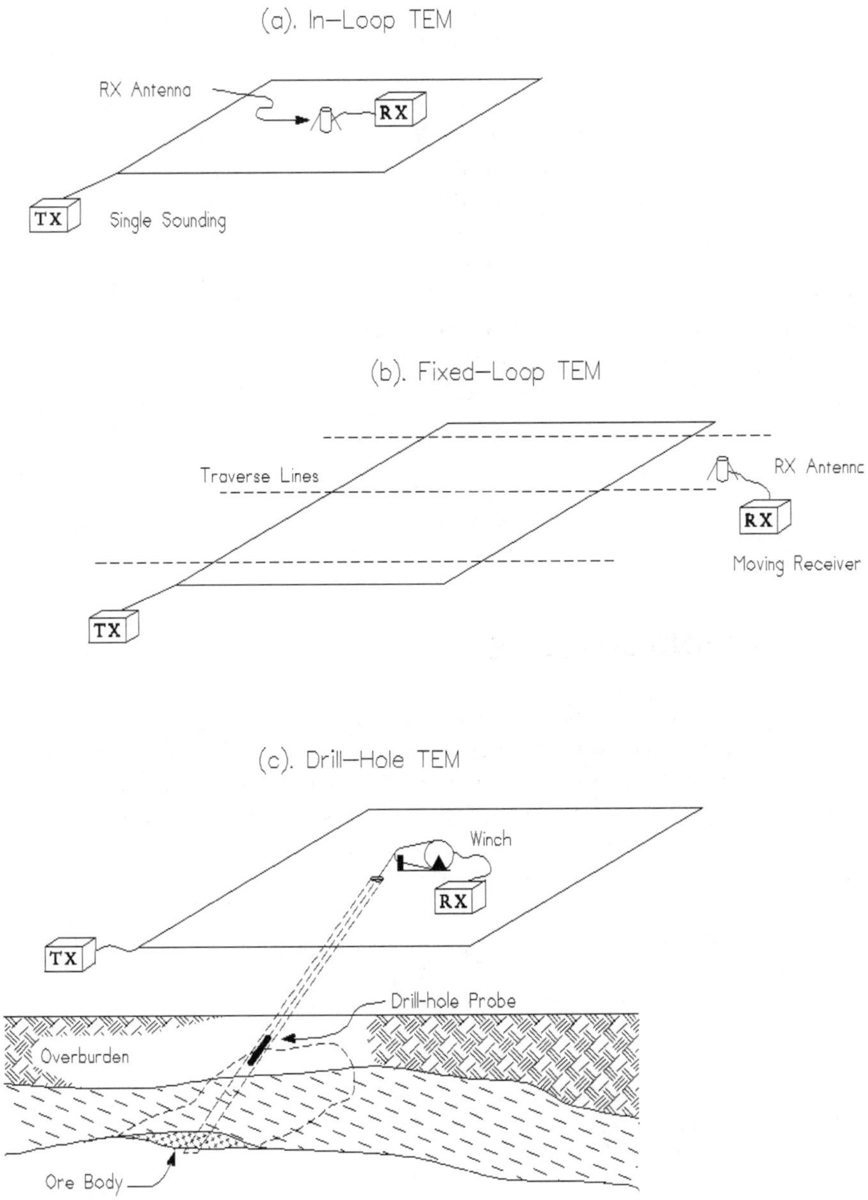

FIGURE 6.64
Field setup for (a) vertical sounding, (b) profiling and (c) downhole TEM measurements

Vertical Sounding Arrays

The vertical sounding method uses a large square loop, typically on the order of 100 m to 300 m on a side, but it can range from tens of meters to more than 1000 m, depending upon the depth of investigation required. Large loops allow a greater depth of investigation but are more cumbersome to set up. There are three commonly used configurations: single-loop, where the transmitter and receiver use the same large loop (this configuration can be used only with receiver/transmitter combinations that permit connecting the receiver directly to the transmitter output); coincident-loop, for which the transmitter and receiver use different large loops placed with sides parallel and offset by a small amount; and in-loop, which uses a small air-core or mumetal/ferrite-cored antenna placed at the center of the large transmitter loop. Figure 6.65 shows these three configurations in schematic form.

FIGURE 6.65
Vertical sounding configurations for TEM receivers (Rx) and transmitters (Tx): (a) single loop, (b) coincident loop, and (c) central or in-loop

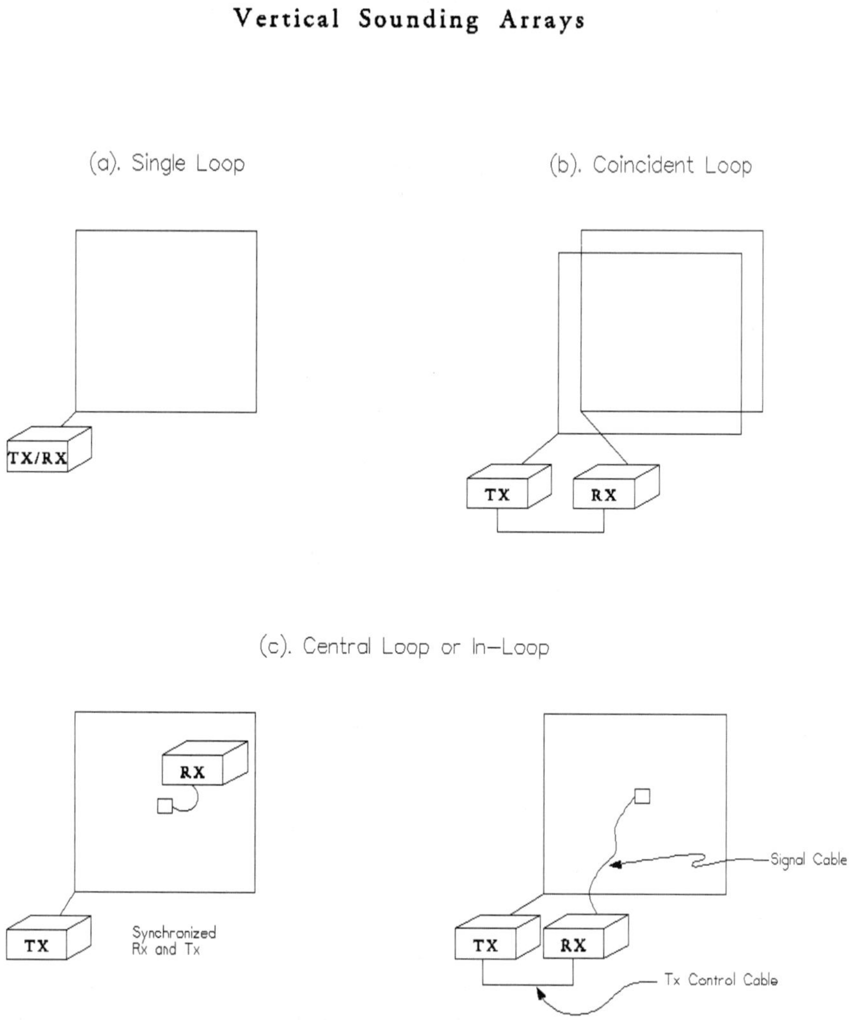

The vertical field component, Hz, is normally measured, and the data are usually presented as amplitude versus time decay plots, depth versus resistivity sections, or profiles of amplitudes at fixed delay times.

In the vertical sounding mode, the TEM depth of investigation is approximately 28 ρ_at meters, where ρ_a is the apparent resistivity in ohm-meters and t is time in milliseconds. Table 6.2 lists the basic equations used for vertical sounding applications. These equations hold for a layered earth and are independent of transmitter loop size.

There is no theoretical limit to the maximum depth of exploration. The practical limit, however, is dictated by the magnitude of the smallest detectable signal. The size of the transmitter loop, the current used (energy put into the ground), the sensitivity of the receiver system and the ambient noise background all contribute to the maximum depth of exploration. Figure 6.66 summarizes these parameters in a nomograph for a typical background noise level of 1 nanovolt per square meter.

As a general rule, the depth of exploration is between one and two times the diameter of the transmitter loop. The depth of

Table 6.2
Useful equations for vertical sounding TEM

Equations for Vertical Sounding TEM Methods

Apparent resistivity, ρ_a, (late time):

$$\rho_a = \left(\frac{I\ A_T\ A_R}{V}\right)^{\frac{2}{3}} \left(\frac{1}{t}\right)^{\frac{5}{3}} \times 6.3219E-3 \quad \text{(ohm-meters)}$$

Diffusion depth, d:

$$d = 40\sqrt{\frac{t}{\sigma}} = 40\sqrt{\rho \cdot t} \quad \text{(meters)}$$

Equivalent depth of investigation, D:

$$D = 28\sqrt{\frac{t}{\sigma}} = 28\sqrt{\rho \cdot t} \quad \text{(meters)}$$

Where: A_T = Transmitter moment in square meters

A_R = Receiver moment in square meters

I = Transmitter current in amperes

V = Recieved voltage in microvolts

t = Time in milliseconds

σ = Conductivity in siemen/meter = $\frac{1}{\rho}$

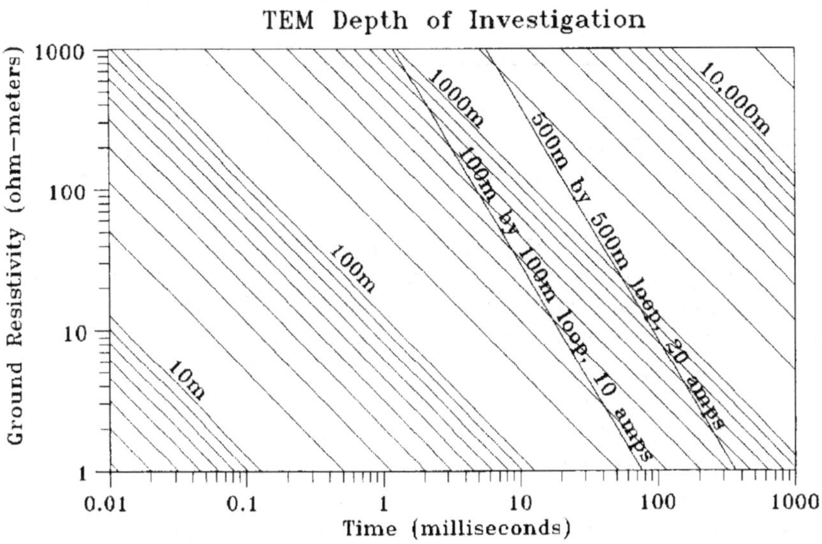

Adapted from Fitterman, 1989, Geophysics v54 using a noise level of 1 nV/m².

FIGURE 6.66
Depth of investigation plot for TEM assuming a noise level of 1 nanovolot per square meter, showing the absolute limits for 100 by 100 and 500 by 500 m loops. Adapted from Fitterman, 1989

exploration is approximately two times the loop diameter in resistive ground and closer to one loop diameter over very conductive ground. All of this assumes nominal transmitter currents and background noise.

Vertical sounding arrays are most useful in a layered-earth environment, since they do not couple well into 3-D objects that have a small cross section compared to loop diameter. Although other loop geometries can be used for special problems, the configurations mentioned above are the most common.

Profiling Arrays

There are three basic types of profiling arrays: fixed-loop, which uses a fixed, large transmitting loop with the receiver moving along lines oriented perpendicular to one side of the loop; moving-loop (also called separated-loop or loop-loop), which is usually a transmitter loop and receiver loop of similar size, spaced at least one loop diameter apart, moving along a profile line together; and LOTEM (long-offset TEM), which uses a grounded dipole transmitter placed a large distance from the receiver lines (very similar to CSAMT). Figure 6.67 displays these three different configurations.

As in the sounding arrays discussed above, the profiling arrays always measure the vertical component of the magnetic field, Hz. Often a horizontal component of the magnetic field (Hx) oriented parallel to the direction of the profile line is also measured to give estimates of 2 dimensionality and 3 dimensionality.

One standard configuration for the fixed-loop survey consists of a rectangular transmitter loop, say 300 by 600 m, with the long side oriented parallel to the expected strike of the target conductors. Survey lines are oriented perpendicular to the long side of the transmitter loop and geologic strike as shown in figure 6.67a. This provides maximum coupling into narrow, steeply dipping conductive features outside of the transmitter loop.

The receiver lines run from inside the transmitter loop out to twice the width of the transmitter loop. Transmitter loop size is a compromise between the largest loop possible for maximum power and area coverage and the need for efficient layout logistics.

The moving-loop configuration shown in figure 6.67b also couples well into vertical dike-like features when running perpendicular to strike. This array is most easily used with small loops and in open areas.

A method pioneered by Newmont uses a very large loop, perhaps 1000 m on a side or larger, to encircle an area of interest. Then all three components of the H-field (Hx, Hy, Hz) are measured over a grid, inside and outside of the loop, to map the 3 dimensionality of targets at depth. To maximize coupling, the long sides of the loop are oriented parallel to the suspected strike of conductors. Data can be plotted in both profile and contoured plan-map form.

The LOTEM technique is logistically similar to the CSAMT method with the transmitter dipole placed 5 km to 20 km away from the receiver area. The vertical magnetic field is always acquired, and the horizontal electric field (parallel to the transmitter dipole) is sometimes measured for better lateral definition. This method is used in both sounding and profiling modes.

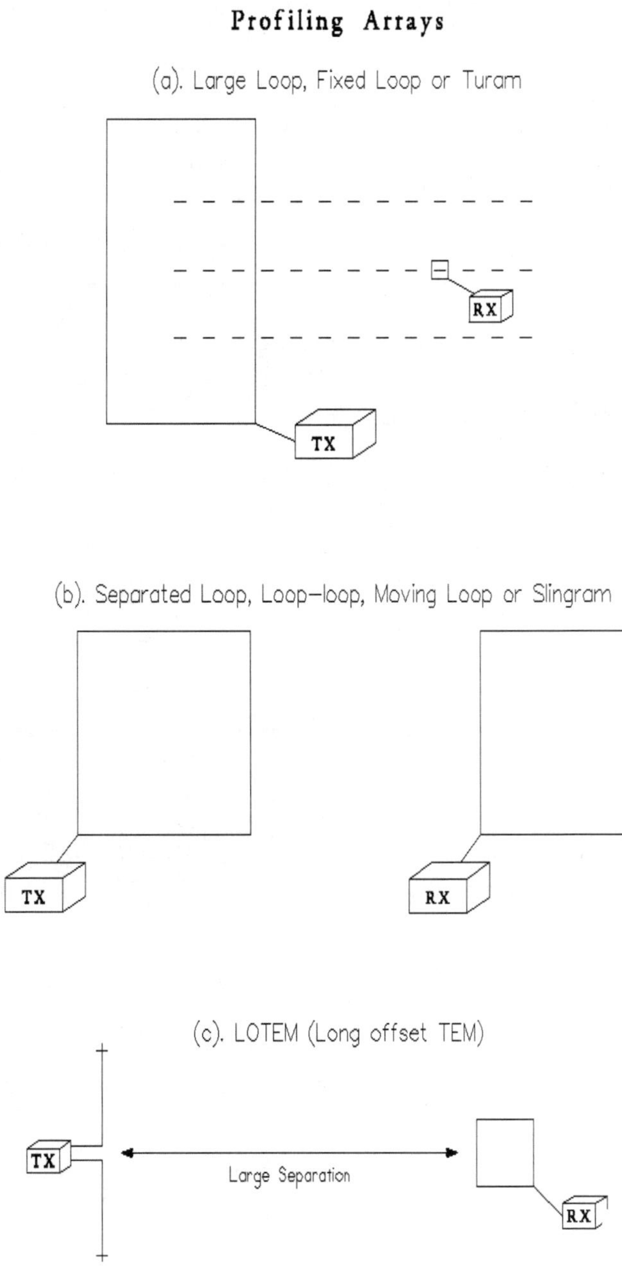

FIGURE 6.67
Profiling arrays for TEM: (a) large loop or fixed loop, (b) separated loop or moving loop, and (c) long offset TEM (LOTEM)

Drill-Hole Arrays

The most commonly used drill-hole array consists of one or more transmitter loops placed on the surface around a drill hole, with the receiver antenna placed down hole. (See figure 6.68.) Often multiple loops are used to try to locate conductive features that may occur off to the side of a drill hole. Offset loops may also be necessary to cover the target when angle holes are used.

Figure 6.68a shows the configuration for a single-loop placed symmetrically around a vertical drill hole. Figure 6.68b shows a typical setup for determining on which side of a drill hole a missed orebody can be located. In figure 6.68c, the transmitter loop is offset from an angle drill hole and is used to excite anomalous areas in the vicinity of the probe at depth.

In practice, 100-m by 100-m loops can be used in resistive environments for surveying close up drill holes to about 400 m in depth. For deeper holes, a 1-3 loop-size to depth ratio is commonly used. For example, to log a 1000-m hole, one would chose to use a 300-m, or larger, surface loop. In a conductive environment, you might want to drop to a 1-2 ratio.

Many different configurations could be conceived for special conditions. The main consideration is to orient the transmitter loop for maximum coupling to the feature of interest and to the down-hole probe.

FIGURE 6.68
Drill-hole arrays: (a) single-loop, (b) multiple loop, and (c) offset loop

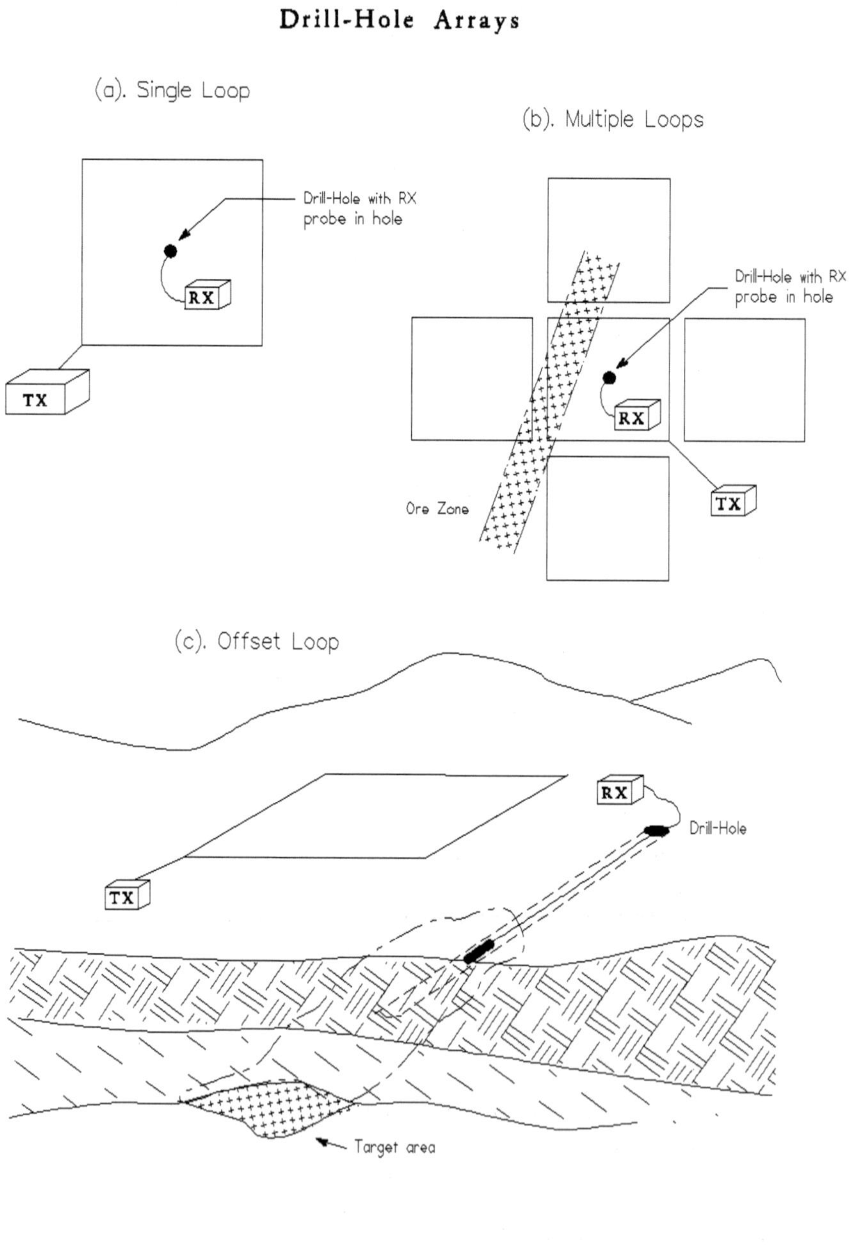

DESCRIPTION OF MEASURED FIELDS

With the exception of LOTEM, all TEM systems couple energy into the ground by magnetic induction. That is, energy is put into the ground and coupled into conductive features by rapidly changing current in a loop of wire. The resultant induced electromotive force (EMF) is proportional to the time rate of change of the primary magnetic field. This EMF in turn generates currents in conductive materials that decay at a rate proportional to its conductivity. Induced currents in poor conductors (moderate resistivity) decay quickly, currents in good conductors (very low resistivity) decay slowly, and very poor conductors (highly resistive silicified dikes, for example), will not sustain any measurable induced currents.

The parameter normally measured for TEM is the time rate of change of the magnetic field, or the voltage induced in a loop or coil antenna. This type of antenna has a frequency domain response that is directly proportional to frequency; e.g., if the frequency of the magnetic field is increased by a factor of two, the signal magnitude coming out of the antenna increases by a factor of two. Most of the plots in this chapter use volts/ampere for the magnitude axis, which is the received voltage normalized by the current in the transmitter loop.

Some TEM systems, notably UTEM and LOTEM, also acquire electric field data for special types of surveys. Since we are considering only magnetic induction effects in this discussion, this is just mentioned here as a point of interest.

Figure 6.69 is a rough schematic showing amplitude versus time plots of idealized transmitted and received waveforms. Notice that you get pulses of energy during turn-on and turn-off of the current in the transmitter loop. Most instruments are designed to turn the current on more slowly than it is turned off. It is important to control the turn-off portion of the current to better analyze and model the data. The generated EMF is proportional to the maximum amplitude of the current and how quickly it is turned off. Also, the faster the current is turned off, the shallower one can sound, assuming that the receiver can sample equally fast at the early times.

FIGURE 6.69
Generalized schematic of TEM system waveforms

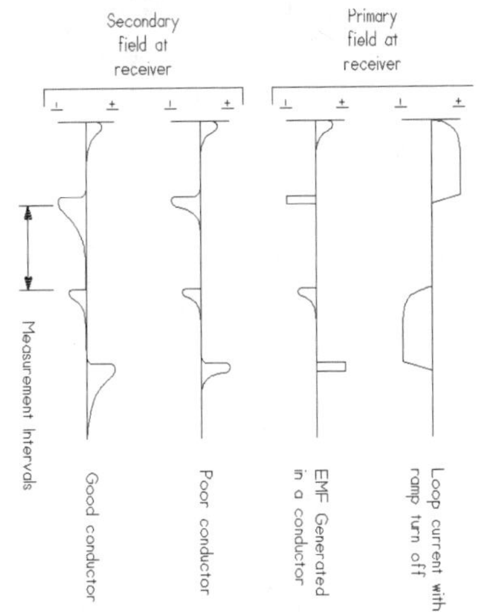

Data Presentation

Data presentations are usually selected from five basic formats: log amplitude versus log time plots of decay waveforms and apparent resistivity, log amplitude versus station profile plots, linear amplitude versus station profile plots, resistivity versus time or depth pseudosections (similar to the CSAMT resistivity plots), and contour plots in plan map form.

The following data set demonstrates four different presentation formats. It was acquired over a test site near Tombstone, Arizona, using an in-loop setup with a 600 by 600 ft (183 by 183 m) transmitter loop and 600-ft station spacing.

Figure 6.70a is a plot of decay waveforms for nine stations, and 70(b) shows the corresponding late time apparent resistivity. Notice that the delay times approximate logarithmic time intervals to give a uniform data presentation. Figure 6.71 shows the same data plotted in log amplitude profile form, and figure 6.72 presents the data in four linear groupings. Figure 6.73a is a plot of the data in apparent resistivity versus depth section form, and 73b displays the results of smooth-model inversion. Notice that the basic low-high-low resistivity layering is evident in both plots.

FIGURE 6.70
Tombstone test data: (a) Log amplitude vs. log time plot of the decay waveforms and (b) corresponding late-time resistivity plots

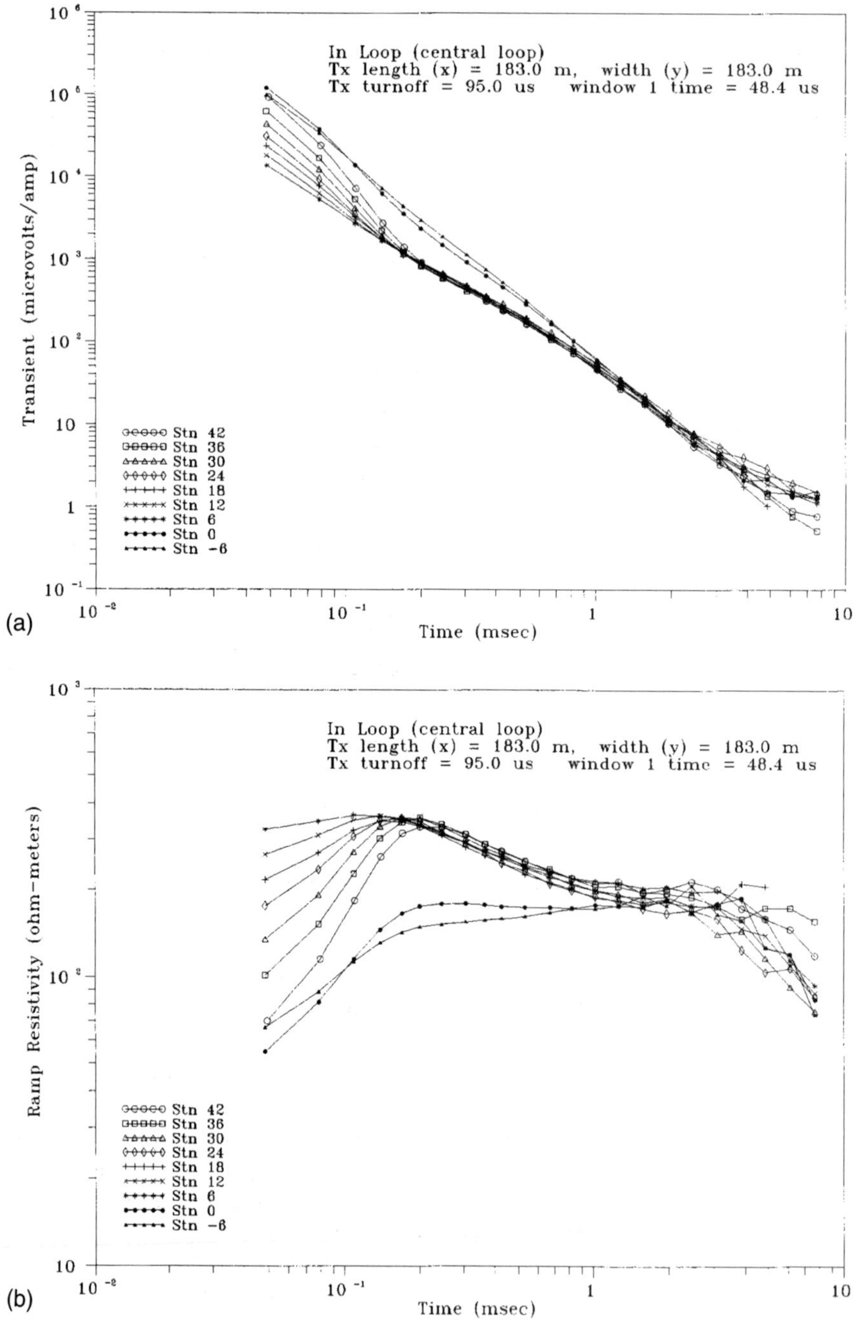

The in-loop array permits easy calculation of apparent resistivity and the generation of resistivity versus depth sections. Resistivity calculations for separated-loop and fixed-loop systems are not quite as straightforward, since the mathematical description of the field behavior is much more complicated. Depth and apparent resistivity for this in-loop example are calculated using the equations in table 6.2.

The contoured plan map approach is useful when a large amount of gridded data are available, e.g., from fixed-loop surveys. Contour plan maps are based on plotting values for a single time delay for each plot. Time delays are chosen which represent the target depths of interest for the survey. See figure 6.101 for an example of contoured field data from a fixed-loop survey in Australia.

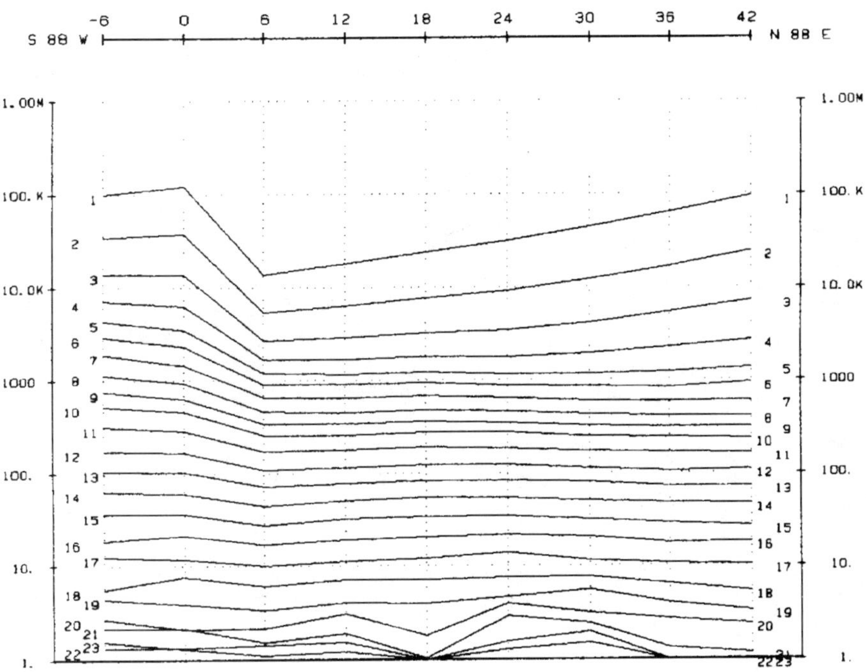

FIGURE 6.71
Tombstone TEM test data: log amplitude vs. station profile plot

Window NUMBER and TIME (seconds)

W 1: 48.00u	W11: 534.0u	W21: 4.868m
W 2: 79.00u	W12: 653.0u	W22: 6.119m
W 3: 109.0u	W13: 820.0u	W23: 7.706m
W 4: 140.0u	W14: 1.017m	
W 5: 170.0u	W15: 1.258m	
W 6: 200.0u	W16: 1.561m	
W 7: 245.0u	W17: 1.952m	
W 8: 306.0u	W18: 2.465m	
W 9: 367.0u	W19: 3.101m	
W10: 442.0u	W20: 3.886m	

Generation of Measured Magnetic Fields

The received voltage is made up of two components: fields from the vortex current that is induced into and contained within conductive bodies, and the galvanic current that is induced into the host rock or host material.

Conductive bodies in conductive hosts tend to channel or gather the galvanic currents in their vicinity. This can result in a stronger response from conductive bodies than expected. Predictions based on a free-space model will make the anomaly more conductive and deeper than it actually is.

Figure 6.74 is a schematic showing the electromagnetic fields and currents about a conductive plate in a conducting half space. In figure 6.74a, vortex currents (Jv) flow in the conductor in a direction that generates a secondary magnetic field (Bv) which opposes the primary magnetic field B. Galvanic currents (figure 6.74b) will flow in a direction parallel to the transmitter loop wire and perpendicular to the primary magnetic field. Notice that these galvanic currents channel through the conductor and generate a secondary magnetic field (Bg) that also opposes the primary field. The galvanic response adds to the response from the vortex currents providing an enhanced anomalous response for this plate. This is one example of current gathering.

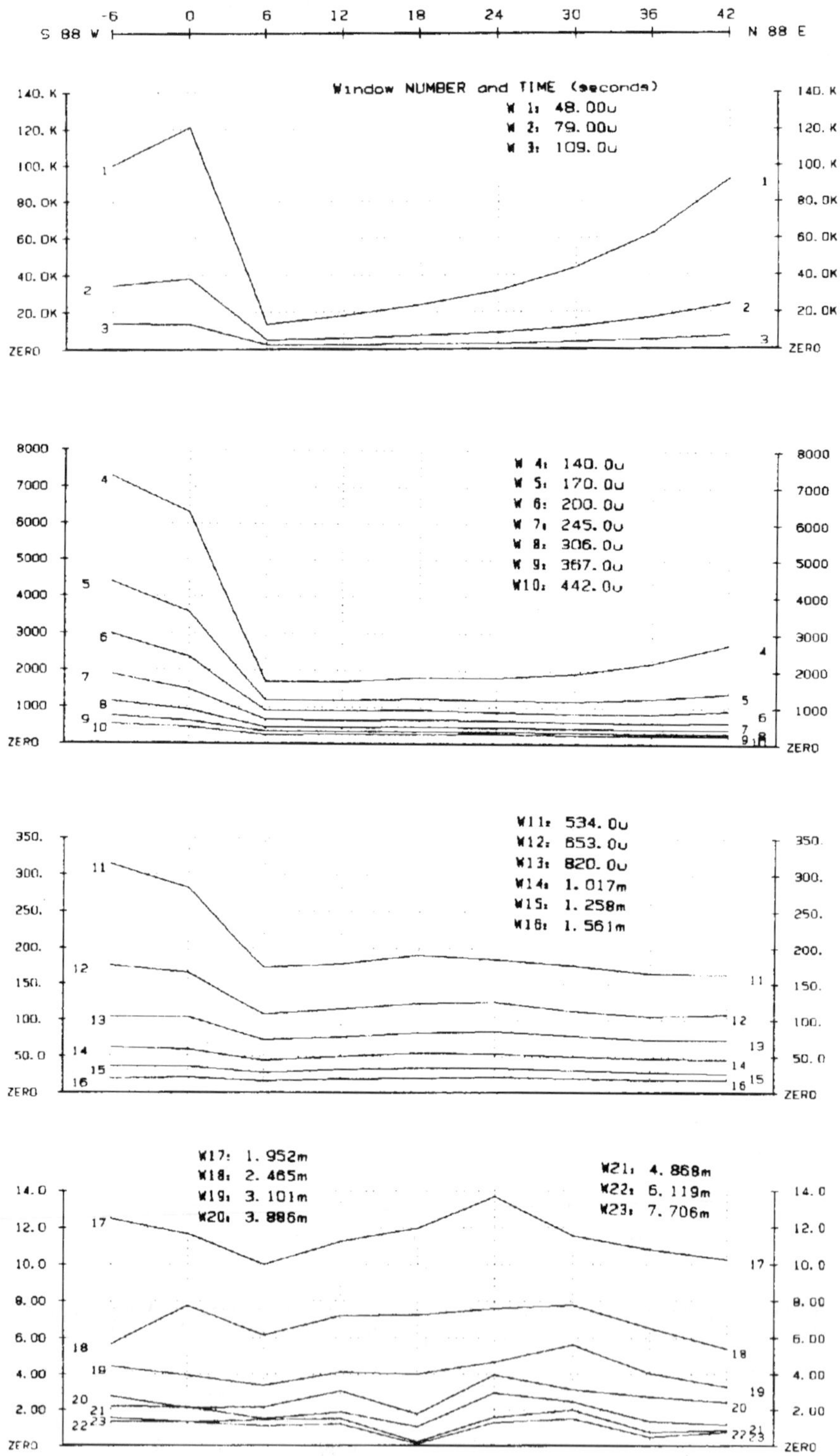

FIGURE 6.72
Tombstone test data: linear amplitude vs. station profile plots with the 23 delay curves divided into four linear scale groupings

FIGURE 6.73
Tombstone test data: (a) ramp-corrected apparent resistivity and (b) smooth-model inversion of the same data

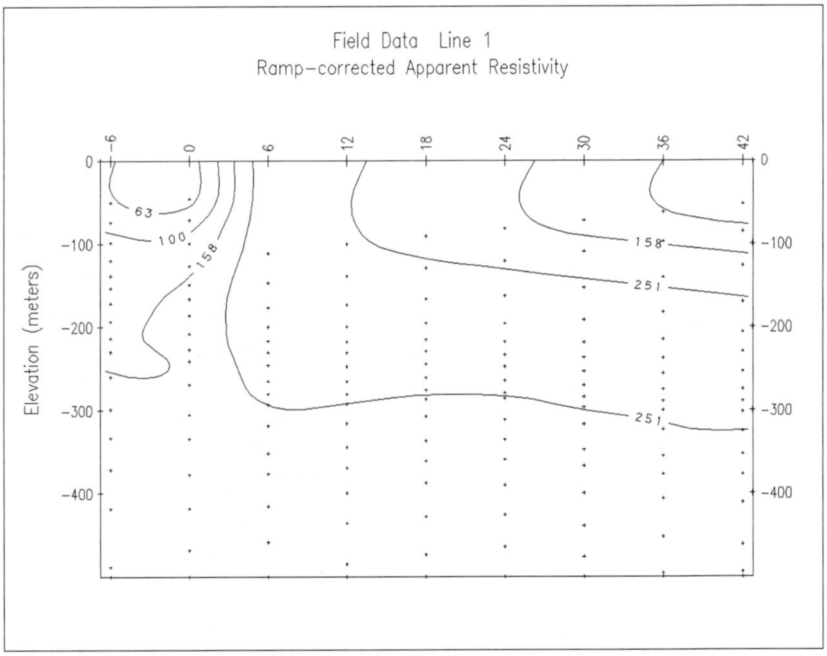

(a)

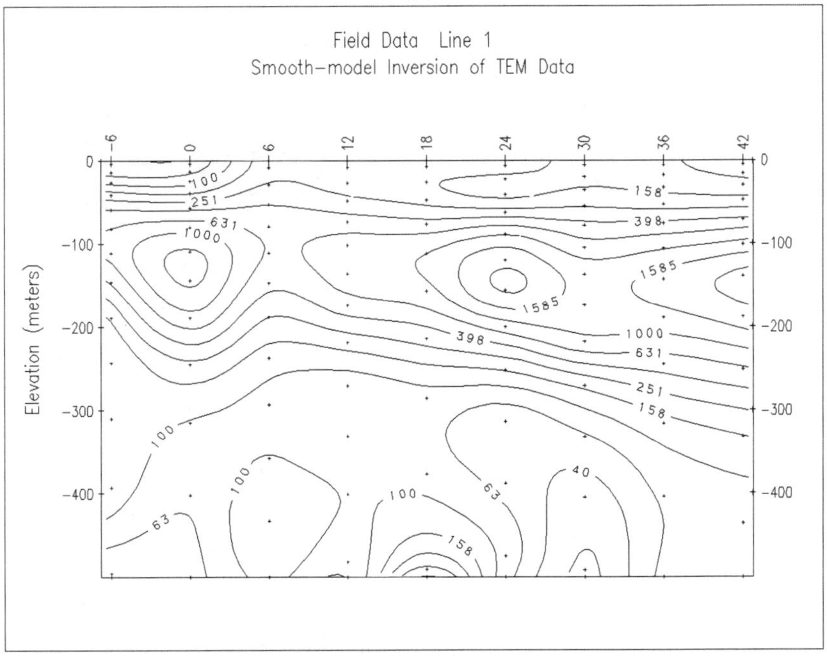

(b)

An aid to understanding the time variation of magnetic fields is the smoke-ring analogy first discussed by Nabighian, 1979. Figure 6.75 shows the propagation of equivalent current filament rings downward into the earth after current turn-off in the loop on the surface. These current filaments propagate downward and outward at a rate proportional the square root of the resistivity of the ground and at a 47-degree angle from the center of the loop on the surface. The rings will propagate quickly in resistive ground and more slowly in conductive ground.

Figure 6.76 shows the magnetic field lines from one of these current filaments. Notice the three vertical receiver antenna

FIGURE 6.74
EM fields around a conductive plate in a conductive half-space. (a) Vortex currents Jv in the plate induced by the changing magnetic field B. Bv is the secondary magnetic field generated by these vortex currents. (b) Galvanic currents Jg generated in the plate by galvanic coupling with the component of the primary electric field in the plate. Bg is the secondary magnetic field generated by the galvanic current. Notice, on the surface, both Bv and Bg are oriented in the same direction. After McNeill et al, 1984

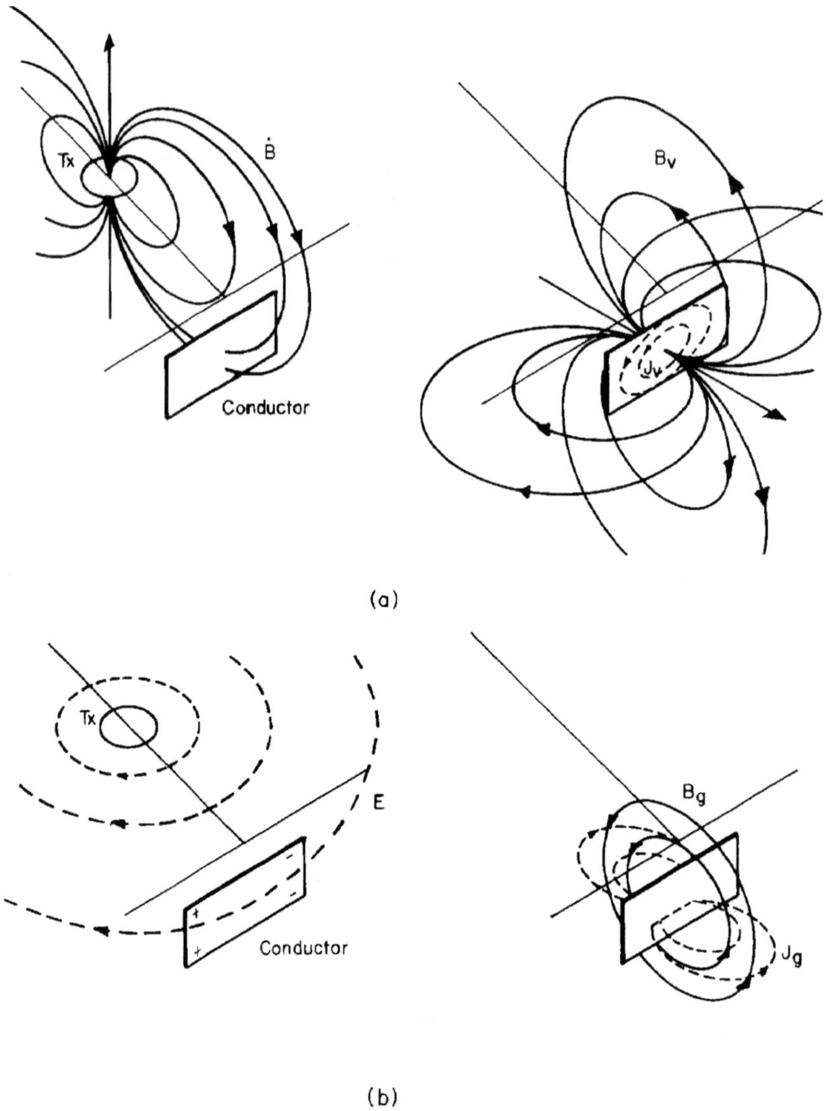

FIGURE 6.75
Smoke ring analogy of a system of equivalent current filiments at various times after current turn-off. Filiments propagate downward and outward with time, generating the magnetic field as seen in figure 6.76. After Nabighian, 1979

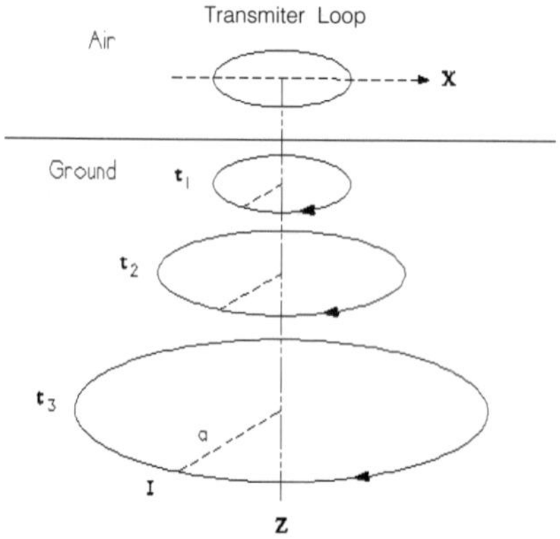

positions. The one closest to the transmitter loop is measuring the signal from a positive vertical magnetic field, the antenna in the middle is at zero vertical field and the antenna farthest away is measuring a negative vertical component. Using this picture, one can see that for very early times, all these surface loops will give negative values (picture the smoke ring near the surface and about the same size as the transmitter loop), and for very late times, all three receiver loops will measure a positive-going magnetic response.

By combining the results of figures 6.75 and 6.76, we can see that an observer on the surface will see a sign change in the magnetometer output at a time that is dependent upon the conductivity of the ground and separation from the transmitter. Conductive ground and/or large separation will extend the time to the sign change, whereas resistive ground and/or close proximity to the transmitter loop will shorten the time, perhaps even to the extent that it cannot be observed if the receiver cannot sample the very early time windows.

Figure 6.77 shows graphically how the time for the polarity change to take place increases as the separation between the transmitter and receiver increases over a one ohm-meter (one siemen/meter)

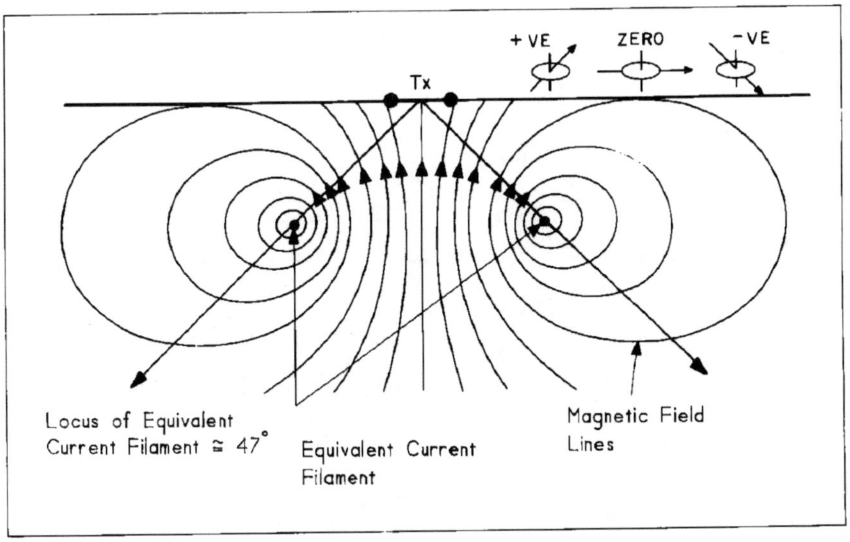

FIGURE 6.76
Secondary magnetic field lines from equivalent current filament showing coupling with surface receiver coils. Adapted from Spies, 1979

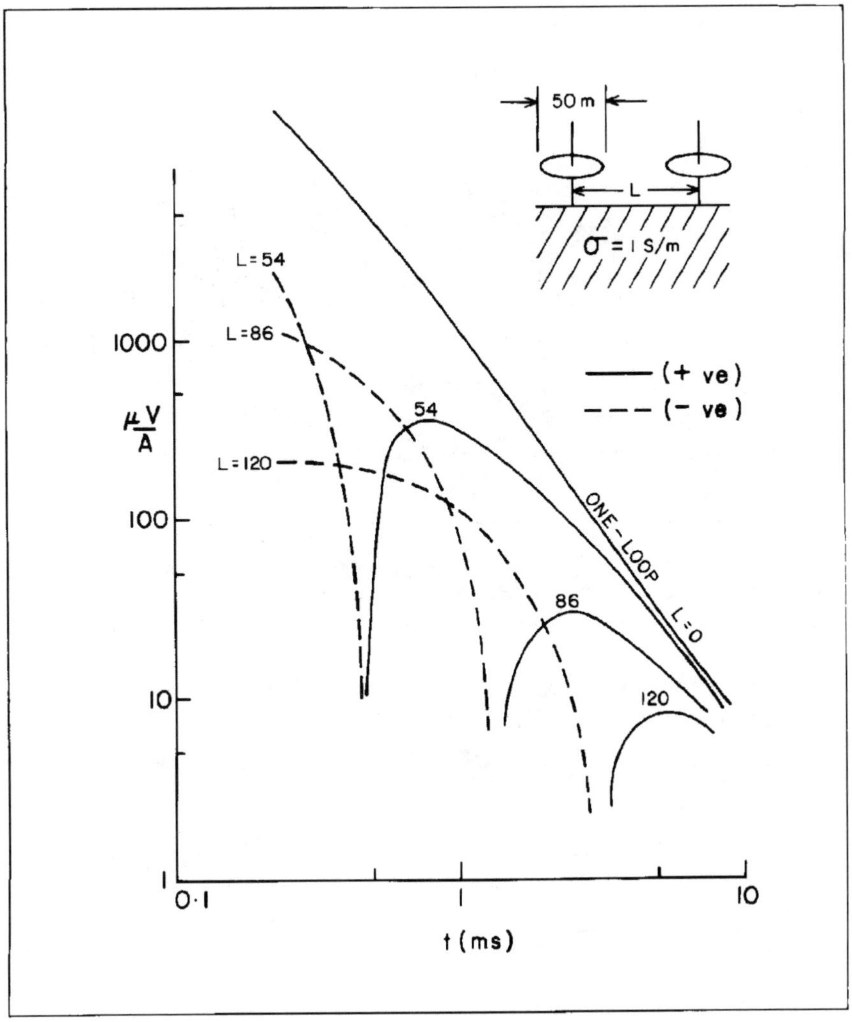

FIGURE 6.77
One-loop and two-loop system responses over a homogeneous half space. After Spies, 1980

half space. Notice that it takes roughly 500 microseconds for the polarity change to take place with a 54-m separation, and 3 milliseconds at 120 m.

FIGURE 6.78
Delay-time profiles over a homogeneous earth of: (a) 1000Ω m, (b) 100 Ωm, (c) 10 Ω m and (d) 1Ω m

Window NUMBER and TIME (seconds)

W 1: 165.0u	W11: 2.107m	W21: 19.44m
W 2: 287.0u	W12: 2.585m	W22: 24.45m
W 3: 409.0u	W13: 3.251m	W23: 30.80m
W 4: 530.0u	W14: 4.039m	W24: 38.65m
W 5: 652.0u	W15: 5.005m	W25: 48.55m
W 6: 773.0u	W16: 6.215m	W26: 61.11m
W 7: 953.0u	W17: 7.779m	W27: 76.88m
W 8: 1.197m	W18: 9.832m	W28: 96.75m
W 9: 1.441m	W19: 12.37m	
W10: 1.742m	W20: 15.51m	

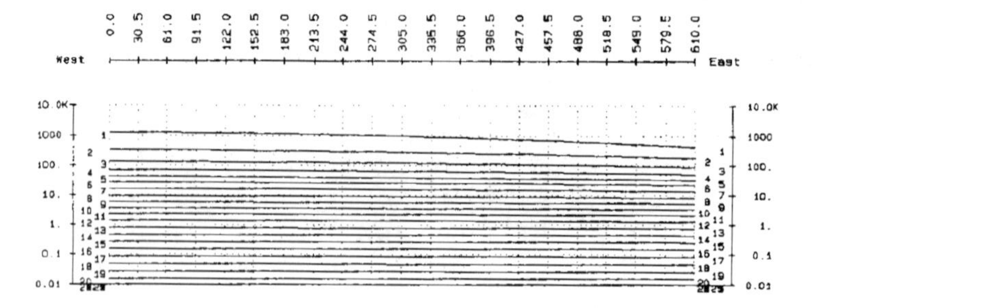

(a)

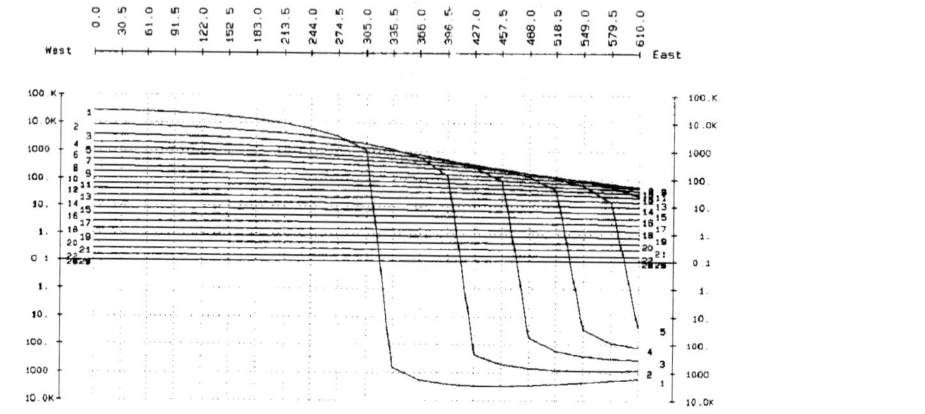

(b)

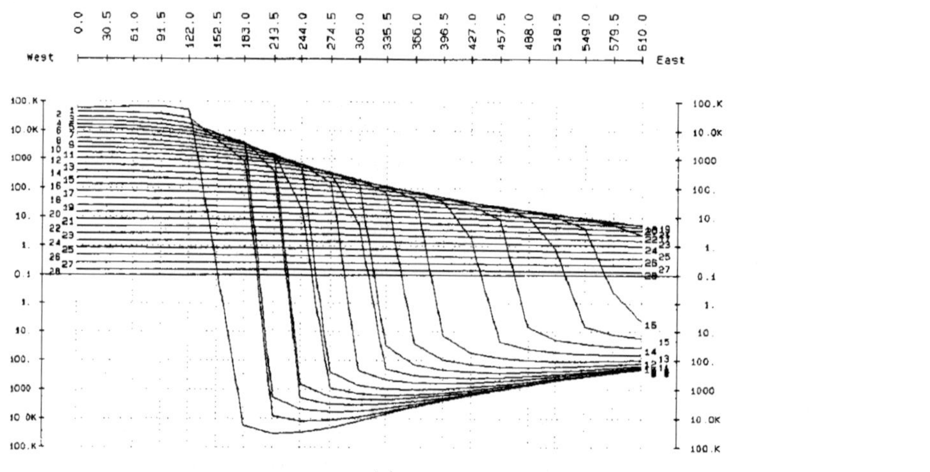

(c)

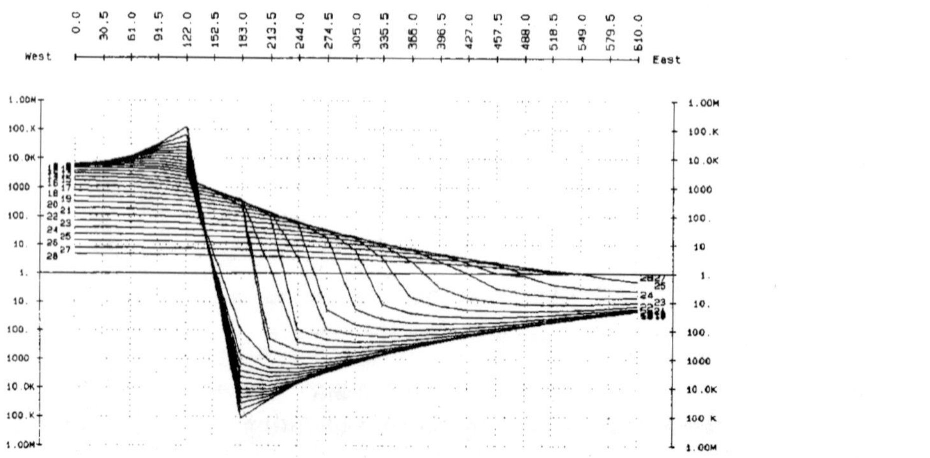

(d)

Figure 6.78 contains four profile sections for a fixed-loop system over a homogeneous half space, each containing 28 profiles on a logarithmic scale, from 165 microseconds to 96.75 milliseconds. A 305-m (1000-ft) loop is centered at station 0.0 and crosses the line at station 152.5. Data are obtained every 30.5 m (100 ft). The data collected at station 0.0, the center of the transmitting loop, is the same as in-loop or one-loop data (vertical sounding). Profile sections are presented for (a) 1000 Ωm, (b) 100 Ωm, (c) 10 Ωm and (d) 1 Ω m half space. Notice the two extremes: the data gathered over the 1000 Ωm half space do not have a sign change over the entire line, even at station 610, which is more than 450 m away from the edge of the transmitter loop. This behavior is typical of Canadian shield and other high-resistivity environments. The early time data collected over the 1 Ωm section (a worst-case Australian environment) at station 183, the first station outside of the loop, however, cross over at 2.6 milliseconds and earlier. The sign change gets progressively later in time as you move away from the transmitter loop. By the time station 610 is reached, the data remain negative for all time channels (out to 96 milliseconds). The results over the 10 Ωm and 100 Ωm sections show intermediate behavior and are typical of what to expect in the southwest United States and Australia.

FIGURE 6.79
Generalized sketches of a coincident-loop response over tabular features of various sizes and dip angles

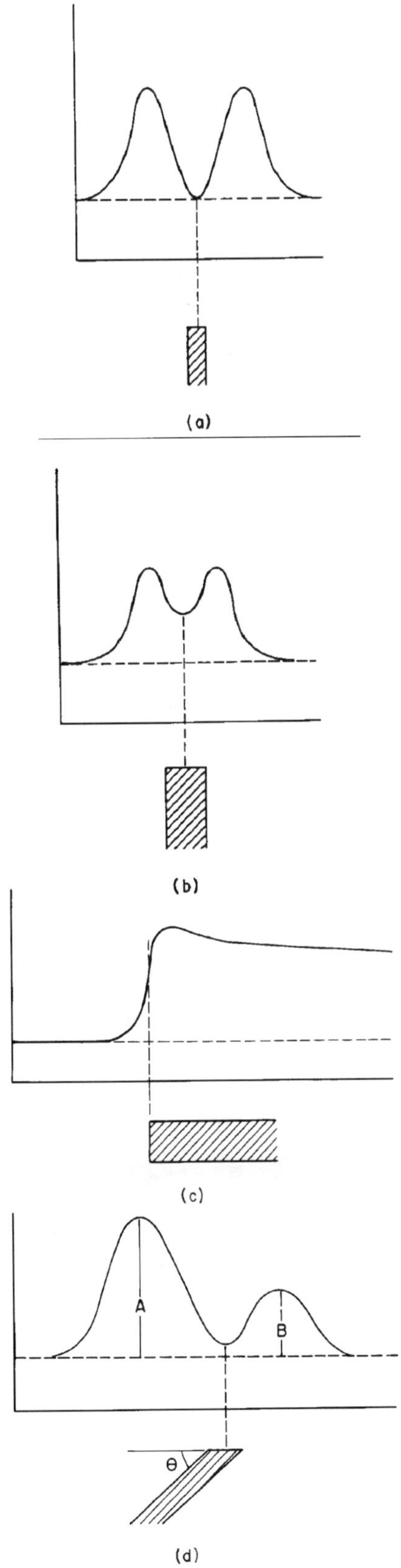

APPLICATIONS, TYPES OF TARGETS

For best coupling into conductive targets, the magnetic lines of flux from the transmitter should intersect the conductor as close to a right angle as possible. Therefore, the coincident-loop type of system couples best into flat layers directly beneath the loop, and the separated-loop systems couple best into high-angle conductive bodies that occur between the transmitter and receiver loops and perpendicular to a line connecting the centers of the two loops.

In the following discussion figures, 6.79, 6.81 and 6.83 are presented to give the reader a general feel for the shapes of responses obtained from different loop configurations run over several target orientations. The actual field response will be a function of loop size, loop separation, window (delay) time, size and depth of the target, conductivity of the host rock and target, etc.

Single-Loop Systems

Figure 6.79 shows a delay-time profile for a single-, coincident-, or in-loop system over a tabular body of different sizes and orientations. This type of loop configuration usually produces a symmetrical side-lobe response with a null over the center of thin, vertical bodies (figure 6.79a). If the loop size is comparable to the depth of burial and the width of the tabular body, the central null will diminish (figure 6.79b) or disappear (see figure 6.97). Figure 6.79c shows the response over the edge of a horizontal, thin, tabular body. This configuration has the same type of response as a traverse over a low/high resistivity contact (see figure 6.81), with the exception of the slight peak in the response just before the edge of the conductor.

The asymmetry of the side-lobes over a dipping, tabular body can be used to estimate the dip. In figure 6.79d, the ratio B/A gives an estimate of the dip angle .

Single-loop systems do not couple well into thin, vertical bodies. Although the response shown here over a vertical dike is typical, the actual amplitude of the response is quite small, as seen in figure 6.80. These modeled results show the profiles of a coincident-loop system

FIGURE 6.80
Model data for a coincident-loop system as a function of the dip of a half-plane conductor (After Spies, 1974)

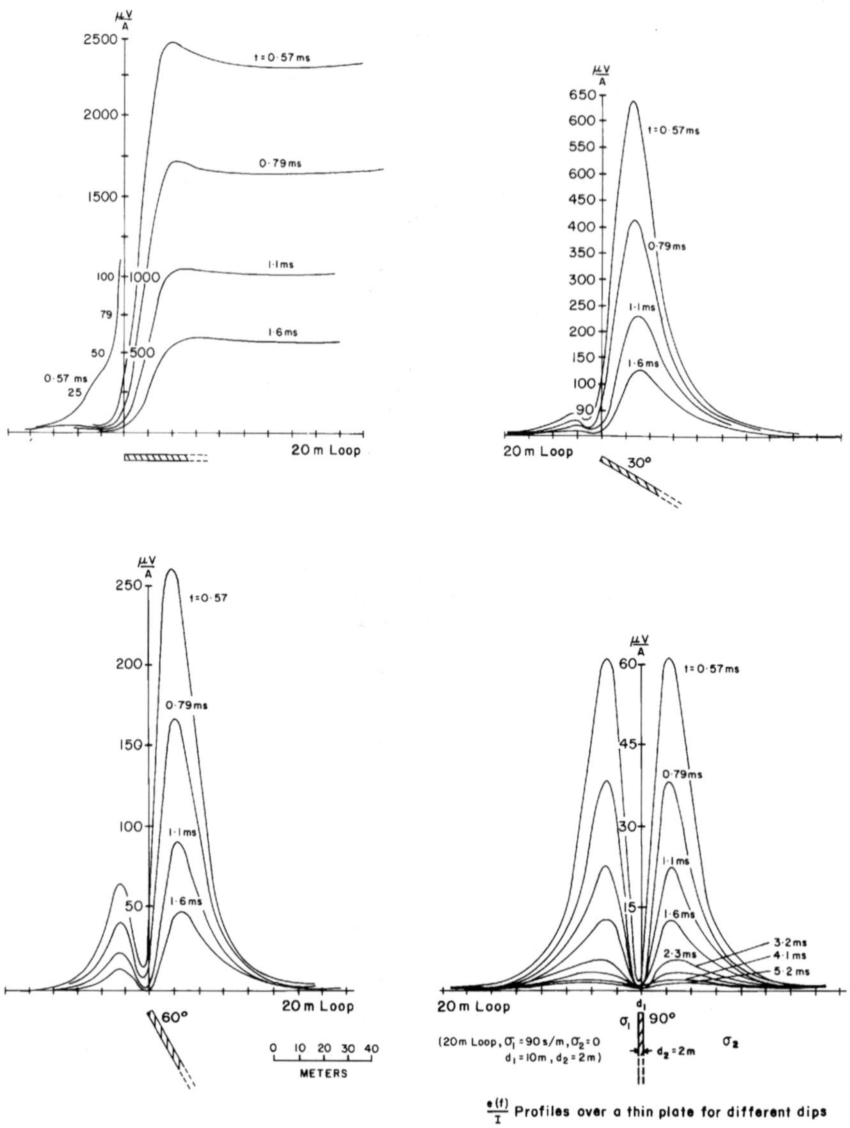

as a function of the dip of a half-plane conductor in free space. Notice the very strong response when the conductor is horizontal, and how quickly the response diminishes with increasing angle. At 90 degrees, the maximum response of the side-lobes is a factor of 40 less than the response over the horizontal conductor.

The modeled response of a single-loop system running across a lateral contact is shown in figure 6.81. Notice that lateral resolution is a function of loop size and station spacing. Single-loop sounding systems do not provide the sharp lateral resolution that is found with systems that measure the electric field; but TEM systems which measure only the magnetic field are not troubled by strong responses from near-surface inhomogeneities (static effects, as discussed in the section on CSAMT).

Separated-Loop Systems

The separated-loop, moving-loop, or loop-loop system couples best into high-angle conductors occurring between the receiver and transmitter loops. Figure 6.82 displays single delay time profiles

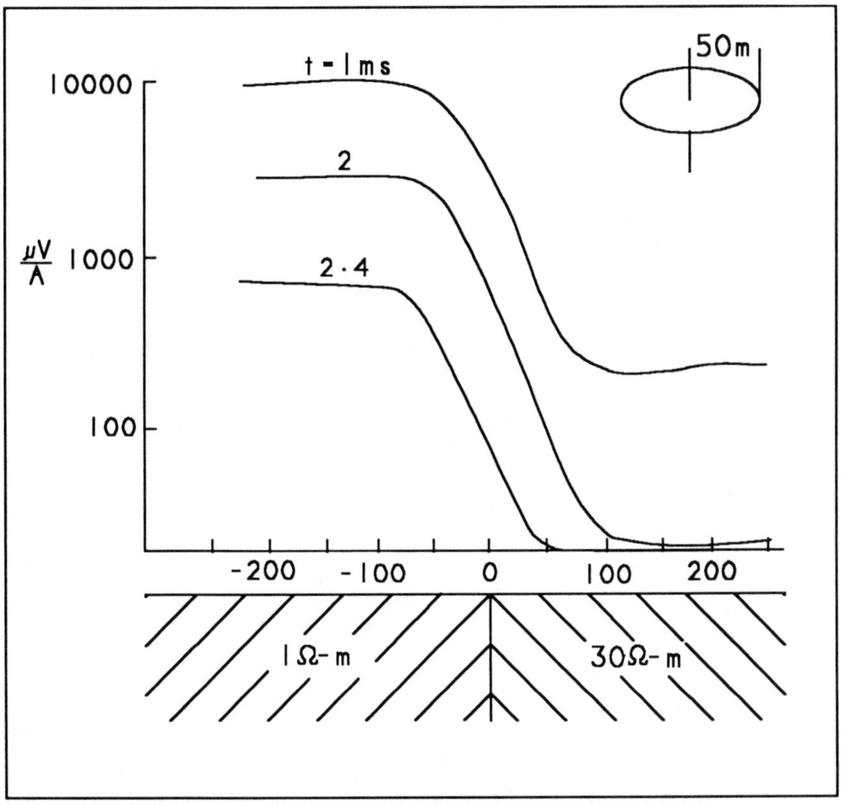

FIGURE 6.81
TEM coincident-loop profiles over a vertical contact model (After Spies, B.R. 1980)

FIGURE 6.82
Generalized sketches of a separated-loop system response over tabular features as a function of dip

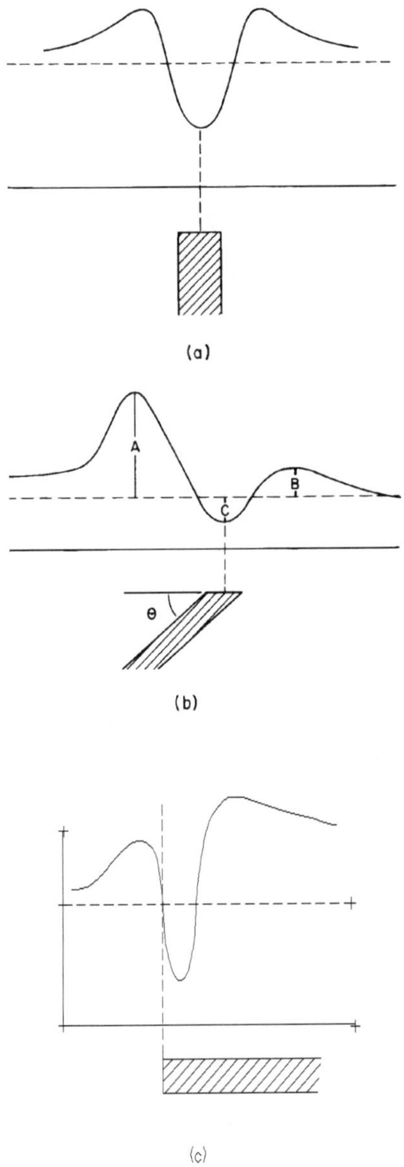

over several orientations of a tabular body. Notice that the main difference between the single-loop and separated-loop systems is that the peak of the response occurs directly over a vertical tabular conductor for the separated-loop system, (figure 6.82a) compared with a null for the single-loop, and this peak is accompanied by side-lobes of opposite sign. Figure 6.82b is the response of a loop-loop system over a thin dipping dike and is similar to the single-loop system with the exception of the negative lobe over the top of the structure. Figure 6.82c is a sketch of the response over the edge of a thin plate near the surface. This response is similar to that of a thin vertical conductor and could be difficult to differentiate in the presence of geologic noise.

Figure 6.83 shows the modeled profile results for a loop-loop system over a conducting plate in free space for three orientations of the plate: (a) 90 degrees, (b) 45 degrees and (c) 15 degrees. Notice that the width of the negative peak at the cross-over points is equal to the spacing between the receiver and transmitter loops in this thin plate example. Dip of the conductor can be estimated from the ratio of the side-lobe amplitudes.

Now, compare these results with figure 6.84, which is a loop-loop model across a lateral contact. Again, with geologic noise included, it could be difficult to differentiate between a surface weathered zone (a horizontal conductive slab), a lateral contact, and a steeply dipping conductive tabular feature. This is not the case for the coincident-loop system in figures 6.79, 6.80 and 6.81, and demonstrates one reason that it is often good to run several types of arrays over anomalies of interest.

FIGURE 6.83
Model data for separated-loop TEM profiles over a conducting sheet in free-space for dip angles of (a) 90 degrees, (b) 45 degrees and (c) 15 degrees (from Frischknecht et al, 1991)

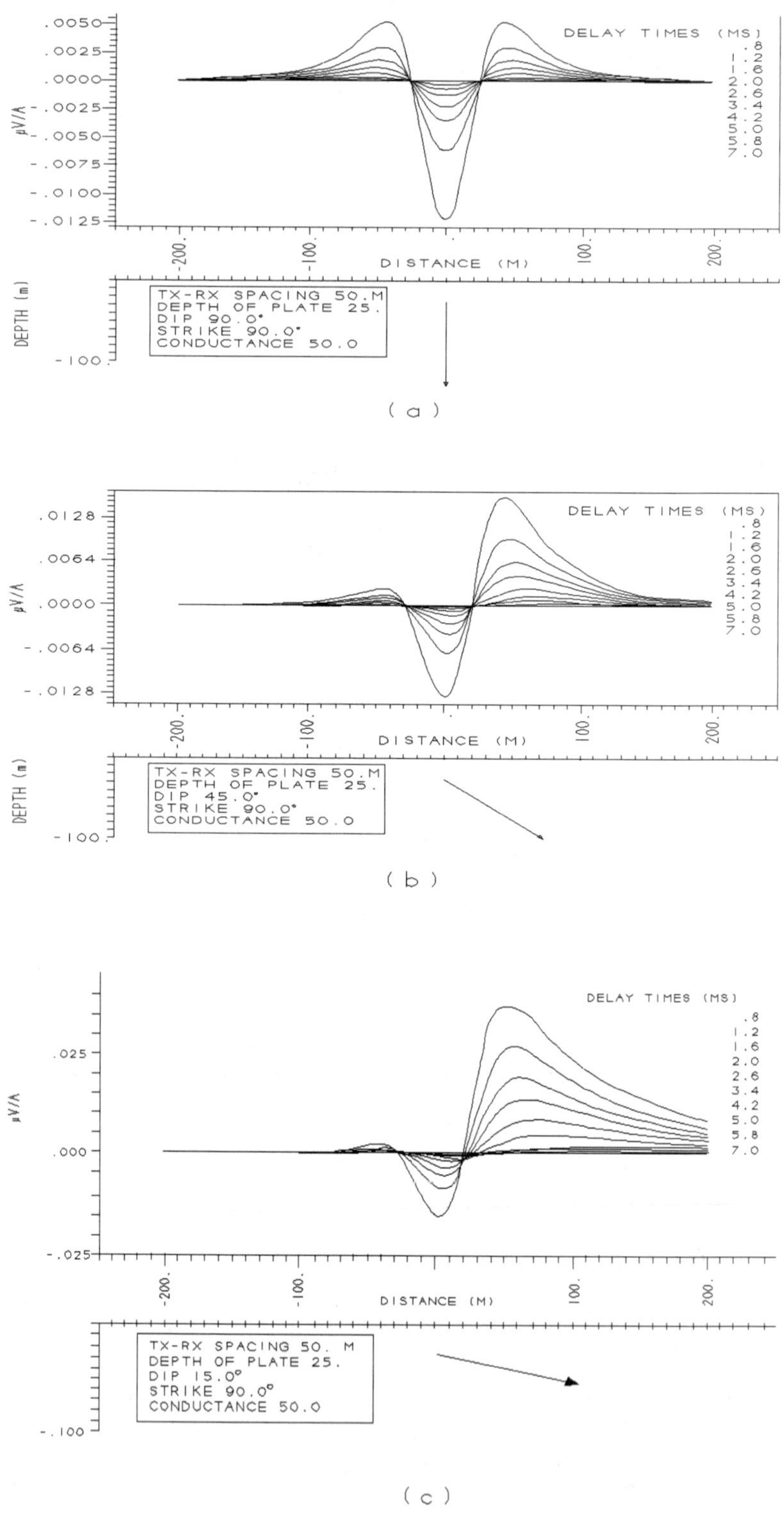

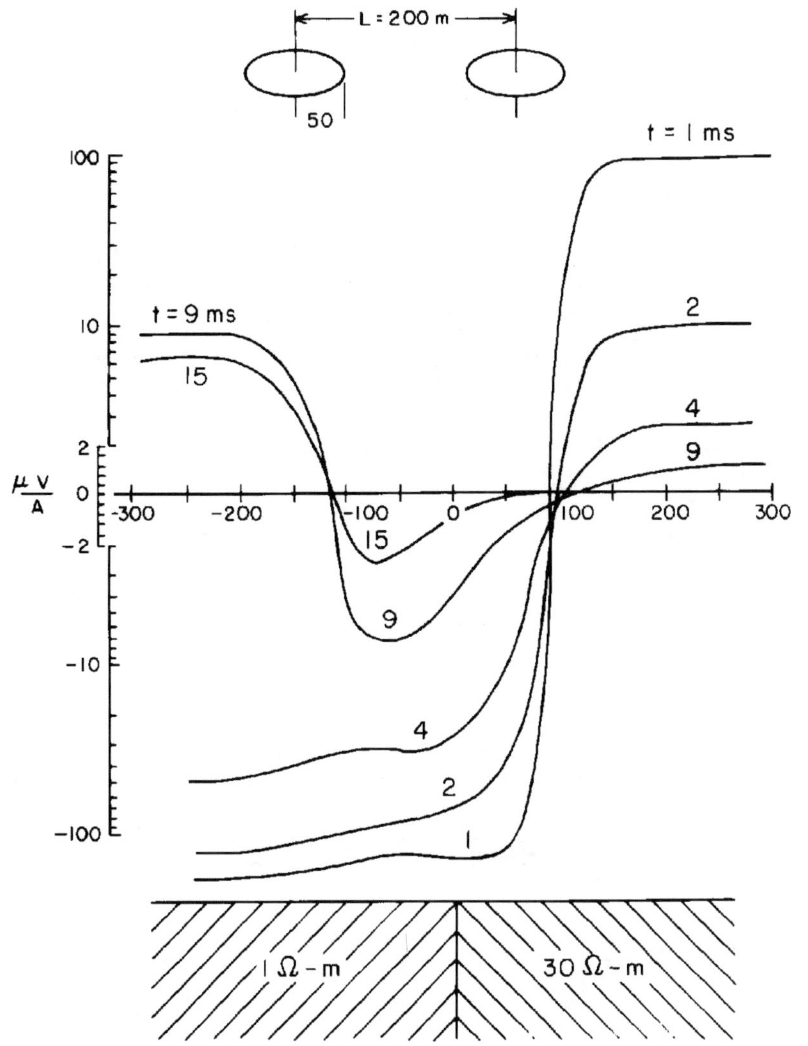

FIGURE 6.84
Model data for a separated-loop system over a vertical contact (After Spies, B.R. 1980)

Nomograms are available for calculating the approximate depth, conductivity and dip of tabular bodies using the loop-loop technique. See for example Ogilvy, 1986.

Fixed-Loop Systems

Fixed-loop systems couple best into steeply dipping tabular bodies that run parallel to the long side of the transmitter loop with receiver lines oriented as shown in figure 6.64. Figure 6.85 shows single delay anomaly profiles over a rectangular body with three orientations. Both the vertical (Z) and horizontal (H) components are shown. Notice that the response of the two components interchange when going from a flat-lying to a vertical body. For fixed-loop surveys, the vertical and horizontal components are routinely measured to help in the interpretation process.

Figure 6.86 shows the fixed-loop response of a half-plane conductor in free space at various dip angles. This modeled, point-normalized UTEM data is very similar to other TEM system responses and provides a good picture of the effect of dip angle on the overall response. Although not shown here, it is possible to have a conductive body dipping toward the transmitter loop that becomes parallel to the magnetic field and which will have an anomaly response near zero. This is called the "angle of extinction".

When running fixed-loop surveys in areas where target dip or orientation is unknown, it is usually best to repeat traverses with a second transmitter location. This can be done by moving the transmitter loop to the opposite end of the traverse and repeating the measurements.

The fixed-loop system can also have problems in differentiating between surface conductors (e.g. weathered zones) and vertical conductors if only a single traverse direction is run. With the transmitter loop set up over the surface conductor and the receiver run across the conductor to the resistive zone, the edge of the conductor is defined by a sharp, stationary crossover in the vertical magnetic field response. When the transmitting loop is set up over the resistive host and the traverse is run across the conductive zone in the opposite direction, however, a crossover in the vertical field response still occurs near the edge of the conductor at early times, but the crossover point will migrate into the conductive feature. See the Brittania field example beginning page 517.

FIGURE 6.85
Generalized sketches of the response of a tabular body as a function of dip for a fixed loop system (After Nabighian, M.N. and Macnae, J.C., 1991)

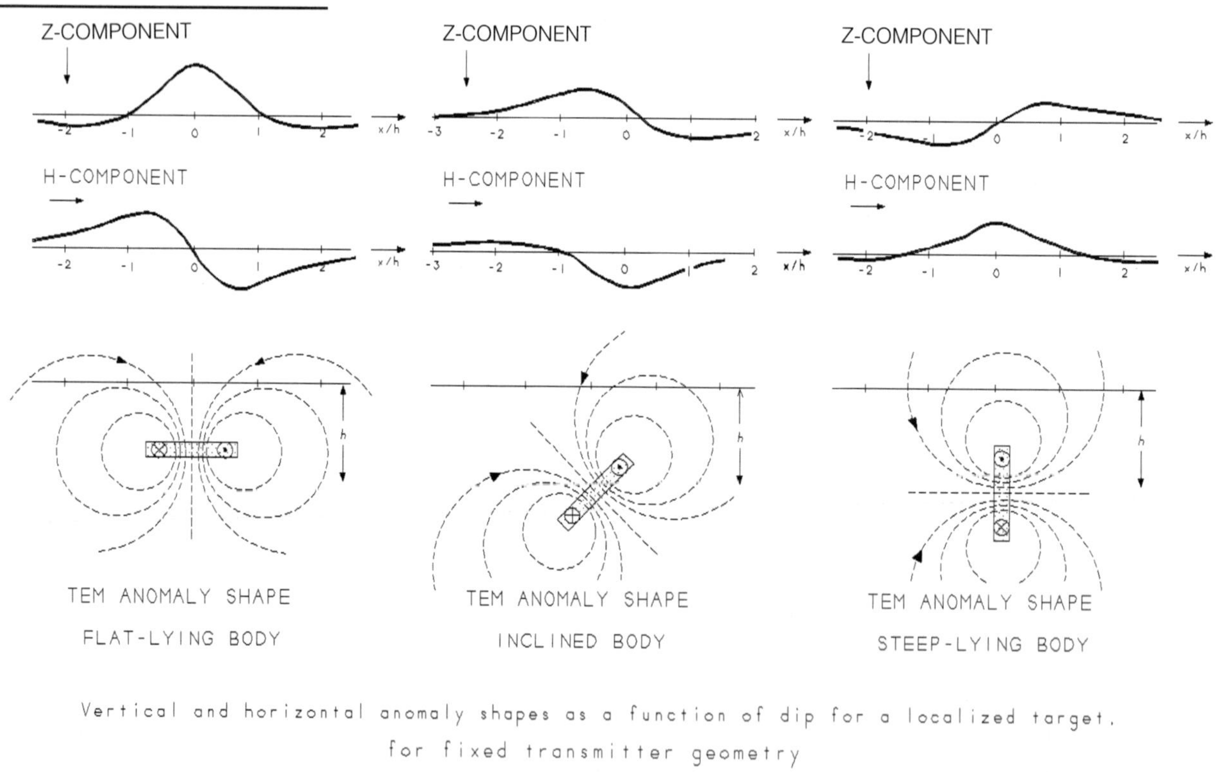

Drill-Hole TEM

Since drill holes sample only the small cylinder occupied by the drill bit, it is desirable to re-occupy drill holes with a geophysical tool to make sure off-hole targets were not missed. The most commonly used arrays for TEM surface to drill-hole logging are shown in figure 6.68. Use of multiple holes for cross-hole tomography is now under

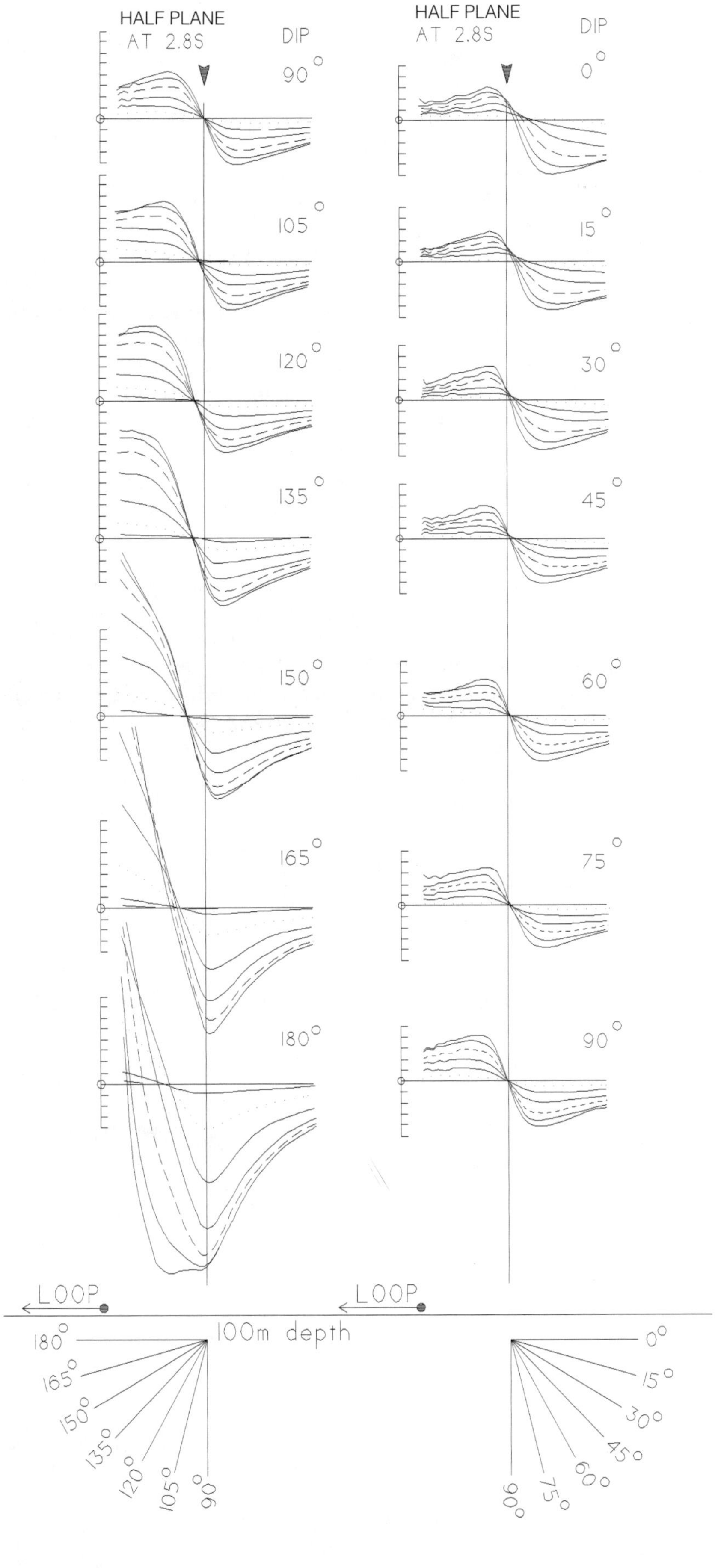

FIGURE 6.86
Point-normalized, fixed-loop TEM response of a half-plane conductor for various dips: (a) toward the transmitter loop, and (b) away from the transmitter loop (after Irvine and Staltari, 1984, courtesy Lamontagne Geophysics)

FIGURE 6.87
Generalized sketches of TEM response for four different down-hole situations: Type A - intersection of the orebody; Type B - contact with the edge; Type C - a near miss; and Type D - a far miss (After Crone, 1980)

BOREHOLE PULSE EM
TYPE RESPONSE CURVES

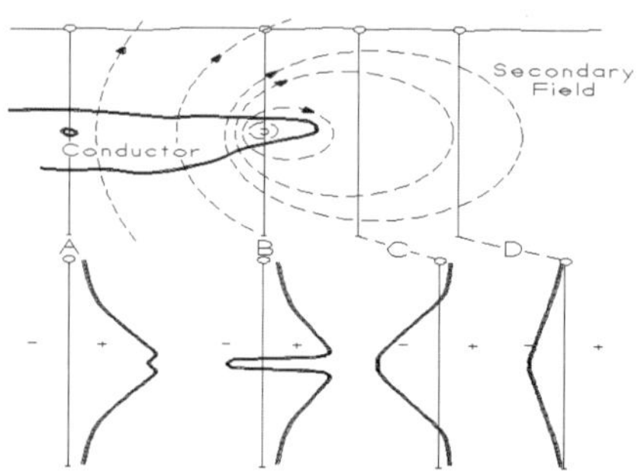

development but will become more common in the future as new equipment and techniques become available for economically acquiring large amounts of data. We will concern ourselves at this time with single-hole measurements.

Figure 6.87 displays the four basic drill-hole responses obtained when a conductor is energized from a surface loop. The type-A response is generated when the drill hole penetrates the central portion of a conductive body. The strong positive response with the small indentation at the peak is typical of the type-A curve. Type-B responses are observed when the TEM probe goes through the outer edge of a conductive body and passes near the induced vortex currents flowing near the outside edge. The strong negative lobe is induced by magnetic fields in proximity of the vortex currents. Just missing the edge of a conductor with a drill hole will result in a type-C response. This response has small positive lobes and a broad negative lobe opposite the conductor. Notice that the zero crossings for both the type-B and type-C curves occur where the induced magnetic fields from the conductive body are perpendicular to the hole. A broad type-D anomaly occurs when the drill hole is distant from the conductor. The farther away, the broader and shallower the anomaly.

Figure 6.88 is a field example from a massive sulfide body in Manitoba, Canada, showing examples of type-A, type-B and type-C responses. Although all three holes intersected sulfides, the type-C response in hole number 3 indicates that these sulfides are not connected to the massive lens encountered in drill holes 1 and 2.

When drill holes intersect lens-like conductive bodies close to 90-degree angles and the drill hole and target are within the vertical projection of the transmitter loop on the surface, drill-hole interpretation is relatively straightforward. If the downhole probe or the target move outside of the surface loop, then interpretation is less direct. In this case, the best thing to do is to sketch out the secondary field lines from a suspected anomaly, determine how they will align with

FIGURE 6.88
Field data from a massive sulfide body in Manitoba, Canada, exhibiting type A, B, and C behavior for drill holes 1, 2 and 3 respectively (after Crone 1980)

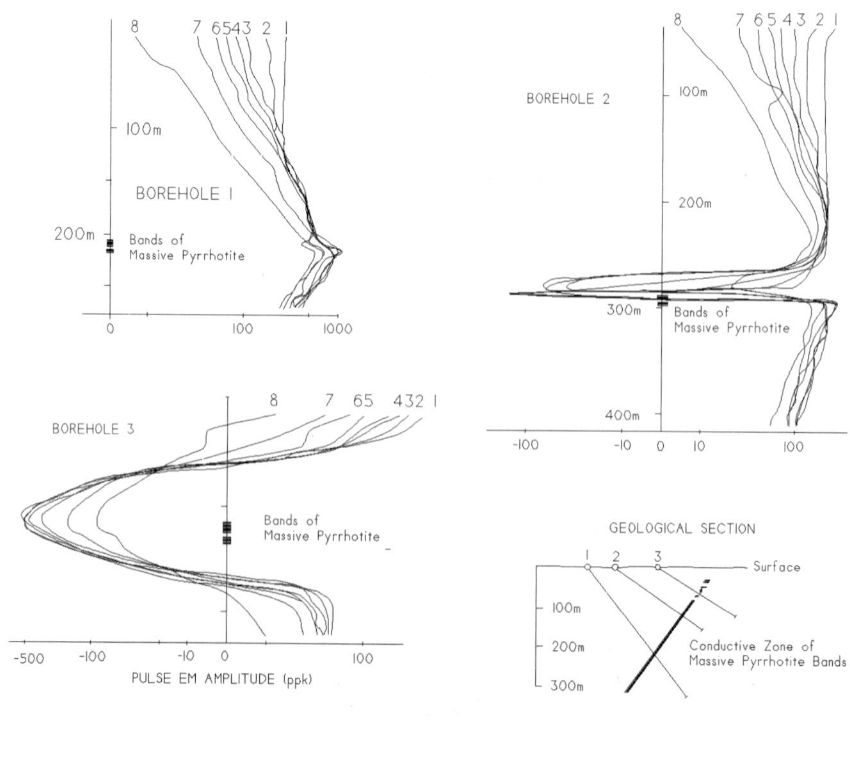

the probe and sketch out the expected result. Some examples are available from the literature, but it is impractical to make a catalog of all possible 3-D configurations for conductive bodies, drill holes and transmitter loop positions. Several good modeling programs are available for a conductive plate and sphere in free space.

LOOP SIZE AND LOOP SEPARATION

Figure 6.20 in the CSAMT section displayed the effects of receiver/transmitter separation. This parameter variation applies only to separated- and fixed-loop TEM systems in our present discussion. Figure 6.89 shows that the vertical magnetic field drops off as the inverse fifth power of the distance between the receiver and transmitter for large separations. Hence the limitation is encountered in fixed-loop surveys of traverses being no more than two or three loop diameters away from the edge of the loop.

For CSAMT, increasing the length of the transmitter dipole increased the signal strength at the receiver. For TEM, increasing the transmitter loop size also increases the signal strength at the receiver,

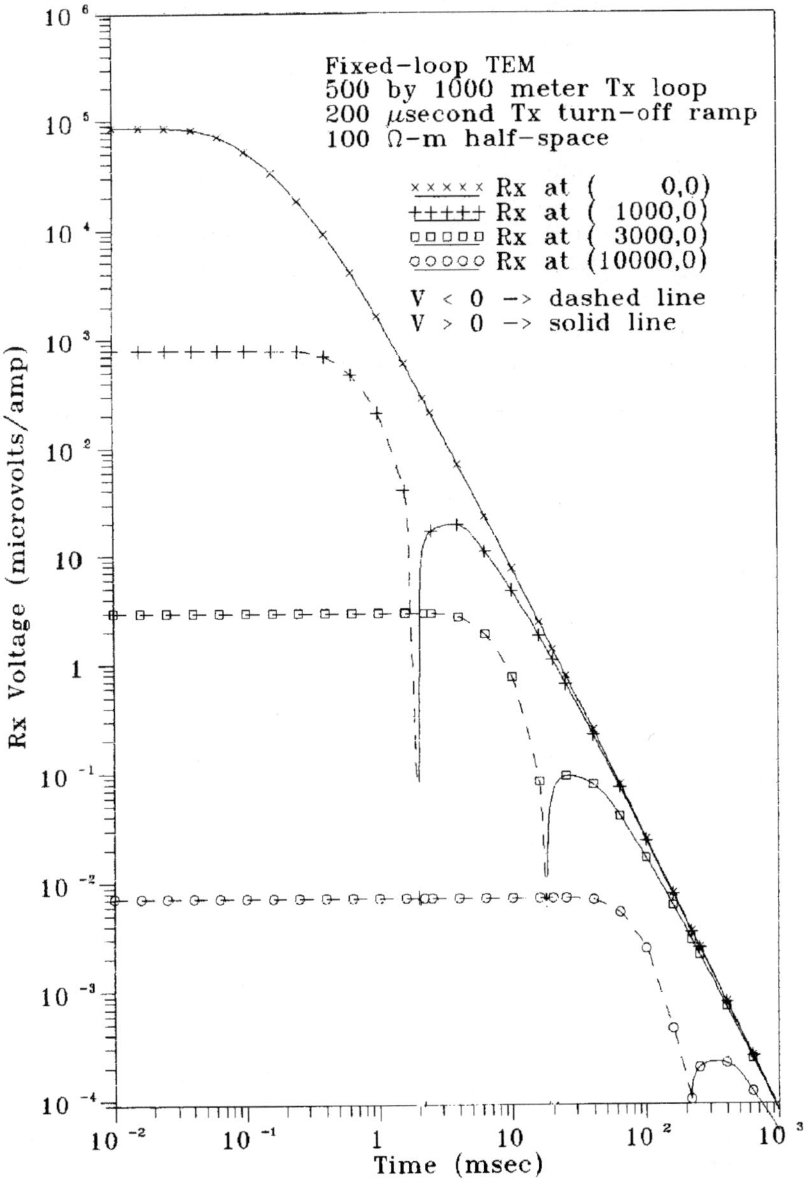

FIGURE 6.89
Received voltage drop-off as a function of receiver/transmitter separation for a fixed-loop system

but it also slows the transition to late-time behavior, which puts a limit on the minimum depth one can obtain with a particular loop size. This means that to make shallow soundings, one should use small loops; to make deep soundings, large loops are in order. This is usually not a problem, as it is infrequent that large loops will be used to make shallow soundings. Figure 6.90 shows the early time behavior for three different loop sizes over a 10 Ωm earth. Using the depth equation from table 6.2 and noticing the late time decay begins around 20 microseconds for the 100-m by 100-m loop, the minimum depth for late time sensitivity for that loop is about 12 m.

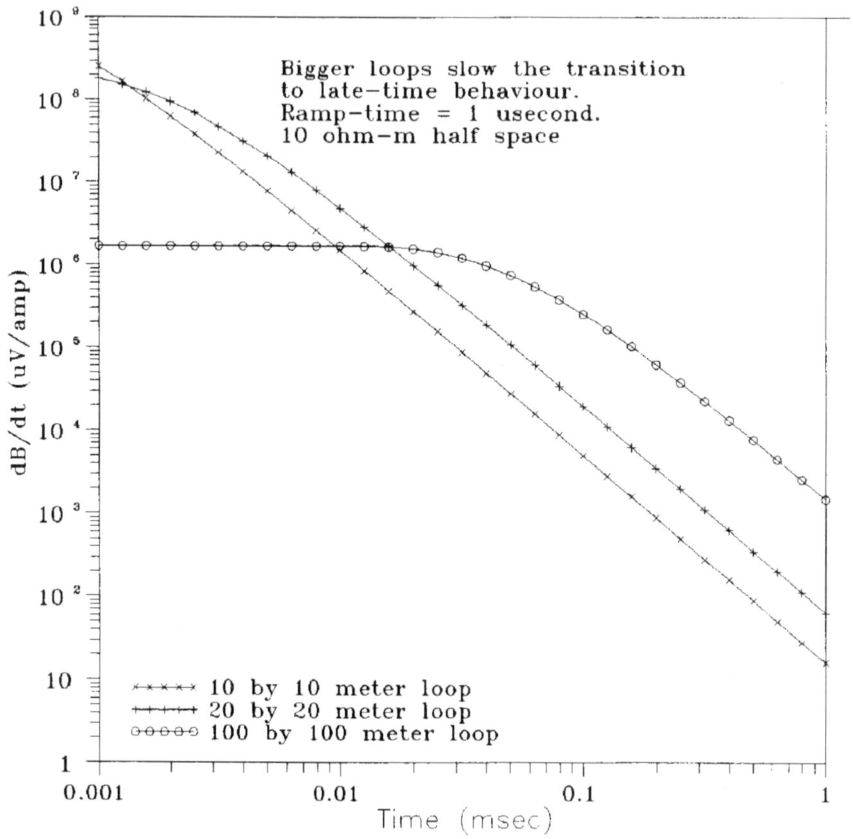

FIGURE 6.90
Early-time behavior for a single-loop TEM system, as a function of loop size

SINGLE-LOOP SOUNDINGS OVER A LAYERED EARTH

Single-loop (in-loop, coincident-loop) TEM and CSAMT give almost identical vertical sounding data over layered-earth environments, which should be expected. One can anticipate a difference in coupling to surrounding culture (fences, pipelines, power lines, etc.) because of differences in transmitter locations, however, and a smoother response to ground resistivity changes for the TEM system.

To stress the similarity between vertical sounding TEM and CSAMT, we will use the same layered-earth models that were used for demonstrating changes in CSAMT values in figures 6.20 through 6.25 and observe how these variations affect the TEM- decay waveform.

Figures 6.91 through 6.95 present TEM data in two formats: decay waveforms in the form of received voltage versus time and apparent resistivity versus $1/(2\pi t)$. The apparent resistivity plots are included for comparison with the equivalent CSAMT plots. Plotting the horizontal axis in the form of $1/(2\pi t)$ (seconds)-1 normalizes the data to a frequency axis so that the TEM resistivity curves will directly overlay the CSAMT curves. For the inverse time axis, 1/sec is equal to 160 milliseconds (ms), 10/sec equals 16 ms, 100/sec equals 1.6 ms, 1000/sec equals 0.16 ms or 160 microseconds and 10000/sec is 16 microseconds. Notice that this axis matches the frequency values on the CSAMT plots.

Decay curves for a homogeneous half space with resistivity varying from 100 to 3200 Ω m are shown in figure 6.91a. This basically shows the decreasing time response (fast decay) for increasing resistivities. Figure 6.91b shows the corresponding apparent resistivity curves for comparison with figure 6.21 in the CSAMT section. The only difference in the curves is in the low-frequency behavior of the CSAMT data when it gets into the near field.

FIGURE 6.91
Effect of ground resistivity on the decay waveform. and (b) the late-time apparent resistivity. Compare (b) with figure 6.21 (a)

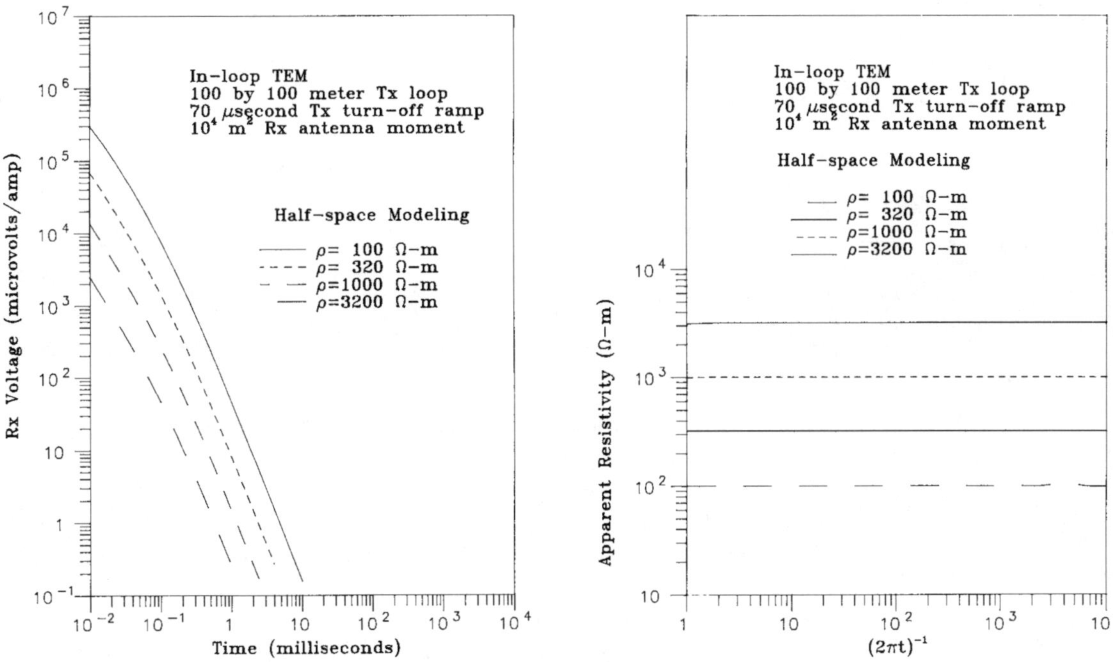

(a) (b)

Figure 6.92 is a two-layer earth, 10 Ω m over 100 Ω m. The top layer has thicknesses of 1, 3, 10, 32 and 100 m. Notice in figures 6.92a and 6.92b that all curves are converging at late times as the major influence becomes the bottom 100 Ωm layer. The apparent resistivity curves of figure 6.92b will overlay the resistivity curves in figure 6.22a. In both plots, the influence of the 3-m layer is seen in the early time behavior and at the highest frequencies, but the 1-m layer has little affect on the curves.

Figure 6.93 demonstrates the effect of varying surface resistivity while holding the thickness of the surface layer constant at 10 m. The surface layer resistivity varies between 1, 10, 100 and 1000 Ω m, with the lower layer held constant at 100 Ω m. The influence of the 1 Ωm surface layer is extreme, and the 10 Ωm layer is strong, whereas the 1000 Ω m layer is barely evident in the early time data. Figure 6.93b will overlay figure 6.23a.

FIGURE 6.92
Effect of the thickness of a conductive surface layer on (a) the decay waveform and (b) the late-time apparent resistivity. Compare (b) with figure 6.22 (a)

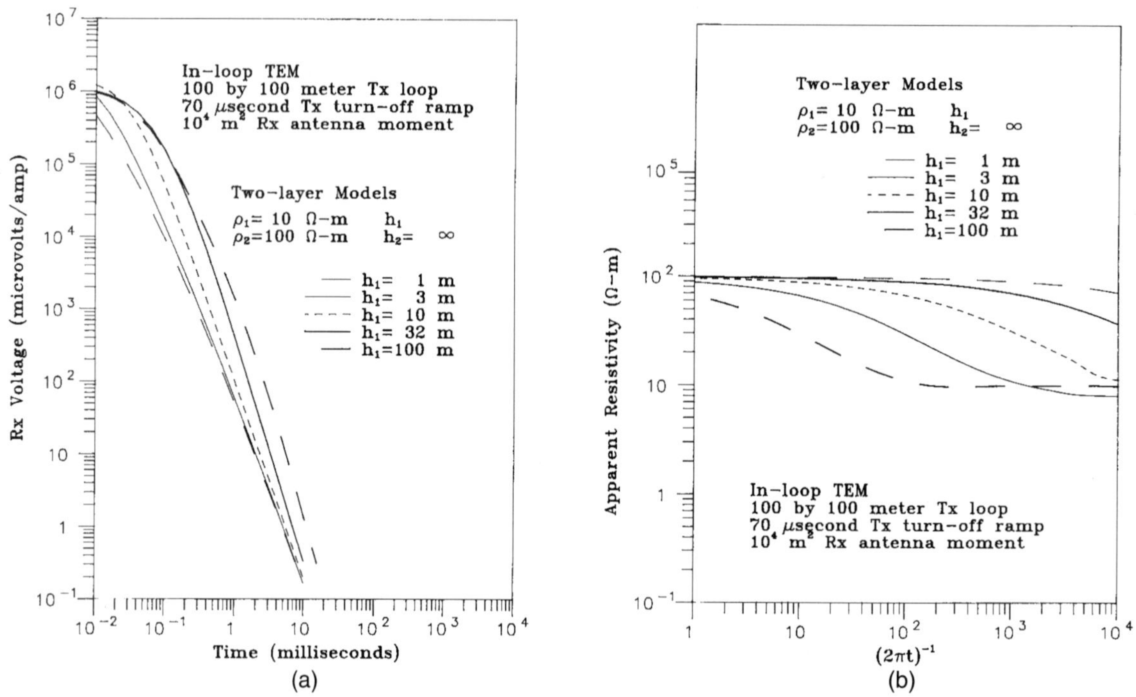

Figure 6.94 displays the effect of a conductive mid-layer in a three-layer model. The top layer is held at 100 Ωm and is 100 m thick, while the middle layer is 10Ω m and varies in thickness between 1, 10, 100 an 1000 m. The bottom layer is also held at 100Ω m. Note that the central 1-m layer is not resolved in any of the data, the 10-m layer is barely resolved, the 100-m layer is fully resolved, and the 1000-m layer influences most of the late-time data. The bottom of the 1000-m layer cannot be seen in the late-time data or in the low-frequency data of figure 6.94b. These data will compare with figure 6.24a. Notice that the bottom of the 1000-m mid-layer does show up in the transition-zone notch of figure 6.24a.

Figure 6.95 demonstrates the effect of a 100-m mid-layer with resistivities that vary from 1 Ω m to 1000 Ω m. As is the case with CSAMT, the curves are less responsive to resistive layers as compared to conductive layers. The curves of figure 6.95b will overlay the curves of figure 6.25a, with the main variations coming from the near-field/transition zone behavior of the CSAMT data at low frequencies.

EFFECT OF CONDUCTIVE OVERBURDEN AND HOST ROCK

We have observed the effect of conductive host rock in figure 6.78 and how it complicates the interpretation of a homogeneous half space. Conductive overburden and weathered surface rock (in arid

FIGURE 6.93
Effect of varying the resistivity of a 10-m thick surface layer on (a) the decay waveform and (b) the late-time apparent resistivity. Compare (b) with figure 6.23 (a)

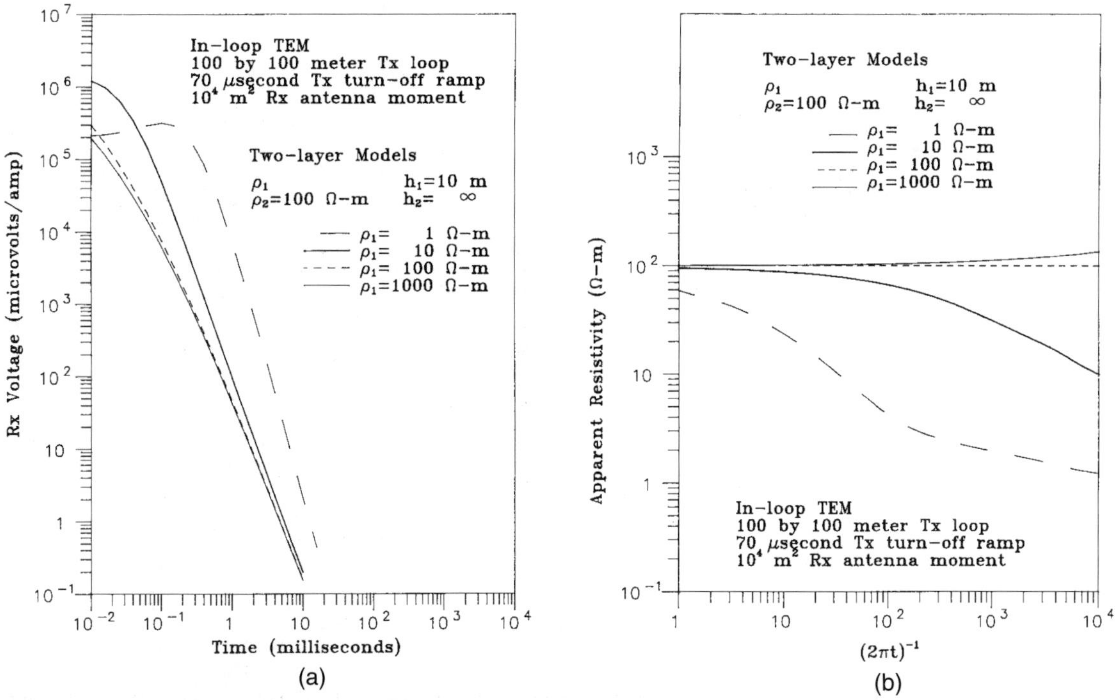

environments such as the southwest USA and Australia) can complicate TEM measurements, however, even if the target of interest is in relatively resistive host rock.

The equivalent depth equation from table 6.2 can be used to calculate a rule of thumb for determining the earliest time that you can expect to detect a target under conductive overburden. Solving this equation for t gives:

$$t \approx 1.3\, \sigma\, d^2\, 10^{-3}$$

where σ is the average conductivity of the overburden (siemen/m), d is the depth of burial (meters) and t is the delay time (milliseconds).

As a first approximation for interpretation of data in a conductive host, you can treat the host response and anomaly response as two separate but additive features. By doing this, the host response can be "stripped" from the total response leaving only the anomalous response. One method is to calculate a homogeneous earth response using the average host resistivity and then remove this from the field data. Layered-earth responses can also be removed in a similar manner. Another approach is to curve-fit the profile data with a smoothed approximation and then treat the residual as the anomalous response. See figure 6.96 for an example of the residual component approach.

FIGURE 6.94
Effect of a conductive mid-layer with variable thickness on (a) the decay waveform and (b) the late-time apparent resistivity. Compare (b) with figure 6.24 (a)

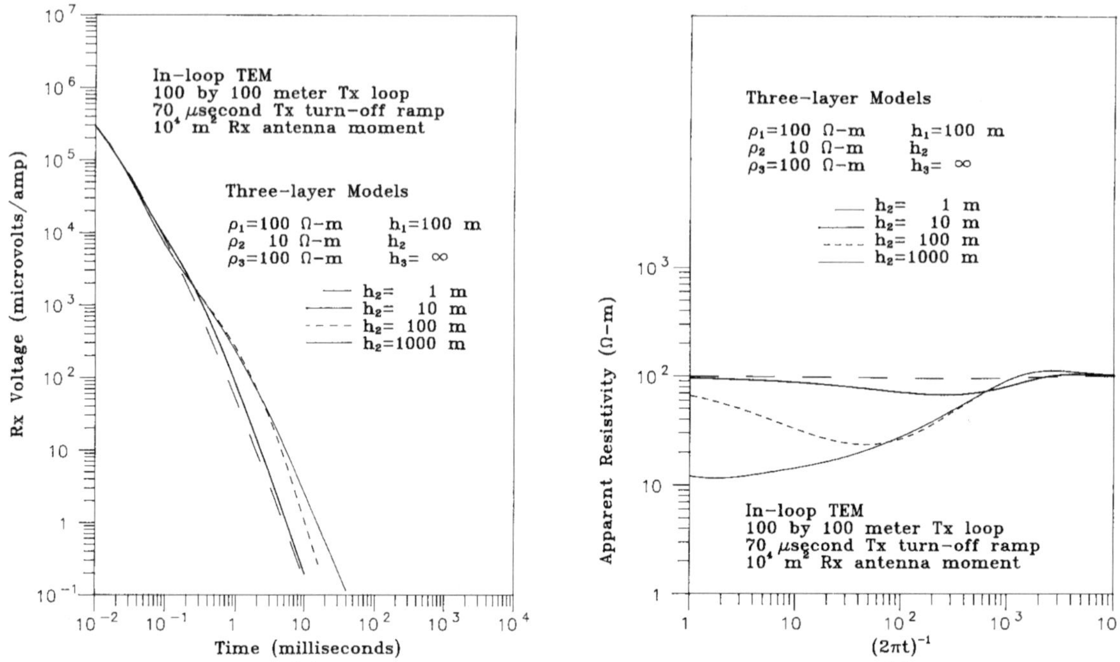

(a) (b)

Elura Deposit, Australia

One of the most widely published examples of orebody detection beneath a conductive overburden is the Elura deposit near Cobar, New South Wales, Australia. (See Emerson, 1980.) This cigar-shaped lead-silver-zinc orebody is about 100 m by 200 m in cross section and has an average resistivity of 0.3Ω m. It is covered by 100 m of 10Ω m weathered overburden. Figure 6.97 shows profiles of 100-m coincident-loop data from line 2500E, across the western side of the orebody. The strong early time values result from the conductive overburden, and the orebody does not begin to be seen until almost 2 milliseconds. This time correlates well with the time estimate of the above equation to see through the overburden.

Figure 6.98 shows a physical model configured to approximate the Elura environment. These data are for a 130-m coincident-loop survey comparing the free-space ($\sigma = 0$ for the host rock) anomalous response with the response of the same orebody imbedded in a conductive host rock ($\sigma = 0.16$). In the actual field situation, the resistivity of the host rock beneath the weathered overburden becomes gradually higher with depth, so these models are not exactly correct but give a good approximation of the expected response. Notice the strong early time values resulting from the conductive host and the weak anomalous response imbedded in the profiles for the later times; which are very similar to the field results. Figure 6.99 shows the same model but using a two-loop system with loop separations of 77 and 142 m. The weak anomalous values are similar to the coincident-loop results. The two-loop system does not provide any additional anomaly discrimination.

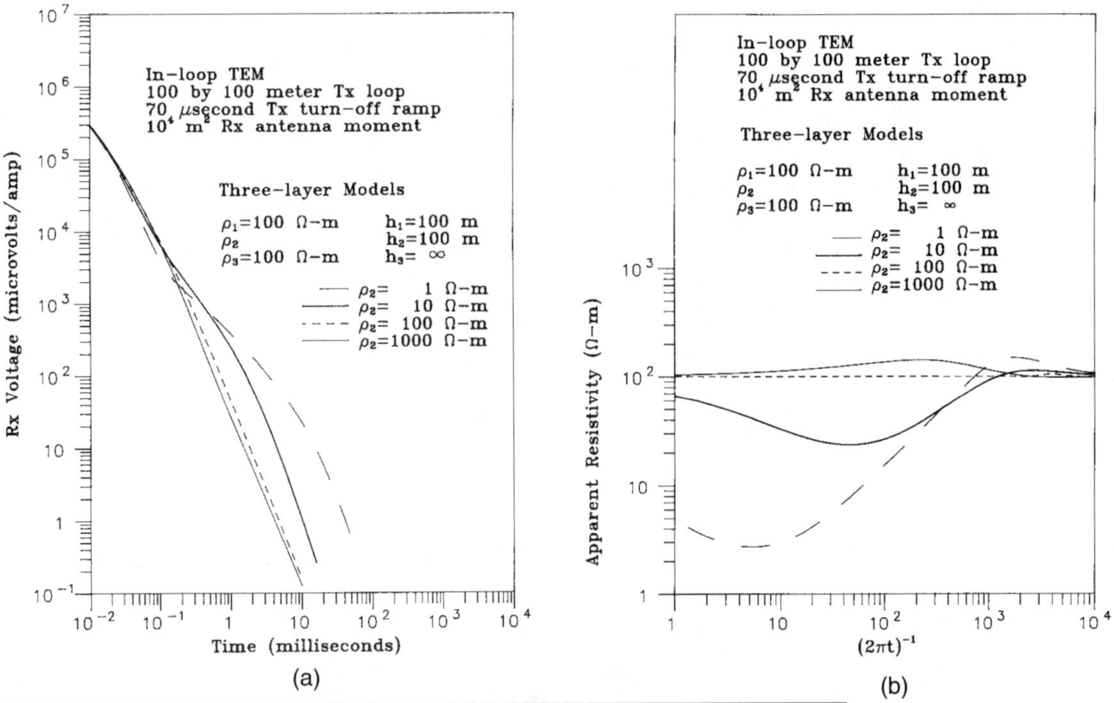

FIGURE 6.95
Effect of a 100 m thick mid-layer with variable resistivity on (a) the decay waveform and (b) the late-time apparent resistivity. Compare (b) with figure 6.25 (a)

Figure 6.100a is a resistivity/depth section obtained from the profile data in figure 6.97; figure 6.100b displays image resistivity (similar to Bostick resistivity discussed in the CSAMT section) versus depth, and 6.100c shows the result of smooth-model inversion. Notice that the orebody is well defined in all plots, and the depth of burial is relatively well delineated.

Mutooroo Prospect, Australia

As you can see, conductive overburden or host rock can complicate interpretation for TEM surveys, but the method is still viable and is used extensively in conductive environments. An example of fixed-loop EMP data acquired in a conductive environment is shown in figure 6.101. This survey was run over a long massive pyrrhotite-pyrite-chalcopyrite body at Newmont's Motooroo prospect situated 65 km southwest of Broken Hill, Australia. The orebody varied from 6 m to 12 m in thickness, dipped westerly at roughly 50 degrees and was covered by 100 m of 20 Ωm weathered overburden. Figure 6.101 summarizes the field data gathered. Sample profiles (b) and a contour map (c) are shown for the 5.63-ms delay window.

CSAMT AND TEM JOINT SURVEYS

Brittania Prospect, Australia

Data obtained over the Brittania prospect in northeast Queensland, Australia, provide a direct comparison between TEM and CSAMT.

FIGURE 6.96
Residual component of a TEM profile obtained by fitting a background response curve to the data (after Nabighian and Macnae, 1991)

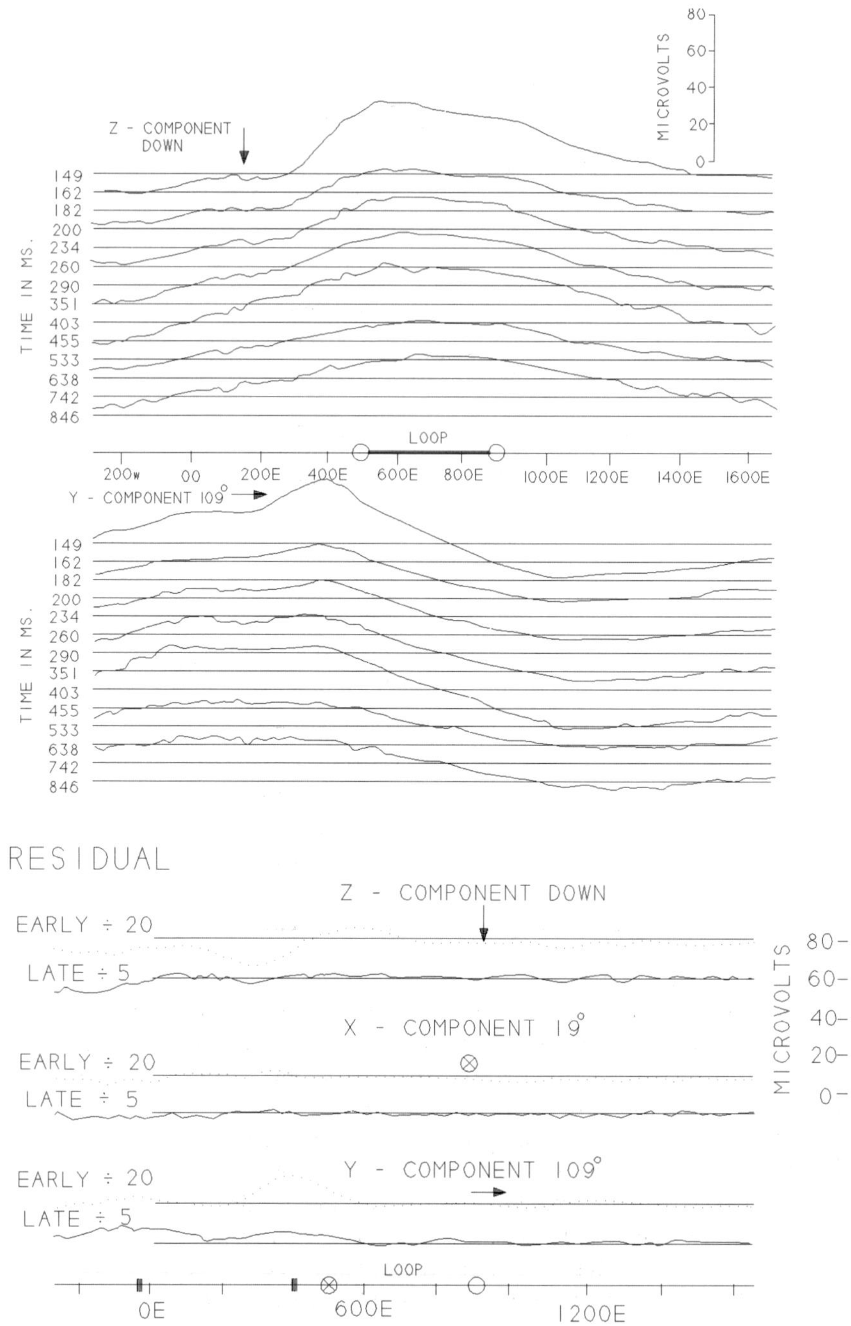

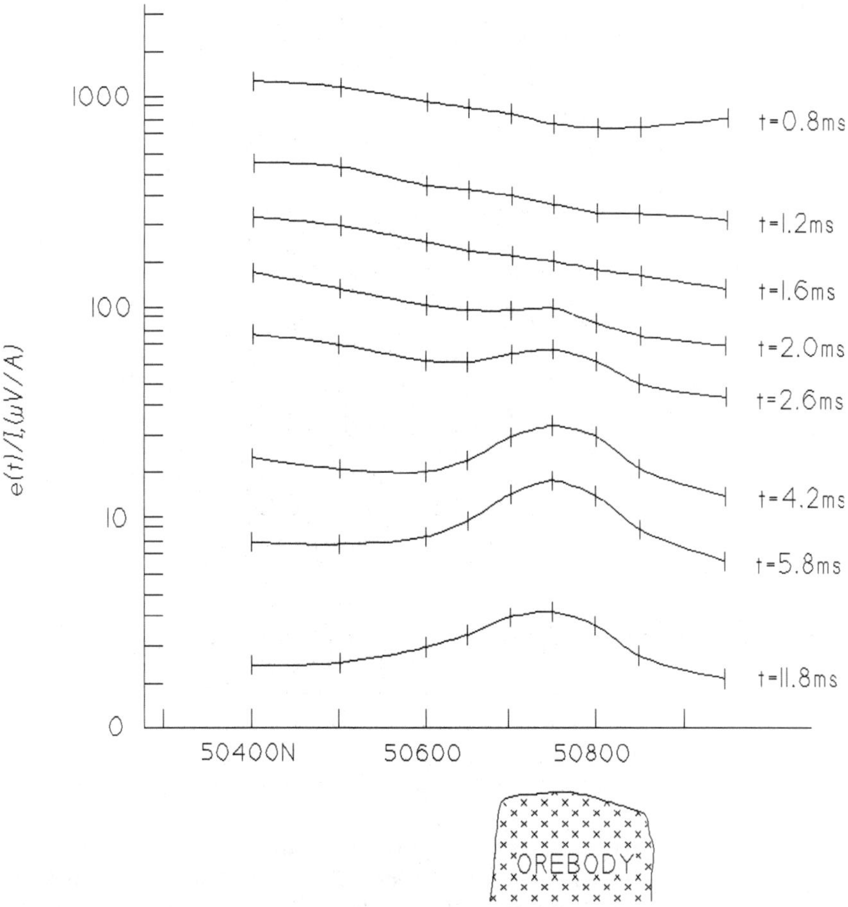

FIGURE 6.97
Coincident-loop TEM profile across the Elura orebody, line 20150E (old grid), using 100-m loops (from Spies, 1980)

FIGURE 6.98
Coincident-loop TEM model approximating the Elura orebody with the modeled body in free space (σ=0) and in a conductive host (σ=0.16). After Spies, 1980

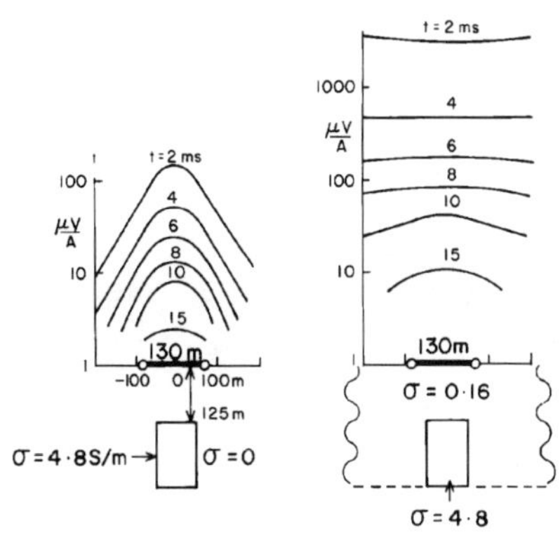

The plan map (figure 6.102) and the generalized geologic cross section (figure 6.103a) shows the geology over Brittania. The area is overlain by the Campaspe Sediments, a wedge of conductive alluvial sediments which thickens towards the south. Beneath these sediments, a block of andesites and metasediments abuts a unit of older, hydrothermally altered, weakly mineralized rhyolitic pyroclastics. The exploration target was sulfides associated with the high-angle andesite/rhyolite contact. Earlier percussion drilling had indicated anomalous concentrations of copper, lead and zinc near this contact.

A series of TEM tests at Brittania is described by Irvine and Staltari (1984). Separated-loop SIROTEM showed a strong anomaly which was interpreted as a steeply dipping bedrock conductor near the rhyolite/andesite contact. The plan-view trace of the interpreted conductor is shown in figure 6.102, and the data for line 10,000E are shown in figure 6.103b. EM37 soundings using transmitter loop 1 (located over alluvium) showed similar results (figure 6.103c), although Irvine and Staltari question the thoroughness of the original interpretation. The interpreted conductor was then drilled (drill hole Brit-1), but only weak sulfide mineralization was encountered. A downhole SIROTEM survey showed no evidence of further mineralization near the drill hole, casting doubt upon the original interpretation of a bedrock conductor.

To investigate the possibility that the TEM responses were caused by an edge effect from conductive sediments at the surface, three more surveys were run: coincident-loop SIROTEM, a second EM37 survey using a new transmitting loop over bedrock, and UTEM

FIGURE 6.99
Separated-loop TEM model approximating the Elura orebody with the modeled body in free space (σ=0) and in a conductive host (σ=0.16), using two loop separations: 77 m and 142 m. After Spies, 1980

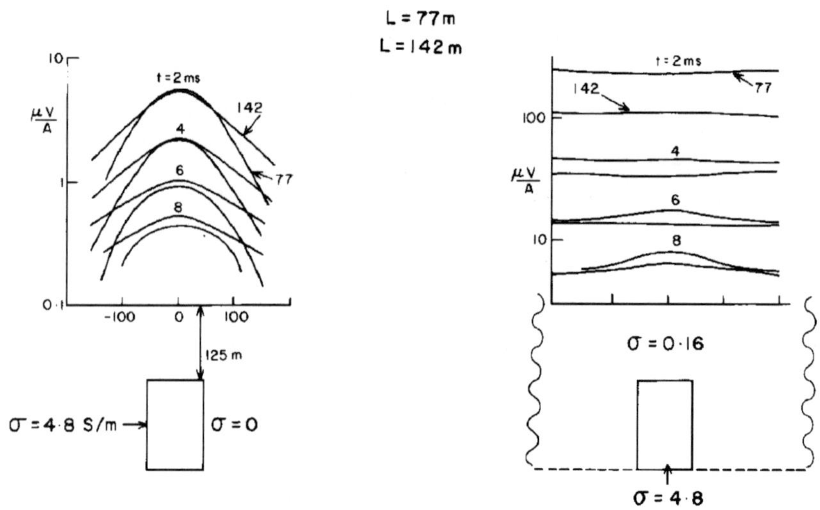

employing loops 1 and 2. The coincident-loop SIROTEM data are similar to figure 6.81. Data from the second EM37 work are shown in figure 6.103d. These additional surveys confirmed that surface conductors had caused a "false" anomaly in the original data. Follow-up Schlumberger and Wenner array resistivity work strengthened this conclusion.

A CSAMT survey was run over the same line to see how it would respond to the conductive surface layer. The survey used a 25-m E-field dipole and a 1100-m, north/south-oriented transmitter dipole located 3.6 km east of the survey line. Figure 6.104b shows the Cagniard resistivity data. South of station 9800, a strong conductive region is observed.

Comparison of the Cagniard resistivity and phase difference (figure 6.104a, b) indicates that some small static effects are present in the resistivity data. Static corrections were performed on the data set. The results, plotted with calculated image-depth on the vertical axis, are shown in figure 6.104d.

The static-corrected data show a low-over-high resistivity environment. The surficial low is especially pronounced on the south half of the line because of the combined effects of conductive alluvium and an even more conductive weathered zone in the top 20 m of the andesite bedrock. The CSAMT agrees with the known geology and the final TEM interpretation by Irvine and Staltari.

This case history substantiates the contention that multiple types and configurations of TEM surveys should be run over attractive anomalies. Once the fixed-loop EM-37 surveys were run in both directions and the data showing the sharp cross-over with the transmitter loop on the weathered bedrock and alluvium were compared with the migrating cross-over data obtained with the transmitter antenna on the rhyolites, the high-angle conductor interpretation was put in serious doubt. The additional surveys just confirmed the more accurate conductive surface layer interpretation. The CSAMT data, by virtue of the electric field data measured, provided a more straightforward interpretation and would not have been misunderstood in this instance.

FIGURE 6.100
Resistivity/depth sections obtained from the profile data in figure 6.97. (a) Late-time resistivity and (b) Image resistivity, with depths calculated from the depth equation (table 6.2). (c) Smooth-model inversion of the same data set

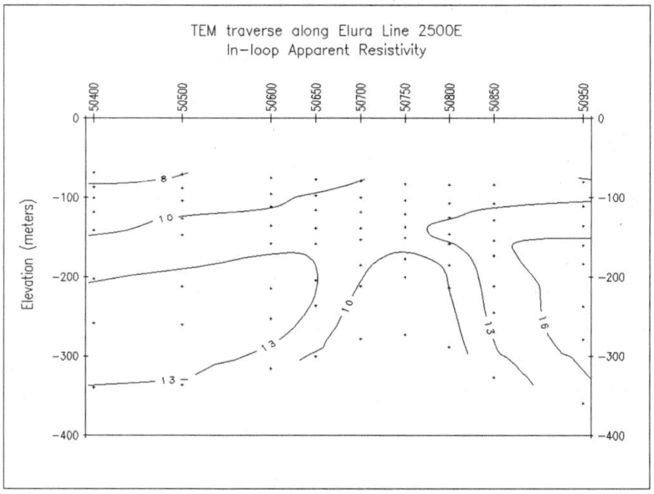

(a)

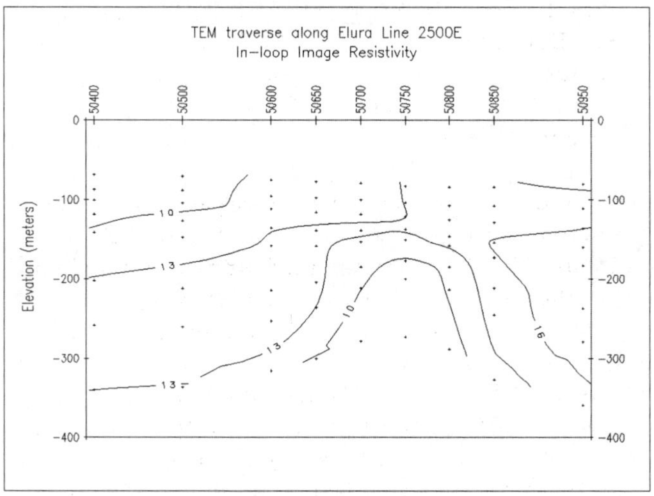

(b)

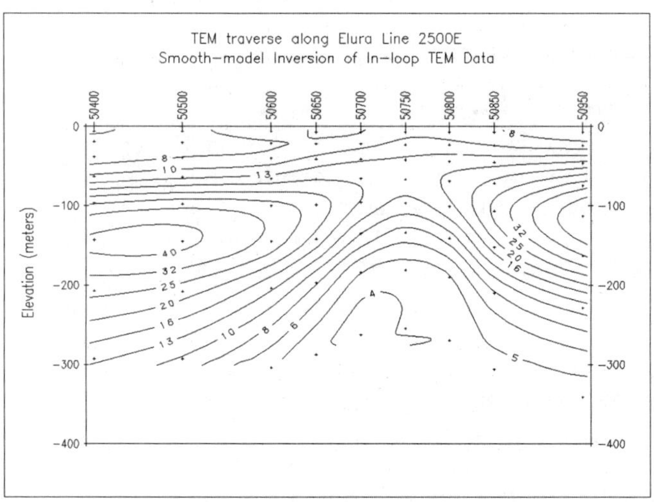

(c)

FIGURE 6.101
Motooroo EMP survey (from Nabighian and Macnae, 1991)

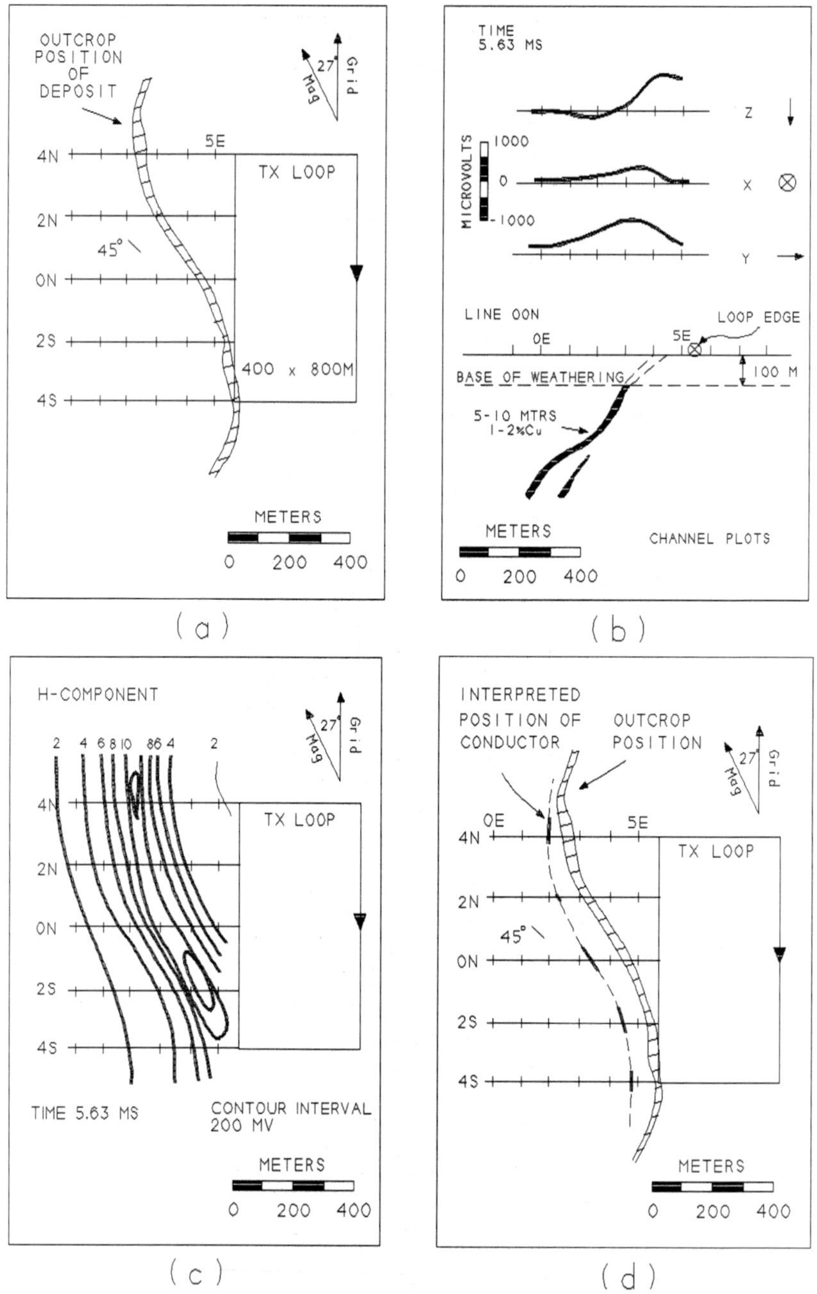

Woodchester, Australia

Fixed-loop UTEM survey data and CSAMT survey data over the Woodchester prospect in South Australia provide a case history with excellent correlation between the two methods. For the CSAMT portion of the survey, the transmitter dipole was offset from the line 6.7 km to the east and station spacing of 25 m with 25-m E-field dipoles were used along the line. Figure 6.105 shows the Cagniard resistivity and impedance phase data in pseudosection format.

For the UTEM survey, an 800-m by 800-m transmitter loop was set up on the east end of the line with the closest transmitter wire passing through station 29. Vertical (Hz) and horizontal (Hx, parallel to line) magnetic fields were measured on 25-m station centers. These results are shown in figures 6.106 and 6.107.

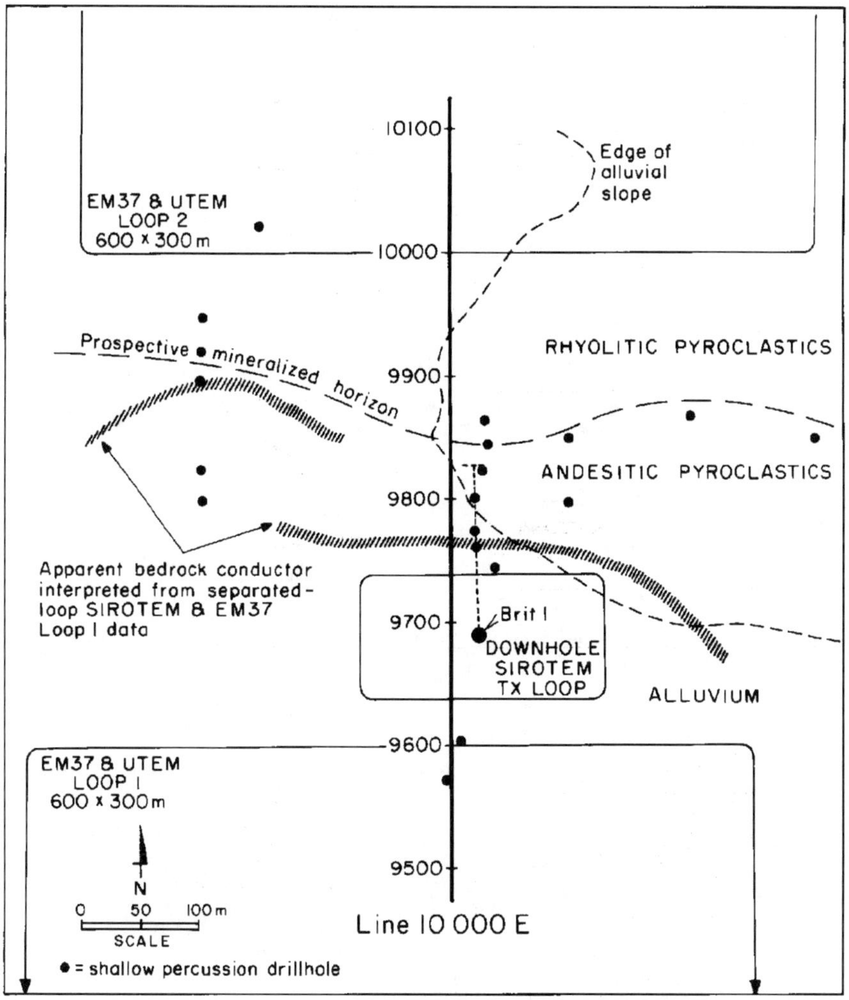

FIGURE 6.102
Map of the TEM and CSAMT tests over the Brittania prospect, Queensland, Australia. Data presented here are from line 10,000E. Adapted from Irvine and Staltari, 1984

Geology along the line consists of steeply dipping, interbedded quartzites and schists at stations to the west of station 16 and fine-grained quartzites to the east. A 40- to 50-m weathered layer of moderate resistivity overlies fresh bedrock.

Conductor B between stations 22 and 23 shows up in the CSAMT data as a near-surface feature with some deep roots, which are also observed in the phase section. In the UTEM data, the conductor is seen in a short time-constant, rapid decay response, indicating that it is a near-surface feature. Since this anomaly was not drilled, the existence of deep roots is not verified.

Conductor A, which is seen between stations 15 and 17 in the CSAMT resistivity data, also appears to have some deep roots and is accompanied by anomalous phase behavior at depth. This feature is seen in the UTEM data in both the early and late times. The migration of the Hz data in the early times toward the west tends to indicate that this near-surface response is caused by the weathered layer, but the consistent, sharp cross-over of the late-time data indicate that there is a near-vertical conductor in the same vicinity.

FIGURE 6.103
Geology and selected TEM results, Brittania prospect, from Irvine and Staltari, 1984. (a) Geology. (b) Separated-loop SIROTEM. (c) Fixed-loop EM 37 with the transmitter loop over alluvium. (d) Fixed-loop EM-37 with the transmitter loop over bedrock

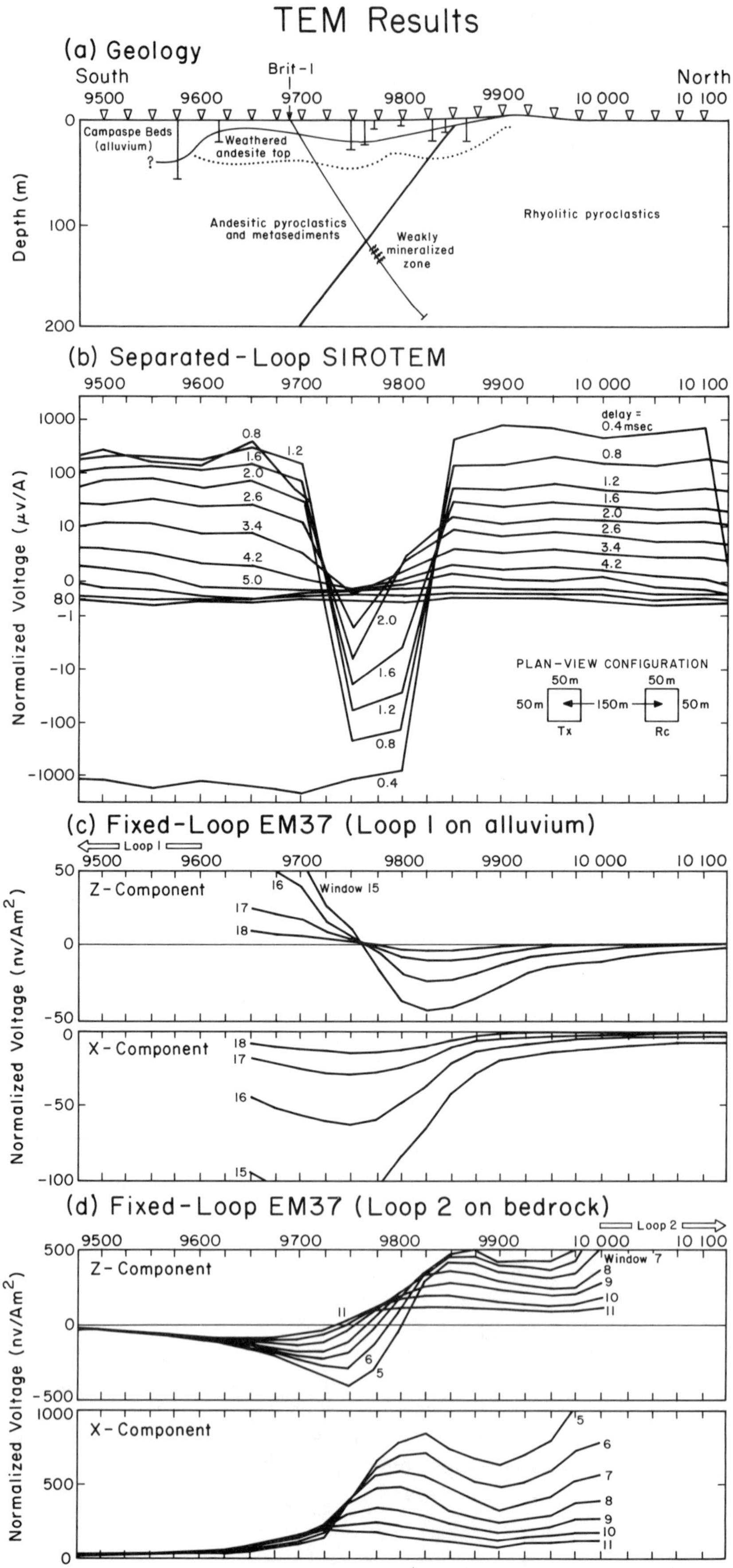

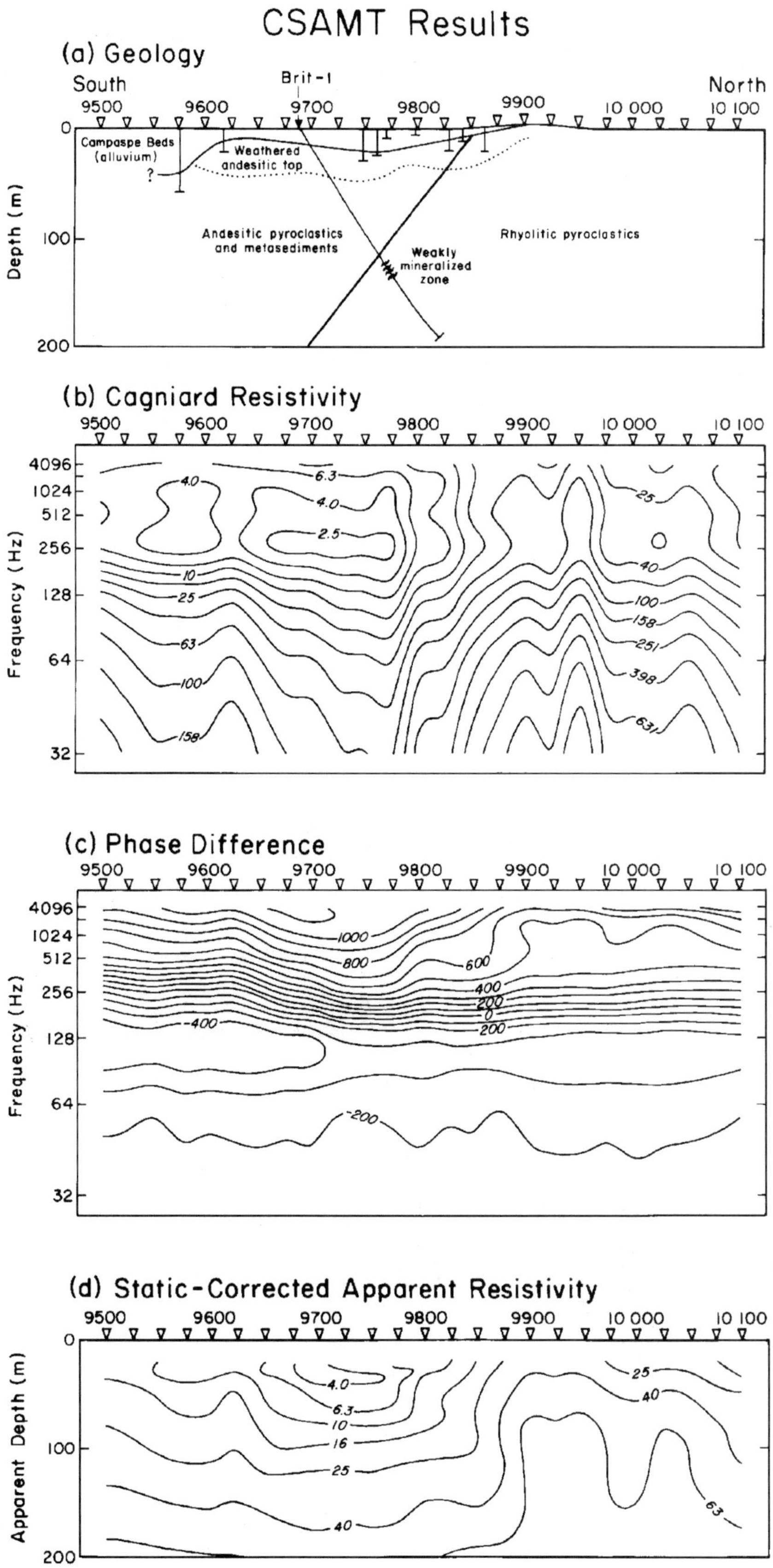

FIGURE 6.104
CSAMT survey results, Brittania prospect. (a) Geology. (b) Cagniard resistivity. (c) Impedance phase data. (d) Static corrected resistivity/depth section

FIGURE 6.105
CSAMT survey results over the Woodchester prospect, South Australia. (a) Cagniard resistivity. (b) Impedance phase

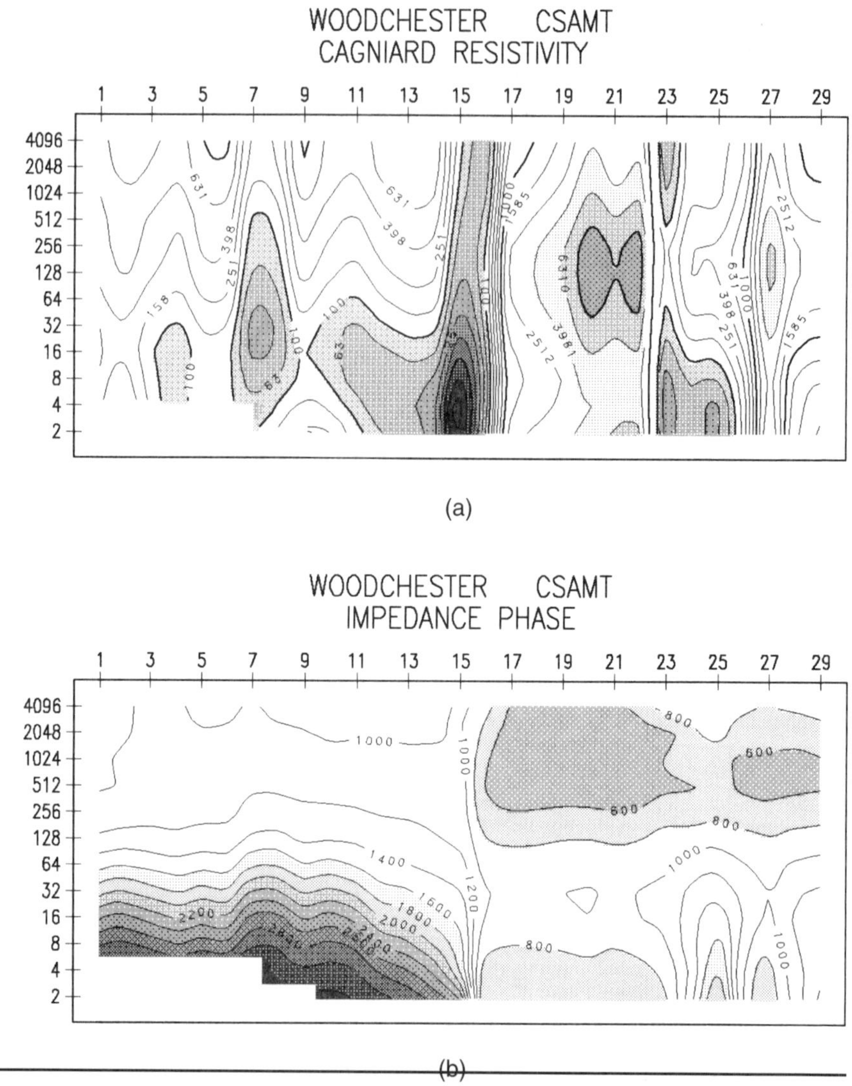

FIGURE 6.106
UTEM channel 1 normalized vertical magnetic field data. Notice the cross-overs near conductors A and B. The transmitter loop wire crossed station 29 (1200E)

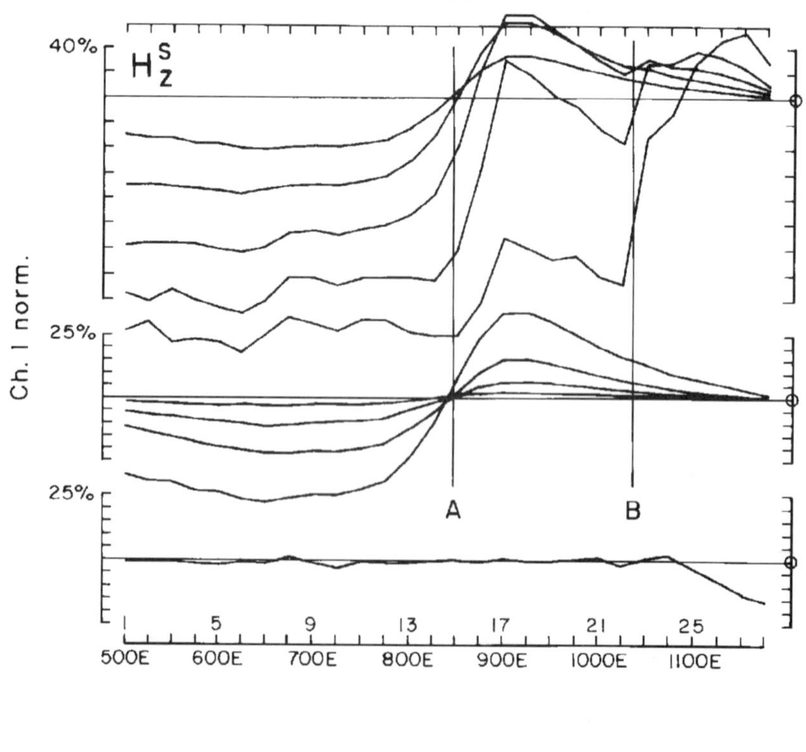

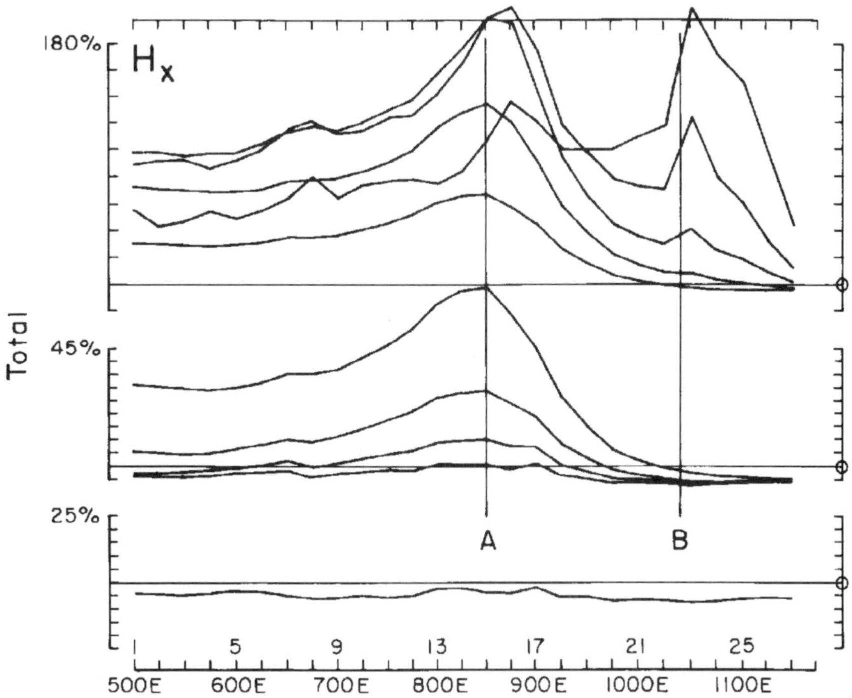

FIGURE 6.107
UTEM magnetic field data, horizontal component. Notice the peaks over conductors A and B

Figure 6.108 is a geologic cross section derived from a drill hole as shown. Conductor A is confirmed to be a narrow, near-vertical, bedded sulfides zone. The interbedded quartzites and biotite schists (with sulfides?) toward the west may account for the deep anomalous phase response of the CSAMT data. This type of behavior is typical of the response obtained over vertically-bedded sediments containing conductive material, when running with the electric field dipoles perpendicular to the bedding planes. This effect disappears when the E-fields are oriented parallel to bedding planes.

FIGURE 6.108
Drilling results near conductor A

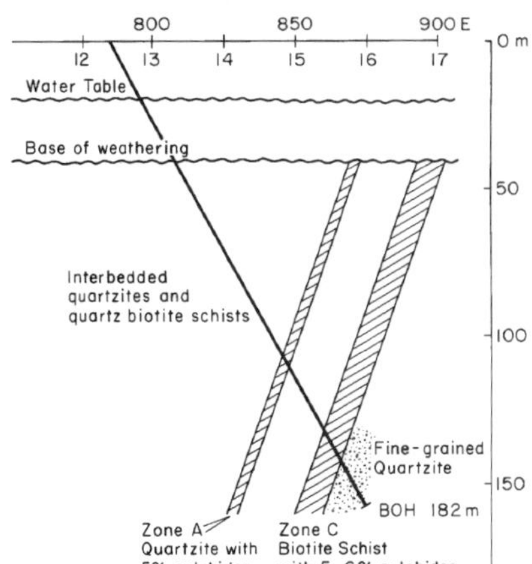

CONCLUSIONS

This discussion of the various facets of TEM and CSAMT is not intended to be a thorough treatment of the subject, but we tried to cover most of the basic concepts. I hope the readers will better understand the strong and weak points of both techniques and will be able to apply TEM and CSAMT successfully to their exploration programs.

For those interested in a much more detailed description of CSAMT and TEM, I recommend "Electromagnetic Methods in Applied Geophysics", Volume 2, edited by Misac Nabighian and published by the Society of Exploration Geophysicists, Tulsa, Oklahoma.

ACKNOWLEDGEMENTS

Completion of this chapter would not have been possible without the help of Dr. Scott MacInnes, Norm Carlson, Ms. Liu De-Xin, Bart Black, and others, who helped acquire, review, model and draft the data, and review the manuscript.

REFERENCES

Butterworth, N.A., 1988, Controlled-source audio frequency magnetotelluric responses of three dimensional bodies: M.S. thesis, University of Utah, Salt Lake City.

Cagniard, L., 1953, Basic theory of the magnetotelluric method of geophysical prospecting: Geophysics, 18, 605-635.

Crone, J.D., 1980, Field results using borehole pulse EM methods: Case History Note, Crone Geophysics Ltd.

Emerson, D.W. (Ed.), 1980, The geophysics of the Elura orebody, Cobar, New South Wales: Bull. Austr. Soc. Expl. Geophys., 2, 143-347.

Fitterman, D.V., 1989, Detectability levels for central induction transient soundings: Geophysics, 54, 127-129.

Frischknecht, F.C., Labson, V.F., Spies, B.R., and Anderson, W.A., 1991, Profiling methods using small sources, in Nabighian, M.N., Ed., Electromagnetic methods in applied geophysics, Vol. 2, Soc. Expl. Geophys., 105-270.

Goldstein, M.A., 1971, Magnetotelluric experiments employing an artificial dipole source: Ph.D. thesis, University of Toronto, Toronto, Ontario, Canada.

Hay, K.R., 1989, Exploration case history of the Golden Cross Prospect, Waihi, New Zealand, Aust.I.M.M., Melbourne.

Irvine, R.J. and Staltari, G., 1984, Case history illustrating interpretation problems in transient electromagnetic surveys: Exploration Geophysics, 15, 155-167.

McNeill, J.D., Edwards, R.N., and Levy, G.M., 1984, Approximate calculations of the transient electromagnetic response from buried conductors in a conductive half-space: Geophysics, 49, 918-924.

Nabighian, M.N., 1979, Quasi-static transient response of a conducting half-space — An approximate representation: Geophysics, 44, 1700-1705.

Nabighian, M.N., and Macnae, J.C., 1991, Time domain electromagnetic prospecting methods: in Nabighian, M.N., Ed., Electromagnetic methods in applied geophysics, Vol 2, Soc. Expl. Geophys., 427-520.

Ogilvy, R.D., 1986, Theoretical EM response curves over a thin dipping dyke in free space — separated inline loop configuration: Geophys. Prosp.,34, 769-788.

Schmidt, P.G., Dolence, J.D., Lluria, M.R., and Parsons, G. III, 1978, Geology of Crandon massive sulfide deposit in Wisconsin: Skillings Mining Review, May 6, 8-11.

Spies, B.R., 1974, Transient EM model studies, 1973: Bur. Mineral Res. Austra. Record 1974, 156.

Spies, B.R., 1979, Experience with time domain EM methods in Australia: (abs) Geophysics, 553.

Spies, B.R., 1980, Interpretation and design of time domain EM surveys in areas of conductive overburden: Bull. Austr. Soc. Expl. Geophys., 2, 272-281.

Zonge, K.L., and Hughes, L.J., 1991, Controlled source audio-frequency magnetotellurics: in Nabighian, M.N., Ed., Electromagnetic methods in applied geophysics, Vol. 2, Soc. Expl. Geophys., 713-809.

CHAPTER 7
IMAGE PROCESSING
THE BASICS OF IMAGE PROCESSING

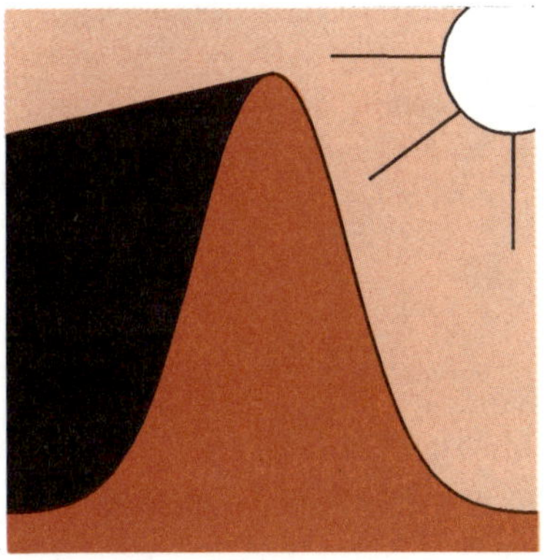

Modern image processing had its start in the early days of space exploration at the Jet Propulsion Laboratory in California. The techniques now used to improve the limited resolution of satellite pictures are basically the same as those employed by medical, machine vision and the earth resources industries. The computers used in the early days of image processing had limited capabilities and low performances, so software development focused on techniques for the rapid analysis of the high-volume digital images. Medical uses for image processing soon appeared for the enhancement of low-contrast and blurred X-rays. As computer hardware advanced, image processing moved from the expensive mainframes to minicomputers. Today, low-cost microprocessors and continually improving graphics performance have brought image processing to the PC microcomputer.

Douglas C. Fraser, Chairman, Dighem Surveys & Processing Inc.

Stephen L. Fogarasi, Scientific Programmer, Dighem Surveys & Processing Inc.

Douglas Fraser obtained bachelors and masters degrees in geology at the University of New Brunswick and a Ph.D. in geophysics at the University of California at Berkeley. He has published papers on copper geochemistry, electromagnetics, VLF, induced polarization, ground and airborne resistivity, digital filtering and correlation techniques, image processing and new interpretation methods. He has been responsible for the design and development of DIGHEM airborne electromagnetic systems.

Dr. Fraser has been employed in geological mapping and geophysical exploration in both the mining and petroleum industries and in geochemical prospecting for base metals. He is chairman of Dighem Surveys & Processing Inc., and consulting geophysicist for Dighem's I-POWER division. Dr. Fraser is a member of the Association of Professional Engineers of Ontario, the Society of Exploration Geophysicists, the European Association of Exploration Geophysicists and is a member and past-president of the Canadian Exploration Geophysical Society.

Stephen L. Fogarasi is a scientific programmer at Dighem Surveys & Processing Inc. He is responsible for the design, development and continued maintenance of Dighem's PC-based image processing, data display and enhancement software as well as the continual development of frequency and spatial domain filtering of acquired geophysical data. Further, he is responsible for the development of presentation software for a field processing system and supervision of programmers for I-POWER workstation activity.

High resolution and a large number of available colors, or gray scales, are required to take advantage of the fastest image processing techniques. A PC microcomputer, with graphics controller and color graphics screen, can be purchased for less than $3,000, with software costing as little as $1,000. Such a system is not powerful but is suitable for manipulating a ground geophysical dataset or an airborne geophysical dataset comprising 1,000 line-miles of data. Larger datasets can also be manipulated with the above simple system, but much greater productivity and detail are obtainable with more powerful systems.

Image processing techniques allow rapid manipulation of data. The techniques can identify folding, faulting and lithologic changes in far greater detail than is possible with alternative methods of display such as contour maps.

Color Perception

Colors and how we perceive them are a complex interaction of light, eye and brain. Three kinds of cones within the retina of the eye respond to wavelengths of light, centering on what we recognize as blue, green and red (figure 7.1).

FIGURE 7.1
Frequency response of the human eye

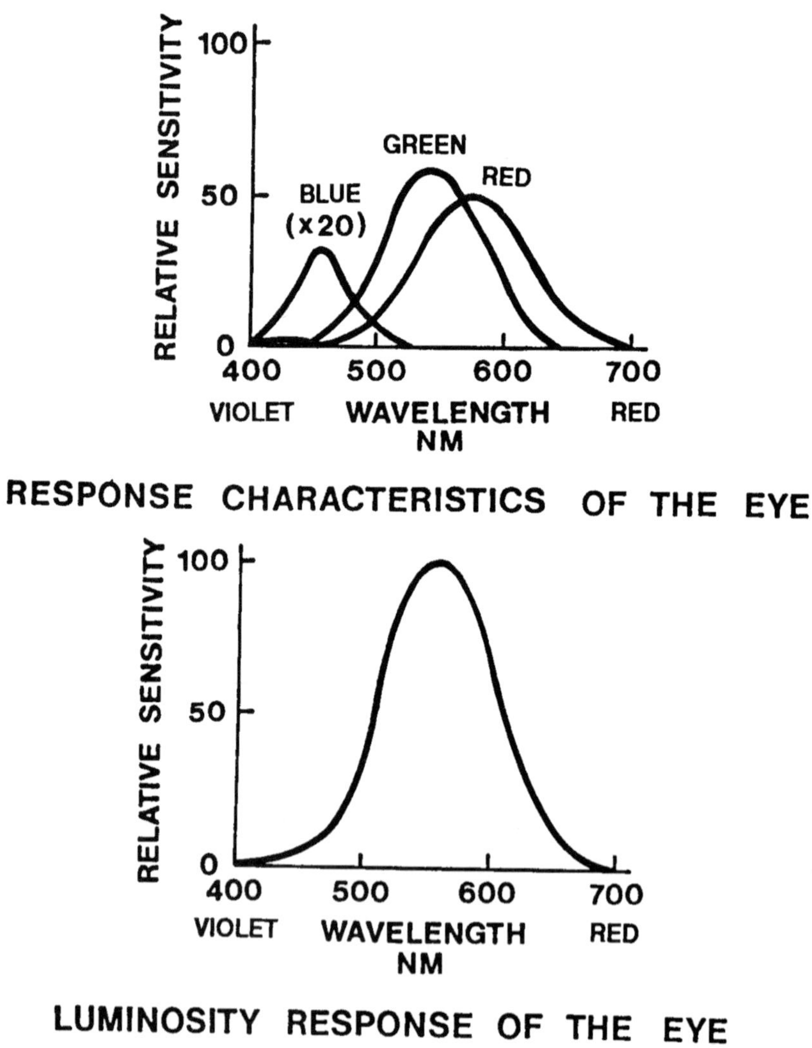

There are also rods within the retina that are sensitive to a broader range of wavelengths but are usually saturated at low intensities and are believed not to contribute to normal color perception. How the brain decodes these three red, green and blue "channels" is not well understood, but suffice it to say that all that is required to produce any recognizable color is the contribution of only these three primary colors. Something to note, however, is the high sensitivity of the eye to the yellow-green part of the spectrum. This information is pertinent to the color selection of map backgrounds.

The primary colors -- red, green and blue -- are termed additive colors and represent what is known as the RGB system (figure 7.2). When equal intensities of each are added together, they produce the familiar gray scale. White occurs when the three are at maximum intensity. A computer graphics screen (i.e., color monitor, or CRT) is the most familiar RGB device. Every picture element, or pixel, on a screen is composed of three smaller cells, each with their own red, green or blue color sensitivity. Each pixel is small enough that the three cells appear to our eye as a single color cell, and the color we see is the combined or additive result.

FIGURE 7.2
The RGB and CMY color systems (from Gonzalez and Wintz, 1977)

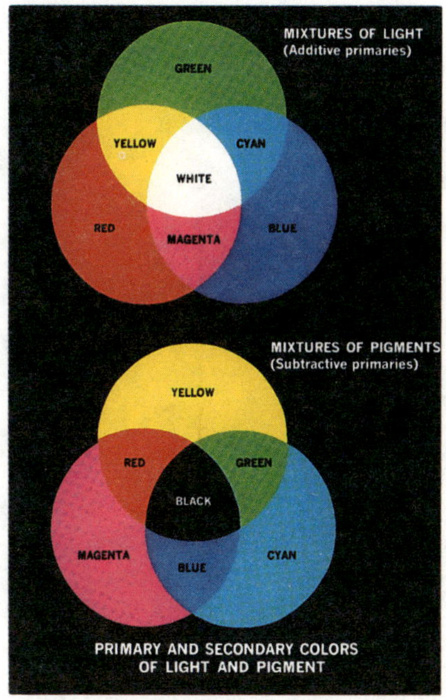

A paper image cannot generate any light of its own but must rely on removing or subtracting colors from some incident light source. An alternate coloring system can be used to describe this secondary or subtractive color scheme. Its fundamental colors are called cyan, magenta and yellow, or the CMY system (figure 7.2). They are subtractive in nature. For example, removing red from white light leaves cyan, removing green from white produces magenta, and removing blue from white leaves yellow.

It is a simple process to convert from RGB to CMY using equation 7.1, and it is required in any color conversion of graphics screen images to paper copies and vice versa. The conversion equation in dimensionless units is the simple matrix operation:

$$\begin{bmatrix} C \\ M \\ Y \end{bmatrix} = 1 - \begin{bmatrix} R \\ G \\ B \end{bmatrix} \tag{7.1}$$

Within the computer, a graphics controller employs a palette to generate a color scheme (figure 7.3). This is simply a table of RGB or CMY intensities that are used to generate the colors. The palette is a hardware device through which every pixel is first numerically defined before activating the RGB or CMY channels. It has the advantage of being independent of the size of the data grid or image while being readily accessible to a user. As a digital file, color palettes are typically much smaller and can be updated faster than the data grid itself. Clearly, manipulation of the palette would be more desirable than redefining the image pixels in the data grid if time is a critical factor.

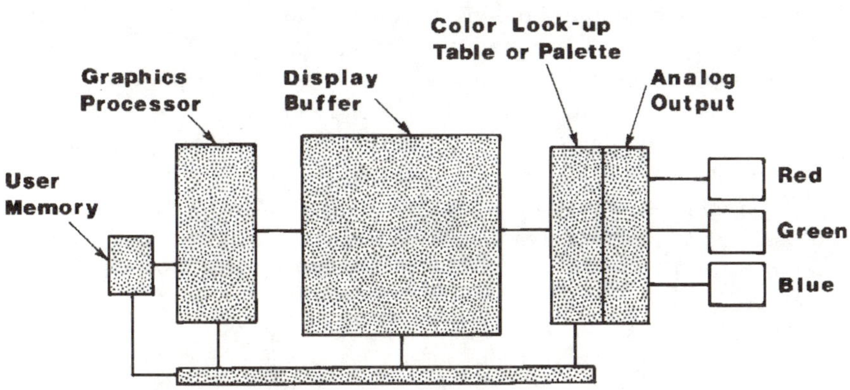

FIGURE 7.3
Simple graphics controller

Assume, for example, that the 55000 nT level of a magnetic image is colored yellow. This yellow color would result if the 55000 nT is replaced in the data grid with the code number, say, 32, if this is the code number in the color palette which identifies the color as yellow. If it is desired to recolor the map so that 55000 nT becomes orange, this may be accomplished by recoding 55000 from 32 (yellow) to 48 (orange), if 48 is the code number which identifies orange in the color palette. This process requires a complete recoding of the entire grid in order to recolor the map.

Alternatively, the grid coding in the data grid can be left untouched. Rather, the color palette can be altered, e.g., the code 32 can be moved up to the orange area of the palette. In both of the above cases, what was yellow before now becomes orange.

The advantage of manipulating the color palette instead of the data grid lies in speed. In this example, recoloring the magnetic map by operating on the palette will involve only 64 manipulations if 64 colors are defined in the palette. In comparison, recoding a 1000 x 1000 magnetic grid would involve one million manipulations.

Dynamic Range and Compression

Geophysical data collected in the field typically have several orders of magnitude of dynamic range and several digits of precision. Prior to image processing, these data must be compressed to a limited dynamic range and precision suitable for digital storage and manipulation. In airborne magnetic data, suitable precision may range from tenths of gammas to thousands of gammas. In airborne resistivity, the dynamic range may encompass six orders of magnitude. In the case of magnetics, a simple linear transformation is often sufficient to translate the precision of field information to storable two-byte format. This allows a dynamic range of about 65,000, i.e., a sensitivity of 1 part in 65,000. Resistivity, on the other hand, typically requires a log transformation to compress the huge dynamic range into a two-byte format.

Two bytes often are sufficient to retain all the significant information from the field measurements of geophysical data for purposes of image processing. Two-byte storage is usually sufficient, because image processing is a qualitative tool for the recognition of geometric patterns rather than a quantitative tool to determine physical properties of the rocks. There is no pictorial way to display even this limited dynamic range to its full sensitivity in any single product, however, so some other process must be applied that can selectively enhance any particular feature or group of features. Image processing achieves contrast enhancement at the cost of dynamic range. A goal of designing software for such systems is to achieve the desired enhancements in a manner which overcomes dynamic range limitations.

Commonly available imaging hardware has memory restrictions which preclude even two-byte (16 bit) pixel storage. In the case of the extended VGA, which is now commonly available as a "low-end" image processing platform, only 1 byte or 256 distinct levels are simultaneously available. Very careful manipulation of the data must be followed to avoid losing any potentially significant information.

No single image processing technique can instantly provide all the desired information in a single product. Rather, dynamic, fast, interactive techniques are used to enhance single features or classes of features.

Histograms

One of the most important tools of image processing is the data histogram (figure 7.4). Rather than manipulating the individual grid cells, many interactive image processing systems manipulate the grid's histogram. In a histogram, the data values are pigeonholed into separate classes. In a color map, this classification is the familiar color bar or legend, where each color represents a range of values.

The number of classes that can be used is usually determined by the hardware employed to display the image. For example, the original Applicon color palette was commonly limited by the user to provide 40 or so pleasing colors. In current graphics controllers, the complete 256 simultaneously available colors are often displayed. Obviously, the more classes which are available, the more the histogram will match the original data.

FIGURE 7.4
The histogram of the data grid is commonly used to design the color scheme

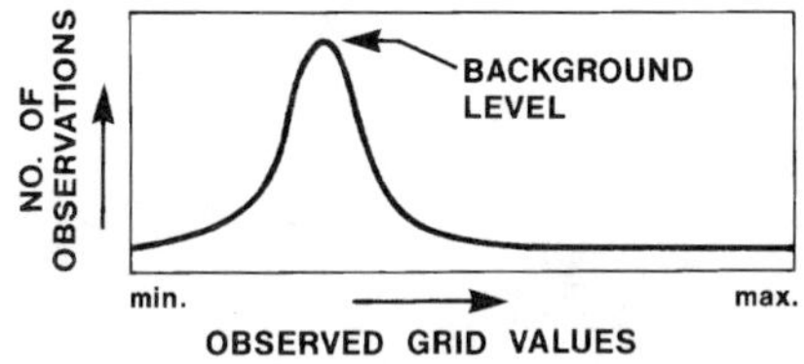

ORIGINAL DATA RANGE: -32000 TO +32000 LEVELS

PROBLEM: How to reduce the range to 1 to 256 levels with minimal loss of information

SOLUTION: Histogram equalize, or histogram specify

Color Palettes and Lookup Tables

The color palette of the graphics controller is a hardware device that can quantify how much of the primary colors (red, green and blue for color graphics screens) or secondary colors (cyan, magenta and yellow for paper plots) are used to generate the colors for each class of the histogram. For the inexpensive image processing systems commonly used in geophysics, the graphics controller in the computer is an 8-bit device, displaying 256 colors. The lookup table is the set of instructions in software, which controls the color palette, to provide the desired color scheme. Lookup tables may be generated in many forms to provide, for example, a vibrant color scheme, pastel colors, or any variety of colors limited only by the imagination of the programmer. The color palette/lookup table combination is commonly referred to simply as the "color palette". This is the context which will be followed here.

From the above description, it is evident that a one-to-one correspondence between the histogram and the color palette exists. Therefore, manipulation of the color palette is identical to manipulating the histogram which, in turn, is a means of manipulating the color image of the data grid.

Histogram Transformation

The eye is much better at discerning contrast than dynamic range.

The intensity of the gray scale in each band of figure 7.5 is constant, even though it does not appear to be so to most viewers. The human eye preferentially enhances the zone change by brightening along the dark side and darkening along the bright side. This optical illusion, called the Mach-band effect, is a characteristic that may be exploited in image processing, as a means of enhancing contrasts. The histogram can be modified to increase contrast by manipulating the boundaries of the different color classes of data. This requires the development of a transformation function that converts an image with a particular intensity distribution to a different intensity distribution. To describe the process, the notation of Gonzales and Wintz (1977) will be followed.

FIGURE 7.5
The intensity of the gray scale in each band is constant, but appears to vary in intensity due to the Mach-band effect (Gonzales and Wintz, 1977)

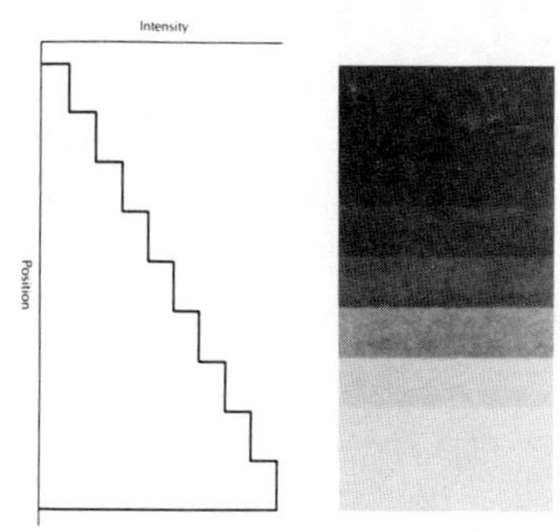

First, consider the continuous case with all pixel values normalized to lie within the range:

$$0 \le r \le 1 \tag{7.2}$$

A distribution of the gray levels would then be a probability density function $p_r(r)$.

Figure 7.6 illustrates probability density functions for a "dark" and "light" image. A great deal can be surmised about general characteristics of an image solely from these probability densities. Now assume there is a transformation of the form,

$$s = T(r) \qquad 0 \le s \le 1 \tag{7.3}$$

that maps to a level s every pixel value r in the original image. Assume also that T(r) is single valued, monotonically increasing in $0 \le r \le 1$ and,

$$0 \le T(r) \le 1 \qquad 0 \le r \le 1 \tag{7.4}$$

In this case, we can expect a transform of the reverse situation to exist, i.e.,

$$r = T^{-1}(s) \qquad 0 \le s \le 1 \tag{7.5}$$

where $T^{-1}(s)$ will also satisfy equation 7.4.

Therefore, from elementary probability theory, if $p_r(r)$ and T(r) are known, we have,

$$p_s(s) = p_r(r)\frac{dr}{ds} \qquad r = T^{-1}(s) \tag{7.6}$$

Thus, a transformation from one distribution to another is possible knowing only the original and final distribution functions, although dr/ds may not be easily obtainable.

Now, if we take, as a particular case,

$$s = T(r) = \int_0^r p_r\,(w)dw \qquad 0 \le r \le 1 \tag{7.7}$$

(where the right side of equation 7.7 is recognized at the cumulative distribution function, or CDF of r), we solve for ds/dr:

$$\frac{ds}{dr} = p_r(r) \tag{7.8}$$

and equation 7.6 becomes,

$$p_s(s) = p_r(r)\frac{1}{p_r(r)} = 1 \tag{7.9}$$

What we see is that using a transformation function equal to the CDF of r produces an image whose gray levels have a uniform (equal) density.

In a usable form, we must take the discrete case. Here,

$$p_r\,(r_k) = \frac{nk}{k} \qquad \begin{array}{l} 0 \le r_k \le 1 \\ k = 0,\ 1,\ \ldots,\ L-1 \end{array} \tag{7.10}$$

where $p_r(r_k)$ is the probability of the kth gray level, L is the number of levels, n_k is the number of times this level appears in the image and n is the total number of pixels in the image. A plot of $p_r(r_k)$ vs r_k is the familiar histogram.

The discrete form of equation 7.7 is,

$$\begin{aligned} s_k = T(r_k) &= \sum_{j=0}^{k} \frac{n_j}{n} \\ &= \sum_{j=0}^{k} p_r(r_j) \qquad 0 \le r_k \le 1 \\ &\qquad\qquad\qquad k = 0,\ 1,\ \ldots,\ L-1 \end{aligned} \tag{7.11}$$

The above process allows a dark or light image to be altered by stretching the histogram to fill the available range. If this process is designed to yield a flat histogram, it is referred to as histogram equalization.

The process of histogram equalization is quite straightforward:

- generate a histogram for an image
- determine the CDF of the histogram
- determine the new levels using equation 7.11, and reset the histogram origin
- map the original classes to the equalized classes.

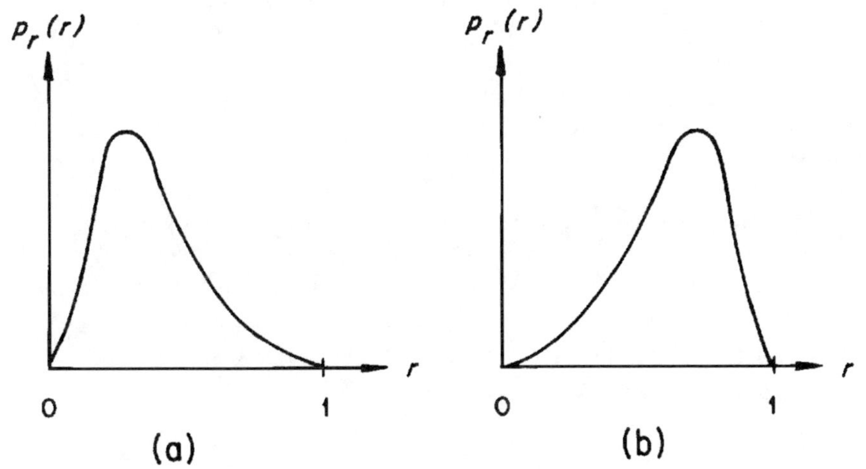

FIGURE 7.6
Probability density functions for, (a) a dark image, and (b) a light image

The result of this operation is illustrated in figure 7.7 where equalization has been applied to stretch out the original histogram.

The equalized histogram is, however, rarely flat as a result of the discrete nature of the original data. Also, the number of levels in the resultant histogram is always less than or equal to the original histogram. Levels tend to be lost in favour of contrast enhancement.

The same process can be applied to generate a histogram of any shape. This process, histogram specification, can be used to tailor an image to provide the desired contrast. In all cases, the CDF of the desired, rather than the original, histogram is used to generate the transformation table.

The example of figure 7.8(a) illustrates a rather dark image. It is from the Getchell area of Nevada which was surveyed with the DIGHEM^{IV} helicopter-borne EM/Magnetic/VLF system. Figure 7.8(a) displays the ratio of 7200 Hz resistivity to the 900 Hz resistivity. The ratio map may be used as an indicator of bedrock conductors. Ratios are notoriously difficult to handle when either of the terms tend to zero. This image of figure 7.8(a) was originally colored with a linear distribution with arbitrary cutoff points to avoid problems of low-resistivity amplitudes in the ratio. Some potential targets are evident as red highs. By stretching out the histogram (which involves moving histogram levels across to adjacent colors), however, contrast is increased [figure 7.8(b)]. Instantly, much more detail is brought out in the background.

The above histogram techniques are not restricted to whole images. Smaller sections can be windowed out for histogramming, and then the resulting modified histogram can be applied to the entire data grid.

The histogram equalization technique has one flaw. By its nature, all classes appear at equal frequency. A background or base level is not, therefore, readily apparent. A somewhat better approach is to produce a normal histogram distribution with a peak corresponding

FIGURE 7.7
Adjacent histogram levels are shifted and combined to produce a histogram with equal levels. Due to the discrete nature of the data, Histogram Equalization rarely produces a flat histogram, but a reasonable approximation can be made

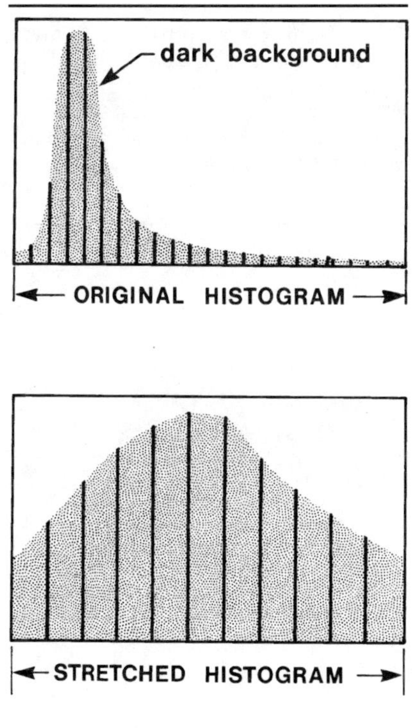

FIGURE 7.8
Color map of the ratio of airborne resistivities at 7200 Hz to 900 Hz for (a) linear histogram and (b) equalized histogram

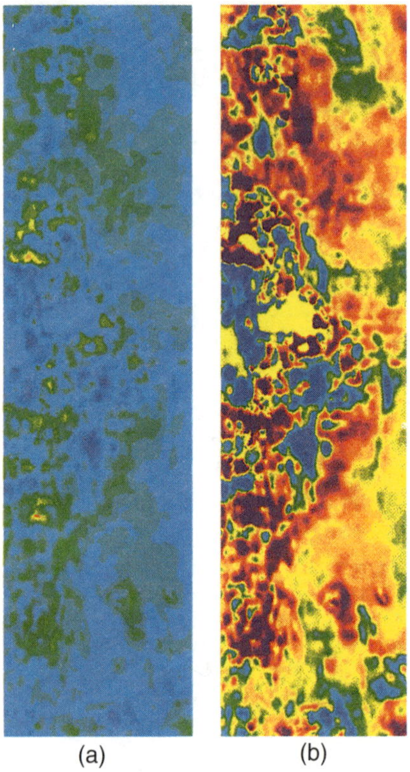

to the background level. A better regional picture can be revealed with this distribution.

In the above cases, the histogram has been modified while the color classification has remained the same. Alternatively, the histogram can be left static and the color classification modified. A simple but effective tool for interactive image processing is to rotate through the color palette by shifting all components of the RGB lookup table one unit at a time.

Figure 7.9b shows a snapshot of a revolving color bar as it moves from left to right. This palette rotation, performed interactively on the graphics screen, provides a dynamic picture that highlights features which a static image or map could not produce.

Enhancing this concept further, a "variable sensitivity" color classification can be used (e.g., figure 7.9c) where a more sensitive high-contrast color bar is embedded within the existing color bar.

This dual sensitivity technique has little value in a conventional static image but, on palette rotation, the high-contrast bar can highlight low-amplitude features in different parts of the histogram.

Radiometric Enhancement Techniques

The image enhancement techniques described previously are performed individually on each pixel. The neighbouring pixels have no influence on the operations being performed on that pixel. Such operations are referred to as point- or pixel-specific operations. Collectively, all such operations are called radiometric enhancement techniques. These techniques can all be represented as a table or graph that expresses the relationship between old and new color or brightness values for the pixels. All of the techniques described above are radiometric enhancement techniques.

FIGURE 7.9
Examples of color bars showing, (a) the usual static single color distribution, (b) a snapshot of the color bar as it revolves from left to right through the histogram, and (c) a snapshot of a revolving dual sensitivity color bar

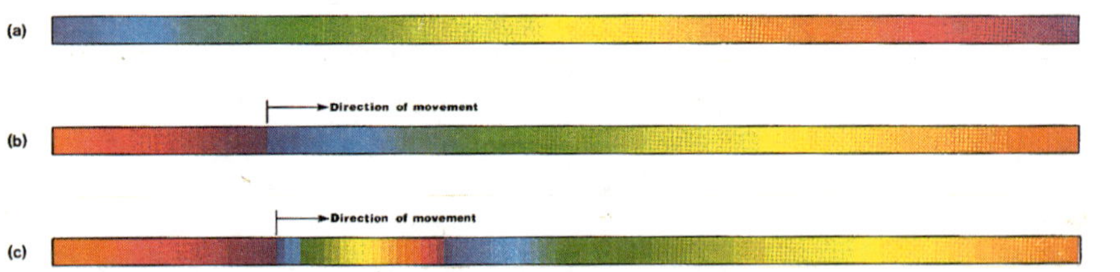

Shaded Relief Maps

While radiometric enhancement techniques can be used to yield a considerable amount of lithologic information from simple color images, dynamic range limitations do not allow flexible structural analysis. First and second vertical derivative maps for potential field data, as well as other spectral filters, help to identify structure, but interpretation can sometimes be ambiguous, misleading and difficult. Shaded relief maps (sometimes referred to as reflectance or shadow maps), on the other hand, provide immediate structural recognition. In such techniques, the operation on a pixel depends on its neighbours' values, and these techniques are fundamentally different from the radiometric enhancement techniques described earlier.

A reflectance map is basically an enhanced directional gradient product. The enhancement is accomplished by treating the grid "surface" as a reflecting body, where the properties of the reflecting body are used in determining the enhancement coefficients.

The apparent brightness of a surface grid cell or element depends on the orientation of the surface as expressed by gradients or surface normals, the position of the light source and the properties of the reflecting surface.

The surface normal is a vector perpendicular to the local surface. The direction of the surface normal **n** is found by taking the cross-product of any two vectors parallel to lines tangent to the surface, i.e.,

$$n = (1, 0, p) \times (0, 1, q) = (-p, -q, 1) \quad (7.12)$$

where (p, q) is the gradient along the x and y direction.

FIGURE 7.10
Geometry for determining shaded relief images

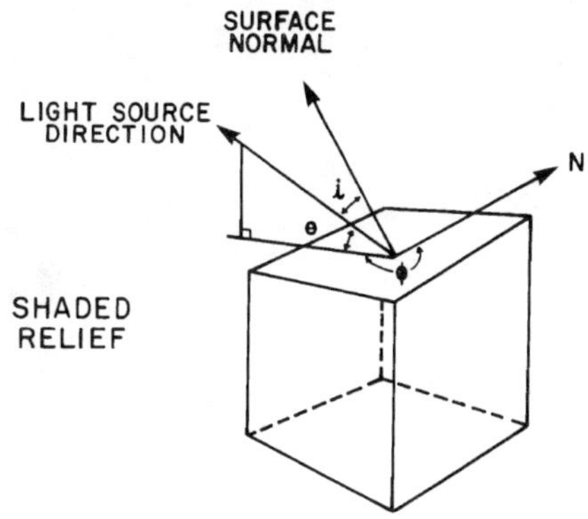

Figure 7.10 shows,

$$p = -\cos\phi \tan\theta \quad (7.13)$$
$$q = -\sin\phi \tan\theta$$

Similarly, for the light source,

$$p_o = -\cos\phi_o \tan\theta_o \quad (7.14)$$
$$q_o = -\sin\phi_o \tan\theta_o$$

where ϕ_b is the azimuth and θ_o is the inclination of the incident light source.

The amount of light captured by a surface will depend on its inclination relative to the incident beam. Looking from the illumination source, the surface projection is foreshortened by a factor equal to the cosine of the incident angle.

If one postulates an ideal surface that reflects all the incident light on it and appears equally bright from all viewing directions, an ideal diffuser or Lambertian reflector results. Horn (1981) describes numerous other reflecting surfaces, each with its own merit. The simple Lambertian model, however, will suffice for most geophysical applications. From figure 7.10, the cosine of the incident angle is seen to be the dot product of the surface normal and the illumination direction. Normalized, this becomes,

$$\cos(i) = \textit{Reflectance} = \frac{1 + p_o + q_o q}{(1 + p^2 + q^2)^{1/2} \, (1 + p_o^2 + q_o^2)^{1/2}} \quad (7.15)$$

When $1 + p_o p + q_o q < 0$, the surface element is turned away from the source and is self shadowed, i.e., reflectance = 0. Also, a scale factor is usually applied to provide an effective vertical exaggeration to this map to enhance the illusion of a relief map. Note that no actual shadow projections are computed. This eliminates the intensive computations required of the typical ray tracing algorithms familiar to three-dimensional graphics work. A gray scale plot of the reflectance values of equation 7.15 provides the shadow map.

The typical shadow map or, more precisely, reflectance map, provides an excellent single image describing the linear features of an area. It is highly sensitive to features which strike at right angles to the illumination direction and totally insensitive to features oriented parallel to the illumination direction. A means of seeing all features is to produce two static maps with different illumination directions.

Fast Shadowing

The shadow calculation is computationally slow with large datasets. It is unsuited for interactive processing on a graphics screen. Equation 7.15 illustrates that the gradients p and q must be calculated for each grid point. For a different choice of the azimuth or inclination of the light source, the gradients would have to be re-computed. A fast shading algorithm is available, however, which bypasses this repeated step.

A review of equation 7.15 for deriving the reflectance map reveals that it can be separated into two major processes:

- the generation of orthogonal (e.g., east and north) gradient pairs at each data point
- the calculation of reflectance normals for these gradients.

Figure 7.11 provides a pictorial view of the fast shadow algorithm. Only one orthogonal direction is shown for simplicity. The first of the above two processes requires the input of the original data grid [i.e., step (a) of figure 7.11], yielding the gradients as the output [figure 7.11, step (b)]. The second of the above two processes only operates on the values of these gradients. If the gradient pairs are first compressed down to acceptable limits, the reflectance calculations need only be performed on this compressed data set. Specifically, for a color palette with 256 levels, the two orthogonal gradients (x and y direction) must first be compressed to 16 levels each [step (d) of figure 7.11]. The product of this compression uses all the available histogram levels (i.e., 16 x 16 = 256). Then only 256 reflectance calculations are needed, independent of the size of the grid. For instance, in a grid of dimension 1024 x 1024, the compression allows the shadows to be computed 4,000 times faster than when using noncompressed, non-gradient data.

FIGURE 7.11
Beginning with the input data of (a), there are five steps in the fast shadow algorithm. The primary steps are (b), (e) and (f). Steps (c) and (d) yield the guidance necessary to perform step (e)

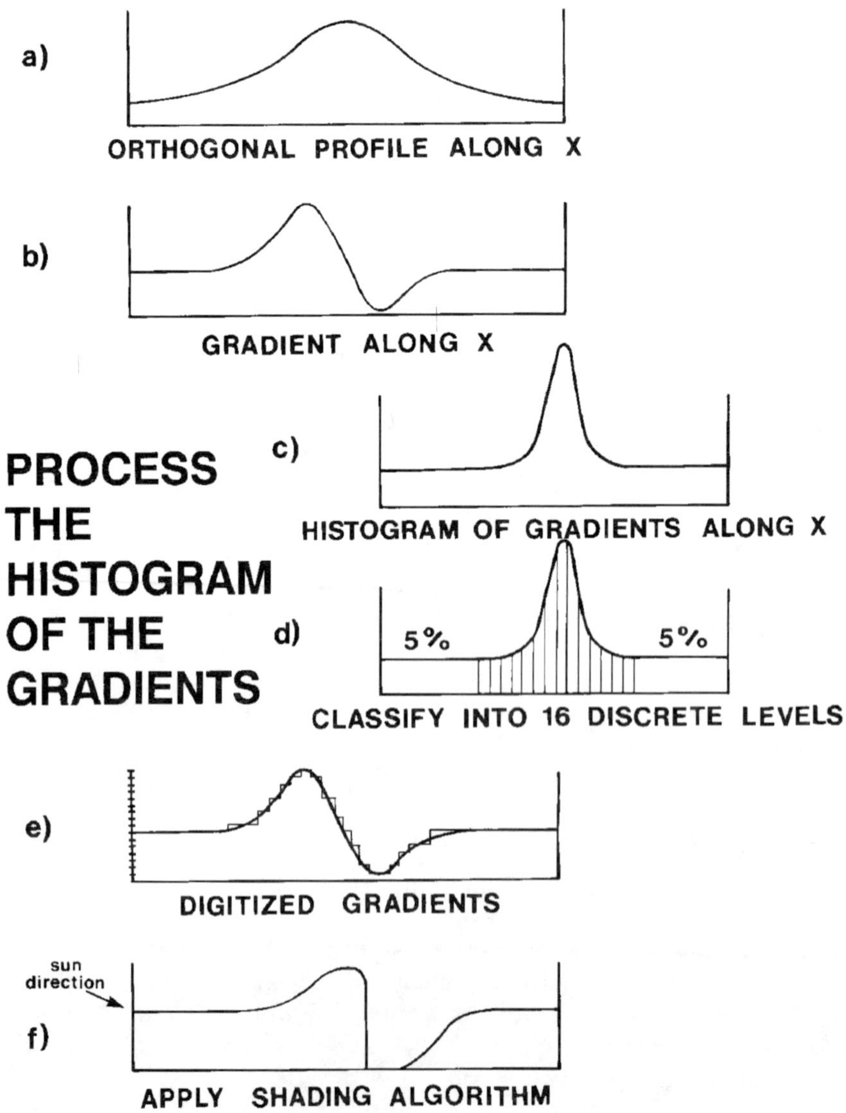

The compression of the processed gradient grid must first take into account the nature of the original orthogonal gradients. Gradients, in theory, can have a range approaching infinity. Obviously, a straight linear distribution of the histogram is not a constructive approach to performing the compression. Histogram equalization of the gradients is also not a good approach, since the reflectance calculation is not a strictly continuous function. A look at a typical histogram of gradients reveals the problem [cf. step (c) of figure 7.11]. A very few gradient values in the extremes stretch the histogram. These end gradients can be grouped into a single class and the remainder simply linearly distributed [step (e) of figure 7.11]. Categorizing the extreme 5 percent of the gradients in a single saturated color commonly provides acceptable results, although this percentage can be case dependent. Finally, the shading is performed on the 16 digitized gradients [step (f)] for each of the two orthogonal gradients.

Images with low gradient backgrounds, but having major high gradient features, tend to distort the gradient histogram from the ideal normal distribution. Typically in these cases, the image must first be "linearized". This can involve either some compression of highs and lows or, in extreme cases, the gradient calculations can be performed on the logarithms of the original data.

A disadvantage of the fast shadowing technique is that the resolution is degraded by the need to compress the two orthogonal gradients into 16 levels each.

Ternary Maps and Multi-layered Images

All of the above examples refer to single component images. It is possible to analyze multicomponent or multiparameter images. Even if the hardware limits the resolution of both the data grid and palette levels, some interesting effects can still be shown.

Figure 7.12 shows a high performance graphics controller used in some remote sensing applications. Instead of one-byte (8-bit) display buffers, three-byte buffers are employed. This allows 256^3 or 16 million simultaneous colors to be displayed. Such a range of colors in itself is rarely useful. This range permits three parameter sets, each having a full range of 256 colors, to be displayed simultaneously.

Figure 7.13 shows the generation of a ternary, or three-component image. It is commonly used in the display of spectrometer data but can also be used with potential field data. In this example, three resistivities, for the frequencies of 900 Hz, 7200 Hz and 56000 Hz, are each channelled into one of the three yellow, cyan and magenta ink

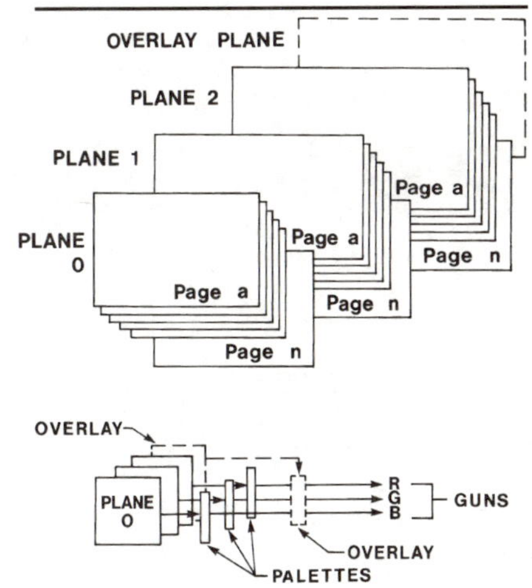

FIGURE 7.12
The high-performance graphics controller has 3 planes vs the single plane for the simple controller of figure 7.3. For either type of controller, a plane comprises a number of pages where each page has 64 KB of memory. For a graphics display of 1024 x 1024 pixels, 16 pages would therefore be required, or 1 MB per plane. The 3-plane high-performance graphics controller therefore requires 3 MB of memory. In addition to the above, both simple and high-performance controllers typically have an overlay plane of 2 or 3 bits per pixel for overlaying line vectors and text

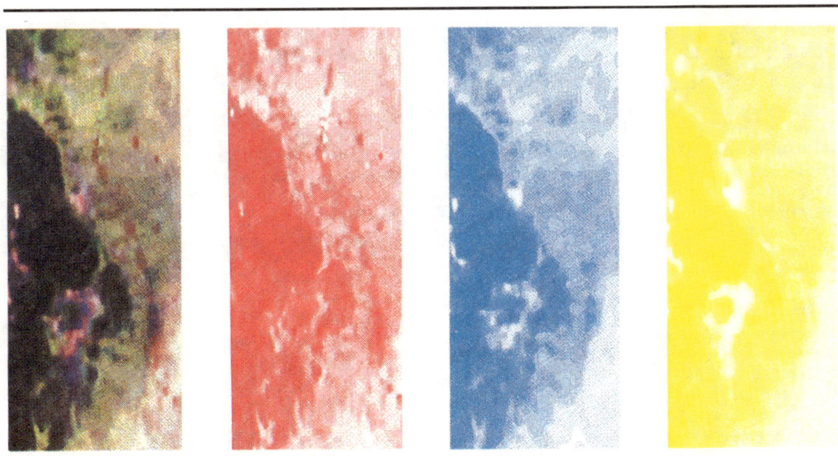

(a) (b) (c) (d)

FIGURE 7.13
Generation of a ternary image. One color is used for each parameter, where (a) yellow is the resistivity for the 900 Hz frequency, (b) cyan is the resistivity for 7200 Hz, and (c) magenta is the resistivity for 56,000 Hz. The low resistivities are white for each of these maps and the high resistivities are colored. The three maps are combined in (d) such that yellow indicates a surficial conductor and magenta a deep conductor

jets. Where they superimpose with high resistivities, the area is black. Where the resistivities are all low (or very conductive), the area is white. In this case, the image provides information on the three dimensionality of the resistivity, since each frequency explores to a different depth, e.g., the magenta (i.e., reddish) colors identify conductors at depth. These data are from the Getchell area of Nevada (Taylor, 1990).

Multicomponent images can be generated in two ways on graphics screens. In the first method, the individual components must be compressed down to accommodate the maximum number of available levels in the color palette. In a two-component image, this would be the square root of the number of available colors; in a three-component image, the cube root, etc. Adding many components rapidly limits the resolution of the individual components.

The second approach uses a dithering scheme. In this case, a two-by-two pixel array is used as the fundamental color element of the multicomponent image. Each of the four cells of the pixel array can then represent a different component. The various components must each share the available palette colors, however. As an example, assume that, (1) the number of color levels in the color palette is 256, (2) the graphics screen can display 800 columns by 600 rows of pixels (e.g., an extended super VGA graphics controller is used) and (3) four components are to be displayed on the screen. For the first method, the basic resolution is a single pixel and each component has the fourth root of the color levels available for it, i.e., the resolution remains at 800 x 600 pixels but the number of color levels per component are only four. In the second method, the basic resolution is 2 x 2 pixels and each component has 1/4 of the color levels available for it, i.e., the resolution is degraded to 400 x 300 "new pixels" but the number of color levels per component is 64.

While the second method yields a useful color range, the resolution of the dataset is seriously impaired. This loss of resolution may be unacceptable, particularly with a graphics controller having less than 1024 x 1024 pixels. This loss of resolution often is less of a limitation, however, than the major reduction in the color levels per component which is a consequence of the first method.

Printed versions of multicomponent images usually are in accordance with the CMY color system. Each component is overlayed directly onto the previous component. In these cases, high correlation between components is indicated by dark areas, which is the opposite to the RGB graphics screen displays.

IMAGE PROCESSING IN EXPLORATION

Image processing in mining exploration has forced improvements in data acquisition technology and in quality control. Prior to the use of computers in exploration, geophysical data commonly were displayed on hand-drawn contour maps. When computers and plotters began to be used, they replaced the hand-drawn contour maps with machine-drawn ones. In both cases, the result was a relatively low-resolution means of displaying the geophysical data. Such data could contain low-level flaws which could not be detected by the low-resolution display.

Low-level problems imbedded in the geophysical data became evident immediately when exploration data began to be image processed. This led to the observation that there is more to image processing than the imaging. The integrity of the data required for image processing is far greater than the integrity required to produce good quality contour maps. This has led to improvements in geophysical hardware and data acquisition technology as well as quality control techniques in both the field and data processing centers.

The benefits from the improvements in data quality and from the use of image processing are dramatic. Image processing highlights folding, faulting and lithologic changes to an extent not seen on contour maps. It also highlights flight path or traverse errors, which can be called positional errors. Levelling errors in the geophysical data are immediately obvious. Therefore, image processing is important in quality control. This is a positive feature. On the negative side, poor data can look terrible when it is image processed.

FIGURE 7.14
Pixels used for imaging may be identical in dimensions to grid cells used for contouring

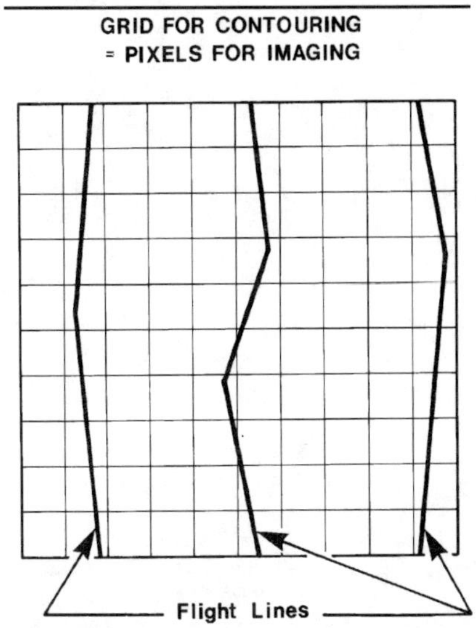

Preparation of Data for Imaging

Image processing is performed on gridded data. Figure 7.14 shows a typical distribution of grid cells. The grid cells each contain a geophysical data value, such as gammas, which would be interpolated from the magnetic survey flight lines shown in this figure.

Gridded data are used to produce a contour map by threading contours through the grid cells. Alternatively, if we apply colors or shading to the grid cells, we have image processing, and each grid cell is then called a pixel. On a color graphics screen, the pixel size is often identical to the grid cell size used to produce contour maps.

Figure 7.14 illustrates that contouring and image processing are performed mainly on interpolated data rather than on measured data. In image processing, as in contouring, there are commonly three-to-four times more pixels or grid cells than measured data points. Therefore, a well-designed gridding program is a primary factor in the quality of the resulting images.

A situation may arise where the form of data is unsuited to image processing. One such example is the typical EM anomaly map generated from an airborne survey. The anomalies are plotted on maps as points or symbols, and such a presentation is unsuited to image processing because points cannot be gridded. The EM profile data can be altered to yield continuous resistivity or conductivity values in place of point-like anomalies, however, and these continuous values can be gridded and imaged.

EM Data as Images

Figure 7.15 shows a schematic of a helicopter geophysical system comprising a long EM sensor or bird, and shorter magnetometer and VLF sensors. Both the magnetometer and VLF sensors yield data which are suitable for gridding and, therefore, can be contoured or imaged. The sensor outputs can be displayed after only a minimal amount of manipulations, e.g., levelling, diurnal and/or spike removal and, perhaps, a light brush with a lowpass filter. The treatment of EM data, however, requires a degree of sophistication, if the data are to be recast as continuous resistivity values which are suitable for imaging.

Figure 7.16 illustrates a computer-produced flight record showing many EM anomalies (Fraser, 1990). Of interest to image processing is the resistivity channel which is shown in this figure. The resistivity channel is calculated from the inphase and quadrature EM channels from one of the transmitter-receiver coil-pairs (Fraser, 1978).

Figure 7.17 presents a typical EM anomaly map on the left and, on the right, the corresponding resistivity contour map.

FIGURE 7.15
Helicopter geophysical system showing VLF, magnetometer and EM sensors

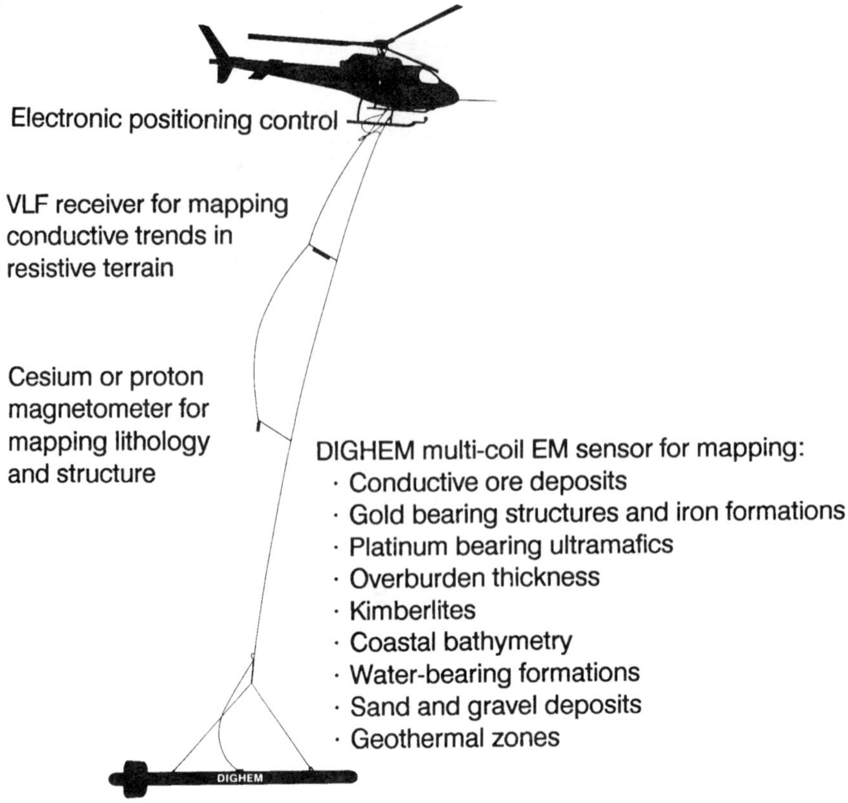

The advantage of the airborne resistivity map is that much more EM information is shown on the map. The resistivity map is free of any prejudices of an interpreter and, therefore, provides a totally unbiased picture of the electrical properties of the survey area. Further, the resistivity data are suitable for image processing because they yield a continuous flight profile (e.g., figure 7.16), rather than the point-like EM anomalies of the classical EM map. In any situation where data are shown as anomalies, e.g., geochemical maps, image processing can be performed provided the data can be recast as values which are suitable for gridding.

The Imaging of Amplitudes

Image processing typically yields products of two classes, i.e., the imaging of the amplitudes of the data, and the imaging of the gradients of the data.

The image processing of amplitudes may display the amplitudes directly, or they may be truncated, linearly compressed into classes, log compressed, etc., as discussed earlier. In any event, some function of the amplitudes is eventually colored and the pixels displayed on a graphics screen or color map.

Figure 7.18 shows a standard contour map from a magnetic survey, and the right side shows the corresponding color map. The color map is the result of image processing, and yet this is only the machine equivalent of a hand colored map.

A close comparison of the two maps would cause most people to agree that the magnetic features are more noticeable on the colored map than on the contour map. Because the colored map can be generated by hand, however, this is hardly impressive as a demonstration of the value of image processing.

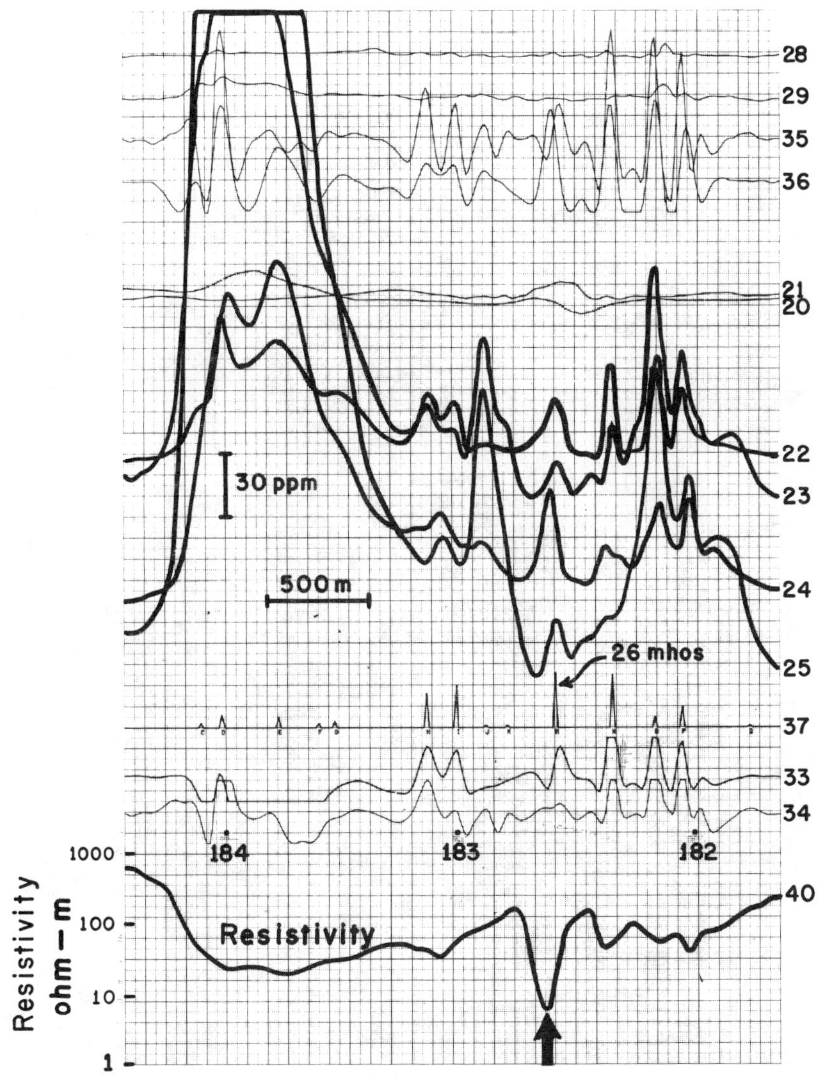

FIGURE 7.16
The resistivity channel for this digital flight record identifies the Woodlawn massive sulfide deposit, Australia (see arrow)

The Imaging of Gradients

An increasingly popular process is the shading of geophysical data to produce shadow maps. Figure 7.19 shows a magnetic profile which can be treated as if it were a topographic profile. An artificial "sun" is positioned to shine upon this profile. Obviously, the sun will reflect best off the profile on the facing side of the magnetic high or hill, whereas the back side of the hill will be in shade.

On the bottom of the figure, a swath of ground is shown which corresponds to the magnetic traverse. The magnetic anomaly is seen to be bright on the facing side and dark on the shaded side. Elsewhere, shades of gray occur. This shadowing process can be seen to be very simple conceptually.

The shaded images of figure 7.20 are from the same dataset as figure 7.18. Figure 7.20 shows the image processed color magnetic map on the left. The right side of figure 7.20 presents the image processed shadow map in gray tones. There are many features which are evident on this gray-tone shadow map which cannot be seen on the color amplitude map on the left side of figure 7.20. This gray-tone shadow example illustrates the potential of image processing.

FIGURE 7.17
Comparison of the DIGHEM^{IV} resistivity map (right), which is generated from EM data, with the corresponding EM anomaly map (left)

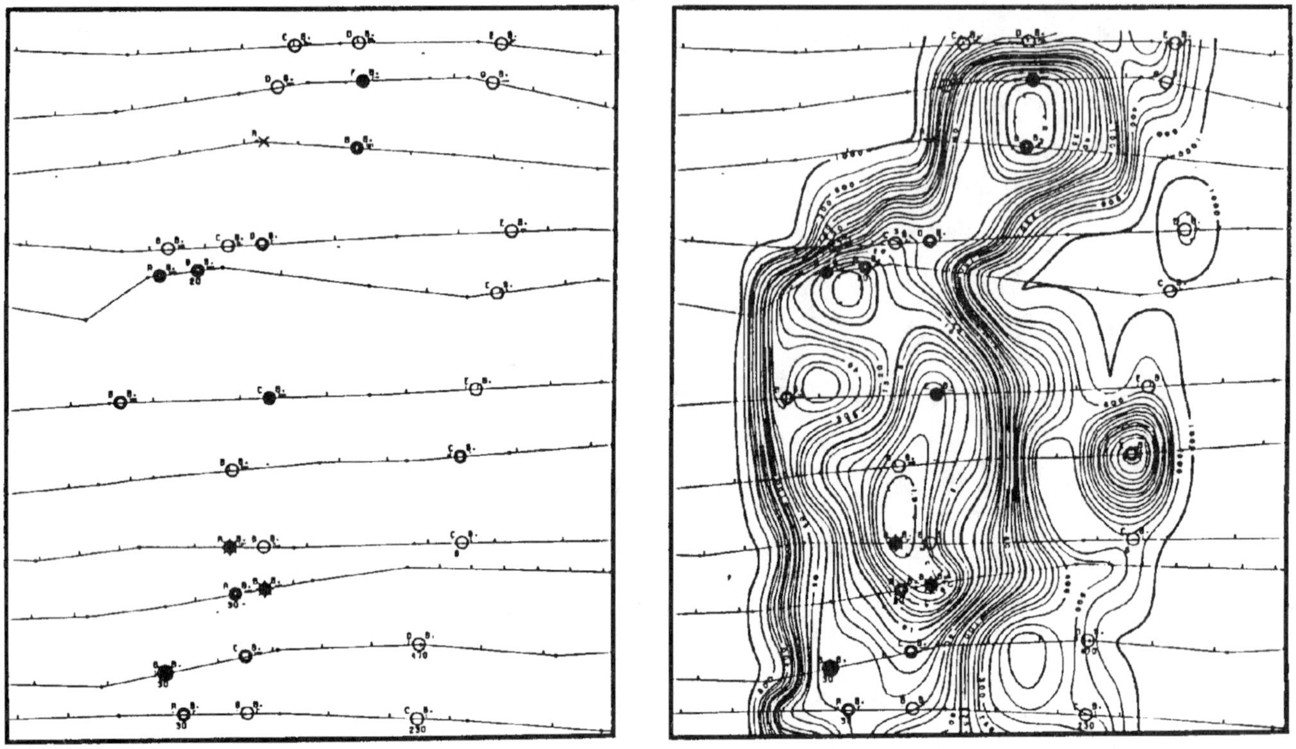

Unlike the color amplitude map on the left, there is no practical way to produce a shadow map by hand, and so image processing is required to obtain the information shown in the shadow map.

The color amplitude map on the left side of figure 7.20 and the gray-tone shadow map on the right can be combined to produce the shaded color image shown in the center. This combined shaded color map is useful for those who wish to see both color amplitudes and shadows in a single presentation. The shaded color map is great for displays, board meetings and so on. Be aware, however, that detail is sacrificed when color and shadow are combined in a single presentation. This is because the shadows lose dynamic range and sensitivity. Also, dark shadows can obscure some color features.

The color map and the shadow map highlight different features of the geology. The color is more sensitive to lithology. The shadow is more sensitive to structure. Clearly, there is considerable overlap in the information content of color and shadow. From the viewpoint of highlighting lithology and lithologic changes, however, the color map should be used as the basic tool. To highlight structure, the shadow map should be used.

FIGURE 7.18
Comparison of contoured and colored magnetic maps

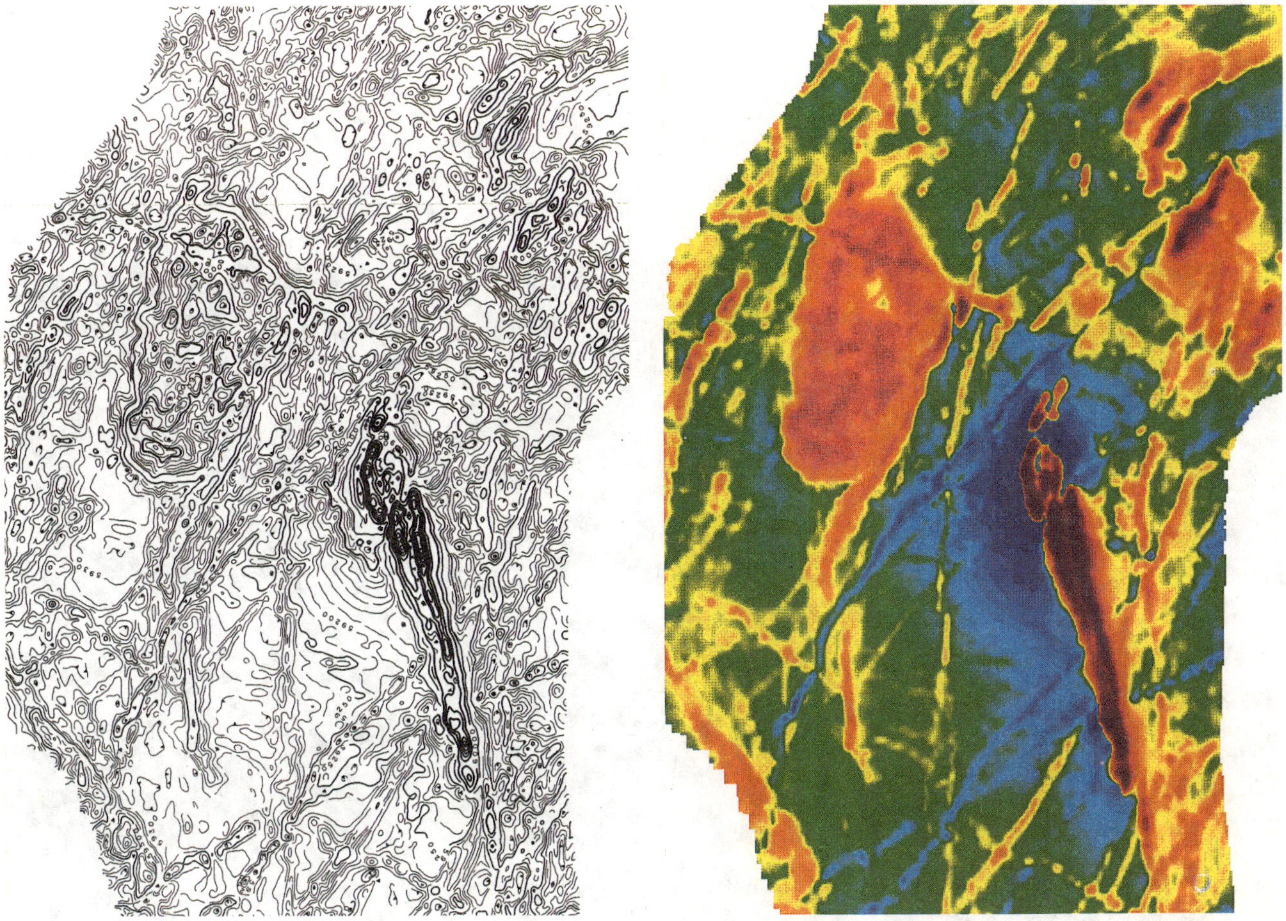

The shadow technique involves the following steps, using a magnetic survey as an example.

- The magnetic data are gridded.
- Bearing in mind that shadows are cast by magnetic "hills" and "ridges" in the magnetic grid, a vertical exaggeration factor must be selected. Shadows tend to become darker as the vertical exaggeration is increased, for a fixed inclination of the artificial sun.
- An inclination (dip) is selected for the artifical sun, e.g., 30° above the horizon as in the shadow map of figure 7.20.
- A declination (azimuth) is selected for the sun.

The vertical exaggeration may be determined by trial and error, to provide a pleasing image, or by a statistical analysis of the data. Too much exaggeration yields a choppy appearance which can mask subtle structures. Too little exaggeration results in loss of useful information through insensitivity.

FIGURE 7.19
An artificial sun is used to shine on a magnetic surface. Shadows develop on the back side of magnetic highs or "hills"

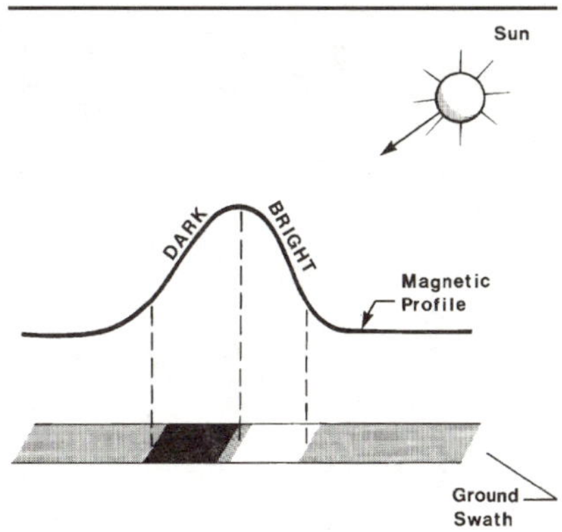

FIGURE 7.20
The magnetic amplitudes are shown in the color map on the left, and the magnetic gradients are highlighted in the gray-tone shadow map on the right. The artificial sun is shining from the upper left. Both maps are combined to yield the shaded color map in the center

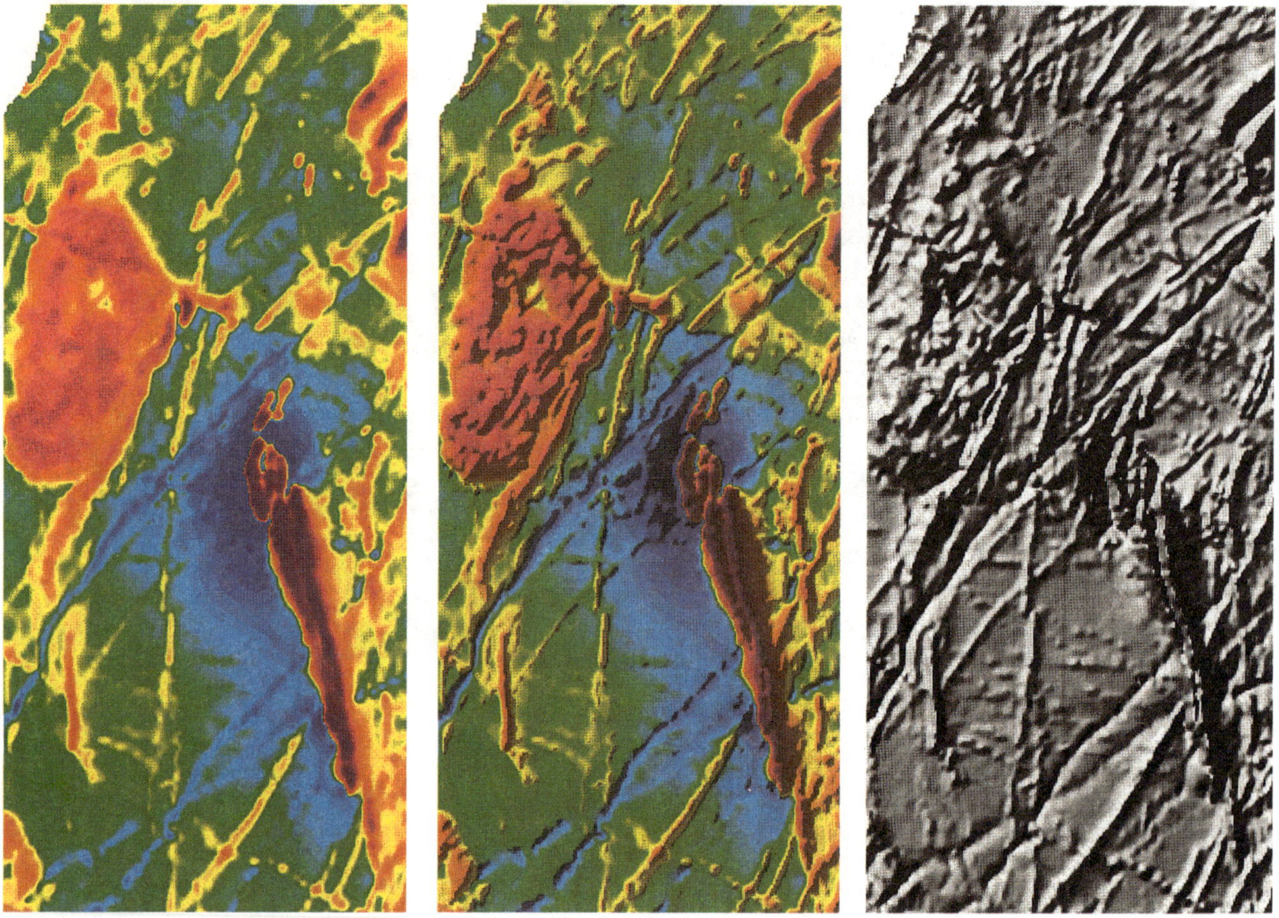

The inclination of the sun may be set at 30° for most applications, based on empirical results. The desired shadowing sensitivity is typically obtained by varying the vertical exaggeration rather than the sun's inclination.

The selection of the azimuth of the sun is the most rewarding variable in the shadowing process.

Returning to the profile of figure 7.19, it is obvious that the sun will be reflected off the facing side of the hill. In plan view, the same is true for magnetic features which are orthogonal to the sun's azimuth. Such features will be bright on the side which faces the sun and will be in shadow on the back side. If the sun were to be rotated so that its azimuth was parallel to a linear magnetic feature, that feature would not cast a shadow and would, in effect, be extinguished. A magnetic diabase dike striking east-west, for example, will cast a shadow when the sun's azimuth is north or south and will cast no shadow when the sun's azimuth is east or west.

The shadow technique is often thought of as a means of highlighting strata or structures of a given strike by setting the sun's azimuth at right angles to the desired strike. Such a conception may be flawed. Often, desired strata are masked by undesirable features such as diabase dikes which cut the desired lithology. The best means of defining the desired strata may be to shine the sun parallel to the dikes so that they are extinguished, rather than orienting the artificial sun perpendicular to the desired strata or structure. The resulting shadows then will reflect all features which are not extinguished, i.e., the desired strata. An analogy may be made with a wooden fence. Regardless of the sun's azimuth, the fence casts a noticeable shadow unless the strike of the fence is within a few degrees of the sun's azimuth. When selecting a sun azimuth for preparing a shadow map, one rule is to pay close attention to those undesirable geologic features which distract from the desired structures and select an azimuth parallel to these undesired features. Clearly, there is an advantage to producing two (or more) shadow maps with different sun azimuths. Depending on the strike angles of undesirable features, the two sun azimuths may not be mutually perpendicular.

Figure 7.21 illustrates two magnetic shadow maps over Precambrian geology in the Wawa greenstone belt of Ontario, Canada.

Diabase dikes dominate the map of figure 7.21a; the direction of the artificial sun is from the upper right of this map.

The diabase dikes are largely extinguished on the map of figure 7.21b because the sun's azimuth is parallel to the strike of the dikes. Note how details of the structure of the greenstone belt are more easily seen on the map of figure 7.21b.

The Rotating Shadow Technique

The vertical exaggeration of the data grid as well as the declination and inclination of the sun can be varied rapidly in interactive processing using a color graphics screen. The technique described earlier called fast shadowing is employed.

The fast shadowing technique has a number of advantages. First, the trial-and-error approach of selecting a pleasing vertical exaggeration requires only a few seconds of interactive manipulation of the image. Second, the sun's azimuth can be varied in fractions of a second. The sun in fact can be rotated through 360 degrees of azimuth, in increments of 1°, in less than a minute of time on a 386-based PC. This allows the viewer to select exactly those sun azimuths which best define geologic features of interest. Shadow map plots then can be generated with those selected azimuths.

The rotating shadow technique is an interactive process. The sun is rotated automatically and continuously through 360 degrees with a full sweep every half minute or so. Geologic features appear, disappear and reappear as the sun sweeps around the compass. The moving sun casts moving shadows which light up features of every strike. Subtle structures may only be noticeable with this moving image, much like a rabbit may only be noticed in the bush when it

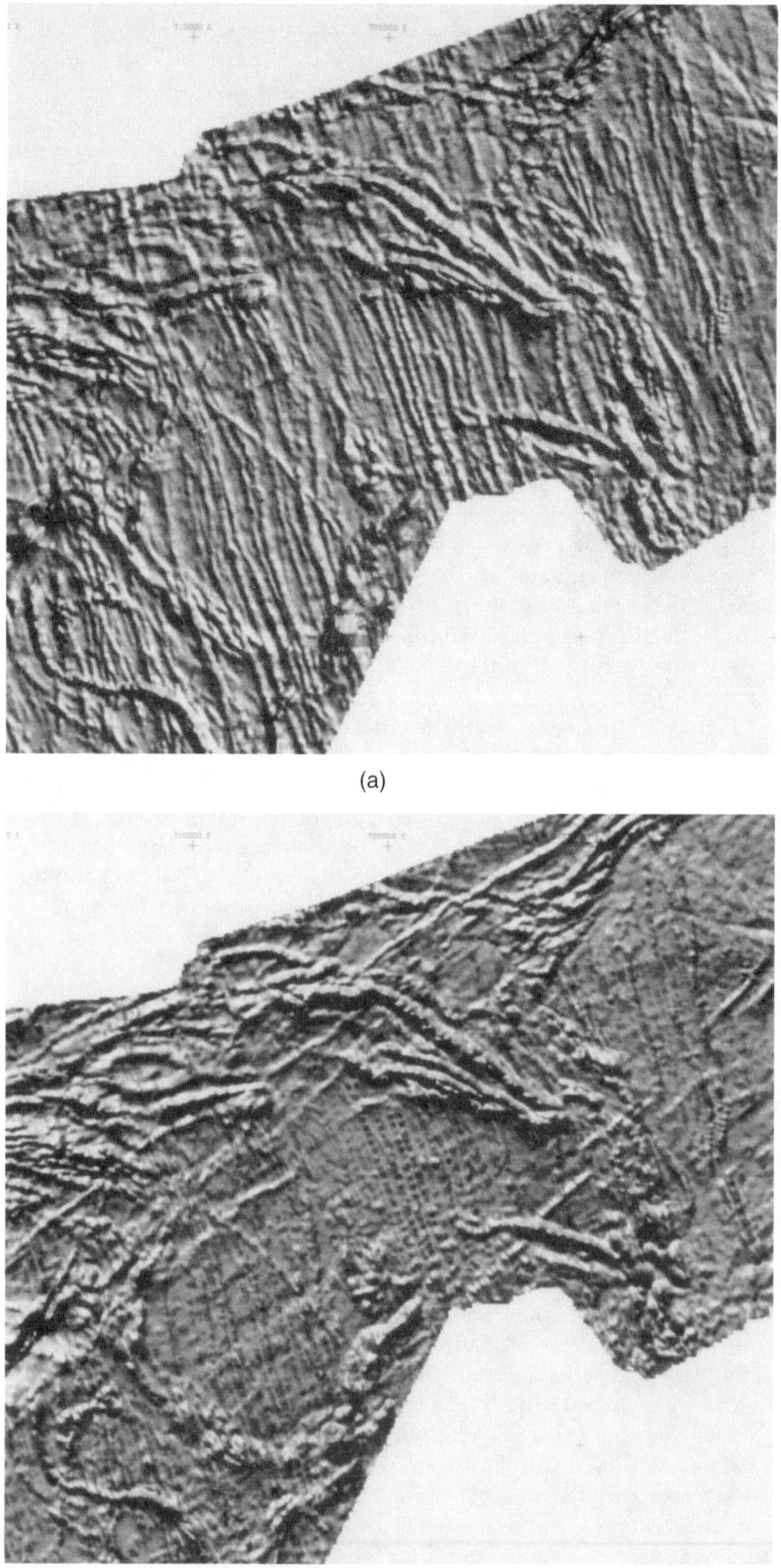

(a)

(b)

FIGURE 7.21
Different sun directions were used to produce these two shadow magnetic maps of the Wawa greenstone belt in Ontario, Canada. Diabase dikes dominate on the top map but are largely extinguished on the bottom map

moves. The rotating shadow technique is an extremely powerful method for identifying geologic structure. It requires an imaging workstation because the effect cannot be duplicated using a set of shadow plots.

The Revolving Color Technique

The value of image processing for the imaging of gradients is indisputable. This is because shadow maps cannot be produced by hand. The value of image processing for the imaging of amplitudes can be arguable, however, because colored maps can be prepared by hand. The imaging of amplitudes only becomes impressive when data are manipulated interactively on a color graphics screen.

The color spectrum of a map, such as in figure 7.22, can be revolved to produce a moving image. The revolving color technique is an interactive process so an imaging workstation is required.

Figure 7.22 attempts to provide the sense of the interactive processing by portraying four steps in the process. This figure shows a portion of the magnetic map from a helicopter-borne DIGHEM^{IV} survey flown for the U.S. Geological Survey in the basin and range area of northern Nevada known as the Getchell trend (Taylor, 1990). The magnetic high which dominates the center of figure 7.22a reflects the Cretaceous granodiorite of the Osgood Mountains.

Figure 7.22a shows the standard color magnetic map which has been produced using the standard color bar. The poorly magnetic rocks are portrayed by cold colors (i.e., blue) and the highly magnetic rocks by hot colors (i.e., red). The background is greenish-yellow.

The color bar in figure 7.22a can be revolved so that blue (rather than greenish-yellow) defines the background, for example. This recoloring may allow subtle lithologic changes to be identified. It generally is more rewarding, however, to redefine the color bar before rotation.

The color bar of figure 7.22b shows a highly resolved narrow color bar imbedded in the left (or weakly magnetic) side of the standard color bar. Note that the highly resolved color bar on the left side contains all the colors from blue to red. The main color bar extends over the remainder, also containing all the colors from blue to red. Thus, the color bar of figure 7.22b contains two areas of blue, two of yellow, two of red, etc. Figure 7.22b is therefore of no value with respect to the use of color to define amplitude, e.g., the red color appears for both weakly and strongly magnetic rocks. The value of figure 7.22b lies in its high definition of the poorly magnetic rocks, since the poorly magnetic part of the color bar (i.e., the left side) contains a full color spectrum.

It is a simple matter to revolve interactively the color bar on an imaging workstation. If the desire is to obtain high definition in the background region, the color bar is simply moved to the right until the narrow, highly resolved, imbedded color bar is in the center (figure 7.22c). If, instead, the goal is to obtain high definition in the highly magnetic rocks, the narrow imbedded color bar is moved to the extreme right side of the color bar (figure 7.22d).

The purpose of revolving the color bar is analogous to the purpose of generating rotating shadow maps. In both cases, the geometry of the image patterns is used to gain geologic insight. Compare, for example, the dual color bar of figure 7.22d with the single, standard color bar of figure 7.22a. In the map of figure 7.22a, the granodiorite is seen to be distributed in two reddish-purple lobes.

Otherwise, the intrusive appears to be but poorly differentiated. The map of figure 7.22d, however, tells a different story. It indicates that differentiation within the granodiorite indeed exists. This type of analysis provided by the dual color bar on an imaging workstation can be very helpful in the magnetic study of carbonatites and other commonly differentiated intrusives.

In revolving the color bar, note that the technique does not operate on the data grid. It does not even manipulate the data histogram. It is simply the color palette which revolves, and this process is exceedingly fast.

It was shown above that the rotating shadow technique yields a hands-free method of producing a continuously moving image which aids in the identification of structure. Similarly, the revolving color technique is an interactive process (whereby the colors are automatically and continuously revolved) with a full revolution every half minute or so. Just as subtle structures may only be noticeable with a rotating shadow image, so subtle lithologic changes may only be noticeable with a revolving color image. Both of these techniques are extremely powerful tools for the geologic study of an area. The techniques are applicable at virtually any scale. For example, single images in color and shadow have been generated for the magnetic map of North America. On the other extreme, images have been produced for soil geochemical data involving 10 lines totalling 300 samples.

FIGURE 7.22
(a)Color magnetic map of the Getchell area, Nevada, showing data histogram and color bar.
(b)Dual spectrum color map and color bar, where a highly resolved color bar is imbedded in the weakly magnetic part of the main color bar.
(c)The highly resolved color bar has been moved to the center of the histogram.
(d)The highly resolved color bar has been moved to the highly magnetic part of the histogram

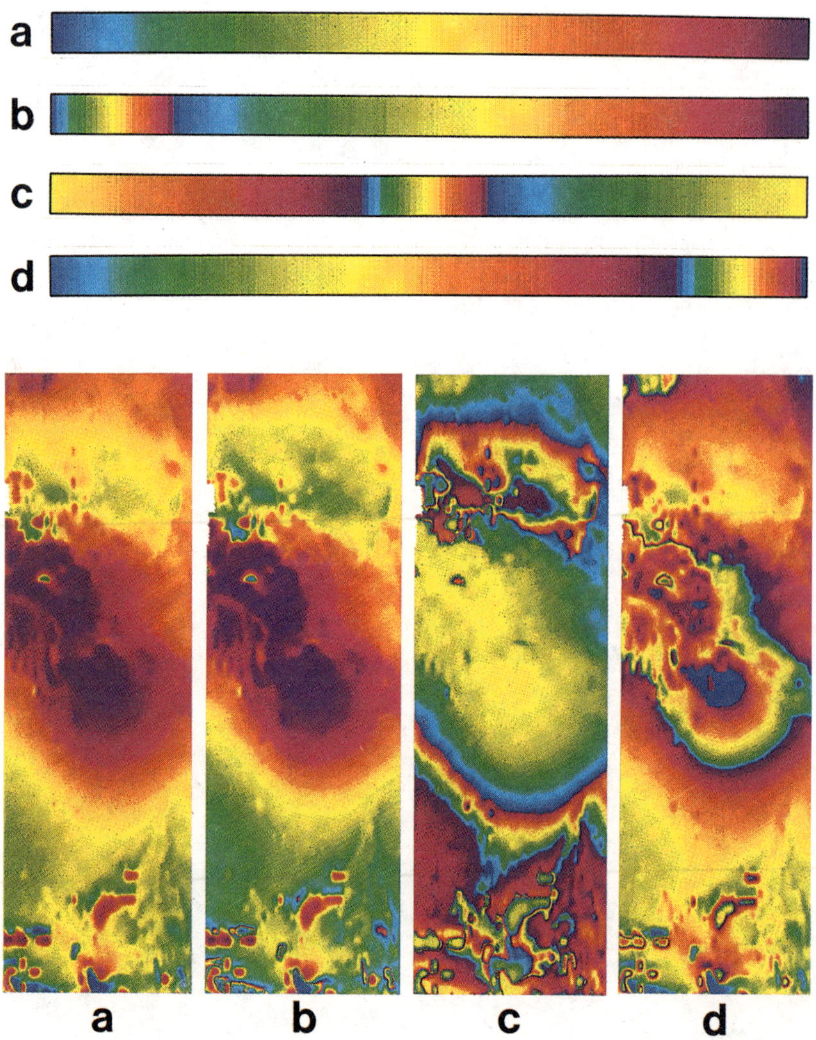

CONCLUSIONS

The products of image processing are either supplementing or replacing contour maps as a means of displaying gridded data in exploration geophysics. The imaging of data also provides information which cannot be obtained by contouring. As such, image processing is destined to become the primary means of displaying gridded datasets.

The end product of imaging still tends to be the colored or shaded map. Far more information can be obtained, however, by interactively manipulating imaged data on a color graphics screen. The revolving color and rotating shadow techniques are exceedingly powerful in this regard. As a result, some geologists and geophysicists request that their geophysical data be delivered on floppy disks because they wish to manipulate the images themselves. This evolution from contour map to color map to interactive computer display is expected to continue as geoscientists become more familiar with the tools and utility of image processing.

The technology of image processing is advancing rapidly as more powerful microcomputer systems become available at reasonable prices. Many of the limitations discussed in this chapter, for example with respect to multicomponent images, will cease to be a problem as hardware technology continues to advance.

REFERENCES

Foley, J.D. and Van Dam, A., Fundamentals of interactive computer graphics: Addison-Wesley Publishing Co., Reading, Massachusetts, 1984.

Fraser, D.C., 1978, Resistivity mapping with an airborne multicoil electromagnetic system: Geophysics, v. 43, pp. 144-172.

Fraser, D.C., 1990, Electromagnetic anomaly recognition within geological and cultural noise; in Fitterman, D.V., ed., Developments and applications of modern airborne electromagnetic surveys: U.S. Geol. Surv. Bull. 1925, pp. 53-64.

Gonzales, R.C. and Wintz, P., Digital image processing: Addison-Wesley Publishing Co., Reading, Massachusetts, 1977.

Horn, B.K.P., Hill shading and the reflectance map: Proceedings of the IEEE, Vol. 69, No. 1, January, 1981.

Pavlidis, T., Algorithms for graphics and image processing: Computer Science Press, Rockville, MD, 1982.

Richards, J.A., Remote sensing digital image analysis, an introduction: Springer-Verlag, Burkin, 1986.

Taylor, R.S., 1990, Airborne EM resistivity applied to exploration for disseminated precious metal deposits: Geophy. Lead. Edge, Feb. 1990, pp. 34-41.

Teskey, D. and Broome, J., Computer programs for production of shaded relief maps: Current Research, Part B, Geological Survey of Canada, Paper 84-1B, pp. 375-379, 1984.

COMPILER

Richard "Dutch" Van Blaricom is the Chief Geophysicist for Cominco American Resources Inc. in Spokane Washington. He is responsible for all geophysical work done by the company in the continental United States and Alaska. Dutch has been involved in mining geophysics since 1963. He has also worked for McPhar Geophysics Ltd., ASARCO, and has also worked an independent geophysical consultant. Dutch has experience in all phases of geophysical exploration in most geological environments, having done surveys for virtually all types of mineral deposits, most industrial minerals, and some contamination plumes.

Dutch was the Course director for the first short course in Geophysics in 1978, and again in 1989. He is widely published and has given many papers on geophysical exploration for mineral deposits. He is a member of the Society of Exploration Geophysicists, American Geophysical Union, Northwest Mining Association, Society for Mining Metallurgy and Exploration Inc., and the National Ground Water Association. He is a registered geophysicist in the State of California and a registered geologist in the state of Idaho. Dutch has a B.S. in geology from Portland State University, a M.S. in geophysics from the University of Arizona, and is currently working toward his Ph.D. at the University of Idaho.

NOTES

Application of Integrated Geophysical Methods

This section indicates which methods are recommended for use on individual deposits, with helpful hints on how to determine which methods one should choose to use on other unlisted deposits. The data are presented in outline form because the "why and how" have been enumerated in previous sections of this manual. A modified version of a cost information graph designed by Geoterex Limited of Canada is included; this information is extremely helpful when planning any geophysical surveys.

Deposits	Direct Detection	Indirect Detection	Geologic Studies, Structures, Etc.
Disseminated sulfide	IP Resistivity SP EM	Magnetics Radiometric (alteration studies) Gravity	Gravity (basin-and-range studies Magnetics (detection of intrustive, etc., helpful in geologic mapping) MT-AMT (basement studies)
Massive sulfide	IP Resistivity Magnetics Gravity SP AMT	Gravity Magnetics	Magnetics VLF, EM, etc.
Uranium	Radiometric studies Radon cups	Radiometric studies Radon cups VLF and ELF IP and resistivity Gas studies (helium and radon) Magnetics	VLF and ELF (structure) Magnetics (geologic mapping) IP and resistivity (rollfront delineation) Magnetics (rollfront studies)
Water	Resistivity EM Seismics AMT	Seismic (aquifer location) Gravity (basement studies)	Magnetics
Contamination Plume	Resistivity EM Seismics AMT	Seismic (aquifer location) Gravity (basement studies)	Magnetics
Asbestos	IP and resistivity Seismics	Magnetics EM	Magnetics EM
Stream channel (placer)		Magnetics (black sand) IP and resistivity (black sand)	Seismic (stream channel Location) Resistivity (clay detection)
Lateritic	Seismic soundings Resistivity soundings	Gravity Magnetics	Magnetics (geologic mapping)
Geothermal studies "hot water"	Resistivity Passive seismics EM Geothermal gradient AMT	SP Radiometric studies	Magnetics Gravity
Mississippi Valley-type	IP and resistivity EM AMT	Gravity Magnetics Seismic	Gravity (basement topography) Magnetics (basement topography) Seismics

Nondestructive material testing	Dam site evaluation and concrete testing: seismics, ripability studies, etc.; resistivity
Overburden depth studies	Seismics; resistivity; gravity
Gravity	Direct detection: seismic studies; resistivity measurements
Location of pipes or other buried features	Applied potential; magnetics; EM; gravity (detection of voids or caverns)
Other deposits or objects	General consideration: Determine the physical parameters of the deposit or object and its host. This includes magnetic susceptibility, remnant magnetism, density, resistivity, induced polarizability, color, seismic velocity, radioactivity, temperature, and structural implications as to the location of the deposit or object itself. Choose parameters that would be most likely to differentiate the deposit or object from its host. Run test survey on known area. Run survey to detect feature.

Application of geophysical techniques

	Method (A) Active (P) Passive	Parameter Measured		Characteristic Physical Property	Main Causes of Anomalies
		Designation	Unit		
ELECTRICAL	RESISTIVITY (A)	Apparent resistivity	Ω-m	Resistivity Conductivity	Conductive veins, sedimentary layers, volcanic intrusions, shear zones, faults, weatherings, hot waters
ELECTRICAL	INDUCED POLARIZATION (A)	Time domain: chargeability polarizability Frequency domain: frequency effect Phase domain: phase shift	ms pfe pfe mr	Ionic-electronic Over voltage	Conductive mineralizations: disseminated or massive (graphite, sulfides), clay
ELECTRICAL	SELF POTENTIAL (P)	Natural potential	mV	Conductivity Oxydability	Massive conductive ores Graphite Electro-filtration Faults
ELECTRICAL	MISE-A-LA-MASSE (A)	Applied potential	mV	Conductivity	Extension of previously located conductive orebodies
ELECTRICAL	TELLURIC (P)	Relative ellipse area Ratios—apparent resistivity	Ω-m	Conductance	Basin and Range studies Conductance of sedimentary series Salt domes, geothermal
ELECTRICAL	MAGNETOTELLURIC (P) MT, AMT (A) CS-AMT	Apparent impedance (resistivity and phase)	Ω-m degrees	Resistivity Conductivity	Conductive veins, sedimentary layers, shear zones, faults, weatherings, resistive basements, bedded ores
ELECTRICAL	ELECTROMAGNETIC (A)	Phase difference Tilt angle Amplitude ratio Sampling of decay curve induced in receiving coil by eddy currents In-phase, out-of-phase components	degrees Ω-m σt	Electrical conductivity	Conductive mineralizations Surficial conductors Shear zones
	MAGNETIC (P)	Earth magnetic field Vertical component Z, Total intensity Horizontal gradient Vertical gradient	$1\gamma = 10^{-5}$ gauss	Magnetic susceptibility	Contrasts of magnetization Magnetite content of the materials
	GRAVITY (P)	Gravity field	milligal (1 gal = 1 cm/s^2)	Density	Deposits of heavy ores Salt domes (light) Basement rocks
	RADIOACTIVITY (P)	Intensity and spectral composition of gamma rays	cps	Radioactivity	Radioactive elements Uranium-Thorium-K_{40}
SEISMIC	REFRACTION (A) REFLECTION (A)	Traveling time of seismic waves	milliseconds feet per second	Seismic wave velocity Dynamic modulus	Contrasts of velocity: markers at variable depth Fissured rocks
	THERMOMETRY (P)	Temperature	°C	Geothermal gradient and temperature	Abnormal flux of heat Thermal inertia of rocks

Applications		Geophysical Crew		
Direct Detection	**Indirect Detection**	**Production**	**Survey Cost**	**Interpretation**
Massive sulfides Oil shales Clays Geothermal reservoirs	Bulk material Base metals Coal & natural steam	Depending upon depth of investigation El. soundings: 20 to 250 E.S./month Resistivity profiling: 20 to 100 line km/month Resistivity maps: 5 to 20 sq km/month	650/day	Curve matching Computer modeling
Conductive: sulfides oxides mn oxides	Associated minerals (zinc, gold, silver, tin, uranium, etc.)	800 to 2000 stations/month 20 to 70 line km/month	750/day	Curve matching Computer modeling
Sulfides: pyrite pyrrhotite copper Mn ore	Associated minerals (lead, gold, silver, zinc, nickel)	2500 to 3500 stations/month 60 to 200 line km/month	200/day	Rules of thumb Curve matching Visual interpretation
Conductive ores	Associated minerals (zinc, tin)	5 to 15 sq km/month	400/day	Computer reduced, computer modeling, visual interpretation
Structural studies Steam	Regional exploration	20 to 500 sq km/month	450/day	Computer reduced Curve matching Visual
Massive sulfides Clays Natural steam	Shear zones General tectonics General structure	Up to 3 stations/day Up to 20 stations/day	1200/day to 1000/station	Computer reduced Computer inversion Visual interpretation
Conductive: sulfides oxides Mn oxides	Kimberlites Associated minerals Ground follow up (lead, nickel), shear zones, weathered zones, conductivity maps	50 to 150 line km/month 6000 to 10,000 line km/month	350/day (ground) 35-65 km air	Curve matching Visual interpretation Computer modeling
Magnetite Pyrrhotite Titano—magnetite	Molybdenum Iron ore Chromite Copper ore Kimberlites Asbestos Geological mapping in terms of magnetic changes (basic rocks . . .) and/or discontinuities inventory of mineral resources	4000 to 8000 stations/month 80 to 300 line km/month (on shore) 20 to 80 line km/working day (off shore) 1000 to 15,000 line km/month	200/day (ground) 25-30 km air	Computer reduced Rule of thumb Computer modeling Computer inversion Visual interpretation
Chromite Pyrite Chalcopyrite Lead	Placer configuration Karstic cavities Basement topography Structure	400 to 1500 stations/month 20 to 50 stations/working day Surface meter combined with seismic turnover	450/day	Computer reduced Computer modeling Computer inversion Curve matching
Uranium Thorium Phosphates	Ground follow up Geological structural mapping (differentiation in granites)	80 to 400 line km/month (ground) 1000 to 15,000 line km/month (air)	350/day 25-60 km	Computer reduced Rations of responses Geologic correlation
Buried channels Faults Morphological traps Basement topography	Tin, diamonds Heavy minerals Natural steam Uranium	10 to 50 line km/month (on shore) 10 to 40 line km/working day (off shore)	350/day 1200/day	Computer interpreted Computer inverted
Thermal springs	Natural steam Boron, sulfur, Subterranean volcanism		1000/day	Computer reduced

Index